AF588396

Green Chemistry and Sustainable Technology

Aims and Scope

The series *Green Chemistry and Sustainable Technology* aims to present cutting-edge research and important advances in green chemistry, green chemical engineering and sustainable industrial technology. The scope of coverage includes (but is not limited to):

- Environmentally benign chemical synthesis and processes (green catalysis, green solvents and reagents, atom-economy synthetic methods etc.)
- Green chemicals and energy produced from renewable resources (biomass, carbon dioxide etc.)
- Novel materials and technologies for energy production and storage (biofuels and bioenergies, hydrogen, fuel cells, solar cells, lithium-ion batteries etc.)
- Green chemical engineering processes (process integration, materials diversity, energy saving, waste minimization, efficient separation processes etc.)
- Green technologies for environmental sustainability (carbon dioxide capture, waste and harmful chemicals treatment, pollution prevention, environmental redemption etc.)

The series *Green Chemistry and Sustainable Technology* is intended to provide an accessible reference resource for postgraduate students, academic researchers and industrial professionals who are interested in green chemistry and technologies for sustainable development.

More information about this series at http://www.springer.com/series/11661

Wai-Yeung Wong
Editor

Organometallics and Related Molecules for Energy Conversion

Editor
Wai-Yeung Wong
Department of Chemistry
Hong Kong Baptist University
Hong Kong, China

ISSN 2196-6982 ISSN 2196-6990 (electronic)
Green Chemistry and Sustainable Technology
ISBN 978-3-662-46053-5 ISBN 978-3-662-46054-2 (eBook)
DOI 10.1007/978-3-662-46054-2

Library of Congress Control Number: 2015936685

Springer Heidelberg New York Dordrecht London

Printed on acid-free paper

Springer-Verlag GmbH Berlin Heidelberg is part of Springer Science+Business Media (www.springer.com)

Preface

With our growing energy needs and increasing environmental concern, there is considerable research interest in developing renewable energy sources as alternatives to the non-renewable fossil fuels as well as improving the technologies for energy conversion.

Solar energy is an inexhaustible and freely available energy source. More energy comes from the sun in 1 hr each day than is used by all humankind activities in one year. The challenge today is to capture and utilize solar energy for sustainable development on a grand scale. There are different manifestations of solar energy conversion, of which one relies fundamentally on chemistry for its scientific underpinnings, i.e., the conversion of light into electricity. Photovoltaic devices are available for decades for the direct conversion of sunlight into electrical energy. The technological promise of the characteristics of OLEDs puts them at the forefront of research in the past two decades. OLEDs possess a number of advantages over conventional dominant display devices, such as high luminous efficiency, high brightness and contrast, fast response time, wide viewing angle, low power consumption and light weight. These new technologies offer a virtually unlimited choice of colors and good potential of low manufacturing cost. While OLED displays can be fabricated on large area and flexible substrates, a stream of new OLED products has reached the marketplace. Therefore, the transformations of light into electricity (solar energy conversion) and electricity into light (light generation in light-emitting diode) are two important interrelated areas in energy conversion. Organometallic molecules and related compounds hold great promise as versatile functional materials for use in these light-electricity transformations.

In addition, development of new molecular systems for converting carbon dioxide to useful chemicals using solar light, i.e., photocatalytic CO_2 reduction systems, is gaining increasing attention for solving the energy shortage and global warming problems. Hydrogen has also attracted great interest because of its potential to serve as an energy storage medium, i.e., an energy carrier. There is great interest in using metal complexes to store hydrogen at higher densities and metal-based catalysts will certainly play major roles in its safe and efficient production and utilization.

As the research fields on energy conversion are growing rapidly and their impacts are both far-reaching and pervasive, other fundamental challenges still remain which would require further multidisciplinary studies. Therefore, to bring a focus on the recent research developments in this direction, a collection of chapters from leading scientists is presented in this book. Many aspects of the field by utilizing a vast number of organometallics and related molecules and their device tactics are covered, which provide readers with a good source of information in solving many of the critical issues on energy conversion. These include the investigation of new light-harvesting and OLED materials (by J.A. Gareth Williams, Etienne Baranoff, Tao Chen, Tsuyoshi Michinobu, Qiang Zhao, Jianzhang Zhao, Suning Wang, Di Liu, Zhen-Tao Yu and Shi-Jian Su), the studies on carbon dioxide reduction (by Shunichi Fukuzumi) and water oxidation (by Khurram Saleem Joya, Kwok-Yin Wong, Luca Gonsalvi, Qiang Xu and Torsten Beweries) as well as other related topics (Toshikazu Hirao and myself).

Clearly, the challenge confronting the twenty-first century is tied to energy crisis and it is encouraging to see that organometallic compounds and related materials are playing important roles in these frontier areas. What organometallic materials can do to address this challenge will have significant impacts on our society, and we are optimistic to witness more successful stories to come in the upcoming future.

Hong Kong, China Wai-Yeung Wong

Contents

Contributors

Toru Amaya Department of Applied Chemistry, Graduate School of Engineering, Osaka University, Suita, Osaka, Japan

Etienne Baranoff School of Chemistry, University of Birmingham, Birmingham, UK

Federica Bertini Consiglio Nazionale delle Ricerche, Istituto di Chimica dei Composti Organometallici (CNR-ICCOM), Firenze, Italy

Torsten Beweries Leibniz-Institut für Katalyse e.V. an der Universität Rostock, Rostock, Germany

Shuai Chang Department of Physics, The Chinese University of Hong Kong, Hong Kong, China

Dongcheng Chen State Key Laboratory of Luminescent Materials and Devices and Institute of Polymer Optoelectronic Materials and Devices, South China University of Technology, Guangzhou, China

Tao Chen Department of Physics, The Chinese University of Hong Kong, Hong Kong, China

Gemma R. Freeman Department of Chemistry, Durham University, Durham, UK

Shunichi Fukuzumi Department of Material and Life Science, Division of Advanced Science and Biotechnology, Graduate School of Engineering, Osaka University, ALCA, Japan Science and Technology Agency (JST), Suita, Osaka, Japan

Luca Gonsalvi Consiglio Nazionale delle Ricerche, Istituto di Chimica dei Composti Organometallici (CNR-ICCOM), Firenze, Italy

Antonella Guerriero Consiglio Nazionale delle Ricerche, Istituto di Chimica dei Composti Organometallici (CNR-ICCOM), Firenze, Italy

Jian He Department of Physics, The Chinese University of Hong Kong, Hong Kong, China

Toshikazu Hirao Department of Applied Chemistry, Graduate School of Engineering, Osaka University, Suita, Osaka, Japan

Cheuk-Lam Ho Department of Chemistry and Partner State Key Laboratory of Environmental and Biological Analysis, Institute of Molecular Functional Materials, Hong Kong Baptist University, Hong Kong, China

Wei Huang Key Laboratory for Organic Electronics and Information Displays and Institute of Advanced Materials, Jiangsu National Syngerstic Innovation Center for Advanced Materials (SICAM), Nanjing University of Posts and Telecommunications (NUPT), Nanjing, China

Key Laboratory of Flexible Electronics (KLOFE) and Institute of Advanced Materials (IAM), Jiangsu National Synergistic Innovation Center for Advanced Materials (SICAM), Nanjing Tech University (NanjingTech), Nanjing, China

Zachary M. Hudson Department of Chemistry, Queen's University, Kingston, ON, Canada

Khurram Saleem Joya Leiden Institute of Chemistry, Gorlaeus Laboratory, Leiden University, Leiden, The Netherlands

Division of Physical Sciences and Engineering, KAUST Catalysis Center (KCC), King Abdullah University of Science and Technology (KAUST), Thuwal, Saudi Arabia

Department of Chemistry, University of Engineering and Technology, Lahore, Punjab, Pakistan

Lawrence Yoon Suk Lee Department of Applied Biology and Chemical Technology, The Hong Kong Polytechnic University, Hong Kong SAR

Di Liu State Key Laboratory of Fine Chemicals, School of Chemistry, Dalian University of Technology, Dalian, China

Shujuan Liu Key Laboratory for Organic Electronics and Information Displays and Institute of Advanced Materials, Jiangsu National Syngerstic Innovation Center for Advanced Materials (SICAM), Nanjing University of Posts and Telecommunications (NUPT), Nanjing, China

Poulomi Majumdar State Key Laboratory of Fine Chemicals, School of Chemical Engineering, Dalian University of Technology, Dalian, China

Irene Mellone Consiglio Nazionale delle Ricerche, Istituto di Chimica dei Composti Organometallici (CNR-ICCOM), Firenze, Italy

Tsuyoshi Michinobu Department of Organic and Polymeric Materials, Tokyo Institute of Technology, Tokyo, Japan

Toshiyuki Moriuchi Department of Applied Chemistry, Graduate School of Engineering, Osaka University, Suita, Osaka, Japan

Shi-Jian Su State Key Laboratory of Luminescent Materials and Devices and Institute of Polymer Optoelectronic Materials and Devices, South China University of Technology, Guangzhou, China

Tomoyoshi Suenobu Department of Material and Life Science, Division of Advanced Science and Biotechnology, Graduate School of Engineering, Osaka University, ALCA, Japan Science and Technology Agency (JST), Suita, Osaka, Japan

Baohua Wang Department of Physics, The Chinese University of Hong Kong, Hong Kong, China

Suning Wang Department of Chemistry, Queen's University, Kingston, ON, Canada

Xiang Wang Department of Chemistry, Queen's University, Kingston, ON, Canada

J.A. Gareth Williams Department of Chemistry, Durham University, Durham, UK

Kwok-Yin Wong Department of Applied Biology and Chemical Technology, The Hong Kong Polytechnic University, Hong Kong SAR

Wai-Yeung Wong Department of Chemistry and Partner State Key Laboratory of Environmental and Biological Analysis, Institute of Molecular Functional Materials, Hong Kong Baptist University, Hong Kong, China

HKBU Institute of Research and Continuing Education, Hong Kong Baptist University, Shenzhen, China

Qiang Xu National Institute of Advanced Industrial Science and Technology (AIST), Ikeda, Osaka, Japan

Yusuke Yamada Department of Material and Life Science, Division of Advanced Science and Biotechnology, Graduate School of Engineering, Osaka University, ALCA, Japan Science and Technology Agency (JST), Suita, Osaka, Japan

Tianshe Yang Key Laboratory for Organic Electronics and Information Displays and Institute of Advanced Materials, Jiangsu National Syngerstic Innovation Center for Advanced Materials (SICAM), Nanjing University of Posts and Telecommunications (NUPT), Nanjing, China

Zhen-Tao Yu Collaborative Innovation Center of Advanced Microstructures, College of Engineering and Applied Sciences, Nanjing University, Nanjing, Jiangsu, China

Xinglin Zhang Key Laboratory for Organic Electronics and Information Displays and Institute of Advanced Materials, Jiangsu National Syngerstic Innovation Center for Advanced Materials (SICAM), Nanjing University of Posts and Telecommunications (NUPT), Nanjing, China

Key Laboratory of Flexible Electronics (KLOFE) and Institute of Advanced Materials (IAM), Jiangsu National Synergistic Innovation Center for Advanced Materials (SICAM), Nanjing Tech University (NanjingTech), Nanjing, China

Jianzhang Zhao State Key Laboratory of Fine Chemicals, School of Chemical Engineering, Dalian University of Technology, Dalian, China

Qiang Zhao Key Laboratory for Organic Electronics and Information Displays and Institute of Advanced Materials, Jiangsu National Syngerstic Innovation Center for Advanced Materials (SICAM), Nanjing University of Posts and Telecommunications (NUPT), Nanjing, China

Yanying Zhao National Institute of Advanced Industrial Science and Technology (AIST), Ikeda, Osaka, Japan

Department of Chemistry, Zhejiang Sci-Tech University, Hangzhou, China

About the Editor

Wai-Yeung Wong is currently Chair Professor and Head of the Department of Chemistry at Hong Kong Baptist University. He received his B.Sc. (1992) and Ph.D. (1995) degrees from the University of Hong Kong. After postdoctoral training with Prof. F. Albert Cotton at Texas A&M University in 1996 and Profs. The Lord Lewis (FRS) and Paul R. Raithby at the University of Cambridge in 1997, he joined Hong Kong Baptist University as an Assistant Professor in 1998, rising through the academic ranks to become Chair Professor in 2011 at the age of 40. His research interests lie in the areas of metallopolymers and metallo-organic molecules with energy functions and photofunctional properties. His research activities are documented in more than 440 scientific articles, 2 edited books, 16 book chapters and 2 US patents. Professor Wong was recently named in the list of Highly Cited Researchers 2014 in the Materials Science category published by Thomson Reuters. He becomes the first Chinese scientist to be presented with the Chemistry of the Transition Metals Award by the Royal Society of Chemistry in 2010. He has also won the FACS Distinguished Young Chemist Award from the Federation of Asian Chemical Societies in 2011, Ho Leung Ho Lee Foundation Prize for Scientific and Technological Innovation in 2012 and State Natural Science Award (Second-class) of China in 2013. Recently, he was awarded the Japanese Photochemistry Association Lectureship Award for Asian and Oceanian Photochemist (Eikohsha Award) in 2014. Professor Wong is currently the Regional Editor of *Journal of Organometallic Chemistry* and Associate Editor of *Journal of Materials Chemistry C*. At present, he is the Chairman of the Hong Kong Chemical Society.

Chapter 1
Organometallic Versus Organic Molecules for Energy Conversion in Organic Light-Emitting Diodes and Solar Cells

Cheuk-Lam Ho and Wai-Yeung Wong

Abstract With the rapid growth of population and urbanization, the energy demand is increasing annually. Due to the major problems concerning the rapid depleting nature of the extraction of fossil resources, energy conservation and transition to renewable energy supplies have been a hot topic worldwide. Organic light-emitting diodes and solar cells represent two important techniques to allow the efficient utilization of energy resources in energy-saving devices and exploration of using renewable energy in energy-producing devices. In this chapter, the importance of using organometallic and organic molecules in both areas of research is discussed.

Keywords Organic light-emitting diodes • Organic solar cells • Organometallic • Energy conversion • Fluorescence • Phosphorescence • Light harvesting

1.1 Introduction

To face the threat to the planet posed by climate change, we need to develop new technologies that can alter the way we generate and use electricity. Organic solar cells and organic light-emitting diodes (OLEDs) represent two sides of this coin, via the transformations of light into electricity and electricity into light. In recent years, these two interrelated areas have become increasingly important, and there is

---

C.-L. Ho
Department of Chemistry and Partner State Key Laboratory of Environmental and Biological Analysis, Institute of Molecular Functional Materials, Hong Kong Baptist University, Waterloo Road, Kowloon Tong, Hong Kong, China

W. Y. Wong (✉)
Department of Chemistry and Partner State Key Laboratory of Environmental and Biological Analysis, Institute of Molecular Functional Materials, Hong Kong Baptist University, Waterloo Road, Kowloon Tong, Hong Kong, China

HKBU Institute of Research and Continuing Education, Hong Kong Baptist University, Shenzhen Virtual University Park, Shenzhen 518057, China
e-mail: rwywong@hkbu.edu.hk

W.-Y. Wong (ed.), *Organometallics and Related Molecules for Energy Conversion*, Green Chemistry and Sustainable Technology, DOI 10.1007/978-3-662-46054-2_1

great interest in developing new active materials for organic solar cells and OLEDs to replace those expensive highly crystalline inorganic semiconductors [1, 2]. Semiconductive organic materials (both organometallic and organic compounds) have shown great promise in these two areas because of their ability to tailor the molecular properties and versatility of the production methods, so that their chemical and physical properties can be easily fine-tuned to fit particular applications and reduce the manufacturing process and cost. In this chapter, we will highlight the recent development on light-emitting and light-absorbing photoactive organometallic and organic compounds for OLEDs and organic solar cells.

1.2 Organic Light-Emitting Diodes

1.2.1 Fluorescence and Phosphorescence

The reports of Tang and VanSlyke and later of Burroughes on electroluminescence (EL) from thin organic films made of small molecular weight molecules and conducting polymers, respectively, opened up a new field of research in light-emitting materials science [3]. Over the past two decades, OLEDs have been widely studied, and great strides have been made in the development of new materials and device architectures for both display and solid-state lighting applications [4]. To have an efficient OLED, it is necessary to have balanced charge injection (equal numbers of holes and electrons from opposite electrodes) and transport and capture of all the injected charges to form excitons in the emissive layer and radiative decay of all the excitons. In the process of injecting charges, it is not possible to control the spins of the electrons, and hence, there is a certain probability that they form a singlet or triplet exciton when they meet on the photoactive material. Theoretically, the ratio of singlets to triplets is 1:3 for small molecules. If the material is fluorescent, only the singlets can emit light, while if it is phosphorescent, both singlets and triplets can be captured for light emission. Therefore, phosphorescent materials inherently have the potential to form the most efficient light-emitting devices. Currently, blue, green, and red OLEDs with excellent performance and high color purity have been achieved by employing phosphorescent complexes of iridium(III). It is believed that 100 % harvesting of the electrogenerated singlet and triplet excitons is caused by the strong spin–orbit coupling of heavy Ir(III) atom [5]. Their high photoluminescence efficiency, relatively short lifetimes, and flexible color tunability render them to be the most successful phosphorescent materials for OLED application to date [6–8].

Recent progress in the development of Ir(III) phosphors will be highlighted as follows. High external quantum efficiency (η_{ext}) greater than 30 % was demonstrated in blue phosphorescent OLEDs using **FIrpic** as the sky-blue emitter [9]. The current (η_L) and power efficiencies (η_P) of such device were improved to 53.6 cd/A and 50.6 lm/W, respectively, by adapting suitable high-triplet energy host materials. **Ir(ppy)$_3$** has been the most widely used green phosphor [10–14], for which the best device performance with a high η_P of 133 lm/W was reported by Kido and his coworkers [12]. The green OLEDs based on its heteroleptic counterpart **Ir (ppy)$_2$(acac)** were also reported to show attractive performance with η_{ext} of 23.7 % and η_P of 105 lm/W [15]. By using **Ir(tfmppy)$_2$(tfmtpip)** as the green phosphor, a maximum η_P and η_L of 113.23 lm/W and 115.39 cd/A, respectively, were achieved at a doping level of 5 wt% [16], in which the introduction of CF_3 moiety to the Ir(III) complex was believed to increase the electron mobility of the complex and thus enhance the overall efficiencies of the device. **Ir(piq)$_3$** is a well-known red phosphorescent emitter [17]. Deep red **Ir(piq)$_3$**-based OLEDs doped in host materials 1,3,5-tris(3-(carbazol-9-yl)phenyl)-benzene, 2,4,6-tris(3-(carbazol-9-yl)phenyl)-pyridine, or 2,4,6-tris(3-(carbazol-9-yl)phenyl)-pyrimidine showed superior efficiency and suppressed efficiency roll-off [18]. All the devices show very high η_{ext} of over 18 % and η_P of around 19 lm/W at low current density due to the bipolar nature of the host materials. For **Ir(piq)$_2$(acac)**, comparable OLED performance as **Ir (piq)$_3$** has been reported in the literature. A deep red phosphorescent **Ir(piq)$_2$(acac)**-based OLED hosted by bipolar triphenylamine/oxadiazole hybrid gave a η_{ext} up to 21.6 %, η_L of 15.9 cd/A, and η_P of 16.1 lm/W [19]. Generally, for both homoleptic and heteroleptic Ir(III) complexes, the cyclometalating C^N ligands play essential roles in determining the phosphorescent energy and the emission color of the complexes, although occasionally the ancillary ligand is also important. Accordingly, the typical strategy utilized to tune the emissive color of the iridium complexes is to incorporate different substituents in either C-related arenes or N-related heterocycles while fixing the C^N ligand skeleton. For example, incorporating different substituents at different positions of the 2-phenylpyridine ligand framework has resulted in a continuous tuning of the phosphorescence of their homoleptic iridium(III) complexes from sky blue (468 nm) to orange-red (595 nm) [20, 21].

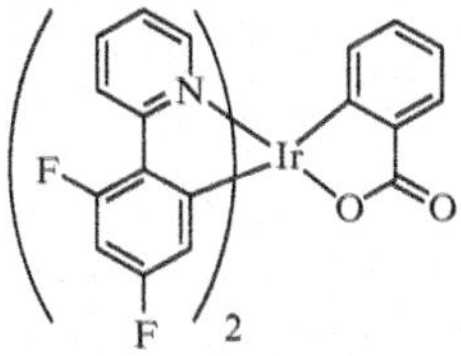

FIrpic 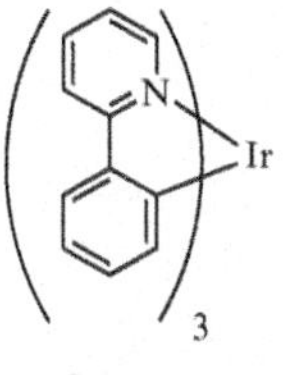 **Ir(ppy)$_3$** 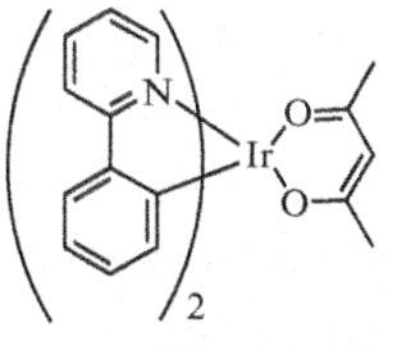 **Ir(ppy)$_2$(acac)**

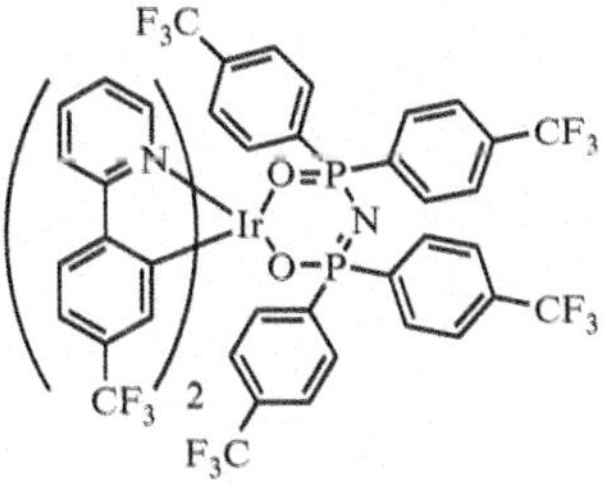

Ir(tfmppy)$_2$(tfmtpip) 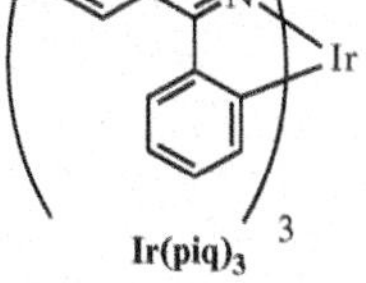 **Ir(piq)$_3$** 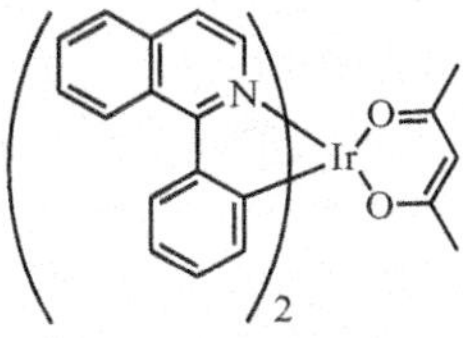 **Ir(piq)$_2$(acac)**

White OLED (WOLED) technology has attracted much attention because of its great potential in solid-state lighting and displays [22]. The first WOLEDs comprising small organic molecules were reported by Kido [23–25]. Later on, the incorporation of exciplex formation and emission from organic molecules may generally be used for the realization of more practical WOLEDs; however, the overall efficiency of such devices seems to be rather low. The only exception so far is the one reported by Tang et al., discussing WOLEDs based on a single emissive material **TPyPA**, where the emission originates from its singlet and exciplex states [26]. This device reached a maximum luminous efficacy of 9.0 lm/W with CIE coordinates of (0.31, 0.36). In 2006, Tsai and Jou reported a highly efficient fluorescent WOLEDs based on a mixed host/emitting layer doped with two fluorophores. By optimizing the blend layer of hole-transporting *N,N′*-Di (1-naphthyl)-*N,N′*-diphenyl-(1,1′-biphenyl)-4,4′-diamine (NPB) and electron-transporting 9,10-di-(2-naphthyl)-2-*t*-butyl-anthracene (TBADN) to a 1:1 ratio, a maximum luminous efficacy of 11.2 lm/W was obtained [27]. Yang et al. discussed a hybrid combination of simultaneous host and dopant emission, forming a two-color fully fluorescent white-emitting device. They mixed the red emitter **DCM** with the blue-emitting host bis[2-(2-hydroxyphenyl)pyridine]beryllium ($Bepp_2$); maximum values of 5.6 % in η_{ext}, 9.2 lm/W in η_P with color coordinates of (0.332, 0.336), and color rendering index (CRI) of 80 were achieved [28]. Many innovative device architectures have been devised using various material systems to achieve high-performance WOLEDs [29]. A large part of the research conducted to achieve white light used the fluorescence and phosphorescence hybrid WOLED (F-P hybrid WOLED), which can separately utilize the singlet excitons with a blue fluorescent emitter and the triplet excitons with green, orange, and red phosphorescent dopants. The main reason for this device concept is that blue phosphorescent emitters are poor in long-term stability as compared to blue fluorescent emitters [30]. Also, while blue phosphorescent materials call for host materials with even larger bandgaps, the operating voltage based on phosphorescent blue emitters will increase as the luminous efficacy decreases [31, 32]. Following this approach, the efficiency of single-emitting layer F-P hybrid WOLED based on blue organic fluorescent emitter and green/red Ir(III) phosphors showed up to a maximum η_P of 59.8 lm/W and a maximum η_{ext} of 24.7 %, as reported by Lee and Zhang in 2013 [33]. Among the various concepts for WOLEDs, by far, most effort has been spent on research dealing with devices based solely on phosphorescent materials. This is probably due to the fact that phosphors inherently offer internal efficiencies of unity, so that in general, the only remaining task in device engineering is the distribution of excitons to different emitters for white emission. In order to increase the color quality and luminous efficacy of phosphorescent OLEDs, three primary colors need to be employed (red, blue, and green, RBG). In 2004, D'Andrade, Holmes, and Forrest reported on the first three-color WOLED devices, based on **FIr6**, **Ir(ppy)$_3$** and **PQ2Ir** [34]. The highest efficiency of all phosphorescent RBG WOLEDs was recently reported by Lee et al. Efficiency values η_P of 45.9 lm/W and η_{ext} of 22 % were demonstrated without the use of any out-coupling enhancement techniques while still maintaining high CRI value of 84 [35]. Sasabe et al. also reported a highly efficient WOLED with a blue-emitting iridium(III) carbene complex **Ir(dbfmi)$_3$** and additional ultrathin layers of red and green emitters, giving a η_{ext} of 21.5 % and η_P of 43.3 lm/W at 1000 cd/m^2 with CRI of 80.2 and CIE of (0.43, 0.43) [36]. The highest CRI to date for a three-emitter system was reported by Chang et al. The emission of their optimized device is close to equi-energy white point and shows a very high CRI of 94 at 1,000 cd/m^2 with color coordinates of (0.322, 0.349) [37]. White light can also be obtained using two complementary colors. Most of the research in this field used the archetype phosphorescent blue-emitter **FIrpic** in connection with various emitters. Yellow-emitting complex **Ir-1**-based device produced high efficiencies of 75.9 cd/A, 48.2 lm/W, and 23 % with CIE of (0.46, 0.53), which represents the highest efficiencies for yellow OLEDs up to now [38]. Furthermore, when **Ir-1** was used to fabricate two-element WOLEDs in combination with **FIrpic**, high efficiencies of 57.9 cd/A and 21.9 % were achieved. A two-element WOLED with a peak η_L of 68 cd/A was also demonstrated using a related derivative as an orange emitter [22]. Another two-color WOLEDs with good performance based on blue-emitting **FIrpic** and **PQ2Ir** as a complementary red emitter was reported by Su and coworkers, in which the η_{ext} of 25 % and η_P of 46 lm/W were achieved at 1,000 cd/m^2 [30]. One of the drawbacks of two-element WOLEDs is the lack of green emission in the spectrum which results in poor color quality, and the CRI typically is limited to values of ~70 [30]. The highest color quality of two-color phosphorescent WOLEDs up to now was obtained by using an iridium complex **Ir(dfbppy)(fbppz)$_2$** as a blue phosphorescent emitter combined with a red-emitting osmium heavy metal complex **Os(bptz)$_2$(dppee)** [39]. The blue emitter with PL maximum at 450 nm and strong vibronic side bands at approximately 480 and 520 nm can alleviate the lack of green emission, resulting in a high CRI = 79. At 100 cd/m^2, a white-emitting device based on these emitters reached η_{ext} of 6.8 % and η_P of 0.0 lm/W with color coordinates of (0.324, 0.343), closely matching the equi-energy white point. One should note that with its rather light-blue emission corresponding to CIE coordinates of (0.17, 0.34), the **FIrpic** spectrum is typically mixed with the emission of a red emitter [30, 40]. If yellow phosphorescent emitters were applied in a two-color approach, deep-blue emitters with CIE coordinates ($x < 0.2$, $y < 0.2$) should be employed in order to get a reasonable white emission [41].

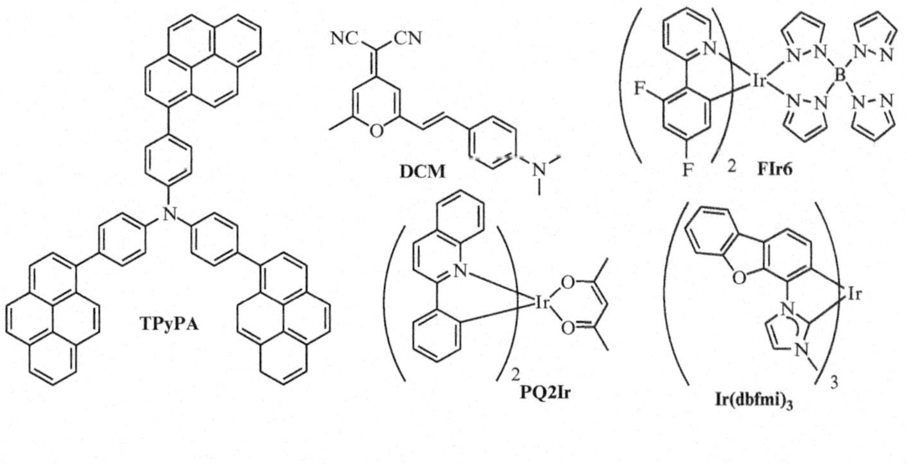

DCM

FIr6

TPyPA

PQ2Ir

Ir(dbfmi)$_3$

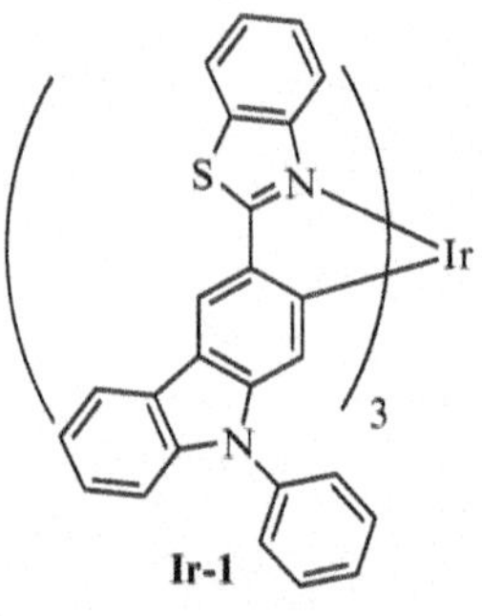

Ir-1

While efficient phosphorescent OLEDs have typically employed Ir-based emitters, recent reports of nearly 100 % internal quantum efficiency for blue and green emission in devices utilizing Pt(II) emitters demonstrate their usefulness in the future commercialization of OLEDs [42, 43]. **Pt-1** bearing tetradentate O^N^C^N ligands displays high emission quantum efficiency of 90 % and is an excellent green phosphorescent dopant for OLEDs with excellent efficiency of 66.7 cd/A and low efficiency roll-off [43]. The blue-emitting device employing **Pt-2** demonstrated a peak η_{ext} of 26.3 % and CIE coordinates of (0.12, 0.24) [44]. The excimer formation from **Pt-2** resulted in white emission from OLEDs which exhibited a peak η_{ext} of 25.7 % with a CRI of 70. The stabilities of both devices were proven by their impressive device operational lifetimes. From the device-engineering point of view, it is always desirable to simplify the OLED structure. Efficient electrophosphorescent excimer emission at a longer wavelength from planar Pt(II) complexes allows simplification of the architecture of WOLED. Cocchi et al. introduced Pt (II) complex, **PtL^{22}Cl**, which emits excimeric phosphorescence at ca. 650 nm; in WOLEDs, the η_{ext} reached 16.6 % with η_P of 9.6 lm/W at 500 cd/m^2. The corresponding CIE coordinates are (0.42, 0.38) [45]. The highest CRI based on the monomer excimer approach was reported by Zhou et al. using **Pt-Ge** doped into CBP host material [46]. At the emitter concentration of 10 wt% for **Pt-Ge**, the CIE coordinates of (0.354, 0.360) were obtained with a very high CRI of 97. The corresponding peak η_{ext} value is 4.13 %.

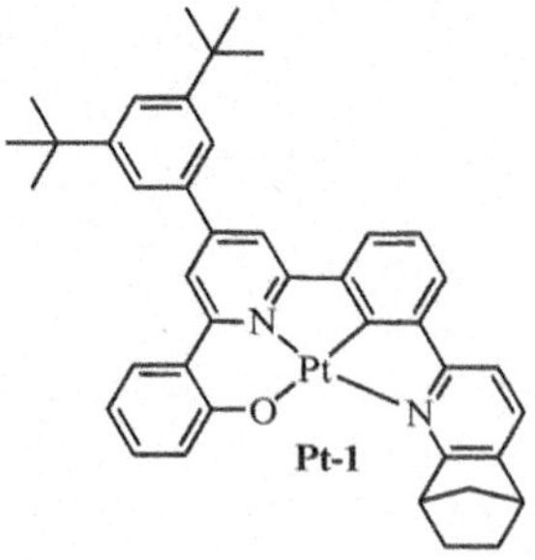

Pt-1

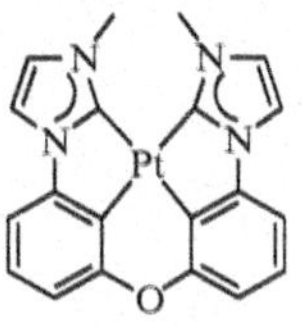

Pt-2

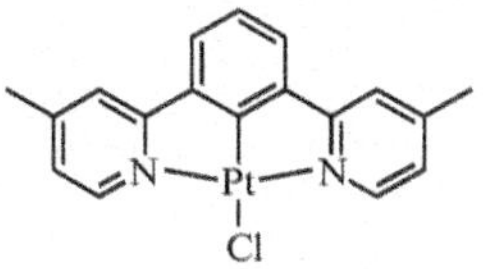

PtL²²Cl 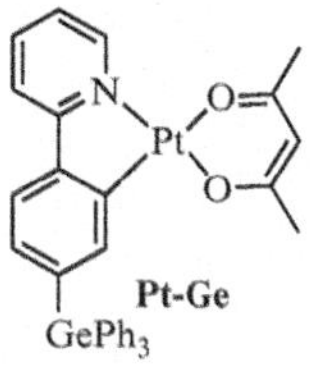 **Pt-Ge**

As compared to Ir(III) and Pt(II) complexes, Os(II) phosphors are less explored in the literature. In comparison, Os(II) complexes may gain certain advantages, such as the reduction of radiative lifetime and hence a possibility of higher luminescent efficiency. A shorter radiative decay lifetime may improve the OLED device efficiencies by minimizing the unwanted triplet–triplet annihilation that occurs at a higher operating current density [47–49]. The EL properties of Os (II) complexes were first reported by Ma and coworkers, in which a stable and uniform red EL emission was observed when the voltage was applied to the device [50]. The maximum η_L of **Os-1**-based device reached 7.0 cd/A with CIE coordinates (0.650, 0.347) [51]. Until 2013, the efficiency of red OLEDs based on Os (II) complexes is up to 9.8 % by using tetradentate bis(pyridylpyrazolate) chelates (**Os-2**) [52]. This OLED exhibits saturated red emission with a maximum brightness of 19,540 cd/m^2 at 11.6 V. Attempts have been made by employing orange-emitting **Os-3** in the fabrication of WOLEDs in combination with blue emitters. By doping with a blue light emitting amino-substituted distyrylarylene fluorescent dye, the device exhibited an intense white emission having CIE coordinates of (0.33, 0.34), a peak η_{ext} of 6.12 %, η_L of 13.2 cd/A, and a maximum brightness of 11,306 cd/m^2 [53]. The efficiencies of WOLED based on **Os-3** climbed up to 40.4 cd/A and 25.6 lm/W when it was doped with blue phosphor **FIrpic** [54]. Shu, Chi, and Chou synthesized a series of electrophosphorescent copolymers, in which a red-emitting Os(II) complex was embedded in the backbone of polyfluorene, giving a highly efficient PLED with maximum η_{ext} of 18.0 % and maximum brightness of 38,000 cd/m^2 with an emission centered at 618 nm [55]. Furthermore, through covalent attachment of a low concentration of a green-emitting benzothiadiazole unit in the same polymers, efficient WOLED with η_L of 10.7 cd/A and η_{ext} of 5.4 % was achieved.

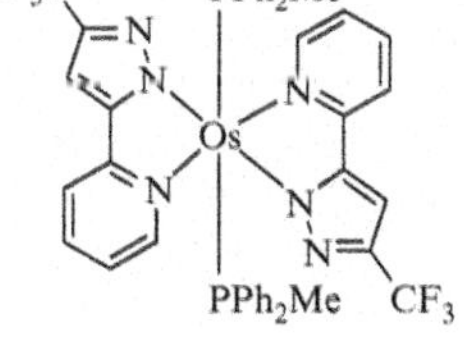

Os-1 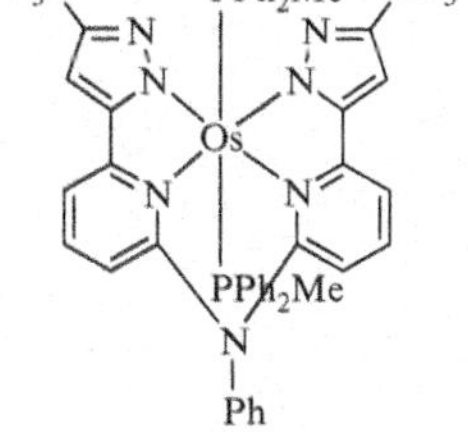 **Os-2**

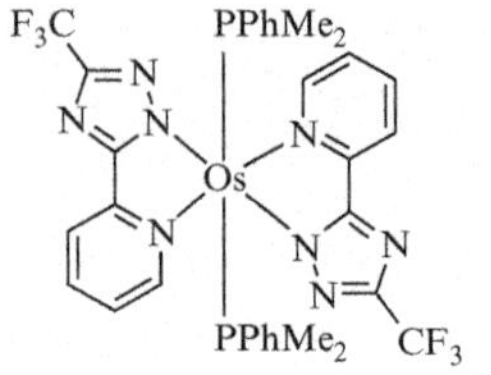

Os-3

1.2.2 *Delayed Fluorescence*

Very recently, efficient OLEDs can be developed by using the idea of thermally activated delayed fluorescence (TADF) from organic molecules to achieve device performance comparable to that of phosphorescent OLEDs and much higher than conventional fluorescent-based OLEDs. TADF originates from the contributions of triplet excitons, which are accessed through efficient upconversion from triplet excited state (T_1) to singlet excited state (S_1) by thermal activation. To obtain efficient TADF, a small energy gap between the S_1 and T_1 states is required, which is realized in molecules with a small overlap between their highest occupied molecular orbital (HOMO) and lowest unoccupied molecular orbital (LUMO). A series of highly efficient TADF emitters based on carbazolyl dicyanobenzene was reported by Uoyama [56] and Nakanotani [57]. Their emission wavelengths were easy to tune by alternating the electron-donating ability of the peripheral groups, which can be achieved by changing the number of carbazolyl groups or introducing relevant substituents. For the green OLED based on **TADF-2**, a η_{ext} of 19.3 % was achieved, while the orange (**TADF-3**) and sky-blue (**TADF-1**) OLEDs had η_{ext} of 11.2 % and 8.0 %, respectively. In 2013, the η_{ext} of the device based on sky-blue-emitting **TADF-1** reached up to 13.6 % [58]. OLED based on green-emitting **TADF-4** exhibited comparable efficiencies of $\eta_L = 50.0$ cd/A, $\eta_{ext} = 17.0$ %, and η_P of 35.7 lm/W. An orange-red OLED incorporating **TADF-5** exhibited high EL performance with a maximum η_{ext} of 17.5 %, a maximum η_P of 22.1 lm/W, and a maximum η_L of 25.9 cd/A. Relative to fluorescent materials, TADF materials have greater potential for OLED application in the upcoming future.

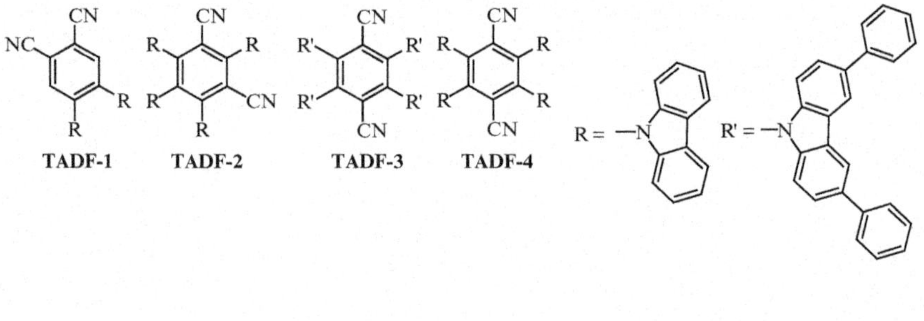

TADF-1 **TADF-2** **TADF-3** **TADF-4**

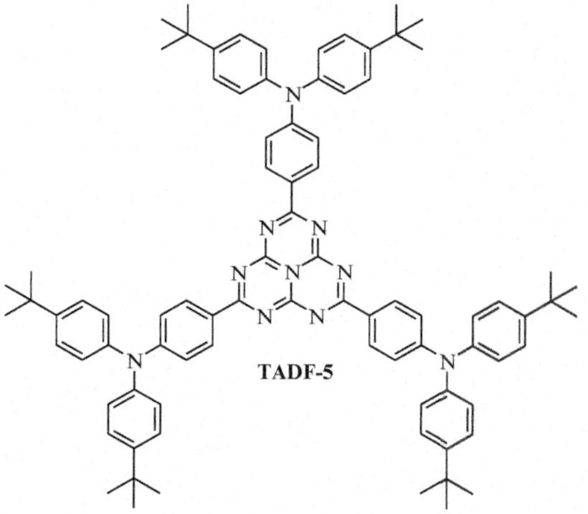

TADF-5

1.3 Photovoltaic Solar Cells

Photovoltaic devices act in the opposite way to OLEDs, in which the light is absorbed, the formed exciton is dissociated, and then the separated charges have to migrate to the electrodes without being quenched to generate electricity. The ideal material must have a high absorption coefficient and be able to absorb light with wavelengths that correspond to the solar absorption profile. Two types of device architectures are generally used, namely, dye-sensitized solar cells (DSSCs) and bulk heterojunction solar cells (BHJSCs) [59, 60].

DSSCs are photoelectrochemical devices that rely on the sensitization of a wide bandgap n-type semiconductor by a light-harvesting molecule or particle to generate electricity from incident sunlight. In DSSCs, a dye is absorbed onto an inorganic semiconductor such as TiO_2. On excitation, the electron from the exciton formed on the dye is transferred to the inorganic semiconductor and then hops to the electrode and travels around the circuit. The oxidized dye is then reduced through an electrolyte and ready for another excitation. The performance of DSSCs highly depends on the molecular structure of the photosensitizers. Nowadays, the majority of successful photosensitizers in DSSC applications are constituted by transition metal Ru(II) complexes since the pioneering report from O'Regan and Grätzel [61]. One of the first and well-known Ru(II) dyes reported by Grätzel is **N3** complex $[Ru(dcbpy)_2(NCS)_2]$ (dcbpy = 4,4′-dicarboxy-2,2′-bipyridine), which is a rare example of a molecule satisfying several criteria that an efficient sensitizer has to fulfill [62–68]. These include broad and intense absorption profile that arises from a series of MLCT transitions; oxidation and reduction potentials which are ideally positioned for dye generation and electron injection into TiO_2; and strong electronic coupling between its excited state and the semiconductor conduction band and high chemical stability. DSSCs based on **N3** gave an overall efficiency over 10 % and a certified stability of ten years [69]. By changing the number of acidic protons of **N3**, **N719** renders an impressive η of 11.2 % [70]. Efficient performance has also been predicted for **black dye** [Ru(tcterpy)$(NCS)_3$] (tcterpy = 4,4′,4″-tricarboxy-2,2′:6′,2″-terpyridine), in which the absorption of this dye is superior to those with **N3** especially in the near-infrared region which results in an efficiency of 10.4 % [71]. Many research groups have attempted to modify the structures of coordinated chromophoric and ancillary ligands with the goal of improving the photovoltaic performance, though only in few cases, the results have been better than those observed with the **N3** dye. By replacing one of the dcbpy ligands in **N3** with a modified 2,2′-bipyridine ligand in order to block the electrolyte from interacting with the surface and absorb more light, a series of dyes such as **C101** and **C106** were designed [72]. Here we only highlight those champion dyes that hit the efficiencies over 11 %. **C101** features an extended conjugation using thiophenes and long alkyl chain to prevent interfacial recombination, resulting in an efficiency of 11 % with an attractive FF of 0.785 [73]. By simply modifying the alkyl chain in **C101** using thiophene fragment, a very encouraging η of 11.4 % was achieved for **C106** [74]. It was proved that **C106** is more applicable to make efficient solar cells with very thin photoactive layers in comparison to its counterpart **C101**. This is mainly attributed to the merit of enhancing the optical absorptivity of stained titania films with **C106**, especially in the weak absorption red region as well as the blue region. The long-term photochemical stability of this dye has also been demonstrated due to the alkylthiophene fragment. Placement of extra thiophene rings in **CYC-B11** increases the molar extinction coefficients, and hence, the J_{sc} value (20.05 mA/cm^2) was enhanced as compared to **C106**, resulting in a higher η of 11.5 % [75]. While there are a number of fundamentally new approaches to dye design that have recently emerged based on the paradigmatic **N3**, one that has gained traction is the substitution of the monodentate NCS− ligands with anionic cyclometalating C^N ligands [76–78]. These chelating ligands have the potential to circumvent degradation pathways and optimize the light-harvesting capacity of DSSC dye [79–81]. An important finding for this approach was provided in 2009 when Grätzel and his coworkers reported a high DSSC efficiency of 10.1 % for **YE05**, which is the highest efficiency for a photosensitizer devoid of NCS group [82]. The high efficiency is ascribed to the suitable reduction potential for **YE05** that results in a larger thermodynamic driving force for dye regeneration. **Ru-1** with strong electron-withdrawing CF_3 groups installed on the phenyl ring of ppy and the two COOH groups on one of the dcbpy ligands being replaced by 2-hexylthiophene units can extend the absorption spectrum down to 600 nm. This paradigm provides the unique opportunity to control the thermodynamic position of the HOMO energy level and a way of manipulating the thermodynamic driving force for the regeneration of the photooxidized Ru(II) dye. DSSCs prepared with **Ru-1** produced a cell efficiency of 7.3 % at full sun [76]. These results may provide an advantage for commercialization because Ru(II) dyes without labile Ru−NCS bonds have potential to increase the overall stability of DSSC devices.

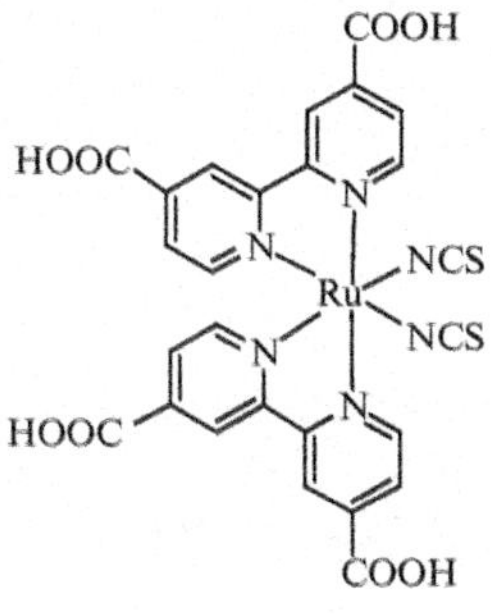

N3 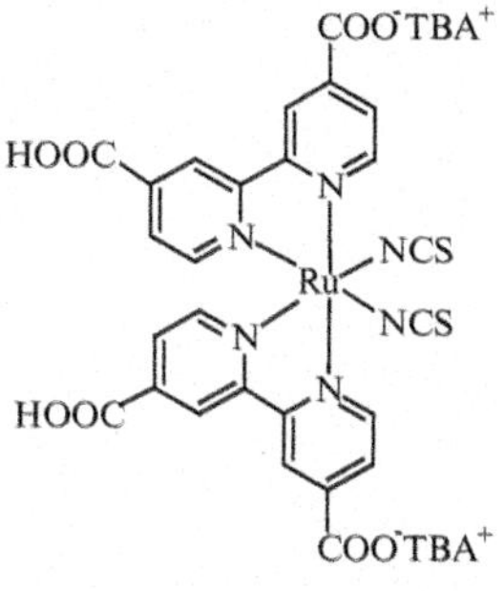 **N719** 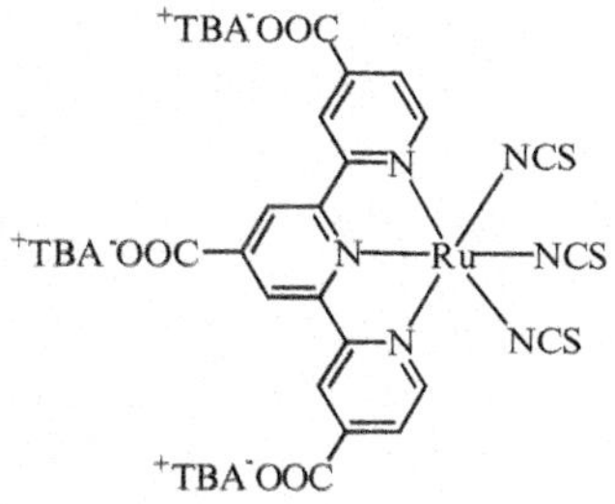 **Black dye**

TBA+ = tetrabutylammonium ion

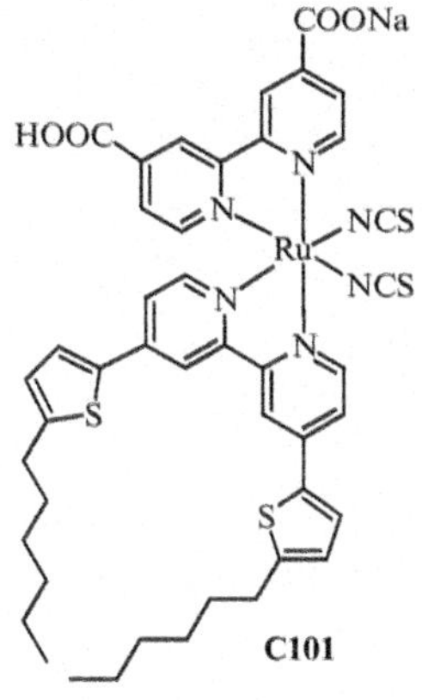

C101 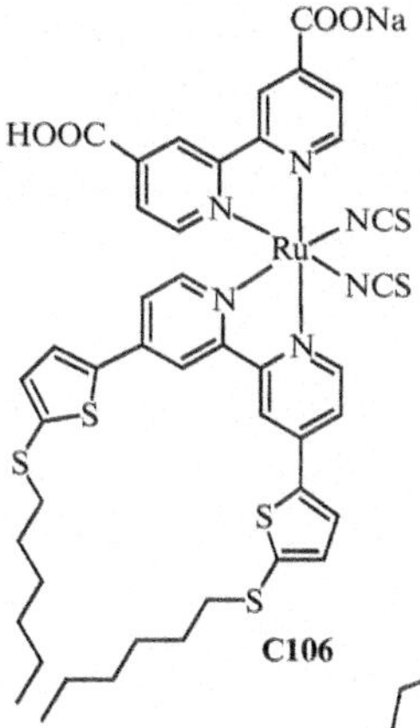 **C106** 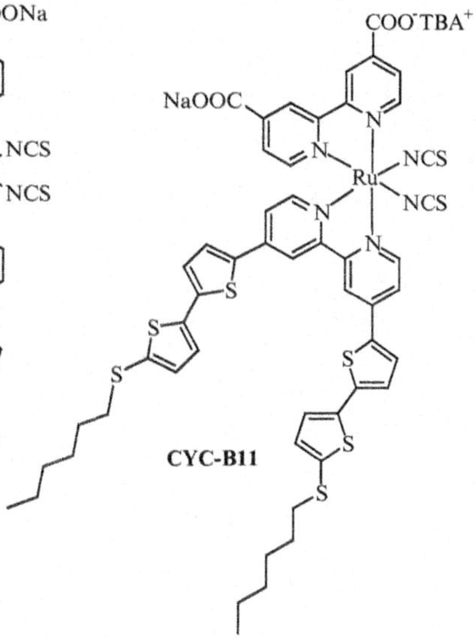 **CYC-B11**

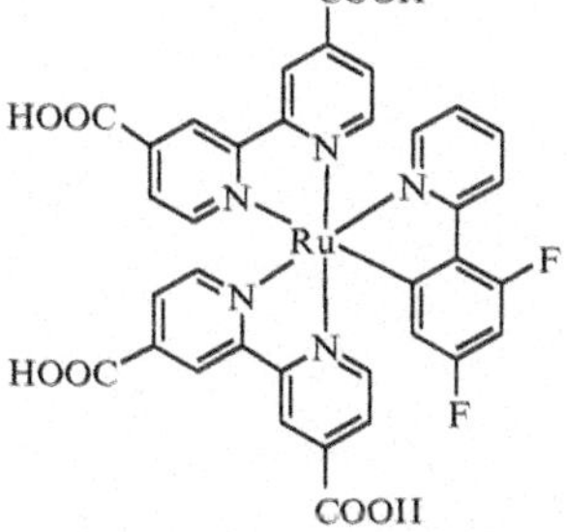

YE05 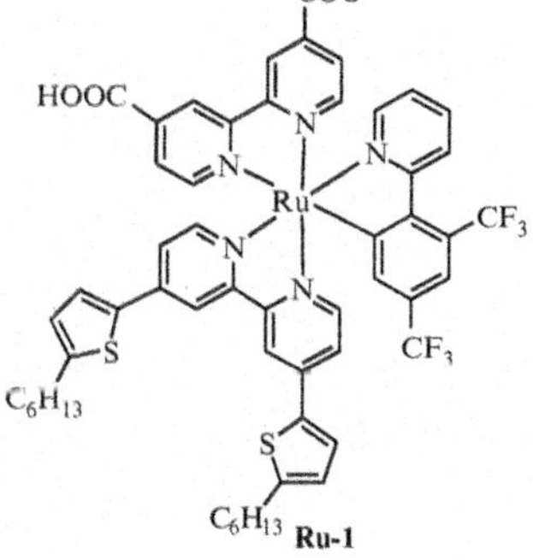 **Ru-1**

In addition, for optimizing Ru(II) photosensitizers, it has been reported that upon introduction of 4,4′,4″-tricarboxy-2,2′:6,2″-terpyridine, the lowest energy transition could be extended toward the near-IR region, thus affording the panchromatic sensitizer known as black dye. Chi and coworkers reported Ru(II) complexes **TF-2**–**TF-4** bearing tridentate anionic ligand and 2,6-bis(5-pyrazolyl)pyridine ligand in the DSSC with efficiencies in excess of 10 % [78]. Thiophene derivatives were tethered to the central pyridyl group in an attempt to increase the light-harvesting capability. The best cell showed a η as high as 10.7 % for **TF-3**, with $J_{sc} = 21.39$ mA/cm^2, $V_{oc} = 760$ mV, and FF = 0.66. These dyes with multidentate coordination improve the long-term stability, as evidenced by the great lifespan of solar cells fabricated with these dyes. This result completely rules out the need for thiocyanate ancillary ligands. Another mixed dentate system with efficiency exceeding 10 % was recently reported for **JK-206** [83]. The high efficiency of 10.39 % and excellent stability may be attributed to the intrinsic stability of the cyclometalated ruthenium(II) complex with the C^N^N ligand and the broad and red-shifted absorption property of the cyclometalated complex.

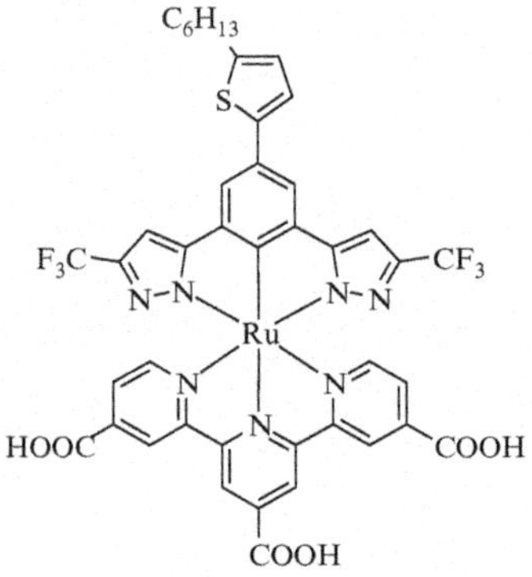

TF-2 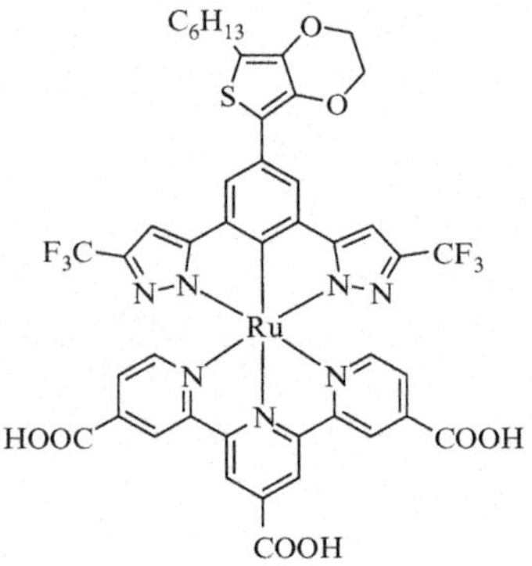 **TF-3**

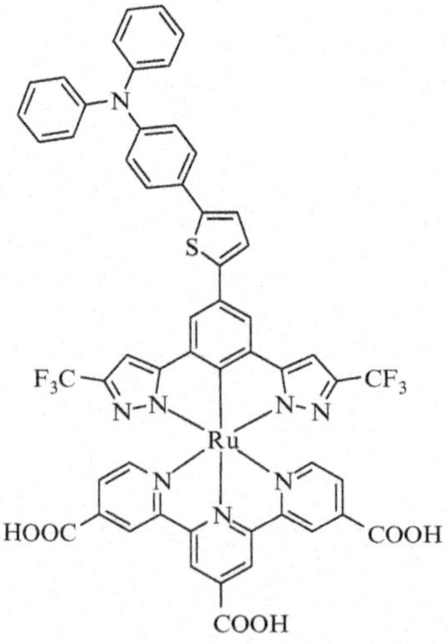

TF-4 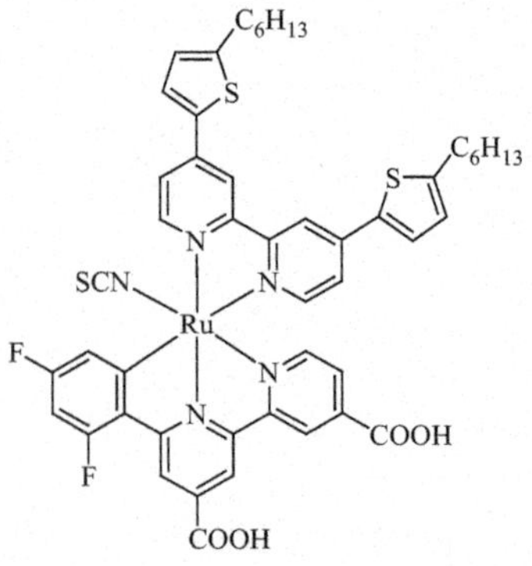 **JK-206**

Although record efficiency and stability have been achieved with Ru(II)-based photosensitizers, the high cost and scarcity of ruthenium have necessitated the consideration of other options. Porphyrins are an obvious choice as a class of simple, robust, efficient, and economically viable photosensitizers due to their well-known role as light absorbers and charge separators in natural photosynthetic systems. Recently, Zn(II) porphyrins with a push–pull system (D-π-A) have become popular because of their ease of modulation on electron-donating and anchoring sites, through which their optical, electrochemical, and photochemical properties can be tuned to afford high efficiencies. A very exciting accomplishment by using Zn(II) porphyrin **YD2** [84] as the photosensitizer in DSSCs was reported by Bessho et al.. It is the first champion dye not to contain Ru(II) ion that can hit the 10 % benchmark in DSSCs. **YD2** possesses high molar extinction coefficient because of the judicious installation of a light-harvesting donor group juxtaposed to the acceptor group to promote charge transfer toward the surface upon light absorption via intramolecular charge transfer transitions. A further elaboration of this dye with **YD2-o-C8** in the DSSC application has resulted in an efficiency over 12 %, which currently stands as the world record for the DSSC efficiency [85]. The IPCE of **YD2-o-C8** was approaching 90 % in the 400–700 nm range together with high V_{oc} (~1 V). The unprecedented efficiency of this dye may arise from the suppression of recombination and the reduction of the over potential for dye regeneration.

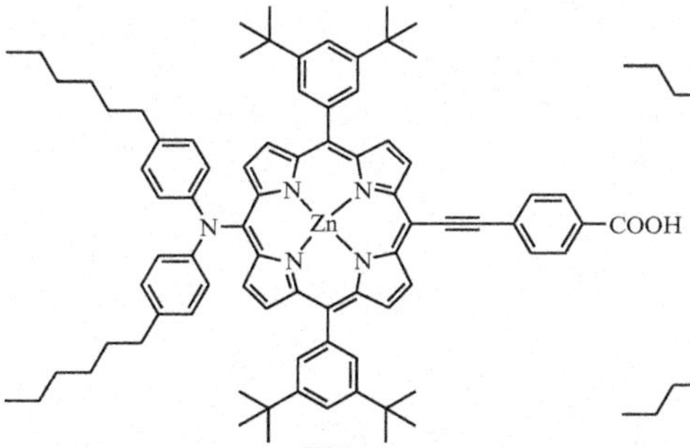

YD2 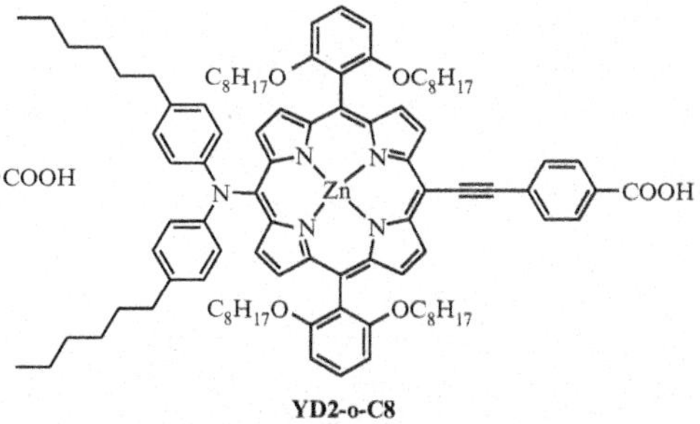 **YD2-o-C8**

An increasing attention is devoted to metal-free donor–acceptor organic dyes in DSSC application due to their advantageous features, especially their high molar extinction coefficients and tunable absorptions achieved through variation of the molecular structure. In general, the photoconversion efficiencies observed with organic dyes are lower as compared to the Ru(II)-based photosensitizers; however, their efficiencies and performances can be improved by the proper selection or tuning of the molecular components. A major breakthrough in the development of DSSC based on organic photosensitizers came with the incorporation of dyes with a D-π-A structure which facilitates photoinduced charge separation. Many molecular compounds such as fluorene, coumarin, indoline, cyanine, carbazole, perylene, triarylamine, etc., have been employed as donor in organic dye motif. But, the most successful and common donor units are represented by functionalized triarylamine moieties [86] because they have been shown to be excellent light harvester with strong electron-donating ability and hole-transporting properties

[87]. The simple architect of **TA-St-CA** showed a η of 9.1 % [88]. The most efficient metal-free organic dye reported up to now is the phenyldihexyloxy-substituted triphenylamine **Y123** prepared by Grätzel and his coworkers [89], which exhibits a η of 10.3 % together with a cobalt redox shuttle. Slowing down the interface charge recombination by its bulky phenyldihexyloxy groups and the high electron-donating ability by triarylamine unit both contribute to the good performance of the devices. By optimizing the conjugated bridge based on the substituted triphenylamine dye, Wang et al. synthesized **C219** which gave a η of 10.3 % in the DSSCs with an iodine/iodide redox shuttle [90]. The introduction of multiple electron-donating substituents at the arylamine core to form D-D-π-A structure could lead to a bathochromic shift of the absorption profile, an enhanced absorptivity, and better photochemical and thermal stability over D-π-A framework and will help suppress the recombination due to extended delocalization of the radical cation. Chio and coworkers have reported an efficiency of 9.1 % for the fluorene-incorporated triarylamine **JK113** using a liquid-based electrolyte. Excellent stability under light soaking was shown if solvent-free ionic liquid-based electrolyte was employed based on this dye [91]. Indoline dyes are one of the most efficient types of organic dyes due to their high absorption coefficients that result in high photocurrent. The more powerful electron-donating capacity of indoline also leads to a bathochromic shift of the absorption spectra as compared to the triarylamine-substituted one [92]. In combination with rhodamine network with a long alkyl chain in **D205**, the η was increased up to 9.52 % [93]. However, this dye easily suffers from desorption from TiO_2 and results in poor stability. Conjugated spacer plays a key role in the final performance of the dye. By expansion of the π-conjugation length using thiophene units, enhanced spectral response and charge transporting ability were obtained. Dye **C229** consisting of a 4,4-dihexyl-4*H*-cyclopenta[2,1-*b*:3,4-*b*′]dithiophene unit as the linker with substituted alkyl chain has proved to be very efficient. A η of 9.4 % was obtained for **C229**-based cells in combination with a cobalt-based redox shuttle [94]. Another strategy to improve the extinction coefficient to get a panchromatic light harvesting is to increase the planarity of π-bridged systems. **C217** consisting of a binary π-conjugated spacer leads to a significant increase of the absorption spectrum of the photosensitizer to give a η of 9.8 %. The 3,4-ethylenedioxythiophene unit connected to the arylamine donor lifts the energy of the HOMO, and the thienothiophene unit conjugated with the acceptor leads to a suitable LUMO energy. Additionally, this dye illustrates an excellent stability and high η with a solvent-free electrolyte [95]. Incorporation of a low bandgap chromophore unit into the dye, which is expected to act as an electron trap to separate charges and facilitate migration to the final acceptor, would provide better long-term stability under light soaking of the sensitizer and often leads to panchromatic absorption. Zu et al. obtained a η of 9.04 % for high stability indoline **WS9** which incorporates a benzothiazole unit in the conjugated space [96].

While the parts above only summarize the organic dyes with efficiencies hitting up to 9 %, there are a number of dyes reported to date, and there is still room for further improvement by judicious molecular design. However, it should be noted that it is not always clear what will be the most appropriate strategy to follow in the molecular engineering of organic dyes in order to obtain competitive efficiency.

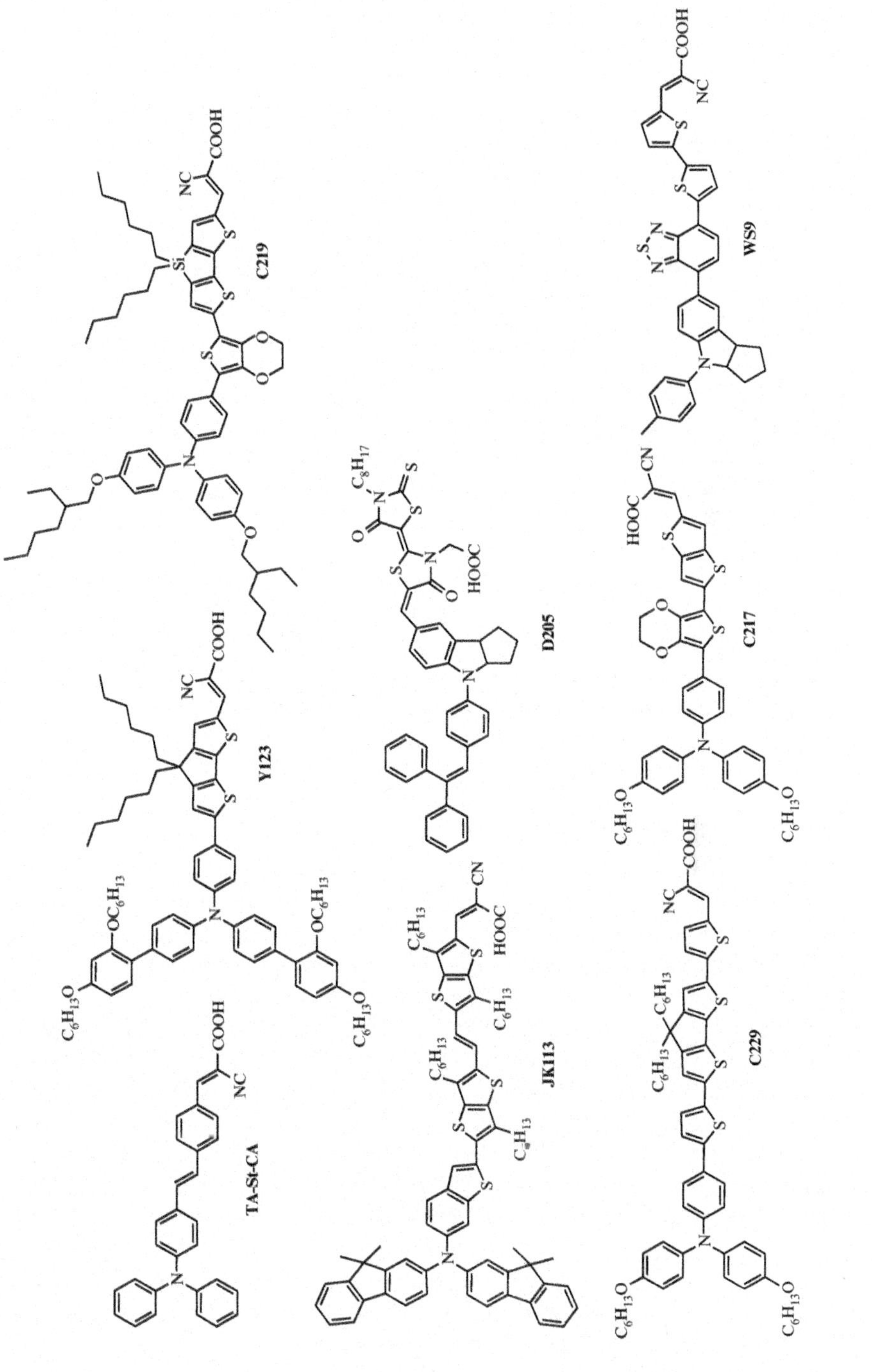

TA-St-CA

Y123

C219

JK113

D205

C229

C217

WS9

In a BHJSC, instead of having layered device structure, the materials with different electron and hole affinities are blended. When an exciton is formed in the hole-transporting material, the excited electrons hop onto the high electron-affinity material. The two charges then have to move through the layer by hopping between regions of their respective materials to the electrodes. The fundamental knowledge of BHJSCs was obtained through a long journey of research. The first successful photovoltaic device was reported as early as 1986 by Tang et al. using a bilayer structure of p-type copper phthalocyanine and n-type perylene diimide derivative [97]. This charge separation at the donor–acceptor interface was found to be very efficient, and an impressive η of ~1 % and a high FF of 65 % were demonstrated. Since the first report of electron transfer from polymer to fullerene by Heeger et al. in the early 1990s [98], much efficiency improvement was achieved by using the bulk heterojunction (BHJ) structure in the solar cell. Since then, the research on BHJSC is mainly focused on developing novel organic semiconducting materials, including polymers and small molecules. Conjugated polymer-based electron donor materials are the most studied materials to date for the BHJSCs. Among them, poly(3-hexylthiophene) (**P3HT**) is the most commonly used material due to the advantages of easy synthesis, high charge carrier mobility, good processability, etc. [99]. Yoshino et al. reported photoconductivity enhancement when polythiophene is blended with C_{60} buckyball [100]. After modifying the morphology and device architecture by numerous research group, efficiencies of BHJSCs based on **P3HT** can now go up to 7 % with external quantum efficiency of around 70 % [101–105]. To address the main issue of large bandgap (>1.9 eV) with **P3HT** and high HOMO level, which lead to insufficient NIR photon absorption and low V_{oc} of BHJSCs, numerous new organic polymers have been designed for photovoltaic applications. The main strategies for designing new materials are to harvest a greater part of the solar spectrum and provide as high V_{oc} as possible, narrow the bandgap, optimize the HOMO and LUMO levels, and ensure good charge mobility. The most successful strategy for harvesting more photons through tuning of the energy levels of conjugated polymers involves the incorporation of electron push–pull molecular units in the conjugated main chain. Alternating push and pull units allows internal charge transfer process along the conjugated chain to increase the effective resonance length of the π electrons, leading to smaller bandgaps as a result of facilitated π electron delocalization. Nowadays, the efficiencies of polymer-based BHJSCs are generally over 7 % in single-junction device architecture. For example, **PDTS-TPD** with bandgap of 1.73 eV gave a η of 7.3 % [106]. With further optimization on this structure by Reynolds, the polymer **PDTG-TPD** shows similar performance of 7.3 % [107]. Using an inverted device structure and modified Zn electrode, an enhanced η of 8.1 % was recorded on a **PDTG-TPD**-based device [108]. One of the most effective push units with a planar structure is benzo[1,2-*b*:4,5-*b*′]dithiophene (BDT). Members of push–pull polymers incorporating BDT units display low bandgaps ($E_g < 1.8$ eV) and low HOMO energy levels, resulting in good photovoltaic performance [109–111]. A η of around 8.2 % was achieved in **PBDTT-DPP**-based device with a tandem configuration [111]. **PBDTT-SeDPP** showed a broader photoresponse and demonstrated excellent photovoltaic performance in single-junction device with η of 7.2 % [112]. When this polymer was applied into a tandem device, a η of 9.5 % was achieved with improved J_{sc}. Very recently, linear side chains in benzo[1,2-b:4,5-b′] dithiophene-thieno[3,4-c]pyrrole-4,6-dione polymer **PBDTTPD** exhibit good thin film morphology which benefits solar cell application [113]. Its η reached 8.5 % with $V_{oc} = 0.97$ eV, $J_{sc} = 12.6$ mA/cm^2, and FF = 0.7 in the standard device structure. A certified η of 9.2 % for **PTB7**-based device with inverted structure was reported by Wu in 2012 [114]. Because of a remarkably improved J_{sc} of 17 mA/cm^2, the inverted polymer solar cells exhibit a superior overall device performance when compared to regular devices, as well as good ambient stability. Up to now, the highest η for a single-junction cell is provided by Mitsubishi Chemical with a record of 10 % that has been certified by the National Renewable Energy Laboratory (NREL), but no detailed information on either the active layer composition or the device structure was given [115]. The highest certified efficiency record on tandem polymer solar cells was obtained with **P3HT** and **PDTP-DFBT**, which produced a η of 10.6 % [116]. In addition to the conventional regular and tandem devices, there are some attempts to make homogeneous tandem solar cells. A η of 9.64 % in a triple junction solar cell was achieved by applying a wide bandgap polymer and two identical low-bandgap-based sub-cells [117].

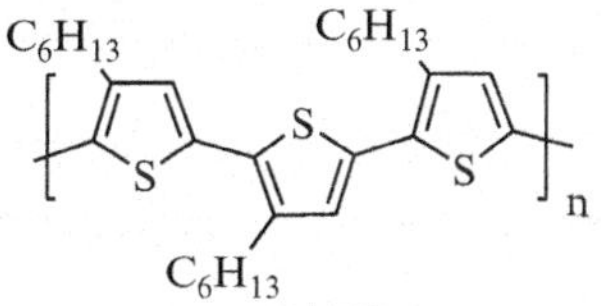

P3HT

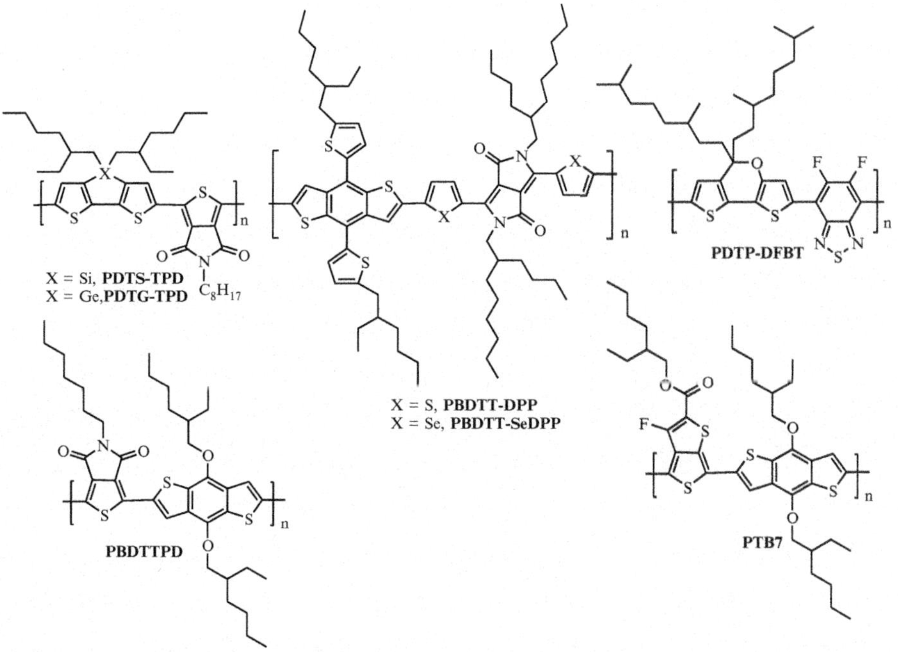

X = Si, **PDTS-TPD**
X = Ge, **PDTG-TPD**

X = S, **PBDTT-DPP**
X = Se, **PBDTT-SeDPP**

PDTP-DFBT

PBDTTPD

PTB7

As an alternative to the organic polymer, solution-processed conjugated organic small-molecule-based solar cells have attracted much attention in the past few years [118, 119]. Compared to the polymeric counterpart, small molecules are expected to have higher molecular precision relative to the statistically determined nature of synthetic polymers and less batch-to-batch variations [120, 121]. Early efforts on solution-processed small-molecule solar cells showed that the efficiency was limited by the low photocurrent and FF, and the initial η were only 1–3 %. Until 2009, an encouraging discovery was reported by Nguyen et al. A small molecule of **DPP (TBFu)$_2$** can give η up to 4.4 % in a BHJ device [122]. Encouraged by this work, there is a rapid progress in the solution-processed small-molecule solar cells. **DR3TBDT** was published with a high η of 7.4 % with J_{sc} of 12.2 mA/cm^2, V_{oc} of 0.93 V, and FF of 0.65 [123]. Recently, Gupta and Heeger et al. reported a material, **DTS(FBTTh$_2$)$_2$**, which exhibited an amazing η of 7.88 % with a J_{sc} of 15.2 mA/cm^2, a V_{oc} of 0.77 V, and a FF of 0.67 [124].

In parallel to the polymer version of tandem solar cell, small-molecule tandem solar cells, which are mainly based on vacuum fabrication process, have also made significant progress. A German company, Heliatek, announced a η of 10.7 % from dual-junction tandem cells based on small molecules in 2012. Recently, they have achieved a η of 12 % based on triple junction OPV cells by thermal evaporation technology [125]. Efficiency of 8 % for single junction from solution-processed small molecules has been achieved by several groups [123, 126]. These results are important and attractive toward large area roll-to-roll printing of organic solar cells.

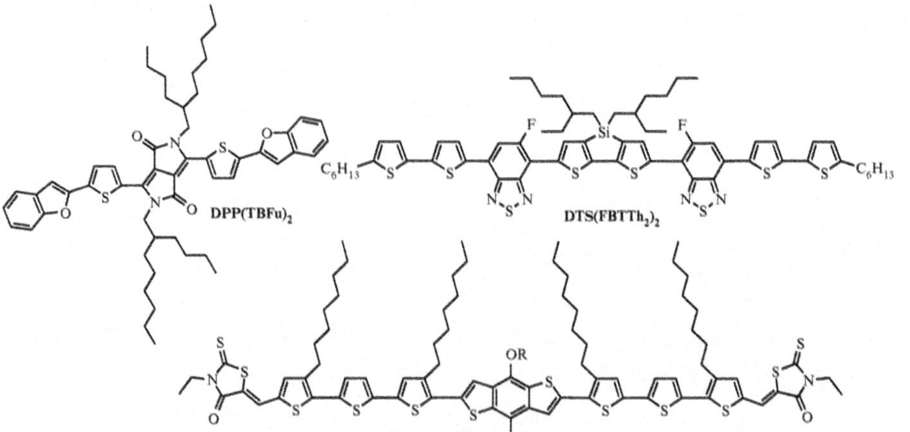

In addition to the general strategies of using organic donor and acceptor segments in the main chain, platinum conjugated polymers have attracted a great deal of interest [127–130]. The incorporation of Pt fragments into the polymers would enhance the electron conjugation and delocalization along the polymer chain as a result of the overlap between d-orbital of the Pt and p-orbital of the alkyne unit. **Pt-P1** contains the thiophene ring as the donor and benzothiadiazole unit as the acceptor to afford a bandgap of 1.85 eV [131]. The best cell obtained by this polymer gave V_{oc} of 0.82 eV with a very high J_{sc} of 15.43 mA/cm^2, which resulted in a η of 4.93 % without thermal annealing. By changing the acceptor to bithiazole rings in **Pt-P2**, the highest η value achieved was 2.5 % [132]. The photovoltaic performance of the **Pt-P2** series showed an increasing trend for η values due to the increasing optical absorbance and hole mobility. Pt-metalated conjugated polymer **Pt-P3** using thieno[3,2-*b*]thiophene connected benzothiadiazole as the central core exhibited high hole mobility due to the more rigid structure by enhancing the electron coupling between the donor and acceptor units along the polymer backbone. The solar cell devices based on **Pt-P3** and $PC_{71}BM$ resulted in a high η of 4.3 % without the need of post-thermal annealing. Although the overall efficiencies of these metalated conjugated polymers are poorer as compared to their organic counterpart, the potential of such system for the application in BHJSCs cannot be neglected. The contribution of organometallic species in the conjugated polymers for BHJSCs mainly came from the strong spin–orbit coupling and efficient intersystem crossing, which facilitates the formation of longer-lived triplet excited states and thus allows extended exciton diffusion length.

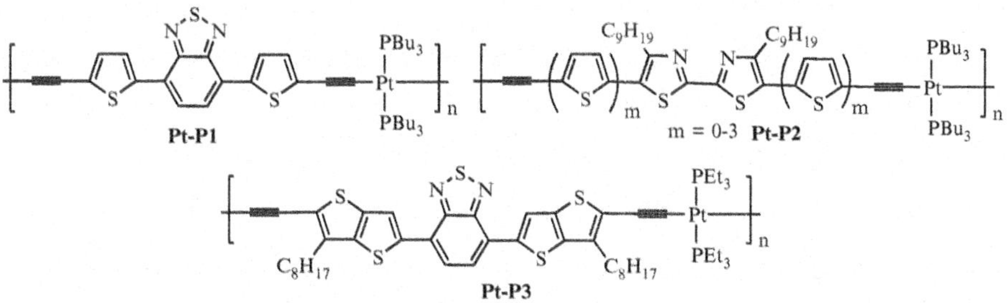

Pt-P1 m = 0-3 **Pt-P2**

Pt-P3

Yang and coworkers pioneered the employment of triplet excitons by using small-molecule 2,3,7,8,12,13,17,18-octaethyl-21*H*,23*H*-porphineplatinum(II) (**Pt-S1**) as the electron-donating material in organic solar cells [133]. By using a simple bilayer structure in OSCs composed of evaporated **Pt-S1** and C_{60}, a promising η of 2.1 % was obtained. For small-molecular platinum(II) bis(aryleneethynylene) complexes **Pt-S2** consisting of benzothiadiazole as the electron acceptor and triphenylamine and/or thiophene as the electron donor, OSCs reached the best η of 2.37 % with the V_{oc} of 0.83 V, J_{sc} of 7.10 mA/cm^2, and FF of 0.40. These works illuminate the potential of well-defined organometallic complexes in developing light harvesting small molecules for efficient power generation in organic photovoltaics implementation. Further improvement in the device efficiency could be achieved by tuning the intramolecular charge transfer absorption and energy levels as well as the oligomeric chain length and thin film morphology.

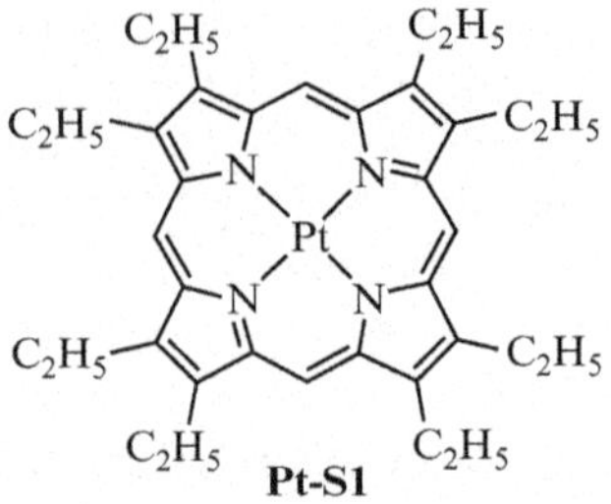

Pt-S1

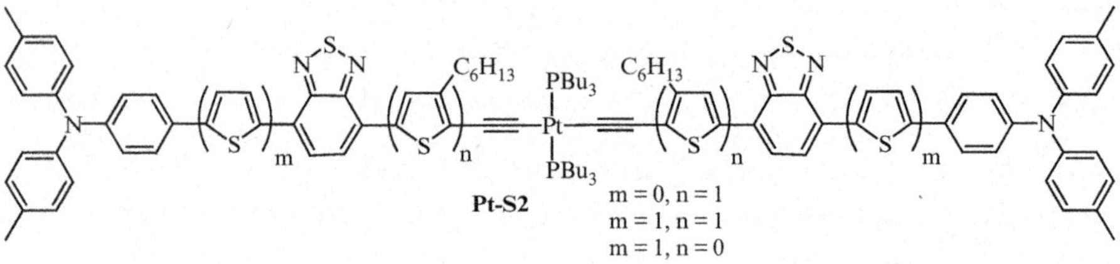

Pt-S2

m = 0, n = 1
m = 1, n = 1
m = 1, n = 0

As the most popular phosphorescent materials in OLEDs, cyclometalated iridium(III) complexes can also be potential candidates as sole donors in BHJSCs, because this class of materials can have favorable photophysical and electrochemical stability, versatility in the modification of molecular properties, tunable energy levels, long exciton lifetime, and potentially long exciton diffusion length. Through a judicious material design, these properties make a compelling case that cyclometalated Ir(III) complexes can be an excellent material candidate for photovoltaic applications. **Ir-S1** has a low HOMO level of 5.5 eV, which shows the potential to gain high V_{oc} [134]. The highest V_{oc} value can be up to 0.94 V if a 30 nm thick of photoactive **Ir-S1** layer was applied. However, the low FF value of **Ir-S1**-based device limited the device efficiency (2.10 %). The low FF was found to be attributed to the insufficient extraction of holes due to the low hole mobility of **Ir-S1**. Zhen reported a solution-processed BHJSCs by employing **Ir-S2** as the electron donor which gave a η of 2.0 % [135]. A bilayer device with **Ir-S3** and C_{60} resulted in a η of 2.8 % with a high V_{oc} value of 1 V [136]. Generally, although the overall efficiencies of Ir(III) complexes are lower than organic active materials, this class of materials tends to give high V_{oc} value as compared to the organic one. Future development of Ir(III) complexes with improved mobility and wider absorption range will enable this class of materials to make an even significant impact.

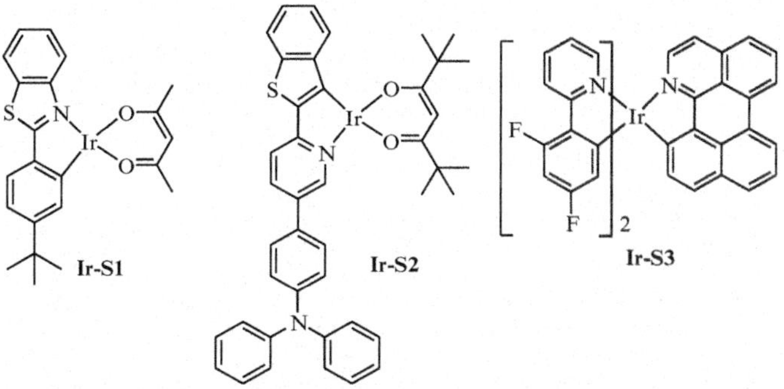

Ir-S1 Ir-S2 Ir-S3

It is believed that introducing appropriate organic materials with long exciton lifetime is a very promising way to improve photovoltaic performance. High-mobility phosphorescent materials with improved absorption matching the solar spectrum are desired for next-generation devices. With the help of phosphorescent materials purposely designed for photovoltaics, the device performance could be greatly improved in the future.

1.4 Conclusion

The field of organic electronics is progressing rapidly. There are numerous new materials being developed, and the research works in both OLEDs and organic solar cells are actively underway in terms of material design and engineering of nanostructures to optimize the optical properties (emission/absorption), carrier mobility and efficiency, operational stability, and large-scale manufacturability. Both organometallic and organic materials take important roles in both areas and have the potential to significantly change the way that OLEDs and organic solar cells are manufactured. They should open up a new revolution in the materials and device aspects for energy conservation in the coming centuries.

Acknowledgments We thank the National Basic Research Program of China (973 Program) (2013CB834702); the National Natural Science Foundation of China (project number 51373145); the Science, Technology and Innovation Committee of Shenzhen Municipality (JCYJ20120829154440583); Hong Kong Baptist University (FRG2/12-13/083 and FRG1/13-14/053); Hong Kong Research Grants Council (HKBU203011); and Areas of Excellence Scheme, University Grants Committee of HKSAR, China (project No. AoE/P-03/08). The work was also supported by Partner State Key Laboratory of Environmental and Biological Analysis (SKLP-14-15-P011) and Strategic Development Fund of HKBU.

References

1. Murawsk C, Leo K, Gather MC (2013) Efficiency roll-off in organic light-emitting diodes. Adv Mater 25:6801–6827
2. Ameri T, Li N, Brabec CJ (2013) Highly efficient organic tandem solar cells: a follow up review. Energy Environ Sci 6:2390–2413

3. Tang CW, Vanslyke SA (1987) Organic electroluminescent diodes. Appl Phys Lett 51:913–915
4. Kalinowski J, Fattori V, Cocchi M, Williams JAG (2011) Light-emitting devices based on organometallic platinum complexes as emitters. Coord Chem Rev 255:2401–2525
5. Ho CL, Wong WY (2013) Charge and energy transfers in functional metallophosphors and metallopolyynes. Coord Chem Rev 257:1614–1649
6. Baldo MA, O'Brien DF, You Y, Shoustikov A, Sibley S, Thompson ME, Forrest SR (1998) Highly efficient phosphorescent emission from organic electroluminescent devices. Nature 395:151–154
7. Gong X, Ostrowski TC, Bazan GC (2003) Electrophosphorescence from a conjugated copolymer doped with an iridium complex: high brightness and improved operational stability. Adv Mater 15:45–49
8. Chen Z, Bian Z, Huang C (2010) Functional Ir^{III} complexes and their applications. Adv Mater 22:1534–1539
9. Lee CW, Lee JY (2013) Above 30 % external quantum efficiency in blue phosphorescent organic light-emitting diodes using pyrido[2,3-b]indole derivatives as host materials. Adv Mater 25:5450–5454
10. King KA, Spellane PJ, Watts RJ (1985) Excited-state properties of a triply ortho-metalated iridium(III) complex. J Am Chem Soc 107:1431–1432
11. Adachi C, Baldo MA, Forrest SR, Thompson ME (2000) High-efficiency red electrophosphorescence devices. Appl Phys Lett 78:170–175
12. Tanaka D, Sasabe H, Li YJ, Su SJ, Takeda T, Kido J (2007) Ultra high efficiency green organic light-emitting devices. Jpn J Appl Phys 46:L10–L12
13. Zhu MR, Ye TL, He X, Cao XS, Zhong C, Ma DG, Qin JG, Yang CL (2011) Highly efficient solution-processed green and red electrophosphorescent devices enabled by small-molecule bipolar host material. J Mater Chem 21:9326–9331
14. Chou HH, Cheng CH (2010) A highly efficient universal bipolar host for blue, green, and red phosphorescent OLEDs. Adv Mater 22:2468–2471
15. Tao YT, Wang QA, Yang CL, Zhong C, Qin JG, Ma DG (2010) Multifunctional triphenylamine/oxadiazole hybrid as host and exciton-blocking material: high efficiency green phosphorescent OLEDs using easily available and common materials. Adv Funct Mater 20:2923–2928
16. Li HY, Zhou L, Teng MY, Xu QL, Lin C, Zheng YX, Zuo JL, Zhang HJ, You XZ (2013) Highly efficient green phosphorescent OLEDs based on a novel iridium complex. J Mater Chem C 1:560–565
17. Tsuboyama A, Iwawaki H, Furugori M, Mukaide T, Kamatani J, Igawa S, Moriyama T, Miura S, Takiguchi T, Okada S, Hoshino M, Ueno K (2003) Homoleptic cyclometalated iridium complexes with highly efficient red phosphorescence and application to organic light-emitting diode. J Am Chem Soc 125:12971–12979
18. Su SJ, Cai C, Kido J (2012) Three-carbazole-armed host materials with various cores for RGB phosphorescent organic light-emitting diodes. J Mater Chem 22:3447–3456
19. Tao Y, Wang Q, Ao L, Zhong C, Qin J, Yang C, Ma D (2010) Molecular design of host materials based on triphenylamine/oxadiazole hybrids for excellent deep-red phosphorescent organic light-emitting diodes. J Mater Chem 20:1759–1765
20. Grushin VV, Herron N, LeCloux DD, Marshall WJ, Petrov VA, Wang Y (2001) New, efficient electroluminescent materials based on organometallic Ir complexes. Chem Commun 1494–1496
21. Tamayo AB, Alleyne BD, Djurovich PI, Lamansky S, Tsyba I, Ho NN, Bau R, Thompson ME (2003) Synthesis and characterization of facial and meridional tris-cyclometalated iridium(III) complexes. J Am Chem Soc 125:7377–7387
22. Wang RJ, Liu D, Ren HC, Zhang T, Yin HM, Liu GY, Li JY (2011) Highly efficient orange and white organic light-emitting diodes based on new orange iridium complexes. Adv Mater 23:2823–2827

23. Kido J, Hongawa K, Okuyama K, Nagai K (1994) White light-emitting organic electroluminescent devices using the poly(N-vinylcarbazole) emitter layer doped with three fluorescent dyes. Appl Phys Lett 64:815–817
24. Kido J, Kimura M, Nagai K (1995) Multilayer white light-emitting organic electroluminescent device. Science 267:1332–1334
25. Kido J, Shionoya H, Nagai K (1995) Single-layer white light-emitting organic electroluminescent devices based on dye-dispersed poly(N-vinylcarbazole). Appl Phys Lett 67:2281–2283
26. Tong QX, Lai SL, Chan MY, Tang JX, Kwong HL, Lee CS, Lee ST (2007) High-efficiency nondoped white organic light-emitting devices. Appl Phys Lett 91:023503-1–023503-3
27. Tsai YC, Jou JH (2006) Long-lifetime, high-efficiency white organic light-emitting diodes with mixed host composing double emission layers. Appl Phys Lett 89:243521-1–243521-3
28. Yang Y, Peng T, Ye KQ, Wu Y, Liu Y, Wang Y (2011) High-efficiency and high-quality white organic light-emitting diode employing fluorescent emitters. Org Electron 12:29–33
29. Wang Q, Ma DG (2010) Management of charges and excitons for high-performance white organic light-emitting diodes. Chem Soc Rev 39:2387–2398
30. Su SJ, Gonmori E, Sasabe H, Kido J (2008) Highly efficient organic blue-and white-light-emitting devices having a carrier- and exciton-confining structure for reduced efficiency roll-off. Adv Mater 20:4189–4194
31. Seidler N, Reineke S, Walzer K, Lüssem B, Tomkeviciene A, Grazulevicius JV, Leo K (2010) Influence of the hole blocking layer on blue phosphorescent organic light-emitting devices using 3,6-di(9-carbazolyl)-9-(2-ethylhexyl)carbazole as host material. Appl Phys Lett 96:093304-1–093304-3
32. So F, Kondakov D (2010) Degradation mechanisms in small-molecule and polymer organic light-emitting diodes. Adv Mater 22:3762–3777
33. Liu XK, Zheng CJ, Lo MF, Xiao J, Chen Z, Liu CL, Lee CS, Fung MK, Zhang XH (2013) Novel blue fluorophor with high triplet energy level for high performance single-emitting-layer fluorescence and phosphorescence hybrid white organic light-emitting diodes. Chem Mater 25:4454–4459
34. D'Andrade BW, Holmes RJ, Forrest SR (2004) Efficient organic electrophosphorescent white-light-emitting device with a triple doped emissive layer. Adv Mater 16:624–628
35. Lee J, Lee JW, Cho NS, Hwang J, Joo CW, Sung WJ, Chu HY, Lee JI (2014) Highly efficient all phosphorescent white organic light-emitting diodes for solid state lighting applications. Curr Appl Phys 14:S84–S87
36. Sasabe H, Takamatsu J, Motoyama T, Watanabe S, Wagenblast G, Langer N, Molt O, Fuchs E, Lennartz C, Kido K (2010) High-efficiency blue and white organic light-emitting devices incorporating a blue iridium carbene complex. Adv Mater 22:5003–5007
37. Chang CH, Tien KC, Chen CC, Lin MS, Cheng HC, Liu SH, Wu CC, Hung JY, Chiu YC, Chi Y (2010) Efficient phosphorescent white OLEDs with high color rendering capability. Org Electron 11:412–418
38. Li J, Wang R, Yang R, Zhou W, Wang X (2013) Iridium complexes containing 2-arylbenzothiazole ligands: color tuning and application in high-performance organic light-emitting diodes. J Mater Chem C 1:4171–4179
39. Chang CH, Chen CC, Wu CC, Chang SY, Hung JY, Chi Y (2010) High-color-rendering pure-white phosphorescent organic light-emitting devices employing only two complementary colors. Org Electron 11:266–272
40. Wang Q, Ding JQ, Ma DG, Cheng YX, Wang LX, Jing XB, Wang FS (2009) Harvesting excitons via two parallel channels for efficient white organic LEDs with nearly 100 % internal quantum efficiency: fabrication and emission-mechanism analysis. Adv Funct Mater 19:84–95
41. Lai SL, Tao SL, Chan MY, Ng TW, Lo MF, Lee CS, Zhang XH, Lee ST (2010) Efficient white organic light-emitting devices based on phosphorescent iridium complexes. Org Electron 11:1511–1513

42. Turner E, Nakken N, Li J (2013) Cyclometalated platinum complexes with luminescent quantum yields approaching 100 %. Inorg Chem 52:7344–7351
43. Kui SCF, Chow PK, Cheng G, Kwok CC, Kwong CL, Low KH, Che CM (2013) Robust phosphorescent platinum(II) complexes with tetradentate O^N^C^N ligands: high efficiency OLEDs with excellent efficiency stability. Chem Commun 49:1497–1499
44. Li G, Fleetham T, Li J (2014) Efficient and stable white organic light-emitting diodes employing a single emitter. Adv Mater 26:2931–2936
45. Cocchi M, Kalinowski J, Murphy L, Williams JAG, Fattori V (2010) Mixing of molecular exciton and excimer phosphorescence to tune color and efficiency of organic LEDs. Org Electron 11:388–396
46. Zhou GJ, Wang Q, Ho CL, Wong WY, Ma DG, Wang LX (2009) Duplicating "sunlight" from simple WOLEDs for lighting applications. Chem Commun 3574–3576
47. Zhen H, Jiang C, Yang W, Jiang J, Huang F, Cao Y (2005) Synthesis and properties of electrophosphorescent chelating polymers with iridium complexes in the conjugated backbone. Chem Eur J 11:5007–5016
48. Kalinowski J, Mezyk J, Meinardi F, Tubino R, Cocchi M, Virgili D (2005) Phosphorescence response to excitonic interactions in Ir organic complex-based electrophosphorescent emitters. J Appl Phys 98:063532-1–063532-3.
49. Baldo MA, Adachi C, Forrest SR (2000) Transient analysis of organic electrophosphorescence. II. Transient analysis of triplet-triplet annihilation. Phys Rev B 62:10967–10977
50. Ma Y, Zhang H, Shen J, Che C (1998) Electroluminescence from triplet metal-ligand charge-transfer excited state of transition metal complexes. Synth Met 94:245–248
51. Lu J, Tao Y, Chi Y, Tung Y (2005) High-efficiency red electrophosphorescent devices based on new osmium(II) complexes. Synth Met 155:56–62
52. Chang SH, Chang CF, Liao JL, Chi Y, Zhou DY, Liao LS, Jiang TY, Chou TP, Li EY, Lee GH, Kuo TY, Chou PT (2013) Emissive osmium(II) complexes with tetradentate bis (pyridylpyrazolate) chelates. Inorg Chem 52:5867–5875
53. Shih PI, Shu CF, Tung YL, Chi Y (2006) Efficient white-light-emitting diodes based on poly (N-vinylcarbazole) doped with blue fluorescent and orange phosphorescent materials. Appl Phys Lett 88:251110-1–251110-3.
54. Liao TC, Chou HT, Juang FS, Tsai YS, Wang SH, Tuan V, Chi Y (2011) Optimizing blue iridium complex and orange-red osmium complex doping concentrations to improve phosphorescent white organic light emitting diodes. Curr Appl Phys 11:S175–S178
55. Chien CH, Liao SF, Wu CH, Shu CF, Chang SY, Chi Y, Chou PT, Lai CH (2008) Electrophosphorescent polyfluorenes containing osmium complexes in the conjugated backbone. Adv Funct Mater 18:1430–1439
56. Uoyama H, Goushi K, Shizu K, Nomura H, Adachi C (2012) Highly efficient organic light-emitting diodes from delayed fluorescence. Nature 492:234–238
57. Nakanotani H, Masui K, Nishide J, Shibata T, Adachi C (2014) Promising operational stability of high-efficiency organic light-emitting diodes based on thermally activated delayed fluorescence. Sci Rep 3:2127
58. Masui K, Nakanotani H, Adachi C (2013) Analysis of exciton annihilation in high-efficiency sky-blue organic light-emitting diodes with thermally activated delayed fluorescence. Org Electron 14:2721–2726
59. Brabec CJ, Sacrificiti NS, Hummelen JC (2001) Plastic solar cells. Adv Funct Mater 11:15–26
60. Brabec CJ (2004) Organic photovoltaics: technology and market. Sol Energy Mater Sol Cells 83:273–292
61. O'Regan B, Grätzel M (1991) A low-cost, high-efficiency solar cell based on dye-sensitized colloidal TiO_2 films. Nature 353:737–740
62. Hagfedlt A, Grätzel M (1995) Light-induced redox reactions in nanocrystalline systems. Chem Rev 95:49–68

63. Nazeeruddin MK, Zakeeruddin SM, Lagref JJ, Liska P, Comte P, Barolo C, Viscardi G, Schenk K, Grätzel M (2004) Stepwise assembly of amphiphilic ruthenium sensitizers and their applications in dye-sensitized solar cell. Coord Chem Rev 248:1317–1328
64. Argazzi R, Iha NTM, Zabri F, Odobel F, Bignozzi CA (2004) Design of molecular dyes for application in photoelectrochemical and electrochromic devices based on nanocrystalline metal oxide semiconductors. Coord Chem Rev 248:1299–1316
65. Polo AS, Itokazu MK, Iha NYM (2004) Metal complex sensitizers in dye-sensitized solar cells. Coord Chem Rev 248:1343–1361
66. Meyer GJ (2005) Molecular approaches to solar energy conversion with coordination compounds anchored to semiconductor surfaces. Inorg Chem 44:6852–6864
67. Robertson N (2006) Optimizing dyes for dye-sensitized solar cells. Angew Chem Int Ed 45:2338–2345
68. Xie P, Guo F (2007) Molecular engineering of ruthenium sensitizers in dye-sensitized solar cells. Curr Org Chem 11:1272–1286
69. Nazeeruddin MK, Kay A, Rodicio I, Humphry-Baker R, Müller E, Liska P, Vlachopoulos N, Grätzel M (1993) Conversion of light to electricity by cis-X_2bis(2,2'-bipyridyl-4,4'-dicarboxylate)ruthenium(II) charge-transfer sensitizers ($X = Cl^-$, Br^-, I^-, CN^-, and SCN^-) on nanocrystalline titanium dioxide electrodes. J Am Chem Soc 115:6382–6390
70. Nazeeruddin MK, De Angelis F, Fantacci S, Selloni A, Viscardi G, Liska P, Ito S, Takeru B, Grätzel M (2005) Combined experimental and DFT-TDDFT computational study of photoelectrochemical cell ruthenium sensitizers. J Am Chem Soc 127:16835–16847
71. Nazeeruddin MK, Pechy P, Renouard T, Zakeeruddin SM, Humphry-Baker R, Comte P, Liska P, Cevey L, Costa E, Shklover V, Spiccia L, Deacon GB, Bignozzi CA, Grätzel M (2001) Engineering of efficient panchromatic sensitizers for nanocrystalline TiO_2-based solar cells. J Am Chem Soc 123:1613–1624
72. Paolo GB, Kiyoshi CDR, Koivisto BD, Berlinguette CP (2012) Cyclometalated ruthenium chromophores for the dye-sensitized solar cell. Coord Chem Rev 256:1438–1450
73. Gao F, Wang Y, Shi D, Zhang J, Wang MK, Jing XY, Humphry-Baker R, Wang P, Zakeeruddin SM, Grätzel M (2008) Enhance the optical absorptivity of nanocrystalline TiO_2 film with high molar extinction coefficient ruthenium sensitizers for high performance dye-sensitized solar cells. J Am Chem Soc 130:10720–10728
74. Cao Y, Bai Y, Yu Q, Cheng Y, Liu S, Shi D, Gao F, Wang P (2009) Dye-sensitized solar cells with a high absorptivity ruthenium sensitizer featuring a 2-(hexylthio)thiophene conjugated bipyridine. J Phys Chem C 113:6290–6297
75. Chen CY, Wang M, Li JY, Pootrakulchote N, Alibabaei L, Ngocle CH, Decoppet D, Tsai JH, Grätzel C, Wu CG, Zakeeruddin SM, Grätzel M (2009) Highly efficient light-harvesting ruthenium sensitizer for thin-film dye-sensitized solar cells. ACS Nano 3:3103–3109
76. Bomben PG, Gordon TJ, Schott E, Berlinguette CP (2011) A trisheteroleptic cyclometalated Ru^{II} sensitizer that enables high power output in a dye-sensitized solar cell. Angew Chem Int Ed 50:10682–10685
77. Bomben PG, Robson KCD, Sedach PA, Berlinguette CP (2009) On the viability of cyclometalated Ru(II) complexes for light-harvesting applications. Inorg Chem 48:9631–9643
78. Chou CC, Wu KL, Chi Y, Hu WP, Yu SJ, Lee GH, Lin CL, Chou PT (2011) Ruthenium (II) sensitizers with heteroleptic tridentate chelates for dye-sensitized solar cells. Angew Chem Int Ed 50:2054–2058
79. Nguyen PT, Lam BXT, Andersen AR, Hansen PE, Lund T (2011) Photovoltaic performance and characteristics of dye-sensitized solar cells prepared with the N719 thermal degradation products $[Ru(LH)_2(NCS)(4\text{-tert-butylpyridine})][N(Bu)_4]$ and $[Ru(LH)_2(NCS)(1\text{-methylbenzimidazole})][N(Bu)_4]$. Eur J Inorg Chem 2533–2541
80. Nguyen PT, Degn R, Nguyen HT, Lund T (2009) Thiocyanate ligand substitution kinetics of the solar cell dye Z-907 by 3-methoxypropionitrile and 4-tert-butylpyridine at elevated temperatures. Sol Energy Mater Sol Cells 93:1939–1945

81. Bomben PG, Koivisto BD, Berlinguette CP (2010) Cyclometalated Ru complexes of type $[RuII(N^N)_2(C^N)]^z$: Physicochemical response to substituents installed on the anionic ligand. Inorg Chem 49:4960–4971
82. Bessho T, Yoneda E, Yum JH, Guglielmi M, Tavernelli I, Imai H, Rothlisberger U, Nazeeruddin MK, Grätzel M (2009) New paradigm in molecular engineering of sensitizers for solar cell applications. J Am Chem Soc 131:5930–5934
83. Kim JJ, Choi H, Paek S, Kim C, Lim K, Ju MJ, Kang HS, Kang MS, Ko J (2011) A new class of cyclometalated ruthenium sensitizers of the type C^N^N for efficient dye-sensitized solar cells. Inorg Chem 50:11340–11347
84. Bessho T, Zakeeruddin S, Yeh CY, Diau EG, Grätzel M (2010) Highly efficient mesoscopic dye-sensitized solar cells based on donor–acceptor-substituted porphyrins. Angew Chem Int Ed 49:6646–6649
85. Yella A, Lee HW, Tsao HN, Yi C, Chandiran AK, Nazeeruddin MK, Diau EGW, Yeh CY, Zakeeruddin SM, Grätzel M (2011) Porphyrin-sensitized solar cells with cobalt (II/III)–based redox electrolyte exceed 12 percent efficiency. Science 334:629–634
86. Kitamura T, Ikeda M, Shigaki K, Inoue T, Anderson NA, Ai X, Lian TQ, Yanagida S (2004) Phenyl-conjugated oligoene sensitizers for TiO_2 solar cells. Chem Mater 16:1806–1812
87. Liang M, Chen J (2013) Arylamine organic dyes for dye-sensitized solar cells. Chem Soc Rev 42:3453–3488
88. Hwang S, Lee JH, Park C, Lee H, Kim C, Park C, Lee MH, Lee W, Park J, Kim K, Park NG, Kim C (2007) A highly efficient organic sensitizer for dye-sensitized solar cells. Chem Commun 4887–4889
89. Tsao HN, Burschka J, Yi C, Kessler F, Nazeeruddin MK, Grätzel M (2011) Influence of the interfacial charge-transfer resistance at the counter electrode in dye-sensitized solar cells employing cobalt redox shuttles. Energy Environ Sci 4:4921–4924
90. Zeng W, Cao Y, Bai Y, Wang Y, Shi Y, Zhang M, Wang F, Pan C, Wang P (2010) Efficient dye-sensitized solar cells with an organic photosensitizer featuring orderly conjugated ethylenedioxythiophene and dithienosilole blocks. Chem Mater 22:1915–1925
91. Choi H, Raabe I, Kim D, Teocoli F, Kim C, Song K, Yum JH, Ko J, Nazeeruddin MK, Grätzel M (2010) High molar extinction coefficient organic sensitizers for efficient dye-sensitized solar cells. Chem Eur J 16:1193–1201
92. Liu B, Zhu W, Zhang Q, Wu W, Xu M, Ning Z, Xie Y, Tian H (2009) Conveniently synthesized isophoronedyes for high efficiency dye-sensitized solar cells: tuning photovoltaic performance by structural modification of donor group in donor–π–acceptor system. Chem Commun 1766–1768
93. Ito S, Miura H, Uchida S, Takata M, Sumioka K, Liska P, Comte P, Pechy P, Grätzel M (2008) High-conversion-efficiency organic dye-sensitized solar cells with a novel indoline dye. Chem Commun 5194–5196
94. Bai Y, Zhang J, Zhou D, Wang Y, Zhang M, Wang P (2011) Engineering organic sensitizers for iodine-free dye-sensitized solar cells: red-shifted current response concomitant with attenuated charge recombination. J Am Chem Soc 133:11442–11445
95. Zhang G, Bala H, Cheng Y, Shi D, Lv X, Yu Q, Wang P (2009) High efficiency and stable dye-sensitized solar cells with an organic chromophore featuring a binary π-conjugated spacer. Chem Commun 2198–2200
96. Qu S, Qin C, Islam A, Wu Y, Zhu W, Hua J, Tian H, Han L (2012) A novel D–A-π-A organic sensitizer containing a diketopyrrolopyrrole unit with a branched alkyl chain for highly efficient and stable dye-sensitized solar cells. Chem Commun 48:6972–6974
97. Tang CW (1986) Two-layer organic photovoltaic cell. Appl Phys Lett 48:183–185
98. Sariciftci NS, Smilowitz L, Heeger AJ, Wudl F (1992) Photoinduced electron transfer from a conducting polymer to buckminsterfullerene. Science 258:1474–1476
99. McCullough RD, Lowe RD (1992) Enhanced electrical conductivity in regioselectively synthesized poly(3-alkylthiophenes). J Chem Soc Chem Commun 70–72

100. Morita S, Zakhidov AA, Yoshino K (1992) Doping effect of buckminsterfullerene in conducting polymer: Change of absorption spectrum and quenching of luminescence. Solid State Commun 82:249–252
101. Irwin MD, Buchholz DB, Hains AW, Chang RPH, Marks TJ (2008) p-Type semiconducting nickel oxide as an efficiency-enhancing anode interfacial layer in polymer bulk-heterojunction solar cells. PNAS 105:2783–2787
102. Chang CY, Wu CE, Chen CY, Cui C, Cheng YJ, Hsu CS, Wang YL, Li Y (2011) Enhanced performance and stability of a polymer solar cell by incorporation of vertically aligned, cross-linked fullerene nanorods. Angew Chem Int Ed 50:9386–9390
103. Sun Y, Cui C, Wang H, Li Y (2011) Combinatorial screening of polymer:fullerene blends for organic solar cells by inkjet printing. Adv Energy Mater 1:105–108
104. Guo X, Cui C, Zhang M, Huo L, Huang Y, Hou J, Li Y (2012) High efficiency polymer solar cells based on poly(3-hexylthiophene)/indene-C_{70} bisadduct with solvent additive. Energy Environ Sci 5:7943–7949
105. Liao SH, Li YL, Jen TH, Cheng YS, Chen SA (2012) Multiple functionalities of polyfluorene grafted with metal ion-intercalated crown ether as an electron transport layer for bulk-heterojunction polymer solar cells: optical interference, hole blocking, interfacial dipole, and electron conduction. J Am Chem Soc 134:14271–14274
106. Chu TY, Lu JP, Beaupre S, Zhang YG, Pouliot JR, Wakim S, Zhou JY, Leclerc M, Li Z, Ding JF, Tao Y (2011) Bulk heterojunction solar cells using thieno[3,4-c]pyrrole-4,6-dione and dithieno[3,2-b:2',3'-d]silole copolymer with a power conversion efficiency of 7.3 %. J Am Chem Soc 133:4250–4253
107. Amb CM, Chen S, Graham KR, Subbiah J, Small CE, So F, Reynolds JR (2011) Dithienogermole as a fused electron donor in bulk heterojunction solar cells. J Am Chem Soc 133:10062–10065
108. Small CE, Chen S, Subbiah J, Amb CM, Tsang SW, Lai TH, Reynolds JR, So F (2012) High-efficiency inverted dithienogermole–thienopyrrolodione-based polymer solar cells. Nat Photonics 6:115–120
109. Jiang JM, Yang PA, Chen HC, Wei KH (2011) Synthesis, characterization, and photovoltaic properties of a low-bandgap copolymer based on 2,1,3-benzooxadiazole. Chem Commun 47:8877–8879
110. Huo L, Hou J, Zhang S, Chen HY, Yang Y (2010) A polybenzo[1,2-b:4,5-b']dithiophene derivative with deep HOMO level and its application in high-performance polymer solar cells. Angew Chem Int Ed 49:1500–1503
111. Dou L, You J, Yang J, Chen CC, He Y, Murase S, Moriarty T, Keith E, Li G, Yang Y (2012) Tandem polymer solar cells featuring a spectrally matched low-bandgap polymer. Nat Photonics 6:180–185
112. Dou L, Chang WH, Gao J, Chen CC, You J, Yang Y (2013) A selenium-substituted low-bandgap polymer with versatile photovoltaic applications. Adv Mater 25:825–831
113. Cabanetos C, Labban AE, Bartelt JA, Douglas JD, Mateker WR, Frechet JMJ, McGehee MD, Beaujuge PM (2013) Linear side chains in benzo[1,2-b:4,5-b']dithiophene-thieno[3,4-c] pyrrole-4,6-dione polymers direct self-assembly and solar cell performance. J Am Chem Soc 135:4656–4659
114. He Z, Zhong C, Su S, Xu M, Wu H, Cao Y (2012) Enhanced power-conversion efficiency in polymer solar cells using an inverted device structure. Nat Photonics 6:591–595
115. Su YW, Lan SC, Wei KH (2012) Organic photovoltaics. Mater Today 15:554–562
116. Dou L, Chen CC, Yoshimura K, Ohya K, Chang WH, Gao J, Liu Y, Richard E, Yang Y (2013) Synthesis of 5H-dithieno[3,2-b:2′,3′-d]pyran as an electron-rich building block for donor-acceptor type low-bandgap polymers. Macromolecules 46:3384–3390
117. Li WW, Furlan A, Hendriks KH, Wienk MM, Janssen RA (2013) Efficient tandem and triple-junction polymer solar cells. J Am Chem Soc 135:5529–5532
118. Roncali J (2009) Molecular bulk heterojunctions: An emerging approach to organic solar cells. Acc Chem Res 42:1719–1730

119. Walker B, Kim C, Nguyen TQ (2011) Small molecule solution-processed bulk heterojunction solar cells. Chem Mater 23:470–482
120. Lin Y, Li Y, Zhan X (2012) Small molecule semiconductors for high-efficiency organic photovoltaics. Chem Soc Rev 41:4245–4272
121. Shen S, Jiang P, He C, Zhang J, Shen P, Zhang Y, Yi Y, Zhang Z, Li Z, Li Y (2013) Solution-processable organic molecule photovoltaic materials with bithienyl-benzodithiophene central unit and indenedione end groups. Chem Mater 25:2274–2281
122. Walker B, Tamayo AB, Dang XD, Zalar P, Seo JH, Garcia A, Tantiwiwat M, Nguyen TQ (2009) Nanoscale phase separation and high photovoltaic efficiency in solution-processed, small-molecule bulk heterojunction solar cells. Adv Funct Mater 19:3063–3069
123. Zhou JY, Wan XJ, Liu YS, Zuo Y, Li Z, He GG, Long GK, Ni W, Li CX, Su XC, Chen YS (2012) Small molecules based on benzo[1,2-b:4,5-b']dithiophene unit for high-performance solution-processed organic solar cells. J Am Chem Soc 134:16345–16351
124. Kyaw AKK, Wang DH, Gupta V, Zhang J, Chand S, Bazan GC, Heeger AJ (2013) Efficient solution-processed small-molecule solar cells with inverted structure. Adv Mater 25:2397–2402
125. Heliatek consolidates its technology leadership by establishing a new world record for organic solar technology with a cell efficiency of 12 % (2013). http://www.heliatek.com/. Accessed Mar 2013
126. Sun Y, Welch GC, Leong WL, Takacs CJ, Bazan GC, Heeger AJ (2012) Solution-processed small-molecule solar cells with 6.7 % efficiency. Nat Mater 11:44–48
127. Wu PT, Bull T, Kim FS, Luscombe CK, Jenekhe SA (2009) Organometallic donor-acceptor conjugated polymer semiconductors: tunable optical, electrochemical, charge transport, and photovoltaic properties. Macromolecules 42:671–681
128. Liu L, Ho CL, Wong WY, Cheung KY, Fung MK, Lam WT, Djurišic AB, Chan WK (2008) Effect of oligothienyl chain length on tuning the solar cell performance in fluorene-based polyplatinynes. Adv Funct Mater 18:2824–2833
129. Mei J, Ogawa K, Kim YG, Heston NC, Arenas DJ, Nasrollahi Z, McCarley TD, Tanner DB, Reynolds JR, Schanze KS (2009) Low-band-gap platinum acetylide polymers as active materials for organic solar cells. ACS Appl Mater Interfaces 1:150–161
130. Baek NS, Hau SK, Yip HL, Acton O, Chen KS, Jen AKY (2008) High performance amorphous metallated π-conjugated polymers for field-effect transistors and polymer solar cells. Chem Mater 20:5734–5736
131. Wong WY, Wang XZ, He Z, Djurišic AB, Yip CT, Cheung KY, Wang H, Mak CSK, Chan WK (2007) Metallated conjugated polymers as a new avenue towards high-efficiency polymer solar cells. Nat Mater 6:521–527
132. Wong WY, Wang XZ, He Z, Chan KK, Djurisic AB, Cheung KY, Yip CT, Ng AMC, Xi YY, Mak CSK, Chan WK (2007) Tuning the absorption, charge transport properties, and solar cell efficiency with the number of thienyl rings in platinum-containing poly(aryleneethynylene)s. J Am Chem Soc 129:14372–14380
133. Shao Y, Yang Y (2005) Efficient organic heterojunction photovoltaic cells based on triplet materials. Adv Mater 17:2841–2844
134. Yu J, Zang Y, Li H, Huang J (2012) Fill factor enhancement of organic solar cells based on a wide bandgap phosphorescent material and C_{60}. Thin Solid Films 520:6653–6657
135. Zhen H, Hou Q, Li K, Ma Z, Fabiano S, Gaob F, Zhang F (2014) Solution-processed bulk-heterojunction organic solar cells employing Ir complexes as electron donors. J Mater Chem A 2:12390–12396
136. Fleetham TB, Wang Z, Li J (2013) Exploring cyclometalated Ir complexes as donor materials for organic solar cells. Inorg Chem 52:7338–7343

Chapter 2
Density Functional Theory in the Design of Organometallics for Energy Conversion

Gemma R. Freeman and J.A. Gareth Williams

Abstract Theoretical methods based on density functional theory (DFT) and time-dependent density functional theory (TD-DFT) are increasingly used to rationalise the excited-state properties of metal complexes and to help guide the design of new materials. This chapter provides a brief introduction of the background to such methods, highlighting some of the features that need to be considered, such as the ability of functionals to deal with charge-transfer states and the challenges associated with triplet-state calculations. Examples are drawn from recent studies on (1) the ground-state and light absorption properties of ruthenium(II) complexes as sensitisers for dye-sensitised solar cells (DSSCs) and (2) the triplet excited states of luminescent platinum(II) complexes that are potential phosphors for organic light-emitting diodes (OLEDs).

Keywords Time-dependent density functional theory (TD-DFT) • Metal complexes • Phosphorescence • Iridium • Platinum • Ruthenium • Organic light-emitting diode • OLED • Dye-sensitised solar cell (DSSC)

2.1 Introduction

Metal complexes and organometallics play a key role in many devices for energy conversion, particularly those where light is involved. For example, in the new generation of display screen equipment and energy-efficient lighting – organic light-emitting diodes (OLEDs) – the use of complexes of heavy transition metals as phosphorescent emitters allows large gains in efficiency through the harvesting of otherwise wasted triplet states [1]. Meanwhile, in the reverse process of light-to-electrical energy conversion, metal complexes that absorb light to generate charge-transfer states are at the forefront of research into dye-sensitised solar cells (DSSCs) [2]. Metal complexes with long excited-state lifetimes and that are strong excited-state oxidants or reductants also attract attention in the field of 'artificial photosynthesis' (AP) – the conversion of light-to-chemical energy, for example, in

G.R. Freeman • J.A.G. Williams (✉)
Department of Chemistry, Durham University, Durham DH1 3LE, UK
e-mail: j.a.g.williams@durham.ac.uk

W.-Y. Wong (ed.), *Organometallics and Related Molecules for Energy Conversion*,
Green Chemistry and Sustainable Technology, DOI 10.1007/978-3-662-46054-2_2

photocatalysed 'water splitting' to generate H_2 for subsequent use as a non-polluting fuel [3].

All of these applications require certain properties of the metal complexes to be optimised. For example, in DSSCs and AP, the compounds need to absorb light very strongly, preferably across the whole of the visible region of the spectrum and into the near infrared (NIR). Moreover, they need to form the appropriate kind of excited state (e.g. metal-to-ligand charge transfer, MLCT, in the case of DSSCs) and with the correct directionality (i.e. charge transfer to one particular ligand, normally the one bound to the semiconductor). Complexes for both applications must also satisfy quite stringent requirements in terms of ground- and excited-state redox potentials, since electron transfer is a key step. Similarly, in the field of OLEDs, metal complexes are required that display phosphorescent emission from triplet states with very high efficiency (i.e. high luminescence quantum yield) and with well-defined and tunable emission maxima for use in red-green-blue (RGB) displays [4].

These requirements have led to a large number of labour-intensive synthetic projects around the world, whereby potential molecular materials are prepared in the laboratory and their photophysical and electrochemical properties then studied. Such projects have led to extensive empirical data. But theoretical methods are increasingly important in allowing such data to be interpreted, making it more useful in the informed *design* of new materials. The goal of such theoretical methods should be to allow the properties of a possible molecule to be predicted accurately and rapidly, so that synthetic effort can be directed specifically to those systems likely to show the best results.

Whilst empirical molecular mechanics [5] and semiempirical methods [6] have found utility in calculations on transition-metal complexes, the requirement for suitable parametrisation has limited their use. More common are ab initio techniques, in particular Hartree-Fock (HF) methods, which can provide reasonable results for second and third row transition metals [7]. These, however, become increasingly impracticable with increasing number of atoms, and density functional theory (DFT) calculations are favoured for their similar accuracy at a reduced computational cost as well as their inclusion of electronic correlation effects [8].

DFT is based on the Hohenberg-Kohn theorem, namely, that all ground-state properties – including the ground-state energy – are uniquely determined by the electron density $\rho(r)$, where r represents spatial coordinates [9]. The density and hence energy are optimised using variational procedures to determine the ground-state geometry and properties. DFT can be used to calculate both closed- and open-shell systems, although the molecules typically of interest in the field of this chapter are closed shell, and we shall limit the ensuing discussion to these cases.

The electron density $\rho(r)$ is obtained by summing over all occupied Kohn-Sham (KS) orbitals $\phi_i(r)$:

$$\rho(r) = \sum_i n_i |\phi_i(r)|^2 \tag{2.1}$$

where n_i is the orbital occupancy. The KS orbitals can then be expressed in terms of a linear combination of atomic orbitals χ_j or basis set, as in the wave function-based techniques:

$$\phi_i = \sum_j c_{ij}\chi_j \tag{2.2}$$

The orbitals $\phi_i(r)$ and their energies are obtained by solving the KS equations, which incorporate terms for the kinetic and electrostatic potential energy, as well as a term V_{XC} that describes electron exchange and correlation. Since an exact mathematical expression for V_{XC} is unknown, approximate functionals have to be used. Indeed, the availability of increasingly complex but accurate exchange-correlation functionals has been a key reason behind the rapid development and wide uptake of DFT methods over recent years. Excited states are dealt with using time-dependent DFT (TD-DFT), wherein the molecule is considered to be subject to a time-dependent perturbation arising from an oscillating electrical field associated with incident radiation. The electron density $\rho(r, t)$ becomes time dependent.

Accuracy has now reached the stage where DFT calculations may be used almost routinely in combination with experimental studies of transition-metal complexes. For example, in the field of organic light-emitting diodes (OLEDs), the widely used emissive material aluminium tris(8-hydroxyquinoline) (Alq_3), used in the landmark report of Tang and VanSlyke [10], was studied by a number of groups around the turn of the century, with reasonable correlation with experiment [11–13]. Similarly, cyclometallated iridium(III) complexes – phosphorescent emitters in contemporary commercial OLED display screens used in some mobile phones, for example – have been extensively investigated using TD-DFT methods [14]. The same methods have also been applied to related complexes with other metal ions such as platinum(II) and gold(III) [15, 16]. There is also an increasing interest in modelling DSSC dyes using theoretical methods, not just the light absorption properties of the isolated molecules but also as semiconductor-adsorbed species and their redox potentials [17].

In this chapter, following a brief overview of the principles, a selection of examples will be presented of the application of DFT and TD-DFT methods to metal complexes for use in DSSCs and OLEDs. This is a very active area, and the aim is – of course – not to be comprehensive but rather to convey a flavour of the activity and some of the challenges. A number of cyclometallated complexes will be used as examples, reflecting not only the attractiveness and increasing importance of such compounds but also the authors' own research interests.

2.2 DFT Methods for Metal Complexes and Organometallics

The way in which DFT is carried out on metal-containing molecules is essentially no different from its application to small organic molecules. Nevertheless, there are some commonly encountered issues, such as the choice of basis set for heavy metal ions and the importance of charge-transfer states in the field, which raise challenges. Such points will be highlighted in specific cases in subsequent sections. At this stage, it may be useful to briefly consider the approach in more general terms.

2.2.1 *Choice of Functional*

Since the ability to probe excited states is key to the applications in mind (absorption or emission of light), the chosen functionals should be able to deal well with excitation energies. Functionals that employ generalised gradient approximations (GGAs), such as PBE, systematically underestimate excitation energies and are typically inappropriate [18]. The introduction of exact exchange helps to deal with these deficiencies. The exact orbital exchange in DFT is the HF exchange energy expression, evaluated using KS orbitals. Global hybrid (GH) functionals employ a fixed amount of exact orbital exchange, α. Examples include B3LYP ($\alpha = 20$ %) [19, 20] and PBE0 ($\alpha = 25$ %), which have emerged as the most popular of the functionals [21].

Nevertheless, such GH functionals do systematically underestimate excitation energies associated with through-space charge-transfer transitions. Ironically, it is precisely such charge-transfer states that are particularly important in energy conversion applications. In DSSCs, the formation of a charge-transfer state by the sensitiser upon the absorption of light is the first step in charge separation, prior to injection of an electron into the semiconducting material such as TiO_2 [22]. Meanwhile, in phosphorescent OLEDs and LECs, it is well established that a high degree of metal-to-ligand charge-transfer (MLCT) character is typically required in the excited state, in order to promote the radiative decay of the triplet states by introducing the spin-orbit coupling effect of the metal ion [23]. Range-separated hybrid (RSH) or Coulomb-attenuated functionals are of interest in this connection [24]. They have been shown to improve calculation of long-range, charge-transfer-type excitation energies, but maintain a good approximation of localised excitations. They work by varying the amount of the exact orbital exchange as a function of the interelectron distance, r_{12}; for example, CAM-B3LYP has an initial α of 19 %, increasing to 65 % at high r_{12} [25].

2.2.2 *Choice of Basis Sets*

For complexes of low-atomic-weight metals, such as those of Al^{3+} and Zn^{2+}, for example, it is usual to use a standard, double-ζ quality, basis set such as 6-31G or cc-PVDZ for all atoms in the molecule, as typically used for organic molecules. In the case of complexes of heavy metals, particularly third row transition-metal ions such as Ir(III) or Pt(II), it is more common to treat the metal using a different basis set from the rest of the atoms, replacing the large number of core electrons of the metal by an 'effective core potential' (ECP). Since these inner electrons are not involved in bonding, very little loss of accuracy is incurred, but there is a large decrease in calculation time. The more significant outer core [e.g. $(5s)^2(5p)^6$ in the case of Ir(III)] and valence electrons [$(5d)^6$ for Ir(III)] are still included. The Los Alamos National Laboratory 2-double-ζ (LANL2DZ) basis set is popular for such metals, which incorporates an ECP. Other variants include the Stuttgart 1997 ECP.

2.2.3 *Potential Energy Surfaces and Optical Transitions*

The first step in applying DFT methods to metal complexes is normally to optimise the ground-state geometry. This process involves the evaluation of the energy of the molecule by DFT and minimisation according to structure (bond lengths, bond angles, torsions, etc.). For most molecules of interest in the field, the ground states are normally singlet states with no unpaired electrons. The large ligand-field splitting associated with second and third row transition metals ensures low-spin configurations. Where a crystal structure of the molecule of interest is available, the use of the atomic coordinates within the crystal as a starting point can help to reduce the computational time. In some cases, however, crystal packing effects and intermolecular interactions can influence molecular geometry, so some care should be exercised when using this approach, in case a local rather than global minimum is determined. Calculation of the vibrational frequencies of the optimised structure normally provides evidence if this is the case, as revealed by negative (imaginary) values, and this check is strongly recommended. Restarting the calculation from a modified geometry will typically lead to the correct structure.

Having obtained the optimised (i.e. global energy-minimised) S_0 structure, TD-DFT can then routinely be applied to determine the energies and oscillator strengths of spin-allowed ($S_0 \rightarrow S_n$) transitions and the energies of spin-forbidden transitions ($S_0 \rightarrow T_n$) at the S_0 geometry, corresponding to the process of light absorption. (Note that the successful incorporation of spin-orbit coupling into such calculations remains an elusively challenging task, such that oscillator strengths for $S_0 \rightarrow T_n$ transitions are unlikely to be very meaningful in commercial packages.) Jacquemin and co-workers have outlined the different methods that are commonly used for the calculation of $S_0 \rightarrow S_1$ excitation energies, which are illustrated in Fig. 2.1 [26, 27]. TD-DFT calculation of $E^{\text{vert-abso}}$ gives the energy

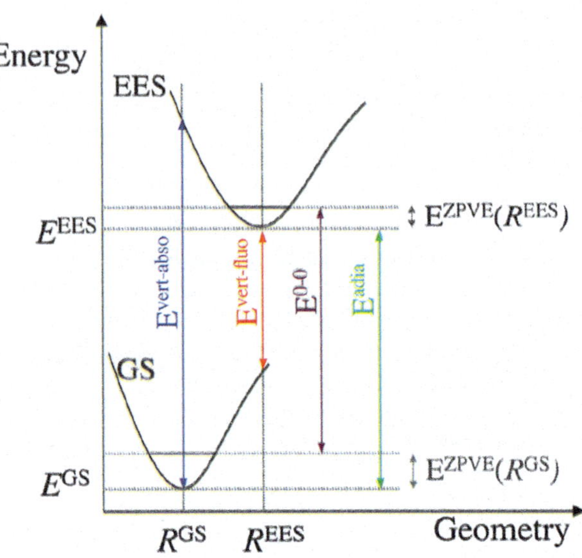

Fig. 2.1 Jablonski diagram representing a ground state (GS) and an excited electronic state (EES) and the different possible methods for calculating excitation energies (Reprinted with permission from [27]. © 2012 American Chemical Society)

of the lowest-energy absorption process: in principle, this should correspond to the longest wavelength band in the absorption spectrum. The same calculation at the S_1 excited-state geometry, $E^{\text{vert-fluo}}$, corresponds to fluorescence. This quantity is smaller than $E^{\text{vert-abso}}$ due to the relaxation of the excited state (typically on a timescale of picoseconds) prior to emission (typically a nanosecond timescale). The difference between the two minima in the scheme of Fig. 2.1 is the adiabatic energy, E^{adia}. It is occasionally used to calculate emission energies [28] but will generally lead to values that are too high, owing to the typically shallower potential energy surface (PES) of the excited state and smaller zero-point energy.

In order to model phosphorescence from the triplet state, the $S_0 \rightarrow T_1$ excitation energy at the T_1 geometry should be calculated. It is important to note that DFT applies rigorously to the lowest state of *any* spin multiplicity; thus, the geometry of the T_1 state can be obtained directly, just as for the S_0 state, and the TD-DFT calculation then performed. It is, however, more likely than for the S_0 that local minima may inadvertently be found as opposed to the true minimum, owing to the typically flatter profile of the PES. It is common practice to model triplet emission by calculating the $S_0 \rightarrow T_1$ excitation energy at the S_0 (as opposed to T_1) geometry [29, 30], since it saves on computational time, requiring only one geometry optimisation. Such values will, however, normally be too high in energy, and good correlation with experiment may be fortuitous through cancellation of errors.

Calculations of the $S_0 \rightarrow T_1$ excitation at the T_1 geometry can, however, sometimes produce excitation energies that are implausibly low in energy or even imaginary. This phenomenon has recently been discussed in small molecules and ascribed to 'triplet instabilities' [31, 32]. It is a well-known deficiency of HF theory [33, 34], and thus it is perhaps unsurprising that DFT functionals that incorporate some degree of exact orbital exchange will suffer from similar difficulties with triplet states. The use of the Tamm-Dancoff approximation (TDA) is recommended for calculating excitation energies with low triplet stabilities [35, 36]. The TDA corresponds to setting $B = 0$ in the generalised eigenvalue equations (2.3); i.e. allowing only excitation between occupied-virtual orbital pairs (given by the eigenvector X) as opposed to conventional TD-DFT, where virtual-occupied de-excitation contributions (Y) are also permitted. Since A is Hermitian, the occurrence of imaginary excitation energies is precluded. To date, however, the TDA has been little applied to metal complexes [37]:

$$\begin{pmatrix} A & B \\ B & A \end{pmatrix}\begin{pmatrix} X \\ Y \end{pmatrix} = \begin{pmatrix} 1 & 0 \\ 0 & -1 \end{pmatrix}\begin{pmatrix} X \\ Y \end{pmatrix}. \quad (2.3)$$

2.2.4 *Modelling Solvent*

DFT calculations are normally performed in the gas phase. Since such conditions are difficult to reproduce experimentally and different solvents can have a profound effect on the optical properties, it is often desirable to introduce solvent into the calculations. The two main approaches for the estimation of solvent effects are explicit and implicit models. The former treats every solvent molecule individually, calculating their interaction with one another and the compound of interest. Though occasionally applied to small molecule systems, it is computationally demanding and not used routinely for larger systems such as metal complexes [38].

Implicit solvent models, on the other hand, describe the volume around the compound of interest as a structureless continuum. For example, the polarisable continuum model (PCM) [39] is often used in calculations of large molecules, including third row transition-metal complexes [40–42]. This approach omits specific solvent interactions but describes the polarity of the environment.

2.3 Examples in Light-to-Electrical Energy Conversion: DSSCs

2.3.1 *Background and Brief Guide to What Calculations Can Offer*

In a conventional inorganic semiconductor solar cell or a bulk heterojunction cell, the materials are responsible for absorbing light as well as participating in charge transport. In a DSSC, the two functions of light absorption and charge transport are separated. A dye coated onto a semiconductor is used to absorb light and initiates the transfer of an electron into the semiconductor, which functions as the charge transporter. The basic design and components of a DSSC are shown in Fig. 2.2. After the absorption of light to generate the excited state, the dye injects an electron

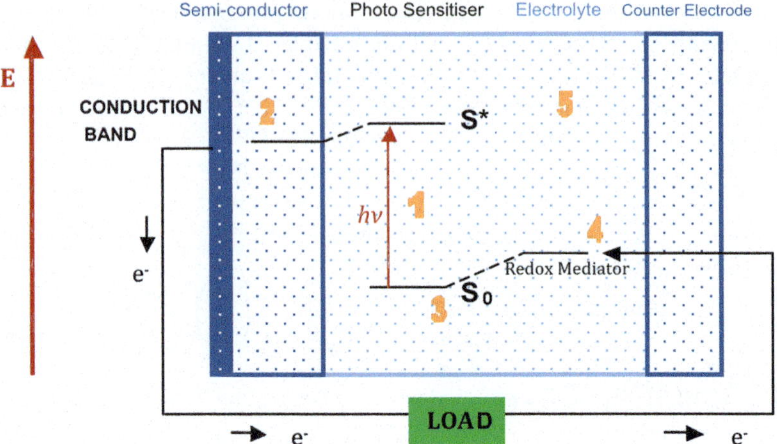

Fig. 2.2 Simplified diagram showing the mode of action of a dye-sensitised solar cell (DSSC). Key processes are (1) excitation of the sensitiser by light absorption, (2) injection of an electron into the semiconductor and hence circuit, (3) reduction of the oxidised sensitiser by the reduced form of the redox mediator and (4) reduction of the oxidised form of the mediator by incoming electrons from the counter electrode

into the semiconductor which subsequently flows around the circuit. The oxidised dye is in turn reduced by electrons from the counter electrode, through the intermedicacy of a redox electrolyte, typically I^-/I_3^-, which acts as an electron shuttle.

The most successful and widely investigated dyes are based on ruthenium-bipyridine complexes related to $[Ru(bpy)_3]^{2+}$ (bpy = 2,2′-bipyridine). This is because their lowest-energy excited states are archetypal MLCT states: the absorption of light has already induced an initial charge separation in a well-defined direction, with transient oxidation of the Ru(II) centre, and reduction of one of the ligands (normally that on which the LUMO is based). If the ligand to which the electron transfer occurs is attached to the semiconductor surface, then the electron can move from the ligand onto the semiconductor, before the regeneration of the ground state occurs though radiative or other non-radiative decay processes. Other reasons for the success of such complexes relate to the suitability of their energy levels (redox potentials) in the ground and excited state: the HOMO and LUMO should be higher in energy than the valence band edge and conduction band edge, respectively, of the semiconductor. The so-called N3 dye (Fig. 2.3) developed by Grätzel and co-workers has become a benchmark compound in the area. The carboxylate groups allow for binding to a TiO_2 surface, and the σ-donating thiocyanate groups help to raise the energy of the metal d-orbitals and thus lower the energy of the MLCT transition compared to $[Ru(bpy)_3]^{2+}$, shifting the absorption band towards the red.

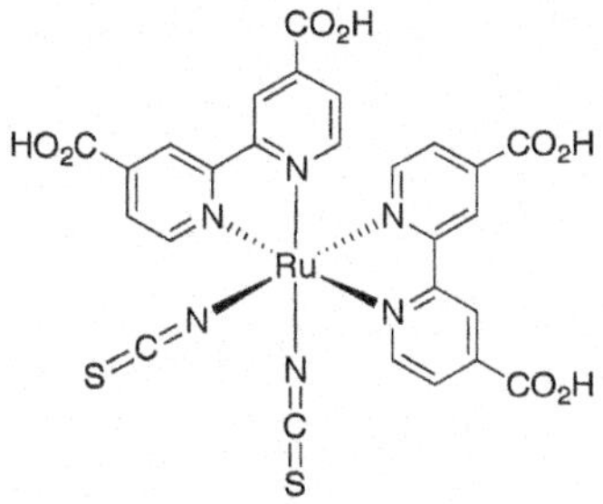

Fig. 2.3 The structure of the benchmark Ru (II) complex known as the N3 dye

Even from this very brief discussion, it is clear that TD-DFT calculations should be able to help rationalise the performance of known dyes and inform the design of new dyes by allowing:

1. The nature of the lowest-energy excited state to be deduced; e.g. is it MLCT as required, or does it have undesirable ligand-centred π–π^* or metal-centred (d–d) character?
2. The location of the acceptor in the CT process (often the LUMO) to be identified; e.g. is it based on the ligand carrying the anchoring groups – through which the complex will bind to the semiconductor – as desired, or on one of the other ligands, which will lead to poorer performance?
3. The absorption spectrum to be deduced; e.g. does the spectrum extend to long wavelengths and with the high extinction coefficients required for good efficiencies (determined by oscillator strengths)?
4. An assessment of other non-radiative decay pathways open to the excited state that may compete with electron injection.
5. Frontier orbital energy levels/electrochemical potentials to be determined.
6. Interaction of the dye with the electrolyte to be assessed.

2.3.2 *Redshifting the Absorption and Localisation of the LUMO*

Liao and co-workers have recently considered the simulation of the absorption spectrum of the N3 dye, comparing results obtained using B3LYP and CAM-B3LYP, and with a range of basis sets, either with or without a solvent included [43]. They found that the inclusion of solvent (DMF) using a PCM led to better agreement with the experimental data (Fig. 2.4). The lowest-energy bands were found to have mixed MLCT/LLCT character, with excitations originating from MOs localised mainly on the Ru and NCS ligands into the π^* orbitals of the bipyridines, as expected. Notably, B3LYP appears to underestimate the energy of these bands, in line with the long-range CT assignment, whereas CAM-B3LYP deals with this issue, but leads to an overestimation of the energy. Using CAM-B3LYP with a PCM, and LANL2DZ for the Ru(II) ion, the use of basis

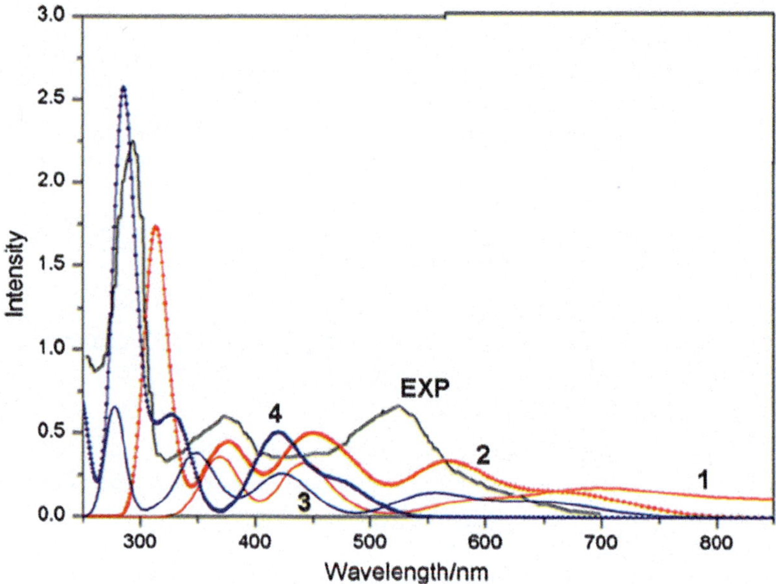

Fig. 2.4 Experimental UV-visible absorption spectra of N3 dye in DMF solvent (EXP) and simulated spectra using TD-DFT: (*1*) B3LYP gas phase, (*2*) B3LYP with PCM, (*3*) CAM-B3LYP gas phase and (*4*) CAM-B3LYP with PCM (basis set = LANL2DZ for Ru, 3-21G* for other atoms in each case) (Reprinted with permission from [43]. © 2013 Elsevier)

sets that add a set of diffuse functionals on other atoms (e.g. 6-311 + G*) was found to give improved results.

One of the limitations of dyes such as N3 is that the absorption (and hence the incident photon conversion efficiency or IPCE) falls off rapidly in the red region of the spectrum. Researchers are actively seeking to extend the absorption of such dyes into the red and NIR regions. One way to achieve this is to increase the extent of conjugation in one of the donor ligands and/or to append electron-rich substituents. For example, a number of complexes incorporating thiophene-appended bipyridine ligands have been explored. Liu et al. carried out TD-DFT calculations and an orbital analysis on a series of complexes including those of Fig. 2.5 [*op. cit*]. Despite the redshift induced by thiophene pendants, the amount of photoinduced charge transfer (Δq) towards the bipyridine dicarboxylate (dcbpy) acceptor from the rest of the molecules in the S_1 state does not significantly increase as the conjugation length in the ancillary ligand increases from T to TT to TTT. However, a notable observation is that in some higher excited states S_n ($n > 1$), the thiophene-appended ancillary ligand (AL) actually acts as an acceptor: increased conjugation lowers the vacant orbitals on these units. Such an effect will clearly be detrimental to the charge injection efficiency, as the electron should be transferred to the dcbpy. Interestingly, the change of one of the two carboxylates to a CN unit ensures that the bpy remains the acceptor, as shown in Fig. 2.6.

R =	R' =	Compound name
thiophene–C_8H_{17}	$-CO_2H$	T
	$-CN$	T-CN
thienothiophene–C_8H_{17}	$-CO_2H$	TT
	$-CN$	TT-CN
dithienothiophene–C_8H_{17}	$-CO_2H$	TTT
	$-CN$	TTT-CN

Fig. 2.5 Structures of thiophene-appended variants of the N3 dye investigated by Liu et al. [43]

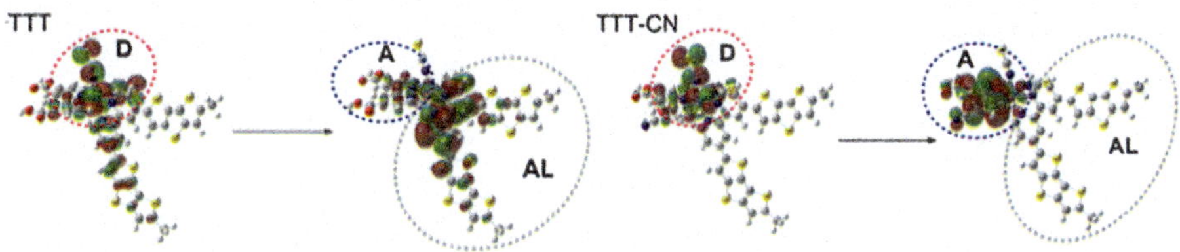

Fig. 2.6 Natural transition orbitals for the $S_0 \rightarrow S_3$ transition of TTT (*left*) and TTT-CN (*right*), determined by TD-DFT using CAM-B3LYP. The circled groups D and A are the donor (metal/thiocyanate) and acceptor (dcbpy), respectively, whilst AL indicates the thiophene-appended 'ancillary' ligands (Reprinted with permission from [43]. © 2013 Elsevier)

Another interesting example comes from the recent work by Gros and colleagues, who have been exploring the utility of dithienylpyrrole (DTP) pendants for redshifting the absorption of N3-like dyes [44, 45]. Complexes such as those shown in Fig. 2.7 have been prepared. They show significantly enhanced extinction coefficients compared to N3 and good coverage over the visible spectrum. The tris-heteroleptic complex **C4** with the thienyl-linked DTP, in particular, displays significant absorption to long wavelengths > 700 nm. Nevertheless, these complexes were found to display only low IPCE values, using either TiO_2 or SnO_2 as the semiconductor, indicating that charge injection is inefficient. The possible explanation is that this might be due to the localisation of the electron in the MLCT state in the 'wrong' ligand – i.e. the DTP-appended bipyridine as opposed to the dcbpy ligand was subsequently confirmed by TD-DFT study, supported by transient absorption spectroscopy. The virtual natural transition orbitals of **C1**, for example (Fig. 2.8), show the confinement of the excited electron on the metal- and DTP^1-appended ligand, with no participation of the dcbpy ligand. In the case of **C4**, although the lowest-energy transition does show the desired localisation of the excited electron on the dcbpy ligand, other higher-energy transitions with much higher oscillator strengths again suffer from the problem of localisation of the electron on the DTP^2 ligand, remote from the semiconductor surface.

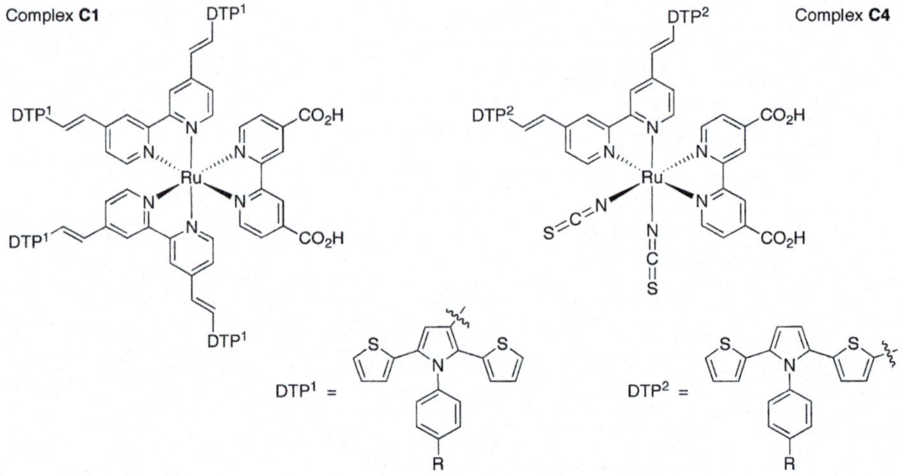

Fig. 2.7 Ru(II) complexes **C1** and **C4** incorporating dithienylpyrrole units studied by Gros and co-workers [45]

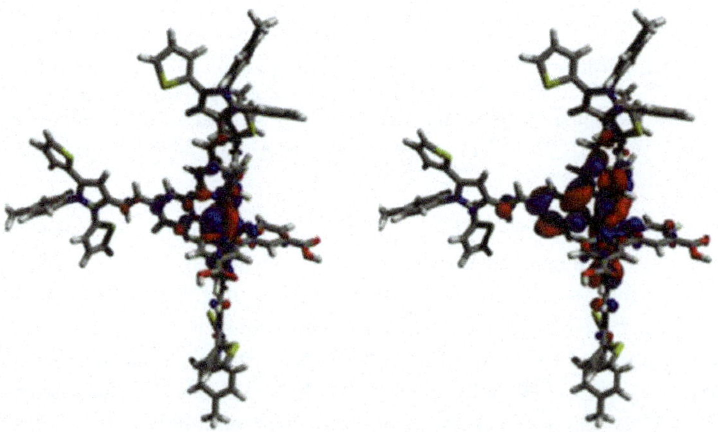

Fig. 2.8 Natural transition orbitals for complex **C1** at 488 nm, occupied (*left*) and virtual (*right*) orbitals. A similar picture is obtained at 512 nm (Reprinted with permission from [45]. © 2014 Elsevier)

2.3.3 Cyclometallation Versus Monodentate Thiocyanate Ligands

Many of the most thoroughly investigated ruthenium-based DSSC dyes contain monodentate thiocyanate ligands. As noted above, the thiocyanates raise the energy of filled metal orbitals compared to $[Ru(bpy)_3]^{2+}$ and also lower the symmetry, leading to a broader absorption band. Nevertheless, the NCS ligands have drawbacks: they make it difficult to further tune the HOMO, and they can compromise the long-term chemical stability of the complex, one of the biggest issues limiting widespread uptake of the technology.

The past few years has witnessed increasing interest in cyclometallated ligands in place of the thiocyanates. The archetypal cyclometallating ligand in the field is 2-phenylpyridine (ppyH), which binds as an anionic N^C ligand, forming a 5-membered chelate ring, isoelectronic with neutral bpy. The resemblance between the electronics/MO energy levels of such complexes and their thiocyanate analogues has been highlighted by Berlinguette and co-workers, using a combination of experimental and DFT data [46]. It can be seen (Fig. 2.9) that the introduction of the ppy ligand in place of a bpy in $[Ru(bpy)_3]^{2+}$ has a similar effect to thiocyanate ligand, actually destabilising the filled metal orbitals to a slightly greater extent and leading to a pronounced long-wavelength tail in the absorption spectrum (Fig. 2.10) [47]. A similar analysis has been carried out by Grätzel and co-workers [48]. Notably, modification of the phenyl ring of the cyclometallating ligand (e.g. introduction

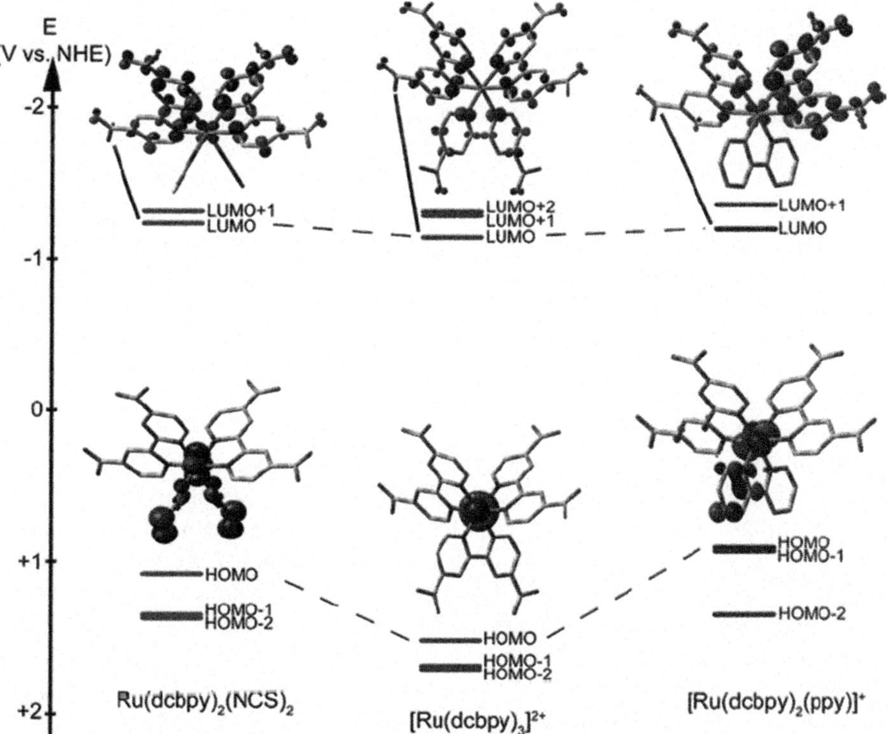

Fig. 2.9 Energy level diagram and selected MOs for the N3 dye (left), $[Ru(dcbpy)_3]^{2+}$ (*centre*) and the cyclometallated complex $[Ru(dcbpy)_2(bpy)]^{+}$. The positions of HOMO and LUMO energy levels are from experimental measurements; other energy levels calculated by TD-DFT and adjusted relative to the HOMO and LUMO (Reprinted with permission from [46]. © 2012 Elsevier)

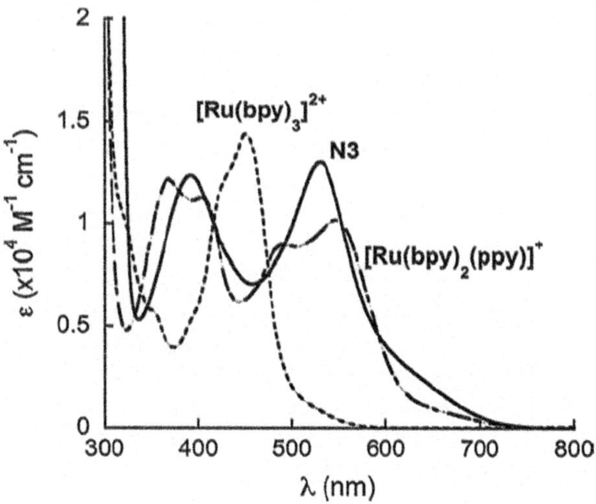

Fig. 2.10 Experimental UV-visible absorption spectra of the three complexes of Fig. 2.9 in MeCN at 298 K (Reprinted with permission from [46]. © 2012 Elsevier)

of different substituents) can be used to provide subtle control over the energy of the HOMO [49]. Chen et al. have used a similar strategy with a 'pseudo-cyclometallate' – an anionic N^N$^-$-binding pyrazolylpyridine ligand [50].

There has been considerable interest in applying DFT methods to dyes bound to the semiconductor surface, as opposed to studies on the isolated dye molecules in vacuo. The methods employed for this purpose can be divided mainly into two types. In one type, periodic DFT calculations with plane-wave methods are carried out, where the TiO_2 101 surface is created by cleaving the TiO_2 anatase crystal [51]. The other method treats the semiconductor as a 'molecular' species – a nanoparticle $(TiO_2)_n$, with n typically in the range 16–82 (e.g. [52]). Although the smaller nanoparticles do not necessarily allow all binding modes of the dye to be identified, they offer computational advantages. Calculations on dye–$(TiO_2)_n$, including TD-DFT calculations, are then carried out as isolated molecules in the gas phase or in solvent.

Singh and co-workers have very recently applied such methods to a series of cyclometallated Ru(II) complexes incorporating tridentate ligands (Fig. 2.11) [53]. Complex **M3** showed the best DSSC performance with an experimental overall energy conversion efficiency, η, of 7.1 %. TD-DFT calculations on the isolated complex showed that the presence of the cyclometallating ligand on one side of the molecule leads to high directionality in the charge-transfer process; for example, the most intense singlet transition in the low-energy region is mainly HOMO–1 → LUMO + 1 in character: the former spans the metal (56 % contribution) and the two ligands, whilst the latter is almost exclusively on the terpyridine ligand (Fig. 2.12). It is thanks to the electron-donating substituents on the cyclometallated ligand, which raise HOMO–1, that **M3** has the most redshifted absorption bands and the highest η amongst the four complexes.

The complex was modelled on $(TiO_2)_{38}$ using plane-wave methods [*op. cit*]. The bidentate bridging mode of binding via two carboxylates (Fig. 2.13b) was found to be energetically more favoured than that through only one carboxylate group (Fig.

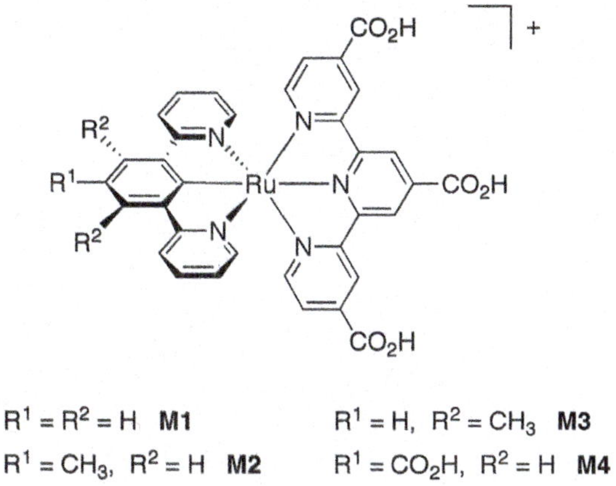

Fig. 2.11 Structures of the cyclometallated, bis-terdentate Ru (II) complexes studied by Singh and co-workers [53]

$R^1 = R^2 = H$ **M1** $R^1 = H$, $R^2 = CH_3$ **M3**

$R^1 = CH_3$, $R^2 = H$ **M2** $R^1 = CO_2H$, $R^2 = H$ **M4**

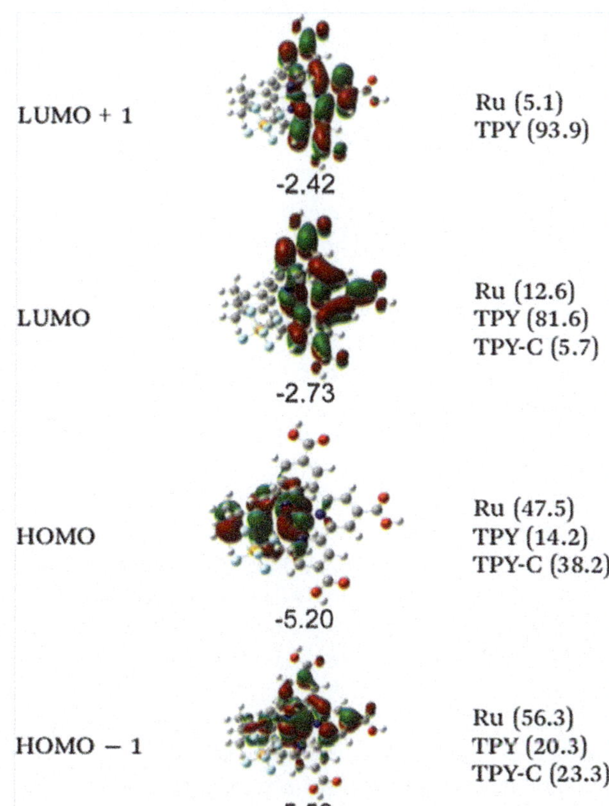

Fig. 2.12 Frontier orbitals of the complex **M3** (Fig. 2.11), their energies in eV and molecular orbital composition (%) (Reproduced from [53] with permission of the PCCP Owner Societies)

2.13a). TD-DFT calculations on this assembly gave absorption data that were in good agreement with those measured experimentally. Interestingly, the results suggest that the complexes exhibit some degree of direct-charge transfer to TiO_2 upon excitation.

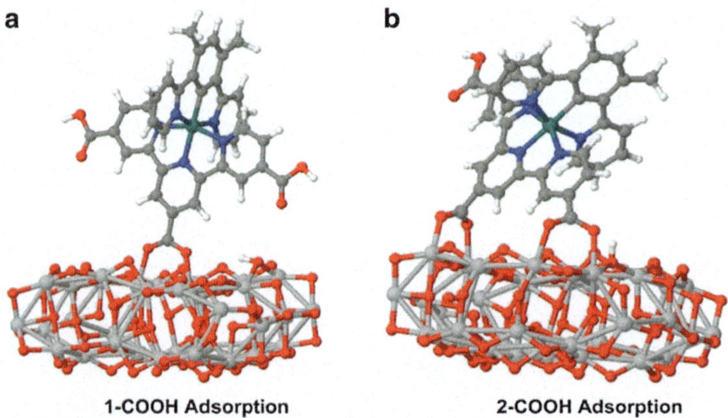

Fig. 2.13 Possible absorption configurations of the complex **M3** on $(TiO_2)_{38}$ (Reproduced from [53] with permission of the PCCP Owner Societies)

2.3.4 Electrochemical Potentials and Interaction with the Electrolyte

Theoretical methods are increasingly being applied to other aspects and processes involved in the DSSC. The nature of the interaction of the metal-complex dye with the redox electrolyte is of particular interest, since it determines the energetics and kinetics of dye regeneration. On the basis of experimental results, Clifford et al. proposed that dyes such as N3 – when used in DSSCs with the I^-/I_3^- redox couple – undergo regeneration via a transient [dye$^+$–iodide] intermediate complex, formed by the reaction of the photogenerated dye cation with I^- [54]. Subsequent reaction with further I^- forms I_2^-. DFT methods have been applied to probe such reactions by Schreckenbach and co-workers [55] and, independently, by Privalov et al. [56]. The former authors used SDD small-core ECPs and VDZ basis sets for the halogen with solvation effects included, whilst the latter study employed a large-core ECP for iodine. Briefly, the theoretical results rule out the formation of inner-sphere, 7-coordinate Ru(II) species containing I^-, but do reveal that a number of outer-sphere complexes are feasible. The results also indicate that the subsequent reaction $N3^+I^- + I^- \rightarrow N3I_2^-$ is the rate-limiting step, in line with experimental conclusions.

Finally, we note that the ability to model accurate ground- and excited-state oxidation potentials of dyes for DSSCs is also clearly valuable. Excited-state oxidation potentials require accurate optimisation of the excited-state geometry. In a study on donor-acceptor triphenylamine-based organic dyes, De Angelis and co-workers have highlighted the need to use functionals with a large amount (~50 %) of HF exchange for this purpose [57]. Such functionals are required to avoid the formation of artificial minima in twisted geometries with a high degree of charge transfer [58]. It is likely that the conclusions of this study will be equally applicable to many metal-based dyes.

2.4 Emission Properties of Metal Complexes

Phosphorescent organometallic complexes have been widely applied in the field of phosphors for organic light-emitting devices (OLEDs), including white-light OLEDs (WOLEDs) [1, 4, 59–62]. A wide range of different metals and ligands in various combinations have been synthesised and studied both in solution and in devices. Most of these synthetic targets are time consuming and potentially costly to make, such that it is often desirable to be able to predict their emissive properties prior to synthesis as well as to rationalise experimental properties in retrospect. The knowledge of the nature of excitations taking place then allows for tailored design of new complexes. For the design of a device of a specific colour, it is important to know the energy of emission. The efficiency of emission must also be considered since compounds exhibiting a very low quantum yield in solution, or no room-temperature emission, are unlikely to be viable candidates for OLED applications. DFT and TD-DFT are increasingly used to predict and rationalise these properties, selected cases of which will be given in this section with platinum-based examples, after a brief overview of the principles involved.

2.4.1 *Calculation of Emission Energy*

Generation of orbital plots for the $S_0 \rightarrow T_1$ excitation of a complex can be extremely informative as to the nature of the excitation. They are also useful for suggesting what effect substituent groups at certain positions of the ligand can have on the emission energy. Such calculations for organometallic complexes incorporating the arylheterocycle unit in Fig. 2.14 – common to many of the most successful phosphors – show that the HOMO is generally located primarily on the metal and cyclometallated aryl ring, with the LUMO based on the heterocyclic ring. This was illustrated for a series of iridium(III) complexes by Hay [14]. Alteration of substituents on the aryl ring will therefore typically affect the energy of the HOMO much more than the LUMO, the opposite being true of the heterocyclic ring. Naturally, the observed effect will differ according to different positions of substitution within the rings, particularly at the 5-position where conjugation through the two rings may affect both frontier orbitals [63].

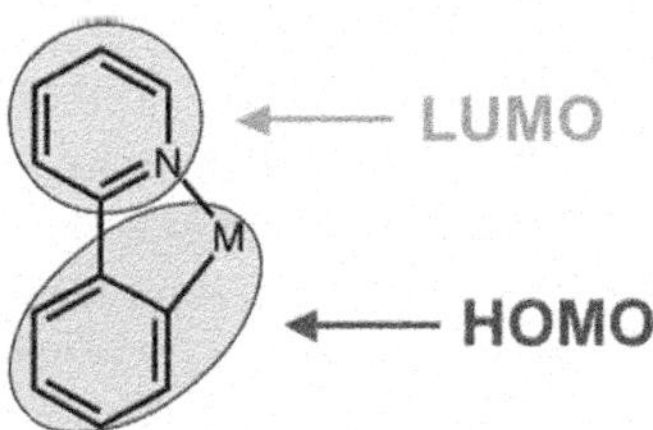

Fig. 2.14 Schematic diagram of the typical location of the frontier orbitals in cyclometallated arylpyridine complexes {e.g. where M = Ir(III) or Pt(II)}

Electron-withdrawing groups usually stabilise molecular orbitals whilst electron-donating substituents cause destabilisation. Knowledge of the position of the frontier orbitals, which can be derived from TD-DFT calculations, therefore allows control over the energy of emission obtained. For example, a compound with a higher energy of emission could be produced by addition of an electron-withdrawing group to the HOMO and/or an electron-donating group to the LUMO (a redshift is achieved by the reverse) [64]. Whilst excitation energies themselves can be produced by TD-DFT calculations, orbital plots are often more informative since they allow deliberate design of complexes tuned towards a particular wavelength of emission. Where excitation energies can be adversely affected by triplet instabilities leading to an incorrect emission energy (vide supra), orbital plots often provide a more reliable – albeit qualitative – way to predict how the HOMO and LUMO will be affected by various substituent groups.

The emission of light accompanying the $T_1 \rightarrow S_0$ transition does not usually involve purely HOMO and LUMO orbitals, but is rather made up of many small contributions from various other orbitals. For this reason, when considering orbital plots, those plots with the highest contribution to the excitation are normally presented alongside an indication as to the degree of their involvement. Alternatively, density difference plots can be used, which are generated by considering all the transitions, combining them and taking occupied from unoccupied to give an idea of the net movement of the electron upon excitation. Some examples follow in subsequent sections.

2.4.2 *Efficiency of Emission*

Experimentally, the efficiency of a compound's luminescence is measured by its quantum yield Φ_{lum}. This is determined by the relative rate constants for radiative (k_{r}) and non-radiative (Σk_{nr}) decay, as outlined in Equation 3 (where n_{E} is the number of photons emitted and n_{A} is the number absorbed), assuming that the emitting state is formed with unitary efficiency upon the absorption of light:

$$\Phi_{\text{lum}} = \frac{n_{\text{E}}}{n_{\text{A}}} = \frac{k_{\text{r}}}{k_{\text{r}} + \Sigma k_{\text{nr}}}. \tag{2.4}$$

In contrast to many purely organic molecules, for complexes with small ligands and metals with high spin-orbit coupling (SOC) constants, intersystem crossing is much faster than the rate of emission from the singlet excited state. Any observed emission will then normally emanate from the triplet excited state, giving phosphorescence. The exception to this rule of thumb is if the excited state is isolated away from the metal centre (e.g. when 'extended' ligands are used) [65–67].

Equation 2.4 makes clear that luminescence can be promoted either by increasing the rate of radiative decay or decreasing the rate of non-radiative decay. In phosphorescent metal complexes, the degree of metal character in the excited state becomes important in determining the efficiency of emission through its effect on

k_r, since it is thanks to the SOC effect of the metal that the spin selection rule is relaxed to promote triplet radiative decay. The rate constant for radiative decay will therefore normally be higher for excited states comprising significant metal character. Meanwhile, the Franck-Condon principle states that transitions with a high degree of orbital overlap will be more favourable than those with a lower degree of overlap, thus proceeding at an increased rate [68, 69].

Non-radiative decay of the excited states of organometallic complexes can occur through a number of different routes. The 'energy gap law' states that as the excited-state energy decreases (within a series of structurally similar compounds having a mutually common type of excited state), the rate of non-radiative decay of an excited-state complex will increase exponentially through intramolecular energy transfer into vibrations [70, 71]. Geometrical distortion of a compound in the excited state relative to the ground state also facilitates non-radiative decay, as illustrated schematically in Fig. 2.15. Diagram (a) illustrates the case where there is little or no distortion between the S_0 ground state and the T_1 excited state and (b) that where there is a large degree of distortion between the two states. Where there is no distortion, relaxation of the excited state to the ground state results in the emission of light (unless there is some other deactivating process). In (b), the excited-state PES is shifted with respect to that of the ground state, giving a crossing point between the two curves through which non-radiative decay can occur.

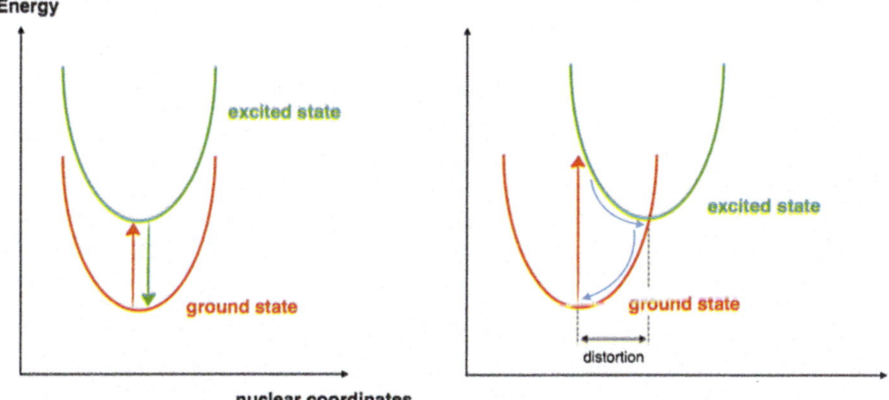

Fig. 2.15 Simplified schematic diagram highlighting how non-radiative decay is facilitated by distortion in the excited state

2.4.3 Selected Examples for Electrical-to-Light Energy Conversion (OLEDs)

2.4.3.1 Energy of Emission

A recent example of the utility of TD-DFT for understanding emission energies is provided by the work of Nisic et al. on a pair of isomeric, styryl-appended Pt (II) complexes (Fig. 2.16) [72]. It was found that irradiation of the complex with UV light resulted in isomerisation of the *trans* (E) isomer to the *cis* (Z), a process which was accompanied by a dramatic change in the absorptive and emissive properties. Whilst the *trans* isomer was non-emissive at room temperature with weak red emission at 77 K ($\lambda_{max} = 634$ nm), the *cis* showed bright green room-temperature emission. The analysis by TD-DFT (at the DFT-optimised T_1 geometry in each case) showed a marked difference in the density difference plots of the $S_0 \rightarrow T_1$ excitation between the two isomers. Whilst the *cis* isomer showed almost no involvement of the pendant group in the excitation – and indeed displays emission very much like that of the parent unsubstituted complex [73] – the plots

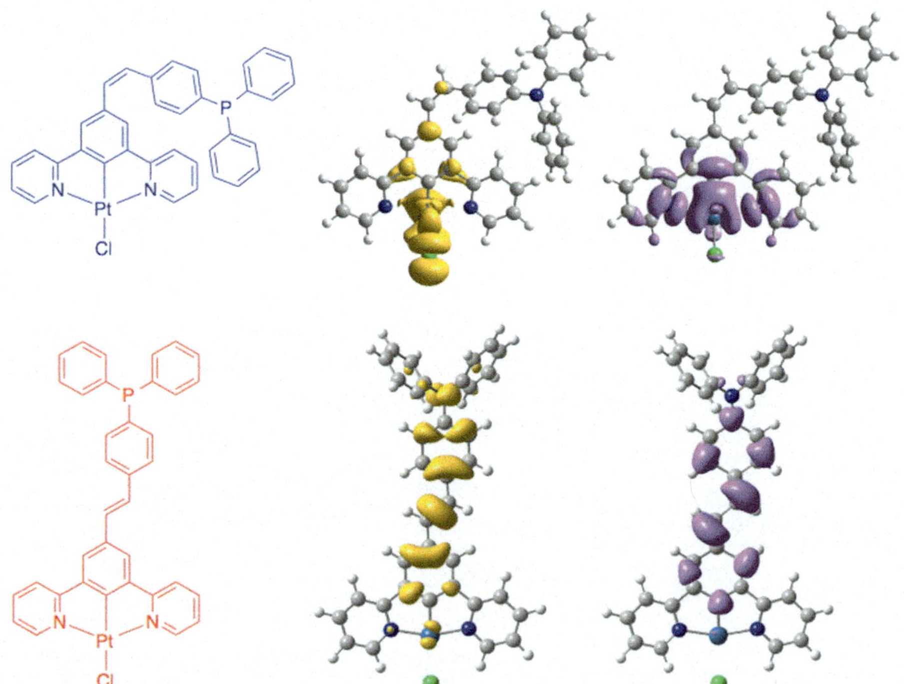

Fig. 2.16 Structures of the *cis* (*top*) and *trans* (*bottom*) isomers of the platinum complex studied by Nisic et al. [72]. Density difference plots for the $S_0 \rightarrow T_1$ transition are shown at the T_1 geometry, calculated by TD-DFT, using PBE0 in DCM. Published by The Royal Society of Chemistry

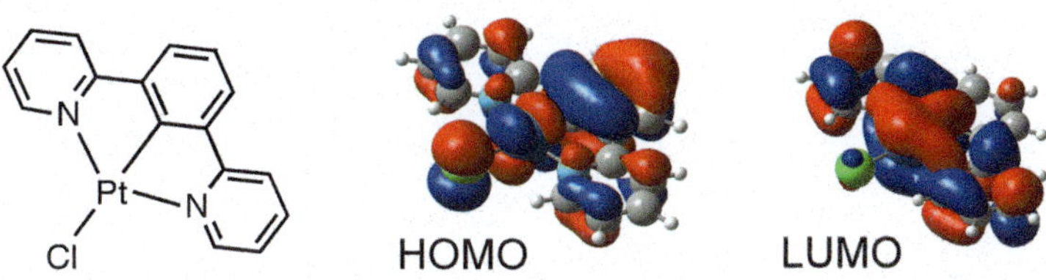

Fig. 2.17 Frontier orbitals for the Pt(II) complex of 1,3-di(2-pyridyl)benzene obtained using B3LYP and a PCM for dichloromethane solvent

for the *trans* isomer show that the movement of electron density is predominantly confined within the organic pendant itself (Fig. 2.16). Moreover, it is visually apparent from the plots that the amount of metal character in the excited state of the *trans* isomer is smaller than in the *cis*, providing an intuitively reasonable explanation for the former's low efficiency of emission.

As previously discussed, knowledge of the location of frontier orbitals allows for the rational targeting of an emission energy and hence colour: electron-donating groups destabilise molecular orbitals, whilst electron-withdrawing groups stabilise them. Farley et al. reported the synthesis and TD-DFT studies of the platinum (II) complex of 1,3-di(2-pyridyl)benzene and substituted derivatives (Fig. 2.17) [74]. The frontier orbitals generated by TD-DFT clearly show that the 4-position of the phenyl ring plays a large role in the HOMO, but almost none in the LUMO [75], and indeed, increasingly electron-donating substituents at this position increasingly redshift the emission. The reverse is true of the 4-positions of the pyridyl rings, which show involvement in the LUMO but not in the HOMO. Experimentally, electron-donating groups at these positions were found to *blueshift* the emission, in line with the prediction [64]. Overall, by judicious choice of substituents at both of these positions, the emission maxima in these systems can be tuned very simply over a wide range from about 450 to 600 nm.

2.4.3.2 Efficiency of Emission

The inclusion of spin-orbit coupling (SOC) in TD-DFT calculations is very complex and time intensive and so is not routinely attempted. Unlike singlet-singlet transitions, SOC is necessary when considering the oscillator strengths of phosphorescent processes since this is essentially the factor which is making the formally forbidden $T_1 \rightarrow S_0$ transition allowed. For this reason, the evaluation of k_r using TD-DFT remains far from routine. Tong and Che employed some of these techniques in an attempt to understand the emission efficiencies of the five Pt (II) complexes shown in Fig. 2.18 [15]. Experimental studies have shown that C^N^C-coordinated Pt(II) complexes are generally weakly emissive at room temperature and N^N^C analogues are moderately emissive, whereas N^C^N-coordinated isomers are often highly intense emitters. The experimental quantum yields for the five compounds are included in Fig. 2.18 [74, 76, 77]. Tong and Che

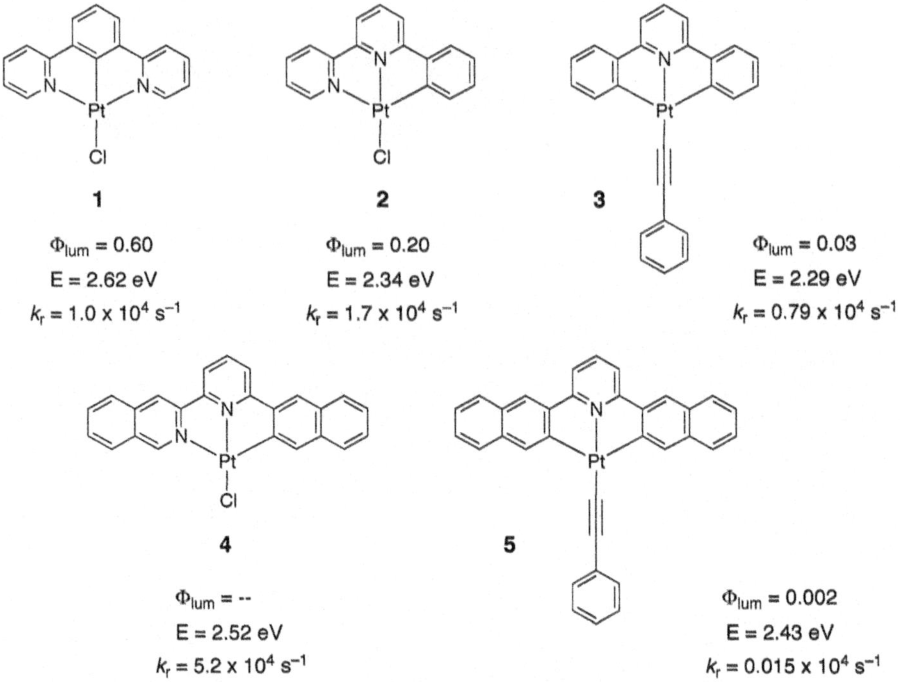

1

$\Phi_{lum} = 0.60$
$E = 2.62$ eV
$k_r = 1.0 \times 10^4\ s^{-1}$

2

$\Phi_{lum} = 0.20$
$E = 2.34$ eV
$k_r = 1.7 \times 10^4\ s^{-1}$

3

$\Phi_{lum} = 0.03$
$E = 2.29$ eV
$k_r = 0.79 \times 10^4\ s^{-1}$

4

$\Phi_{lum} =$ --
$E = 2.52$ eV
$k_r = 5.2 \times 10^4\ s^{-1}$

5

$\Phi_{lum} = 0.002$
$E = 2.43$ eV
$k_r = 0.015 \times 10^4\ s^{-1}$

Fig. 2.18 Structures of the complexes studied by Tong and Che, together with experimental quantum yields Φ_{lum}, calculated phosphorescence energies E (in DCM) and calculated rate of radiative decay k_r for the $T_1 \rightarrow S_0$ process

sought to calculate the rates of radiative decay by looking at a combination of three factors: (1) SOC matrix elements between the emissive triplet and singlet excited states, (2) the energy ratio between those two states and (3) the oscillator strength of the $S_n \rightarrow S_0$ transition with which the emissive triplet state undergoes SOC. The value of k_r calculated for each complex is shown in Fig. 2.18. For compounds **2**, **3** and **5**, the calculated values for k_r correlate with the experiment, whereas **1** is a better emitter than expected on the basis of the calculated k_r. The authors reasoned that the emission from **1** was instead emanating from the T_2 state since k_r was much larger for that transition ($109 \times 10^4\ s^{-1}$) and the Stokes shift for the $S_0 \rightarrow T_2$ excitation at the T_1 optimised geometry was more like the value obtained experimentally than for the $S_0 \rightarrow T_1$ excitation. Consideration of the excited-state geometries of compounds **2** and **3** showed that these compounds undergo significant distortion at the T_1 excited state, suggesting efficient non-radiative decay. The complexity of considering d-orbital splittings was also underlined in this study, which described how a compromise must be made between the need for large splitting between occupied and unoccupied d-orbitals necessary to make deactivating d–d excited states thermally inaccessible and the need for occupied d-orbitals to be close in energy for efficient SOC [15].

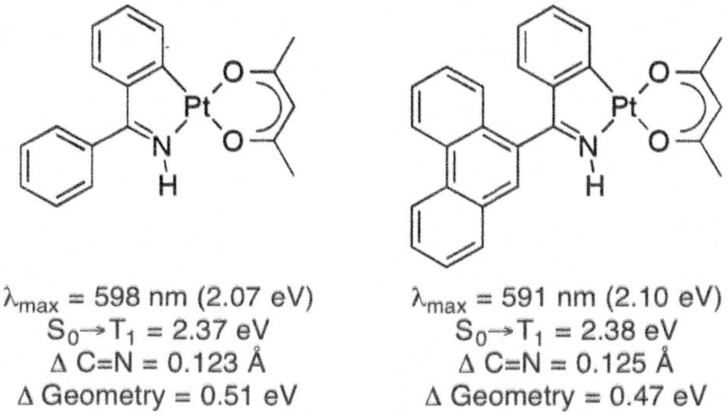

λ_{max} = 598 nm (2.07 eV)
$S_0 \rightarrow T_1$ = 2.37 eV
Δ C=N = 0.123 Å
Δ Geometry = 0.51 eV

λ_{max} = 591 nm (2.10 eV)
$S_0 \rightarrow T_1$ = 2.38 eV
Δ C=N = 0.125 Å
Δ Geometry = 0.47 eV

Fig. 2.19 Structures and properties of two imine-based N^C-coordinated Pt(II) complexes studied by Pandya et al. λ_{max} represent the experimentally measured emission in DCM at 298 K, $S_0 \rightarrow T_1$ the calculated excitation energy at the ground-state geometry, ΔC=N the change in the C=N bond length between the S_0 and T_1 geometries and Δ geometry the difference between the energy of the ground state at the S_0 geometry and the single-point energy at the T_1 geometry

The influence of excited-state distortion has also been considered for a series of Pt(II) complexes with imine-based N^C-coordinating ligands, of which two examples are shown in Fig. 2.19 [78]. Both complexes showed almost no emission in solution at 298 K, despite the structural resemblance to numerous successful emitters such as Pt(N^C-ppy)(acac). DFT and TD-DFT calculations were employed in an attempt to rationalise the processes taking place that result in such efficient non-radiative deactivation of the excited state. The $S_0 \rightarrow T_1$ excitation energy at the S_0 and T_1 geometries was calculated. These showed very different energies from those obtained experimentally, suggesting that the T_1 excited-state geometries were very different from the ground-state geometries. The calculated T_1 geometry revealed that the C = N bond is substantially elongated compared to the ground state. It was also noted that single-point singlet calculations at the T_1 geometry gave quite different SCF energies from the values calculated at the ground-state geometries (Fig. 2.19), which again points to excited-state distortion. The application of this technique does, however, rely heavily on small differences in computed energies and so should be approached with some caution: inherent errors of up to about 0.3 eV are considered quite normal for TD-DFT work, and the presence of triplet stabilities can lead to a detrimental effect on the accuracy of triplet-state energies.

The synthesis, liquid crystal and luminescent properties of a series of ortho-platinated complexes incorporating N^C-coordinated arylpyridines in conjunction with β-diketonate ligands have been reported by Spencer et al. [79]. Such complexes (i.e. those of the form [Pt(N^C)(O^O)]) generally form a lowest triplet excited state located on the metal and cyclometallated ligand (as in Fig. 2.14), with the diketonate acting as an 'innocent' ancillary ligand, with little effect on the emission properties. However, it was observed that, in this instance, whilst Pt(ppy)(acac) displayed the expected phosphorescence in solution at room temperature, Pt(ppy)(hfac) showed no emission under the same conditions (the structures are shown in Fig. 2.20). Moreover, two isomers of the tfac complex were formed, of

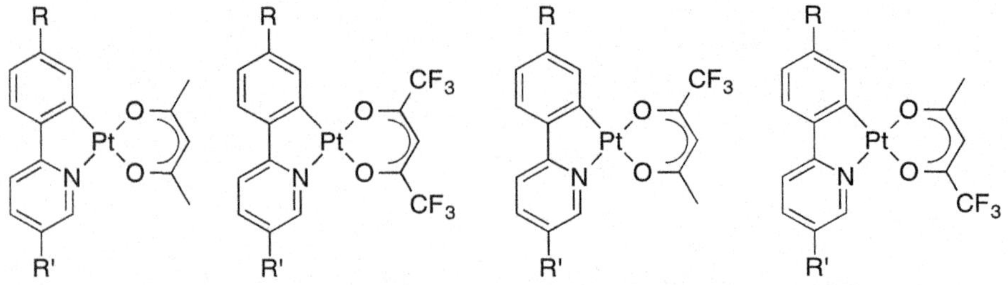

[Pt(ppy)(acac)] [Pt(ppy)(hfac)] *trans*-[Pt(ppy)(tfac)] *cis*-[Pt(ppy)(tfac)]

R = H	R' = H	DFT studies
R = OC_6H_{13}	R' = $PhOC_6H_{13}$	Experimental measurements

Fig. 2.20 The structures of the four Pt(N^C)(O^O) complexes studied by Spencer et al. [79]

which the *trans* isomer emits at a higher energy than the *cis*, but with a significantly lower quantum yield (here, *trans/cis* refers to the relative disposition of the CF_3 group and the pyridyl ring). S_0ground-state geometries of the four complexes were optimised using DFT, and the density difference plots for the $S_0 \rightarrow T_1$ excitation of each complex at this geometry are shown in Fig. 2.21. It can be seen that the non-emissive Pt(ppy)(hfac) shows a marked difference from the other three compounds. Its density difference plot involves electron depletion from the aryl ring of the ppy ligand and augmentation on the hfac ligand. In contrast, the other three complexes display rearrangement of electron density on the ppy ligand upon excitation, in line with excited states normally calculated in Pt(N^C)(O^O) complexes, i.e. $d_{Pt} / \pi_{N^{\wedge}C} \rightarrow \pi^*_{N^{\wedge}C}$. The change in excitation shown for Pt(ppy)(hfac) was attributed to the stabilisation of the orbitals by the electron-withdrawing CF_3 groups to such an extent that the LUMO is positioned on the O^O ligand instead. It was also shown that this excitation has a lower orbital overlap, showing that the reduced quantum yield could – at least in part – be due to a lower rate of radiative decay.

Moreover, by evaluating the first triplet excited-state geometry (T_1) of each of the four complexes, the researchers revealed another likely contribution to the variation of emission efficiency amongst the four complexes. The comparison of the S_0 and T_1 geometries (both calculated by DFT) for each of the four complexes shows more significant excited-state distortion for some complexes than others. Figure 2.22 shows the T_1 geometry of each complex superimposed upon its respective ground-state geometry, S_0. Whilst Pt(ppy)(acac) undergoes virtually no geometrical distortion upon excitation, all of the other three complexes do so, to varying degrees. The largest change in geometry is shown by Pt(ppy)(hfac). The significant distortion in the hfac complex is a likely reason for the increase in non-radiative decay, which, combined with a decrease in radiative decay noted above, accounts for the lack of emission displayed by this complex. Similar reasoning may explain the difference in emission efficiencies between *cis*- and

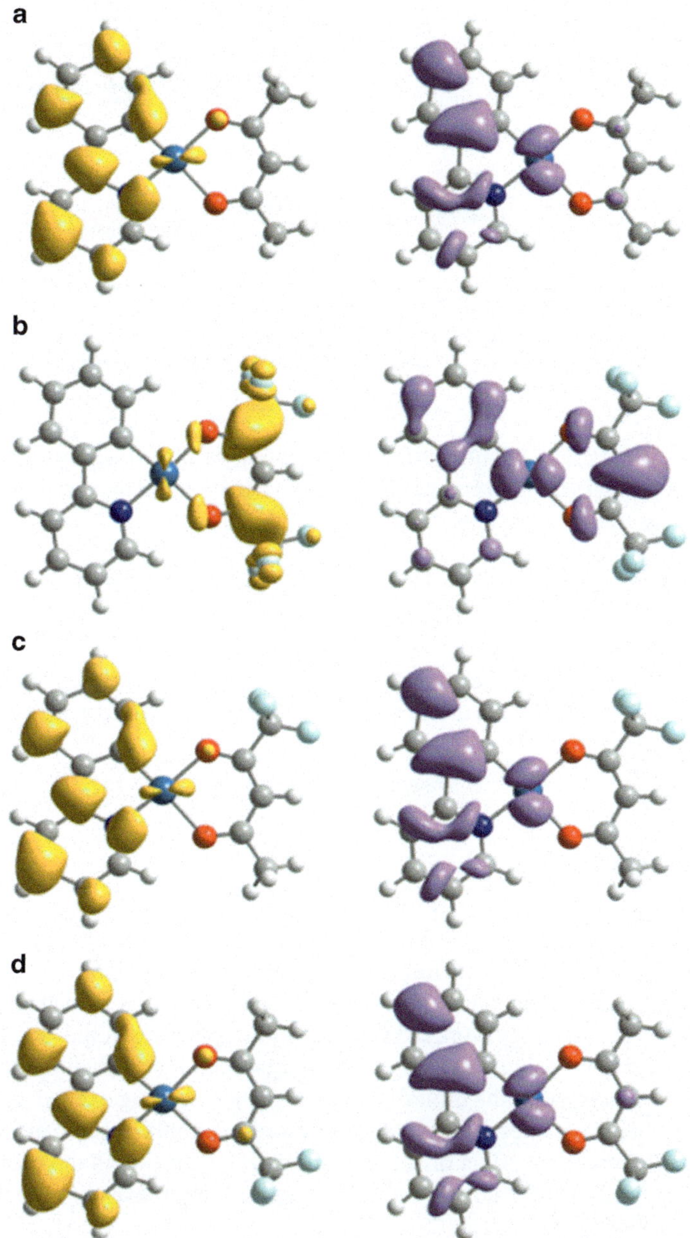

Fig. 2.21 Density difference plots for the $S_0 \rightarrow T_1$ excitation of the four complexes of Fig. 2.20 at the ground-state geometry: (**a**) Pt(ppy)(acac), (**b**) Pt(ppy)(hfac), (**c**) *trans*-Pt(ppy)(tfac) and (**d**) *cis*-Pt(ppy)(tfac). Electron depletion shown on the *left* and accretion on the *right* (Reproduced from [79] with permission from The Royal Society of Chemistry)

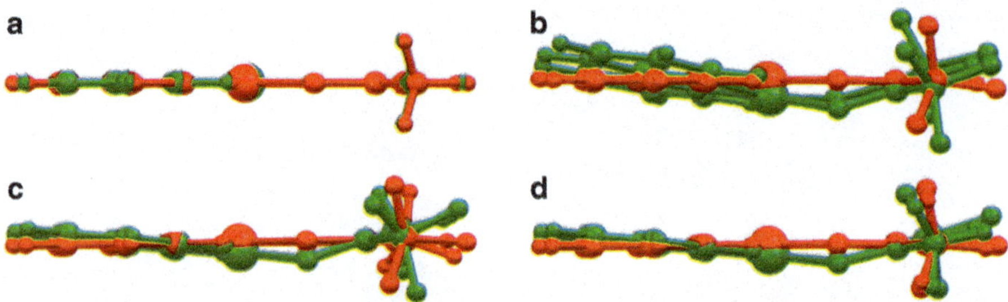

Fig. 2.22 Superimposed structures of the S_0 ground state (planar) and T_1 excited state (distorted) for the complexes of Fig. 2.21: (**a**) Pt(ppy)(acac), (**b**) Pt(ppy)(hfac), (**c**) *trans*-Pt(ppy)(tfac) and (**d**) *cis*-Pt(ppy)(tfac) (Reproduced from [79] with permission from The Royal Society of Chemistry)

trans-Pt(ppy)(tfac). The quantification of the extent of distortion by calculation of the root-mean-square displacement of the atoms between the two states (S_0 and T_1) shows that the extent of distortion increases in the order:

$$[\mathrm{Pt(ppy)(acac)}] < cis\text{-}[\mathrm{Pt(ppy)(tfac)}] < trans\text{-}[\mathrm{Pt(ppy)(tfac)}] < [\mathrm{Pt(ppy)(hfac)}].$$

Greater distortion in the T_1 state of the *trans* isomer compared to the *cis* may lead to a greater rate of non-radiative decay and thus to the observed less efficient emission at room temperature.

2.5 Conclusions

In this short contribution, we have sought to provide an indication of how theoretical methods based on density functional theory are being increasingly used to probe ground- and excited-state properties of metal complexes relevant to their use in energy conversion. Early work in this area at the beginning of the century tended to focus primarily on using TD-DFT to obtain qualitative information about trends in absorption energies and the orbital parentage of lowest-lying excited states. More recently, studies have expanded to consider more explicitly factors such as triplet states, phosphorescence, effect of solvent, structural distortion in the excited states compared to the ground state and immobilisation of complexes onto semiconductors, to name but a few.

There remain major challenges: methods are still mostly used retrospectively to rationalise behaviour and rarely to predict properties in advance of synthesis. In particular, being able to predict the efficiency of emission is a complicated problem: a detailed analysis would require a full treatment of spin-orbit coupling pathways and evaluation of the coupling mechanisms between excited electronic states and all vibrational modes – utterly unfeasible for molecules of this size. Nevertheless, by using simple concepts such as those relating deactivation pathways to the degree of structural distortion in the excited state, valuable deductions can be made.

As far as excited-state energies are concerned, there is often a good agreement between theory and experiment for singlet states, and many studies have been able to show a satisfactory match between experimental absorption spectra and those simulated using TD-DFT. On the other hand, the analysis of emission is much less common, as it requires knowledge of the relaxed excited-state geometry. For phosphorescence from the triplet state, the situation should be aided by the fact that DFT can be used to directly determine the lowest state of any multiplicity, thus including the T_1. But triplet instabilities can cause unexpected problems, particularly in those excited states which have high orbital overlap. The recent implementation of the Tamm-Dancoff approximation into some of the commercial DFT packages is likely to go some ways to dealing with these issues.

Clearly, the continual development of new and improved functionals and basis sets, as well as faster computer power, renders the area one which is bound to enjoy major advances over the next decade.

Acknowledgements We thank our colleagues Prof David Tozer (Durham University) and Dr Michael Peach (University of Lancaster) for stimulating discussions and collaborative work on theory applied to excited states of metal complexes. Our work in the field has been supported by a studentship from EPSRC (to G.R.F., grant ref. EP/G06928X/1) and facilitated by Durham University through access to its high-performance computing facility.

References

1. Yersin H (ed) (2007) Highly efficient OLEDs with phosphorescent materials. Wiley-VCH, Berlin
2. Hagfeldt A, Boschloo G, Sun L, Kloo L, Pettersson H (2010) Dye-sensitized solar cells. Chem Rev 110:6595–6663
3. Young KJ, Martini LA, Milot RL, Snoeberger RC, Batista VS, Schmuttenmaer CA, Crabtree RH, Brudvig GW (2012) Light-driven water oxidation for solar fuels. Coord Chem Rev 256:2503–2520
4. Gildea LF, Williams JAG (2013) Iridium and platinum complexes for OLEDs. In: Buckley A (ed) Organic light-emitting diodes (OLEDs): materials, devices and applications. Woodhead, Cambridge
5. Hay BP, Clement O (1998) Metal complexes. In: Schleyer PR (ed) Encyclopedia of computational chemistry. Wiley, Chichester
6. Holder AJ (1998) Semiempirical methods: Transition Metals. In: Schleyer PR (ed) Encyclopedia of computational chemistry. Wiley, Chichester
7. Frenking G, Wagener T (1998) Transition metal chemistry. In: Schleyer PR (ed) Encyclopedia of computational chemistry. Wiley, Chichester
8. Vlček A Jr, Záliš S (2007) Modeling of charge-transfer transitions and excited states in d^6 transition metal complexes by DFT techniques. Coord Chem Rev 251:258–287
9. Koch W, Holthausen MC (2001) A chemist's guide to density functional theory, 2nd edn. Wiley-VCH, Weinheim
10. Tang CW, VanSlyke SA (1987) Organic electroluminescent diodes. Appl Phys Lett 51:913–915

11. Martin RL, Kress JD, Campbell IH, Smith DL (2000) Molecular and solid-state properties of tris-(8-hydroxyquinolate)-aluminium. Phys Rev B 61:15804–15811
12. Halls MD, Schlegel HB (2001) Molecular orbital study of the first excited state of the OLED material tris(8-hydroxyquinoline)aluminium(III). Chem Mater 13:2632–2640
13. Curioni A, Andreoni W (1999) Metal-Alq$_3$ complexes: the nature of the chemical bonding. J Am Chem Soc 121:8216–8220
14. Hay PJ (2002) Theoretical studies of the ground and excited electronic states in cyclometalated phenylpyridine Ir(III) complexes using density functional theory. J Phys Chem A 106:1634–1641
15. Tong GSM, Che CM (2009) Emissive or nonemissive? A theoretical analysis of the phosphorescence efficiencies of cyclometalated platinum(II) complexes. Chem Eur J 15:7225–7237
16. Costa PJ, Calhorda MJ (2006) A DFT and MP2 study of luminescence of gold(I) complexes. Inorg Chim Acta 359:3617–3624
17. Labat F, Le Bahers T, Ciofini I, Adamo C (2012) First-principles modeling of dye-sensitized solar cells: challenges and perspectives. Acc Chem Res 45:1268–1277
18. Perdew JP, Burke K, Ernzerhof M (1996) Generalized gradient approximation for the exchange-correlation hole of a many-electron system. Phys Rev B 54:16533–16539
19. Becke AD (1993) Density functional thermochemistry 3. The role of exact exchange. J Chem Phys 98:5648–5652
20. Lee C, Yang W, Parr RG (1988) Development of the Colle-Salvetti correlation-energy formula into a functional of the electron-density. Phys Rev B 37:785–789
21. Perdew JP, Burke K, Ernzerhof M (1996) Generalized gradient approximation made simple. Phys Rev Lett 77:3865–3868
22. Grätzel M (2005) Solar energy conversion by dye-sensitized solar cells. Inorg Chem 44:6841–6851
23. Yersin H, Rausch AF, Czerwieniec R, Hofbeck T, Fischer T (2011) The triplet state of organo-transition metal compounds, Triplet harvesting and singlet harvesting for efficient OLEDS. Coord Chem Rev 255:2622–2652
24. Gill PMW, Adamson RD, Pople JA (1996) Coulomb-attenuated exchange energy density functionals. Mol Phys 88:1005–1009
25. Yanai T, Tew DP, Handy NC (2004) A new hybrid exchange-correlation functional using the Coulomb-attenuating method (CAM-B3LYP). Chem Phys Lett 393:51–57
26. Adamo C, Jacquemin D (2013) The calculations of excited-state properties with time-dependent density functional theory. Chem Soc Rev 42:845
27. Jacquemin D, Planchat A, Adamo C, Mennucci B (2012) TD-DFT assessment of functionals for optical 0–0 transitions in solvated dyes. J Chem Theory Comput 8:2359–2372
28. Zhan H, Lamare S, Ng A, Kenny T, Guernon H, Chan WK, Djurisic AB, Harvey PD, Wong WY (2011) Synthesis and photovoltaic properties of new metalloporphyrin-containing polyplatinyne polymers. Macromolecules 44:5155–5167
29. Wu W, Wu W, Ji S, Guo H, Song P, Han K, Chi L, Shao J, Zhao J (2010) Tuning of the emission properties of cyclometalated platinum(II) complexes by intramolecular electron-sink/arylethynylated ligands and its application for enhanced luminescent oxygen sensing. J Mater Chem 20:9775
30. Wu W, Wu W, Ji S, Guo H, Zhao J (2011) Accessing the long-lived emissive ^{3}IL triplet excited states of coumarin fluorophores by direct cyclometallation and its application for oxygen sensing and upconversion. Dalton Trans 40:5953–5963
31. Sears JS, Koerzdoerfer T, Zhang CR, Brédas JL (2011) Orbital instabilities and triplet states from time-dependent density functional theory and long-range corrected functionals. J Chem Phys 135:151103
32. Peach MJG, Williamson MJ, Tozer DJ (2011) Influence of triplet instabilities in TD-DFT. J Chem Theory Comput 7:3578–3585
33. Seeger R, Pople JA (1977) Self-consistent molecular-orbital methods 18. Constraints and stability in Hartree-Fock theory. J Chem Phys 66:3045–3050

34. Hirata S, Head-Gordon M (1999) Time-dependent density functional theory within the Tamm-Dancoff approximation. Chem Phys Lett 314:291–299
35. Tamm I (1941) Theory of the mesotron and of nuclear forces. Usp Fiz Nauk 25:136–143
36. Dancoff SM (1950) Non-adiabatic meson theory of nuclear forces. Phys Rev 78:382–385
37. Hanson K, Roskop L, Djurovich PI, Zahariev F, Gordon MS, Thompson ME (2010) A paradigm for blue- or red-shifted absorption of small molecules depending on the site of π extension. J Am Chem Soc 132:16247–16255
38. Barone V, Bloino J, Monti S, Pedone A, Prampolini G (2011) Fluorescence spectra of organic dyes in solution: a time-dependent multilevel approach. Phys Chem Chem Phys 13:2160–2166
39. Tomasi J, Mennucci B, Cammi R (2005) Quantum mechanical continuum solvation models. Chem Rev 105:2999–3093
40. Kataoka Y, Kitagawa Y, Kawakami T, Okumura M (2013) Photophysical properties of mono- and di-nuclear platinum(II) complexes with the tridentate ligand 2-phenyl-6-(1H-pyrazol-3-yl) pyridine: A DFT and TDDFT study. J Organomet Chem 743:163–169
41. Li Z, Badaeva A, Ugrinov A, Kilina S, Sun W (2013) Platinum chloride complexes containing 6-[9,9-di(2-ethylhexyl)-7-R-9H-fluoren-2-yl]-2,2'-bipyridine ligand ($R = NO_2$, CHO, benzothiazol-2-yl, n-Bu, carbazol-9-yl, NPh_2): tunable photophysics and reverse saturable absorption. Inorg Chem 52:7578–7592
42. Zhang R, Liang Z, Han A, Wu H, Du P, Lai W, Cao R (2014) Structural, spectroscopic and theoretical studies of a vapochromic platinum(II) terpyridyl complex. CrystEngComm 16:5531–5542
43. Liu P, Fu JJ, Guo MS, Zuo X, Liao Y (2013) Effect of the chemical modifications of thiophene-based N3 dyes on the performance of dye-sensitized solar cells: A density functional theory study. Comput Theor Chem 1015:8–14
44. Noureen S, Caramori S, Monari A, Assfeld X, Argazzi R, Bignozzi CA, Beley M, Gros PC (2012) Strong π-delocalisation and substitution effect on electronic properties of dithienylpyrrole-containing bipyridine ligands and corresponding ruthenium complexes. Dalton Trans 41:4833–4844
45. Noureen S, Argazzi R, Monari A, Beley M, Assfeld X, Bignozzi CA, Caramori S, Gros PC (2014) Novel Ru-based sunlight harvesters bearing dithienylpyrrolo (DTP)-bipyridine ligands: synthesis, characterization and photovoltaic properties. Dyes Pigments 101:318–328
46. Bomben PG, Robson KCD, Koivisto BD, Berlinguette CP (2012) Cyclometalated ruthenium chromophores for the dye-sensitized solar cell. Coord Chem Rev 256:1438–1450
47. Bomben PG, Robson KCD, Sedach PA, Berlinguette CP (2009) On the viability of cyclometalated Ru(II) complexes for light-harvesting applications. Inorg Chem 48:9631–9643
48. Yoneda E, Nazeeruddin MK, Grätzel M (2012) Cyclometalated ruthenium dyes for DSSC. J Photopolym Sci Technol 25:175–181
49. Bomben PG, Koivisto BD, Berlinguette CP (2010) Cyclometalated Ru complexes of type $[RuII(N^\wedge N)_2(C^\wedge N)]^z$: physicochemical response to substituents installed on the anionic ligand. Inorg Chem 49:4960–4971
50. Chen BS, Chen K, Hong YH, Liu WH, Li TH, Lai CH, Chou PT, Chi Y, Lee GH (2009) Neutral, panchromatic Ru(II) terpyridine sensitizers bearing pyridine pyrazolate chelates with superior DSSC performance. Chem Commun 45:5844–5846
51. Srinivas K, Yesudas K, Bhanuprakash K, Rao VJ, Giribabu L (2009) A combined experimental and computational investigation of anthracene based sensitizers for DSSC: comparison of cyanoacrylic and malonic acid electron withdrawing groups binding onto the TiO_2 anatase (101) surface. J Phys Chem C 113:20117–20126
52. Lundqvist MJ, Nilsing M, Persson P, Lunell S (2006) DFT study of bare and dye-sensitised TiO2 clusters and nanocrystals. Int J Quantum Chem 106:3214–3234
53. Chitumalla RK, Gupta KSV, Malapaka C, Fallahpour R, Islam A, Han L, Kotamarthi B, Singh SP (2014) Thiocyanate-free cyclometalated ruthenium(II) sensitizers for DSSC: a combined experimental and theoretical investigation. Phys Chem Chem Phys 16:2630–2640

54. Clifford JN, Palomares E, Nazeeruddin MK, Grätzel M, Durrant JR (2007) Dye dependent regeneration dynamics in dye sensitized nanocrystalline solar cells: evidence for the formation of a ruthenium bipyridyl cation/iodide intermediate. J Phys Chem C 111:6561–6567
55. Hu CH, Asaduzzaman AM, Schreckenbach G (2010) Computational studies of the interaction between ruthenium dyes and X^- and X_2^-, X = Br, I. At. Implications for dye-sensitized solar cells. J Phys Chem 114:15165–15173
56. Privalov T, Boschloo G, Hagfeldt A, Svensson PH, Kloo L (2009) A study of the interactions between I^-/I_3^- redox mediators and organometallic sensitizing dyes in solar cells. J Phys Chem C 113:783–790
57. Pastore M, Fantacci S, De Angelis F (2010) Ab initio determination of ground and excited state oxidation potentials of organic chromophores for dye-sensitized solar cells. J Phys Chem C 114:22742–22750
58. Wiggins P, Williams JAG, Tozer DJ (2009) Excited state surfaces in density functional theory: a new twist on an old problem. J Chem Phys 131:091101
59. Williams JAG, Develay S, Rochester DL, Murphy L (2008) Optimising the luminescence of platinum(II) complexes and their application in organic light emitting devices (OLEDs). Coord Chem Rev 252:2596–2611
60. Kalinowski J, Fattori V, Cocchi M, Williams JAG (2011) Light-emitting devices based on organometallic platinum complexes as emitters. Coord Chem Rev 255:2401–2425
61. Wong WY, Ho CL (2009) Functional metallophosphors for effective charge carrier injection / transport: new robust OLED materials with emerging applications. J Mater Chem 19:4457–4482
62. Chi Y, Chou PT (2010) Transition-metal phosphors with cyclometalating ligands: fundamentals and applications. Chem Soc Rev 39:638–655
63. Frey J, Curchod BFE, Scopelliti R, Tavernelli I, Rothlisberger U, Nazeeruddin MK, Baranoff E (2014) Structure-property relationships based on Hammett constants in cyclometalated iridium(III) complexes: their application to the design of a fluorine-free FIrPic-like emitter. Dalton Trans 43:5667–5679
64. Murphy L, Brulatti P, Fattori V, Cocchi M, Williams JAG (2012) Blue-shifting the monomer and excimer phosphorescence of tridentate cyclometallated platinum(II) complexes for optimal white-light OLEDs. Chem Commun 48:5817–5819
65. Wong WY, He Z, So SK, Tong KL, Li Z (2005) A multifunctional platinum-based triplet emitter for OLED applications. Organometallics 24:4079–4082
66. Chen YL, Li SW, Chi Y, Cheng YM, Pu SC, Yeh YS, Chou PT (2005) Switching luminescent properties in osmium-basedβ-diketonate complexes. ChemPhysChem 6:2012–2017
67. Kozhevnikov DN, Kozhevnikov VN, Shafikov MZ, Prokhorov AM, Bruce DW, Williams JAG (2011) Phosphorescence vs fluorescence in cyclometalated platinum(II) and iridium(III) complexes of (oligo)thienylpyridines. Inorg Chem 50:3804–3815
68. Franck J, Dymond EG (1926) Elementary processes of photochemical reactions. Trans Faraday Soc 21:536–542
69. Condon E (1926) A theory of intensity distribution in band systems. Phys Rev 28:1182–1201
70. Englman R, Jortner J (1970) The energy gap law for radiationless transitions in large molecules. Mol Phys 18:145–164
71. Whittle CE, Weinstein JA, George MW, Schanze KS (2001) Photophysics of diimine platinum (II) bis-acetylide complexes. Inorg Chem 40:4053–4062
72. Nisic F, Colombo A, Dragonetti C, Roberto D, Valore A, Malicka JM, Cocchi M, Freeman GR, Williams JAG (2014) Platinum(II) complexes with cyclometallated 5-π-delocalised-donor-1,3-di(2-pyridyl)benzene ligands as efficient phosphors for NIR-OLEDs. J Mater Chem C 2:1791–1800
73. Williams JAG, Beeby A, Davies ES, Weinstein JA, Wilson C (2003) An alternative route to highly luminescent platinum(II) complexes: cyclometalation with N^C^N-coordinating dipyridylbenzene ligands. Inorg Chem 42:8609–8611

74. Farley SJ, Rochester DL, Thompson AL, Howard JAK, Williams JAG (2005) Controlling emission energy, self-quenching, and excimer formation in highly luminescent N^C^N-coordinated platinum(II) complexes. Inorg Chem 44:9690–9703
75. Rochester DL, Develay S, Záliš S, Williams JAG (2009) Localised to intraligand charge-transfer states in cyclometalated platinum complexes: an experimental and theoretical study into the influence of electron-rich pendants and modulation of excited states by ion binding. Dalton Trans 1728–1741
76. Kui S, Sham I, Cheung C, Ma CW, Yan B, Zhu N, Che CM, Fu WF (2007) Platinum (II) complexes with π-conjugated, naphthyl-substituted, cyclometalated ligands (R-C^N^N): structures and photo- and electroluminescence. Chem Eur J 13:417–435
77. Lu W, Chan MCW, Cheung KK, Che CM (2001) π–π interactions in organometallic systems. Crystal structures and spectroscopic properties of luminescent mono-, bi-, and trinuclear trans-cyclometalated platinum(II) complexes derived from 2,6-diphenylpyridine. Organometallics 20:2477–2486
78. Pandya SU, Moss KC, Bryce MR, Batsanov AS, Fox MA, Jankus V, Al Attar HA, Monkman AP (2010) Luminescent platinum(II) complexes containing cyclometallated diaryl ketimine ligands: synthesis, photophysical and computational properties. Eur J Inorg Chem 1963–1972
79. Spencer M, Santoro A, Freeman GR, Diez A, Murray PR, Torroba J, Whitwood AC, Yellowlees LJ, Williams JAG, Bruce DW (2012) Phosphorescent, liquid-crystalline complexes of platinum(II): influence of the β-diketonate co-ligand on mesomorphism and emission properties. Dalton Trans 41:14244–14256

Chapter 3
First-Row Transition Metal Complexes for the Conversion of Light into Electricity and Electricity into Light

Etienne Baranoff

Abstract Ruthenium and iridium complexes have been widely used as sensitizer for dye-sensitized solar cells (conversion of light into electricity) and as highly phosphorescent emitters for organic electroluminescence (conversion of electricity into light). The high costs and limited availability of these platinoid metals have motivated the search for alternatives based on first-row transition metals. First-row transition metal complexes have also been used as alternatives to existing materials as redox mediator. This chapter provides an overview of such materials used as an active component of the aforementioned devices.

Keywords First-row transition metal complexes • Dye-sensitized solar cells • Redox mediator • Photovoltaic • OLEDs • Light-emitting electrochemical cells • Electroluminescence

Abbreviations

Dcbpy	4,4′-dicarboxy-2,2′-bipyridine
DSC	dye-sensitized solar cell
ff	fill factor
HOMO	highest occupied molecular orbitals
IPCE	incident monochromatic photon-to-current conversion efficiency
iTMCs	ionic transition metal complexes
Jsc	short-circuit current
LEC	light-emitting electrochemical cells
LF	ligand field
LUMO	lowest unoccupied molecular orbitals
OLEDs	organic light-emitting diodes
MLCT	metal-to-ligand charge transfer
MPCT	metal-to-particle charge transfer

E. Baranoff (✉)
School of Chemistry, University of Birmingham, Edgbaston, B15 2TT Birmingham, UK
e-mail: e.baranoff@bham.ac.uk

W.-Y. Wong (ed.), *Organometallics and Related Molecules for Energy Conversion*, Green Chemistry and Sustainable Technology, DOI 10.1007/978-3-662-46054-2_3

ppy	2-phenylpyridine
TADF	thermally activated delayed fluorescence
TCO	transparent conductive oxide
Voc	open-circuit potential

3.1 Introduction

Our society is craving for energy. The current yearly global energy consumption is roughly 570 exajoules, about 100 times the consumption of a century ago [1]. This vast increase in energy use has been implicated in the fabulous technological developments and overall significant improvement of our quality of life society has witnessed over the same time period, at least from a material point of view. However, this hides important disparities as half the population of the world lives on less than $2.50 per day with no access to clean and reliable energy [2].

The key to this recent phenomenal technological development is the access to a cheap source of energy, namely, the fossil fuels. This apparent golden energy glut age is coming to an end. Not that the fossil fuels have disappeared, but they are becoming less accessible. Consequently, they tend to become more expensive and disproportionately so to the poorest people.

It is then clear that an affordable and plentiful energy source is required. An obvious source of energy is the Sun, and research about the conversion of sunlight into electricity, photovoltaics, is extremely important, both fundamentally and technologically. About 89 petajoules of solar energy is absorbed by the Earth's surface every second, to be compared to the 18 terajoules utilized by human beings in the same time. As such, a million square kilometres (0.2 % of the Earth's surface) of 10 % efficient solar panels in areas of good insolation would be sufficient to meet the world's present and near-future energy needs.

It is also necessary to improve the processes using energy; our society consumes effectively *only* half of the energy produced [1]. One important area for improvement is lighting. Artificial lighting accounts for about 20 % of the worldwide electricity consumption. Lighting is mainly produced with incandescent sources, where a heated material emits light as a blackbody radiator. However, most of the emitted wavelengths are infrared, which is of little use for our visual perception. Compact fluorescent bulbs are an improvement, yet they raise issues at the end of life of the bulb due to the use and requirement to dispose of hazardous materials. Organic electroluminescence has been shown to have internal efficiency of unity and therefore constitutes a promising technology for future lighting systems.

In this chapter, we discuss the dye-sensitized solar cells as a low-cost photovoltaic technology and the organic electroluminescence as an efficient lighting technology. These technologies have mainly used materials based on platinum-group metals such as ruthenium, iridium and platinum. The high costs and limited availability of these metals have stimulated research for alternatives based on the more available first-row transition metals. First-row transition metal complexes have also been used as alternatives to existing materials for other roles such as redox mediators for solar cells. We have limited ourselves to first-row transition metal complexes used as an active component of the devices; therefore, this chapter does not cover such materials as a starting material to obtain the active component.

3.2 Dye-Sensitized Solar Cells

3.2.1 Brief History

Among all the photovoltaic technologies, the dye-sensitized solar cell (DSC) is very attractive due to low cost of fabrication and low embodied energy cost [3].

Moser reported the first sensitized photoelectrode in 1887 [4], but the operating principle by injection of electrons from the photo-excited dye into the conduction band of the n-type semiconductor substrates was proposed only in the 1960s [5]. Chemisorption of the dye on the surface of the semiconductor followed soon after [6, 7]. The key moment in the history of the DSC is the use of mesoporous thin films to improve the surface area enabling high dye loading [8]. At the moment, champion cells using ruthenium complexes exhibit around 11 % power conversion efficiency under AM1.5 conditions [9].

3.2.2 Operating Principles

The dye-sensitized solar cell is made mainly of five components: (1) a mechanical support coated with transparent conductive oxides (TCO), (2) a semiconductor film of TiO_2, (3) a sensitizer adsorbed onto the surface of the semiconductor, (4) an electrolyte containing a redox shuttle and (5) a counter electrode to regenerate the redox shuttle [10, 11]. Of interest for this chapter are the dye and the redox shuttle. Polyimine ruthenium complexes fuelled the early successes of the DSCs and are still commonly used, although they are being replaced with more efficient materials such as organic dyes [12] and, more recently, perovskites [13]. For high-efficiency liquid cells, the main redox couple used has been iodide/triiodide, I^-/I_3^-.

The operating principles of the DSCs are shown schematically in Fig. 3.1. First, the sensitizer S absorbs a photon, resulting in the excited sensitizer S*, which injects an electron into the conduction band of the semiconductor. At the cathode, the redox mediator is reduced and then regenerates the oxidized dye, completing the circuit. Under illumination, the cell is a regenerative and stable photovoltaic energy conversion system.

All components should be compatible and optimized to obtain high conversion efficiency. The conversion efficiency (η) of the device is function of the photocurrent density (J_{ph}), the open-circuit potential (V_{OC}), the fill factor (*ff*) of the cell and the intensity of the incident light (P_{irr}) (Eq. 3.1):

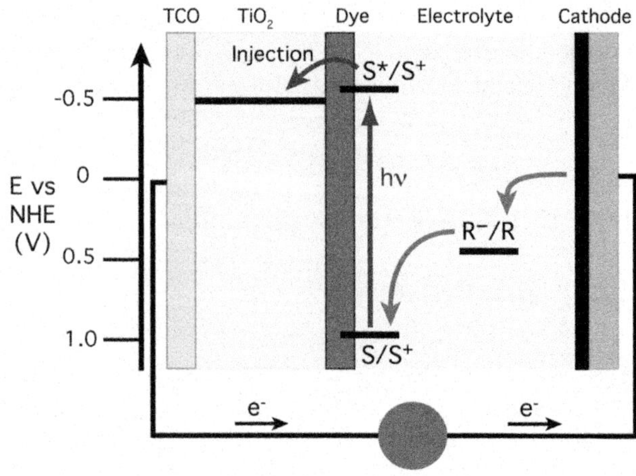

Fig. 3.1 Operating principles and energy-level diagram of the DSC; S/S$^+$/S* = sensitizer in the ground, oxidized and excited state; R$^-$/R = redox mediator

$$\eta_{\text{global}} = \frac{J_{\text{ph}} \cdot V_{\text{OC}} \cdot ff}{P_{\text{irr}}} \tag{3.1}$$

V_{OC} is related to the energy difference between the Fermi level of the solid under illumination and the Nernst potential of the redox couple in the electrolyte, while J_{ph} reflects in part the light-harvesting ability of the sensitizing dye.

3.2.3 *Archetypal Ruthenium Dyes*

The first reported efficient dye was a trinuclear ruthenium complex [8], and, until recently, the best photovoltaic performances (conversion efficiencies and long-term stability) have been obtained with polypyridyl ruthenium complexes based on the structure of **N719** (Fig. 3.2), a doubly deprotonated version of the **N3** dye [14]. The role of the carboxylic acid groups is to bind strongly to the semiconductor surface, while the –NCS groups finely tune the oxidation potential of the complex to optimize the regeneration by the redox shuttle and to increase the light absorption in the visible [15]. An optimized device gives a conversion efficiency of 11.18 % (Table 3.1) [9], that is, >30 % improvement from earlier reports [14, 15].

As a single junction solar cell, the DSC should ideally absorb all photons below a 920 nm threshold for optimum conversion efficiency. **N719** results in photon collection only up to about 780 nm, leaving ample room for further improvement of the device performances. **N749** (black dye) was obtained by molecularly engineering **N719** to respond to this requirement [16]. **N749** can harvest light up to 920 nm resulting in an enhanced photocurrent density, which translates into improved conversion efficiency compared to other champion cells at that time. In 2006, Chiba et al. reported an optimized device using the black dye with improved efficiency of 11.1 % [17].

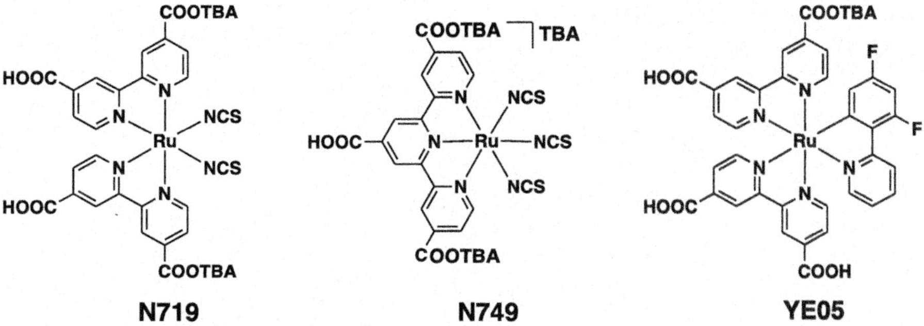

Fig. 3.2 Archetypal ruthenium complexes for DSC

Table 3.1 Device performances of archetype ruthenium dyes

Dye	J_{ph} mA cm^{-2}	V_{oc} V	*ff*	η %	Reference
N719	17.73	0.846	0.75	11.18	[9]
N719	17.0	0.73	0.68	8.4	[14]
N3	19.0	0.60	0.65	7.4	[14]
N3	18.2	0.72	0.73	10.0	[15]
N749	20.5	0.72	0.704	10.4	[16]
N749	20.9	0.736	0.722	11.1	[17]
YE05	17.0	0.80	0.74	10.1	[18]

Stability issues may arise from the monodentate thiocyanate ligands in the previous dyes. The two –NCS of **N719** have been replaced by a bidentate cyclometallated ligand diFppy = 2-(2,4-difluorophenyl)pyridine to give the **YE05** dye [18]. This modification does not significantly change the oxidation potential (E_{ox} = 1.08 V/NHE compared to 1.12 V/NHE for **N719**) allowing efficient regeneration by the I^-/I_3^- redox shuttle. The excited state oxidation potential is also well placed for favourable injection of electrons into the TiO_2 conduction band (E_{ox}* = −0.76 V/NHE). Furthermore, **YE05** possesses panchromatic absorption with an additional absorption band at 490 nm and overall increased molar extinction coefficient compared to **N719** (Fig. 3.3). This shift in design paradigm has motivated a surge of research into cyclometallated ruthenium dyes [19] away from traditional designs [10].

3.2.4 First-Row Transition Metal Complexes as Sensitizing Dyes

There are multiple examples reported of dyes based on first-row transition metal complexes. As most of them have been recently reviewed [20–22] and many are

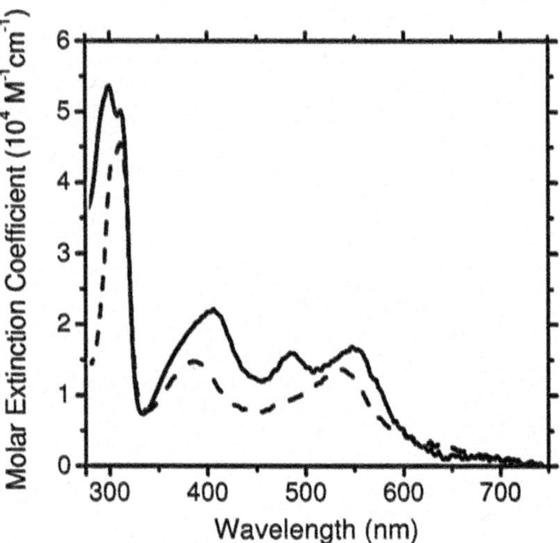

Fig. 3.3 Absorption spectra of **YE05** (*solid line*) and **N719** (*dashed line*) (Adapted with permission from Ref. [18]. Copyright 2009 American Chemical Society)

resulting in very poor device performance, only few selected examples are discussed.

3.2.4.1 Iron Complexes

Iron is the most abundant transition metal in the Earth's crust and is in the same group as ruthenium, which makes it particularly attractive as an alternative to ruthenium.

The first sensitization of TiO_2 with iron complex was carried out in 1987 by Vrachnou et al. [23]. The colourless ferrocyanide, **1** (Fig. 3.4), was adsorbed onto a TiO_2 electrode resulting in a pronounced orange absorption band with a maximum at 420 nm and extending up to 700 nm. This transition is attributed to an Fe(II) to Ti (IV) intervalence charge transfer (ICT). The incident monochromatic photon-to-current conversion efficiency reaches 37 % at 420 nm, which was attributed to the high roughness factor of the TiO_2 layer. Due to the ICT character of the transition, sensitization of the TiO_2 occurs through a direct injection of an electron from the iron centre to the Ti(IV) acceptor sites [24]. This direct injection mechanism, termed metal-to-particle charge transfer (MPCT) [25], was later confirmed by density functional theory (DFT) and time-dependent DFT (TDDFT) calculations [26].

The complex $[Fe^{II}(2,2'\text{-bipyridine-4,4'-dicarboxylic acid})_2(CN)_2]$, **2** (Fig. 3.4), as a sensitizer of TiO_2 was reported by Ferrere and Gregg in 1998 [27]. Its structure is similar to **N3** with the two thiocyanate ligands replaced by two cyanide ligands resulting in an oxidation potential of 0.213 V/ferrocene [28]. The UV-visible absorption spectra of **2** (Fig. 3.5) resembles the one of **N719** (Fig. 3.3) with an intense bipyridyl-based π- π^* transition at 318 nm and two main metal-to-ligand

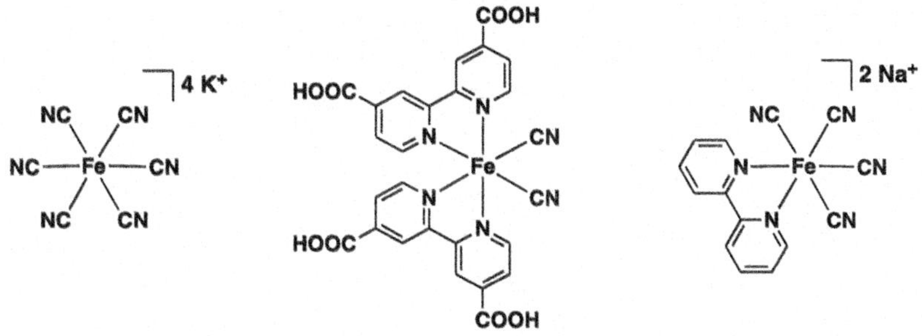

Fig. 3.4 Iron(II) dyes with cyanide ancillary ligands

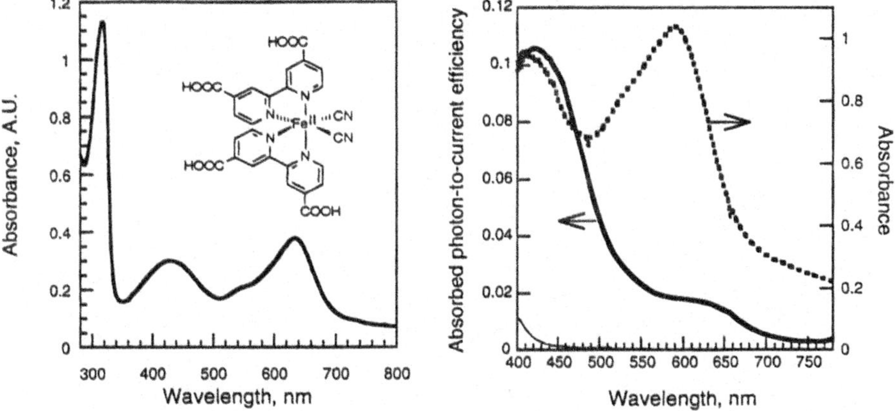

Fig. 3.5 *Left*: Absorption spectrum of **2** in DMSO (concentration $= 8 \times 10^{-5}$ M in a 1 cm path length cell). *Right*: (---) Absorbance spectrum of **2** adsorbed on TiO_2 in 0.5 M LiI in acetonitrile; (—) photocurrent action spectrum of the same film; (—) is the action spectrum of an undyed TiO_2 film (Adapted with permission from Ref. [27]. Copyright 1998 American Chemical Society)

charge-transfer (MLCT) bands at 430 and 635 nm with a shoulder at about 550 nm. The MLCT bands are redshifted compared to **N719**, but their intensity is also significantly lower. When adsorbed onto nanocrystalline films of TiO_2 in the presence of chenodeoxycholic acid (cheno) with an iodine-based liquid electrolyte and platinum as counter electrode, under illumination a short-circuit photocurrent of 290 μA cm^{-2} and an open-circuit voltage of 0.36 V are obtained without optimization of the device [27]. Interestingly, the photocurrent action spectrum (Fig. 3.5) shows "band-selective" sensitization. While the IPCE mirrors the absorption spectra of TiO_2 films sensitized with ruthenium complexes, in the case of **2**, the injection of electrons into the TiO_2 conduction band is much less efficient from the lower-energy MLCT band.

Meyer and co-workers have observed a similar phenomenon using Fe(bpy)$(CN)_4^{2-}$, **3** in Fig. 3.4 [25]. When adsorbed onto TiO_2, **3** displays both MLCT absorption band around 520 nm and MPCT absorption band at higher energy. As such, **3** is a structurally intermediate complex between **1** and **2** and displays absorption characteristics of both. However, based on the absorptance spectra, injection from the MLCT band is less efficient than injection from the MPCT band.

Recent theoretical calculations shed some light on these results [29]. In general, electron injection is in competition with the deactivation of the MLCT excited state to the ligand field (LF) state. In contrast to ruthenium, the e_g orbitals of iron are lower in energy than the π^* orbitals of the ligands due to the weak LF of iron. Consequently, relaxation to the LF state is very favourable. Furthermore, there is a lower driving force for the injection of electrons from the lower-energy MLCT states, which favours the non-injecting LF pathway even more. It is then proposed to improve the efficiency of iron-based dyes by increasing the energy of the LUMO and LUMO + 1 states to have better matching energy levels between the dye and the TiO_2. This modification should also blueshift the absorption, possibly resulting in a ruthenium-like absorption spectrum.

3.2.4.2 Copper Complexes

Copper is about 10^5 times more abundant than ruthenium in the Earth's crust and can easily form photoactive polyimine complexes, which make them particularly attractive as an alternative to ruthenium complexes.

Before the advent of the mesoscopic solar cell, copper complexes have been used to sensitize high bandgap semiconductors [30]. A decade after this initial work, the same group used complex **4** (Fig. 3.6) as a sensitizer for mesoporous titania [31]. The complex uses two 2,9-diphenyl-1,10-phenanthroline substituted by carboxylate groups, which act as anchoring groups. It displays a maximum absorption at 440 nm in methanolic solution with absorption coefficient of ~3,000 M^{-1} cm^{-1}, extending up to about 620 nm, similar to the complex without carboxylate groups. Such characteristics are much below ruthenium complexes, both in terms of panchromaticity and intensity of absorption; this is one important inconvenience of polyimine copper complexes for solar cell applications. Using a 475 nm cut-off filter ($P_{irr} \sim 300$ mW cm^{-2}), $J_{ph} \sim 0.6$ mA cm^{-2}, $V_{OC} \sim 0.62$ V and *ff* of 0.6 have been obtained, resulting in an overall efficiency of 7.4×10^{-2} %.

Considering the position of the carboxylate, complex **5** was proposed as an improvement of **4** [32]. The direct grafting of the anchoring groups onto the polyimine ligand insures favourable interactions of the complex with the semiconductor to give improved charge injection. Interestingly, the absorption characteristics of **5** are also very much improved, with a maximum absorption at 450 nm and an absorption coefficient of 6,400 M^{-1} cm^{-1}. An optimized cell resulted in $J_{ph} \sim 3.9$ mA cm^{-2}, $V_{OC} \sim 0.63$ V, for a claimed energy conversion efficiency of 2.5 % at $P_{irr} = 100$ mW cm^{-2}. (Using Eq. 3.1, it is apparent that these values imply a fill

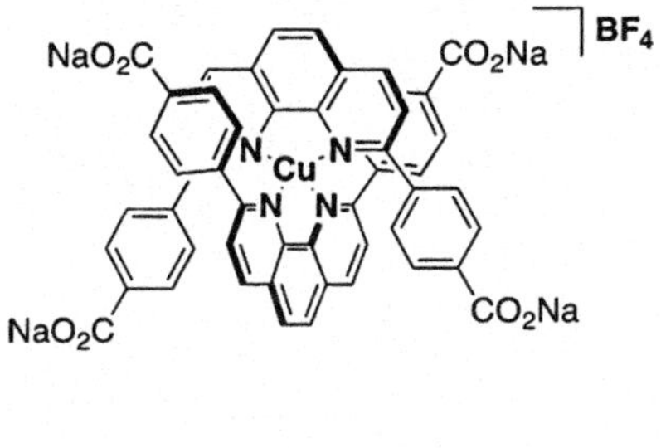

Fig. 3.6 Copper(I) dyes for DSC

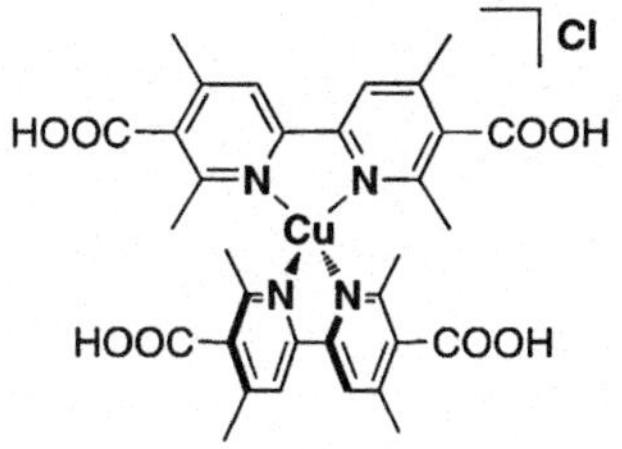

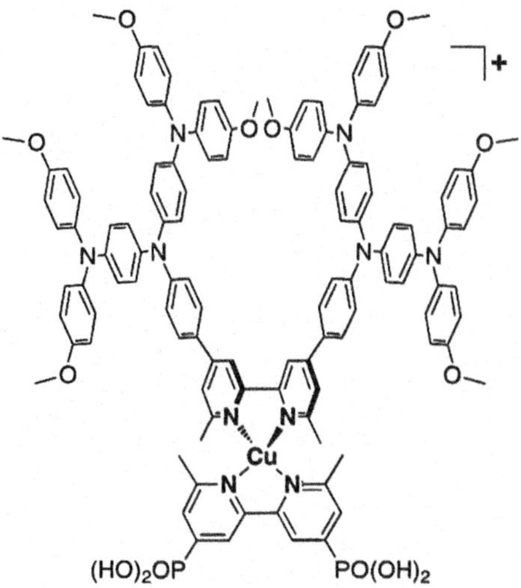

factor *ff* ~ 1, pointing to possible issues with the measurements or calculation of the efficiency.)

From the structures of dyes **4** and **5**, it is clear that the lability of copper polyimine complexes is an issue as only complexes with two identical bidentate ligands are easily accessible [33–35] unless bidentate bisphosphine ligands are used [36]. This apparent disadvantage has been cleverly harnessed to prepare heteroleptic copper complexes. The principle is based on an in situ stepwise synthesis: The first step is to cover the semiconductor surface with the free anchoring ligand, followed by ligand exchange with a homoleptic copper complex (Fig. 3.7). With this strategy in hand, a whole new class of copper(I) sensitizers was accessible and has been tested [37, 38]. Complex **6** (Fig. 3.6) shows very attractive efficiencies from 1.57 up to 2.61 %, depending on measurement methodology. Importantly, these efficiencies are compared with efficiencies from devices using **N719** as the sensitizer that are close to the best reported to date.

Theoretical calculations have been used to compare copper(I) and ruthenium (II) sensitizers [39]. In particular, these studies anticipate that as copper (I) complexes can be optimized to exhibit similar optical properties to ruthenium (II) sensitizers, copper(I)-based sensitizers are promising sensitizers. It should be noted that copper(I) complexes experience significant distortion in the excited state, possibly influencing the dynamics of the interfacial electron transfer processes [40]. It will be important to control this distortion to improve further the device efficiencies.

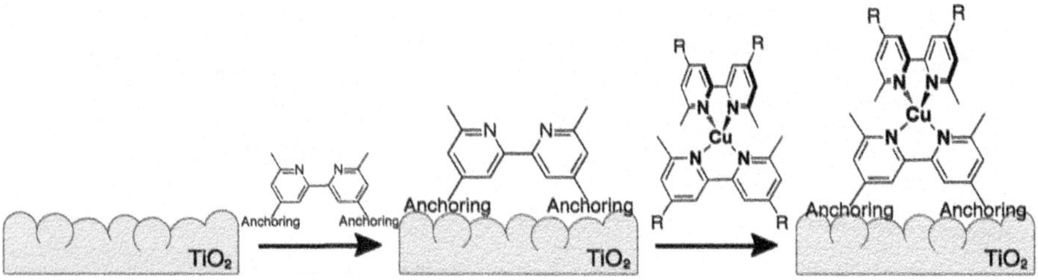

Fig. 3.7 In situ stepwise synthesis of heteroleptic copper(I) complexes

3.2.5 *Iodide/Triiodide as the Archetypal Redox Mediator*

The redox mediator is another key component of the DSC for which transition metal complexes have been explored.

Iodide/triiodide, I^-/I_3^-, has been the preferred redox couple as it results in the overall best performing devices [41]. This is due to its peculiar properties. First, it has generally very good solubility in organic solvent, which makes it suitable for liquid electrolytes. Second, although having yellow to brown colour, its visible light absorption has limited impact on the efficiency of the DSC [42]. Third, its standard potential is well placed to provide the necessary driving force for rapid regeneration of the sensitizer. Fourth, and arguably most important, are the very slow kinetics for the recombination of electrons in TiO_2 and triiodide.

A unique property of I^-/I_3^- is that regeneration of the sensitizer occurs schematically in a two-step mechanism. Initially the dye is regenerated by I^- and produces the diiodide radical $I_2^{\bullet-}$, which disproportionate into I^- and I_3^-. The redox potential of $I^-/I_2^{\bullet-}$ is at most 0.93 V vs NHE [43], compared to 1.1 V for **N3**, resulting in driving force for dye regeneration >0.17 V. Electron transfer from the TiO_2 to I_3^- is then a slow two-electron process, while it is catalyzed at the metallic counter electrode due to dissociative chemisorption of iodine species.

Despite its interesting properties as redox shuttle, the couple I^-/I_3^- suffers from some drawbacks. Mainly, the redox potential of I^-/I_3^- is 0.35 V vs NHE and is the redox couple to be taken into account for the resulting V_{OC}. This significant loss of potential has considerable impact on the efficiency of the device, and increasing the oxidizing strength of the electrolyte has been an important focus of the research of alternative electrolyte. Also I^-/I_3^- results in corrosion of the metallic counter electrode.

Redox shuttles for DSC, including iodine-free alternatives, have been reviewed recently [41, 44–48]. Therefore, only selected examples are discussed.

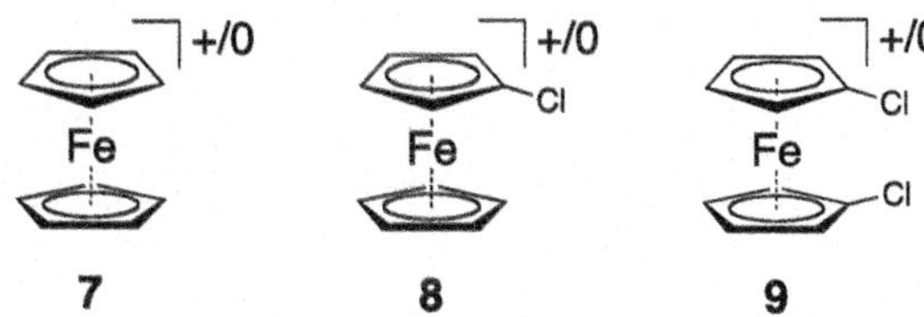

Fig. 3.8 Ferrocenium/ferrocene-based redox shuttles

3.2.6 First-Row Transition Metal Complexes as Redox Mediator

3.2.6.1 Iron Complexes

Ferrocene/ferrocenium (Fc/Fc^+), **7** (Fig. 3.8), is an example of one-electron, outer-sphere redox mediator. Regeneration is usually very fast, as, unfortunately, is the recombination between electrons in the TiO_2 and ferrocenium. Passivation of the semiconductor interface significantly decreases the recombination rates [49, 50].

The redox potential of Fc/Fc^+ is 0.63 V vs NHE, significantly lower than **N3**. Ferrocene-based redox shuttles with more positive oxidation potential have been developed through the introduction of chlorine substituents [50]. Compounds **8** and **9** display reversible oxidation at 0.81 and 0.94 V vs NHE, respectively. Interestingly, **9** has similar oxidation potential to $I^-/I_2^{\bullet -}$. This shift in potential results in an increase of V_{OC}, but device efficiency remains very low, below 1 %.

A 7.5 % efficiency energy-converting DSC using Fc/Fc^+ as redox mediator was reported in 2011 [51]. Developments underpinning this breakthrough are twofold. First, the design of the organic sensitizer, Carbz-PAHTDTT, appears crucial with a large donor unit and a long spacer unit. It is believed to significantly reduce the kinetics of the recombination process by keeping the ferrocenium away from the TiO_2 surface; the dye itself acts as a passivation layer. Second, the electrolyte was prepared in a glove box, excluding the presence of oxygen. Because of its instability towards oxygen, ferrocenium decomposes into various ferric species, modifying the composition of the electrolyte. It is possible that low efficiencies reported earlier are due to such reaction with oxygen.

Following up on this promising result, systematic studies have been reported with series of ferrocene derivatives having a range of oxidation potentials from 0.088 to 0.935 V vs NHE [52] and combining them with dyes having oxidation potential ranging from 0.897 to 1.136 V vs NHE [53]. It was found that the optimum driving force should be $\approx$ 0.20–0.25 V.

3.2.6.2 Cobalt Complexes

Polyimine cobalt complexes, Co^{2+}/Co^{3+}, have been the most widely explored family of transition metal complexes for use as redox mediator. As with ferrocene-based materials, their redox potential can be easily tuned by modification to the ligands coordinated to the central metal. Furthermore, Co^{3+} is generally in a

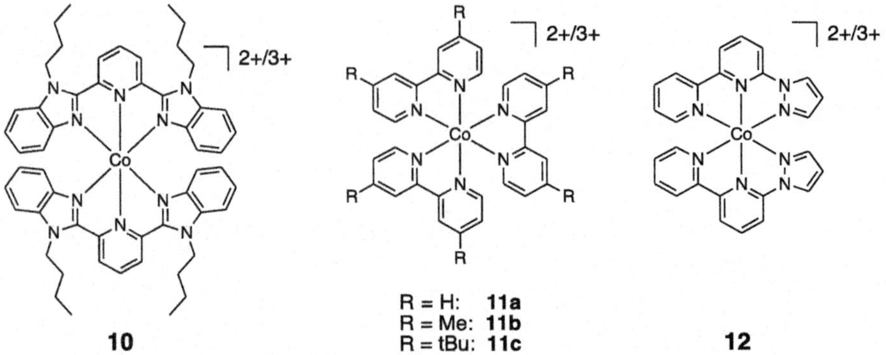

Fig. 3.9 Cobalt complexes **10–12**

low spin state, while Co^{2+} is in high spin state, resulting in slow self-exchange rates. Co^{2+} would effectively regenerate the oxidized dye, while Co^{3+} will be less efficient in electron recombination. Finally, cobalt complexes have very low absorption coefficient in the visible region, about 100 M^{-1} cm^{-1}, which allows for an increase of the useful light-harvesting capability of the device.

The first efficient cobalt-based redox shuttle for use in DSC, complex **10** in Fig. 3.9, was reported in 2001 [54]. Using **Z316** as the sensitizer, it resulted in efficiency of 5.2 % at light intensity of 94 W cm^{-2}. At standard AM1.5 irradiance (1,000 W m^{-2}), the efficiency drops to 2.2 %, with a poor fill factor (0.46 compared to 0.68 at low light intensity) pointing to issues with the diffusion of the large complex compared to the iodide system [55]. The redox potential of **10** is 0.36 V vs SCE, and the V_{OC} at high light intensity is 0.67 V for a photocurrent $J_{ph} = 6.8$ mA cm^{-2}. A detailed electrochemical study of **10** has been reported [56].

A series of 14 cobalt complexes has been studied as possible alternatives to the iodide system; variations were made to the core of the ligand (terpyridine, bipyridine and phenanthroline) and on a range of substituents to tune the redox potential [57]. The best system was tris(4,4′-di-*tert*-butyl-2,2′-dipyridyl)cobalt (II/III) perchlorate, **11c**. The efficiencies observed are however low, up to 1.6 % only, which can now be ascribed to the sensitizer used, **N3**. It was recently shown that insulation by bulky groups is necessary to obtain high-efficiency devices based on ruthenium dyes with cobalt complexes as redox mediators [58].

This principle of insulating the dye by introducing bulky groups on the dye originates from the development of organic dyes showing good efficiencies with cobalt complexes as redox shuttles. In 2010, the D-π-A dye **D35** with butoxyl chains on the donor groups was used with simple tris-bipyridyl cobalt complex **11a**, and energy conversion efficiency of the DSC reached 6.4 % under full sunlight with an important V_{OC} of 0.9 V [59]. An increase of the pore size of the TiO_2 film was also important to improve the diffusion of the redox mediator.

While the redox potential of the cobalt complexes can be tuned simply by grafting suitable substituents onto the ligands, it would increase the size of the complex, resulting in lower diffusion. An approach to tune the properties of the

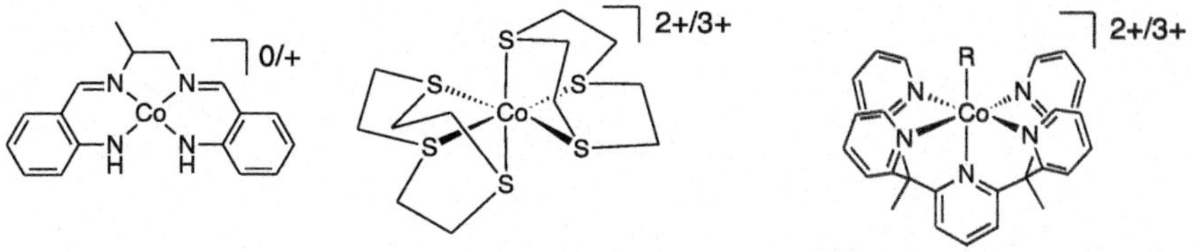

13 14

R = 4-*tBu*-pyridine: **15a**
R = *N*-methylbenzoimidazole: **15b**

Fig. 3.10 Cobalt complexes **13–15**

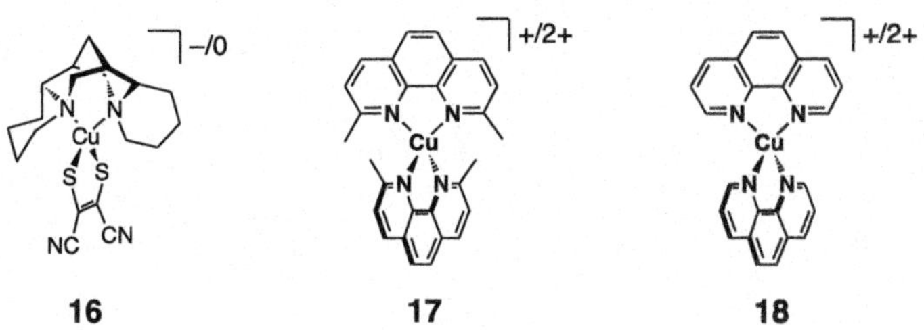

Fig. 3.11 Copper complexes **16–18**

complexes was therefore to change a pyridine ring to a pyrazole one [60]. As such, the redox potential can be increased to 0.86 V vs NHE, resulting in V_{OC} over 1 V, with efficiency approaching 11 %.

Other designs with particular features have been explored in cobalt complexes (Fig. 3.10). Complex **13** is neutral in its reduced state [61], while **14** is a low spin cobalt(II) complex for fast dye regeneration [62]. Complexes **15** have a pentadentate pyridyl-based ligand, which leaves one position available on the cobalt [63]. This position can be used by components of the electrolyte such as 4-*tBu*-pyridine or *N*-methylbenzimidazole to fine-tune the properties of the complexes.

3.2.6.3 Copper Complexes

Copper complexes offer a possible strategy to improve the electron transfer kinetics, that is, increase the regeneration rate while slowing the recombination. Upon oxidation of the copper(I) complex during the regeneration of the dye, the geometry of the complex will significantly distort from the initial pseudo-tetrahedron.

The one-electron reduction potentials of $\mathbf{16}^{0}$, $\mathbf{17}^{2+}$ and $\mathbf{18}^{2+}$ (Fig. 3.11) are 0.29, 0.66 and −0.10 V vs SCE, respectively. As such, they are all more negative than the potential of **N719** and have been tested as redox shuttles for DSC [64]. Devices based on **17** are the most efficient ($\eta = 1.4$ %) among this series. High efficiency of 7 % has been obtained with **17** as the redox mediator in a DSC using an organic dye, **C218**, as the sensitizer [65]. Importantly, the efficiency is higher than the efficiency from

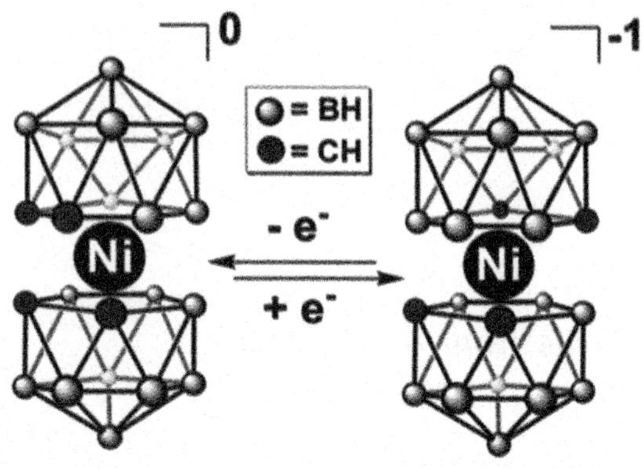

Fig. 3.12 Redox-induced rotation in nickel bis (dicarbollide) complexes (Adapted with permission from Ref. [66]. Copyright 2010 American Chemical Society)

devices using iodine redox shuttle, despite the very poor *ff* observed with the copper complex, which bodes well for future improvements. This low *ff* was attributed to the very low electron transfer rates of the complex on the counter electrode [65].

3.2.6.4 Nickel Complexes

Recently, Ni(III)/(IV) bis(dicarbollide) complexes (Fig. 3.12) have been reported and offer an alternative strategy to gated electron transfer [66, 67]. As shown in Fig. 3.12, upon oxidation and reduction, there is a rotation of one dicarbollide group relative to the other one, resulting in the two carbon atoms of each dicarbollide group to be either in a *cis* or *trans* conformation. To further limit the recombination process, alumina was used to passivate the TiO_2 electrode. While the oxidation process (corresponding to regeneration of the sensitizer) is nevertheless very fast, slightly faster than regeneration by I^-, it was demonstrated that interception of the injected electron by the Ni(IV) complex is 10 to 100 times slower than dye regeneration [66]. Remarkably, with only one passivation cycle, V_{OC} above 0.64 V are obtained.

A series of complexes with various substituents to tune the redox potential have been prepared and studied [67]. Optimization of the redox potential resulted in V_{OC} of 0.77 V. However, in every case, the power conversion efficiencies are low, only up to 2 %, due to low photocurrent density. One reason for such low J_{ph} could be due to the sensitizer used, **N719**. As seen with other transition metal-based redox shuttle, the best results have been obtained with organic dyes or largely modified ruthenium complexes, while **N719** appears incompatible with these redox shuttles.

3.2.6.5 Manganese Complexes

As the latest addition to the list of redox shuttles based on first-row transition metal complexes, the couple tris(acetylacetonato)manganese(III)/(IV), $[Mn(acac)_3]^{0/1+}$, was tested with both organic dyes and **N719** [68]. Best efficiencies in this case were obtained with **N719** at 10 % Sun intensity (4.5 % versus 3.8 % using **MK2** dye). At full Sun, the efficiencies are identical (4.4 %). Recombination losses appear to limit the conversion efficiency, and improvement is expected using bulkier substituents and/or passivation of the TiO_2 electrode.

3.3 Electroluminescence from Small Molecules

3.3.1 Brief History

Almost 50 years after the observation of electroluminescence from silicon carbide, Bernanose and co-workers reported the first electroluminescence from organic materials [69, 70]. They used acridine derivatives, and light emission was observed upon application of an alternating current with voltage >400 V. A decade later, Pope and co-workers reported electroluminescence from anthracene and from tetracene (as an impurity of anthracene) [71].

In 1987, Tang and Van Slyke demonstrated electroluminescence from a two-layer heterojunction using 1,1-bis((di-4-tolylamino)phenyl)cyclohexane (**TAPC**) and tris(8-hydroxyquinolinato)aluminium (**Alq$_3$**) (Fig. 3.13) [72]. For the first time, devices could operate at a voltage below 10 V.

A second key development was made in 1998 by Forrest, Thompson and co-workers with the use of platinum porphyrin **PtOEP** as a phosphorescent dopant [73] followed by the use of the cyclometallated iridium complex **Ir(ppy)$_3$** (ppy = 2-phenylpyridine) in 2000 [74]. The use of phosphorescent dopant enables 100 % internal efficiency by harvesting both singlet and triplet excitons.

3.3.2 Operating Principles

In its simplest form, an OLED is made of a single film of organic material sandwiched between two electrodes.

When the device is operated (Fig. 3.14), electrons are injected in the LUMO of the material and holes in the HOMO. The charges then drift towards the other

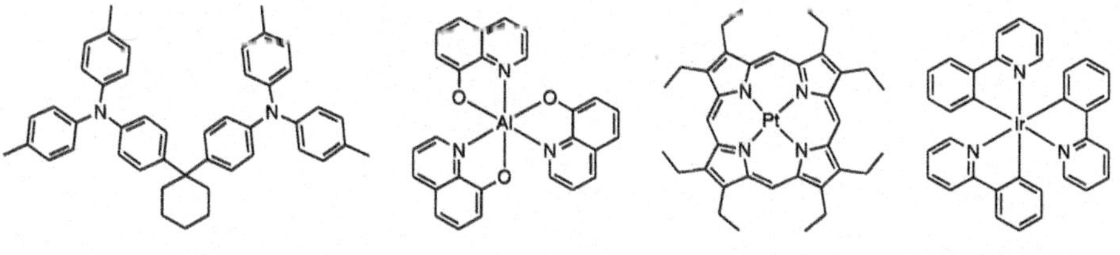

Fig. 3.13 Historical compounds for modern OLEDs

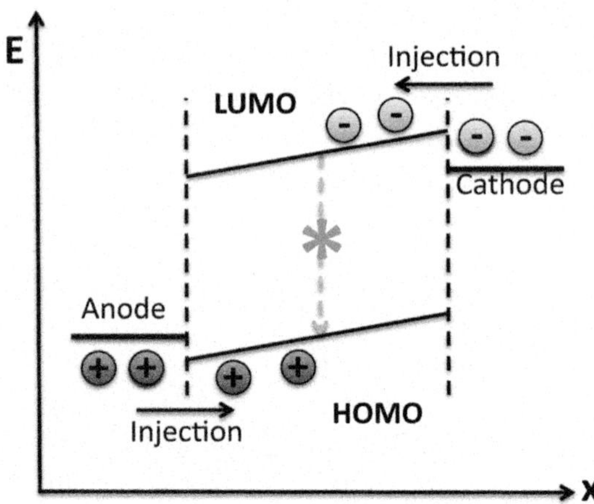

Fig. 3.14 Operating principles of electroluminescent devices

electrode. A hole and an electron may interact and recombine to create an exciton, which will deactivate by emitting light.

To improve efficiency and stability of devices, multilayer architectures are generally used. Additional layers can be injection layers (to facilitate the injection of charges), charge-transporting layers (to balance charges in the emissive layer) and charge blocking layers (to confine charges in the emissive layer).

The external conversion efficiency (η_{ext}) of the device is function of the outcoupling (η_{out}), the ratio of electrons to holes (γ), the fraction of luminescent excitons (χ) and the photoluminescent quantum yield of the emitter (Φ_{lum}) (Eq. 3.2):

$$\eta_{ext} = \eta_{out} \cdot \gamma \cdot \chi \cdot \Phi_{lum} \tag{3.2}$$

χ is a very important parameter. The recombination of holes and electrons statistically produces 25 % of singlet excitons and 75 % of triplet excitons. With a fluorescent emitter, only the singlet excitons can undergo radiative deactivation and χ will be 0.25, resulting in low efficiency. It is possible to have fusion of two non-radiative triplet states, resulting in a radiative singlet state and a ground state; however, the maximum χ will be 0.625. Phosphorescent emitters can efficiently produce light from triplet excitons, enabling the maximum value for $\chi = 1$. Efficient utilization of triplet excitons can also be achieved with processes such as thermally activated delayed fluorescence (TADF) [75–77].

3.3.3 *Archetypal Phosphorescent Materials*

3.3.3.1 Cyclometallated Iridium Complexes

Cyclometallated iridium(III) complexes possess unique photophysical properties such as high phosphorescence quantum yields, relatively short triplet excited

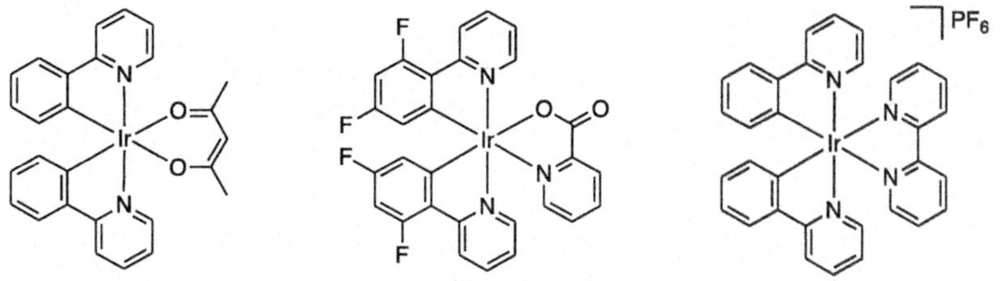

19 $Ir(ppy)_2(acac)$ **20** FIrPic **21** $Ir(ppy)_2(bpy)$

Fig. 3.15 Standard bis-heteroleptic cyclometallated iridium complexes

state lifetime and wide emission colour tunability [78, 79]. As a result of these characteristics, they have attracted considerable interest for virtually any application requiring highly phosphorescent materials [80–83].

Tris-homoleptic cyclometallated iridium complexes are neutral complexes of the type $Ir(C^\wedge N)_3$. The archetypal complex is $Ir(ppy)_3$ (Fig. 3.13), where the coordination of the ppy ligand to the metal centre is as that found in 2,2′-bipyridine except that a nitrogen is replaced by an anionic carbon. The electronic absorption spectrum of $Ir(ppy)_3$ displays strong ligand-centred (LC, π- π^*) and metal-to-ligand charge-transfer (MLCT) transitions in the UV and visible, respectively. In degassed solution in 2-MeTHF at room temperature, the excited triplet state shows strong green phosphorescence at around 508 nm, with an excited state lifetime of 1.6 μs and a photoluminescence quantum yield of 0.97 [84].

Bis-heteroleptic cyclometallated iridium complexes are of the form Ir $(C^\wedge N)_2(L_{anc})$, where L_{anc} is an ancillary ligand system, which can be a bidentate ligand or a combination of two monodentate ligands and can be neutral or charged. Bis-heteroleptic complexes are advantageous due to their ease of synthesis and due to the extended possibilities to control the excited state properties of the complexes compared to their tris-homoleptic counterparts [85]. L_{anc} such as acac (acetylacetonate) and Pic (2-picolinate) are commonly used (Fig. 3.15).

When a neutral L_{anc} system is used, such as bpy (2,2′-bipyridine), a charged complex is obtained (Fig. 3.15). Ionic transition metal complexes (iTMCs) are attractive for their use in light-emitting electrochemical cells (LECs) [86]. LECs constitute an alternative to OLEDs due to their potential low-cost production. The presence of mobile ions in the emissive layer allows for devices with simpler architecture and the possibility of using air-stable electrodes allowing non-rigorous encapsulation processes.

3.3.3.2 Cyclometallated Platinum Complexes

Cyclometallated platinum complexes are very similar to cyclometallated iridium complexes as they show as well high phosphorescence quantum yields, relatively short triplet excited state lifetime and wide emission colour tunability.

Fig. 3.16 Examples of cyclometallated platinum complexes

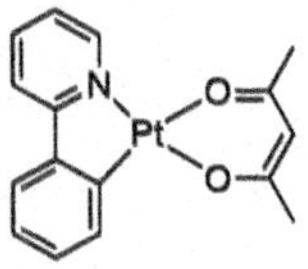

22
Pt(ppy)(acac)

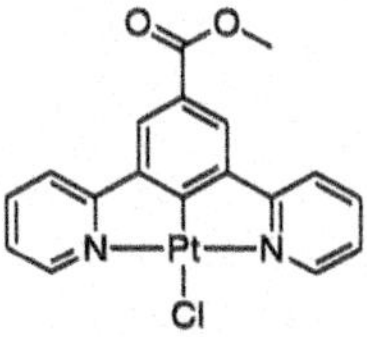

23

Their structure and design rules are also similar to their iridium counterparts and the same ligands can be used (complex **22** in Fig. 3.16) [87]. The main difference is their square planar geometry leading to the possible formation of dimers through Pt-Pt interactions. The emission of the dimeric species being redshifted compared to the one of the monomeric complex and the amount of dimers formed being function of the concentration of complex in the emissive layer, platinum complexes can be used to prepare WOLEDs from a single complex. For example, with **23**, white emission is obtained from the combination of blue emission from the monomers and orange emission from the dimers [88].

3.3.4 *First-Row Transition Metal Complexes as Emitter for Electroluminescence*

3.3.4.1 Scandium

Since the report of the heterojunction for electroluminescence, **Alq_3** has attracted interest as a basic molecular design to be modified in order to improve device performance and vary the colour of emission. One research direction was to change the central aluminium atom to other metal ions, keeping the hydroxyquinoline (or similar N^O chelates) as ligand (Fig. 3.17).

Scq_3 was used in a study of metal quinolate complexes, Mq_3, where the impact of the central atom on the performances of EL devices was compared [89]. The device architecture was ITO/TPD/Mq_3/Mg-Ag(10:1)/Ag, with TPD: *N,N′*-diphenyl-*N,N′*-bis(3-methylphenyl)1,1′-biphenyl-4,4′-diamine. Along the series with M = Al, Ga, In and Sc, **Scq_3** was found to be the least efficient with EL efficiency about 10 % of **Alq_3**. One reason for low efficiency could be the structure of **Scq_3**, which forms mixture of monomers and oligomers in solution, resulting in films with extremely low photoluminescence efficiency.

Scq_3 was recently re-examined, and it was found that devices are now about twice as efficient as **Alq_3**-based devices with current efficiencies at 300 cd m^{-2} of 4.6 and 2.7 cd A^{-1} and power efficiencies of 2.6 and 1.2 lm W^{-1}, respectively [90]. Reasons behind the improved performances have been attributed to three factors. First, **Scq_3** was prepared accordingly to a new synthetic methodology resulting in anhydrous mononuclear complexes, which improved the photoluminescence quantum yield [91]. Second, the cathode was changed to ytterbium as a

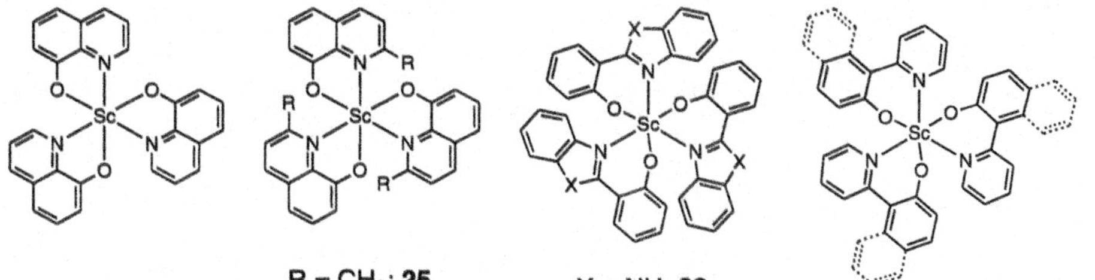

R = CH_3: **25**
R = CN: **26**
R = NH_2: **27**

X = NH: **28**
X = O: **29**
X = S: **30**

24
Scq_3

Phenol: **31**
Naphthol: **32**

Fig. 3.17 Scandium complexes used as electroluminescent emitters

metal with a lower work function of 2.59 eV than Mg-Ag alloy (3.7 eV [92]), resulting in improved injection of charges. Third, it was claimed that **Scq_3** has improved electron mobility over **Alq_3** [90].

Following these results, various scandium complexes have been studied using an ytterbium cathode. Using donor and acceptor substituents on the quinolinate ligand, complexes **25–27**, the emitted colour could be tuned from blue to orange [93]. Replacing the quinolate ligand with phenolate bearing benzimidazole, **28**; benzoxyazole, **29**; and benzothiazole, **30**, resulted in whitish emission (emitting wavelength from about 400 nm to 700 nm) with various hues [94]. The broad emission can be attributed to aggregation effects, as a pure layer of complex was used as the emissive layer. Surprisingly, the replacement of the phenol fragment of **31** with a naphthol fragment, **32**, does not result in a redshift of the emission, and both complexes have very similar emission in the blue-green region of the visible spectrum [95]. On the other hand, the naphthol group appears to have a drastic detrimental effect on the performance of the device, which drops from power efficiency of 6.49 lm W^{-1} for **31** to 0.16 lm W^{-1} for **32** [95].

3.3.4.2 Copper

An attractive alternative to iridium and platinum is copper. As early as 1999, Ma and co-workers have reported the use of a tetranuclear copper(I) complex as the active emitting species in an OLED [96]. The properties of the complex in solution appeared promising ($\lambda_{em} = 522$ nm, $\Phi = 0.42$ and $\tau = 9.8$ μs in CH_2Cl_2 at RT); however, an EL device with a simple bilayer architecture using polymeric PVK host performed poorly, with a reported efficiency of 0.1 %, mainly due to the long lifetime of the excited state. Although copper(I) readily forms coordination complexes, which show phosphorescence in the visible [97, 98], their use in EL applications has been hampered by their limited spectral coverage, mainly in the green region of the spectrum, and the low quantum yield of photoluminescence yielding poorly efficient devices.

Interest in copper complexes was renewed in 2004 when the Cu(N^N)(P^P)$^+$ complexes **33** (Fig. 3.18) were reported to yield OLEDs with improved

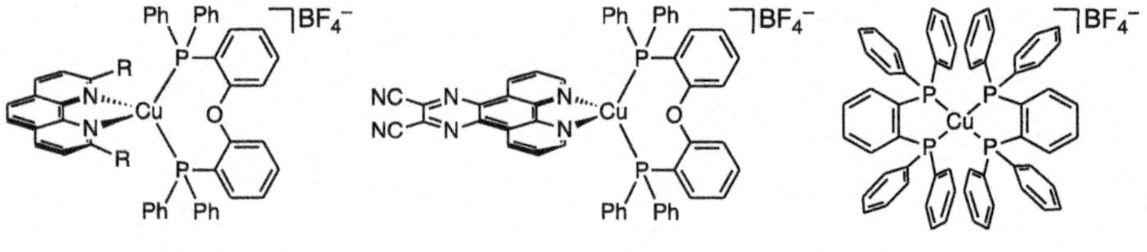

Fig. 3.18 Charged copper complexes used as emitter for electroluminescence

performances, that is, current efficiencies >10 cd A^{-1} at current densities below 1 mA cm^{-2} and luminance >1,000 cd m^{-2} at about 30 V [99]. Various concentrations of **33**, with R = *n*-butyl, in PVK were used with the overall device architecture being ITO/PEDOT/**33**:PVK/BCP/Alq_3/LiF/Al. With deeper HOMO and higher LUMO energy levels than **33**, PVK was a suitable host, which explains, in part, the good performance. The same group also reported a study on the effect of the counterion and of the cathode [100]. The same complex was used in a simpler device architecture based on a single layer of a 75:25 % w/w mixture of the complex in PMMA as the emissive layer sandwiched between ITO and silver electrodes [101].

Complex **34** was stable enough to be sublimed under vacuum, and a multilayer vacuum-processed OLED was prepared using **34** doped in CBP as the emissive layer [102]. At a 6 wt% doping concentration, the device shows orange emission with current efficiencies of 1.9 and 9.2 cd A^{-1} at 100 and 1 mA cm^{-2}, respectively. Furthermore, this complex has been used as the orange-emitting component for white OLED using FIrPic as the sky-blue component [103, 104].

Complex **35** is an interesting homoleptic P^P complex. When doped at a concentration of 12.5 wt% in PVK, whitish electroluminescence is observed, with the EL spectrum covering 400 nm up to 750 nm. However, the turn-on voltage is still very high (15 V) and the brightness poor (below 500 cd m^{-2}) [105].

Charged copper(I) complexes have been the most utilized design for copper OLED complexes, and multiple examples are found in the literature [106–113].

The rich coordination chemistry of copper(I) and its ability to bind to anionic groups lead to the investigation of some interesting multinuclear complexes (Fig. 3.19), in addition to the aforementioned first example of copper complex for EL. Complex **36** is a dinuclear "diamond-core" copper complex using a terdentate P^N^P ligand [75]. It resulted in a device with maximum external quantum efficiency (EQE) of 16.1 %, which is very promising when it comes to challenging the high efficiency of devices using iridium complexes (>20 %). Uniquely, this complex displays E-type delayed fluorescence, or thermally activated delayed fluorescence (TADF): with triplet and singlet energy levels close in energy, the triplet excited state is thermally activated to the higher energy singlet excited state, and radiative deactivation occurs efficiently from the singlet state. This approach allows for harvesting efficiently both the singlet and triplet excitons, even in cases

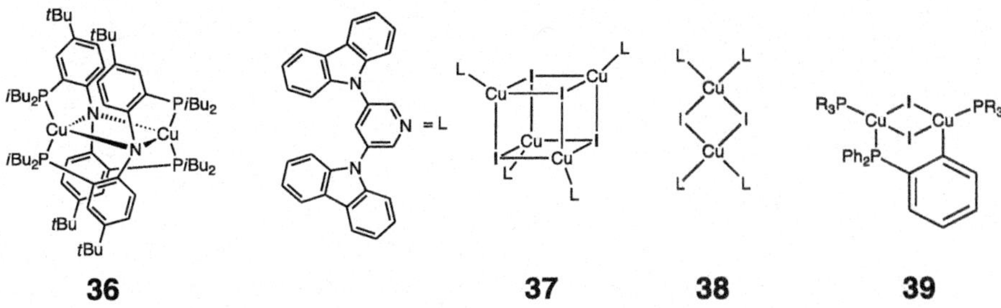

Fig. 3.19 Neutral multinuclear copper complexes

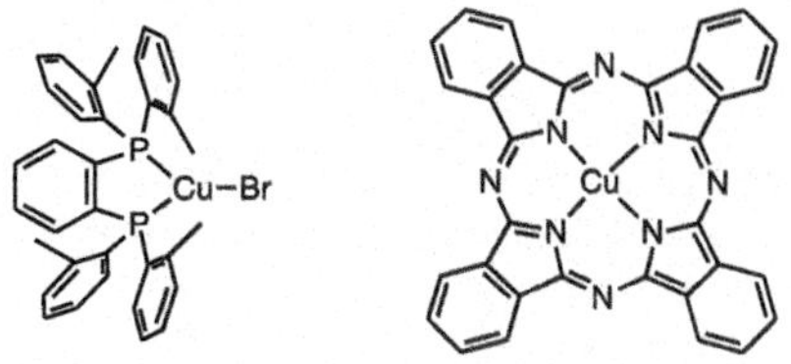

Fig. 3.20 Other copper complexes for electroluminescence

where the radiative deactivation of the triplet state is not an efficient process and would otherwise result in poor device efficiency.

Tetranuclear complexes such as **37** can be easily synthesized, isolated and crystallized [114]. However, when the components used to make **37** are co-evaporated to make the emissive layer of an OLED, the dinuclear complex **38** is formed in situ and acts as the emissive material to give a green emission [115].

A series of complexes based on **39**, where PR_3 represents various phosphine ligands, has been found to be excellent emitters in solid state, up to 99 % photoluminescence quantum yield for PR_3 = triphenylphosphine, with the maximum of emission wavelength ranging from 500 to 555 nm [116]. The complexes also show TADF. Solution-processed devices with an emissive layer composed of 45 wt% of complex, 45 wt% of *m*CPy and 10 wt% PVK display high turn-on voltage (>8 V) and limited performance.

A mononuclear three-coordinated copper(I) complex, (dtpb)CuBr with dtpb = 1,2-bis(*o*-ditolylphosphino)benzene, **40** (Fig. 3.20), was reported to give high luminescence quantum yield of 71 % when doped in amorphous films of *m*CP. It seems that small distortion of the complex in the excited state limits the non-radiative deactivation for this particular complex, when compared to other three-coordinated complexes. A green OLED was prepared, and the maximum current efficiency of 65.3 cd A^{-1} was measured at a current density of 0.01 mA cm^{-2} [117]. The efficiency quickly drops at higher current densities.

Finally, copper phthalocyanine, **CuPc**, was used as an emitter for infrared electroluminescence. To obtain improved efficiencies, the strategy is to use electroluminescence from another material to sensitize the IR emission of **CuPc**.

An ytterbium complex [118] and an iridium complex [119] have been used as the sensitizer, and electroluminescence centred at 1,120 nm was observed from the device.

3.3.4.3 Zinc

Zinc complexes for electroluminescence are generally based on N^O ligands, similar to aluminium (and scandium, see above) complexes. An important difference is the possibility for the oxygen of the N^O ligand to bridge two zinc atoms, resulting in multinuclear (usually binuclear) complexes in the solid state [120–123]. The coordination sphere can also accommodate water molecules [124] or the complex be found as a co-crystal of a mono-ligand dichloro zinc complex and a protonated ligand [125]. This variety of structure likely has an impact on some device performance. Unfortunately, the exact material in the emissive layer is not always characterized.

The complex **Znq$_2$** was reported as early as 1993 [126]. Using a simple device structure ITO/TPD/complex/MgIn, series of **Mq$_x$** emitters were tested and a device using the zinc complex was found to emit at 568 nm (yellow, compared to the device using **Alq$_3$** emitting at 519 nm, green) with high luminance of 16,200 cd m^{-2} at 20 V (compared to 15,800 cd m^{-2} at 22 V for **Alq$_3$**).

Using the same device structure, the azomethine-zinc complex **43** (Fig. 3.21) results in blue electroluminescence with EL peak at 462 nm [127]. Modification of the alkyl bridge between the two nitrogen atoms leads to minor variations in colour and maximum luminance.

The complex **44** was reported as a ZnL$_2$ mononuclear complex resulting in a greenish-white emission [128]. The CIE coordinates are (0.246, 0.363) and luminance reaches 10,190 cd m^{-2} at 8 V for a maximum luminous efficiency of 0.89 lm W^{-1} at 173 cd A^{-1}. The low driving voltage and, consequently, relatively high efficiency have been attributed to excellent electron-transporting properties of the complex. Although the emission is slightly blueshifted and the efficiency of about 1 cd A^{-1} at high luminance is similar to a recently reported single-centre

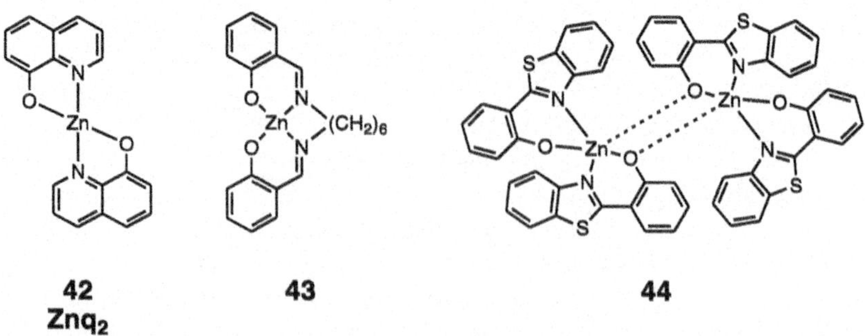

Fig. 3.21 Examples of Zn complexes for electroluminescence

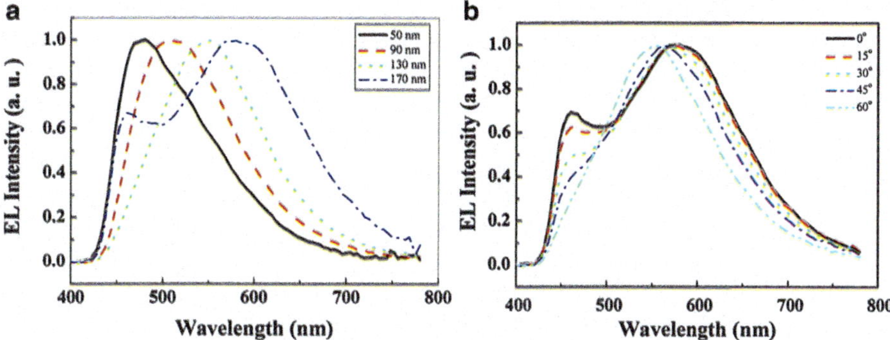

Fig. 3.22 (**a**) EL spectra from devices using different thickness of **44** emissive layer; (**b**) EL spectra from the device with **44** emissive layer 170 nm thick as a function of the viewing angle (Adapted with permission from Ref. [132]. Copyright 2007 American Chemical Society)

iridium-based white emitter [129], the luminance of the device using **44** is 10 times higher. The binuclear structure of **44** was reported few years later, and its electron transport properties were studied [130]. In particular, it was found that **44** has better electron transport properties than **Alq3**, which is primarily attributed to close intermolecular π-π interactions of the ligands in the solid state.

The reasons behind the unusual whitish electroluminescence of **44** have been explored recently [131]. It was found that the EL spectra vary significantly with the thickness of the emissive layer (Fig. 3.22 a) and that, for a given thickness, the EL spectra also depend on the viewing angle (Fig. 3.22 b). Photophysical studies also show that blue fluorescence at room temperature and orange phosphorescence at 77 K can be observed in thin film. The whitish EL emission was then attributed to microcavity effects and to dual emission fluorescence/phosphorescence. A white OLED was also made using **44** doped with the red DCJTB (4-(dicyanomethylene)-2-*t*-butyl-6-(1,1,7,7-tetramethyljulolidyl-9-enyl)-4*H*-pyran)[132].

Besides these key complexes, examples of zinc complexes with N^O ligands for electroluminescence are multiple. Modifications encompass changes of the core ligand and series of substituents on a given core ligand [133–140]. A study of the mechanism of degradation in the device was also reported, pointing to reversible (charge trapping) and irreversible (changes at the interfaces) [141].

3.4 Conclusions

Described herein are some of the most promising first-row transition metal complexes to be used as low-cost alternatives of the ruthenium, iridium and platinum complexes in DSC and electroluminescence arenas. Their use as redox mediators in DSCs as potential replacements for the iodide/triiodide system was also described. The complexes used in electroluminescence with a function different from the emissive dopant were not described due to the limited number of examples in the literature, although dopant for charge-transporting layer has a rich patent literature.

In general, it can be seen that the device performance using such metal complexes still has room for improvement in particular in terms of device efficiency converting light into energy and energy into light. Furthermore, the stability of the devices has not been widely addressed. Due to the lability of many of the complexes presented, it is expected that this may be limited.

Nevertheless, the field of first-row transition metal complexes for energy applications has made tremendous progress over the past decade. Although it is very unlikely that these materials will become the next champions of high-efficiency devices, they are promising alternative to platinoid metals for low cost energy applications and possibly when a single use is desired, such as point-of-care diagnostic devices and on-site testing.

Acknowledgements The financial support from the European Union (HetIridium, CIG322280) is greatly appreciated. I am very grateful to Dr John Fossey for the discussion and time for proofreading the manuscript.

References

1. http://eia.gov. Accessed 21 Dec 2013
2. http://econ.worldbank.org/. Accessed 21 Dec 2013
3. Veltkamp AC, de Wild-Scholten MJ (2006) In: Proceedings of renewable energy conference, 2006, Makuhari, Chiba, Japan, October 2006
4. Moser J (1887) Notiz über Verstärkung photoelektrischer Ströme durch optische Sensibilisirung. Monatsh Chem 8:373
5. Gerischer H, Tributsch H (1968) Elektrochemische Untersuchungen zur spektralen Sensibilisierung von ZnO-Einkristallen. Ber Bunsenges Phys Chem 72:437–445
6. Dare-Edwards MP, Goodenough JB, Hamnett A et al (1980) Sensitisation of semiconducting electrodes with ruthenium-based dyes. Faraday Discuss Chem Soc 70:285–298
7. Tsuborama H, Matsumura M, Nomura Y et al (1976) Dye sensitised zinc oxide: aqueous electrolyte: platinum photocell. Nature 261:402–403
8. O'Regan B, Grätzel M (1991) A low-cost, high-efficiency solar cell based on dye-sensitized colloidal TiO_2 films. Nature 353:737–740
9. Nazeeruddin MK, De Angelis F, Fantacci S et al (2005) Combined experimental and DFT-TDDFT computational study of photoelectrochemical cell ruthenium sensitizers. J Am Chem Soc 127:16835–16847
10. Nazeeruddin MK, Baranoff E, Grätzel M (2011) Dye-sensitized solar cells: a brief overview. Sol Energy 85:1172–1178
11. Hagfeldt A, Boschloo G, Sun L et al (2010) Dye-sensitized solar cells. Chem Rev 110:6595–6663
12. Yella A, Lee HW, Tsao HN et al (2011) Porphyrin-sensitized solar cells with cobalt (II/III)-based redox electrolyte exceed 12 percent efficiency. Science 334:629–634
13. Snaith H (2013) Perovskites: the emergence of a new era for low-cost, high-efficiency solar cells. J Phys Chem Lett 4:3623–3630
14. Nazeeruddin MK, Humphry-Baker R, Liska P et al (2003) Investigation of sensitizer adsorption and the influence of protons on current and voltage of a dye-sensitized nanocrystalline TiO_2 solar cell. J Phys Chem B 107:8981–8987

15. Nazeeruddin MK, Kay A, Rodicio I et al (1993) Conversion of light to electricity by cis-X2bis(2,2'-bipyridyl-4,4'-dicarboxylate)ruthenium(II) charge-transfer sensitizers ($X = Cl^-$, Br^-, I-, CN^-, and SCN^-) on nanocrystalline titanium dioxide electrodes. J Am Chem Soc 115:6382–6390
16. Nazeeruddin MK, Péchy P, Renouard T et al (2003) engineering of efficient panchromatic sensitizers for nanocrystalline TiO_2-based solar cells. J Am Chem Soc 123:1613–1624
17. Chiba Y, Islam A, Watanabe Y et al (2006) Dye-sensitized solar cells with conversion efficiency of 11.1%. Jpn J Appl Phys 45:L638–L640
18. Bessho T, Yoneda E, Yum JH et al (2009) New paradigm in molecular engineering of sensitizers for solar cell applications. J Am Chem Soc 131:5930–5934
19. Bomben PG, Robson KCD, Koivisto BD et al (2012) Cyclometalated ruthenium chromophores for the dye-sensitized solar cell. Coord Chem Rev 256:1438–1450
20. Bozic-Weber B, Constable EC, Housecroft CE (2013) Light harvesting with Earth abundant d-block metals: development of sensitizers in dye-sensitized solar cells (DSCs). Coord Chem Rev 257:3089–3106
21. Ragoussi ME, Ince M, Torres T (2013) Recent advances in phthalocyanine-based sensitizers for dye-sensitized solar cells. Eur J Org Chem 29:6475–6489
22. Li LL, Diau EWG (2013) Porphyrin-sensitized solar cells. Chem Soc Rev 42:291–304
23. Vrachnou E, Vlachopoulos N, Grätzel M (1987) Efficient visible light sensitization of TiO_2 by surface complexation with $Fe(CN)_6{}^{4-}$. J Chem Soc Chem Commun 12:868–870
24. Yang M, Thompson DW, Meyer GJ (2000) Dual pathways for TiO_2 sensitization by $Na_2[Fe(bpy)(CN)_4]$. Inorg Chem 39:3738–3739
25. Yang M, Thompson DW, Meyer GJ (2002) Charge-transfer studies of iron cyano compounds bound to nanocrystalline TiO_2 surfaces. Inorg Chem 41:1254–1262
26. De Angelis F, Tilocca A, Selloni A (2004) Time-dependent DFT study of $[Fe(CN)_6]^{4-}$ sensitization of TiO_2 nanoparticles. J Am Chem Soc 126:15024–15025
27. Ferrere S, Gregg BA (1998) Photosensitization of TiO2 by [FeII(2,2'-bipyridine-4,4'--dicarboxylic acid)2(CN)2]: band selective electron injection from ultra-short-lived excited states. J Am Chem Soc 120:843–844
28. Ferrere S (2002) New photosensitizers based upon $[Fe^{II}(L)_2(CN)_2]$ and $[Fe^{II}L_3]$, where L is substituted 2,2'-bipyridine. Inorg Chim Acta 329:79–92
29. Bowman DN, Blew JH, Tsuchiya T et al (2013) Elucidating band-selective sensitization in iron(II) polypyridine-TiO_2 assemblies. Inorg Chem 52:8621–8628
30. Alonso-Vante N, Ern V, Chartier P et al (1983) Spectral sensitization of semiconductors by copper(I) complexes in photoelectrochemical systems. Nouv J Chim 7:3–5
31. Alonso-Vante N, Nierengarten JF, Sauvage JP (1994) Spectral sensitization of large-band-gap semiconductors (thin films and ceramics) by a carboxylated bis(1,10-phenanthroline) copper(I) complex. J Chem Soc Dalton Trans 11:1649–1654
32. Sakaki S, Kuroki T, Hamada T (2002) Synthesis of a new copper(I) complex, $[Cu(tmdcbpy)_2]+$ (tmdcbpy = 4,4',6,6'-tetramethyl-2,2'-bipyridine-5,5'-dicarboxylic acid), and its application to solar cells. J Chem Soc Dalton Trans 2:840–842
33. Bessho T, Constable EC, Graetzel M et al (2008) An element of surprise-efficient copper-functionalized dye-sensitized solar cells. Chem Commun 32:3717–3719
34. Constable EC, Hernandez Redondo A, Housecroft CE et al (2009) Copper(I) complexes of 6,6'-disubstituted 2,2'-bipyridine dicarboxylic acids: new complexes for incorporation into copper-based dye sensitized solar cells (DSCs). Dalton Trans 7:6634–6644
35. Yuan YJ, Yu ZT, Zhang JY et al (2012) A copper(I) dye-sensitised TiO_2-based system for efficient light harvesting and photoconversion of CO_2 into hydrocarbon fuel. Dalton Trans 41:9594–9597
36. Linfoot CL, Richardson P, Hewat TE et al (2010) Substituted [Cu(I)(POP)(bipyridyl)] and related complexes: synthesis, structure, properties and applications to dye-sensitised solar cells. Dalton Trans 39:8945–8956

37. Bozic-Weber B, Constable EC, Housecroft CE et al (2011) The intramolecular aryl embrace: from light emission to light absorption. Dalton Trans 40:12584–12594
38. Bozic-Weber B, Chaurin V, Constable EC et al (2012) Exploring copper(I)-based dye-sensitized solar cells: a complementary experimental and TD-DFT investigation. Dalton Trans 41:14157–14169
39. Lu X, Wei S, Wu CML et al (2001) Can polypyridyl Cu(I)-based complexes provide promising sensitizers for dye-sensitized solar cells? a theoretical insight into Cu(I) versus Ru(II) sensitizers. J Phys Chem C 115:3753–3761
40. Huang J, Buyukcakir O, Mara MW et al (2012) Highly efficient ultrafast electron injection from the singlet MLCT excited state of copper(I) diimine complexes to TiO2 nanoparticles. Angew Chem Int Ed 51:12711–12715
41. Boschloo G, Hagfeldt A (2009) Characteristics of the iodide/triiodide redox mediator in dye-sensitized solar cells. Acc Chem Res 42:1819–1826
42. Kubo W, Kambe S, Nakade S et al (2003) Photocurrent-determining processes in quasi-solid-state dye-sensitized solar cells using ionic gel electrolytes. J Phys Chem B 107:4374–4381
43. Wang X, Stanbury DM (2006) Oxidation of iodide by a series of Fe(III) complexes in acetonitrile. Inorg Chem 45:3415–3423
44. Hamann TW, Ondersma JW (2011) Dye-sensitized solar cell redox shuttles. Energy Environ Sci 4:370–381
45. Tian H, Sun L (2011) Iodine-free redox couples for dye-sensitized solar cells. J Mater Chem 21:10592–10601
46. Wang M, Grätzel C, Zakeeruddin SM et al (2012) Recent developments in redox electrolytes for dye-sensitized solar cells. Energy Environ Sci 5:9394–9405
47. Cong J, Yang X, Kloo L, Sun L (2012) Iodine/iodide-free redox shuttles for liquid electrolyte-based dye-sensitized solar cells. Energy Environ Sci 5:9180–9194
48. Yu Z, Vlachopoulos N, Gorlov M et al (2011) Liquid electrolytes for dye-sensitized solar cells. Dalton Trans 40:10289–10303
49. Gregg BA, Pichot F, Ferrere S et al (2001) Interfacial recombination processes in dye-sensitized solar cells and methods to passivate the interfaces. J Phys Chem B 105:1422–1429
50. Hamann TW, Farha OK, Hupp JT (2008) Outer-sphere redox couples as shuttles in dye-sensitized solar cells. Performance enhancement based on photoelectrode modification via atomic layer deposition. J Phys Chem C 112:19756–19764
51. Daeneke T, Kwon TH, Holmes AB et al (2011) High-efficiency dye-sensitized solar cells with ferrocene-based electrolytes. Nat Chem 3:211–215
52. Daeneke T, Mozer AJ, Kwon TH et al (2012) Dye regeneration and charge recombination in dye-sensitized solar cells with ferrocene derivatives as redox mediators. Energy Environ Sci 5:7090–7099
53. Daeneke T, Mozer AJ, Uemura Y et al (2012) Dye regeneration kinetics in dye-sensitized solar cells. J Am Chem Soc 134:16925–16928
54. Nusbaumer H, Moser JE, Zakeeruddin SM et al (2001) $Co^{II}(dbbip)_2^{2+}$ complex rivals tri-iodide/iodide redox mediator in dye-sensitized photovoltaic cells. J Phys Chem B 105:10461–10464
55. Wang H, Sun Z, Zhang Y et al (2014) Charge transport limitations of redox mediators in dye-sensitized solar cells: investigation based on a quasi-linear model. J Phys Chem C 118:60–70
56. Cameron PJ, Peter LM, Zakeeruddin SM et al (2004) Electrochemical studies of the Co(III)/Co(II)$(dbbip)_2$ redox couple as a mediator for dye-sensitized nanocrystalline solar cells. Coord Chem Rev 248:1447–1453
57. Sapp SA, Elliott CM, Contado C et al (2002) Substituted polypyridine complexes of cobalt (II/III) as efficient electron-transfer mediators in dye-sensitized solar cells. J Am Chem Soc 124:11215–11222

58. Liu Y, Jenning JR, Huang Y et al (2011) Cobalt redox mediators for ruthenium-based dye-sensitized solar cells: a combined impedance spectroscopy and near-IR transmittance study. J Phys Chem C 115:18847–18855
59. Feldt SM, Gibson EA, Gabrielsson E et al (2010) Design of organic dyes and cobalt polypyridine redox mediators for high-efficiency dye-sensitized solar cells. J Am Chem Soc 132:16714–16724
60. Yum JH, Baranoff E, Kessler F et al (2012) A cobalt complex redox shuttle for dye-sensitized solar cells with high open-circuit potentials. Nat Commun 3:631
61. Nakade S, Makimoto Y, Kubo W et al (2005) Roles of electrolytes on charge recombination in dye-sensitized TiO2 solar cells (2): the case of solar cells using cobalt complex redox couples. J Phys Chem B 109:3488–3493
62. Xie Y, Hamann TW (2013) Fast low-spin cobalt complex redox shuttles for dye-sensitized solar cells. J Phys Chem Lett 4:328–332
63. Kashif MK, Axelson JC, Duffy NW et al (2012) A new direction in dye-sensitized solar cells redox mediator development: in situ fine-tuning of the cobalt(II)/(III) redox potential through Lewis base interactions. J Am Chem Soc 134:16646–16653
64. Hattori S, Wada Y, Yanagida S et al (2005) Blue copper model complexes with distorted tetragonal geometry acting as effective electron-transfer mediators in dye-sensitized solar cells. J Am Chem Soc 127:9648–9654
65. Bai Y, Yu Q, Cai N et al (2011) High-efficiency organic dye-sensitized mesoscopic solar cells with a copper redox shuttle. Chem Commun 47:4376–4378
66. Li TC, Spokoyny AM, She C et al (2010) Ni(III)/(IV) Bis(dicarbollide) as a fast, noncorrosive redox shuttle for dye-sensitized solar cells. J Am Chem Soc 132:4580–4582
67. Spokoyny AM, Li TC, Farha OK et al (2010) Electronic tuning of nickel-based Bis (dicarbollide) redox shuttles in dye-sensitized solar cells. Angew Chem Int Ed 49:5339–5343
68. Perera IR, Gupta A, Xiang W et al (2014) Introducing manganese complexes as redox mediators for dye-sensitized solar cells. Phys Chem Chem Phys 16:12021–12028
69. Bernanose A, Comte M, Vouaux P (1953) A new method of emission of light by certain organic compounds. J Chim Phys Phys Chim Biol 50:64–68
70. Bernanose A, Vouaux P (1953) Organic electroluminescence: type of emission. J Chim Phys Phys Chim Biol 50:261–263
71. Pope M, Kallmann HP, Magnante P (1963) Electroluminescence in organic crystals. J Chem Phys 38:2042–2043
72. Tang CW, VanSlyke SA (1987) Organic electroluminescent diodes. Appl Phys Lett 51:913–915
73. Baldo MA, O'Brien DF, You Y et al (1998) Highly efficient phosphorescent emission from organic electroluminescent devices. Nature 395:151–154
74. Baldo MA, Thompson ME, Forrest SR (2000) High-efficiency fluorescent organic light-emitting devices using a phosphorescent sensitizer. Nature 403:750–753
75. Deaton JC, Switalski SC, Kondakov DY et al (2010) E-type delayed fluorescence of a phosphine-supported $Cu_2(\mu\text{-}NAr_2)_2$ diamond core: harvesting singlet and triplet excitons in OLEDs. J Am Chem Soc 132:9499–9508
76. Uoyama H, Goushi K, Shizu K et al (2012) Highly efficient organic light-emitting diodes from delayed fluorescence. Nature 492:234–238
77. Dias FB, Bourdakos KN, Jankus V et al (2013) Triplet harvesting with 100% efficiency by way of thermally activated delayed fluorescence in charge transfer OLED emitters. Adv Mater 25:3707–3714
78. Flamigni L, Barbieri A, Sabatini C et al (2007) Photochemistry and photophysics of coordination compounds: iridium. Top Curr Chem 281:143–203
79. You Y, Park SY (2009) Phosphorescent iridium(III) complexes: toward high phosphorescence quantum efficiency through ligand control. Dalton Trans 8:1267–1282
80. Lowry MS, Bernhard S (2006) Synthetically tailored excited states: phosphorescent, cyclometalated iridium(III) complexes and their applications. Chem Eur J 12:7970–7977

81. Ulbricht C, Beyer B, Friebe C et al (2009) Recent developments in the application of phosphorescent iridium(III) complex systems. Adv Mater 21:4418–4441
82. Baranoff E, Yum JH, Graetzel M et al (2009) Cyclometallated iridium complexes for conversion of light into electricity and electricity into light. J Organomet Chem 694:2661–2670
83. Lo KK, Choi AW, Law WH (2012) Applications of luminescent inorganic and organometallic transition metal complexes as biomolecular and cellular probes. Dalton Trans 41:6021–6047
84. Sajoto T, Djurovich PI, Tamayo AB et al (2009) Temperature dependence of blue phosphorescent cyclometalated Ir(III) complexes. J Am Chem Soc 131:9813–9822
85. Li J, Djurovich PI, Alleyne BD et al (2005) Synthetic control of excited-state properties in cyclometalated Ir(III) complexes using ancillary ligands. Inorg Chem 44:1713–1727
86. Costa RD, Ortì E, Bolink HJ et al (2012) Luminescent ionic transition-metal complexes for light-emitting electrochemical cells. Angew Chem Int Ed 51:8178–8211
87. Brooks J, Babayan Y, Lamanski S et al (2002) Synthesis and characterization of phosphorescent cyclometalated platinum complexes. Inorg Chem 41:3055–3066
88. Cocchi M, Kalinowski J, Virgili D et al (2007) Single-dopant organic white electrophosphorescent diodes with very high efficiency and its reduced current density roll-off. Appl Phys Lett 90:163508
89. Burrows PE, Sapochak LS, McCarty DM et al (1994) Metal ion dependent luminescence effects in metal tris-quinolate organic heterojunction light emitting devices. Appl Phys Lett 64:2718–2720
90. Katkova MA, Ilichev VA, Konev AN et al (2008) Electroluminescent characteristics of scandium and yttrium 8-quinolinolates. J Appl Phys 104:053706
91. Katkova MA, Kurskii YA, Fukin GK et al (2005) Efficient synthetic route to anhydrous mononuclear tris(8-quinolinolato)lanthanoid complexes for organic light-emitting devices. Inorg Chim Acta 358:3625–3632
92. Chen BJ, Sun XW (2005) The role of MgF_2 buffer layer in tris-(8-hydroxyquinoline)aluminium-based organic light-emitting devices with Mg:Ag cathode. Semicond Sci Technol 20:801–804
93. Shestakov AF, Katkova MA, Emel'yanova NS et al (2010) Experimental and theoretical study of the effect of the substituent nature on the luminescent properties of scandium complexes with substituted 8-hydroxyquinolines. High Energy Chem 44:503–510
94. Katkova MA, Balashova TV, Lichev VA et al (2010) Synthesis, structures, and electroluminescent properties of scandium N, O-chelated complexes toward near-white organic light-emitting diodes. Inorg Chem 49:5094–5100
95. Burin ME, Ilichev VA, Pushkarev AP et al (2012) Synthesis and luminescence properties of lithium, zinc and scandium 1-(2-pyridyl)naphtholates. Org Electron 13:3203–3210
96. Ma YG, Chan WH, Zhou XM et al (1999) Light-emitting diode device from a luminescent organocopper(I) compound. New J Chem 23:263–265
97. Min J, Zhang Q, Sun W et al (2011) Neutral copper(I) phosphorescent complexes from their ionic counterparts with 2-(2'-quinolyl)benzimidazole and phosphine mixed ligand. Dalton Trans 40:686–693
98. Babashkina MG, Safin DA, Klein A et al (2010) Synthesis, characterisation and luminescent properties of mixed-ligand copper(I) complexes incorporating N-thiophosphorylated thioureas and phosphane ligands. Eur J Inorg Chem 2010:4018–4026
99. Zhang Q, Zhou Q, Cheng Y et al (2004) Highly efficient green phosphorescent organic light-emitting diodes based on Cu^{I} complexes. Adv Mater 16:432–436
100. Zhang Q, Zhou Q, Cheng Y et al (2006) Highly efficient electroluminescence from green-light-emitting electrochemical cells based on CuI complexes. Adv Funct Mater 16:1203–1208
101. Armaroli N, Accorsi G, Holler M et al (2006) Highly luminescent Cu^{I} complexes for light-emitting electrochemical cells. Adv Mater 18:1313–1316

102. Che G, Su Z, Li W et al (2006) Highly efficient and color-tuning electrophosphorescent devices based on Cu^I complex. Appl Phys Lett 89:103511
103. Su Z, Li W, Che G et al (2007) Enhanced electrophosphorescence of copper complex based devices by codoping an iridium complex. Appl Phys Lett 90:143505
104. Su Z, Li W, Chu B et al (2008) Efficient white organic light-emitting diodes based on iridium complex sensitized copper complex. J Phys D Appl Phys 41:085103
105. Moudam O, Kaeser A, Delavaux-Nicot B et al (2007) Electrophosphorescent homo- and heteroleptic copper(I) complexes prepared from various bis-phosphine ligands. Chem Commun 29:3077–3079
106. Jia WL, McCormick T, Tao Y et al (2005) New phosphorescent polynuclear Cu (I) compounds based on linear and star-shaped 2-(2′-pyridyl)benzimidazolyl derivatives: syntheses, structures, luminescence, and electroluminescence. Inorg Chem 44:5706–5712
107. Zhang Q, Ding J, Cheng Y et al (2007) Novel heteroleptic CuI complexes with tunable emission color for efficient phosphorescent light-emitting diodes. Adv Funct Mater 17:2983–2990
108. Si Z, Li J, Li B et al (2009) High light electroluminescence of novel Cu(I) complexes. J Lumin 129:181–186
109. Zhang L, Li B, Su Z (2009) Realization of high-energy emission from $[Cu(N-N)(P-P)]^+$ complexes for organic light-emitting diode applications. J Phys Chem C 113:13968–13973
110. Sun W, Zhang Q, Qin L et al (2010) Phosphorescent cuprous complexes with N, O ligands – synthesis, photoluminescence, and electroluminescence. Eur J Inorg Chem 2010:4009–4017
111. Costa RD, Tordera D, Ortí E et al (2011) Copper(I) complexes for sustainable light-emitting electrochemical cells. J Mater Chem 21:16108–16118
112. Wada A, Zhang Q, Yasuda T et al (2012) Efficient luminescence from a copper(I) complex doped in organic light-emitting diodes by suppressing C–H vibrational quenching. Chem Commun 48:5340–5342
113. Igawa S, Hashimoto M, Kawata I et al (2013) Highly efficient green organic light-emitting diodes containing luminescent tetrahedral copper(I) complexes. J Mater Chem C 1:542–551
114. Parmeggiani F, Sacchetti A (2012) Preparation and luminescence thermochromism of tetranuclear copper(I)–pyridine–iodide clusters. J Chem Educ 89:946–949
115. Liu Z, Qayyum MF, Wu C et al (2011) A codeposition route to CuI–pyridine coordination complexes for organic light-emitting diodes. J Am Chem Soc 133:3700–3703
116. Volz D, Zink DM, Bocksrocker T et al (2013) Molecular construction kit for tuning solubility, stability and luminescence properties: heteroleptic MePyrPHOS-copper iodide-complexes and their application in organic light-emitting diodes. Chem Mater 25:3414–3426
117. Hashimoto M, Igawa S, Yashima M et al (2011) Highly efficient green organic light-emitting diodes containing luminescent three-coordinate copper(I) complexes. J Am Chem Soc 133:10348–10351
118. Yan F, Li WL, Chu B et al (2007) Sensitized electrophosphorescence of infrared emission diode based on copper phthalocyanine by an ytterbium complex. Appl Phys Lett 91:203512
119. Yan F, Li W, Chu B et al (2009) Sensitized infrared electrophosphorescence based on divalent copper complex by an iridium(III) complex. Org Electron 10:1408–1411
120. Yuan GZ, Huo YP, Nie XL et al (2012) Structure and photophysical properties of a dimeric Zn(II) complex based on 8-hydroxyquinoline group containing 2,6-dichlorobenzene unit. Tetrahedron 68:8018–8023
121. Czugler M, Neumann R, Weber E (2001) X-ray crystal structures and data bank analysis of Zn(II) and Cd(II) complexes of 2- and 7-nonyl substituted 8-hydroxyquinoline and 8-hydroxyquinaldine extractive agents. Inorg Chim Acta 313:100–108
122. Xu HB, Wen HM, Chen ZH et al (2010) Square structures and photophysical properties of Zn2Ln2 complexes (Ln = Nd, Eu, Sm, Er, Yb). Dalton Trans 39:1948–1953
123. Huo YP, Zhu SZ, Hu S (2010) Synthesis and luminescent properties of Zn complex based on 8-hydroxyquinoline group containing 3,5-bis(trifluoromethyl) benzene unit with unique crystal structure. Tetrahedron 66:8635–8640

124. Palenik GJ (1964) The structure of coordination compounds. III. A refinement of the structure of zinc 8-hydroxyquinolinate dehydrate. Acta Crystallogr 17:696–700
125. Najafi E, Amini MM, Ng SW (2011) 8-Hydroxy-2-methylquinolinium dichlorido (2-methylquinolin-8-olato-^{2}N,O) zincate acetonitrile disolvate. Acta Crystallogr E67:m1280
126. Hamada Y, Sano T, Fujita M et al (1993) Organic electroluminescent devices with 8-hydroxyquinoline derivative-metal complexes as an emitter. Jpn J Appl Phys 32:L514–L515
127. Hamada Y, Sano T, Fujita M et al (1993) Blue electroluminescence in thin films of azomethin-zinc complexes. Jpn J Appl Phys 32:L511–L513
128. Hamada Y, Sano T, Fujii H et al (1996) White-light-emitting material for organic electroluminescent devices. Jpn J Appl Phys 35:L1339–L1341
129. Bolink HJ, De Angelis F, Baranoff E et al (2009) White-light phosphorescence emission from a single molecule: application to OLED. Chem Commun 31:4672–4674
130. Yu G, Yin S, Liu Y et al (2003) Structures, electronic states, and electroluminescent properties of a zinc(II) 2-(2-hydroxyphenyl)benzothiazolate complex. J Am Chem Soc 125:14816–14824
131. Xu X, Liao Y, Yu G et al (2007) Charge carrier transporting, photoluminescent, and electroluminescent properties of Zinc(II)-2-(2-hydroxyphenyl)benzothiazolate complex. Chem Mater 19:1740–1748
132. Kim DE, Kim WS, Kim BS et al (2008) Improvement of color purity in white OLED based on $Zn(HPB)_2$ as blue emitting layer. Thin Solid Films 516:3637–3640
133. Son HJ, Han WS, Chun JY et al (2008) Generation of blue light-emitting zinc complexes by band-gap control of the oxazolylphenolate ligand system: syntheses, characterizations, and organic light emitting device applications of 4-coordinated Bis(2-oxazolylphenolate) Zinc (II) complexes. Inorg Chem 47:5666–5676
134. Wang P, Hong Z, Xie Z et al (2003) A bis-salicylaldiminato Schiff base and its zinc complex as new highly fluorescent red dopants for high performance organic electroluminescence devices. Chem Commun 14:1664–1665
135. Wang R, Deng L, Fu M et al (2012) Novel ZnII complexes of 2-(2-hydroxyphenyl) benzothiazoles ligands: electroluminescence and application as host materials for phosphorescent organic light-emitting diodes. J Mater Chem 22:23454–23460
136. Aleksanyan DV, Kozlov VA, Petrov BI et al (2013) Lithium, zinc and scandium complexes of phosphorylated salicylaldimines: synthesis, structure, thermochemical and photophysical properties, and application in OLEDs. RSC Adv 3:24484–24491
137. Xie J, Qiao J, Wang L et al (2005) An azomethin-zinc complex for organic electroluminescence: Crystal structure, thermal stability and optoelectronic properties. Inorg Chim Acta 358:4451–4458
138. Yu T, Su W, Li W et al (2006) Synthesis, crystal structure and electroluminescent properties of a Schiff base zinc complex. Inorg Chim Acta 359:2246–2251
139. Zeng HP, Wang GR, Zeng GC et al (2009) The synthesis, characterization and electroluminescent properties of zinc(II) complexes for single-layer organic light-emitting diodes. Dyes Pigments 83:155–161
140. Bagatin IA, Legnani C, Cremona M (2009) Investigation on Al(III) and Zn(II) complexes containing a calix[4]arene bearing two 8-oxyquinoline pendant arms used as emitting materials for OLEDs. Mater Sci Eng C 29:267–270
141. Lepnev L, Vaschenko A, Vitukhnovsky A et al (2009) OLEDs based on some mixed-ligand terbium carboxylates and zinc complexes with tetradentate Schiff bases: mechanisms of electroluminescence degradation. Synth Met 159:625–631

Chapter 4
Ruthenium-Based Photosensitizers for Dye-Sensitized Solar Cells

Jian He, Baohua Wang, Shuai Chang, and Tao Chen

Abstract Dye-sensitized solar cell (DSSC) is a type of excitonic solar cells with photoanode sensitized by organic molecules, which serve as light harvester. Ruthenium complex-based dyes are a significant class of molecules in the development of DSSCs since 1991. Basically, ruthenium complex is composed of a ruthenium metal center and ancillary ligands. They have been intensively studied because of their excellent photovoltaic properties. In this chapter, we extensively summarize the efforts made to the development of Ru-based dye molecules, such as the approaches to change ancillary ligands of ruthenium complexes and tune the energy levels of the lowest unoccupied molecular orbital and the highest occupied molecular orbital. We also summarize the stability improvement of the dye molecules by introducing hydrophobic ligands. Finally, alternative approaches to expand light response of the Ru dye-sensitized devices including co-sensitization and plasmon-enhanced light harvesting are discussed.

Keywords Ruthenium-based complex • Solar cells • Anchoring group • Co-sensitization • Charge recombination • Electron injection • Photosensitizer • Ancillary ligand • Deprotonation • Titanium dioxide

4.1 Introduction

Solar energy is undoubtedly one of the most attractive sustainable energy sources. The conversion of solar energy into electricity by solar cell is featured as clean and safe. Therefore, a great deal of efforts have been devoted to the development of solar devices, among which dye-sensitized solar cell (DSSC) is one of the most promising low-cost alternatives to the conventional silicon solar cells. In addition, the production of colorful and semitransparent DSSCs is especially attractive in the so-called building integrated photovoltaics (BIPV).

J. He • B. Wang • S. Chang • T. Chen (✉)
Department of Physics, The Chinese University of Hong Kong, Shatin, N.T., Hong Kong, China
e-mail: taochen@phy.cuhk.edu.hk

W.-Y. Wong (ed.), *Organometallics and Related Molecules for Energy Conversion*, Green Chemistry and Sustainable Technology, DOI 10.1007/978-3-662-46054-2_4

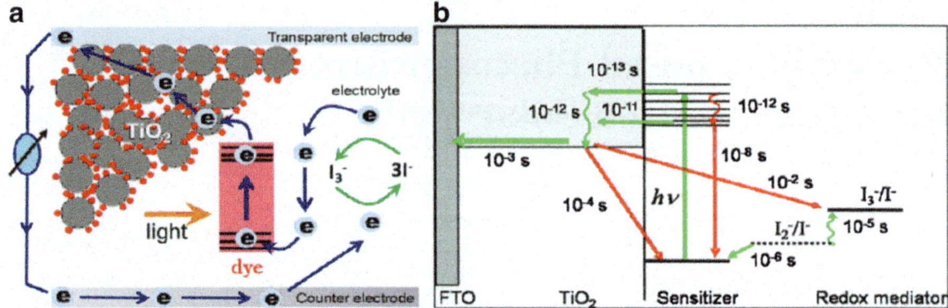

Fig. 4.1 (**a**) Schematic illustration of typical dye-sensitized solar cell structure; (**b**) working processes and typical time constant of DSSC with Ru-based sensitizers; green lines stand for charge transfer process with timescale for illustration, red lines stand for recombination process (Reproduced with permission from Ref. [2]. Copyright © 2010, American Chemical Society)

Typically, DSSC possesses a sandwich structure with electrolyte sealed in between the two electrodes (Fig. 4.1a). The photoanode is consisted of dye-sensitized TiO_2 nanoparticle network on FTO glass. Platinum sputtered on FTO is usually used as cathode. When light illuminates from the photoanode side and irradiates the dyes anchored on the TiO_2 surface, electrons could be excited from the HOMO (highest occupied molecular orbitals) of the dyes to the LUMO (lowest unoccupied molecular orbitals) and injected to the conduction band of TiO_2, leaving the dyes oxidized. Then electrons will diffuse across the TiO_2 nanocrystal mesoporous film and reach the photoanode FTO. The oxidized dye will be regenerated by redox couple in the electrolyte, usually I^-/I_3^- redox couple in nitrile-based solvents. The I^- ions will reduce the dye to its ground state, followed by transformation back to I_3^-. On the other hand, the counter electrode transfers electrons to the electrolyte reducing I_3^- to I^-.

The mesoporous TiO_2 photoanode is critical in the device which can dramatically improve the DSSC performance since the mesoporous TiO_2 film can provide much larger surface area than that of bulk TiO_2 film for sufficient amount of dye uptake. There are two common phases of TiO_2 used in DSSCs: rutile and anatase. Anatase TiO_2 is a preferred photoanode material in DSSCs. Electron transport process is slower in the rutile phase than that in anatase TiO_2, which leads to low PCE [1]. However, even in the anatase TiO_2 nanoparticle network, the inherent resistance of the mesoporous film is considerably large. The charge transfer process in TiO_2 mesoporous is usually regarded as a hopped process from one individual TiO_2 crystallite to another [2].

From the kinetic perspective, DSSC operates in a mechanism called electrons competitive injection and transport (Fig. 4.1b). There are many charge transfer processes in a working DSSC. Different transfer processes happen in different time constants. The injection from dyes to TiO_2 is the fastest and charge collection is fast as well compared to the recombination processes. As a consequence, electrons are generated and transferred efficiently from the photoexcited dyes to the TiO_2 and finally to the FTO, driving DSSC working in a favorable way. This lays the ground for the operation of DSSCs.

There are several parameters that are used to characterize DSSCs. The open-circuit voltage (V_{oc}) is determined by the difference between the Fermi level of TiO_2 and the redox couple potential of the electrolyte. The light-harvesting ability of the dye molecules is a determinant factor on the final short-circuit photocurrent density (J_{sc}). In addition, the device quality is reflected by fill factor (*FF*), which is affected by the cell's internal resistance. The internal resistance is composed of resistance of electrolytes, electrodes, and different interfaces such as FTO/TiO_2, TiO_2/electrolyte, TiO_2/dye, dye/electrolyte, and electrolyte/counter electrode.

The current density versus voltage (*J-V*) curve can be obtained from experimental photoelectric measurement. Thus, one can get the J_{sc} and V_{oc}. The maximum output power can also be extracted by calculating the maximum area of the inscribed square of the *J-V* curve. Since the intensity of the incident light (P_{in}) is known, the overall light-to-electrical power conversion efficiency (η) can be obtained. The *FF* of a cell can be calculated based on Eq. (4.1). The *FF* depends highly on the quality of the solar cell:

$$\eta = \frac{J_{sc}{}^{*}V_{oc}{}^{*}FF}{P_{in}} \tag{4.1}$$

The method to evaluate the photoresponse of the solar cell is the incident photon-to-current conversion efficiency (IPCE) spectrum. IPCE measures the percent of incident photons that are converted into electrons as a function of the wavelength of monochromatic illumination on the cell (Eq. 4.2). The photocurrent value can also be calculated based on the IPCE result (Eq. 4.3). The IPCE of DSSC sensitized with general Ru-based complex is about 80 % for wavelength from 300 to 650 nm and tails off at about 750 nm. This situation can be improved by modifying ancillary ligands of the dye molecules as discussed in the following sections:

$$\text{IPCE} = \frac{J_{sc}(\lambda)}{e\phi(\lambda)} = 1,240\frac{J_{sc}(\lambda)[A^{*}\text{cm}^{-2}]}{\lambda[\text{nm}]P_{in}(\lambda)[W^{*}\text{cm}^{-2}]} \tag{4.2}$$

$$J_{sc} = q_e \int_{\lambda_{min}}^{\lambda_{max}} J_{photon}(\lambda)\text{IPCE}(\lambda)\text{d}\lambda \tag{4.3}$$

To date, several classes of dye molecules have been developed to improve the device efficiency [2–4]. Notably, solar cells sensitized with porphyrin- and Ru-based molecules have reached efficiency around 12 % [5, 6]. Most recently, perovskite-based light sensitizers have also been applied in the DSSCs to replace the traditional organic molecules [7]. In this chapter, we pay special attention to the Ru-based dye molecules for the solar cells, and the device structures are based primarily on liquid electrolyte with $I^{-}/I_3{}^{-}$ as redox couples.

4.2 Fundamentals and Brief History of Ruthenium Complex

From chemistry perspective, ruthenium-based dyes belong to a class of metal complexes (organometallics) [2, 8]. Generally, metal complex dyes consist of a central metal ion with ancillary ligands. There is at least one anchor group on the ligand for linking the dye with the semiconductor surface and for facilitating electron injection. The light absorption is usually attributed to the metal-to-ligand charge transfer process (MLCT). The MLCT excited states of $d\pi^6$ coordination compounds are the most efficient for light harvesting. An electron will be excited from the metal *d* orbitals to the ligand π^* orbitals when light irradiates the dye. HOMO is commonly on the ruthenium center and electron-donating ligands [9]. LUMO is usually on the polypyridyl ligand. Figure 4.2 presents a typical calculated HOMO and LUMO distribution of the **N749** (also known as black dye) [10]. Ancillary ligands of Ru-based dyes can be tuned by different substituents like heterocycle and aryl to improve the dye performance in DSSC. For efficient charge injection, generally, the electron cloud should situate close to the supporting semiconduction oxide in excited state.

In 1979, Ru-based dyes with carboxylated bipyridine ligands were used to sensitize TiO_2 for the first time [11]. Later in 1985, Desilvestro et al. used Ru-complexes with three carboxylated bipyridine ligands to obtain the first efficient DSSC with IPCE over 40 % in the wavelength of 450–500 nm [3]. However, the breakthrough of DSSC is that Grätzel achieved an efficiency of 7.9 % in 1991 by using a trinuclear Ru-complex and a new mesoporous TiO_2 film as photoanode [12].

By improving the trinuclear Ru-complex, Grätzel and coworkers synthesized a series of mononuclear Ru-complexes including cis-$(SCN)_2$ bis(2,2′-bipyridyl-4,4-′-dicarboxylate) ruthenium(II), also known as **N3**, which displays outstanding photovoltaic properties [13]. DSSC with **N3** (Fig. 4.3) can achieve an overall

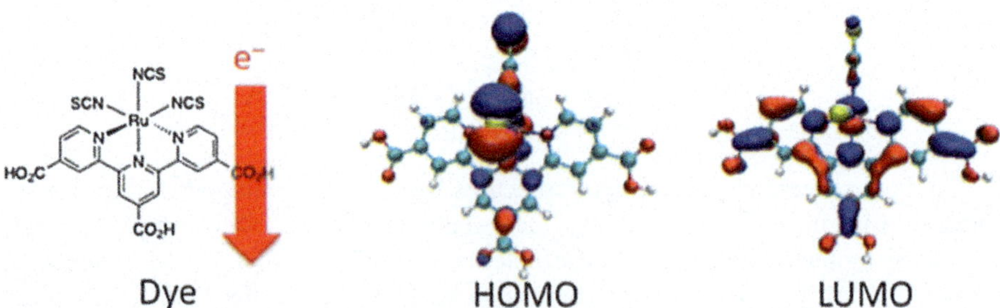

Fig. 4.2 Electron flow and the HOMO and LUMO distributions of black dye (**N749**) (Reproduced from Ref. [10] with permission from The Royal Society of Chemistry)

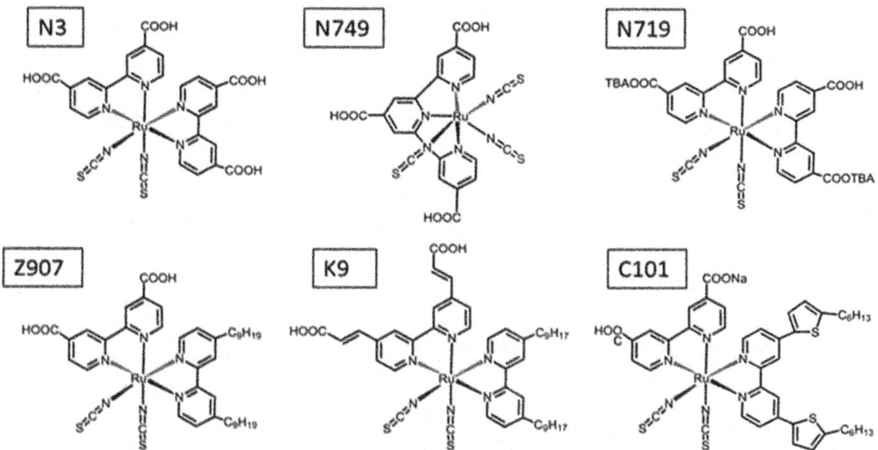

Fig. 4.3 Typical ruthenium-based dyes (**a**) **N3**; (**b**) **N749**; (**c**) **N719**; *TBA tetra*-n-butylammonium (**d**) **Z907**; (**e**) **K9**; (**f**) **C101**

solar-to-electric energy conversion efficiency of 10 %. After that, most Ru-based dyes have only one ruthenium metal ion center with various ancillary ligands. Changing ancillary properly can extend the absorption region of IPCE from visible to near infrared (NIR) [14], increase effective electron injection, suppress dye aggregation, tune HOMO and LUMO levels, increase thermal stability of DSSC [15], and decrease recombination rate of electrons and holes.

To extend absorption region of NIR to 920 nm, Grätzel and coworkers designed a kind of dye known as **N749** by changing the ligands of Ru-complexes on the basis of **N3** [14]. The **N749** has a ruthenium center attached with three thiocyanato ligands and one terpyridine ligand substituted with three carboxyl groups. The π^* level of the terpyridine ligand decrease and the t_{2g} metal orbital level increase, which leads to the redshift in the MLCT band comparing to **N3** [14]. Finally, DSSC device with **N749** has achieved efficiency of 10.4 % under AM 1.5G [16].

Similar to the synthesis of **N749**, **N719** is designed on the basis of **N3** as well. **N719**, $(Bu_4N)_2[Ru(dcbpyH)_2(NCS)_2]$, is the doubly deprotonated form of **N3** exhibiting better power conversion efficiency [17]. This improvement shows the deprotonation of the carboxylic acid groups in **N3**, which contributes to the production of **N719**, enables the oxidation and reduction potentials of the dye in electrolyte moving toward a more negative level.

The **N3** and **N719** dyes are widely used as reference dyes for DSSC in the literature, and they are considered as standard dyes for designing new Ru-based dyes. A variety of Ru-complexes have been synthesized by changing **N3** or **N749** ancillary ligands. Figure 4.3 shows the molecular structures of several classical ruthenium-based dyes.

4.3 Structural Optimization of Ru-Complex for High-Performance DSSCs

Ancillary ligands attaching to the ruthenium center can significantly affect performance of Ru-complex dyes when used as sensitizers in DSSCs. **N3** and **N719** dyes are widely used as the basic dyes for optimization. Most of the developed ruthenium complexes were obtained by changing **N3** or **N749** ancillary ligands.

4.3.1 Extend Absorption Spectrum with New Ligands

To harvest the light in the longer wavelength, ruthenium complex with a tetradentate ligand was fabricated, which was coded as **N886** (trans-[Ru(L)$(NCS)_2$], L = 4, 4″-di-tertbutyl-4′, 4″-bis(carboxylic acid) -2, 2′ : 6′, 2″ : 6″, 2‴-quarterpyridine). The **N886** dye exhibits a panchromatic absorption spectrum up to 900 nm. But the overall efficiency is rather low by 5.9 % because of the low extinction coefficient, though the spectral response is spanned [18].

The dye **HRS-q**, formed by introduction of 2-thiophene-2-yl-vinyl into the ancillary ligand of **N719**, increased the absorption spectrum by a 10 nm redshift compared to **N719**. There was a 30 % increase in the extinction coefficient as well. As a result, the dye **HRS-q** achieved 9.5 % efficiency while **N719** was 8.9 % as a reference [19].

When one of the bipyridyl ligands in Ru-complex is substituted with alkylthiophene, ethylenedioxythienyl (EDOT), or carbazole, energy levels of ruthenium center and LUMO of the ligands will change [20–22]. Substitution of bipyridyl ligands with thiophene ligands raised ruthenium center energy level and LUMO level of the ligands.

Changing ligands can tune the HOMO and LUMO levels of the dye to extend absorption region to NIR. The HOMO must be low enough for redox couple to regenerate the oxidized dye, while the LUMO should be high enough for efficient electron injection into the conduction band of TiO_2. Meanwhile, HOMO-LUMO gap should be as narrow as possible to absorb longer wavelength light. Successful example is **N749** as mentioned. Nazeeruddin reported that dyes **K9** and **K23** can generate high J_{sc} in DSSC when compared with **Z907**, which was due to the increased absorption in NIR regions [23].

4.3.2 Increase Stability of the Dyes

Tridentate bipyridine-pyrazolate ancillary ligands and elongated π-conjugated system in Ru-complexes was also investigated. One of these improved dyes showed 5.7 % efficiency as the reference dye **N3** shows PCE of 6.0 %. However, the dye is rather stable in superior thermal and light soaking conditions. It was concluded in the report that "the concomitant tridentate binding properties offered by the bipyridine-pyrazolate ligand may render a more stable complication, such that extending the lifespan of DSCs is expected" [24].

In most of the highly efficient ruthenium-based dyes, thiocyanate ligand (SCN) was widely used. However, thiocyanate is chemically unstable comparing to other parts in Ru-complexes. Many efforts have been made to replace thiocyanate with more stable ligands. Cyclometalated ruthenium complexes without thiocyanate ligands, [Ru(C^N^N)-(N^N^N)] type, were an example achieving comparable photocurrent and absorption spectrum to **N719**. It was reported by Grätzel, which exhibited an IPCE of 83 % and conversion efficiency of 10.1 % at AM 1.5 G [25].

Attachment of hydrophobic ligands to Ru-complexes may increase the stability as well. The anchoring states of dyes to TiO_2 tend to break in trace amount of water, and the dye molecules will desorb from the surface. Attaching hydrophobic chains to ruthenium center as ligands can retard desorption of the dye from TiO_2 surface. Thus, it can make DSSC more stable in a long term [26, 27]. Introduction of two hydrophobic alkyl chains on the bipyridyl ligand will increase the thermal stability, which was demonstrated by **Z907**, an amphiphilic heteroleptic Ru-complex. It was reported that DSSC with **Z907** and hexadecyl phosphonic acid coadsorbing maintained 7 % efficiency for as long as 1,000 h aging test at 80 °C [15]. In another report, the **Z907** dye was compared with the reference dye **N719** for a long-term test in a DSSC device. **N719** showed much poorer stability than **Z907** [27]. In addition, the hydrophobic chains enhance the interaction between the dye and the hole-conducting polymer. Consequently, the wettability of the TiO_2 increases [28].

4.3.3 Facilitate Dye Regeneration by Cyclodextrin (CD) Ligand

Cyclodextrin (CD) unit attaching the ancillary ligand of ruthenium tris-bipyridyl core may facilitate dye regeneration by CD moiety binding I^-/I_3^- redox couple into CD cavity. A supermolecular complex [Ru(dcb)$_2$(a-CD-5-bpy)]Cl$_2$(1-a-CD) (dcb = 4,4′-dicarboxyl-2,2′-bipyridine, a-CD-5-bpy = 6-mono [5-methyl (5′-methyl-2,2′-bipyridyl)]-permethylated a-CD) was fabricated. The device characterization showed larger J_{sc}, V_{oc} and PCE than the dye without CD unit [29]. Thus, it was concluded that CD unit may facilitate dye regeneration.

4.3.4 Retard Charge Recombination

Proper modification of ancillary ligands of ruthenium complexes can decrease recombination rate effectively. We will introduce the mechanism to retard recombination first. As the excited electrons in the LUMO of the dye are injected to TiO_2 rapidly, the dye is left with positive charge density. The positive charge density is distributed over the metal ruthenium, even to the thiocyanate (NCS) ligand. So there is a charged space in the interface between dye and TiO_2, retarding charge recombination between injected electrons and the left positive charges [30].

It was observed that the triarylamine moiety connecting with a conjugated link helped to increase the extinction coefficient. This is the result of faster intramolecular hole transfer from the ruthenium center to the donor and longer-lived charged separation between the injected electrons in TiO_2 and oxidized donor group of the dye. Dyes with triarylamine moiety achieved 3.4 % overall efficiency in solid-state dye-sensitized solar cells while **N719** was only 0.7 % in the same condition [31]. Liquid electrolyte DSSC with this kind of dye achieved efficiency of 10.3 % [32]. The results showed that charge recombination between electrons in TiO_2 and holes in dye can be retarded by introduction of triarylamine donor groups.

4.3.5 Extend the π-Conjugated System to Enhance Light Harvesting

Increasing the conjugation length of the ligand helped to extend the π-conjugated system of the bipyridine to enhance light harvesting. An example is that introducing 3-methoxystyryl into the ligand obtained the **Z910** dye. Devices sensitized with **Z910** reached an efficiency of 10.2 %. [33].

Introducing the tri(ethylene oxide) methyl ether (TEOME) into the 2,2-′-bipyridine ligand generates a novel ion-coordinating dye, **K51**. The outstanding ion-coordinating ability made the dye suitable for nonvolatile electrolyte or hole-transporting material with efficiency of 7.8 % and 3.8 %, respectively [34]. **K60** dye was obtained by extending the π-conjugated system in ion-coordinating dye, which achieved 8.4 % device efficiency [35].

When Ru-complex is with conjugated electron-rich units in its ancillary ligands, its molar extinction coefficient will increase. One example is the production of **C101**. Conjugated electron-rich units include alkylthiophene, alkyl furan, alkyl selenophene, and alkyl thieno[3,2-b]thiophene. In the DSSC device test, **C101** achieved 11 % overall efficiency under AM 1.5G using acetonitrile-based electrolyte [36].

Organic antenna groups which modified Ru-complexes also increased the extinction coefficient. One of these dyes coded as **JK56** achieved 83 % IPCE and 9.2 % energy conversion efficiency under AM 1.5G [37, 38].

4.4 Anchoring of Dye Molecules Toward the Surface of Semiconducting Oxide Nanostructures

The connection between the dye molecules and semiconducting oxide is critical for both efficient charge injection and device stability in terms of the dissociation of dye molecules. First, the effect of the position and the number of carboxyl groups attached to the phenanthroline ligand as anchors were investigated. It was found that two carboxyl groups attached to phenanthroline ligands are necessary for effective electron injection from dyes to TiO_2 [40, 41]. Phosphonic acid groups were used to substitute carboxyl groups. But the dye performed worse in photovoltaic property, but with better stability.

It is of great importance that the dye attaches to TiO_2 intimately for efficient electron injection. Much effort has been made in the anchoring of the dyes onto the oxide surface. According to the attaching force, anchoring can be classified as six modes: covalent attachment which is formed by directly linking groups of interest or linking agents, electrostatic interactions, hydrogen bonding, hydrophobic interactions, Van der Waals forces, and physical entrapment inside the pores or cavities of hosts. Among these modes, covalent attachment is suitable for dyes applied in DSSC [42]. This mode provides the possibility of the intimate adsorption of the dye. Such mode requires the dye having at least one anchoring group to react with the surface hydroxyl groups of TiO_2 forming chemical bonds.

The standard anchoring group in the dye molecule is carboxylic acid groups, which can be seen in Fig. 4.3 of classical ruthenium-based dyes. In 1979, it was found that dehydrative coupling of carboxylic acid groups with surface TiO_2 formed ester-type linkages [43]. The p^* orbitals of the dcb ligand would promote excited electrons to inject to the conduction band of TiO_2 (Fig. 4.5a). However, the linkage is easily hydrolyzed in the presence of trace water affecting the stability of DSSC.

It is worth noting that electron injection into SnO_2 or ZnO is not as efficient as the injection into the TiO_2. This is because the conduction band of TiO_2 is composed mainly of unfilled d orbitals while that of SnO_2 or ZnO show mainly s orbitals character (Fig. 4.5b) [43].

Generally speaking, the carboxylic acid groups attach on TiO_2 via three different ways: unidentate mode, bidentate chelating mode, and bidentate bridging mode (Fig. 4.4) [39, 44]. FT-IR was used to identify the modes of anchoring. The last two

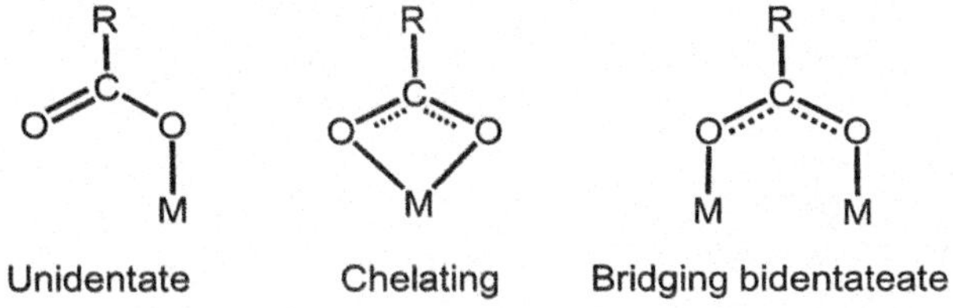

Fig. 4.4 Binding modes for carboxylate acid group on TiO_2 (Reproduce with permission from Ref. [39]. Copyright © 2003, American Chemical Society)

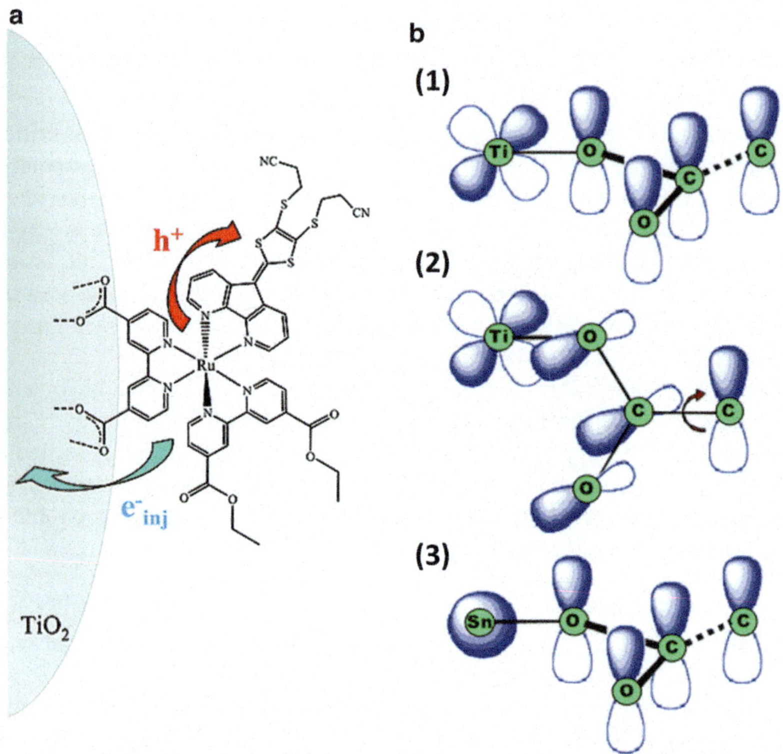

Fig. 4.5 (**a**) A schematic illustration depicting a novel, high extinction coefficient sensitizer bound to a TiO_2 nanocrystallite and photoinduced electron and hole-transfer mechanisms. This sensitizer is unique in that the extended conjugation on the dithiolene-containing ligand is in the 3 and 3′ positions (Reproduced with permission from Ref. [43]. Copyright © 2008, American Chemical Society) (**b**) Orbital diagrams for ester-type binding to the surface of metal oxides. (*1*) For TiO_2, the overlap of the extended *p* system and the Ti *3d* orbitals are thought to aid in electron injection. (*2*) When carboxylates are rotated in such a way as to minimize orbital overlap, the injection yields are thought to suffer. (*3*) Similar effects are proposed for SnO_2 as the Sn *s* orbitals have less efficient orbital mixing with the carboxylate *p* system (Reproduced from Ref. [9] with permission from The Royal Society of Chemistry)

modes, bidentate chelating mode and bidentate bridging mode, perform better than unidentate mode, showing better stability and electron injection efficiency [39].

In addition, there are other anchor groups like sulfonate (-SO_3^-) and silane (SiX_3 or $Si(OX)_3$). Carboxylic acid group derivatives were also adopted including ester, acid chloride, acetic anhydride, carboxylate salt, and carboxylate amide.

Phosphonic acid binding groups were first developed by Pechy and coworkers in 1995. They found that a ruthenium complex with a single phosphonic acid attached to TiO_2 much stronger than **N3** with carboxylic acid groups, and it showed better stability in the presence of trace water, but showed lower PCE [45]. In 2004, dye **Z955** with phosphonic acid linkers achieved high efficiency more than 8 % under AM 1.5G [46].

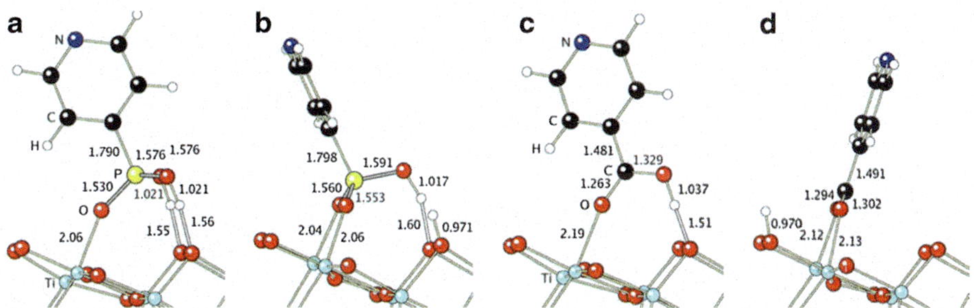

Fig. 4.6 Geometry-optimized structures for (**a**) pyridine phosphonic acid adsorbed on anatase in monodentate binding mode (PM), (**b**) pyridine phosphonic acid in bidentate mode (PB), (**c**) isonicotinic acid in monodentate mode (CM), and (**d**) isonicotinic acid in bidentate mode (CB) (Reproduced with permission from Ref. [47]. Copyright © 2005, Elsevier)

To illustrate the effect of carboxylic acid groups and phosphonic acid groups, we compare **Z955** (with phosphonic acid groups) to **Z907** (with carboxylic acid groups). As mentioned in this chapter, changing ligands affects dye photovoltaic property significantly. Compared to **Z907**, the novel dye **Z955** has a blueshift of the absorption maxima, better stability in device test and slower charge recombination kinetics, but a lower efficiency [46]. Deprotonation of the phosphonic acid groups produced carboxylic acid. **N719** is the product deprotonated from **N3**. And **N719** shows better photovoltaic property, which is a similar case [17].

Density functional theory (DFT) calculations were also used to investigate absorption of pyridine to TiO_2 surface via carboxylic acid and phosphonic acid groups [47]. The result showed that the absorption formed by phosphonic acid groups is stronger than that formed by carboxylic acid groups. As seen from Fig. 4.6, one phosphonic acid has three bonds to TiO_2 surface while carboxylic acid has two bonds. As a consequence, dyes with phosphonic acid groups show better stability when anchoring onto TiO_2. However, the quantity of the electronic interactions of the π^* level (π^* level of the link ligands) with the TiO_2 conduction band decides the electron injection rate from dyes to TiO_2. Carboxylic acid groups have more interaction of the π^* level with the conduction band of TiO_2 than phosphonic acid groups. It was concluded that the rate of surface electron injection via the carboxylic acid groups can be two times as fast as via the phosphonic acid groups.

Besides the protonation of the carboxylic acid groups, their position on the bipyridyl moiety may also affect dye's performance. The majority of ruthenium-based dyes with bipyridyl groups employ 4,4′-sustituted derivatives, while Ru-complex with carboxylic acid groups with 3,3′-substituted bipyridyl shows a decreased efficiency in DSSC [48].

The electron injection from ruthenium dyes to TiO_2 through carboxylic acid groups should be extremely fast in a femtosecond to picosecond. The process is affected by LUMO level of the dye and the conduction band level of TiO_2. And the conduction band level is modified by the redox electrolyte (Fig. 4.7) [49]. Hence the choice of electrolyte is of great importance for efficient electron injection as well.

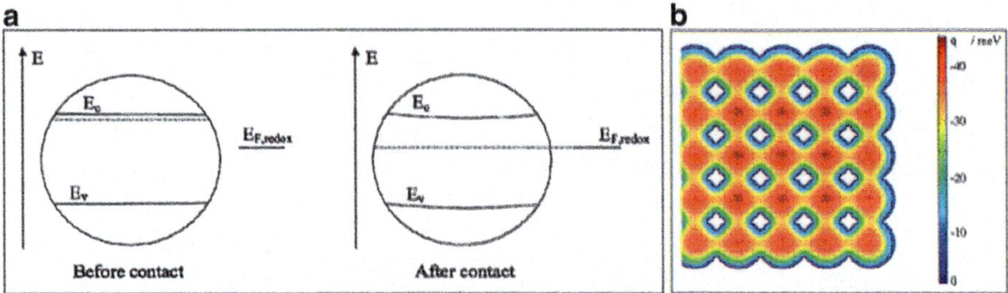

Fig. 4.7 (**a**) Equilibration of Fermi levels between a TiO_2 particle and the I_3^-/I^- redox electrolyte in the dark. (**b**) Band bending in an array of sintered spherical TiO_2 particles. The maximum band bending is 40 meV. The band bending could assist in the channeling of electrons to the contact (Reproduced from Ref. [49] with permission from The Royal Society of Chemistry)

4.5 Alternative Ways to Improve the Light-Harvesting Efficiency of Ru-Complex

4.5.1 Sensitization with Multiple Dye Molecules

An ideal dye in DSSC should absorb all light throughout the solar spectrum efficiently and the excited electrons should inject into the conduction band of TiO_2 swiftly. Unfortunately, such dye is difficult to be obtained. Specifically, the LUMO of the dye approximates the conduction band of TiO_2, which is a way to narrow down the band gap, while the electron injection rate will also decrease in this regard [50]. However, a fact is that there have been different kinds of dyes which possess strong absorption in specific wavelength regions. Thus, combination of different dyes to sensitize TiO_2 anode is a promising way to span the spectral response of the devices. However, co-sensitization is not simply combining different complementary dyes. That won't work because electrons may transfer between the dyes [51]. The combination of different dyes is artistic and the mechanism is worth being investigated. This chapter focuses on the co-sensitization using ruthenium-based dyes.

Experimentally, one method to co-sensitize TiO_2 is using a mixed dye solution. Another way is immersing TiO_2 in a different dye solution sequentially. **N749** and an organic dye (coded as **D131**) were used for co-sensitization in a mixed way and the DSSC device achieved an overall efficiency of 11 % [52].

A new method and structure for co-sensitization was developed in 2004 using a Ru-polypyridyl dye and a Ru-phthalocyanine dye (Fig. 4.8b) [53]. In this device structure, Ru-polypyridyl dye was first used to sensitize TiO_2. Then a thin layer of metal oxide like Al_2O_3 was deposited onto the dye monolayer. Finally, Ru-phthalocyanine dye was attached to the metal oxide layer. The structure can be described as TiO_2/Ru-polypyridyl dye/Al_2O_3/ Ru-phthalocyanine dye. The metal oxide layer is placed in between the two different dye monolayers. Such

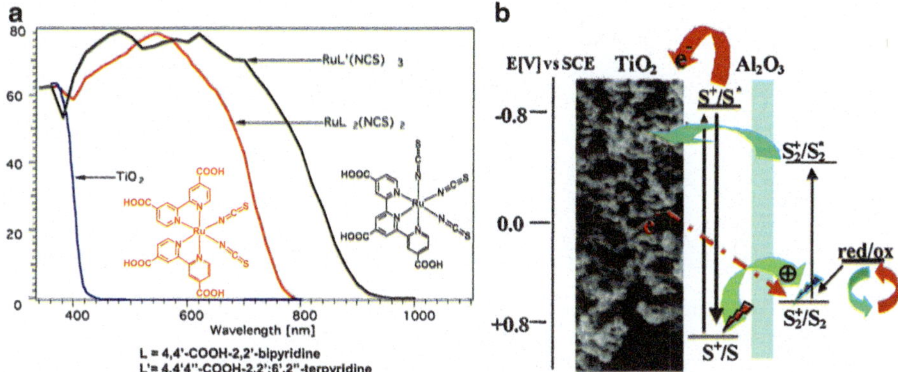

Fig. 4.8 (**a**) Incident photon-to-current conversion efficiency as a function of the wavelength for the standard ruthenium sensitizers **N3** (*red line*), the black dye **N749** (*black curve*), and the blank nanocrystalline TiO_2 film (*blue curve*). The chemical structures of the sensitizers are shown as insets (Reproduced with permission from Ref [54]. Copyright © 2009, American Chemical Society). (**b**) Charge transfer processes in multilayer co-sensitized nanocrystalline TiO_2 films ($S = RuL_2(CN)_2$, $S_2 = RuPc$). Appropriate selection of sensitizer dyes allows vectorial, multistep, electron transfer processes, resulting in a suppression of interfacial charge recombination and a significantly improved photovoltaic device performance relative to single-layer co-sensitization devices (Reproduced with permission from Ref. [53]. Copyright © 2004, American Chemical Society)

structure allowed layered coverage nanoparticles of both dyes rather than a competing absorption on TiO_2 limited surface sites.

There is another unique co-sensitization method with one kind of dye on TiO_2 surface and the other dye with highly photoluminescent chromophores unattached to TiO_2. The dye resides in the electrolyte or the solid hole conductor will transfer energy to the sensitizing dye on TiO_2 as it undergoes Forster resonance [55, 56].

4.5.2 *Plasmon-Enhanced Light Harvesting of Ru-Complex*

Metal NPs, especially Au and Ag, can cause collective oscillation of surface electrons under the external electromagnetic wave. This type of collective oscillation is defined as localized surface plasmon resonance (LSPR); the resonance frequency is dependent on the band structure of the metallic materials and the shape and size of the metal nanoparticles (NPs) [57, 58]. A striking feature of LSPR is that the optical cross sections of the metal NPs are usually much larger than their geometric cross section [59]. For example, the optical cross section of 100 nm AgNP can be ten times more than the geometric cross section at the plasmon resonance frequency. On the other hand, the cross section of Au or Ag is normally 10^5 times of dye molecules. Therefore, the proper use of plasmonic effects is regarded as a promising pathway to increase light absorption in the solar cells. The plasmon-enhanced light harvesting has been demonstrated in various solar

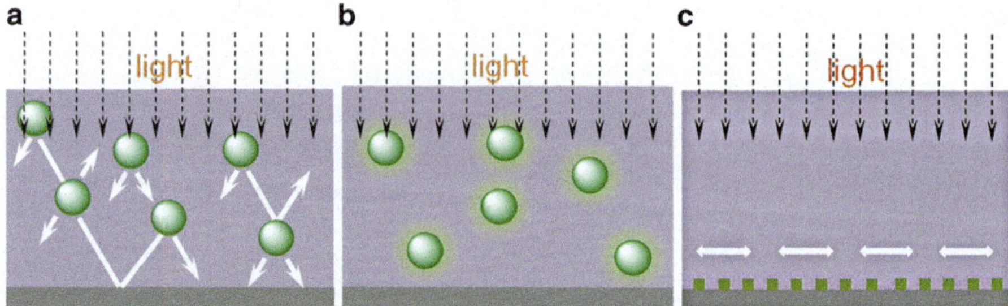

Fig. 4.9 Plasmonic light-trapping mechanisms for solar cells. (**a**) Light trapping by scattering from metal nanoparticles in the solar cell, causing an increase in the effective optical path length in the cell. (**b**) Light trapping by the excitation of localized surface plasmons in metal nanoparticles embedded in the active layer, which increases the optical density around the metal nanoparticles. (**c**) Light trapping by the excitation of surface plasmon polaritons at the electrode, in which a patterned metal nanoparticle structure couples light to surface plasmon polariton or photonic modes that propagates electromagnetic wave to different directions

cells such as amorphous silicon-based solar cells [60, 61], gallium arsenide solar cells [62], organic solar cells [63–65], and DSSCs [66–68]. Generally, there are three mechanisms for the improvement of light utility in DSSCs [69]:

(1) Scattering of light by metallic nanoparticles to increase the effective optical path length. In this approach, metal NPs are usually embedded in the absorber layer of DSSCs, in which light can acquire an angular spread and thus lead to increased optical path length (Fig. 4.9a).
(2) Excitation of localized surface plasmon resonances of metallic nanoparticles. Due to the strong field enhancement of metal NPs, the use of metal NPs can significantly increase the light absorption when dye molecules reside in the vicinity of the metal NPs (Fig. 4.9b).
(3) Coupling to propagating surface plasmon polariton modes. Patterned periodic metal NPs on the electrodes are usually applied to convert light into surface plasmon polariton (SPP) (Fig. 4.9c), which enables the electromagnetic wave to penetrate into or reflect back into the solar cells, depending on the 3D geometric parameters [70]. The generation of SPPs at the electrode can efficiently trap and guide light in the absorber (active) layer of the solar cells, which thus increase the light utilization of DSSCs.

4.5.2.1 Bare Metal Nanoparticles in Ru-Based DSSCs

A straightforward method to introduce plasmon effect into DSSCs is simply incorporating "bare (without additional surface coating layer)" metal NPs into TiO_2 network. Nahm et al. reported that blending AuNPs (diameter at ~100 nm) with commercial P25 NPs as photoanode in DSSCs can lead to the enhancement of light utility [71]. It is found that appropriately incorporating concentration of the metal NPs is vital for the performance improvement. In this case, the mass ratio of 0.05 in the Au/TiO_2 film shows the highest absorption at the plasmon band, indicating that stronger extinction as a result of the LSPR-generated field enhancement prevails over the prolonged optical path lengths from light scattering. And this optimal incorporation of AuNPs significantly improves the absorption of **N719**-sensitized photoanode from 70.1 to 82.6 %. However, a higher addition of AuNPs would be detrimental to absorption at the photoactive materials, and it is proved by experiment that both the IPCE and photocurrent of the device dropped when the Au/TiO_2 mass ratio was increased to 0.07.

The position of metal NPs in the DSSCs is also an important parameter for plasmon-enhanced light harvesting. Hou et al. performed a comparison by embedding AuNPs into anode and on top of TiO_2 anode film [72]. It is observed that the embedded AuNPs in anode show significantly improved light absorption at 500–600 nm, which is corresponding with the plasmon band of spherical AuNPs, while those on top of the anode show less improved light absorption for in that case only a very small amount of dye molecules can utilize the LSPR-increased optical density. This is especially useful in guiding the design of plasmon-enhanced solar cells. Owing to the thin anode films, the PCE of a conventional N719-sensitized DSSCs is 0.94 %, while it was remarkably improved to 2.28 % with AuNPs incorporation.

A similar approach is used by Lin et al. to investigate the plasmon effect of AgNPs in DSSC anode films adsorbed by dye **N3** [73]. It should be noted that AgNPs also possess excellent plasmonic properties, the plasmon band of spherical AgNPs located at round 430 nm [57]. Sandwiched structure composed of TiO_2/AgNPs/TiO_2 for DSSCs has been demonstrated to enhance the light absorption and reduce the reflectance, which leads to a 23 % enhancement in the photocurrent density. A detailed investigation shows that the sandwiched structure is the most efficient one when compared with the incorporation of the Ag on the top of TiO_2 NPs or the anode with only TiO_2/TiO_2 in bilayer structure.

4.5.2.2 Encapsulated Metal Nanoparticles in Ru-Based DSSCs

In the application of naked metal NPs in the anodes, due to the electrical conductivity of metal NPs, they can trap both electrons and holes and thus act as recombination sites. To reduce this negative effect and enable charge carriers to quickly transport to the collecting electrode, anode films are usually fabricated to be relatively thin (no more than 5 μm) in comparison to the typical thickness of DSSC anodes (ca. 12 μm or even thicker); thus only small amounts of dye molecules allow to be adsorbed, which seriously limited the overall performance. Another potential threat is that the NPs may suffer from the corrosion of I^-/I_3^- liquid electrolyte. Therefore, appropriate protections are required. In this regard, coating the metal NPs with semiconducting or insulating materials before employing them into the TiO_2 anode layer of the devices has been demonstrated to be an effective route to make full use of the light trapping by LSPR and light scattering.

The shell thickness is an important parameter since the intensity of LSPR-generated electric field is sensitively dependent on the distance from the metal

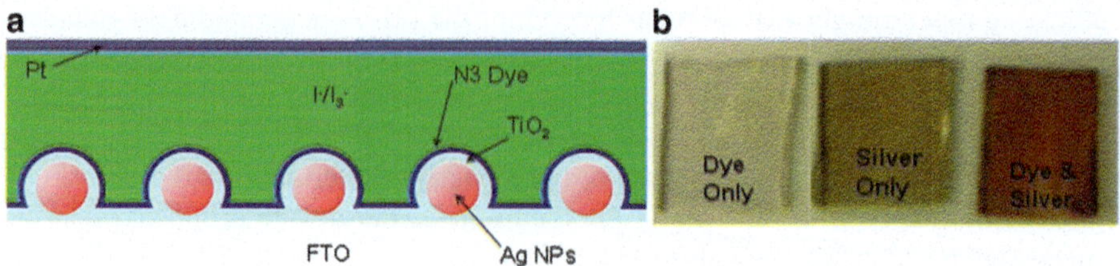

Fig. 4.10 (**a**) Scheme of solar cells composed of AgNPs and atomic layer deposited TiO_2 layer and dye and (**b**) photos of dye on transparent conducting glass substrate, AgNPs on transparent conducting glass substrate, and dye adsorbed on amorphous TiO_2 with 2.0 nm in thickness (Reproduced with permission from ref [68]. Copyright 2009 American Chemical Society)

cores. Pioneering work in this area was done by Hupp's group, who systematically investigated the shell thickness-dependent light harvesting by using atomic layer deposition (ALD) of TiO_2 onto transparent-conductive-oxide-supported AgNPs to finely control the thickness of TiO_2 layer and thus the distance between **N3** dye molecules and Ag cores (Fig. 4.10a) [68]. It is observed that with the thickness increase, the light extinction of dye molecules becomes weaker and weaker. The comparison on the IPCE among the dye only cell, silver only cell, and dye and silver and TiO_2 cell (Fig. 4.10b) was made, and it has been found that the dye- and silver-based DSSCs show greatly enhanced IPCEs. At an optimized distance, the IPCE is increased from 0.2 to 1.4 % in a prototype DSSC.

Although the ALD provides an excellent platform for studying the distance-dependent plasmon-enhanced DSSCs, it is not applicable for 3D anodes of DSSCs. Therefore, a colloidal synthesis of core-shell structures which are then incorporated into TiO_2 anode is more feasible. Qi et al. reported on a very thin shell of TiO_2 (ca. 2 nm) coating on the Ag core for enhancing light harvesting in **N3** dye-based DSSCs; this method was proved to maximize the effect of LSPRs as well as maintain the stability of metal NPs to some extend in the work [66]. Due to the high extinction coefficient of AgNPs, the addition of AgNPs can reduce the thickness of the anode in that the optical density is increased in the device (Fig. 4.11a, b). In addition to the increased optical density and the optical path length, the decreased film thickness can also reduce the recombination probability due to the decreased electron transport path length (Fig. 4.11c, d). The other advantage of using TiO_2 as coating material is that it is compatible with TiO_2 anode and serves as medium for the charge transport (Fig. 4.11e, f).

The Tang group used a simple hydrothermal method to produce Au@TiO_2 hollow submicrospheres with tunable shell for plasmon enhancers [67]. In this method, the AuNPs were firstly synthesized; they were then distributed in the growth solution of TiO_2 spheres. Herein the TiO_2 sphere can also serve as a media for charge transport. Using classic dye **N719**, the PCE is 7.06 % once the TiO_2 hollow submicrospheres were used as anode materials, which is significantly higher than that made of bare TiO_2 electrode with a PCE of 6.25 %.

Besides TiO_2, SiO_2 can also be efficient coating materials. Snaith and coworkers encapsulated AuNPs with a layer of SiO_2 to overcome the major problems (surface

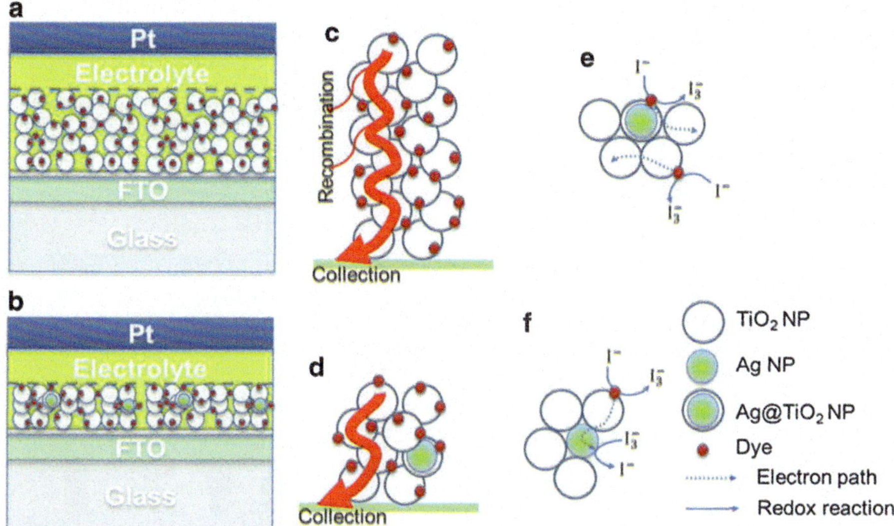

Fig. 4.11 (**a**) Device structures of conventional DSSCs and (**b**) plasmon-enhanced DSSCs, in which plasmon-enhanced DSSCs require thinner film and less material to achieve the same PCE. (**c**) Illustration of photogenerated electron collection in conventional DSSCs and (**d**) plasmon-enhanced DSSCs. (**e**) Mechanisms of plasmon-enhanced DSSCs using Ag@TiO_2 NPs and (**f**) Ag NPs (Reproduced with permission from Ref. [66]. Copyright 2011 American Chemical Society)

corrosion and surface-induced recombination). In this work, I^-/I_3^-electrolyte-based DSSCs and **N719** as sensitizer were used for the DSSCs [74]. Similarly, Sheehan et al. reported that the use of Au@SiO_2@TiO_2 (core-multiple shell structures) can also improve the device efficiency by incorporating them into the TiO_2 anode network [75].

From the above discussion, both SiO_2 and TiO_2 are good coating materials for preventing the corrosion of metal NPs and demonstrating the effectiveness LSPR. Obviously, the electrical conductivity of SiO_2 and TiO_2 is clearly different; SiO_2 is insulator while TiO_2 is wide band gap semiconductor. How the electrical properties influence the device performance is a significant problem that needs to be resolved for rational application of plasmon effect in solar cells. Here, the Kamat group performed a comparative study in the use of Au@TiO_2 and Au@SiO_2 for DSSCs [76]. When Au@SiO_2 was present as intermediary, the influence was mainly limited to LSPR effect, resulting in better charge separation and higher photocurrent. The apparent Fermi level of the composite film was not affected by such LSPR effect. While Au@TiO_2 is different, herein Au is capable of accepting electrons from the neighboring TiO_2/dye particles and undergoing Fermi level equilibration. Such Fermi level equilibration and shift of the apparent Fermi level to more negative potential were reflected as an increase in the open-circuit voltage of the DSSC (Fig. 4.12).

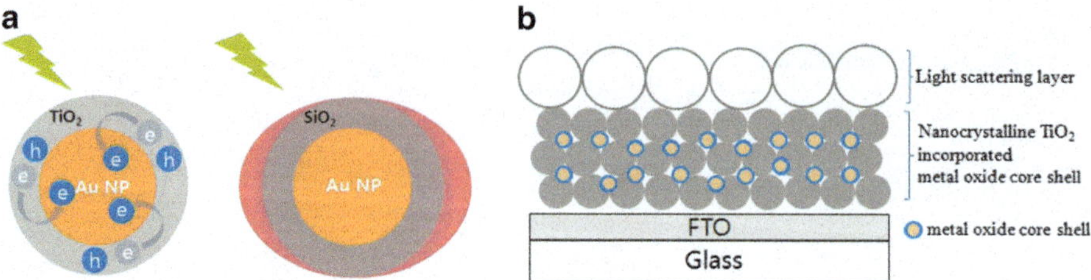

Fig. 4.12 (**a**) Charging effect (Au@TiO_2) versus plasmonic effect (Au@SiO_2) of metal core oxide shell particles. (**b**) Schematic drawing depicting the layered structure of a mesoscopic TiO_2 film incorporating core-shell particles typically employed for high-efficiency DSSC (Reproduced with permission from Ref. [76]. Copyright 2012 American Chemical Society)

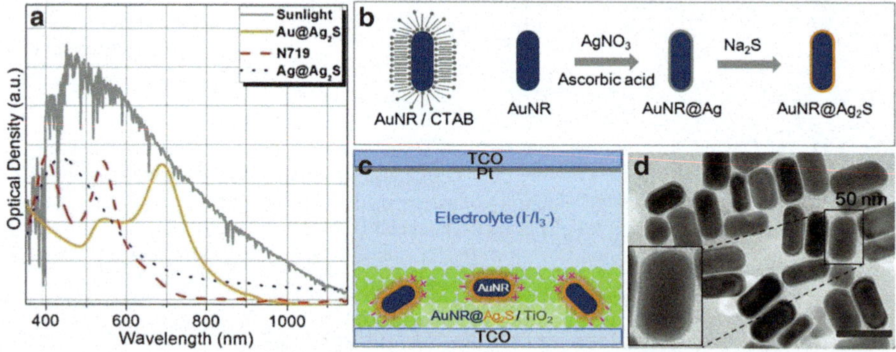

Fig. 4.13 (**a**) Solar irradiance spectrum, absorption of AuNR@Ag_2S (2 nm in shell thickness) and N719 dye solution in acetonitrile and *t*-butanol and Ag_2S-encapsulated Ag nanoparticles, AgNP@Ag_2S; (**b**) schematic illustration of a Au nanorod stabilized by hexadecyltrimethylammonium bromide (CTAB) and a two-step chemical process toward AuNR@Ag_2S; (**c**) device configuration of the plasmon-enhanced DSSCs, in which dye molecules were not included for clearness; (**d**) TEM image of AuNR@Ag_2S with Ag_2S thickness of 2 nm

In literature, most of the studies focus on the utilization of spherical Au and Ag NPs for the improvement of light harvesting in DSSCs. The plasmon bands of them concentrate on 530 and 430 nm, respectively, which are overlapping with (or close to) the absorption peak of conventional Ru-based dye molecules. However, there is plenty of room at the longer wavelengths, for most of the Ru-based dye molecules show significantly reduced absorption coefficient over the longer wavelength (Fig. 4.13a). Rodlike Ag and Au nanostructures have longitudinal plasmon absorption that can be tuned from visible, to NIR, and to the IR region by simply manipulating the aspect ratios of the rod, providing a unique opportunity for utilizing the low energy range of the solar spectrum. Our group has performed the investigation of using encapsulated Au nanorods (NRs) to increase the low-photon energy sunlight harvesting in DSSCs (Fig. 4.13b and c) [77]. Different from the conventional synthesis of spherical Au and Ag NPs, AuNRs are synthesized using cetyltrimethylammonium bromide (CTAB) as a morphological controlling surfactant. Owing to the long aliphatic tail of CTAB, it is difficult to encapsulate AuNR surface with a thin and dense layer of TiO_2 or SiO_2 to introduce AuNRs into DSSCs. We developed a simple method to coat AuNR surface with Ag_2S to introduce plasmonic AuNRs into DSSCs. In the plasmon-enhanced solar cells, the enhancement in the photocurrent generation is more prominent in the plasmonic absorption band of the Au nanorod. To make a detailed comparison, we integrated the IPCE at 600–720 nm and the whole IPCE spectra for the DSSCs with various concentrations of AuNR@Ag_2S (with shell thickness of 2 nm). The value of IPCE (600–720 nm)/IPCE represents the contribution from the 600–720 nm regions to the overall IPCE. As a result, the values significantly change to 9.2 % and 11.7 % with the increase of AuNR@Ag_2S, clearly demonstrating the improved conversion of low energy sunlight into photocurrent. In addition, the Ag@Ag_2S selectively enhances photocurrent generation at around 430 nm, which is an absorption band of Ag@Ag_2S (Fig. 4.13a).

4.6 Conclusions

Ruthenium complex-based dyes play an important role in DSSC. The investigation of the dye molecules not only boosts the device efficiency but also generates some new fundamental understanding of the Ru-complex from both physical and chemical perspectives. In the whole category of the dye molecules, the structures are usually composed of a ruthenium as the metal center and various ancillary ligands, in which **N3** and **N719** are considered as standard dyes for designing new ruthenium-based complexes by changing ancillary ligands. As some general rules for the molecular design, some points about ancillary ligands are highlighted herein:

1. 4,4′-Dicarboxylate-bipyridine ligand on Ru-complexes is favorable for ruthenium complexes anchoring to TiO_2 intimately, because of which electron injection from the dye to TiO_2 is highly efficient and fast. The anchoring mode is the result of reaction between carboxylic acid and TiO_2 surface forming chemical bonds. In addition, there is a charged space between the dye and TiO_2 retarding recombination between injected electrons and dye cations. Thus electrons effectively transfer to FTO.
2. The attachment of cyclodextrin unit to the ancillary ligand of ruthenium trisbipyridyl core could facilitate the dye regeneration.
3. Attachment of hydrophobic chains to ruthenium center can retard desorption of the dye from TiO_2 surface. Otherwise, the anchoring dyes on the TiO_2 surface tend to break in trace amount of water, and the dyes will desorb from the surface.
4. Thiocyanate (SCN), which is chemically unstable compared to other parts in Ru-complexes, is widely used in ruthenium-based dyes. Efforts have been made to replace it with other ligands. Cyclometalated ruthenium complexes without

thiocyanate ligands, such as [Ru(C^N^N)-(N^N^N)] type, are an example of achieving comparable photocurrent and absorption spectrum compared to the state-of-the-art dye molecules.

Acknowledgments This work was substantially supported by a grant from the Research Grants Council of the Hong Kong Special Administrative Region, China, under Theme-based Research Scheme through Project No. T23-407/13-N. The authors also acknowledge the financial support from the CUHK Group Research Scheme and CUHK Focused Scheme B Grant "Centre for Solar Energy Research," CUHK direct grant 4053068 and 4053012.

References

1. Park NG, van de Lagemaat J, Frank AJ (2000) Comparison of dye-sensitized rutile- and anatase-based TiO2 solar cells. J Phys Chem B 104(38):8989–8994. doi:10.1021/Jp994365l
2. Hagfeldt A, Boschloo G, Sun L, Kloo L, Pettersson H (2010) Dye-sensitized solar cells. Chem Rev 110(11):6595–6663. doi:10.1021/cr900356p
3. Desilvestro J, Gratzel M, Kavan L, Moser J, Augustynski J (1985) Highly efficient sensitization of titanium dioxide. J Am Chem Soc 107(10):2988–2990
4. O'Regan B, Gratzel M (1991) A low-cost, high-efficiency solar cell based on dye-sensitized colloidal TiO2 films. Nature 353(6346):737–740
5. Yu QJ, Wang YH, Yi ZH, Zu NN, Zhang J, Zhang M, Wang P (2010) High-efficiency dye-sensitized solar cells: the influence of lithium ions on exciton dissociation, charge recombination, and surface states. ACS Nano 4(10):6032–6038. doi:10.1021/Nn101384e
6. Yella A, Lee HW, Tsao HN, Yi C, Chandiran AK, Nazeeruddin MK, Diau EW, Yeh CY, Zakeeruddin SM, Gratzel M (2011) Porphyrin-sensitized solar cells with cobalt (II/III)-based redox electrolyte exceed 12 percent efficiency. Science 334(6056):629–634. doi:10.1126/science.1209688
7. Burschka J, Pellet N, Moon SJ, Humphry-Baker R, Gao P, Nazeeruddin MK, Gratzel M (2013) Sequential deposition as a route to high-performance perovskite-sensitized solar cells. Nature 499(7458):316–319. doi:10.1038/nature12340
8. Qin YC, Peng Q (2012) Ruthenium sensitizers and their applications in dye-sensitized solar cells. Int J Photoenergy. doi: 10.1155/2012/291579. Artn 291579
9. Ardo S, Meyer GJ (2009) Photodriven heterogeneous charge transfer with transition-metal compounds anchored to TiO2 semiconductor surfaces. Chem Soc Rev 38(1):115–164. doi:10.1039/b804321n
10. Zhang S7, Yang X, Numata Y, Han L (2013) Highly efficient dye-sensitized solar cells: progress and future challenges. Energy Environ Sci 6(5):1443. doi:10.1039/c3ee24453a
11. Anderson S, Constable EC, Dare-Edwards MP, Goodenough JB, Hamnett A, Seddon KR, Wright RD (1979) Chemical modification of a titanium (IV) oxide electrode to give stable dye sensitisation without a supersensitiser. Nature 280(5723):571–573
12. Nazeeruddin MK, Liska P, Moser J, Vlachopoulos N, Grätzel M (1990) Conversion of light into electricity with trinuclear ruthenium complexes adsorbed on textured TiO2 films. Helv Chim Acta 73(6):1788–1803. doi:10.1002/hlca.19900730624
13. Nazeeruddin MK, Kay A, Rodicio I, Humphry-Baker R, Müller E, Liska P, Vlachopoulos N, Grätzel M (1993) Conversion of light to electricity by cis-X2bis (2, 2′-bipyridyl-4, 4′-dicarboxylate) ruthenium (II) charge-transfer sensitizers (X = Cl-, Br-, I-, CN-, and SCN-) on nanocrystalline titanium dioxide electrodes. J Am Chem Soc 115(14):6382–6390
14. Nazeeruddin MK, Pechy P, Gratzel M (1997) Efficient panchromatic sensitization of nanocrystalline TiO2 films by a black dye based on a trithiocyanato-ruthenium complex. Chem Commun 18:1705–1706. doi:10.1039/A703277C

15. Wang P, Zakeeruddin SM, Humphry-Baker R, Moser JE, Gratzel M (2003) Molecular-scale interface engineering of TiO2 nanocrystals: improving the efficiency and stability of dye-sensitized solar cells. Adv Mater 15(24):2101. doi:10.1002/adma.200306084
16. Nazeeruddin MK, Pechy P, Renouard T, Zakeeruddin SM, Humphry-Baker R, Comte P, Liska P, Cevey L, Costa E, Shklover V, Spiccia L, Deacon GB, Bignozzi CA, Gratzel M (2001) Engineering of efficient panchromatic sensitizers for nanocrystalline TiO2-based solar cells. J Am Chem Soc 123(8):1613–1624. doi:10.1021/Ja003299u
17. Nazeeruddin MK, Zakeeruddin SM, Humphry-Baker R, Jirousek M, Liska P, Vlachopoulos N, Shklover V, Fischer C-H, Grätzel M (1999) Acid–base equilibria of (2,2′-bipyridyl-4,4-′-dicarboxylic acid)ruthenium(II) complexes and the effect of protonation on charge-transfer sensitization of nanocrystalline titania. Inorg Chem 38(26):6298–6305. doi:10.1021/ic990916a
18. Barolo C, Nazeeruddin MK, Fantacci S, Di Censo D, Comte P, Liska P, Viscardi G, Quagliotto P, De Angelis F, Ito S, Gratzel M (2006) Synthesis, characterization, and DFT-TDDFT computational study of a ruthenium complex containing a functionalized tetradentate ligand. Inorg Chem 45(12):4642–4653. doi:10.1021/Ic051970w
19. Jiang KJ, Masaki N, Xia JB, Noda S, Yanagida S (2006) A novel ruthenium sensitizer with a hydrophobic 2-thiophen-2-yl-vinyl-conjugated bipyridyl ligand for effective dye sensitized TiO2 solar cells. Chem Commun 23:2460–2462. doi:10.1039/b602989b
20. Chen CY, Lu HC, Wu CG, Chen JG, Ho KC (2007) New ruthenium complexes containing oligoalkylthiophene-substituted 1,10-phenanthroline for nanocrystalline dye-sensitized solar cells. Adv Funct Mater 17(1):29–36. doi:10.1002/adfm.200600059
21. Chen CY, Wu SJ, Li JY, Wu CG, Chen JG, Ho KC (2007) A new route to enhance the light-harvesting capability of ruthenium complexes for dye-sensitized solar cells. Adv Mater 19(22):3888–3891. doi:10.1002/adma.200701111
22. Chen CY, Chen JG, Wu SJ, Li JY, Wu CG, Ho KC (2008) Multifunctionalized ruthenium-based supersensitizers for highly efficient dye-sensitized solar cells. Angew Chem 47(38):7342–7345. doi:10.1002/anie.200802120
23. Jang SR, Yum JH, Klein C, Kim KJ, Wagner P, Officer D, Gratzel M, Nazeeruddin MK (2009) High molar extinction coefficient ruthenium sensitizers for thin film dye-sensitized solar cells. J Phys Chem C 113(5):1998–2003. doi:10.1021/Jp8077562
24. Chen KS, Liu WH, Wang YH, Lai CH, Chou PT, Lee GH, Chen K, Chen HY, Chi Y, Tung FC (2007) New family of ruthenium-dye-sensitized nanocrystalline TiO2 solar cells with a high solar-energy-conversion efficiency. Adv Funct Mater 17(15):2964–2974. doi:10.1002/adfm.200600985
25. Bessho T, Yoneda E, Yum JH, Guglielmi M, Tavernelli I, Imai H, Rothlisberger U, Nazeeruddin MK, Gratzel M (2009) New paradigm in molecular engineering of sensitizers for solar cell applications. J Am Chem Soc 131(16):5930–5934. doi:10.1021/Ja9002684
26. Zakeeruddin SM, Nazeeruddin MK, Humphry-Baker R, Pechy P, Quagliotto P, Barolo C, Viscardi G, Gratzel M (2002) Design, synthesis, and application of amphiphilic ruthenium polypyridyl photosensitizers in solar cells based on nanocrystalline TiO2 films. Langmuir 18(3):952–954. doi:10.1021/La0110848
27. Wang P, Zakeeruddin SM, Moser JE, Nazeeruddin MK, Sekiguchi T, Gratzel M (2003) A stable quasi-solid-state dye-sensitized solar cell with an amphiphilic ruthenium sensitizer and polymer gel electrolyte. Nat Mater 2(6):402–407. doi:10.1038/nmat904
28. Kang MS, Kim JH, Kim YJ, Won J, Park NG, Kang YS (2005) Dye-sensitized solar cells based on composite solid polymer electrolytes. Chem Commun 7:889–891. doi:10.1039/b412129p
29. Faiz J, Philippopoulos AI, Kontos AG, Falaras P, Pikramenou Z (2007) Functional supramolecular ruthenium cyclodextrin dyes for nanocrystalline solar cells. Adv Funct Mater 17(1):54–58. doi:10.1002/adfm.200600188
30. Clifford JN, Palomares E, Nazeeruddin MK, Gratzel M, Nelson J, Li X, Long NJ, Durrant JR (2004) Molecular control of recombination dynamics in dye-sensitized nanocrystalline TiO2

films: free energy vs distance dependence. J Am Chem Soc 126(16):5225–5233. doi:10.1021/Ja039924n
31. Karthikeyan CS, Wietasch H, Thelakkat M (2007) Highly efficient solid-state dye-sensitized TiO2 solar cells using donor-antenna dyes capable of multistep charge-transfer cascades. Adv Mater 19(8):1091–1095. doi:10.1002/adma.200601872
32. Yum J-H, Jung I, Baik C, Ko J, Nazeeruddin MK, Gratzel M (2009) High efficient donor-acceptor ruthenium complex for dye-sensitized solar cell applications. Energy Environ Sci 2(1):100–102. doi:10.1039/B814863P
33. Wang P, Zakeeruddin SM, Moser JE, Humphry-Baker R, Comte P, Aranyos V, Hagfeldt A, Nazeeruddin MK, Gratzel M (2004) Stable new sensitizer with improved light harvesting for nanocrystalline dye-sensitized solar cells. Adv Mater 16(20):1806. doi:10.1002/adma.200400039
34. Kuang DB, Klein C, Snaith HJ, Moser JE, Humphry-Baker R, Comte P, Zakeeruddin SM, Gratzel M (2006) Ion coordinating sensitizer for high efficiency mesoscopic dye-sensitized solar cells: influence of lithium ions on the photovoltaic performance of liquid and solid-state cells. Nano Lett 6(4):769–773. doi:10.1021/Nl060075m
35. Kuang D, Klein C, Ito S, Moser JE, Humphry-Baker R, Zakeeruddin SM, Grätzel M (2007) High molar extinction coefficient ion-coordinating ruthenium sensitizer for efficient and stable mesoscopic dye-sensitized solar cells. Adv Funct Mater 17(1):154–160. doi:10.1002/adfm.200600483
36. Gao F, Wang Y, Shi D, Zhang J, Wang M, Jing X, Humphry-Baker R, Wang P, Zakeeruddin SM, Grätzel M (2008) Enhance the optical absorptivity of nanocrystalline TiO2 film with high molar extinction coefficient ruthenium sensitizers for high performance dye-sensitized solar cells. J Am Chem Soc 130(32):10720–10728. doi:10.1021/ja801942j
37. Jung I, Choi H, Lee JK, Song KH, Kang SO, Ko J (2007) New ruthenium sensitizers containing styryl and antenna fragments. Inorg Chim Acta 360(11):3518–3524. doi:10.1016/j.ica.2007.04.050
38. Choi H, Baik C, Kim S, Kang M-S, Xu X, Kang HS, Kang SO, Ko J, Nazeeruddin MK, Grätzel M (2008) Molecular engineering of hybrid sensitizers incorporating an organic antenna into ruthenium complex and their application in solar cells. New J Chem 32(12):2233. doi:10.1039/b810332a
39. Nazeeruddin MK, Humphry-Baker R, Liska P, Grätzel M (2003) Investigation of sensitizer adsorption and the influence of protons on current and voltage of a dye-sensitized nanocrystalline TiO2 solar cell. J Phys Chem B 107(34):8981–8987. doi:10.1021/jp022656f
40. Hara K, Sugihara H, Tachibana Y, Islam A, Yanagida M, Sayama K, Arakawa H, Fujihashi G, Horiguchi T, Kinoshita T (2001) Dye-sensitized nanocrystalline TiO2 solar cells based on ruthenium(II) phenanthroline complex photosensitizers. Langmuir 17(19):5992–5999. doi:10.1021/La010343q
41. Reynal A, Forneli A, Martinez-Ferrero E, Sanchez-Diaz A, Vidal-Ferran A, Palomares E (2008) A phenanthroline heteroleptic ruthenium complex and its application to dye-sensitised solar cells. Eur J Inorg Chem 2008(12):1955–1958. doi:10.1002/ejic.200800054
42. Kalyanasundaram K, Gratzel M (1998) Applications of functionalized transition metal complexes in photonic and optoelectronic devices. Coord Chem Rev 177:347–414. doi:10.1016/S0010-8545(98)00189-1
43. Staniszewski A, Heuer WB, Meyer GJ (2008) High-extinction ruthenium compounds for sunlight harvesting and hole transport. Inorg Chem 47(16):7062–7064. doi:10.1021/Ic800171h
44. Wang Z-S, Hara K, Dan-oh Y, Kasada C, Shinpo A, Suga S, Arakawa H, Sugihara H (2005) Photophysical and (photo) electrochemical properties of a coumarin dye. J Phys Chem B 109(9):3907–3914
45. Pechy P, Rotzinger FP, Nazeeruddin MK, Kohle O, Zakeeruddin SM, Humphrybaker R, Gratzel M (1995) Preparation of phosphonated polypyridyl ligands to anchor transition-

metal complexes on oxide surfaces – application for the conversion of light to electricity with nanocrystalline Tio2 films (Pg 65, 1995). J Chem Soc Chem Commun 10:1093–1093
46. Wang P, Klein C, Moser JE, Humphry-Baker R, Cevey-Ha NL, Charvet R, Comte P, Zakeeruddin SM, Gratzel M (2004) Amphiphilic ruthenium sensitizer with 4,4 ′-diphosphonic acid-2,2′-bipyridine as anchoring ligand for nanocrystalline dye sensitized solar cells. J Phys Chem B 108(45)):17553–17559. doi:10.1021/Jp046932x
47. Nilsing M, Persson P, Ojamäe L (2005) Anchor group influence on molecule–metal oxide interfaces: periodic hybrid DFT study of pyridine bound to TiO2 via carboxylic and phosphonic acid. Chem Phys Lett 415(4–6):375–380. http://dx.doi.org/10.1016/j.cplett.2005.08.154
48. Xie PH, Hou YJ, Wei TX, Zhang BW, Cao Y, Huang CH (2000) Synthesis and photoelectric studies of Ru(II) polypyridyl sensitizers. Inorg Chim Acta 308(1-2):73—79. doi:10.1016/S0020-1693(00)00214-0
49. Peter LM (2007) Dye-sensitized nanocrystalline solar cells. Phys Chem Chem Phys 9(21):2630–2642. doi:10.1039/b617073k
50. Nazeeruddin MK, Bessho T, Cevey L, Ito S, Klein C, De Angelis F, Fantacci S, Comte P, Liska P, Imai H, Graetzel M (2007) A high molar extinction coefficient charge transfer sensitizer and its application in dye-sensitized solar cell. J Photochem Photobiol A 185 (2–3):331–337. doi:10.1016/j.jphotochem.2006.06.028
51. Ehret A, Stuhl L, Spitler MT (2001) Spectral sensitization of TiO2 nanocrystalline electrodes with aggregated cyanine dyes. J Phys Chem B 105(41):9960–9965. doi:10.1021/Jp011952
52. Ogura RY, Nakane S, Morooka M, Orihashi M, Suzuki Y, Noda K (2009) High-performance dye-sensitized solar cell with a multiple dye system. Appl Phys Lett 94(7):073308. doi:10.1063/1.3086891
53. Clifford JN, Palomares E, Nazeeruddin K, Thampi R, Gratzel M, Durrant JR (2004) Multistep electron transfer processes on dye co-sensitized nanocrystalline TiO2 films. J Am Chem Soc 126(18):5670–5671. doi:10.1021/Ja049705h
54. Grätzel M (2009) Recent advances in sensitized mesoscopic solar cells. Acc Chem Res 42(11):1788–1798. doi:10.1021/ar900141y
55. Hardin BE, Hoke ET, Armstrong PB, Yum JH, Comte P, Torres T, Frechet JMJ, Nazeeruddin MK, Gratzel M, McGehee MD (2009) Increased light harvesting in dye-sensitized solar cells with energy relay dyes. Nat Photonics 3(7):406–411. doi:10.1038/Nphoton.2009.96
56. Yum JH, Hardin BE, Moon SJ, Baranoff E, Nuesch F, McGehee MD, Gratzel M, Nazeeruddin MK (2009) Panchromatic response in solid-state dye-sensitized solar cells containing phosphorescent energy relay dyes. Angew Chem 48(49):9277–9280. doi:10.1002/anie.200904725
57. Wiley B, Sun YG, Xia YN (2007) Synthesis of silver nanostructures with controlled shapes and properties. Acc Chem Res 40(10):1067–1076. doi:10.1021/Ar7000974
58. Wu DJ, Xu XD, Liu XJ (2008) Influence of dielectric core, embedding medium and size on the optical properties of gold nanoshells. Solid State Commun 146(1-2):7–11. doi:10.1016/j.ssc.2008.01.038
59. Bohren CF (1983) How can a particle absorb more than the light incident on it? Am J Phys 51 (4):323–327
60. Mühlschlegel P, Eisler H-J, Martin OJF, Hecht B, Pohl DW (2005) Resonant optical antennas. Science 308(5728):1607–1609. doi:10.1126/science.1111886
61. Ditlbacher H, Krenn JR, Schider G, Leitner A, Aussenegg FR (2002) Two-dimensional optics with surface plasmon polaritons. Appl Phys Lett 81(10):1762–1764
62. Bozhevolnyi SI, Volkov VS, Devaux E, Laluet J-Y, Ebbesen TW (2006) Channel plasmon subwavelength waveguide components including interferometers and ring resonators. Nature 440(7083):508–511
63. Li XH, Choy WCH, Huo LJ, Xie FX, Sha WEI, Ding BF, Guo X, Li YF, Hou JH, You JB, Yang Y (2012) Dual plasmonic nanostructures for high performance inverted organic solar cells. Adv Mater 24(22):3046–3052. doi:10.1002/adma.201200120

64. Chen F-C, Wu J-L, Lee C-L, Hong Y, Kuo C-H, Huang MH (2009) Plasmonic-enhanced polymer photovoltaic devices incorporating solution-processable metal nanoparticles. Appl Phys Lett 95(1):013305–013303
65. Morfa AJ, Rowlen KL, Reilly TH, Iii RMJ, van de Lagemaat J (2008) Plasmon-enhanced solar energy conversion in organic bulk heterojunction photovoltaics. Appl Phys Lett 92(1):013504–013503
66. Qi J, Dang X, Hammond PT, Belcher AM (2011) Highly efficient plasmon-enhanced dye-sensitized solar cells through metal@oxide core–shell nanostructure. ACS Nano 5(9):7108–7116. doi:10.1021/nn201808g
67. Du J, Qi J, Wang D, Tang Z (2012) Facile synthesis of Au@TiO2 core-shell hollow spheres for dye-sensitized solar cells with remarkably improved efficiency. Energy Environ Sci 5(5):6914–6918
68. Standridge SD, Schatz GC, Hupp JT (2009) Distance dependence of plasmon-enhanced photocurrent in dye-sensitized solar cells. J Am Chem Soc 131(24):8407. doi:10.1021/ja9022072
69. Atwater HA, Polman A (2010) Plasmonics for improved photovoltaic devices. Nat Mater 9(3):205–213. doi:10.1038/Nmat2629
70. Ferry VE, Sweatlock LA, Pacifici D, Atwater HA (2008) Plasmonic nanostructure design for efficient light coupling into solar cells. Nano Lett 8(12):4391–4397. doi:10.1021/nl8022548
71. Nahm C, Choi H, Kim J, Jung DR, Kim C, Moon J, Lee B, Park B (2011) The effects of 100 nm-diameter Au nanoparticles on dye-sensitized solar cells. Appl Phys Lett 99(25). doi:10.1063/1.3671087. Article ID 253107
72. Hou WB, Pavaskar P, Liu ZW, Theiss J, Aykol M, Cronin SB (2011) Plasmon resonant enhancement of dye sensitized solar cells. Energy Environ Sci 4(11):4650–4655. doi:10.1039/c1ee02120f
73. Lin SJ, Lee KC, Wu JL, Wu JY (2011) Enhanced performance of dye-sensitized solar cells via plasmonic sandwiched structure. Appl Phys Lett 99(4). doi:10.1063/1.3616139. Article ID 043306
74. Brown MD, Suteewong T, Kumar RSS, D'Innocenzo V, Petrozza A, Lee MM, Wiesner U, Snaith HJ (2011) Plasmonic dye-sensitized solar cells using core-shell metal-insulator nanoparticles. Nano Lett 11(2):438–445. doi:10.1021/Nl1031106
75. Sheehan SW, Noh H, Brudvig GW, Cao H, Schmuttenmaer CA (2013) Plasmonic enhancement of dye-sensitized solar cells using core-shell-shell nanostructures. J Phys Chem C 117(2):927–934. doi:10.1021/jp311881k
76. Choi H, Chen WT, Kamat PV (2012) Know thy nano neighbor. Plasmonic versus electron charging effects of metal nanoparticles in dye-sensitized solar cells. ACS Nano 6(5):4418–4427. doi:10.1021/nn301137r
77. Chang S, Li Q, Xiao X, Wong KY, Chen T (2012) Enhancement of low energy sunlight harvesting in dye-sensitized solar cells using plasmonic gold nanorods. Energy Environ Sci 5:9444–9448

Chapter 5
All-Polymer Solar Cells Based on Organometallic Polymers

Tsuyoshi Michinobu

Abstract Bulk-heterojunction (BHJ) solar cells using n-type semiconducting polymers, instead of the conventional fullerene derivatives, are known as all-polymer solar cells. Their significant advantages include designable structures of both p-type and n-type semiconducting polymers. In this chapter, recent advances in all-polymer solar cells are introduced. Particular attention is focused on the development of high-performance n-type semiconducting polymers. However, organometallic polymers have generally been employed as p-type semiconductors in BHJ solar cells. Therefore, there are a few examples of all-polymer solar cells based on organometallic polymers. In order to solve this problem, a novel method of inverting the semiconducting feature is applied to produce promising Pt-polyyne polymers with potentially n-type energy levels. The main chain alkynes of the precursor Pt-polyyne polymers are modified by the high-yielding transformation into tetracyanobutadiene units through a [2 + 2] cycloaddition-retroelectrocyclization with tetracyanoethylene (TCNE). All-polymer solar cells composed of the p-type poly(3-hexylthiophene) (P3HT) and n-type Pt-polyyne polymer are successfully fabricated, and the photocurrent generation is demonstrated.

Keywords Bulk-heterojunction solar cells • Charge-transfer • Cycloaddition-retroelectrocyclization • Organometallic semiconducting polymers • Pt-polyyne polymers

5.1 Introduction

Bulk-heterojunction (BHJ) solar cells using organic semiconducting polymers offer an emerging technology for alternative energy sources due to many advantages, such as lightweight, excellent flexibility, and processability [1–3]. The active layer of these devices is composed of a mixture of p-type and n-type semiconductors. The

T. Michinobu (✉)
Department of Organic and Polymeric Materials, Tokyo Institute of Technology, 2-12-1 Ookayama, Meguro-ku, Tokyo 152-8552, Japan
e-mail: michinobu.t.aa@m.titech.ac.jp

W.-Y. Wong (ed.), *Organometallics and Related Molecules for Energy Conversion*, Green Chemistry and Sustainable Technology, DOI 10.1007/978-3-662-46054-2_5

most common semiconductor combination is the p-type regio-regular poly (3-hexylthiophene) (P3HT) and n-type fullerene derivatives, as represented by the [6,6]-phenyl-C_{61} or C_{71}-butyric acid methyl ester ([60]PCBM or [70]PCBM) [4]. An initial approach to improve the power conversion efficiency (PCE) from the chemistry viewpoint was based on the pursuit of novel p-type semiconducting polymers. A popular established molecular design was donor-acceptor alternating conjugated polymers, which possess both a longer wavelength absorption and good crystallinity in the thin-film states. To date, the best PCE of BHJ solar cells is approaching 10 % [5, 6]. It is said that this value is close to the threshold for the compensation of commercialization costs [7].

All-polymer solar cells are another approach of which their success mainly relies on the development of high-performance n-type semiconducting polymers [8, 9]. This approach actually dates back to the pioneering work by Friend and Holmes in the 1990s [10, 11]. They initially employed the poly(*p*-phenylenevinylene) derivative (MEH-PPV) **1** as a p-type semiconductor and cyanated PPV derivative (MEH-CN-PPV) **3** as an n-type semiconductor (Fig. 5.1). A mixture of these two polymer blends successfully generated photocurrents. Later, the replacement of MEH-PPV by a polythiophene derivative **2** further improved the photovoltaic properties. Thus, the overall PCE of 1.9 % was reported under a simulated solar spectrum.

Fréchet et al. revisited the initial n-type semiconducting polymer of MEH-CN-PPV in 2009 for the evaluation of all-polymer solar cells [12]. This work stimulated other researchers, and some key components of the n-type semiconducting polymers have been found. For example, it was shown that arylenediimide derivatives are one of the most promising acceptor units. Alkyl chains substituted at the imide nitrogen atoms enable one to control the solubility and crystallinity of the arylenediimide-containing polymers. Marder et al. reported the selection of perylenediimide-containing polymers **4** with oligo(dithienothiophene) structures as the comonomer unit (Fig. 5.2) [13]. These polymers displayed n-type semiconducting performances suitable for application in the BHJ solar cells. In combination with the p-type semiconducting polythiophene derivatives with the extended side chain π-systems, the device based on these two polymer blends exhibited the PCE

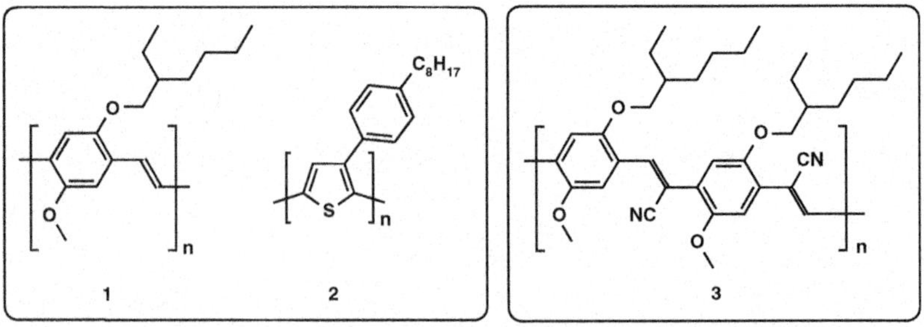

Fig. 5.1 Polymer structures of p-type semiconductors (MEH-PPV**1** or **2**) and an n-type semiconductor (MEH-CN-PPV **3**) employed in initial studies

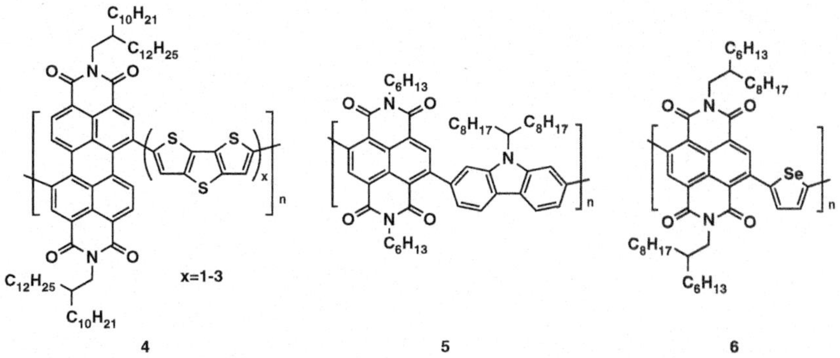

Fig. 5.2 Examples of some high-performance arylenediimide-based n-type semiconducting polymers

of 1.5 %. Tajima and Hashimoto also reported all-polymer solar cells using similar polymer combinations of the naphthalenediimide-carbazole copolymer **5** and high-performance p-type semiconducting polymers [14]. The optimization of processing solvents and additives produced nanoscale phase-separated structures with p-type and n-type semiconducting domains, which eventually led to a dramatic increase in the PCE to 3.68 %. The excellent n-type semiconducting performance in the all-polymer solar cells was also observed for the naphthalenediimide-selenophene copolymer **6** [15]. The best short circuit current density and external quantum efficiency were achieved when a thiazolothiazole copolymer donor was employed.

Another promising electron-accepting structure is the benzothiadiazole unit. The corresponding dibromo derivative, 4,7-dibromo-2,1,3-benzothiadiazole, is usually employed as a synthetic intermediate, and this molecule can be prepared in two steps from the commercially available *o*-phenylenediamine in high yields on a multigram scale. Its potent electron-accepting feature further led to the design of the benzobis(thiadiazole) unit, and these chromophores have been successfully integrated into conjugated polymers for potential applications in polymer solar cells. One of the initial reports about the n-type semiconducting polymer using this unit was the benzothiadiazole-fluorene copolymer **7** (Fig. 5.3) [16]. Lam et al. fabricated polymer nanofibers of the p-type P3HT and n-type polymer **7**. Simple thermal annealing produced the optimized blend morphology, which displayed an enhanced short circuit current by a factor of 10 as compared to the simple blended films. Ito et al. also reported the excellent phase separation of the polymer blends when **8** with the additional thiophene units was employed as an n-type semiconductor [17]. Additionally, they carefully investigated the effect of the molecular weights of **8** on the phase separation. A higher molecular weight **8** exhibited the better PCE of 2.7 % due to the formation of smaller phase-separated structures on the length scale of exciton diffusion. Furthermore, the appropriate selection of comonomer structures realized the control of the polymer energy levels. An n-type polymer **9** with

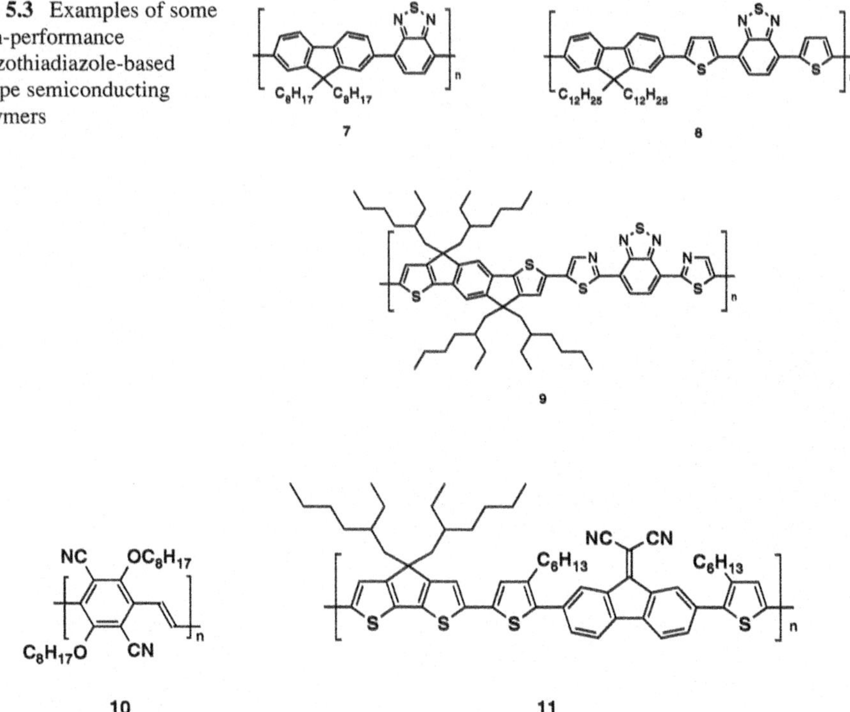

Fig. 5.3 Examples of some high-performance benzothiadiazole-based n-type semiconducting polymers

Fig. 5.4 Examples of some high-performance cyano-containing n-type semiconducting polymers

extended π-conjugated systems possessed a narrow bandgap with elevated highest occupied molecular orbital (HOMO) and lowered lowest unoccupied molecular orbital (LUMO) levels as compared to the counter polymers [18]. Although the blend films of P3HT and **9** showed a sufficient miscibility as deduced from the surface roughness and P3HT crystallinity, the PCE was limited to 1.2 %.

Inspired by MEH-CN-PPV, other cyano-containing acceptor polymers have also been pursued. The substitution pattern of the cyano groups was changed from MEH-CN-PPV. Thus, the use of **10** as an n-type semiconducting polymer in the all-polymer solar cell displayed a PCE of 0.8 % (Fig. 5.4) [19]. The donor-acceptor-type copolymer **11** with the dicyanofluorene unit was prepared, and its space-charge-limited current measurements suggested the electron mobility 1–2 orders higher than the corresponding hole mobility [20]. This result clearly indicated the intrinsic n-type nature of this polymer. Accordingly, the polymer was evaluated as an n-type semiconductor in the all-polymer solar cells, which eventually showed the PCE of 0.07 %. As compared to the cyano-free derivatives, the introduction of cyano acceptors explicitly enhanced the n-type semiconducting characters of the conjugated polymers.

5.2 p-Type Pt-Polyyne Polymers

Organometallic polymers have been a hot topic of contemporary optoelectronic materials due to the combined features of polymers and transition metals [21–23]. The advantages of using polymers include straightforward processing and mechanical properties, whereas the transition metals generally show useful electrical, optical, and magnetic properties. Both kinds of features can be readily integrated into single polymer structures when organometallic polymers are appropriately designed. However, among the many kinds of organometallic polymers, the semi-conducting characters of the Pt-polyyne polymers have mainly been investigated. This was probably due to their high synthetic accessibility. The desired π-chromophores can be readily introduced into the main chain, and the efficient polymerization based on the dehydrohalogenation reaction between the dichloro-platinum monomers and diethynyl comonomers usually produces high molecular weight Pt-polyyne polymers in excellent yields.

The first class of the Pt-polyyne polymers studied for photovoltaic applications was the one with the simple *p*-phenylene spacer **12** (Fig. 5.5) [24]. The blended film with C_{60} revealed the photocurrent generation through photoinduced electron transfer from **12** to C_{60}. It was subsequently found that the replacement of the *p*-phenylene unit by other aromatic spacers significantly improved the photovoltaic performances. For example, Reynolds and Schanze reported the Pt-polyyne polymer with the thiophene spacer **13** [25]. Detailed photophysical observations suggested the involvement of the triplet excited state of **13** for the enhanced PCEs. A further important finding arose from the combination of the donor-acceptor structure and Pt-polyyne polymers. As mentioned above, the thiophene-benzothiadiazole-thiophene unit is an excellent charge-transfer chromophore. When this unit was employed as a π-spacer of the Pt-polyyne polymers, an intensely colored polymer **14** with a low bandgap of 1.85 eV was produced [26]. Despite the absence of any annealing treatment, the blended film of **14** and PCBM showed high PCEs, which are almost comparable to the best value of the annealed P3HT/PCBM systems.

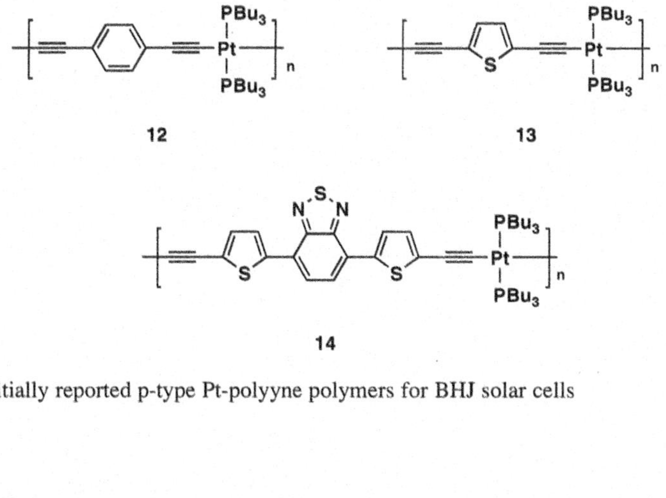

Fig. 5.5 Initially reported p-type Pt-polyyne polymers for BHJ solar cells

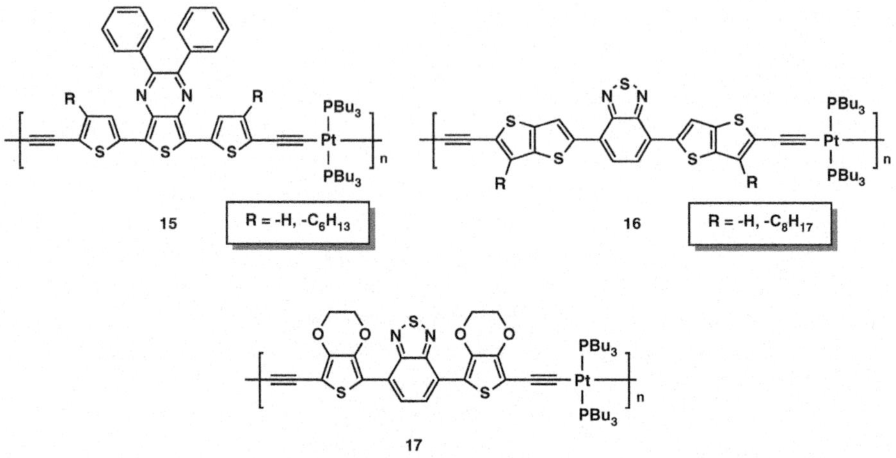

Fig. 5.6 Donor-acceptor-type Pt-polyyne polymers used for p-type semiconductors in BHJ solar cells

The success of the benzothiadiazole-containing polymer **14** initiated the development of new Pt-polyyne polymers as a p-type semiconductor in BHJ solar cells with an n-type semiconductor of PCBM. Luscombe and Jenekhe synthesized a variety of Pt-polyyne polymers with different acceptor units. The pyrido[3,4-*b*]pyrazine and thieno[3,4-*b*]pyrazine moieties were newly examined as an alternative acceptor unit to 2,1,3-benzothiadiazole, as represented by **15** (Fig. 5.6) [27]. However, the careful optimization of the polymer solubilities and device fabrication conditions did not lead to any improvement in the photovoltaic performances. The original Pt-polyyne polymer **14** was a better photovoltaic material. Thus, the other possible molecular design starting from **14** was the exchange of the thiophene units adjacent to the 2,1,3-benzothiadiazole acceptor. It is generally known that fused thiophene structures offer an effective strategy to produce high-mobility semiconductors. The introduction of thieno[3,2-*b*]thiophene units into the Pt-polyyne polymers afforded a more structurally rigid main chain polymer **16**, and accordingly, higher hole mobilities and significantly improved on/off ratios were observed in the thin-film transistor performances [28]. Therefore, the PCEs of **16** with PCBM were more than two times greater than those of the devices based on **14** and PCBM. It is also well known that (ethylenedioxy)thiophene (EDOT) is a stronger donor than thiophene. Consequently, **17** with the EDOT moieties displayed a bathochromically shifted λ_{max} at 621 nm in THF and a narrower optical bandgap of 1.93 eV as compared to **14** [29]. However, due to the difficulties in the energy level match and limited formation of the triplet excited state in **17**, the photovoltaic performances were not remarkable.

Wong et al. further expanded the series of the Pt-polyyne polymers. Various types of aromatic spacers were introduced into the main chain between the alkyne units. New Pt-polyyne polymers containing the bithiazole-oligo(thienyl) units **18** were prepared, and they showed sufficient solubilities in polar organic solvents due to the two nonyl groups (Fig. 5.7) [30]. The photovoltaic activities as p-type

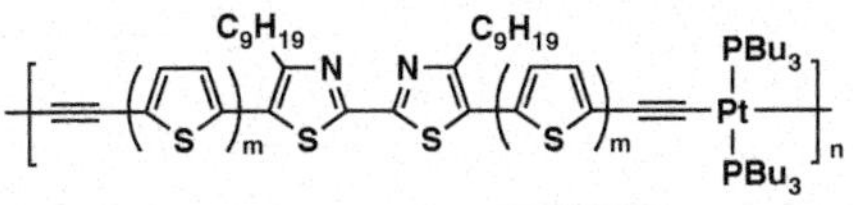

18 m = 0-3

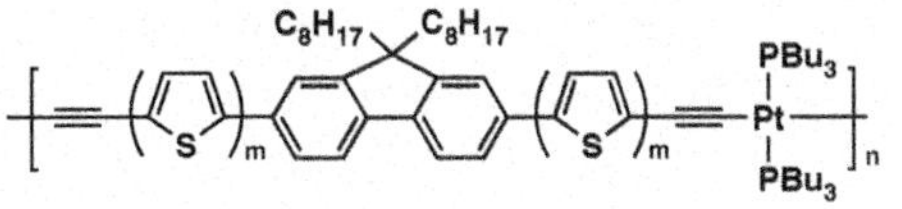

19 m = 0-3

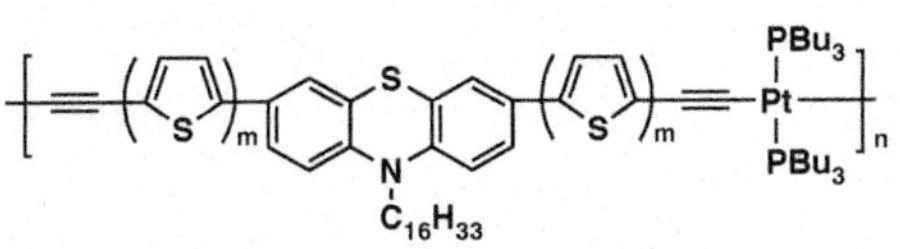

20 m = 0-2

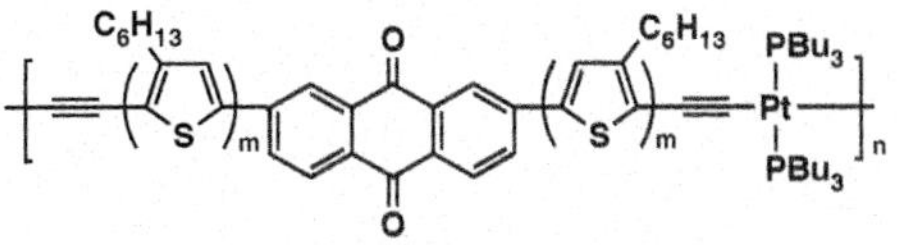

21 m = 0, 1

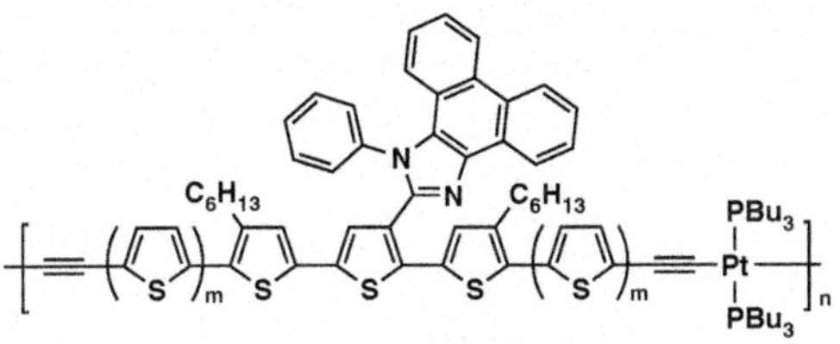

22 m = 0, 1

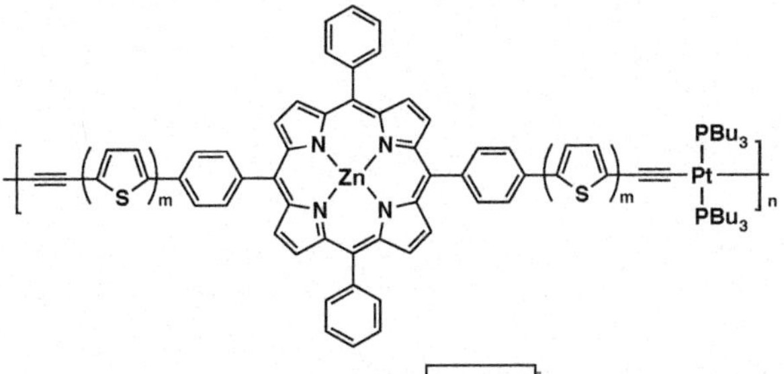

23 m = 0, 1

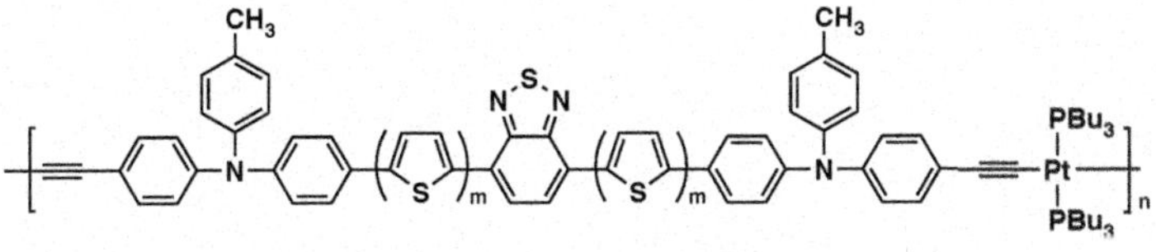

24 m = 0, 1

Fig. 5.7 Pt-polyyne polymers with various aromatic units for p-type semiconductors in BHJ solar cells semiconductors were comprehensively investigated. This study revealed that both the photovoltaic responses and PCEs increased as the number of thienyl rings increased. The optimized PCE was 2.7 %, and the best external quantum efficiency reached 83 %. Interestingly, these performances are comparable to that of the P3HT-based devices. A similar linear dependence of the thienyl ring number on the photovoltaic performances was also reported for the Pt-polyyne polymers containing the fluorene-oligo(thienyl) units **19** [31]. The polymer bandgaps gradually deceased as more thienyl rings were introduced. This feature was directly related to the carrier mobilities. Both the hole and electron mobilities, determined by the space-charge-limited current method, increased with the increasing thienyl ring numbers. Accordingly, **19** ($m = 3$) showed the best PCE of 2.9 % with the external quantum efficiency up to 83 %. In place of the fluorene unit, the electron-rich phenothiazine or electron-deficient 9,10-anthraquinone moieties were examined. The phenothiazine-containing Pt-polyyne polymers **20** served as a good p-type semiconductor as expected from the energy levels [32]. Surprisingly, the 9,10-anthraquinone-containing Pt-polyyne polymers **21** also showed the photocurrent generation, although the PCEs were limited. It was shown that the presence of the thienyl units dramatically improved the p-type semiconducting character [33]. The moderate photovoltaic properties of the Pt-polyyne polymers with the phenanthrenyl-imidazole moiety **22**, with the Zn(II) porphyrin **23**, and with the triphenylamine-benzothiadiazole-triphenylamine (D-A-D) unit **24** were also reported [34–36].

5.3 Other Organometallic Polymers

In addition to the mainstream Pt-polyyne polymers, other types of organometallic polymers were also examined as semiconductors in photovoltaic devices. In this section, these polymers are briefly introduced.

Sargent and Manners reported the photoconductive properties of poly (ferrocenylmethylphenylsilane) **25** when it was mixed with PCBM in the solid state (Fig. 5.8) [37]. However, the authors were suspicious about the semiconducting feature of this polymer. Main chain Ru(II) metallo-polymers have been applied to BHJ solar cell devices. Lin et al. examined three kinds of benzodiazole acceptor units in **26** [38]. As compared to the counter benzoselenodiazole and benzoxadiazole derivatives, the benzothiadiazole-based polymer **26** displayed a better photovoltaic performance mainly because of the higher short circuit currents. Li and Peng also reported similar polymers **27** [39]. It was suggested that the substitution of electron-withdrawing fluorine atoms at the benzothiadiazole unit significantly improved the optical and electrochemical properties. These features led to one of the highest PCE of 2.66 % among the reported metallo-supramolecular polymers.

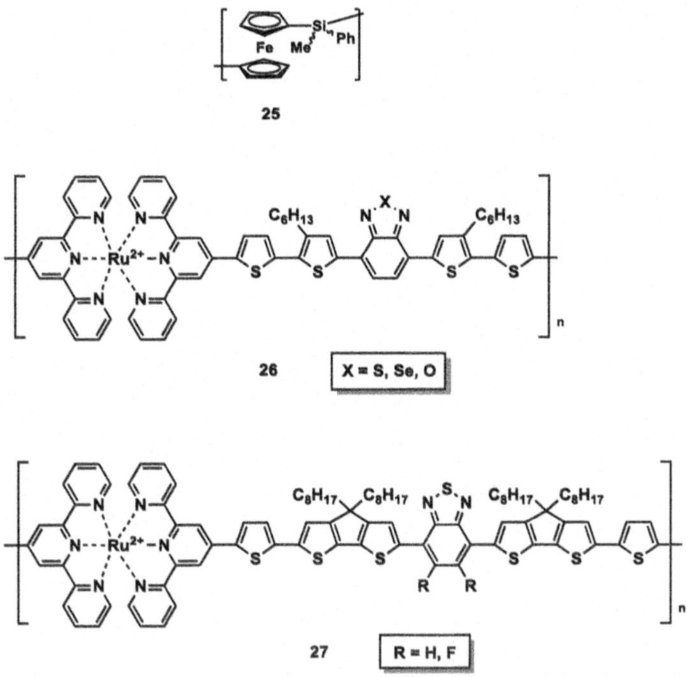

25

26 X = S, Se, O

27 R = H, F

Fig. 5.8 Organometallic polymers containing ferrocene and Ru(II) units in the main chain for p-type semiconductors in BHJ solar cells

The complexation with metal ions at the side chains of conjugated polymers results in a significant change in the optical and electric properties. For example, Chen et al. prepared conjugated polymers comprised of the indacenodithiophene unit [40]. The absorption spectrum of this polymer exhibited a noticeable redshift upon formation of the cyclometalated Pt(II) moieties, producing **28** (Fig. 5.9). This redshift enabled an efficient solar light absorption especially in the longer wavelength. Accordingly, the blended films of **28** and PCBM produced a sufficiently high PCE of 2.9 %. Chan et al. also reported the Re(I) complexation with highly conjugated polymers [41]. The precursor polymers were intrinsically low bandgap polymers, but a further bandgap narrowing occurred upon complexation with Re (I) pentacarbonyl chloride. For example, the precursor polymer film of **29** showed a λ_{max} at 664 nm, while the λ_{max} of the **29** film shifted to 921 nm. Additionally, the hole mobilities of the polymer/PCBM (1:4) thin films were determined by the time of flight measurements. Interestingly, the mobility of **29**/PCBM was in the range of 10^{-6} cm^2 V^{-1} s^{-1}, which is one order of magnitude higher than the precursor polymer blended film. Kimoto et al. investigated the benzothiadiazole-based conjugated polymer as a p-type semiconductor in BHJ solar cells with PCBM [42]. It was found that the open circuit voltage was clearly enhanced from 0.46 to 0.52 V upon complexation with Sn(II), forming **30**.

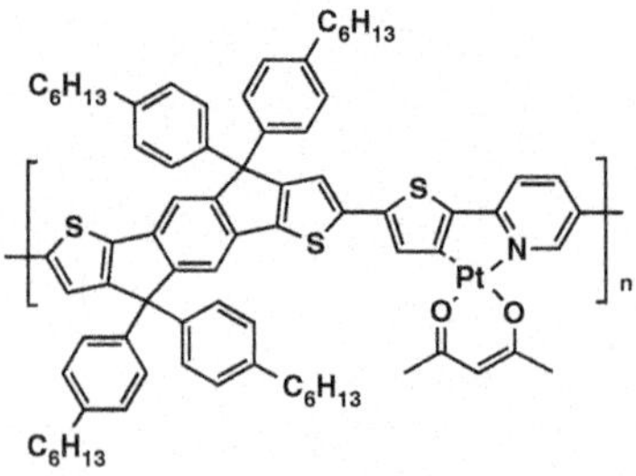

Fig. 5.9 Organometallic polymers complexed with metal ions in the side chain for producing high-performance p-type semiconductors in BHJ solar cells

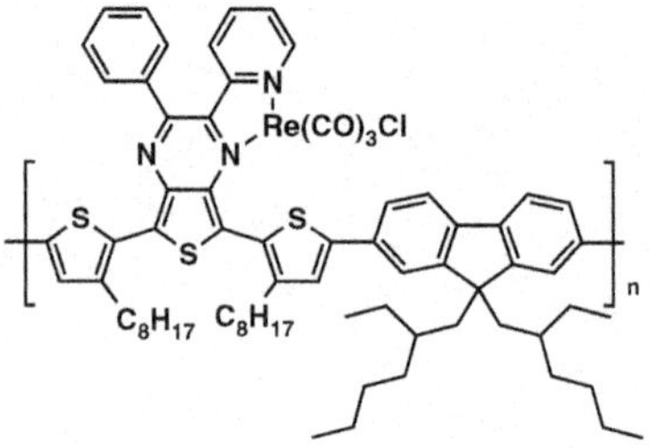

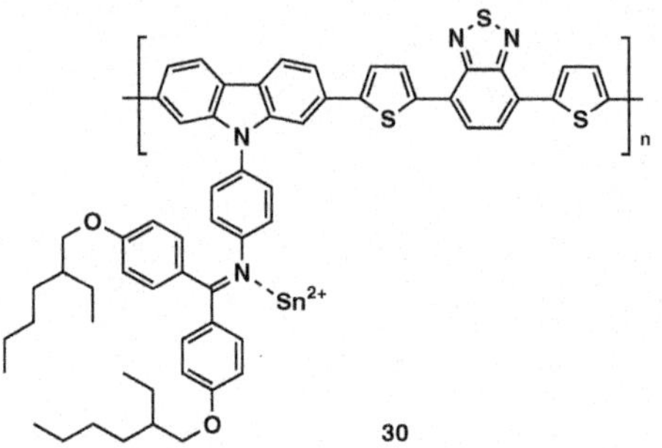

Overall, both the Pt-polyyne polymers and other organometallic polymers reviewed above were used as p-type semiconductors in photovoltaic devices. This fact indicates the intrinsic p-type character of the metal-containing materials. However, there are a few examples of using metal-containing materials in the role of other than p-type semiconductors in electronic devices. Schanze et al. synthesized the C_{60} end-capped Pt-acetylide triad molecule **31** (Fig. 5.10) [43]. Electrochemical and optical spectroscopy measurements suggested that intramolecular photoinduced electron transfer occurs both in solutions at room temperature and in a low-temperature solvent glass. A single-layer photovoltaic device was fabricated by spin coating this molecule onto a PEDOT-PSS deposited ITO followed by LiF/Al deposition. Although the PCE was limited to 0.05 %, the incident photon-to-current efficiency (IPCE) curve agreed with the absorption spectrum, reaching the maximum response in the UV region. Very recently, the Hg-polyyne polymer **32** was examined as a cathode interlayer of inverted polymer solar cells [44]. With this interlayer, the device performances were significantly improved. The authors highlighted the importance of the intermolecular interactions, such as the Hg-Hg interaction and π-π stacking for this polymer. The potent electron transport properties of **32** were demonstrated for the electron-only devices. However, there are, to the best of my knowledge, no reports on n-type semiconductors of metal-containing materials used in photovoltaic devices.

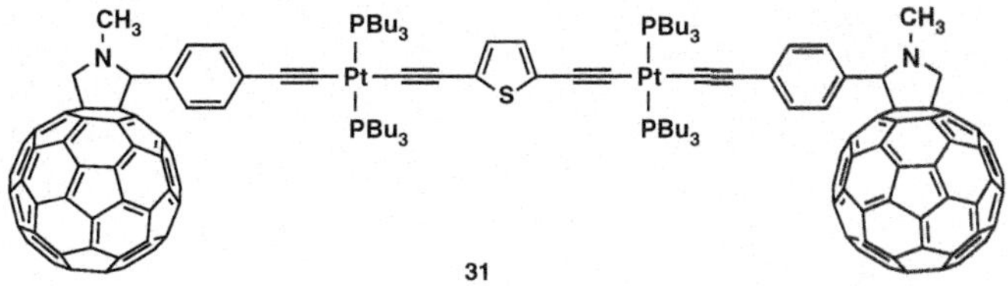

31

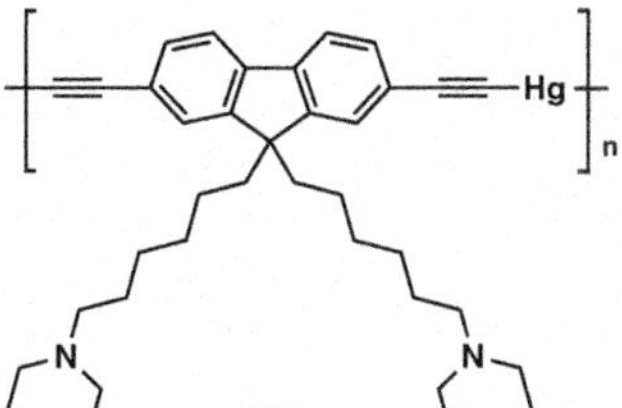

32

Fig. 5.10 Other types of organometallic semiconducting materials

5.4 n-Type Pt-Polyyne Polymers

In this section, a new approach to invert the energy levels or possibly the semiconducting features of Pt-polyyne polymers is introduced. As described above, Pt-polyyne polymers are good p-type semiconductors. In order to construct n-type energy levels, the introduction of electron-accepting units is required.

Cyanated acceptor molecules are known to form charge-transfer complexes, as represented by 7,7,8,8-tetracyanoquinodimethane (TCNQ) with tetrathiafulvalene (TTF). This feature was indeed applied to the semiconducting polymer doping techniques to create metallic polymers in which cyanated acceptor molecules were employed as p-type dopants. The charge-transfer complexes usually have a low energy absorption. Therefore, another reactivity of the cyanated acceptor molecules was overlooked. When the smallest and compact acceptor molecule, tetracyanoethylene (TCNE), was added to the *N,N*-dimethylaniline-substituted buta-1,3-diyne **33**, the solution color quickly changed at room temperature, and the formed intense color was due to the formation of the charge-transfer complex (Fig. 5.11) [45]. However, the X-ray crystal analysis finally revealed the chemical structure of the product, and the proposed reaction mechanism was a [2+2] cycloaddition-retroelectrocyclization reaction, yielding the 1,1,4,4-tetracyanobuta-1,3-diene derivative **34** [46]. A literature search suggested that this reaction significantly depends on the substituents of the alkyne moiety [47–50]. The reactions proceed faster and in higher yields as the electron density of the alkyne increases. From the viewpoint of pure organic chemistry, special attention has been paid to the pursuit of effective electron-donating substituents as well as the extension of the applicable molecular size. Diederich et al. successfully found that the TTF and ferrocene are effective substituents for the reaction with TCNE [51]. Also, linear and dendritic macromolecules were quantitatively functionalized by the acceptor

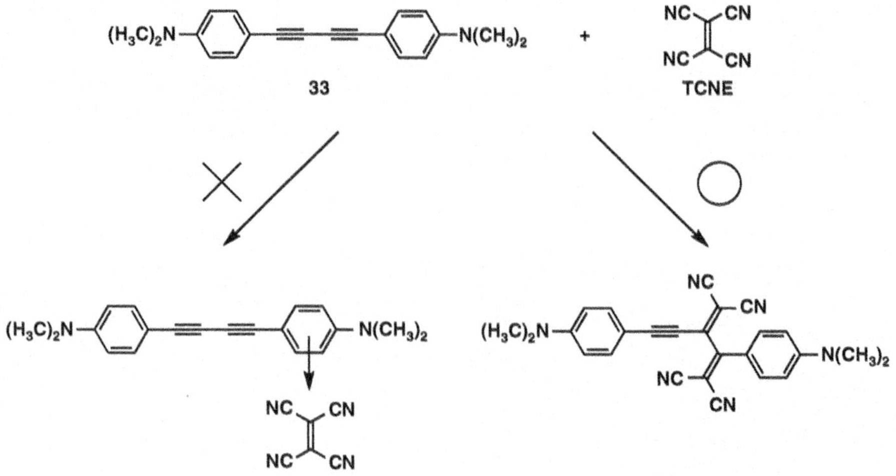

Fig. 5.11 Reaction between 1,4-bis[4-(*N*,*N*-dimethylamino)phenyl]-1,3-butadiyne **33** and TCNE, yielding a charge-transfer chromophore: (left) erroneous charge-transfer complex and (right) donor-substituted 1,1,4,4-tetracyanobuta-1,3-diene **34** obtained in practice

addition [52, 53], and in some cases, super-acceptor molecules rivaling the benchmark compounds of TCNQ and F_4TCNQ were produced [54].

The cyanated products usually show intense charge-transfer bands, which are ascribed to intramolecular donor-acceptor interactions. Consequently, they are promising photovoltaic materials, and there are indeed some reports on the photovoltaic applications of the organic TCNE or TCNQ adducts. Blanchard et al. synthesized the TCNE adduct **35** by a [2+2] cycloaddition-retroelectrocyclization (Fig. 5.12) [55]. This D-A-D-type molecule displayed a well-defined charge-transfer band at 569 nm in CH_2Cl_2 and a relatively low oxidation potential at 1.05 V (vs. SCE). Based on the calculated energy levels, a bilayer heterojunction solar cell using **35** as a p-type semiconductor in combination with C_{60} was fabricated, and the maximum PCE of >1.0 % was achieved. These charge-transfer chromophores were also examined as photosensitizers in dye-sensitized solar cells. Ohshita et al. synthesized **36** by the postfunctionalization of the precursor disilanyleneethynylene polymer [56]. It was found that this polymer was adsorbed onto TiO_2 electrodes, although it does not possess any carboxylic acid groups. The authors postulated that the adsorption was due to the formation of Si–O–Ti bonds. Later, Michinobu and Satoh revealed that the tetracyanate units are effective anchoring groups to the TiO_2 surface [57]. Small molecular weight dipolar D-A molecules **37–40** were synthesized by a [2+2] cycloaddition-retroelectrocyclization between aniline-substituted alkynes and TCNE or TCNQ. All reactions rapidly proceeded at room temperature in a “click chemistry” fashion due to the powerful electron-donating substituents. When TiO_2 electrodes were immersed into the solutions of these D-A molecules, the color of the electrodes

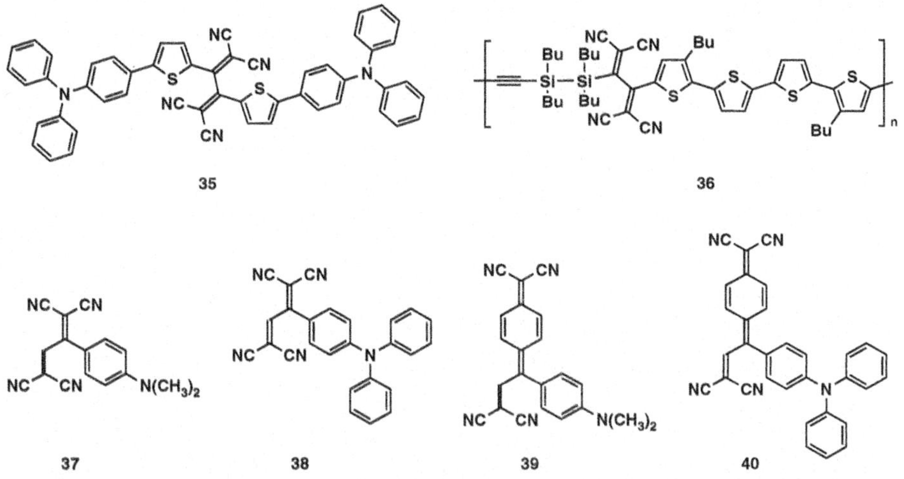

Fig. 5.12 Donor-substituted tetracyanated acceptor chomophores for photovoltaic applications

clearly changed. This fact indicated the dye adsorption onto TiO_2. The photovoltaic properties of the dye-adsorbed TiO_2 electrodes were evaluated, and **40** was the best photosensitizer among the examined dyes.

Metal acetylides are also appropriate precursors for the reaction with cyanated acceptors. Bruce has worked on these reactions of electron-rich alkynyl-metal complexes with TCNE, TCNQ, and some related acceptors. He started this research program in the 1980s using Ru-acetylides [58]. Since then, this chemistry has expanded to a variety of metallo-polyynes (M = Fe, Au, Os, Pd, Pt, Ru, Rh, W) [59]. Among them, the Pt-acetylides have received particular attention due to the synthetic versatility as well as the excellent semiconducting character of the corresponding linear polymers. The reactions of small molecular weight Pt-acetylides with TCNE and TCNQ were comprehensively investigated [60–62]. Most products were unambiguously characterized by X-ray crystallography. The reactions of the Pt-acetylides with additional donor units, such as aromatic amines, gave the corresponding TCNE or TCNQ adducts in satisfactorily high yields (>90 %). In contrast, in the absence of other effective donor groups, the reaction yields were moderate. This result suggested that the Pt moiety definitely activates the alkyne, but it is a weaker activator than aromatic amine units. It should be noted that the reaction of the Pt-acetylides with TCNE or TCNQ did not cause any noticeable side reactions under the optimized conditions.

The systematic studies of the small molecular weight Pt-acetylides raised the applicability of this reaction to Pt-polyyne polymers. A pioneering study by Sonogashira et al. was reported in the proceedings of a domestic conference [63]. The polymer reactions of the precursor Pt-polyyne polymers **12** and **42** were performed by adding TCNE in CH_2Cl_2 at room temperature, yielding the brown-colored polymers **41** and **43**, respectively (Fig. 5.13). The added amount of TCNE (TCNE/Pt) was estimated to be 1 for **41** and 0.6 for **43**. The lower added amount in

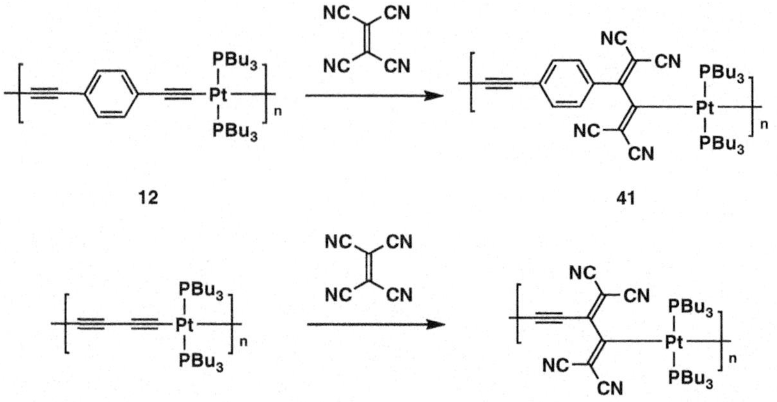

Fig. 5.13 Postfunctional cyanation of Pt-polyyne polymers by a [2+2] cycloaddition-retroelectrocyclization with TCNE

43 was explained by a steric reason. The ^{31}P-NMR, IR, and UV-vis absorption spectra as well as thermal degradation temperatures were measured. However, no further detailed analyses of the products were performed probably because of the limited analytical techniques in the 1980s.

Recently, Michinobu et al. revisited this postfunctionalization method of the Pt-polyyne polymers in order to prepare n-type materials. As an improved structure of **12**, the hexylthiophene unit was introduced into the Pt-polyyne polymer [64]. Since the thiophene unit is a stronger donor than the phenyl group, the reactivity of the main chain alkynes was enhanced. Thus, the thiophene-containing Pt-polyyne polymer **44** was reacted with TCNE or TCNQ (Fig. 5.14). Thanks to the activation of both Pt and the thiophene moieties, the reactions of both TCNE and TCNQ quantitatively proceeded. It should be noted that the TCNE adduct **45** was identified by ^{1}H NMR, while the addition pattern of TCNQ was regio-random. Thus, the ^{1}H NMR of the TCNQ adduct **46** was complicated. However, both products displayed well-defined low-energy absorptions ascribed to the intramolecular charge-transfer. The optical bandgaps of **45** and **46** were 1.83 and 1.22 eV, respectively, which are quite low when compared to that of the precursor polymer **44** (2.78 eV). Additionally, the cyanated Pt-polyyne polymers exhibited very low reduction potentials originating from the cyano-acceptor units. The onset potentials of the first reduction were detected at −0.82 V (vs. Fc/Fc^{+}) for **45** and −0.49 V for **46**. Overall, it was demonstrated that there is a linear correlation between the optical bandgaps and electrochemical bandgaps.

Furthermore, the promising acceptor units of benzothiadiazole or diketopyrrolopyrrole were added to the repeat unit of the thiophene-containing Pt-polyyne polymers. The donor-acceptor-type Pt-polyyne polymers **47** and **48** are typical p-type semiconductors, and they also serve as precursor polymers for the postfunctionalization with TCNE (Fig. 5.15). Due to the conjugative linkage of the

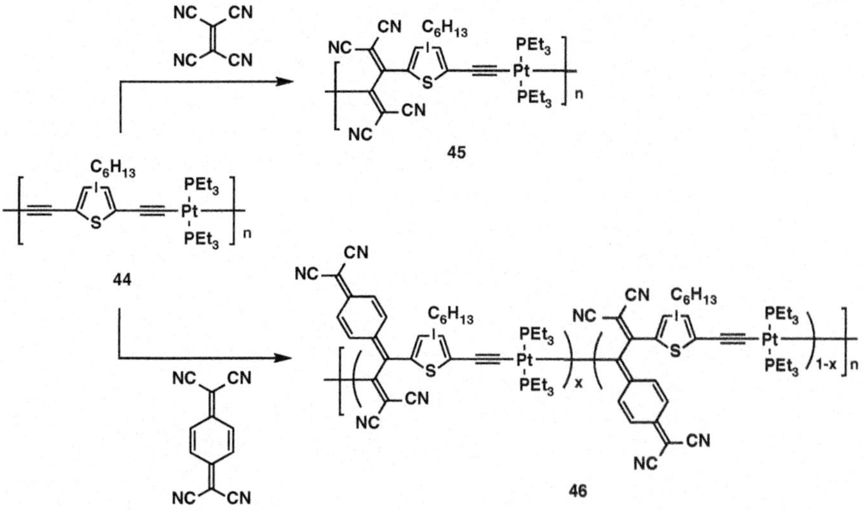

Fig. 5.14 Postfunctional cyanation of Pt-polyyne polymers by a [2+2] cycloaddition-retroelectrocyclization with TCNE and TCNQ

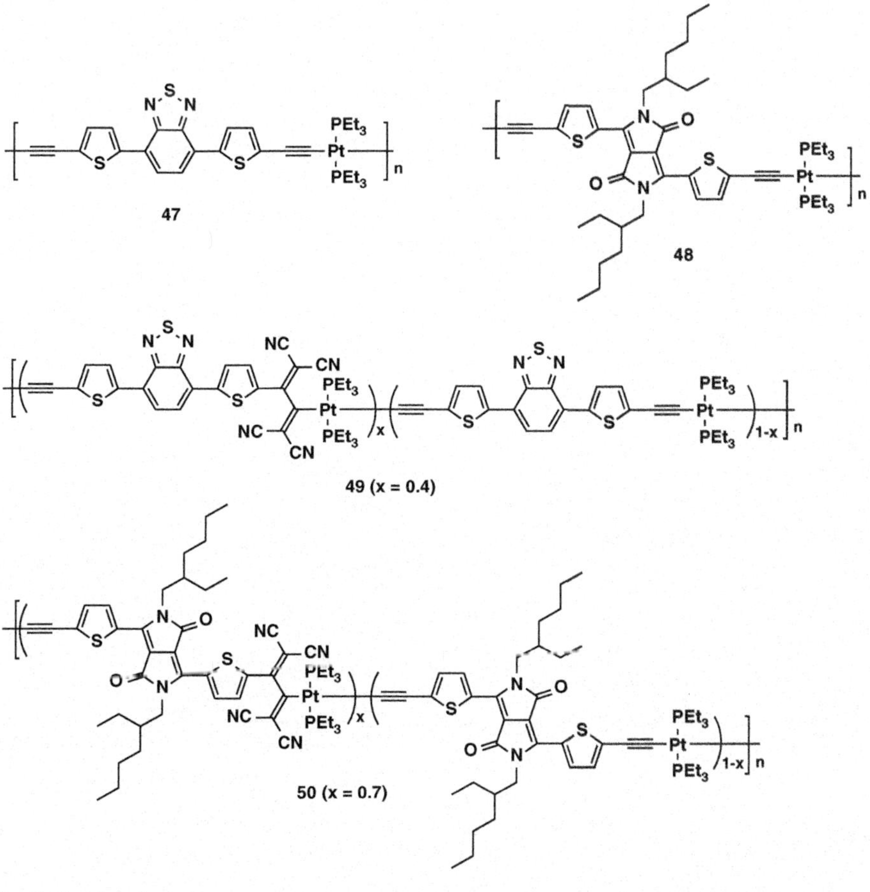

Fig. 5.15 Chemical structures of precursor Pt-polyyne polymers and their TCNE adducts

thiophene ring with electron-accepting benzothiadiazole or diketopyrrolopyrrole moieties, the reactivity of the main chain alkynes of **47** and **48** was reduced [65]. Accordingly, the added amount of TCNE was 40 % for **49** and 70 % for **50**. Despite the limited added amount of TCNE, the resulting Pt-polyyne polymers showed significantly broadened absorption spectra as compared to the corresponding precursor polymers. Therefore, the optical bandgaps of the TCNE-adducted polymers (1.47 eV for **49** and 1.28 eV for **50**) became narrow as compared to those of the precursor polymers (1.87 eV for **47** and 1.79 eV for **48**). The HOMO and LUMO levels associated with the electrochemical redox potentials were also varied. The LUMO levels of the TCNE-adducted polymers (−4.18 eV for **49** and −4.49 eV for **50**) were significantly lower than those of the precursor polymers (−3.14 eV for **47** and −3.36 eV for **48**). This result explicitly indicates the energy level decrease of the Pt-polyyne polymers in terms of the TCNE addition. In other words, **49** and **50** would serve as n-type semiconducting materials in electronic devices. A similar decrease and further control of the energy levels were demonstrated for the purely organic semiconducting polymers [66, 67].

The donor-acceptor-type Pt-polyyne polymers **47–50** were employed as the active components of BHJ solar cells. In order to get insights into the energy level decrease, both precursor and cyanated Pt-polyyne polymers were initially evaluated as p-type semiconductors in the BHJ solar cells. Thus, the device configuration was ITO/PEDOT:PSS/polymer:[70]PCBM(1:2 to 1:4 wt/wt)/TiO_x/Al. The current density-voltage (J-V) and IPCE curves, evaluated under AM 1.5G light illumination at 100 mW cm^{-2}, suggested a clear photocurrent generation for all devices (Fig. 5.16). Both the open circuit voltage and short circuit current of the devices based on the precursor Pt-polyyne polymers were better than those of the devices based on the cyanated Pt-polyyne polymers. For example, the device based on **47**:[70]PCBM showed the open circuit voltage of 0.59 V, short circuit current of 3.97 mA cm^{-2}, and fill factor of 0.40, which lead to the PCE of 0.94 %. In contrast, all the photovoltaic parameters of the device based on the corresponding cyanated derivative **49**:[70]PCBM were reduced. This result reasonably indicates the

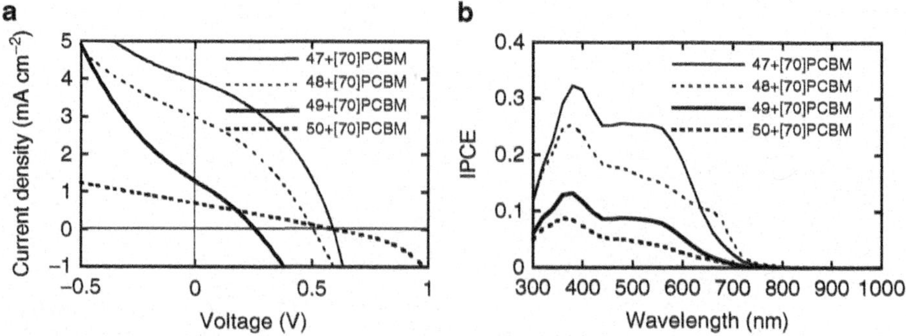

Fig. 5.16 (**a**) J-V and (**b**) IPCE curves of the BHJ solar cells based on Pt-polyyne polymers and [70]PCBM. (Reproduced from ref. [65] by permission of John Wiley & Sons, Ltd)

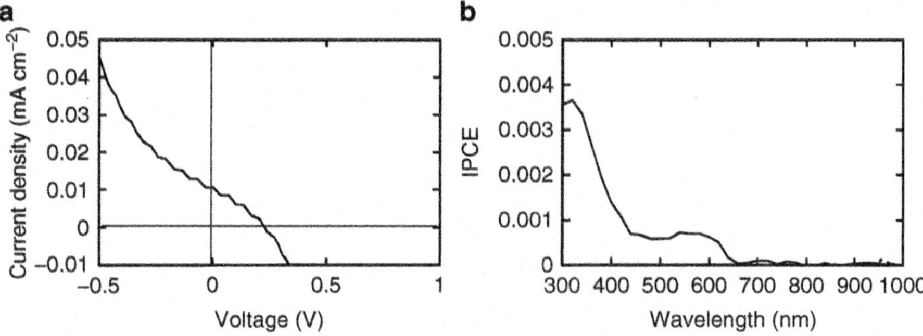

Fig. 5.17 (**a**) *J-V* and (**b**) IPCE curves of all-polymer solar cells based on P3HT and **49**. (Reproduced from ref. [65] by permission of John Wiley & Sons, Ltd)

decrease in the p-type performance of the Pt-polyyne polymers by the TCNE addition, which also agrees with the decrease in the energy levels.

Based on the above result, the all-polymer solar cells based on P3HT as the p-type semiconductor and cyanated Pt-polyyne polymers as the n-type semiconductor were fabricated. The weight ratio of the P3HT/Pt-polyyne polymer was 1:1. Unfortunately, the device based on **50** as the n-type semiconductor did not provide any photocurrents. However, the device containing **49** produced the *J-V* curve with the PCE of 0.00079 % and weak IPCE photocurrents in the range of the absorption bands of this polymer (Fig. 5.17). This result can be explained by the difference in the LUMO levels of the Pt-polyyne polymers. The diketopyrrolopyrrole-containing polymer **50** possesses a deeper LUMO level, resulting in an energy level mismatch with P3HT. The limited performances of the all-polymer solar cells were also suggested by the surface morphologies of the active layers. Atomic force microscopy (AFM) measurements of the active layers comprising P3HT and the cyanated Pt-polyyne polymers (**49** or **50**) provided no nanoscale phase separation images. It is expected that the optimization of the blend ratio of the p-type and n-type semiconducting polymers improves the photovoltaic properties.

5.5 Conclusions

Organometallic semiconducting polymers, which were applied to photovoltaic devices, are described. The construction of n-type semiconducting polymers has been elusive because of the intrinsic p-type character of the organometallic polymers. However, recent efforts in developing purely organic n-type semiconducting materials have opened the way to n-type organometallic polymers. The use of fullerene or other electron-accepting units, such as benzothiadiazole and diketopyrrolopyrrole, is one promising approach, and an efficient photoinduced electron transfer was demonstrated in some cases. In contrast, the postfunctional cyanation of the Pt-polyyne polymers is a new method of constructing n-type materials.

The energy levels of the cyanated Pt-polyyne polymers were significantly low, as demonstrated by the application to the all-polymer solar cells with the p-type semiconducting P3HT. However, the effective n-type semiconducting performances of the cyanated polymers have not yet been achieved. Therefore, the development of high-performance n-type semiconducting polymers containing metal elements is still challenging. In the future, we will see further improvements in the cyanated organometallic polymers for use in all-polymer solar cells. Energy level match and nanoscale phase separation between two types of semiconducting polymers are the next subjects to be solved.

References

1. Thompson BC, Fréchet JMJ (2008) Polymer-fullerene composite solar cells. Angew Chem Int Ed 47:58–77
2. Dennler G, Scharber MC, Brabec CJ (2009) Polymer-fullerene bulk-heterojunction solar cells. Adv Mater 21:1323–1338
3. Brabec CJ, Gowrisanker S, Malls JJM, Laird D, Jia S, Williams SP (2010) Polymer-fullerene bulk-heterojunction solar cells. Adv Mater 22:3839–3856
4. Dang MT, Hirsch L, Wantz G (2011) P3HT:PCBM, best seller in polymer photovoltaic research. Adv Mater 23:3597–3602
5. He Z, Zhong C, Su S, Xu M, Wu H, Cao Y (2012) Enhanced power-conversion efficiency in polymer solar cells using an inverted device structure. Nat Photon 6:591–595
6. You J, Dou L, Yoshimura K, Kato T, Ohya K, Moriarty T, Emery K, Chen CC, Gao J, Li G, Yang Y (2013) A polymer tandem solar cell with 10.6 % power conversion efficiency. Nat Commun 4:1446
7. Zhang Y, Zhou H, Seifter J, Ying L, Mikhailovsky A, Heeger AJ, Bazan GC, Nguyen TQ (2013) Molecular doping enhances photoconductivity in polymer bulk heterojunction solar cells. Adv Mater 25:7038–7044
8. Facchetti A (2013) Polymer donor-polymer acceptor (all-polymer) solar cells. Mater Today 16:123–132
9. Stalder R, Mei J, Graham KR, Estrada LA, Reynolds JR (2014) Isoindigo, a versatile electron-deficient unit for high-performance organic electronics. Chem Mater 26:664–678
10. Halls JJM, Walsh CA, Greenham NC, Marseglia EA, Friend RH, Moratti SC, Holmes AB (1995) Efficient photodiodes from interpenetrating polymer networks. Nature 376:498–500
11. Granström M, Petritsch K, Arias AC, Lux A, Andersson MR, Friend RH (1998) Laminated fabrication of polymeric photovoltaic diodes. Nature 395:257–260
12. Holcombe TW, Woo CH, Kavulak DFJ, Thompson BC, Fréchet JMJ (2009) All-polymer photovoltaic devices of poly(3-(4-*n*-octyl)-phenylthiophene) from Grignard metathesis (GRIM) polymerization. J Am Chem Soc 131:14160–14161
13. Zhan X, Tan Z, Zhou E, Li Y, Misra R, Grant A, Domercq B, Zhang XH, An Z, Zhang X, Barlow S, Kippelen B, Marder SR (2009) Copolymers of perylene diimide with dithienothiophene and dithienopyrrole as electron-transport materials for all-polymer solar cells and field-effect transistors. J Mater Chem 19:5794–5803
14. Zhou E, Cong J, Hashimoto K, Tajima K (2013) Control of miscibility and aggregation via the material design and coating process for high-performance polymer blend solar cells. Adv Mater 25:6991–6996
15. Earmme T, Hwang YJ, Murari NM, Subramaniyan S, Jenekhe SA (2013) All-polymer solar cells with 3.3 % efficiency based on naphthalene diimide-selenophene copolymer acceptor. J Am Chem Soc 135:14960–14963

16. Salim T, Sun S, Wong LH, Xi L, Foo YL, Lam YM (2010) The role of poly(3-hexylthiophene) nanofibers in an all-polymer blend with a polyfluorene copolymer for solar cell applications. J Phys Chem C 114:9459–9468
17. Mori D, Benten H, Ohkita H, Ito S, Miyake K (2012) Polymer/polymer blend solar cells improved by using high-molecular-weight fluorene-based copolymer as electron acceptor. ACS Appl Mater Interfaces 4:3325–3329
18. Cao Y, Lei T, Yuan J, Wang JY, Pei J (2013) Dithiazolyl-benzothiadiazole-containing polymer acceptors: synthesis, characterization, and all-polymer solar cells. Polym Chem 4:5228–5236
19. Sang G, Zou Y, Huang Y, Zhao G, Yang Y, Li Y (2009) All-polymer solar cells based on a blend of poly[3-(10-*n*-octyl-3-phenothiazine-vinylene)thiophene-co-2,5-thiophene] and poly [1,4-dioctyloxyl-*p*-2,5-dicyanophenylenevinylene]. Appl Phys Lett 94:193302-1–193302-3
20. Vijayakumar C, Saeki A, Seki S (2012) Optoelectronic properties of dicyanofluorene-based n-type polymers. Chem Asian J 7:1845–1852
21. Wong WY (2008) Metallopolyyne polymers as new functional materials for photovoltaic and solar cell applications. Macromol Chem Phys 209:14–24
22. Wong WY, Ho CL (2010) Organometallic photovoltaics: a new and versatile approach for harvesting solar energy using conjugated polymetallaynes. Acc Chem Res 43:1246–1256
23. Ho CL, Wong WY (2011) Metal-containing polymers: facile tuning of photophysical traits and emerging applications in organic electronics and photonics. Coord Chem Rev 225:2469–2502
24. Köhler A, Wittmann HF, Friend RH, Khan MS, Lewis J (1996) Enhanced photocurrent response in photocells made with platinum-poly-yne/C_{60} blends by photoinduced electron transfer. Synth Met 77:147–150
25. Guo F, Kim YG, Reynolds JR, Schanze KS (2006) Platinum-acetylide polymer based solar cells: involvement of the triplet state for energy conversion. Chem Commun 42:1887–1889
26. Wong WY, Wang XZ, He Z, Djurišić AB, Yip CT, Cheung KY, Wang H, Mak CSK, Chan WK (2007) Metallated conjugated polymers as a new avenue towards high-efficiency polymer solar cells. Nat Mater 6:521–527
27. Wu PT, Bull T, Kim FS, Luscombe CK, Jenekhe SA (2009) Organometallic donor-acceptor conjugated polymer semiconductors: tunable optical, electrochemical, charge transport, and photovoltaic properties. Macromolecules 42:671–681
28. Baek NS, Hau SK, Yip HL, Acton O, Chen KS, Jen AKY (2008) High performance amorphous metallated π-conjugated polymers for field-effect transistors and polymer solar cells. Chem Mater 20:5734–5736
29. Mei J, Ogawa K, Kim YG, Heston NC, Arenas DJ, Nasrollahi Z, McCarley TD, Tanner DB, Reynolds JR, Schanze KS (2009) Low-band-gap platinum acetylide polymers as active materials for organic solar cells. ACS Appl Mater Interfaces 1:150–161
30. Wong WY, Wang XZ, He Z, Chan KK, Djurišić AB, Cheung KY, Yip CT, Ng AMC, Xi YY, Mak CSK, Chan WK (2007) Tuning the absorption, charge transport properties, and solar cell efficiency with the number of thienyl rings in platinum-containing poly(aryleneethynylene)s. J Am Chem Soc 129:14372–14380
31. Liu L, Ho CL, Wong WY, Cheung KY, Fung MK, Lam WT, Djurišić AB, Chan WK (2008) Effect of oligothienyl chain length on tuning the solar cell performance in fluorene-based polyplatinynes. Adv Funct Mater 18:2824–2833
32. Wong WY, Chow WC, Cheung KY, Fung MK, Djurišić AB, Chan WK (2009) Harvesting solar energy using conjugated metallopolyyne donors containing electron-rich phenothiazine-oligothiophene moieties. J Organomet Chem 694:2717–2726
33. Li L, Chow WC, Wong WY, Chui CH, Wong RSM (2011) Synthesis, characterization and photovoltaic behavior of platinum acetylide polymers with electron-deficient 9,10-anthraquinone moiety. J Organomet Chem 696:1189–1197
34. Zhan H, Wong WY, Ng A, Djurišić AB, Chan WK (2011) Synthesis, characterization and photovoltaic properties of platinum-containing poly(aryleneethynylene) polymers with phenanthrenyl-imidazole moiety. J Organomet Chem 696:4112–4120

35. Zhan H, Lamare S, Ng A, Kenny T, Guernon H, Chan WK, Djurišić AB, Harvey PD, Wong WY (2011) Synthesis and photovoltaic properties of new metalloporphyrin-containing polyplatinyne polymers. Macromolecules 44:5155–5167
36. Wang Q, Wong WY (2011) New low-bandgap polymetallaynes of platinum functionalized with a triphenylamine-benzothiadiazole donor-acceptor unit for solar cell applications. Polym Chem 2:432–440
37. Cyr PW, Klem EJD, Sargent EH, Manners I (2005) Photoconductivity in donor-acceptor polyferrocenylsilane-fullerene composite films. Chem Mater 17:5770–5773
38. Padhy H, Sahu D, Chiang IH, Patra D, Kekuda D, Chu CW, Lin HC (2011) Synthesis and applications of main-chain Ru(II) metallo-polymers containing bis-terpyridyl ligands with various benzodiazole cores for solar cells. J Mater Chem 21:1196–1205
39. Feng K, Shen X, Li Y, He Y, Huang D, Peng Q (2013) Ruthenium(II) containing supramolecular polymers with cyclopentadithiophene-benzothiazole conjugated bridges for photovoltaic applications. Polym Chem 4:5701–5710
40. Liao CY, Chen CP, Chang CC, Hwang GW, Chou HH, Cheng CH (2013) Synthesis of conjugated polymers bearing indacenodithiophene and cyclometalated platinum(II) units and their application in organic photovoltaics. Sol Energy Mater Sol Cells 109:111–119
41. Mak CSK, Cheung WK, Leung QY, Chan WK (2010) Conjugated copolymers containing low bandgap rhenium(I) complexes. Macromol Rapid Commun 31:875–882
42. Kimoto A, Tajima Y (2012) Donor-acceptor-type low bandgap polymer carrying phenylazomethine moiety as a metal-collecting pendant unit: open-circuit voltage modulation of solution-processed organic photovoltaic devices induced by metal complexation. ACS Macro Lett 1:667–671
43. Guo F, Ogawa K, Kim YG, Danilov EO, Castellano FN, Reynolds JR, Schanze KS (2007) A fulleropyrrolidine end-capped platinum-acetylide triad: the mechanism of photoinduced charge transfer in organometallic photovoltaic cells. Phys Chem Chem Phys 9:2724–2734
44. Liu S, Zhang K, Lu J, Zhang J, Yip HL, Huang F, Cao Y (2013) High-efficiency polymer solar cells via the incorporation of an amino-functionalized conjugated metallopolymer as a cathode interlayer. J Am Chem Soc 135:15326–15329
45. Rodríguez JG, Lafuente A, Martín-Villamil R, Martínez-Alcazar MP (2001) Synthesis and structural analysis of 1,4-bis[n-(N,N-dimethylamino)pheny]buta-1,3-diynes and charge-transfer complexes with TCNE. J Phys Org Chem 14:859–868
46. Michinobu T, May JS, Lim JH, Boudon C, Gisselbrecht JP, Seiler P, Gross M, Biaggio I, Diederich F (2005) A new class of organic donor-acceptor molecules with large third-order optical nonlinearities. Chem Commun 41:737–739
47. Kivala M, Diederich F (2009) Acetylene-derived strong organic acceptors for planar and nonplanar push-pull chromophores. Acc Chem Res 42:235–248
48. Kato SI, Diederich F (2010) Non-planar push-pull chromophores. Chem Commun 46:1994–2006
49. Michinobu T (2011) Adapting semiconducting polymer doping techniques to create new types of click postfunctionalization. Chem Soc Rev 40:2306–2316
50. Michinobu T, Li Y, Hyakutake T (2013) Polymeric ion sensors with multiple detection modes achieved by a new type of click chemistry reaction. Phys Chem Chem Phys 15:2623–2631
51. Kato SI, Kivala M, Schweizer WB, Boudon C, Gisselbrecht JP, Diederich F (2009) Origin of intense intramolecular charge-transfer interactions in nonplanar push-pull chromophores. Chem Eur J 15:8687–8691
52. Kivala M, Boudon C, Gisselbrecht JP, Seiler P, Gross M, Diederich F (2007) Charge-transfer chromophores by cycloaddition-retroelectrocyclization: multivalent systems and cascade reactions. Angew Chem Int Ed 46:6357–6360
53. Silvestri F, Jordan M, Howes K, Kivala M, Rivera-Fuentes P, Boudon C, Gisselbrecht JP, Schweizer WB, Seiler P, Chiu M, Diederich F (2011) Regular acyclic and macrocyclic [AB] oligomers by formation of push-pull chromophores in the chain-growth step. Chem Eur J 17:6088–6097

54. Kivala M, Boudon C, Gisselbrecht JP, Enko B, Seiler P, Müller IB, Langer N, Jarowski PD, Gescheidt G, Diederich F (2009) Organic super-acceptors with efficient intramolecular charge-transfer interactions by [2+2] cycloadditions of TCNE, TCNQ, and F_4-TCNQ to donor-substituted cyanoalkynes. Chem Eur J 15:4111–4123
55. Leliège A, Blanchard P, Rousseau T, Roncali J (2011) Triphenylamine/tetracyanobutadiene-based D-A-D π-conjugated systems as molecular donors for organic solar cells. Org Lett 13:3098–3101
56. Ohshita J, Kajihara T, Tanaka D, Ooyama Y (2014) Preparation of poly(disilanylenetetracyanobutadienyleneoligothienylene)s as new donor-acceptor type organosilicon polymers. J Organomet Chem 749:255–260
57. Michinobu T, Satoh N, Cai J, Li Y, Han L (2014) Novel design of organic donor-acceptor dyes without carboxylic acid anchoring groups for dye-sensitized solar cells. J Mater Chem C 2:3367–3372
58. Bruce MI, Rodgers JR, Snow MR, Swincer AG (1981) Cyclopentadienyl-ruthenium and –osmium chemistry. cleavage of tetracyanoethylene under mild conditions: X-ray crystal structures of $[Ru\{\eta^3\text{-}C(CN)_2CPhC=C(CN)_2\}(PPh_3)(\eta\text{-}C_5H_5)]$ and $[Ru\{C[=C(CN)_2]CPh=C(CN)_2\}(CNBu^t)(PPh_3)(\eta\text{-}C_5H_5)]$. J Chem Soc Chem Commun 17:271–272
59. Bruce MI (2013) Reactions of polycyano-alkenes with alkynyl- and poly-ynyl-Group 8 metal complexes. J Organomet Chem 730:3–19
60. Onitsuka K, Takahashi S (1995) Synthesis and structure of *s-cis-* and *s-trans*-μ-butadiene-2,3-diyldiplatinum complexes by the reaction of μ-ethynediyldiplatinum complexes with tetracyanoethylene. J Chem Soc Chem Commun 31:2095–2096
61. Onitsuka K, Ose N, Ozawa F, Takahashi S (1999) Reactions of acetylene-bridged diplatinum complexes with tetracyanoethylene. J Organomet Chem 578:169–177
62. Tchitchanov BH, Chiu M, Jordan M, Kivala M, Schweizer WB, Diederich F (2013) Platinum (II) acetylides in the formal [2+2] cycloaddition-retroelectrocyclization reaction: organodonor versus metal activation. Eur J Org Chem 2013:3729–3740
63. Sonogashira K, Morimoto H, Takai Y, Takahashi S (1981) Reactions of metal poly-yne polymers with free phosphines, coordinated phosphines, carbon monoxide, and tetracyanoethylene. Symp Organomet Chem Jpn 28:64–66
64. Yuan Y, Michinobu T (2012) Construction of donor-acceptor chromophores in platinum polyyne polymer by [2+2] cycloaddition of organic acceptors. Macromol Chem Phys 213:2114–2119
65. Yuan Y, Michinobu T, Oguma J, Kato T, Miyake K (2013) Attempted inversion of semiconducting features of platinum polyyne polymers: a new approach for all-polymer solar cells. Macromol Chem Phys 214:1465–1472
66. Michinobu T (2008) Click-type reaction of aromatic polyamines for improvement of thermal and optoelectronic properties. J Am Chem Soc 130:14074–14075
67. Michinobu T, Seo C, Noguchi K, Mori T (2012) Effects of click postfunctionalization on thermal stability and field effect transistor performances of aromatic polyamines. Polym Chem 3:1427–1435

Chapter 6
Transition-Metal Complexes for Triplet–Triplet Annihilation-Based Energy Upconversion

Xinglin Zhang, Tianshe Yang, Shujuan Liu, Qiang Zhao, and Wei Huang

Abstract In recent years, significant progress has been achieved in the field of triplet–triplet annihilation (TTA)-based energy upconversion, in which transition-metal complexes as the sensitizers play a key role. These complexes are different from organic fluorophores because the triplet excited states, instead of the singlet excited states, are populated with a high intersystem crossing (ISC) efficiency upon photoexcitation. Meanwhile, the long-lived triplet excited states in the microsecond range are observed for these complexes. All these properties are favorable when transition-metal complexes, including Ir(III), Pd(II), Pt(II), Ru(II), Zn(II), Re(I), Cu(I), and Au(III) complexes summarized herein, are used as sensitizers for TTA upconversion. Moreover, some examples of organic sensitizers and the applications of TTA upconversion systems are also summarized.

Keywords Transition-metal complexes • Sensitizers • Triplet–triplet annihilation • Energy upconversion • Solar cells • Bioimaging

X. Zhang • W. Huang (✉)
Key Laboratory for Organic Electronics and Information Displays and Institute of Advanced Materials, Jiangsu National Synergistic Innovation Center for Advanced Materials (SICAM), Nanjing University of Posts and Telecommunications (NUPT), Nanjing 210023, China

Key Laboratory of Flexible Electronics (KLOFE) and Institute of Advanced Materials (IAM), Jiangsu National Synergistic Innovation Center for Advanced Materials (SICAM), Nanjing Tech University (NanjingTech), 30 South Puzhu Road, Nanjing 211816, China
e-mail: wei-huang@njtech.edu.cn

T. Yang • S. Liu • Q. Zhao (✉)
Key Laboratory for Organic Electronics and Information Displays and Institute of Advanced Materials, Jiangsu National Synergistic Innovation Center for Advanced Materials (SICAM), Nanjing University of Posts and Telecommunications (NUPT), Nanjing 210023, China
e-mail: iamqzhao@njupt.edu.cn

W.-Y. Wong (ed.), *Organometallics and Related Molecules for Energy Conversion*, Green Chemistry and Sustainable Technology, DOI 10.1007/978-3-662-46054-2_6

6.1 Introduction

Energy upconversion, by which incident photons of low energy (long wavelength) are converted into photons of high energy (short wavelength), has attracted much attention due to its potential applications in solar cells [1], photocatalysis [2], and bioimaging [3, 4]. Currently, well-established approaches to upconversion include upconversion with two-photon absorption (TPA) dyes [5], lanthanide-doped upconversion nanoparticles [6], and triplet–triplet annihilation (TTA) [7]. TPA upconversion suffers from some fundamental drawbacks. For instance, a coherent laser with a high power density (typically 10^6 W cm^{-2}) is necessary for the excitation, which is far beyond the energy of normal light sources [8]. Moreover, tailoring the molecular structures of TPA dyes to achieve a specific upconversion wavelength and simultaneously maintaining a high TPA cross section are difficult. The upconversion process with rare earth materials has been developed with advantages of excitation at 980 nm, large anti-Stokes shift, high penetration depth, sharp emission bandwidths, as well as high resistance to photobleaching [9]. However, just as TPA upconversion, its excitation requires a coherent laser with a high power density. Meanwhile, the absorption of the rare earth ions (typically, the absorption cross section of Yb^{3+} is 1.2×10^{-20} cm^2 at 980 nm) is usually weak, resulting in a low overall upconversion quantum yield. Another approach is based on TTA, which can be performed by low power (a few mW cm^{-2}) and incoherent excitation sources, even solar light can be used as the excitation source. Though this technology was first introduced by Parker and Hatchard in the 1960s [10], it has developed very slowly for decades due to the low efficiency resulting from the poor intersystem crossing (ISC) yields of the organic sensitizers.

Most recently, the field of TTA-based upconversion has experienced a significant development by introducing various transition-metal complexes as sensitizer conjunct with appropriate acceptors spanning across the visible-to-near-infrared spectral region. And these combinations have successfully generated upconverted photons that can now be easily discerned with the naked eye [11]. Meanwhile, the excitation and emission wavelengths of TTA upconversion are tunable simply by independent selection of the triplet sensitizer and the acceptor.

Herein, the mechanism of TTA-based upconversion has been introduced. Next, the research progress of transition-metal complexes as sensitizers, including iridium(III), palladium(II), platinum(II), ruthenium(II), zinc(II), rhenium(I), copper(I), as well as gold(III) complexes, has been summarized. In addition, some examples of organic sensitizers as well as the applications of TTA-based upconversion are also presented.

6.2 TTA-Based Upconversion

Compared with other upconversion methods, mixing the sensitizer (donor) and the annihilator (acceptor) together is a unique characteristic of the TTA-based upconversion. Tunable excitation and emission wavelengths thus can be achieved simply by independent selection of the sensitizer as well as the annihilator during the upconversion processes. To get a full understanding of these processes, we should clarify the mechanism of TTA-based upconversion.

6.2.1 Mechanism

As illustrated in Fig. 6.1, the singlet excited state of the sensitizer will be firstly populated via selective excitation at long wavelength ($S_0 \rightarrow {}^*S_1$). Then with ISC in which the heavy atom (usually transition-metal atom) effect is often required, the triplet excited state of the sensitizer will be populated (${}^*S_1 \rightarrow {}^*T_1$). Herein, it is worth mentioning that the direct excitation from the ground state into the triplet excited state of the sensitizer ($S_0 \rightarrow {}^*T_1$) is forbidden. Since the lifetime of the triplet excited state is much longer, it is favorable for the energy transfer from the triplet sensitizer to the triplet annihilator via the triplet–triplet energy transfer (TTET) process. The energy transfer between the two triplet states is usually a Dexter process, which requires the contact of two molecules [12].

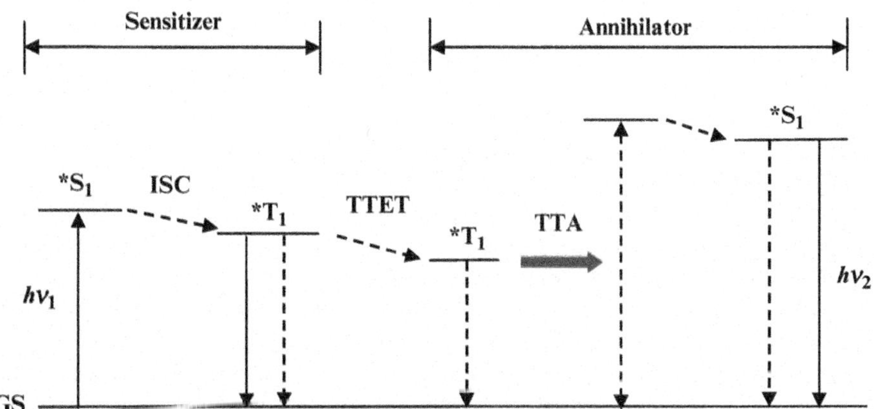

Fig. 6.1 Generalized Jablonski diagram illustrating processes involved in TTA-based upconversion between the sensitizer molecule and the annihilator molecule. Solid lines represent radiative processes. *Dashed lines* represent non-radiative processes. GS is ground state (S_0). ${}^*S_1$ is the singlet excited state. ${}^*T_1$ is the triplet excited state. ISC is intersystem crossing. TTET is triplet–triplet energy transfer, and TTA is triplet–triplet annihilation. $h\nu_1$ is the photon energy of excitation. $h\nu_2$ is the photon energy of emission

The triplet excited annihilator molecules collide with each other and the singlet excited state of the annihilator will be produced. The upconverted radiative fluorescence, the energy of which is higher than the excitation light, can be observed when the singlet excited state of annihilator molecules decays to the ground state. All these processes can be generalized by Eq. (6.1):

$$\begin{aligned} &S(S_0) + h\nu_1 \rightarrow S(^*S_1) \\ &S(^*S_1) \rightarrow S(^*T_1) \\ &S(^*T_1) + A(S_0) \rightarrow S(S_0) + A(^*T_1) \\ &A(^*T_1) + A(^*T_1) \rightarrow A(^*S_1) + A(S_0) \\ &A(^*S_1) \rightarrow A(S_0) + h\nu_2 \end{aligned} \tag{6.1}$$

where $S, A, S_0, {}^*S_1$, and ${}^*T_1$ represent the sensitizer, annihilator, ground state, singlet excited state, and triplet excited state, respectively. h is Planck's constant, and ν is the frequency of light, $\nu_2 > \nu_1$.

6.2.2 Sensitizer and Acceptor

The TTA-based upconversion involves the energy transfer from a sensitizer to an acceptor, the photophysical parameters of which are crucial in these processes. Apart from the ability of light absorption in the visible-to-near-infrared spectral region, which will utilize low energy for excitation, the sensitizer should have a high ISC efficiency near unity as well as a long lifetime of excited triplet state on the order of several microseconds or longer.

The transition-metal complexes can strongly enhance the spin–orbit coupling with singlet–triplet ISC efficiencies near unity and have a long-lived excited triplet state. These properties of transition-metal complexes are desirable for the sensitizer. Long-lived excited triplet state will lead to a further diffusion distance and increase the collision probability between the sensitizer and the acceptor in TTET process, which requires that the energy level of the triplet acceptor must be lower than that of the sensitizer.

In addition, the twofold triplet excited state energy level should be higher than or equal to the singlet excited state energy level of the acceptor, so that TTA process can be achieved. It is favorable to produce intense upconverted fluorescence emission for the efficient radiative decay of the singlet excited state of the acceptor, i.e., the acceptor with near-unity fluorescence quantum yield, because this value ultimately contributes to the overall upconversion quantum efficiency. Moreover, the annihilator molecule is chosen in consideration of its singlet excited state lying above that of the sensitizer's singlet manifold which means "up" in the conversion. Therefore, upconverted fluorescence can be observed from the sample when these design rationales are met and conditions are appropriate.

6.2.3 Upconversion Quantum Yield

TTA-based upconversion consists of the processes of excited-photon absorption, ISC, TTET, TTA, and the fluorescence emission of the acceptor. There are several calculation methods for upconversion quantum yield [7, 13–15].

As we know, TTA plays a key role in the upconversion processes. In order to get a maximum efficiency of TTA, we need to understand the mechanism of this crucial process. The acceptor molecules at the triplet excited state will collide with each other, producing the singlet excited state of the acceptor following the spin–statistics law as shown in Eq (6.2):

$$\begin{aligned} &{}^3A_1{*}+{}^3A_1{*} \leftrightarrow {}^5(AA)_2{*} \leftrightarrow {}^5A_2{*}+{}^1A_0 \\ &{}^3A_1{*}+{}^3A_1{*} \leftrightarrow {}^3(AA)_1{*} \leftrightarrow {}^3A_1{*}+{}^1A_0 \\ &{}^3A_1{*}+{}^3A_1{*} \leftrightarrow {}^1(AA)_0{*} \leftrightarrow {}^1A_0{*}+{}^1A_0 \end{aligned} \tag{6.2}$$

The spin manifold of the triplet encounters of the acceptors is governed by the spin–statistics, which has been reviewed by Saltiel et al. [16, 17]. When two excited triplets interact, the spin multiplicities in a diffusion-controlled environment result in nine encounter-pair spin states which are composed of three distinct sublevels, namely, five quintet, three triplet, and one singlet, and are produced with the equal probability. As we see, spin–statistics forecasts that singlet product represents 1/9 (or 11.1 %) of the TTA events:

$$\Phi_{UC} = \Phi_q \times \Phi_{TTA} \times \Phi_F \tag{6.3}$$

According to the report of Schmidt et al., upconversion quantum yield can be calculated by Eq. (6.3) [14]. In their difinition, two absorbed photons can be converted into one upconverted photon. If the quantum yield is defined as the ratio of output to input photons, the maximum is 50 %. Then the upper limit, based on the spin–statistics probability of forming the singlet excited annihilator (1/9), would be just 5.55 %, given that both the quantum yield of fluorescence and the efficiency of TTET from the sensitizer to the emitter are unity.

However, the triplets (and likely also the quintets) formed upon annihilation can be recycled, which are involved in the TTA again until they decay via the singlet pathway [17, 18]. Furthermore, the quintet state of polyaromatic organic compounds is often inaccessible [19]. The theoretical efficiency of the TTA step increases to 40 % if these recycling effects are taken into account [15]. In the experiments, the upconversion quantum yield (Φ_{UC}) can be determined by Eq. (6.4);

$$\Phi_{UC} = 2\Phi_{std}\left(\frac{A_{std}}{A_{unk}}\right)\left(\frac{I_{unk}}{I_{std}}\right)\left(\frac{\eta_{unk}}{\eta_{std}}\right)^2 \tag{6.4}$$

where Φ_{UC}, A_{unk}, I_{unk}, and η_{unk} represent the upconversion quantum yield, absorbance, integrated photoluminescence intensity of the samples, and the refractive index of the solvents, respectively. Since two photons are required to generate one upconverted photon, the equation is multiplied by a factor of 2 to keep the maximum quantum yield as a unity [7]. To date the highest value was reported to be 27.2 % [20].

Recently, Zhao and coworkers proposed that the light-harvesting ability of the sensitizer should be taken into account. It is more meaningful when overall upconversion yield η is evaluated by Eq. (6.5) [11]:

$$\eta = \varepsilon \times \Phi_{UC} \tag{6.5}$$

where ε is the molar extinction coefficient of the upconversion materials and Φ_{UC} is the upconversion quantum yield determined with Eq. (6.4). TTA-based upconversion with a large η values is more likely to be ideal for practical use.

6.3 Transition-Metal Complexes as Sensitizers

Transition-metal complexes are best candidates as sensitizers in the processes of TTA-based upconversion because the triplet excited states are populated with a high ISC efficiency when excited by light. And long-lived triplet excited states in the microsecond range are favorable for TTET process. Up to now, many kinds of transition-metal complex, such as iridium(III), palladium(II), platinum(II), ruthenium(II), zinc(II), rhenium(I), copper(I), and gold(III) complexes, have been successfully used as the sensitizers.

6.3.1 Ir(III) Complexes

In 2006, Castellano et al. used $Ir(ppy)_3$ (**Ir-1**, ppy = 2-phenylpyridine) as the sensitizer for TTA-based upconversion [21]. The energy level of triplet excited state of **Ir-1** was calculated to be 2.48 eV. Pyrene (**A-1**) and its derivative 3,8-di-tert-butylpyrene (**A-2**) were selected as the triplet annihilators due to the appropriate energy level of T_1 state (ca. 2.09 eV). In deaerated dichloromethane solution, upconverted singlet fluorescence at shorter wavelengths (360–420 nm) from the acceptors was observed with the excitation of 450 nm laser. However, the lifetime of T_1 excited state of **Ir-1** was only 1.55 μs, which was very short for the TTET process and disadvantageous to TTA upconversion.

Therefore, Ir(III) complexes with long-lived triplet excited states are expected. Zhao et al. designed and synthesized a series of visible-light-harvesting cyclometalated Ir(III) complexes (**Ir-2** ~ **Ir-5**) [22]. At room temperature, the emission lifetimes can be up to 75.5 and 73.6 μs for **Ir-4** and **Ir-5**, respectively.

Meanwhile, the energy levels of the T_1 states of **Ir-4** and **Ir-5** (ca. 2.06 eV) are higher than that of the acceptor 9,10-diphenylanthracene (DPA, **A-3**), which is 1.77 eV. With the excitation at 445 nm, the upconversion quantum yields were up to 21.3 % and 23.4 % for **Ir-4** and **Ir-5**, respectively. However, no upconversion was observed under the same experimental conditions for **Ir-2** and **Ir-3**, because of their shorter lifetimes of the T_1 states (0.77 μs for **Ir-2** and 0.68 μs for **Ir-3**) than those of **Ir-4** and **Ir-5** as well as the poor absorption at the excitation wavelength. Moreover, it was demonstrated that transition-metal complexes with a long-lived triplet excited state could be used as triplet sensitizers for TTA upconversion even if the phosphorescence is weak. And these Ir(III) complexes were further used as singlet oxygen sensitizers for photooxidation of 1,5-dihydroxynaphthalene [23].

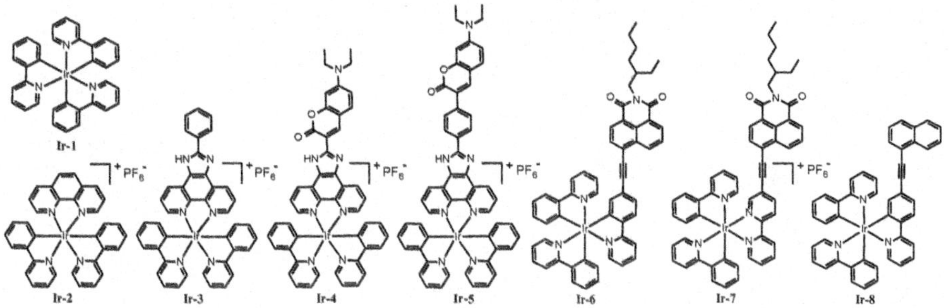

Besides, naphthalimide (NI) moiety was also introduced into the ligands to enhance the absorption of cyclometalated Ir(III) complexes. Two neutral (**Ir-6**) and cationic (**Ir-7**) complexes have been synthesized [24]. By contrast, Ir(III) complex with a naphthal ligand (**Ir-8**) and the model complex (**Ir-1**) were also prepared. In deaerated toluene solution, these complexes were used as triplet photosensitizers and DPA as an acceptor for TTA upconversion ($\lambda_{ex} = 445$ nm) with quantum yields of 14.4 % and 7.1 % for **Ir-6** and **Ir-8**, respectively, whereas the upconversion was negligible for **Ir-1** and **Ir-7**. The high upconversion quantum yield of **Ir-6** was attributed to the strong absorption of the complex at the excitation wavelength and the long-lived T_1 excited state (9.30 μs). The poor absorption of **Ir-1** at 445 nm and the short-lived T_1 excited state (1.34 μs) were responsible for the inefficient TTA upconversion. Similarly, for **Ir-7**, it could not be excited efficiently at 445 nm due to the weak absorption, in spite of a longer lifetime of T_1 excited state (16.45 μs).

The above examples are based on the short excitation wavelength of 445 or 450 nm. TTA upconversion with a longer excitation wavelength is very important. Ir(III) complexes with 3-(2-benzothiazoly)-7-diethylaminocoumarin as the C^N cyclometalated ligands and either 6-piperidine naphthalimide-phenanthroline (**Ir-9**) or 9-aminophenanthroline (**Ir-10**) as the ancillary N^N ligand were prepared and used as sensitizers for TTA upconversion in the presence of DPA under the excitation at 473 nm. Upconversion quantum yields were 19.3 % and 12.7 % for **Ir-9** and **Ir-10**, respectively [25]. Besides, pyreno[4,5-d]imidazole C^N ligand, which was used to access the long-lived T_1 excited state, and a coumarin-derived N^N ligand, which could enhance the absorption in visible range, were used simultaneously to prepare Ir(III) complex (**Ir-11**). In deaerated dichloromethane solution, upconversion quantum yield up to 23.7 % was observed with **Ir-11** as the sensitizer and DPA as the acceptor [26].

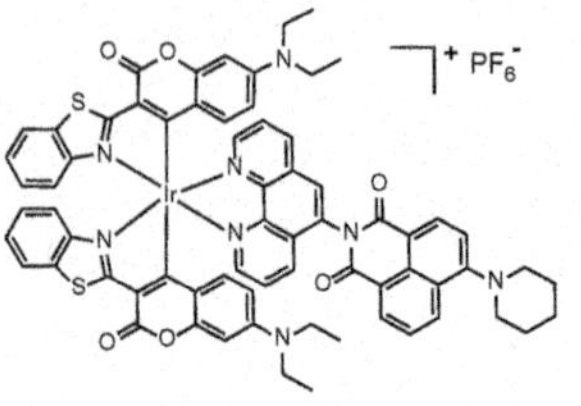

Ir-9 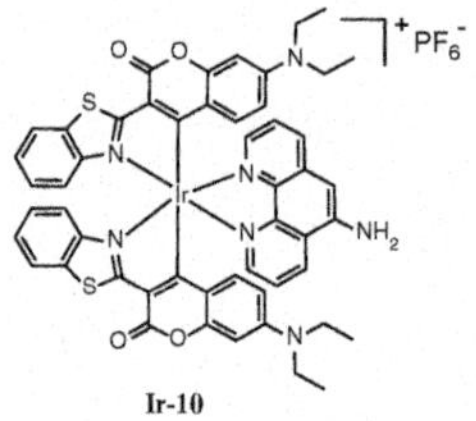 Ir-10 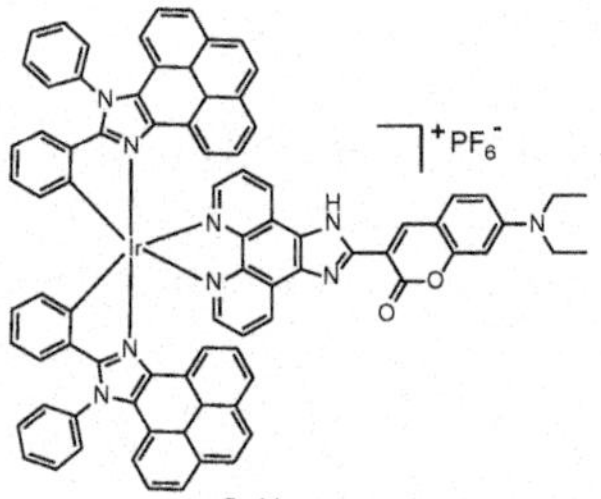 Ir-11

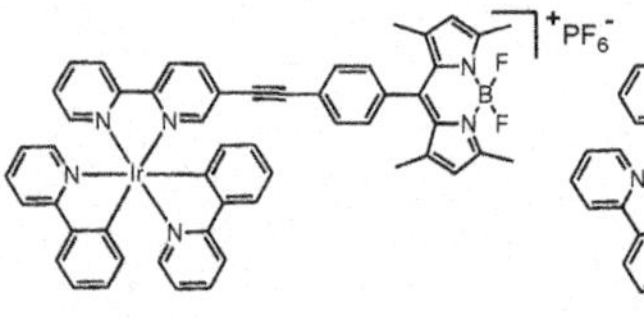

Ir-12 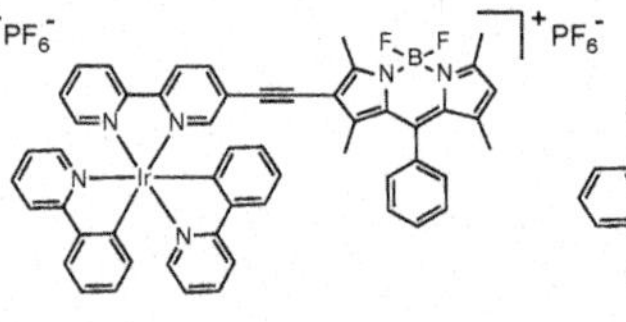 Ir-13 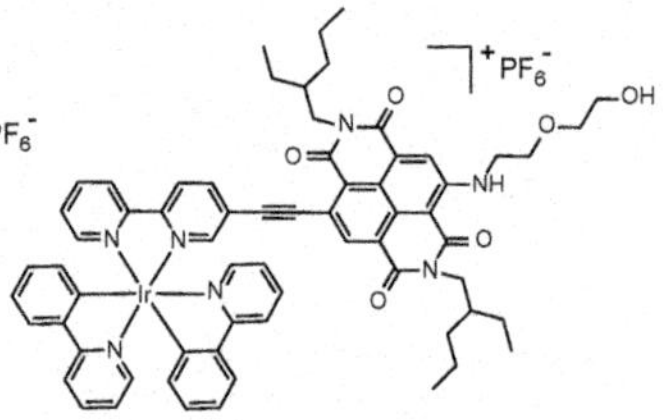 Ir-14

Cationic Ir(III) complexes as sensitizers for TTA upconversion with green excitation wavelength were also obtained, in which boron-dipyrromethene (Bodipy) units were attached to the 2,2′-bipyridine ligand via –C≡C– bonds at either the meso-phenyl (**Ir-12**) or 2-position of the π core of Bodipy (**Ir-13**) [27]. **Ir-12**, with excitation of 473 nm laser (5 mW) and perylene (**A-4**) as an acceptor in deaerated acetonitrile solution, exhibited TTA upconversion emission with $\Phi_{UC} = 1.2$ %. However, for **Ir-13**, the link of the π core of a visible-light-harvesting chromophore (Bodipy) to the Ir(III) coordination center was a very effective way to maximize the ISC effect. With the excitation of green laser ($\lambda = 532$ nm), blue emission in 445–470 nm range of perylene (acceptor) was observed, and the upconversion quantum yield was 2.8 % [27].

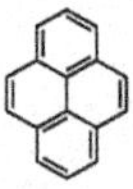

A-1 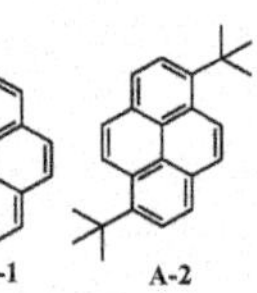 A-2 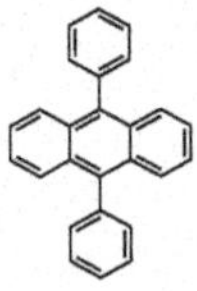 A-3 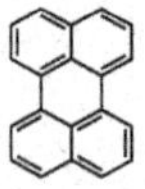 A-4  A-5 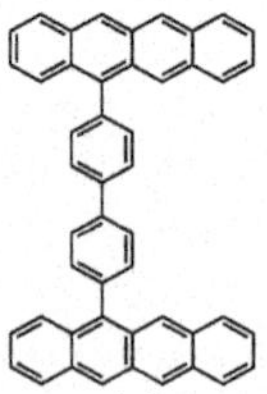 A-6 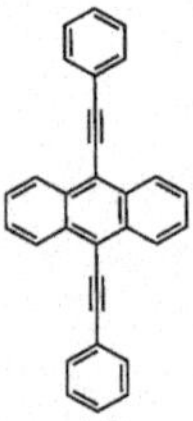 A-7 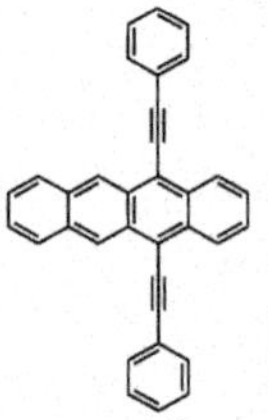 A-8  A-9

Similarly, naphthalenediimide (NDI) was also connected to the ligand of Ir(III) complex (**Ir-14**) via –C≡C– bond to enhance the absorption in the visible region and to access long-lived triplet excited states [28]. **Ir-14** showed strong absorption ($\varepsilon = 11{,}000\ M^{-1}\ cm^{-1}$ at 542 nm), and the lifetime of the NDI-localized triplet excited state was up to 130.0 μs. Using perylene as an acceptor and **Ir-14** as the sensitizer, TTA upconversion was observed upon the excitation at 532 nm (5.6 mW) with an upconversion quantum yield of 6.7 % [28].

6.3.2 Pd(II) Complexes

Pd(II) complexes can be excited with longer wavelength, such as green, red, or even near-infrared (NIR) light. Pd(II) complexes with porphyrin derivatives as ligands are appropriate candidates for TTA upconversion with the absorption in the green or longer wavelength range.

In 2006, Baluschev et al. reported the upconversion photoluminescence excited by ultralow ($10\ W\ cm^{-2}$) intensity from incoherent sunlight for the first time. The fluorescence of DPA was observed when a solution of DPA and 2 wt% PdOEP (**Pd-1**, OEP = octaethylporphyrin) was excited with wavelength of 550 nm and the external efficiency was 1 % [29].

To expand the excitation window for TTA upconversion, the ability to combine sensitizers (**Pd-2** and **Pd-3**) with single acceptor rubrene (**A-5**) was demonstrated. The integral upconverted fluorescence of the acceptor by simultaneous excitation at two wavelengths ($\lambda = 695$ nm for **Pd-2** and $\lambda = 635$ nm for **Pd-3**) was slightly more intense than the sum of the integral fluorescence for each single emitter/sensitizer system [30]. Then, **A-6** as an acceptor with the sensitizer **Pd-2** for TTA upconversion, a single-mode continuous-wave diode laser ($\lambda = 695$ nm) was used as the excitation source, and the anti-Stokes shift was about 0.7 eV. The upconversion process has an external quantum yield of 4 % [31].

The couples PdOEP/DPA ($\lambda_{ex} = 532$ nm, $50\ mW\ cm^{-2}$), **Pd-3/(A-7)** ($\lambda_{ex} = 635$ nm, $70\ mW\ cm^{-2}$), and **Pd-2/(A-8)** ($\lambda_{ex} = 695$ nm, $150\ mW\ cm^{-2}$) were used for TTA upconversion in deaerated toluene solutions, and the corresponding fluorescence of the acceptors was observed [32]. Besides, anthracene(**A-9**) as the energy acceptor and PdOEP as the triplet sensitizer in degassed toluene have been investigated for TTA upconversion recently [33]. The anthryl TTA quantum efficiency was up to 40 %, while the upconversion quantum yield was limited to 10 %, as the singlet fluorescence yield of anthracene is ca. 25 %. It is worth noting that the anthryl fluorescence was observed with incident irradiance as low as $0.6\ mW\ cm^{-2}$ ($\lambda_{ex} = 547$ nm) [33].

In 2008, upconverted yellow fluorescence from rubrene was observed with selective excitation ($\lambda_{ex} = 725$ nm) of the red-light-absorbing triplet sensitizer $PdPc(OBu)_8$ (**Pd-4**, $Pc(OBu)_8$ = octabutoxyphthalocyanine) in deaerated toluene [34]. The energy level of triplet sensitizer is 1.24 eV and that of rubrene is 1.14 eV, which met the requirements of TTET process.

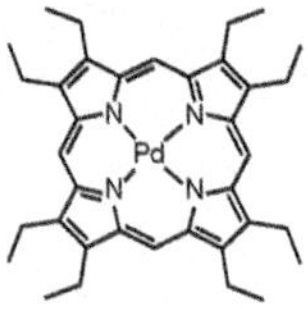

Pd-1 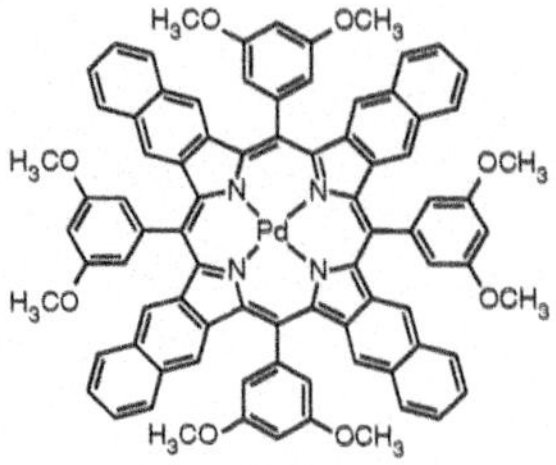 Pd-2  Pd-3 Pd-4

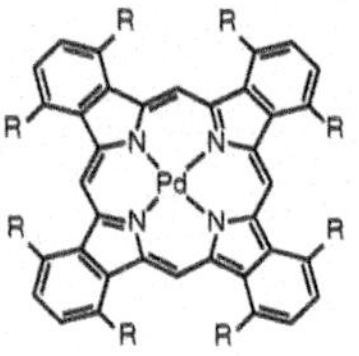

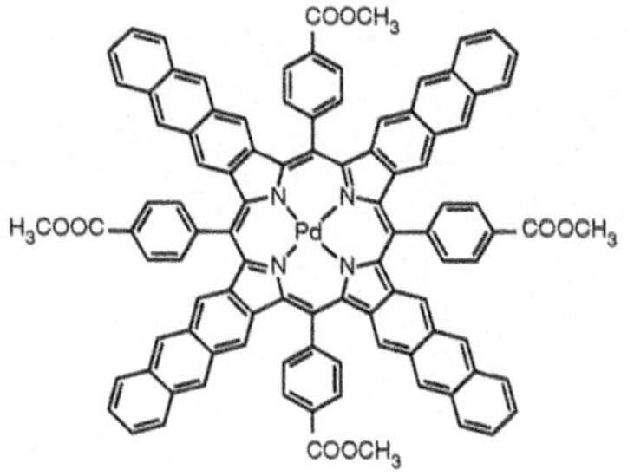

Pd-5 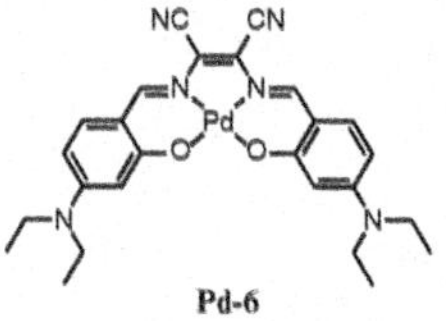 Pd-6 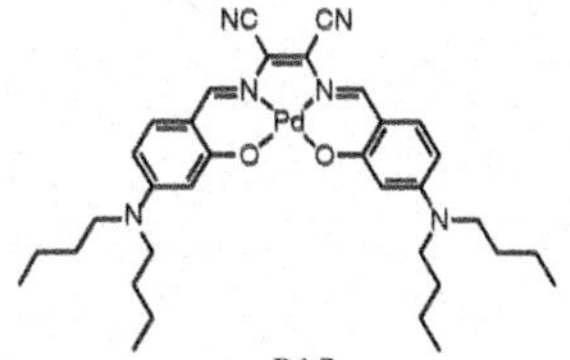 Pd-7

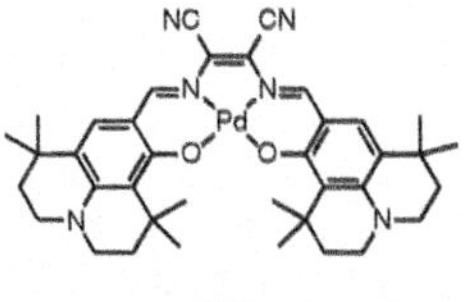

Pd-8 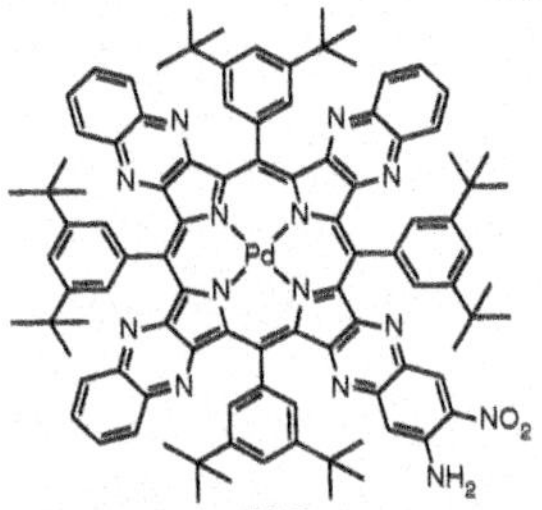 Pd-9

With **Pd-5** as the sensitizer and rubrene as the acceptor in deaerated toluene, the yellow emission of the rubrene was observed under excitation at 785 nm with low intensity (100 mW cm^{-2}), and the upconversion quantum yield was 1.2 % [35]. Moreover, in deoxygenated solutions, red-light-absorbing Pd(II) complexes with Schiff base ligands (**Pd-6**~**Pd-8**) as sensitizer and **A-10** as the acceptor, the upconversion quantum yields were estimated to be 2.0 %, 6.2 %, and 2.0 % for **Pd-6** ($\lambda_{ex} = 635$ nm), **Pd-7** ($\lambda_{ex} = 635$ nm), and **Pd-8** ($\lambda_{ex} = 605$ nm), respectively. For **Pd-8**, with **A-11** as the acceptor, yellow upconverted emission was generated [36].

Recently, TTA-based upconversion with the sensitizer (**Pd-3**)/acceptor (**A-12**) couple was reported [37]. And a new Pd(II) complex (**Pd-9**) was synthesized and can be used as the sensitizer [38]. In addition, TTA upconversion with Pd (II) complex (**Pd-1**) as sensitizer was observed in polyfluorene [39] and P (EO/EP) matrix [40], and the influence of temperature on upconversion in rubbery polymer blends was also investigated [41].

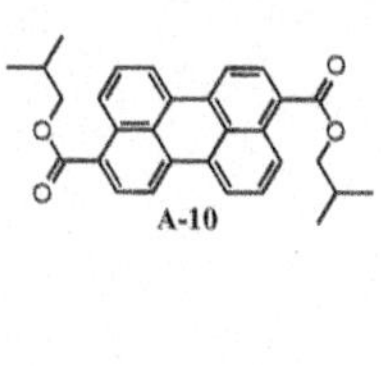

A-10 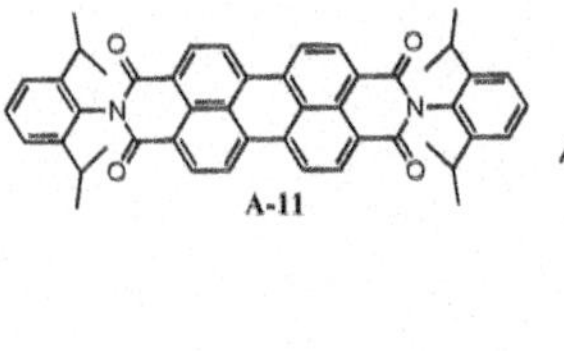 A-11 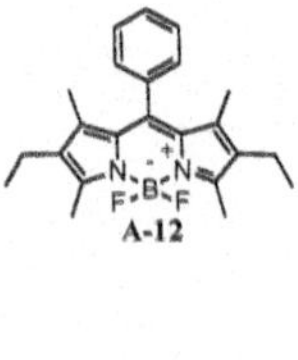 A-12 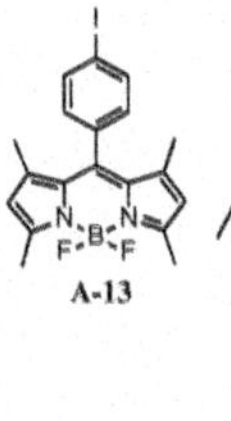 A-13 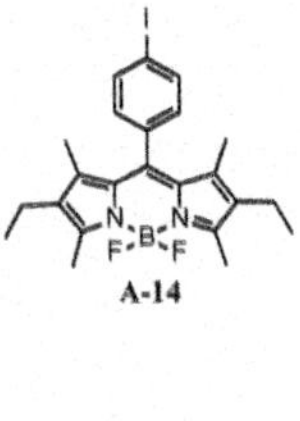 A-14

6.3.3 Pt(II) Complexes

Compared with Ir(III) and Pd(II) complexes, there are a lot of Pt(II) complexes used as triplet sensitizer, which include Pt(II) porphyrin complexes; N^N Pt(II) bisacetylide complexes; C^N Pt(II)(acac) complexes (C^N = cyclometalating ligand; acac = acetylacetonato); Pt(II) Schiff base complexes; N^N^N, N^C^N, and C^N^N cyclometalated Pt(II) complexes with functionalized acetylide ligands; and tributyl-phosphine Pt(II) bisacetylide complexes.

6.3.3.1 Pt(II) Porphyrin Complexes

Pt(II) porphyrin complexes, such as PtOEP (**Pt-1**, OEP = octaethylporphyrin), based on the capability of visible-light absorption and population of triplet excited state upon photoexcitation, have been widely used in luminescent oxygen sensing and photodynamic therapy. PtOEP was firstly used as sensitizer for TTA upconversion in a blue-emitting polymer matrix (PF2/6). When a system consisting of PF2/6 blended with 6 wt% PtOEP excited at 532 nm, the delayed fluorescence from the singlet state of the conjugated polymer matrix was observed. Meanwhile, the upconversion fluorescence was demonstrated to be a result of TTA upconversion [42]. Moreover, upconversion fluorescence was also observed in ladder-type pentaphenylene [43], spirobifluorene-anthracene copolymer [44], when doped with PtOEP. In 2010, PtOEP as a sensitizer and anthracene as an emitter were accumulated into the water-soluble G2 POSS-core dendrimer. TTA upconversion luminescence from visible light to UV light in aqueous solutions was observed with the excitation of 537 nm [45].

Pt(II) porphyrin complex PtTPBP (**Pt-2**, TPBP = tetraphenyltetrabenzoporphyrin) with an absorption in longer wavelength (red light) region is also used as a sensitizer for TTA upconversion. In 2008, with selective excitation of PtTPBP at the wavelength of 635 ± 5 nm, the green (**A-13**) and yellow (**A-14**) upconverted fluorescence of the acceptors were observed in deaerated benzene [46].

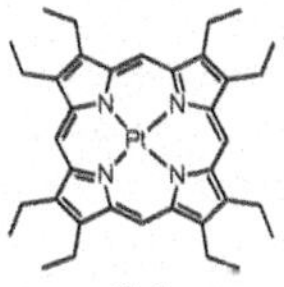

Pt-1 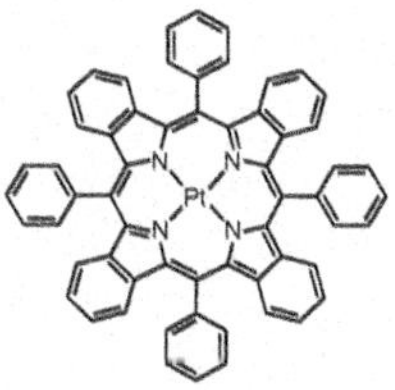 Pt-2 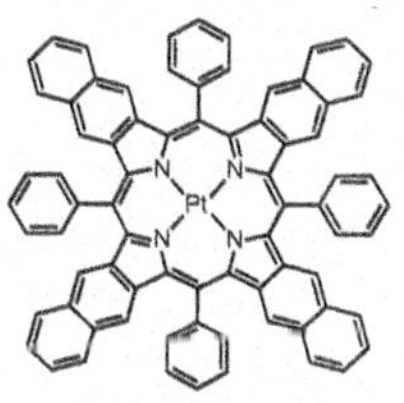 Pt-3 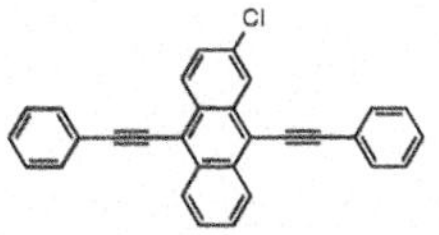 A-15 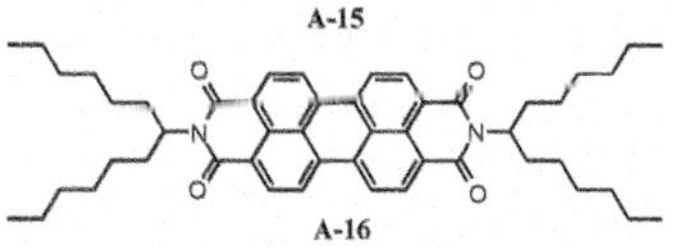 A-16

In the same way, blue-green upconverted emission (peaked at ca. 490 nm) from 2-chloro-bis-phenylethynylanthracene (**A-15**) sensitized by the red-light-absorbing PtTPBP at 635.5 nm was investigated in *N,N*-dimethylformamide (DMF) [47]. With perylene as an acceptor and PtTPBP as the triplet sensitizer, red-to-blue upconversion with anti-Stokes shift of 0.8 eV has been achieved in deaerated benzene. Perylene fluorescence centered at 451 nm was measured with selective excitation at 635 nm of PtTPBP [48]. Similarly, TTA upconversion could be achieved with the combination of PtTPBP (or PdTPBP) and perylene-DIPE (**A-10**) [49].

Recently, a Pt(II) porphyrin complex PtTPTNP (**Pt-3**, TPTNP = tetraphenyltetranaphtho[2, 3]porphyrin) with the longer wavelength absorption was used as sensitizer for TTA upconversion. With rubrene(**A-5**) or PDI (**A-16**, perylenediimide) as the triplet acceptors, yellow fluorescence could be observed with the excitation of 690 nm, and the overall efficiency was in the range of 6–7 % [50].

6.3.3.2 N^N Pt(II) Bisacetylide Complexes

N^N Pt(II) bisacetylide complexes are also used for TTA upconversion because of their efficient ISC process and tunable photophysical properties, such as absorption wavelength and phosphorescence quantum yields, by variation of the acetylide ligands.

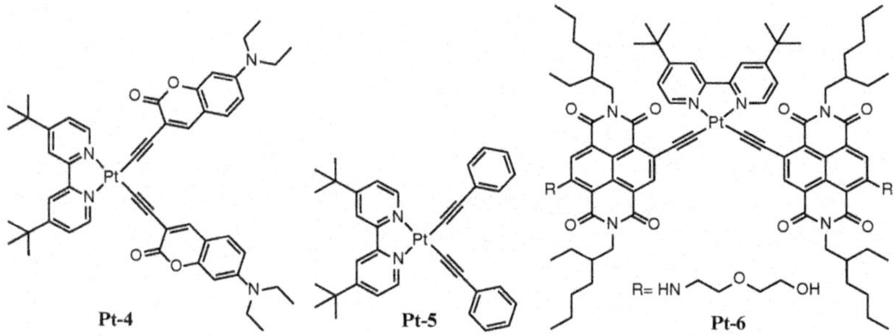

Pt-4 Pt-5 Pt-6

R= HN~O~OH

A dbbpy Pt(II) bis(coumarin acetylide) complex (**Pt-4**, dbbpy = 4,4′-di-*tert*-butyl-2,2′-bipyridine) and dbbpy Pt(II) bis(phenylacetylide) complex (**Pt-5**) were used as sensitizers for TTA upconversion with DPA as the acceptor. The upconversion quantum yields were 14.1 % and 8.9 % for **Pt-4** ($\lambda_{ex} = 474$ nm) and **Pt-5** ($\lambda_{ex} = 403$ nm), respectively [51].

In addition, naphthalenediimide derivative was chosen as acetylide ligand. Connected to Pt(II) center via acetylide, *N^N* Pt(II)bisacetylide NDI complex (**Pt-6**) was excited with 532 nm laser (5 mW) in the presence of perylene. Strong blue emission of perylene was observed in deaerated toluene, and the upconversion quantum yield was determined to be 9.5 % [52].

Moreover, *N^N* Pt(II) bisacetylide complex with the rhodamine fluorophore (**Pt-7**) was used as triplet sensitizer for TTA upconversion with perylene as the triplet acceptor. The upconversion quantum yield was 11.2 % [53]. Difluoroboron (BF_2)-bound phenylacetylide attached to the Pt(II) center of *N^N* Pt(II) bisacetylide (**Pt-8**) was also used as a sensitizer, and the upconversion quantum yield with DPA as the emitter was 8.9 % when excited with 445 nm laser in deaerated toluene [54].

N^N Pt(II) bisacetylide complexes with ethynylpyrene (**Pt-9**) or 4-ethynyl-1,8-naphthalimide (**Pt-10**) that showed long-lived triplet excited states (73.6 μs for **Pt-9** and 118.0 μs for **Pt-10**) were used as triplet sensitizers for TTA upconversion. With DPA as an acceptor, upon excitation with 445 nm laser (5 mW) in deaerated acetonitrile solution, the upconversion quantum yields of 14.2 % and 18.1 % were observed for **Pt-9** and **Pt-10**, respectively [55]. A dbbpy Pt(II) bisfluorescein acetylide complex (**Pt-11**) was used as the triplet sensitizer for TTA-based upconversion with the acceptor DPA. When excited by blue laser $\lambda_{ex} = 473$ nm (5 mW), the upconversion quantum yield was 10.7 % [56].

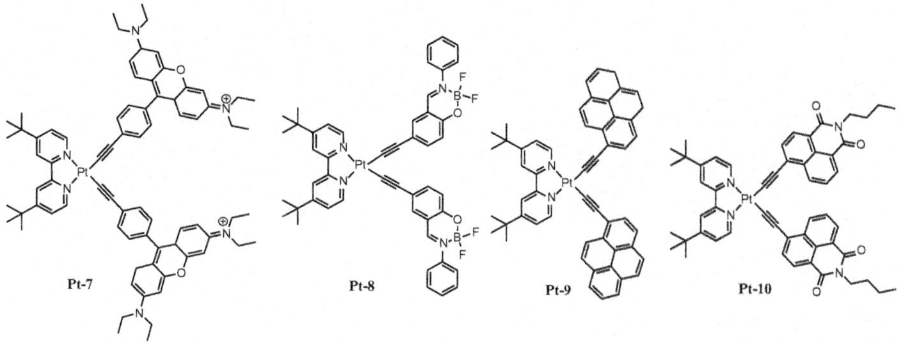

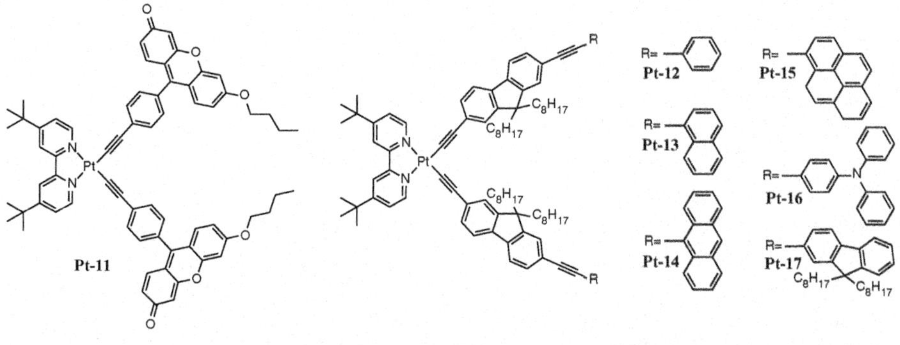

In order to investigate the influence of ligands on photophysical properties of N^N Pt(II) bisacetylide complexes, a series of fluorene-containing aryl acetylide ligands were used to the synthesis of *N^N* Pt(II) bisacetylide complexes, where aryl substituents on the fluorene were phenyl (**Pt-12**), naphthol (**Pt-13**), anthranyl (**Pt-14**), pyrenyl (**Pt-15**), 4-diphenylaminophenyl (**Pt-16**), and 9,9-di-*n*-octylfluorene (**Pt-17**). For **Pt-12**, **Pt-16**, and **Pt-17**, their T_1 excited states were assigned as metal-to-ligand charge-transfer state (^{3}MLCT), whereas those of **Pt-13**, **Pt-14**, and **Pt-15** were assigned as the intraligand state (^{3}IL) [57]. Upon 445 nm laser excitation, the fluorescence of DPA could be observed when **Pt-13** and **Pt-15** were used as sensitizers for TTA upconversion. The upconversion quantum yield was up to 22.4 % with **Pt-15**. The high upconversion quantum yield was attributed to its intense absorption in visible region and long-lived T_1 excited state [57].

In addition, **Pt-18** containing fluorene moiety showed much longer lifetime of triplet excited state (138.1 μs) than **Pt-19** with carbazole moiety (23.0 μs). Upon 473 nm excitation, both of them were used as triplet sensitizers, and upconversion quantum yields were up to 24.3 % and 14.7 % for **Pt-18** and **Pt-19**, respectively [58].

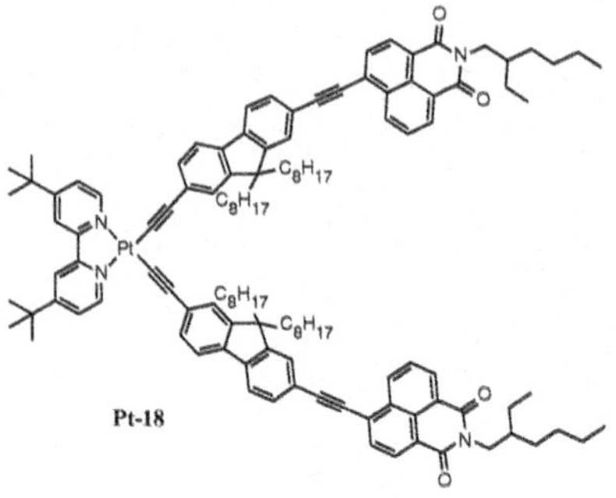

Pt-18 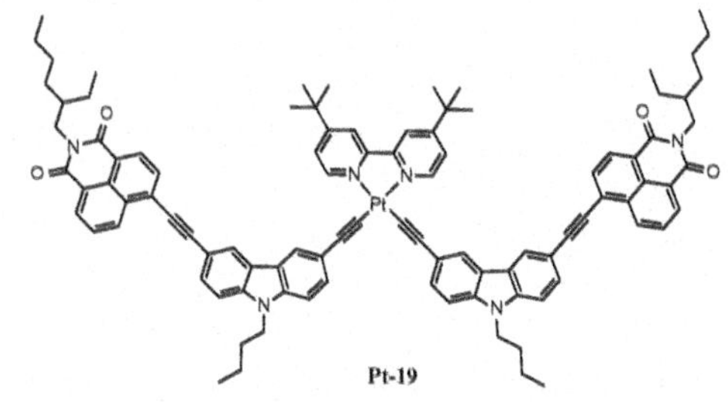 Pt-19

6.3.3.3 C^N Pt(II)(acac) Complexes

C^N Pt(II)(acac) complexes (C^N = cyclometalating ligand, acac = acetylacetonato) with different C^N ligands were used as triplet sensitizers for TTA upconversion. C^N cyclometalated Pt(II) complexes (**Pt-20, Pt-21** and **Pt-22**), in which the thiazocoumarin ligands were directly connected to Pt(II) center, were used as sensitizers for TTA upconversion. With 473 nm laser excitation, the upconversion quantum yields were 15.4 %, 5.3 %, and 2.8 % for **Pt-20, Pt-21**, and **Pt-22**, respectively. The reason of the remarkably high quantum yield was that **Pt-20** showed intense absorption in visible region, while other complexes showed the blue-shifted absorption [59].

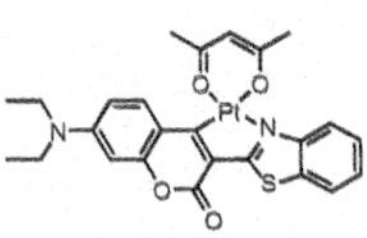

Pt-20 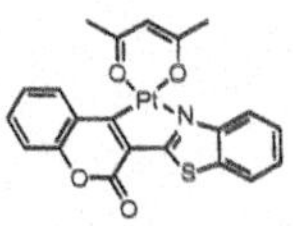 Pt-21 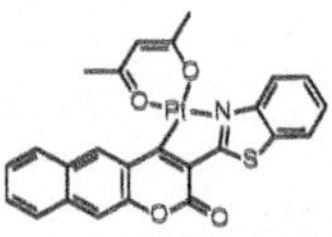 Pt-22 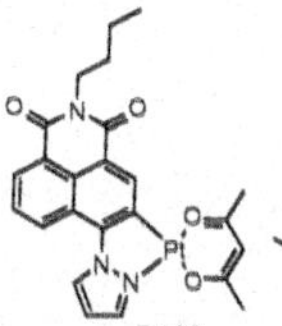 Pt-23 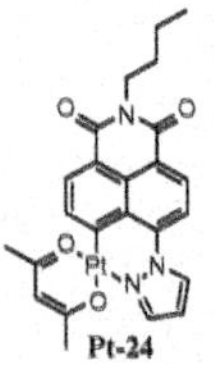 Pt-24 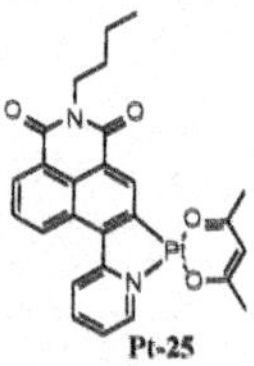 Pt-25

C^N Pt(II)(acac) complexes (**Pt-23 ~ Pt-25**) with naphthalimide (NI) moiety were also used as sensitizers. With DPA in solution, **Pt-24** was an efficient sensitizer with an upconversion quantum yield of 14.1 % due to its light-harvesting ability and long-lived ^{3}IL excited state [60].

6.3.3.4 Pt(II) Schiff Base Complexes

Pt(II) Schiff base complexes containing pyrene subunits (**Pt-26** and **Pt-27**) were used as sensitizers for TTA upconversion. The upconverted blue fluorescence of DPA was observed when excited by 532 nm laser in deaerated acetonitrile solution with the upconversion quantum yield of 9.9 % for **Pt-26** and 17.7 % for **Pt-27** [61]. Moreover, red-light-absorbing Pt(II) complexes with Schiff bases (**Pt-28** and **Pt-29**) were also investigated as sensitizer. With **A-10** as the acceptor in deoxygenated solution, the upconversion quantum yields were estimated to be 1.6 % and 2.0 % for **Pt-28** and **Pt-29** ($\lambda_{ex} = 635$ nm), respectively [36].

Pt-26 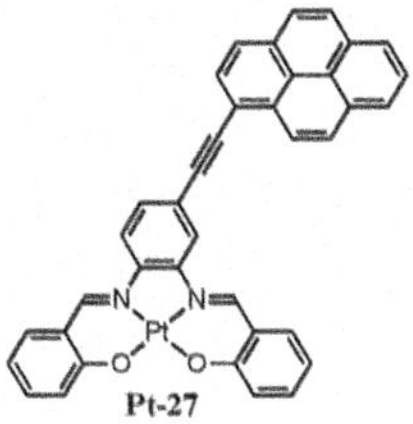 Pt-27 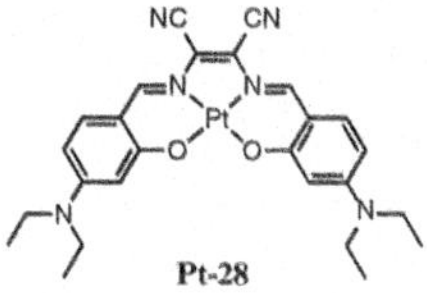 Pt-28 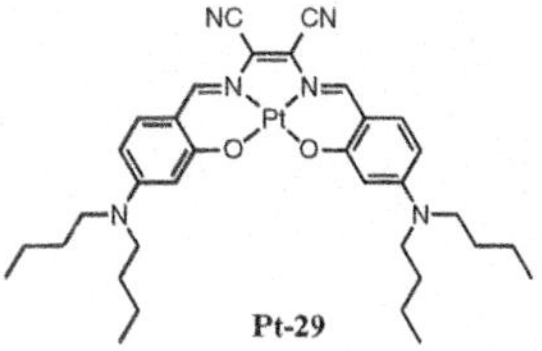 Pt-29

6.3.3.5 N^N^N, N^C^N, and C^N^N Cyclometalated Pt(II) Complexes with Functionalized Acetylide Ligands

Using N^N^N cyclometalated Pt(II) complex (**Pt-30**) with acetylide ligand as the sensitizer and DPA as the acceptor, upconverted blue fluorescence of DPA in the 400–470 nm region ($\lambda_{ex} = 514.5$ nm) was observed in deaerated dichloromethane solution [62]. For a long-wavelength excitation, *N^C^N* cyclometalated Pt (II) complex with acetylide ligands connected to the Bodipy moiety (**Pt-31**) was synthesized and used as triplet sensitizer for TTA upconversion. When excited with a red laser ($\lambda_{ex} = 635$ nm, 20 mW), an upconversion quantum yield of 5.2 % was observed with perylene as the acceptor in deaerated methanol solution [63].

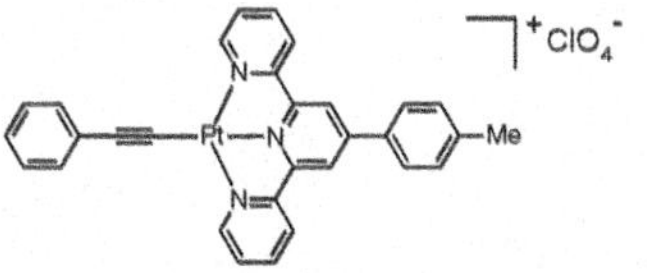

Pt-30 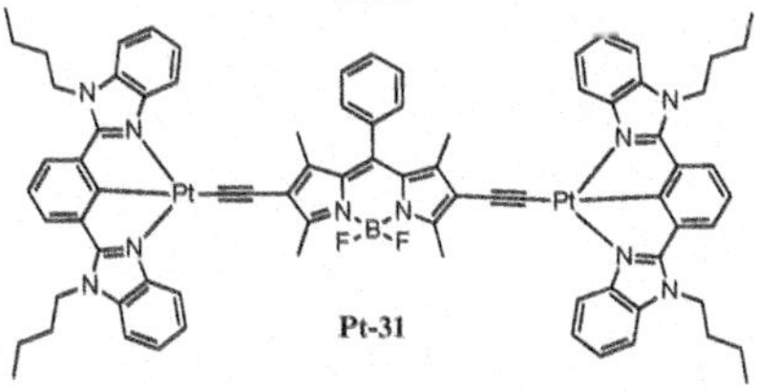 Pt-31 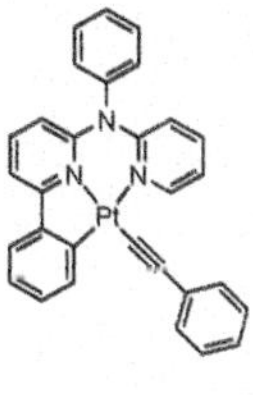 Pt-32 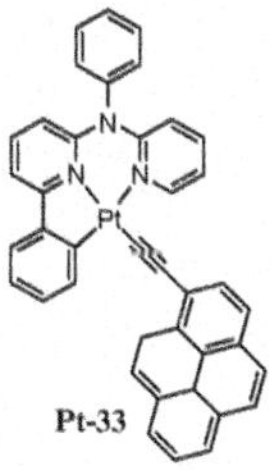 Pt-33 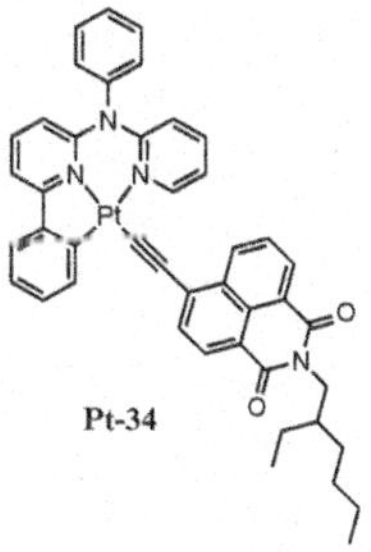 Pt-34

Besides, several C^N^N cyclometalated Pt(II) complexes with different aryl acetylide ligands were prepared, where aryl substituents were phenyl (**Pt-32**), pyrenyl (**Pt-33**), or naphthalimide (**Pt-34**) [64]. Excited at 445 nm in deaerated dichloromethane solution, upconversion quantum yield was 19.5 % with **Pt-34** as the triplet photosensitizer and DPA as the acceptor. No upconverted luminescence of DPA was observed for **Pt-32** and **Pt-33** under the same condition because of their poor absorption at the excitation wavelength [64].

6.3.3.6 Tributyl-Phosphine Pt(II) Bisacetylide Complexes

A series of Pt(II) bis(phosphine) bis(aryleneethynylene) complexes with Bodipy chromophore attached to the Pt(II) centers (**Pt-35** ~ **Pt-40**) were designed for TTA upconversion. With perylene as the acceptor and **Pt-35** ~ **Pt-38** as the sensitizers, the upconversion quantum yields ($\lambda_{ex} = 589$ nm) were determined to be 19.0 %, 3.0 %, 3.5 %, and 14.3 %, respectively. Moreover, with the excitation at 635 nm in the presence of PBI (**A-17**, perylenebisimide, $T_1 = 1.15$–1.2 eV), the TTA upconversion quantum yields were 5.8 % and 5.6 % for **Pt-36** and **Pt-37**, respectively. The higher upconversion quantum yields compared with perylene ($T_1 = 1.53$ eV) as the triplet acceptor may result from the larger driving force of the TTET between the triplet photosensitizer (**Pt-36** and **Pt-37**) and the triplet acceptor PBI. However, **Pt-39** and **Pt-40** showed absorption in the UV range and cannot be used as sensitizers [65].

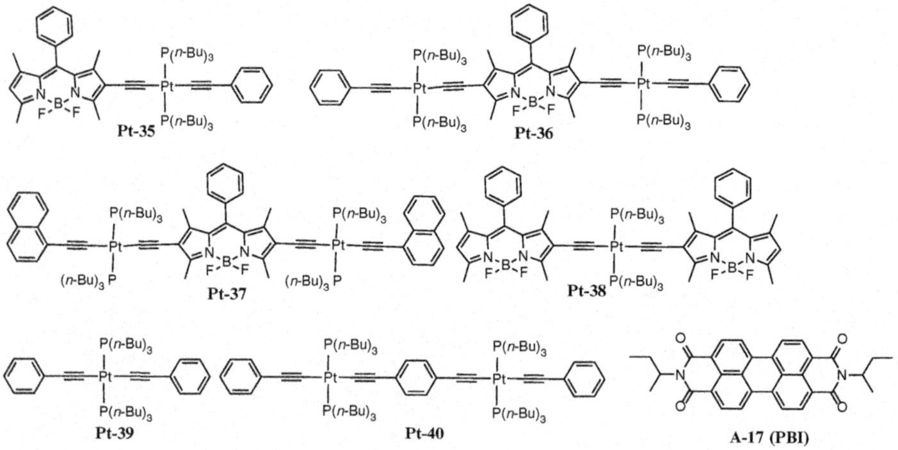

Pt-35 Pt-36 Pt-37 Pt-38 Pt-39 Pt-40 A-17 (PBI)

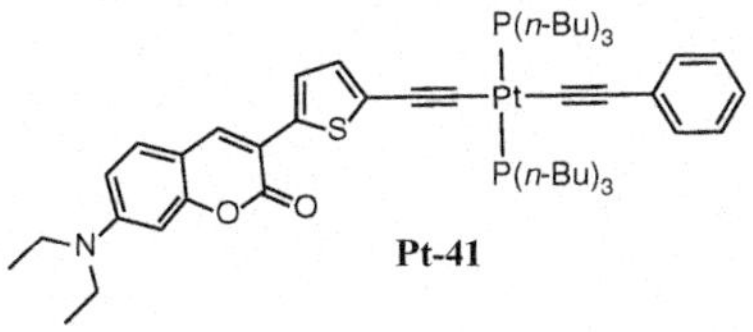

Pt-41 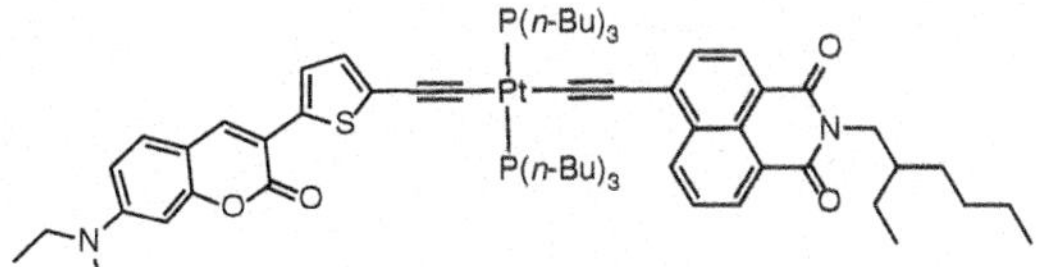 Pt-42

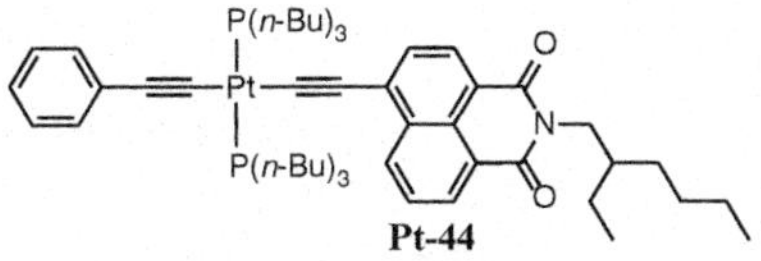

Pt-44 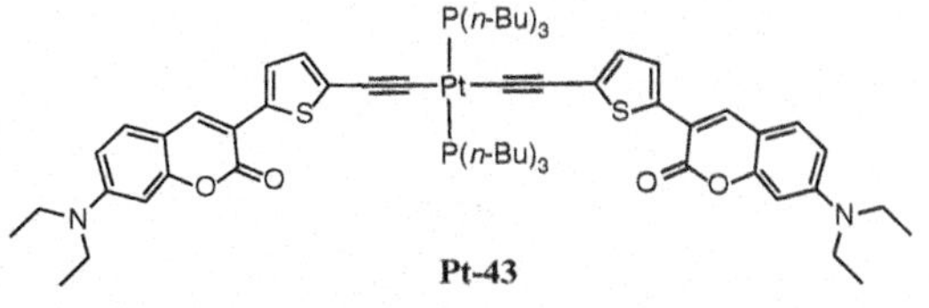 Pt-43

Recently, with DPA as the acceptor, **Pt-44** was used as the sensitizer for TTA upconversion ($\lambda_{ex} = 445$ nm, 5 mW) in deaerated toluene and the upconversion quantum yield was up to 27.2 %. The upconversion quantum yields were 17.1 % and 13.6 % when perylene was used as the acceptor for **Pt-42** and **Pt-43**, respectively. And no upconversion was observed for **Pt-41** under the same condition [20].

6.3.4 Ru(II) Complexes

In 2004, Castellano et al. reported that $[Ru(dmb)_2(bpy\text{-}An)]^{2+}$ (**Ru-1**, dmb is 4,4′-dimethyl-2,2′-bipyridine and bpy-An is 4-methyl-4′-(9-anthrylethyl)-2,2′-bipyridine) produced a reasonable anti-Stokes fluorescence in deaerated acetonitrile solution. However, the quantum yield for the process could not be measured because the method of standardization was unclear. Meanwhile, as a contrast, delayed fluorescence intensity was ~2.9-fold higher in the intermolecular system of $[Ru(dmb)_3]^{2+}$ (**Ru-2**) and anthracene than that for complex **Ru-1** as the sensitizer with the excitation at 450 ± 2 nm [66].

Ru-2 was also used as the sensitizer with two different acceptors DPA and anthracene for TTA upconversion. Excited by a commercial green laser pointer ($\lambda_{ex} = 532$ nm, 5 mW), DPA yielded visible upconverted fluorescence due to its high singlet fluorescence quantum yield (0.95) relative to anthracene (0.27) [67]. Moreover, DMA (**A-18**, 9,10-dimethylanthracene) was selected as the acceptor with the sensitizer **Ru-2** in DMF, and upconverted and downconverted DMA excimer photoluminescence were observed as the white light at the excitation of 514.5 nm [68]. A record anti-Stokes shift of 1.38 eV for sensitized TTA was demonstrated by Castellano et al. with the **Ru-2** and DPA for upconversion, which is the largest anti-Stokes shift, to the best of our knowledge, for every reported TTA-based upconversion. It should be noted that it was the Ti–Sapphire laser where the 860 nm light pulsed that played the critical role in this experiment [69]. Ru(II) polyimine complexes (**Ru-3** ~ **Ru-6**) with a long-lived ^{3}IL excited state or a ^{3}MLCT/^{3}IL equilibrium were used as sensitizers for TTA upconversion. With DPA as the acceptor, the upconversion yields ($\lambda_{ex} = 473$ nm) were 0.9 %, 4.5 %, 9.8 %, and 9.6 % for **Ru-3**, **Ru-4**, **Ru-5**, and **Ru-6**, respectively [70].

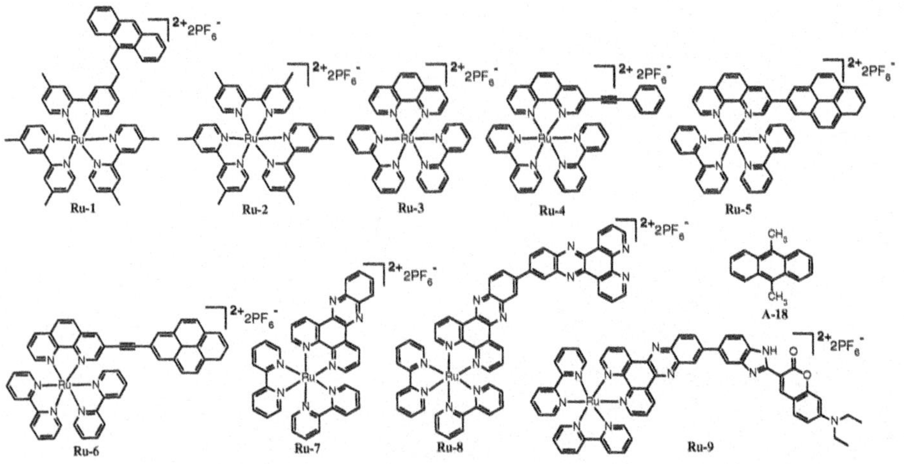

Ru(II) complexes with non-emissive excited state can be used for TTA upconversion. For example, Zhao et al. reported that Ru(II) polyimine–coumarin complexes (**Ru-7** ~ **Ru-9**) with non-emissive ^{3}IL excited state were used as sensitizers and DPA as the acceptor in deaerated acetonitrile solution, and the upconversion yields ($\lambda_{ex} = 473$ nm) were 0.4 %, 1.3 %, and 2.7 % for **Ru-7**, **Ru-8**, and **Ru-9**, respectively [71].

In order to increase the ability of visible-light absorption, a series of Ru (II) complexes (**Ru-10** ~ **Ru-13**) were designed, prepared, and used as triplet sensitizers for TTA upconversion. With DPA as the acceptor, the quantum yield ($\lambda_{ex} = 473$ nm) of **Ru-11** was 15.2 %. Due to the relative low absorption at excitation wavelength, the upconversion quantum yields were 0.95 %, 2.7 %, and 11.3 % for **Ru-10**, **Ru-12**, and **Ru-13**, respectively [72].

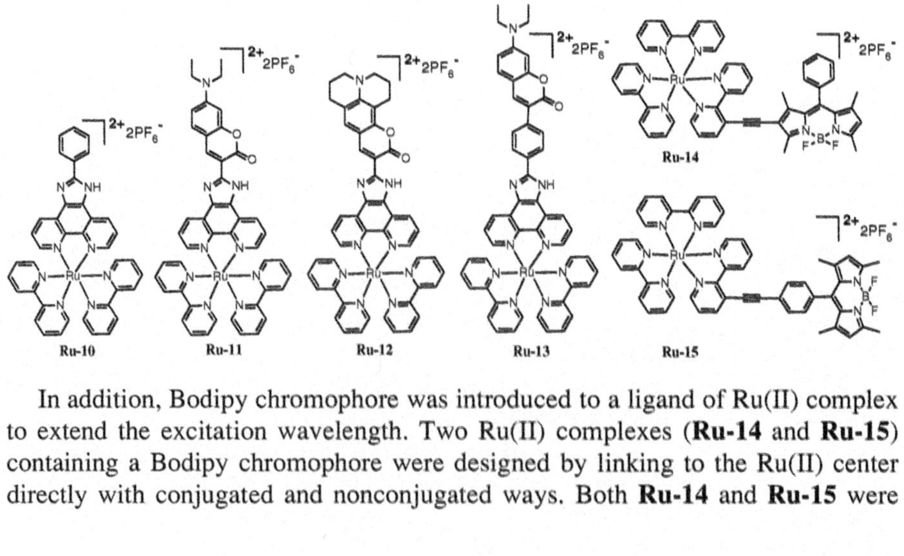

In addition, Bodipy chromophore was introduced to a ligand of Ru(II) complex to extend the excitation wavelength. Two Ru(II) complexes (**Ru-14** and **Ru-15**) containing a Bodipy chromophore were designed by linking to the Ru(II) center directly with conjugated and nonconjugated ways. Both **Ru-14** and **Ru-15** were used as sensitizers with perylene as the acceptor for TTA upconversion, and upconversion quantum yields ($\lambda_{ex} = 532$ nm) were 1.2 % and 0.7 %, respectively. Meanwhile, the $^1O_2$ quantum yields of the triplet photosensitizers were determined to be 0.93 for **Ru-14** and 0.64 for **Ru-15**. The higher quantum yield for **Ru-14** was attributed to the favorable ISC resulting from π-conjugation between the coordination center and Bodipy [73].

6.3.5 Zn(II) Complexes

Zinc tetraphenylporphine (**Zn-1**, ZnTPP) was the first Zn(II) complex used as the triplet sensitizer with perylene and coumarin 343 (**A-19**, C343) as the acceptors for TTA upconversion. For the acceptor perylene, the triplet state (1.53 eV) is lower than that of the sensitizer (1.56 eV). TTET occurred with a high efficiency, and the upconversion was observed when excited at 532 nm in degassed benzene. However, for coumarin 343, the triplet state (2.06 eV) lies at a higher energy than that of the sensitizer **Zn-1**, resulting in a different energy transfer mechanism. Ground-state C343 complexes with the ZnTPP triplet formed a triplet exciplex and then underwent TTA with another ZnTPP triplet to produce the fluorescent state of the acceptor in a three-center process [74].

Zn–Ru complex (**Zn-2**) was synthesized as the sensitizer with different acceptors of PDI (**A-20**) and tetracene (**A-21**) for TTA upconversion. When excited at 780 nm, the fluorescence of PDI and tetracene was observed at 541 nm and 505 nm, respectively [75].

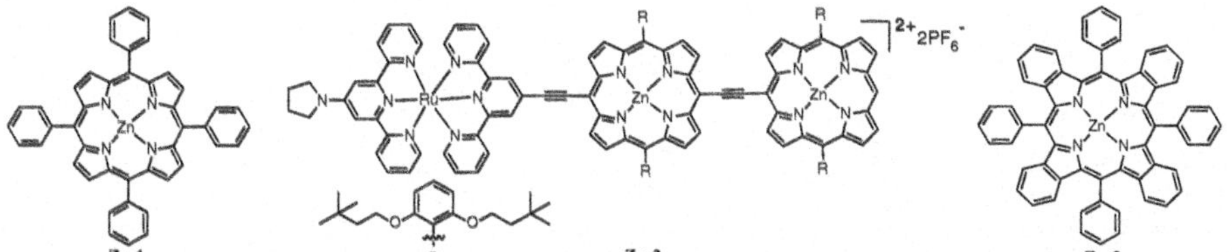

Zn-1 Zn-2 Zn-3

Besides, red-light-absorbing Zn(II) tetraphenyltetrabenzoporphyrin (**Zn-3**, ZnTPTBP) has been used as triplet photosensitizers for TTA upconversion recently. With **A-22** and perylene as the acceptors, the upconversion quantum yields were 0.76 % and 0.32 % when excited at 654 nm in deaerated toluene, respectively [76].

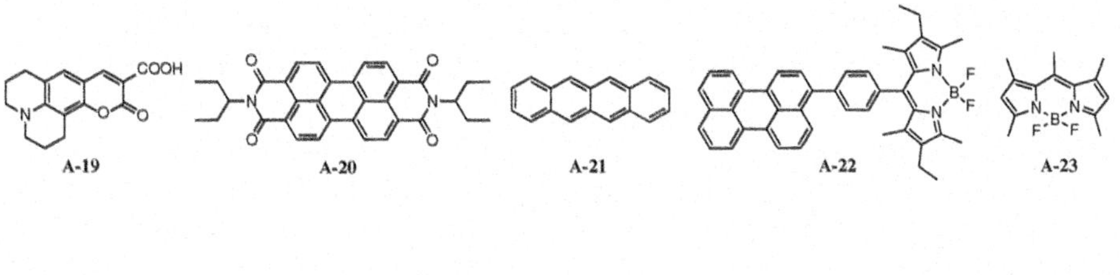

A-19 A-20 A-21 A-22 A-23

6.3.6 Re(I) Complexes

In 2012, three Re(I) complexes, **Re-1**, **Re-2**, and **Re-3**, were used as sensitizers for TTA upconversion. Under excitation at 473 nm and with DPA as the acceptor, the upconversion quantum yields were 17.0 % and 16.9 % for **Re-2** and **Re-3**, respectively, due to their long-lived T_1 states (lifetime was up to 86.0 μs and 64.0 μs, respectively) as well as their strong absorption of visible light. No upconversion was observed for **Re-1** because of the weak absorption and short-lived triplet excited states [77].

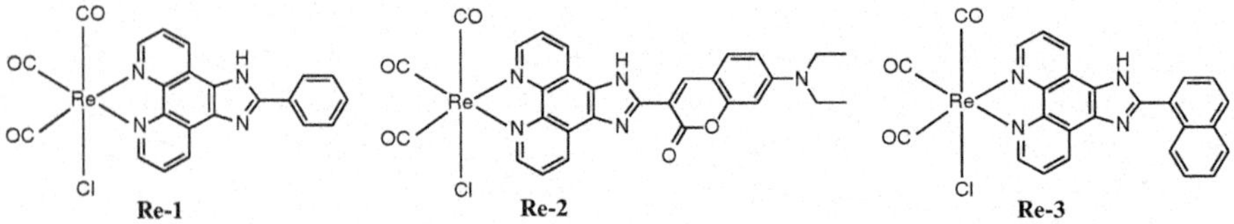

CO OC Re N H N N N OC Cl **Re-1** CO OC Re N H N N N O O OC Cl **Re-2** CO OC Re N H N N N OC Cl **Re-3**

The absorption of Re(I) complexes with coumarin–imidazole phenanthroline ligands is limited in the blue range (<500 nm). **Re-4** (with Bodipy) and **Re-5** (with carbazole-ethynyl Bodipy) showed strong absorption of visible light at 536 nm and 574 nm, respectively. In deaerated toluene, perylene was used as the acceptor, and **Re-4** and **Re-5** were used as sensitizers for TTA upconversion. The upconversion quantum yields ($\lambda_{ex} = 532$ nm) were 8.5 % and 2.0 % for **Re-4** and **Re-5**, respectively [78].

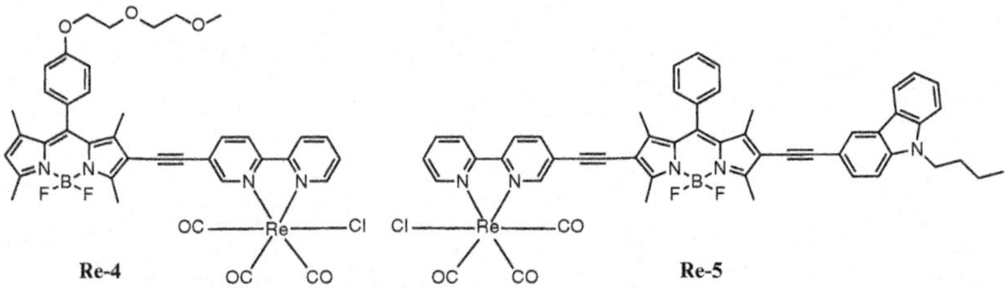

O O O N B N F F N N OC Re Cl OC CO **Re-4** Cl Re CO OC CO N N N B N F F N **Re-5**

6.3.7 Cu(I) Complexes

Cu(I) complexes are a new class of sensitizers. In 2013, bis-(2,9-diphenyl-1,10-phenanthroline) Cu(I) complex (**Cu-1**) was successfully used as a sensitizer with two different acceptors in TTA upconversion. For **Cu-1** and perylene, when excited at 593.5 nm, blue emission of perylene was observed. Meanwhile, **Cu-1** and pyrromethene 546 (**A-23**) accomplished a red-to-green wavelength shift with the excitation of 635 nm. It should be noted that, to access the obvious emission of the acceptors in these experiments, significant incident power density (about or over 1 W/cm^2) was required, which means that it could not be achieved in real-world solar-driven applications. As a result, the upconversion quantum efficiencies were not evaluated [79].

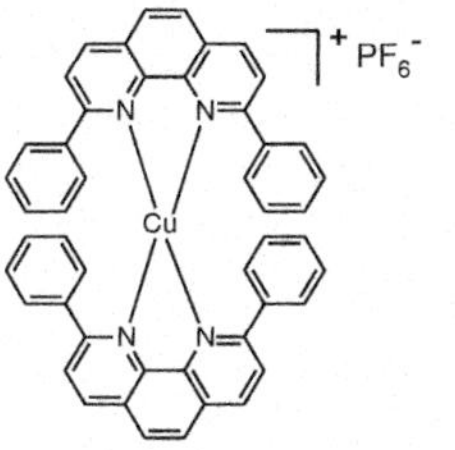

R^1 = *n*Hex
R^2 = 3,5-(*t*Bu)$_2$
R^3 = –≡–*p*-(OMe)C_6H_4

Cu-1 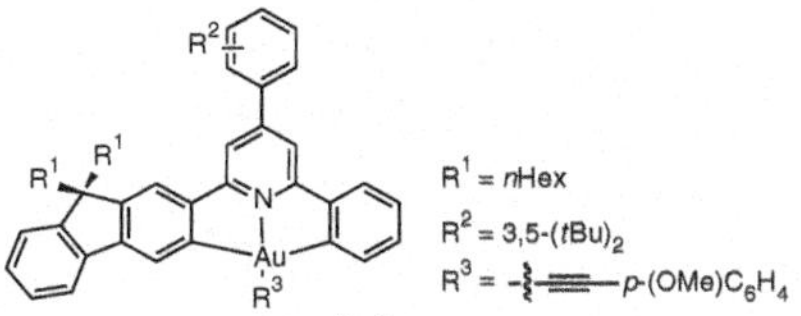 **Au-1**

6.3.8 Au(III) Complexes

A series of Au(III) complexes with tridentate cyclometalated *C^N^C* ligands were prepared with different substituent groups, R^1, R^2, and R^3 [80]. Herein, we take **Au-1** as an example of all these complexes. Firstly, both prompt and delayed fluorescence of DPA and residual phosphorescence of **Au-1** were observed when a mixture of **Au-1** and DPA in deaerated dichloromethane was excited with a Nd: YAG laser (355 nm). Excitation of either **Au-1** or DPA alone under the same conditions did not produce any delayed fluorescence. Moreover, it was demonstrated that the delayed fluorescence was generated by a two-photon process. In addition, the mixture of **Au-1** and DPA in deaerated dichloromethane was excited with a laser light at 476 nm, the fluorescence of DPA was also generated, and TTA upconversion was further confirmed with the upconversion quantum yield of 9.8 %.

Meanwhile, **Au-1** displayed TPA property and two-photon absorption cross section was estimated to be 9.4 GM. Thus, a long-wavelength laser (756 nm) was selected for the excitation of **Au-1** and DPA in deaerated dichloromethane, and the blue fluorescence of DPA was observed. This blue emission could not be observed with the absence of **Au-1** in the control experiment. It should be noted that a remarkable anti-Stokes shift of 1.36 eV was achieved by combining TPA and TTA [80].

6.4 Organic Sensitizers

All the transition-metal complexes discussed above have the advantage of high efficient ISC. Moreover, some organic chromophores can also exhibit ISC with high efficiency. These organic compounds can be used as triplet sensitizers in TTA upconversion. According to the different ways to produce the triplet excited states, organic chromophores as the sensitizers can be divided into three classes, namely, those based on the heavy atom effect, S_{n,π^*} to T_{π,π^*} transitions, and intramolecular spin converter. Some examples of these organic sensitizers (OS) are presented as follows.

6.4.1 *Organic Sensitizers Based on the Heavy Atom Effect*

Organic molecules with the atoms of large atomic number can induce strong spin–orbit coupling, which can enhance ISC process [81]. As a result, iodo- or bromo-substituted organic molecules can be used as triplet sensitizers for TTA upconversion. Some examples are listed as follows: **OS-1** [82], **OS-2** ~ **OS-8** [83], **OS-9** and **OS-10** [84], **OS-11** [85], and **OS-12** ~ **OS-14** [86].

6.4.2 *Organic Sensitizers Based on* S_{n,π^*} *to* T_{π,π^*} *Transitions*

The intersystem crossing rate constant is extremely fast (ca.10^{10} sec^{-1}) in some aromatic carbonyls whose lowest singlet is an n,π* state below which there is a triplet π,π* state, according to El-Sayed rule [87]. The transition of S_{n,π^*} to T_{π,π^*} is allowed in these organic molecules, such as **OS-15** [88], **OS-16** [89], and

OS-17 ~ **OS-19** [90], and these compounds can be used as triplet sensitizers in TTA-based upconversion.

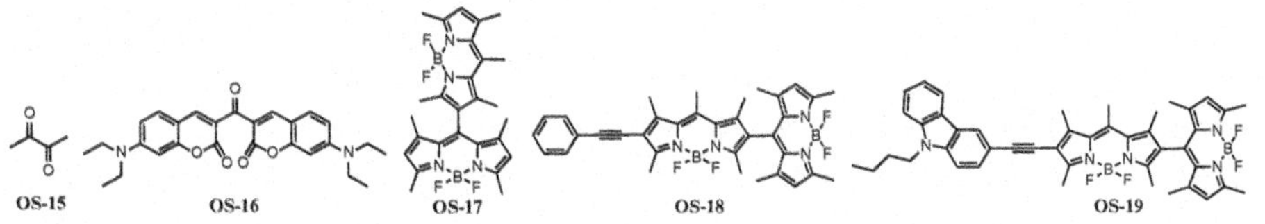

OS-15 OS-16 OS-17 OS-18 OS-19

6.4.3 Organic Sensitizers Based on Intramolecular Spin Converter

As we know, it is difficult to predict the ISC property for organic chromophores without the heavy atom effect. However, the ISC of these heavy atom-free chromophores is predictable when C_{60} is used as a spin converter. Recently, light-harvesting fullerene dyads as organic triplet sensitizers for TTA upconversion were reported. Some examples are listed as follows: **OS-20** and **OS-21** [91], **OS-22** and **OS-23** [92], **OS-24** and **OS-25** [93], and **OS-26** and **OS-27** [94].

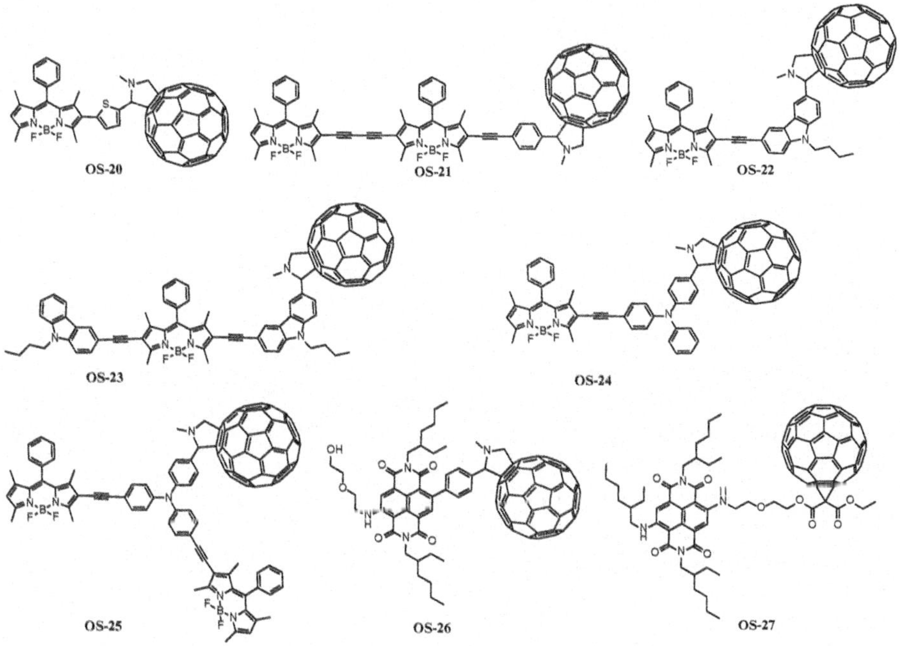

OS-20 OS-21 OS-22

OS-23 OS-24

OS-25 OS-26 OS-27

6.5 Applications

The most significant feature of TTA-based upconversion is that long-wavelength photons can be converted to short-wavelength photons. The produced high-energy photons can be further utilized in solar cells. Alternatively, with long-wavelength photoexcitation, the TTA-based upconversion also exhibits promising application in the field of bioimaging.

6.5.1 *Solar Cells*

It is difficult for single-threshold solar cells to harvest more than about 30 % of the energy available from the sun (unconcentrated). The reason, first derived by Shockley and Queisser [95], is that it cannot harvest photons with energies below the threshold. As a result, harvesting below-threshold photons and reradiating this energy at a shorter wavelength via upconversion would boost the efficiency of such devices.

A hydrogenated amorphous silicon (a-Si:H) thin-film solar cell with the increased light-harvesting efficiency via TTA upconversion was reported in 2012 [96]. Low-energy light in 600–750 nm range was converted to 550–600 nm light when two palladium porphyrins, PQ_4Pd and PQ_4PdNA (**Pd-9**), and rubrene (**A-5**) were dissolved in toluene. And a peak efficiency enhancement of $(1.0 \pm 0.2)\%$ at 720 nm was measured with irradiation equivalent to (48 ± 3) suns (AM1.5) [96].

Based on TTA upconversion with **Pd-9** and **A-5** used as sensitizer and acceptor, respectively, two types of organic solar cells and one amorphous silicon (a-Si:H) solar cell were prepared. It was demonstrated that a solar cell photocurrent was increased up to 0.2 % under a moderate concentration (19 suns), and it was the first report about the behavior of the organic solar cells improved by an upconverting process [1].

Recently, an integrated photovoltaic device combining a dye-sensitized solar cell and TTA upconversion system was reported. After TTA process, the excited singlet emitter returned back to the ground state via emission of a higher energy upconverted photon, which was captured by the D149 dye, and this energy was further utilized by the solar cell (as shown in Fig. 6.2). Thus, the integrated device displayed the enhanced current with sub-bandgap illumination, resulting in a figure of merit (FoM) under low concentration (3 suns), which was competitive, according to the authors, with the best values recorded to date for nonintegrated systems [38].

6.5.2 *Bioimaging*

The TTA-based upconversion bioimaging can definitely open new perspectives for upconversion luminescence materials. The fundamental advantage of the

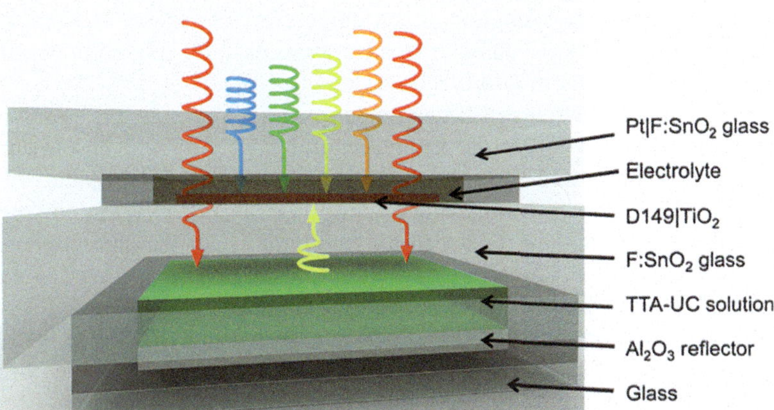

Fig. 6.2 A schematic showing the structure of the integrated device: low-energy photons pass through the active layer of the device and cause TTA upconversion in the TTA upconversion layer. Upconverted photons that are absorbed by the active layer provide extra current to the device (Reprinted with permission from Ref. [38]. Copyright 2013 American Chemical Society)

TTA-based upconversion in the field of the biological applications is the extremely low excitation intensity and high quantum yield compared to the rare earth-based UCL. As a result, the low excitation laser can lead to the harmless to the living organisms. An additional advantage of the TTA-based upconversion is the fact that the absorption spectra of the sensitizer molecules are broad, which is favorable for the selection of the excitation sources. Besides, upconverted luminescence imaging was capable of eliminating background fluorescence from either endogenous fluorophores of biological sample or the colabeled fluorescent probe.

In 2011, the first example of bioimaging based on TTA upconversion was reported [97]. Hydrophilic polymeric nanocapsules embedding a hydrophobic upconversion dye system of a sensitizer (PdOEP) and an acceptor (perylene) showed very efficient upconverted emission in aqueous dispersion under low excitation intensity (0.05 W cm^{-2}). It should be noted that a core/shell system with a liquid core was required instead of complete solid nanoparticles, because the liquid core enabled the high mobility of the emitter and the sensitizer, which was necessary to ensure effective upconversion. With the excitation at 514 nm, an upconverted emission of 450 nm was observed in HeLa cells. And this generated upconversion photons could be served as optical excitation source for consequent light-triggered processes for biological applications [97].

Similarly, Li and coworkers prepared water-soluble upconversion luminescent nanoparticles based on TTA by coloading sensitizer (PdOEP) and annihilator (DPA) into silica nanoparticles [98]. The nanoparticles exhibited excellent photostability as well as low cytotoxicity and were successfully used to label living cells with very high signal-to-noise ratio. Meanwhile, bioimaging with the upconversion nanoparticles could completely eliminate background fluorescence. Upon low-power density excitation of continuous-wave 532 laser (8.5 mW cm^{-2}), blue-emissive upconversion nanoparticles were successfully applied in lymph node imaging in vivo of living mouse [98].

Recently, a general strategy for biocompatible and high-effective upconversion nanocapsules based on triplet–triplet annihilation was proposed by loading both sensitizer and acceptor into BSA–dextran-stabilized soybean oil droplets [99]. Due to soybean oil acted as solvent for dissolving the sensitizer and annihilator, the nanocapsules could maintain high translational mobility of the chromophores and avoid luminescence quenching of chromophore by aggregation. At the same time, the reducibility of soybean oil decreased the O_2-induced quenching of TTA-based upconversion emission. PtTPBP (**Pt-2**) and Bodipy dyes (**A-13** and **A-14** with the maximal fluorescence emission at 528 and 546 nm, respectively) were chosen as sensitizer–acceptor couples to fabricate red-to-green (**A-13**) and red-to-yellow (**A-14**) upconversion luminescent emissive nanocapsules. As shown in Fig. 6.3, these upconversion nanocapsules were successfully applied to lymph node imaging in vivo of living mice without removing the skin, achieving excellent signal-to-noise ratios (>10) upon low-power density excitation by a continuous-wave laser ($\lambda_{ex} = 635$ nm, 12.5 mW cm^{-2}) [99]. This study opened up new perspectives for preparing TTA-based upconversion luminescence materials and their applications in the field of bioimaging in vivo.

Besides, TTA-based upconversion system has been used in the display [100], photoelectrochemistry [101], oxygen sensing materials [49], semiconductor photocatalysis [102], liquid-crystal soft actuators [103], and so on.

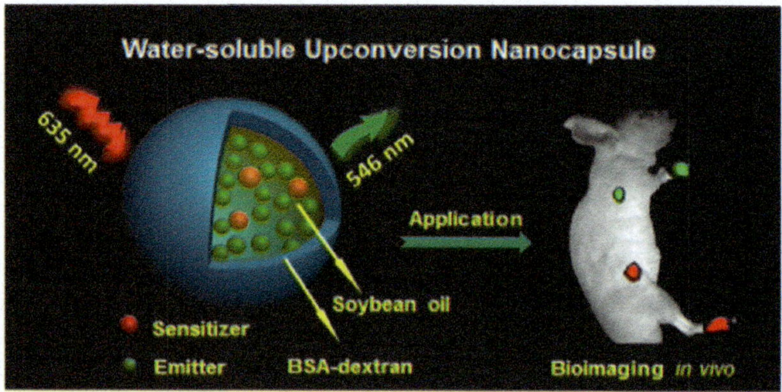

Fig. 6.3 Water-soluble upconversion nanocapsules based on triplet–triplet annihilation for bioimaging in vivo (Reprinted with permission from Ref. [99]. Copyright 2013 American Chemical Society)

6.6 Conclusions and Outlook

Transition-metal complexes play an important role in TTA-based upconversion, which is a promising upconversion process because of the incoherent excitation light sources, low excitation power density, tunable excitation and emission wavelengths, and high upconversion quantum yield. This chapter mainly described fundamental of TTA-based upconversion, summarized transition-metal complex-based sensitizers and some organic sensitizers, and presented the applications in photovoltaics and bioimaging.

Transition-metal complexes as sensitizers with intense absorption of visible light and long-lived T_1 excited state have been synthesized for TTA upconversion. However, most of the currently available sensitizers are synthesized with mono-chromophore and have only one major absorption band in the visible spectral region. Energy funneling resulting from multi-chromophores in the triplet sensitizers is expected for the panchromatic absorption. Furthermore, TTA-based upconversion materials with good water solubility and high quantum yield are still in demand for biological application. The introduction of reductive solvent into nanocapsules entrapping sensitizer and acceptor is feasible. In addition, transition-metal complex with NIR absorption are expected for TTA-based upconversion to meet the demands of the further development for bioimaging and photovoltaics.

Acknowledgments We thank the National Basic Research Program of China (2012CB933301), the National Natural Science Foundation of China (61274018, 61136003 and 21174064), Program for New Century Excellent Talents in University (NCET-12-0740), the Ministry of Education of China (IRT1148), Natural Science Foundation of Jiangsu Province of China (BM2012010, BK20130038 and BK2012835) and Priority Academic Program Development of Jiangsu Higher Education Institutions (YX03001) for financial support.

References

1. Schulze TF, Czolk J, Cheng YY, Fückel B, MacQueen RW, Khoury T, Crossley MJ, Stannowski B, Lips K, Lemmer U, Colsmann A, Schmidt TW (2012) Efficiency enhancement of organic and thin-film silicon solar cells with photochemical upconversion. J Phys Chem C 116:22794–22801
2. Zhang Y, Hong Z (2013) Synthesis of lanthanide-doped $NaYF_4$@TiO_2 core-shell composites with highly crystalline and tunable TiO_2 shells under mild conditions and their upconversion-based photocatalysis. Nanoscale 5:8930–8933
3. Zhou J, Liu Z, Li F (2012) Upconversion nanophosphors for small-animal imaging. Chem Soc Rev 41:1323–1349
4. Liu Y, Chen M, Cao T, Sun Y, Li C, Liu Q, Yang T, Yao L, Feng W, Li F (2013) A cyanine-modified nanosystem for in vivo upconversion luminescence bioimaging of methylmercury. J Am Chem Soc 135:9869–9876
5. Kim HM, Cho BR (2009) Two-photon probes for intracellular free metal ions, acidic vesicles, and lipid rafts in live tissues. Acc Chem Res 42:863–872

6. Wang F, Deng R, Wang J, Wang Q, Han Y, Zhu H, Chen X, Liu X (2011) Tuning upconversion through energy migration in core-shell nanoparticles. Nat Mater 10:968–973
7. Singh-Rachford TN, Castellano FN (2010) Photon upconversion based on sensitized triplet–triplet annihilation. Coord Chem Rev 254:2560–2573
8. Zhu MQ, Zhang GF, Li C, Aldred MP, Chang E, Drezek RA, Li ADQ (2011) Reversible two-photon photoswitching and two-photon imaging of immunofunctionalized nanoparticles targeted to cancer cells. J Am Chem Soc 133:365–372
9. Wang F, Liu X (2009) Recent advances in the chemistry of lanthanide-doped upconversion nanocrystals. Chem Soc Rev 38:976–989
10. Parker CA, Hatchard CG (1962) Sensitised anti-Stokes delayed fluorescence. Proc Chem Soc 386–387
11. Zhao J, Ji S, Guo H (2011) Triplet–triplet annihilation based upconversion: from triplet sensitizers and triplet acceptors to upconversion quantum yields. RSC Adv 1:937–950
12. Monguzzi A, Tubino R, Meinardi F (2008) Upconversion-induced delayed fluorescence in multicomponent organic systems: role of Dexter energy transfer. Phys Rev B 77:155122
13. Ceroni P (2011) Energy up-conversion by low-power excitation: new applications of an old concept. Chem Eur J 17:9560–9564
14. Cheng YY, Khoury T, Clady RGCR, Tayebjee MJY, Ekins-Daukes NJ, Crossley MJ, Schmidt TW (2010) On the efficiency limit of triplet-triplet annihilation for photochemical upconversion. Phys Chem Chem Phys 12:66–71
15. Monguzzi A, Tubino R, Hoseinkhani S, Campione M, Meinardi F (2012) Low power, non-coherent sensitized photon up-conversion: modelling and perspectives. Phys Chem Chem Phys 14:4322–4332
16. Saltiel J, Atwater BW (1988) Spin-statistical factors in diffusion-controlled reactions. Adv Photochem 14:1–90
17. Saltiel J, Marchand GR, Smothers WK, Stout SA, Charlton JL (1981) Concerning the spin-statistical factor in the triplet-triplet annihilation of anthracene triplets. J Am Chem Soc 103:7159–7164
18. Simon YC, Weder C (2012) Low-power photon upconversion through triplet–triplet annihilation in polymers. J Mater Chem 22:20817–20830
19. Bachilo SM, Weisman RB (2000) Determination of triplet quantum yields from triplet–triplet annihilation fluorescence. J Phys Chem A 104:7711–7714
20. Liu L, Huang D, Draper SM, Yi X, Wu W, Zhao J (2013) Visible light-harvesting trans bis (alkylphosphine) platinum(II)-alkynyl complexes showing long-lived triplet excited states as triplet photosensitizers for triplet-triplet annihilation upconversion. Dalton Trans 42:10694–10706
21. Zhao W, Castellano FN (2006) Upconverted emission from pyrene and di-tert-butylpyrene using $Ir(ppy)_3$ as triplet sensitizer. J Phys Chem A 110:11440–11445
22. Sun JF, Wu W, Guo H, Zhao J (2011) Visible-light harvesting with cyclometalated iridium (III) complexes having long-lived ^{3}IL excited states and their application in triplet–triplet-annihilation based upconversion. Eur J Inorg Chem (21):3165–3173
23. Sun J, Zhao J, Guo H, Wu W (2012) Visible-light harvesting iridium complexes as singlet oxygen sensitizers for photooxidation of 1,5-dihydroxynaphthalene. Chem Commun 48:4169–4171
24. Sun J, Wu W, Zhao J (2012) Long-lived room-temperature deep-red-emissive intraligand triplet excited state of naphthalimide in cyclometalated Ir^{III} complexes and its application in triplet-triplet annihilation-based upconversion. Chem Eur J 18:8100–8112
25. Ma L, Guo H, Li Q, Guo S, Zhao J (2012) Visible light-harvesting cyclometalated Ir(III) complexes as triplet photosensitizers for triplet–triplet annihilation based upconversion. Dalton Trans 41:10680–10689
26. Yi X, Yang P, Huang D, Zhao J (2013) Visible light-harvesting cyclometalated Ir(III) complexes with pyreno[4,5-d]imidazole *C^N* ligands as triplet photosensitizers for triplet–triplet annihilation upconversion. Dyes Pigments 96:104–115

27. Sun J, Zhong F, Yi X, Zhao J (2013) Efficient enhancement of the visible-light absorption of cyclometalated Ir(III) complexes triplet photosensitizers with Bodipy and applications in photooxidation and triplet-triplet annihilation upconversion. Inorg Chem 52:6299–6310
28. Ma L, Guo S, Sun J, Zhang C, Zhao J, Guo H (2013) Green light-excitable naphthalenediimide acetylide-containing cyclometalated Ir(III) complex with long-lived triplet excited states as triplet photosensitizers for triplet–triplet annihilation upconversion. Dalton Trans 42:6478–6488
29. Baluschev S, Miteva T, Yakutkin V, Nelles G, Yasuda A, Wegner G (2006) Up-conversion fluorescence: noncoherent excitation by sunlight. Phys Rev Lett 97:143903
30. Baluschev S, Yakutkin V, Wegner G, Miteva T, Nelles G, Yasuda A, Chernov S, Aleshchenkov S, Cheprakov A (2007) Upconversion with ultrabroad excitation band: simultaneous use of two sensitizers. Appl Phys Lett 90:181103
31. Baluschev S, Yakutkin V, Miteva T, Avlasevich Y, Chernov S, Aleshchenkov S, Nelles G, Cheprakov A, Yasuda A, Müllen K, Wegner G (2007) Blue-green up-conversion: noncoherent excitation by NIR light. Angew Chem Int Ed 46:7693–7696
32. Baluschev S, Yakutkin V, Miteva T, Wegner G, Roberts T, Nelles G, Yasuda A, Chernov S, Aleshchenkov S, Cheprakov A (2008) A general approach for non-coherently excited annihilation up-conversion: transforming the solar-spectrum. New J Phys 10:013007
33. Deng F, Blumhoff J, Castellano FN (2013) Annihilation limit of a visible-to-UV photon upconversion composition ascertained from transient absorption kinetics. J Phys Chem A 117:4412–4419
34. Singh-Rachford TN, Castellano FN (2008) Pd(II) phthalocyanine-sensitized triplet-triplet annihilation from rubrene. J Phys Chem A 112:3550–3556
35. Yakutkin V, Aleshchenkov S, Chernov S, Miteva T, Nelles G, Cheprakov A, Baluschev S (2008) Towards the IR limit of the triplet-triplet annihilation-supported up-conversion: tetraanthraporphyrin. Chem Eur J 14:9846–9850
36. Borisov SM, Saf R, Fischer R, Klimant I (2013) Synthesis and properties of new phosphorescent red light-excitable platinum(II) and palladium(II) complexes with Schiff bases for oxygen sensing and triplet-triplet annihilation-based upconversion. Inorg Chem 52:1206–1216
37. Wohnhaas C, Friedemann K, Busko D, Landfester K, Baluschev S, Crespy D, Turshatov A (2013) All organic nanofibers as ultralight versatile support for triplet–triplet annihilation upconversion. ACS Macro Lett 2:446–450
38. Nattestad A, Cheng YY, MacQueen RW, Schulze TF, Thompson FW, Mozer AJ, Fückel B, Khoury T, Crossley MJ, Lips K, Wallace GG, Schmidt TW (2013) Dye-sensitized solar cell with integrated triplet–triplet annihilation upconversion system. J Phys Chem Lett 4:2073–2078
39. Keivanidis PE, Baluschev S, Miteva T, Nelles G, Scherf U, Yasuda A, Wegner G (2003) Up-conversion photoluminescence in polyfluorene doped with metal(II)-octaethyl porphyrins. Adv Mater 15:2095–2098
40. Islangulov RR, Lott J, Weder C, Castellano FN (2007) Noncoherent low-power upconversion in solid polymer films. J Am Chem Soc 129:12652–12653
41. Singh-Rachford TN, Lott J, Weder C, Castellano FN (2009) Influence of temperature on low-power upconversion in rubbery polymer blends. J Am Chem Soc 131:12007–12014
42. Baluschev S, Yu F, Miteva T, Ahl S, Yasuda A, Nelles G, Knoll W, Wegner G (2005) Metal-enhanced up-conversion fluorescence: effective triplet-triplet annihilation near silver surface. Nano Lett 5:2482–2484
43. Baluschev S, Keivanidis PE, Wegner G, Jacob J, Grimsdale AC, Müllen K, Miteva T, Yasuda A, Nelles G (2005) Upconversion photoluminescence in poly(ladder-type-pentaphenylene) doped with metal(II)-octaethyl porphyrins. Appl Phys Lett 86:061904
44. Laquai F, Wegner G, Im C, Büsing A, Heun S (2005) Efficient upconversion fluorescence in a blue-emitting spirobifluorene-anthracene copolymer doped with low concentrations of Pt(II) octaethylporphyrin. J Chem Phys 123:074902

45. Tanaka K, Inafuku K, Chujo Y (2010) Environment-responsive upconversion based on dendrimer-supported efficient triplet-triplet annihilation in aqueous media. Chem Commun 46:4378–4380
46. Singh-Rachford TN, Haefele A, Ziessel R, Castellano FN (2008) Boron dipyrromethene chromophores: next generation triplet acceptors/annihilators for low power upconversion schemes. J Am Chem Soc 130:16164–16165
47. Singh-Rachford TN, Castellano FN (2009) Supra-nanosecond dynamics of a red-to-blue photon upconversion system. Inorg Chem 48:2541–2548
48. Singh-Rachford TN, Castellano FN (2010) Triplet sensitized red-to-blue photon upconversion. J Phys Chem Lett 1:195–200
49. Borisov SM, Larndorfer C, Klimant I (2012) Triplet-triplet annihilation-based anti-Stokes oxygen sensing materials with a very broad dynamic range. Adv Funct Mater 22:4360–4368
50. Deng F, Sommer JR, Myahkostupov M, Schanze KS, Castellano FN (2013) Near-IR phosphorescent metalloporphyrin as a photochemical upconversion sensitizer. Chem Commun 49:7406–7408
51. Sun H, Guo H, Wu W, Liu X, Zhao J (2011) Coumarin phosphorescence observed with *N^N* Pt(II) bisacetylide complex and its applications for luminescent oxygen sensing and triplet-triplet-annihilation based upconversion. Dalton Trans 40:7834–7841
52. Liu Y, Wu W, Zhao J, Zhang X, Guo H (2011) Accessing the long-lived near-IR-emissive triplet excited state in naphthalenediimide with light-harvesting diimine platinum (II) bisacetylide complex and its application for upconversion. Dalton Trans 40:9085–9089
53. Huang L, Zeng L, Guo H, Wu W, Wu W, Ji S, Zhao J (2011) Room-temperature long-lived ^{3}IL excited state of rhodamine in an *N^N* Pt^{II} bis(acetylide) complex with intense visible-light absorption. Eur J Inorg Chem (29):4527–4533
54. Liu Y, Li Q, Zhao J, Guo H (2012) BF_2-bound chromophore-containing *N^N* Pt (II) bisacetylide complex and its applications as sensitizer for triplet-triplet annihilation based upconversion. RSC Adv 2:1061–1067
55. Ji S, Wu W, Zhao J, Guo H, Wu W (2012) Efficient triplet-triplet annihilation upconversion with Platinum(II) bis(arylacetylide) complexes that show long-lived triplet excited states. Eur J Inorg Chem (19):3183–3190
56. Wu W, Zhao J, Wu W, Chen Y (2012) Room temperature long-lived triplet excited state of fluorescein in *N^N* Pt(II) bisacetylide complex and its applications for triplet–triplet annihilation based upconversions. J Organomet Chem 713:189–196
57. Li Q, Guo H, Ma L, Wu W, Liu Y, Zhao J (2012) Tuning the photophysical properties of *N^N* Pt(II) bisacetylide complexes with fluorene moiety and its applications for triplet–triplet-annihilation based upconversion. J Mater Chem 22:5319–5329
58. Guo H, Li Q, Ma L, Zhao J (2012) Fluorene as π-conjugation linker in *N^N* Pt(II) bisacetylide complexes and their applications for triplet–triplet annihilation based upconversion. J Mater Chem 22:15757–15768
59. Wu W, Wu W, Ji S, Guo H, Zhao J (2011) Accessing the long-lived emissive ^{3}IL triplet excited states of coumarin fluorophores by direct cyclometallation and its application for oxygen sensing and upconversion. Dalton Trans 40:5953–5963
60. Wu W, Guo H, Wu W, Ji S, Zhao J (2011) Long-lived room temperature deep-red/near-IR emissive intraligand triplet excited state (^{3}IL) of naphthalimide in cyclometalated platinum (II) complexes and its application in upconversion. Inorg Chem 50:11446–11460
61. Wu W, Sun J, Ji S, Wu W, Zhao J, Guo H (2011) Tuning the emissive triplet excited states of platinum(II) Schiff base complexes with pyrene, and application for luminescent oxygen sensing and triplet-triplet-annihilation based upconversions. Dalton Trans 40:11550–11561
62. Du P, Eisenberg R (2010) Energy upconversion sensitized by a platinum(II) terpyridyl acetylide complex. Chem Sci 1:502–506
63. Wu W, Zhao J, Guo H, Sun J, Ji S, Wang Z (2012) Long-lived room-temperature near-IR phosphorescence of BODIPY in a visible-light-harvesting *N^C^N* Pt(II)-acetylide complex with a directly metalated BODIPY chromophore. Chem Eur J 18:1961–1968

64. Wu W, Huang D, Yi X, Zhao J (2013) Tridentate cyclometalated platinum(II) complexes with strong absorption of visible light and long-lived triplet excited states as photosensitizers for triplet–triplet annihilation upconversion. Dyes Pigments 96:220–231
65. Wu W, Zhao J, Sun J, Huang L, Yi X (2013) Red-light excitable fluorescent platinum(II) bis (aryleneethynylene) bis(trialkylphosphine) complexes showing long-lived triplet excited states as triplet photosensitizers for triplet-triplet annihilation upconversion. J Mater Chem C 1:705–716
66. Kozlov DV, Castellano FN (2004) Anti-Stokes delayed fluorescence from metal-organic bichromophores. Chem Commun (24):2860–2861
67. Islangulov RR, Kozlov DV, Castellano FN (2005) Low power upconversion using MLCT sensitizers. Chem Commun (30):3776–3778
68. Singh-Rachford TN, Islangulov RR, Castellano FN (2008) Photochemical upconversion approach to broad-band visible light generation. J Phys Chem A 112:3906–3910
69. Singh-Rachford TN, Castellano FN (2009) Nonlinear photochemistry squared: quartic light power dependence realized in photon upconversion. J Phys Chem A 113:9266–9269
70. Ji S, Wu W, Wu W, Guo H, Zhao J (2011) Ruthenium(II) polyimine complexes with a long-lived ^{3}IL excited state or a ^{3}MLCT/^{3}IL equilibrium: efficient triplet sensitizers for low-power upconversion. Angew Chem Int Ed 50:1626–1629
71. Ji S, Guo H, Wu W, Wu W, Zhao J (2011) Ruthenium(II) polyimine-coumarin dyad with non-emissive ^{3}IL excited state as sensitizer for triplet-triplet annihilation based upconversion. Angew Chem Int Ed 50:8283–8286
72. Wu W, Ji S, Wu W, Shao J, Guo H, James TD, Zhao J (2012) Ruthenium(II)-polyimine-coumarin light-harvesting molecular arrays: design rationale and application for triplet-triplet-annihilation-based upconversion. Chem Eur J 18:4953–4964
73. Wu W, Sun J, Cui X, Zhao J (2013) Observation of the room temperature phosphorescence of Bodipy in visible light-harvesting Ru(II) polyimine complexes and application as triplet photosensitizers for triplet–triplet-annihilation upconversion and photocatalytic oxidation. J Mater Chem C 1:4577–4589
74. Sugunan SK, Tripathy U, Brunet SMK, Paige MF, Steer RP (2009) Mechanisms of low-power noncoherent photon upconversion in metalloporphyrin organic blue emitter systems in solution. J Phys Chem A 113:8548–8556
75. Singh-Rachford TN, Nayak A, Muro-Small ML, Goeb S, Therien MJ, Castellano FN (2010) Supermolecular-chromophore-sensitized near-infrared-to-visible photon upconversion. J Am Chem Soc 132:14203–14211
76. Cui X, Zhao J, Yang P, Sun J (2013) Zinc(II) tetraphenyltetrabenzoporphyrin complex as triplet photosensitizer for triplet-triplet annihilation upconversion. Chem Commun 49:10221–10223
77. Yi X, Zhao J, Wu W, Huang D, Ji S, Sun J (2012) Rhenium(I) tricarbonyl polypyridine complexes showing strong absorption of visible light and long-lived triplet excited states as a triplet photosensitizer for triplet-triplet annihilation upconversion. Dalton Trans 41:8931–8940
78. Yi X, Zhao J, Sun J, Guo S, Zhang H (2013) Visible light-absorbing rhenium(I) tricarbonyl complexes as triplet photosensitizers in photooxidation and triplet-triplet annihilation upconversion. Dalton Trans 42:2062–2074
79. McCusker CE, Castellano FN (2013) Orange-to-blue and red-to-green photon upconversion with a broadband absorbing copper(I) MLCT sensitizer. Chem Commun 49:3537–3539
80. To WP, Chan KT, Tong GSM, Ma C, Kwok WM, Guan X, Low KH, Che CM (2013) Strongly luminescent gold(III) complexes with long-lived excited states: high emission quantum yields, energy up-conversion, and nonlinear optical properties. Angew Chem Int Ed 52:6648–6652
81. Koziar JC, Cowan DO (1978) Photochemical heavy-atom effects. Acc Chem Res 11:334–341

82. Chen HC, Hung CY, Wang KH, Chen HL, Fann WS, Chien FC, Chen P, Chow TJ, Hsu CP, Sun SS (2009) White-light emission from an upconverted emission with an organic triplet sensitizer. Chem Commun (27):4064–4066
83. Wu W, Guo H, Wu W, Ji S, Zhao J (2011) Organic triplet sensitizer library derived from a single chromophore (BODIPY) with long-lived triplet excited state for triplet-triplet annihilation based upconversion. J Org Chem 76:7056–7064
84. Chen Y, Zhao J, Xie L, Guo H, Li Q (2012) Thienyl-substituted BODIPYs with strong visible light-absorption and long-lived triplet excited states as organic triplet sensitizers for triplet–triplet annihilation upconversion. RSC Adv 2:3942–3953
85. Guo S, Wu W, Guo H, Zhao J (2012) Room-temperature long-lived triplet excited states of naphthalenediimides and their applications as organic triplet photosensitizers for photooxidation and triplet-triplet annihilation upconversions. J Org Chem 77:3933–3943
86. Zhang C, Zhao J, Wu S, Wang Z, Wu W, Ma J, Guo S, Huang L (2013) Intramolecular RET enhanced visible light-absorbing Bodipy organic triplet photosensitizers and application in photooxidation and triplet-triplet annihilation upconversion. J Am Chem Soc 135:10566–10578
87. El-Sayed MA (1968) The triplet state: its radiative and nonradiative properties. Acc Chem Res 1:8–16
88. Singh-Rachford TN, Castellano FN (2009) Low power visible-to-UV upconversion. J Phys Chem A 113:5912–5917
89. Huang D, Sun J, Ma L, Zhang C, Zhao J (2013) Preparation of ketocoumarins as heavy atom-free triplet photosensitizers for triplet-triplet annihilation upconversion. Photochem Photobiol Sci 12:872–882
90. Wu W, Cui X, Zhao J (2013) Hetero bodipy-dimers as heavy atom-free triplet photosensitizers showing a long-lived triplet excited state for triplet-triplet annihilation upconversion. Chem Commun 49:9009–9011
91. Wu W, Zhao J, Sun J, Guo S (2012) Light-harvesting fullerene dyads as organic triplet photosensitizers for triplet-triplet annihilation upconversions. J Org Chem 77:5305–5312
92. Yang P, Wu W, Zhao J, Huang D, Yi X (2012) Using C_{60}-bodipy dyads that show strong absorption of visible light and long-lived triplet excited states as organic triplet photosensitizers for triplet–triplet annihilation upconversion. J Mater Chem 22:20273–20283
93. Huang D, Zhao J, Wu W, Yi X, Yang P, Ma J (2012) Visible-light-harvesting triphenylamine ethynyl C_{60}-BODIPY dyads as heavy-atom-free organic triplet photosensitizers for triplet-triplet annihilation upconversion. Asian J Org Chem 1:264–273
94. Guo S, Sun J, Ma L, You W, Yang P, Zhao J (2013) Visible light-harvesting naphthalenediimide (NDI)-C_{60} dyads as heavy-atom-free organic triplet photosensitizers for triplet–triplet annihilation based upconversion. Dyes Pigments 96:449–458
95. Shockley W, Queisser HJ (1961) Detailed balance limit of efficiency of *p-n* junction solar cells. J Appl Phys 32:510–519
96. Cheng YY, Fückel B, MacQueen RW, Khoury T, Clady RGCR, Schulze TF, Ekins-Daukes NJ, Crossley MJ, Stannowski B, Lip K, Schmidt TW (2012) Improving the light-harvesting of amorphous silicon solar cells with photochemical upconversion. Energy Environ Sci 5:6953–6959
97. Wohnhaas C, Turshatov A, Mailänder V, Lorenz S, Baluschev S, Miteva T, Landfester K (2011) Annihilation upconversion in cells by embedding the dye system in polymeric nanocapsules. Macromol Biosci 11:772–778
98. Liu Q, Yang T, Feng W, Li F (2012) Blue-emissive upconversion nanoparticles for low-power-excited bioimaging in vivo. J Am Chem Soc 134:5390–5397
99. Liu Q, Yin B, Yang T, Yang Y, Shen Z, Yao P, Li F (2013) A general strategy for biocompatible, high-effective upconversion nanocapsules based on triplet-triplet annihilation. J Am Chem Soc 135:5029–5037
100. Miteva T, Yakutkin V, Nelles G, Baluschev S (2008) Annihilation assisted upconversion: all-organic, flexible and transparent multicolour display. New J Phys 10:103002

101. Khnayzer RS, Blumhoff J, Harrington JA, Haefele A, Deng F, Castellano FN (2012) Upconversion-powered photoelectrochemistry. Chem Commun 48:209–211
102. Kim JH, Kim JH (2012) Encapsulated triplet-triplet annihilation-based upconversion in the aqueous phase for sub-band-gap semiconductor photocatalysis. J Am Chem Soc 134:17478–17481
103. Jiang Z, Xu M, Li F, Yu Y (2013) Red-light-controllable liquid-crystal soft actuators via low-power excited upconversion based on triplet-triplet annihilation. J Am Chem Soc 135:16446–16453

Chapter 7
Visible Light-Harvesting Transition Metal Complexes for Triplet–Triplet Annihilation Upconversion

Poulomi Majumdar and Jianzhang Zhao

Abstract Transition metal complexes containing Pt(II), Ir(III), Ru(II), and Re (I) atoms and showing visible light-harvesting ability and long-lived T_1 excited states have been developed and used as photosensitizers for triplet–triplet annihilation (TTA) upconversion, which is a promising upconversion method due to its low excitation power density (solar light is sufficient), high upconversion quantum yield, readily tunable excitation/emission wavelength, and strong absorption of excitation light. TTA upconversion involves triplet energy transfer between a photosensitizer (donor) molecule and an acceptor/annihilator/emitter. This chapter is based on upconversion examples, aiming to cover the challenges that are faced by the developments of TTA upconversion and the molecular structure designing rationales for the triplet photosensitizers and triplet acceptors.

Keywords Visible light • Energy transfer • Intersystem crossing • Transition metal • Triplet–triplet annihilation • Upconversion

7.1 Introduction

Transition metal complexes have attracted much attention due to the applications in electroluminescence and light-harvesting molecular assemblies [1, 2]. These complexes, such as those containing Pt(II), Ir(III), and Ru(II) atoms, are different from organic fluorophores in that the *triplet* excited states, not the *singlet* excited states, are populated upon photoexcitation of these complexes [1–10]. Therefore, long-lived triplet excited states in microsecond range (μs) were observed for these complexes. These properties are potentially useful for application in triplet–triplet annihilation (TTA) upconversion.

TTA upconversion, first introduced by Parker and Hatchard over 40 years [11], is observation of photon emission or, more generally, population of excited state at

P. Majumdar • J. Zhao (✉)
State Key Laboratory of Fine Chemicals, School of Chemical Engineering, Dalian University of Technology, Dalian 116024, China
e-mail: zhaojzh@dlut.edu.cn; http://finechem2.dlut.edu.cn/photochem

W.-Y. Wong (ed.), *Organometallics and Related Molecules for Energy Conversion*, Green Chemistry and Sustainable Technology, DOI 10.1007/978-3-662-46054-2_7

higher energy (shorter wavelength) with excitation at lower energy (longer wavelength). TTA upconversion is of particular interest due to its potential applications for photovoltaics, artificial photosynthesis, photocatalysis, optics, etc. [12–15]. For example, it is difficult for the dye-sensitized solar cell (DSCs) to utilize the solar light in the near-IR region, despite the power of the solar light in this wavelength range being intense. The efficiency of the DSCs can be improved with upconversion materials that can convert the radiation at longer wavelength into radiation at shorter wavelength.

A few methods are available for upconversion, including two-photon absorption dyes (TPA), upconversion with inorganic crystals (such as KDP), rare earth materials, etc. [12–15]. However, these techniques usually suffer from drawbacks of requirement of high excitation power, poor absorption of visible light, low upconversion quantum yield, etc. Therefore, these techniques are unlikely to be used for applications with light source at low excitation power density, such as solar light. Instead, coherent laser with high power density (MW cm^{-2}) is required for excitation of TPA dyes, which is well beyond the power density of normal light source (the power density of the terrestrial solar radiance is ca. 0.10 W cm^{-2}, AM1.5G). Furthermore, from a chemist's perspective, it is difficult to tailor the structure of TPA dyes to achieve a specific upconversion wavelength and, at the same time, to maintain a high TPA cross section. Recently, upconversion with rare earth materials have been investigated. However, the visible light absorption of these materials are usually weak; thus, the overall upconversion capability, or the apparent brightness ($\eta = \varepsilon \times \Phi_{UC}$, ε is the molar absorption coefficient of the upconversion materials at the excitation wavelength and Φ_{UC} is the upconversion quantum yield), is poor.

Recently a new upconversion method based on TTA has been developed [16–27]. TTA upconversion shows advantages over the aforementioned upconversion techniques. For example, the excitation power density required for TTA upconversion is quite low, and the excitation need not be coherent. Excitation with energy density of a few mW cm^{-2} is sufficient [19–21]. Thus, solar light can be used as the excitation source. Furthermore, the excitation wavelength and emission wavelength of TTA upconversion can be readily changed, simply by selection of different triplet photosensitizers and the triplet energy acceptor (annihilator/emitter), given the energy levels of the excited state of the photosensitizers and the acceptors are matched (see Scheme 7.1 and later section for detail) [21]. Thus, the TTA upconversion is promising for applications such as photovoltaics, photocatalysis, and many other light-driven photophysical and photochemical processes.

Different from most of the other upconversion methods, the TTA upconversion is based on mixing the triplet photosensitizer and triplet energy acceptor (annihilator/emitter) together [20, 21]. The excitation energy was harvested by the triplet photosensitizer, and the energy was transferred to the acceptor via triplet–triplet energy transfer (TTET), then the singlet excited state of the energy acceptor will be populated by TTA, which will give emission at higher energy level than the

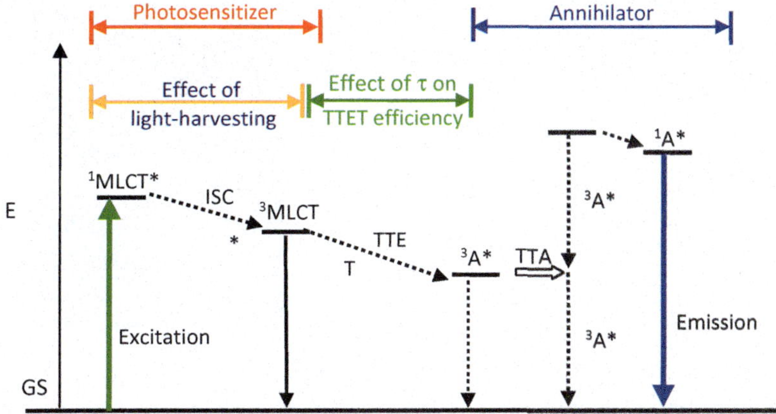

Scheme 7.1 Qualitative Jablonski diagram illustrating the TTA upconversion process between triplet photosensitizer and acceptor (annihilator/emitter). The effect of the light-harvesting ability and the excited state lifetime of the photosensitizer on the efficiency of the TTA upconversion is also shown. E is energy. GS is ground state (S_0). ^{3}MLCT* is the metal-to-ligand-charge-transfer triplet excited state. TTET is triplet–triplet energy transfer. ^{3}A* is the triplet excited state of annihilator. TTA is triplet–triplet annihilation. ^{1}A* is the singlet excited state of annihilator. The emission bands observed for the photosensitizers alone is the ^{3}MLCT emissive excited state. The emission bands observed in the TTA experiment is the simultaneous ^{3}MLCT* emission (phosphorescence) and the ^{1}A* emission (fluorescence) (Adapted from ref. [21] with permission)

excitation. The photophysics of TTA upconversion can be illustrated by a Jablonski diagram (Scheme 7.1).

Firstly the triplet photosensitizer is excited with photo-irradiation (Scheme 7.1). The singlet excited state will be populated ($S_0 \rightarrow S_1$). Then via intersystem crossing (ISC, e.g., $S_1 \rightarrow T_1$), in which the heavy atom effect of transition metal atom is often required, the triplet excited state of the photosensitizer will be populated. Direct excitation into the T_1 state is forbidden (the molar absorption coefficient for $S_0 \rightarrow T_1$ transition is small). Since the lifetime of the triplet excited state is much longer than that of the singlet excited state, thus the energy can be transferred from the triplet photosensitizer to the triplet energy acceptor, via the TTET process. Note the energy transfer between the triplet states is usually a Dexter process and it requires contact of the two components [28]. The triplet acceptor molecules at the triplet excited state will collide with each other and produce the singlet excited state of the acceptor, following the spin–statistic law (Eq. 1). The radiative decay from the singlet excited state of acceptor produces the upconverted fluorescence, for which the energy is higher than the excitation light. One example is illustrated in Fig. 7.1 [29].

Based on the spin–statistic law (Eq. 1), the limit for the efficiency of the TTA upconversion is 11.1 % [20, 21]. However, examples that exceed this limit have been reported [20, 21].

The spin of the triplet encounters of the acceptors was governed by the so-called spin–statistic factors [30, 31]. When two excited triplets ($^3A_1$) interact, nine

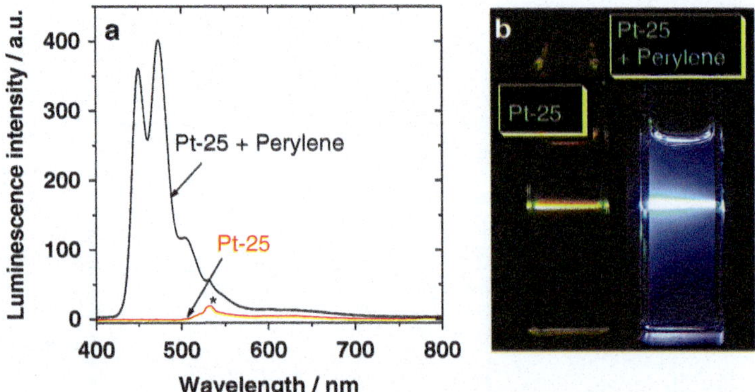

Fig. 7.1 Illustration of the TTA upconversion. (**a**) Emission of the complex **Pt-25** (see Fig. 7.23 for molecular structure) and the upconversion with **Pt-25** as photosensitizer and perylene as acceptor (see Fig. 7.11 for molecular structure). Excited by 532 nm laser. The *asterisk* indicates laser scattering. (**b**) Photographs of the upconversion samples of (**a**) (Reprinted with permission from Ref. [29]. Copyright © 2013, Royal Society of Chemistry)

encounter-pair spin states are produced with equal probability which is composed of three distinct sublevels, five of which are quintet, three are triplet, and one is singlet. Thus, spin statistics predicts that singlet productor represents 1/9 (or 11.1 %) of the annihilation events. Thus, the maximal upconversion quantum yield will be smaller than 11.1 % (Eq. 2), given that both the quenching efficiency Φ_q and the fluorescence quantum yield of acceptor Φ_F are 100 % [30]. While most of the TTA upconversions show quantum yield less than 10 % [14, 15], however, a few recently reported cases (including the results from our laboratory) do show that upconversion quantum yield can be higher than 11.1 % [17, 18]. These results suggest that the quintets are also leading to upconversion.

$$\begin{aligned} &{}^3A_1{*}+{}^3A_1{*} \leftrightarrow {}^5(AA)_2{*} \leftrightarrow {}^5A_2{*}+{}^1A_0 \\ &{}^3A_1{*}+{}^3A_1{*} \leftrightarrow {}^3(AA)_1{*} \leftrightarrow {}^3A_1{*}+{}^1A_0 \\ &{}^3A_1{*}+{}^3A_1{*} \leftrightarrow {}^1(AA)_0{*} \leftrightarrow {}^1A_0{*}+{}^1A_0 \end{aligned} \tag{7.1}$$

$$\Phi_{UC} = \Phi_q \times \Phi_{TTA} \times \Phi_F \tag{7.2}$$

It was proposed that the quintet state ${}^5A^*$ will have a 92 % chance to decay into two molecules at the triplet excited state (${}^3A^*$),[12] which are then involved in the TTA again. Thus, the maximal upconversion quantum yield is definitely higher than the previously thought 11.1 %.

The upconversion quantum yield can be described by Eq. 2, where Φ_q is the energy transfer efficiency (TTET), Φ_{TTA} is the efficiency to produce singlet excited state by the TTA process, and Φ_F is the fluorescence quantum yield of the acceptor.

Several photophysical parameters of the triplet photosensitizer and acceptor are crucial for TTA upconversion. (1) Usually transition metal complexes show weak absorption in visible region and this is detrimental to TTA upconversion [3, 4]. TTA upconversion requires the concentration of the triplet photosensitizers at the triplet excited state to be high; thus, efficient TTET to the acceptor can be ensured. As a result, the acceptor molecules at the triplet excited state will be high, and the upconversion will be more efficient [19, 21] because of the bimolecular feature of the TTET and the TTA processes. With higher concentration of the triplet photosensitizers at the triplet excited state, the TTET process will be more efficient to produce the acceptor's triplet excited state. We propose that the overall upconversion capability (η), or the apparent brightness, of a triplet photosensitizer be better evaluated by $\eta = \varepsilon \times \Phi_{UC}$, i.e., not only the Φ_{UC} value (Eq. 5). (2) The triplet excited state quantum yield of the photosensitizer must be high because it is the triplet excited state, not the singlet excited state that directly produced upon photoexcitation, that is involved in the critical TTET process. Normally the triplet photosensitizers used in TTA upconversion are transition metal complexes, for which the ISC process is with unit efficiency (Φ_{ISC} is close to 100 %) [20, 21]. (3) The lifetime of the triplet excited state of the triplet photosensitizers should be long. Long-lived T_1 excited state of the photosensitizer will lead to more efficient TTET process because TTET process is actually a two-molecular quenching procedure; long-lived T_1 excited state of the photosensitizer will increase the diffusion distance and make the encounter of the photosensitizer and the acceptor more likely [20, 21]. Although some of the triplet photosensitizers show long-lived T_1 excited state, usually the lifetime of the T_1 excited state of the transition metal complexes is short (a few μs) [3, 4]. (4) The energy levels of the triplet photosensitizers and the triplet acceptors must be appropriate to maximize the TTET. (5) The T_1 excited state energy level and the S_1 excited state energy level of the triplet acceptor much fulfill $2 \times E_{T1} > E_{S1}$, where E_{T1} is the energy level of the T_1 excited state and the E_{S1} is the energy level of the S_1 excited state [20, 21]. (6) The radiative decay of the S_1 excited state should be efficient to produce intense upconverted fluorescence emission, i.e., the fluorescence quantum yield of acceptor (Φ_F, Eq. 2) should be high.

Following these lines, the design rationales of the triplet photosensitizers and the energy acceptors can be summarized as follows: (1) triplet photosensitizer should be with strong absorption at the excitation wavelength (large ε values), (2) efficient ISC to produce the T_1 excited state, (3) the lifetime of the T_1 excited state of the photosensitizer must be long, and (4) the energy levels of the excited states of the photosensitizers and the acceptors must be matched in order to enhance the TTET process.

The TTA upconversion can be quantitatively described with at least two parameters, i.e., the efficiency of TTET process and the upconversion quantum yields (Φ_{UC}). The TTET efficiency can be measured by the quenching experiments, with the triplet acceptor as the quencher. Fitting the quenching data with the

Stern–Volmer equation will give the K_{SV} value and the bimolecular quenching constants k_q (Eq. 3):

$$I_0/I = 1 + K_{SV}[Q] \quad K_{SV} = k_q \times \tau_0 \tag{7.3}$$

where τ_0 is the lifetime of the triplet excited state of photosensitizer and [Q] is the concentration of the quenchers at which the I (residual emission of the photosensitizer) is determined. It should be noted that in some cases, the triplet photosensitizer is non-phosphorescent; in this case the quenching can be measured by the variation of the lifetime of the T_1 excited state of the photosensitizer, such as by using time-resolved transient absorption spectroscopy.

$$\Phi_{UC} = 2\Phi_{std}\left(\frac{A_{std}}{A_{unk}}\right)\left(\frac{I_{unk}}{I_{std}}\right)\left(\frac{\eta_{std}}{\eta_{std}}\right)^2 \tag{7.4}$$

The upconversion quantum yield (Φ_{UC}) can be determined by Eq. 4 [20], where Φ_{unk}, A_{unk}, I_{unk}, and η_{unk} represent the quantum yield, absorbance, integrated photoluminescence intensity of the samples, and the refractive index of the solvents, respectively. Since two photons are required to generate one upconverted photon, in order to keep the maximum quantum yield as unit, the equation is multiplied by factor 2 [20, 21].

$$\eta = \varepsilon \times \Phi_{UC} \tag{7.5}$$

Herein we propose to use the $\varepsilon \times \Phi_{UC}$ to evaluate the overall upconversion capability (η) of a triplet photosensitizer (Eq. 5) [21], where ε is the molar absorption coefficient of the upconversion materials and Φ_{UC} is the upconversion quantum yield determined with Eq. 4. Triplet photosensitizers with large η value are more likely ideal for practical applications. On the contrary, materials with large Φ_{UC} value but small ε value are not ideal for application. Furthermore, it should be pointed out that TTA upconversion scheme with excitation at red or near-IR region is ideal for applications, such as photovoltaics.

Although significant efforts have been made by our group to prepare visible light-harvesting transition metal complexes as triplet photosensitizers(PSs) with long-lived triplet states for its crucial application in TTA upconversion, still then much scope is left for the development of the same [3, 4, 20, 21, 25].

7.2 Triplet Photosensitizers for TTA Upconversion

7.2.1 Ru(II) Complexes as Triplet Photosensitizers

7.2.1.1 Ru(II) Polyimine Complexes as Triplet Photosensitizers

Ru(II) polyimine complexes have been investigated for a long history and have attracted much attention owing to the applications in photocatalysis such as photoredox reactions, photocatalytic H_2 production from water, photocatalytic cleavage of DNA, luminescent molecular probes and bioimaging, photovoltaics, and more recently the TTA upconversion [32].

The photophysical features of the Ru(II) polyimine complexes are the efficient population of the triplet excited states upon photoexcitation, moderate absorption in visible range, long-lived T_1 excited state, and efficient ISC (the quantum yield of the $S_1 \rightarrow T_1$ is close to unity). These features are ideal for application of the complexes as triplet photosensitizers for TTA upconversion.

Castellano et al. used $[Ru(dmb)_3][PF_6]_2$**Ru-1** to sensitize the TTA upconversion, with DPA as triplet energy acceptor (Fig. 7.2) [33]. The absorption maxima of the complex is at 450 nm. The energy level of the triplet excited state of the photosensitizer can be derived from the phosphorescence wavelength, at ca. 600 nm (2.07 eV). Thus, 9,10-diphenylanthracene (DPA, **A-1** in Fig. 7.2) is a suitable triplet acceptor, for which the energy level of the T_1 state is 1.77 eV (700 nm). The mixed solution of $[Ru(dmb)_3]^{2+}$ and DPA was excited with green laser ($\lambda_{ex} = 514.5$ nm, 24 mW or $\lambda_{ex} = 532$ nm, <5 mW), and the upconverted blue fluorescence emission of DPA at 430 nm was observed (anti-Stokes shift is ca. 100 nm). This result demonstrated that the TTA upconversion can be achieved with irradiation at low-power density. However, no efficiency of the TTET process and the quantum yield of the upconversion were reported for $Ru(dmb)_3[PF_6]_2$ [33].

Ru(II) complexes with covalently linked anthracene moiety as the integrated photosensitizer/acceptor $[Ru(dmb)_2(bpy\text{-}An)]$ (dmb is 4,4′-dimethyl-2,2′-bipyridine and bpy-An is 4-methyl-4′-(9-anthrylethyl)-2,2′-bipyridine) were reported [34]. The upconversion is more efficient for this intramolecular approach than the intermolecular method (enhanced by 2.9-fold) [34].

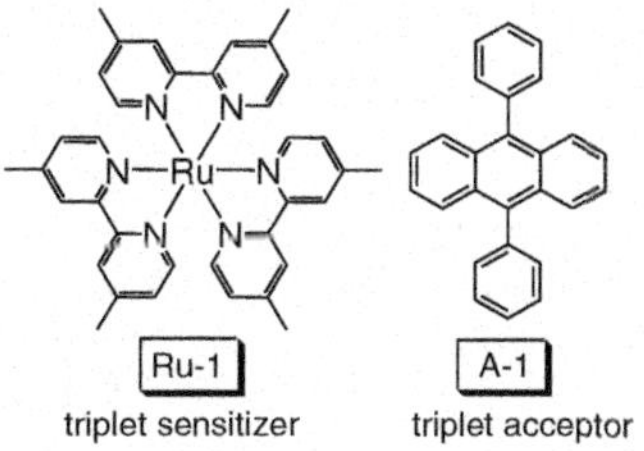

Ru-1
triplet sensitizer
A-1
triplet acceptor

Fig. 7.2 Triplet photosensitizer $[Ru(dmb)_3]^{2+}$ (**Ru-1**, dmb = 4,4′-dimethyl-2,2′-bipyridine) and the triplet acceptor 9,10-diphenylanthracene (DPA, **A-1**) for triplet–triplet annihilation upconversion

TTA upconversion efficiency is dependent on many factors, such as the concentration of triplet photosensitizer and acceptor, the power density of the excitation, etc. The success of using $Ru(dmb)_3[PF_6]_2$ to achieve upconversion with low-power density excitation is attributed to the long lifetime of the ^{3}MLCT excited state ($\tau = 0.87$ μs). The prolonged T_1 excited state lifetime is beneficial to the TTET process; as a result, more DPA molecules at the singlet excited state will be produced.

However, upconversion with $Ru(dmb)_3[PF_6]_2$ suffers from some drawbacks. Firstly, the absorption of the Ru(II) complex is weak in the visible range, and the absorption is limited in the region < 450 nm; this is typical for normal N^N Ru (II) diimine complexes. Triplet photosensitizers with intensive absorption in longer wavelength are desired [4, 21]. Secondly, the typical triplet state lifetimes of Ru (II) complexes are less than 1 μs [4]. Longer lifetime will enhance the TTET process, which has been demonstrated in luminescent O_2 sensing, for which the critical photophysical process is also TTET [35–37]. Therefore, it is highly desired to develop triplet photosensitizers with prolonged T_1 excited state lifetimes.

One method to access the long-lived ^{3}MLCT excited state of Ru(II) polyimine complexes is to optimize the coordination geometry of the Ru(II) center by using ligands with octahedral coordination geometry. For example, the **Ru-2** ($[Ru(tpy)_2]^{2+}$) shows T_1 excited state lifetime of 0.25 ns (Fig. 7.3) [38, 39]. With bpy ligand **Ru-3**, the T_1 state lifetime was extended to 1.0 μs [40]. By optimization of the geometry of N^N^N ligand **Ru-4**, the lifetime of the ^{3}MLCT excited state was extended to 3.0 μs [35]. However, *much*longer-lived T_1 excited state is necessary to enhance the TTET, thus the TTA upconversion [21].

It has been shown that the ^{3}IL (intraligand) excited state of the Ru(II) complexes shows much longer lifetime than the ^{3}MLCT excited state [8, 10, 41, 42]. Previously we demonstrated that O_2 sensing property of the complexes can be significantly improved with the long-lived ^{3}IL excited state [35–37]. Since the critical

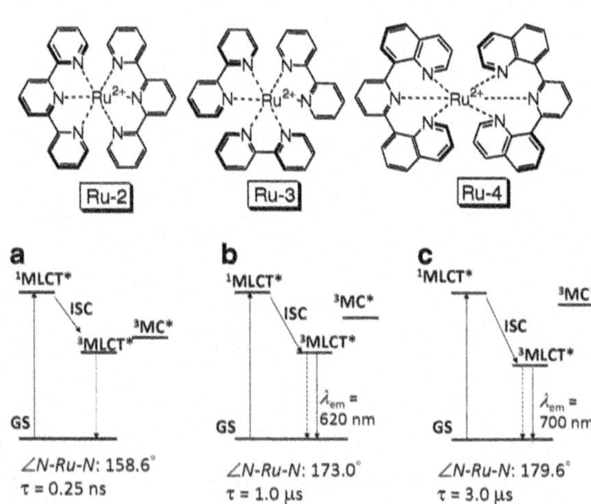

Fig. 7.3 Molecular structures of typical Ru (II) polyimine complexes **Ru-2**, **Ru-3**, and **Ru-4**. *a*, *b*, and *c* are the simplified energy level diagram and the emission states for **Ru-2**, **Ru-3**, and **Ru-4**, respectively (Reprinted with permission from ref. [21]. Copyright © 2011, Royal Society of Chemistry)

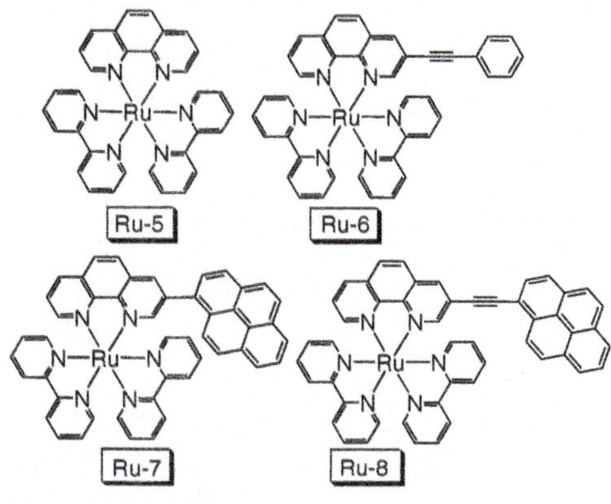

Fig. 7.4 Chemical structures of the photosensitizer Ru^{II} complexes **Ru-5**–**Ru-8**. Note the complexes are dications and the $[PF_6]^-$ ions are omitted for clarity. The compounds are from Ref. [35]

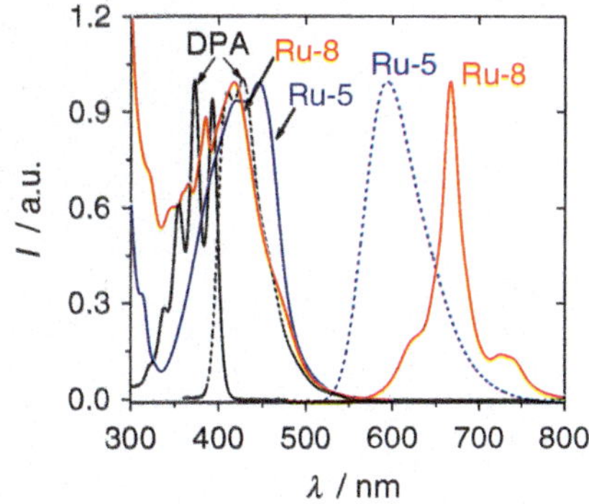

Fig. 7.5 Normalized absorbance (*solid lines*) and emission spectra (*dotted lines*) of DPA, **Ru-5** and **Ru-8** in acetonitrile (1.0×10^{-5} M). DPA, $\lambda_{ex} = 380$ nm; **Ru-5**, $\lambda_{ex} = 446$ nm; **Ru-8**, $\lambda_{ex} = 418$ nm. 25° C. Reprinted with permission from Ref. [43] (Copyright © 2013 WILEY-VCH Verlag Gmbh & Co. KGaA, Weinheim)

photophysical process involved in the luminescent O_2 sensing (TTET) is similar to that of TTA upconversion (Scheme 7.1), thus we envisaged that the TTA upconversion be enhanced with these Ru(II) polyimine complexes showing long-lived ^{3}IL excited state.

The absorption and phosphorescence of **Ru-5** and **Ru-8** were compared (Fig. 7.5). **Ru-8** gives more intense absorption than the model complex **Ru-5**. The typical MLCT emission band was observed for **Ru-5** (structureless). For **Ru-8**, however, the structured ^{3}IL emission band was observed. **Ru-7** also shows prolonged lifetime compared to the model complex **Ru-5**, due to the ^{3}MLCT/^{3}IL excited state equilibrium [35].

The TTA upconversion with **Ru-5**–**Ru-8** as triplet photosensitizers was studied (Fig. 7.6). In the presence of triplet energy acceptor DPA, the phosphorescence of the photosensitizers was quenched (Fig. 7.6b). Concomitantly, the upconverted blue emission of DPA was observed in the region of 400–550 nm. The upconversion is significant for **Ru-7** ($\Phi_{UC} = 9.8$ %) and **Ru-8** ($\Phi_{UC} = 9.6$ %) [43, 44]. The more efficient TTA upconversion with **Ru-7** and **Ru-8** than that with the model

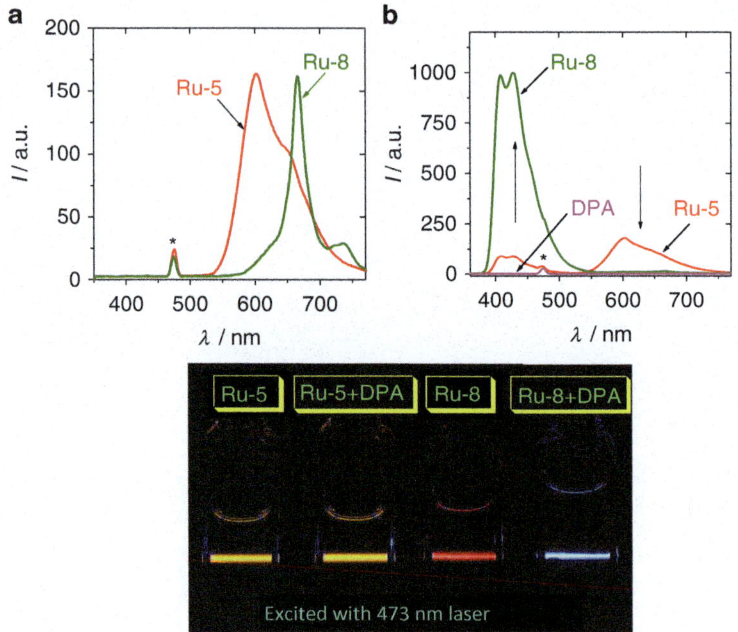

Fig. 7.6 Emission and TTA upconversion with **Ru-5** and **Ru-8** as triplet photosensitizers upon 473 nm laser excitation. (**a**) Emission of the Ru(II) complexes. Excited by blue laser ($\lambda_{ex} = 473$ nm, 5 mW). (**b**) The upconverted DPA fluorescence and the residual phosphorescence of the mixture of DPA (4.3×10^{-5} M) and **Ru-5** or **Ru-8**, respectively. (**c**) The photographs of the upconversion (samples of **a** and **b**). In deaerated CH_3CN solution. The complexes solution are 1.0×10^{-5} M. The *asterisks* in (**a**) and (**b**) indicate the scattered 473 nm excitation laser. 25°C (Reprinted with permission from Ref. [43]. Copyright © 2013 WILEY-VCH Verlag Gmbh & Co. KGaA, Weinheim)

complexes **Ru-5** ($\Phi_{UC} = 0.9$ %) and **Ru-6** ($\Phi_{UC} = 4.5$ %) is attributed to the long-lived T_1 excited state of **Ru-8** and **Ru-7**, with which the critical process of the TTA upconversion, i.e., the TTET process, was enhanced. Under similar conditions, $[Ru(dmb)_3]^{2+}$ gives Φ_{UC} value of only 1.0 %. The TTET process of the TTA upconversion can be evaluated by the quenching of phosphorescence of photosensitizers with acceptor DPA (Fig. 7.7). The largest Stern–Volmer quenching constant was observed for **Ru-8** (9.93×10^5 M^{-1}). The K_{SV} value of **Ru-8** is much larger than that of **Ru-5** (4.45×10^3 M^{-1}) and **Ru-6** (4.59×10^3 M^{-1}). Large value was also observed for **Ru-7** ($\tau = 9.22$ μs. $K_{SV} = 1.70 \times 10^5$ M^{-1}).

The upconversion quantum yield with **Ru-8** as the photosensitizer is much higher than that with **Ru-5** or $Ru(dmb)_3$ as the photosensitizers. We noticed that the upconversion of **Ru-8** is not significantly higher than that of **Ru-7**, despite the much longer T_1 lifetime of **Ru-8** than that of **Ru-7**. We attributed this small upconversion quantum yield to the lower T_1 excited state energy level of **Ru-8**.

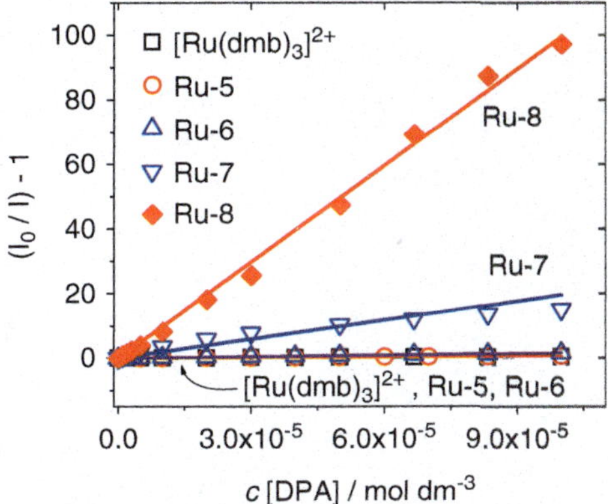

Fig. 7.7 Stern–Volmer plots generated from intensity quenching of complex $[Ru(dmb)_3]^{2+}$ ($\lambda_{ex} = 460$ nm), **Ru-5** ($\lambda_{ex} = 446$ nm), **Ru-6** ($\lambda_{ex} = 450$ nm), **Ru-7** ($\lambda_{ex} = 450$ nm), and **Ru-8** ($\lambda_{ex} = 418$ nm). Phosphorescence measured as a function of DPA concentration in CH_3CN. 1.0×10^{-5} mol dm^{-3}. 25°C. (Reprinted with permission from Ref. [43]. Copyright © 2013 WILEY-VCH Verlag Gmbh & Co. KGaA, Weinheim)

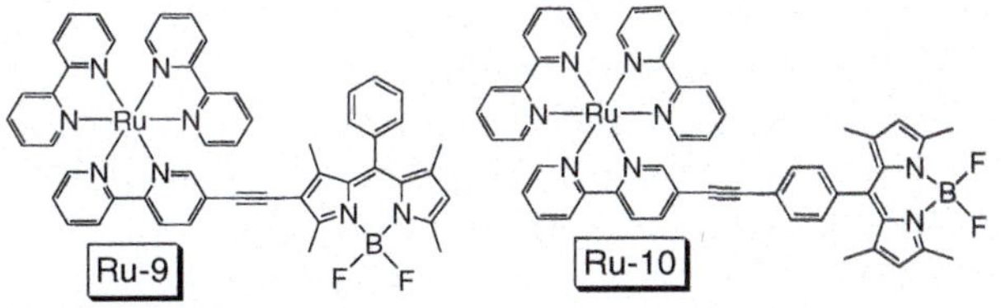

Fig. 7.8 Chemical structures of the photosensitizer Ru^{II} complexes **Ru-9** and **Ru-10**. Note the complexes are dications and the $[PF_6]^-$ ions are omitted for clarity. The compounds are from Ref. [32]

7.2.1.2 BODIPY-Ru(II) Polyimine Complexes as Triplet Photosensitizers

A straightforward approach to prepare a transition metal complex that shows strong absorption in the visible light region is to attach a visible light-harvesting chromophore, typically a fluorophore, such as boron dipyrromethene (BODIPY) [45–48], to the coordination center [49–51].

Till date the visible light-absorbing BODIPY-containing Ru(II) complexes were not studied for TTA applications [49–51]. Recently we reported two Ru (II) polyimine complexes: **Ru-9** and **Ru-10** (Fig. 7.8) [32]; visible light absorbing and exceptionally long-lived triplet excited states were observed. In these complexes, the BODIPY units were either linked by π-conjugation to the coordination center via a C^C bond (**Ru-9**, $\tau_T = 279.7 \mu s$) or tethered on the N^N coordination framework (**Ru-10**, $\tau_T = 246.6$ μs). Application of these complexes as a triplet photosensitizer in triplet–triplet annihilation upconversion has been studied using perylene as the acceptor.

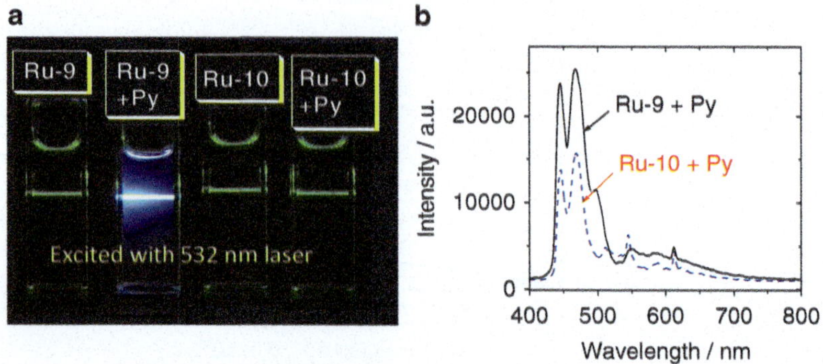

Fig. 7.9 Upconversion with **Ru-9**, **Ru-10** as triplet photosensitizers (1.0×10^{-5} M). (**a**) Photographs of the emission of photosensitizers alone and in the presence of perylene (Py) (4.1×10^{-5} M). (**b**) excited by the OPO laser at the isosbestic point of the UV/Vis absorption of **Ru-9** and **Ru-10** (509 nm). $\lambda_{ex} = 532$ nm, 11.2 mW (MeCN, 20° C) (Reprinted with permission from Ref. [32]. Copyright © 2013, Royal Society of Chemistry)

In the presence of perylene, significant upconverted blue emission was observed in the region of 400 nm−500 nm for **Ru-9** as compared to **Ru-10** (Fig. 7.9b). The TTA upconversion quantum yield of Ru-**9** ($\Phi_{UC} = 1.2$ %) was found to be 2-fold of **Ru-10** ($\Phi_{UC} = 0.7$ %), indicating that direct connection of the π-core of the BODIPY chromophore to the coordination center, i.e., by establishing π-conjugation between the visible light-harvesting chromophore and the metal coordination center, is essential to enhance the effective visible light harvesting of the Ru(II) complexes. To the best of our knowledge, **Ru-9** is the first reported green light excitable Ru(II) complex used for TTA upconversion [21, 32, 52].

7.2.2 Pt(II)/Pd(II)/Zn Porphyrin Complex for TTA Upconversion

Pt(II) porphyrin complexes, such as PtOEP (**Pt-1**, OEP = octaethylporphyrin), have been used in luminescent O_2 sensing and photodynamic therapy; both applications are based on the capability of visible light absorption and population of triplet excited state upon photoexcitation. Different from the typical Ru(II) polyimine complexes, the Pt(II) porphyrin complexes show intense absorption of visible light, and the lifetimes of the T_1 excited state of these complexes are much longer, usually longer than 50 μs [53–55].

Pt(II) porphyrin complexes are triplet emitters with moderate absorption in the green region. For example, the **Pt-1** (Fig. 7.10) shows absorption at ca. 400 nm with $\varepsilon = 1.0\text{-}5.0 \times 10^5\ M^{-1}\ cm^{-1}$. But usually the absorption at longer wavelength, i.e., 530 nm, is much weaker. Triplet state energy of the porphyrin complexes is 1.33–1.93 eV. Thus, **Pt-1** was used with DPA (with T_1 energy of 1.77 eV or 700 nm) for noncoherently excited annihilation upconversion [56]. It should be pointed out that

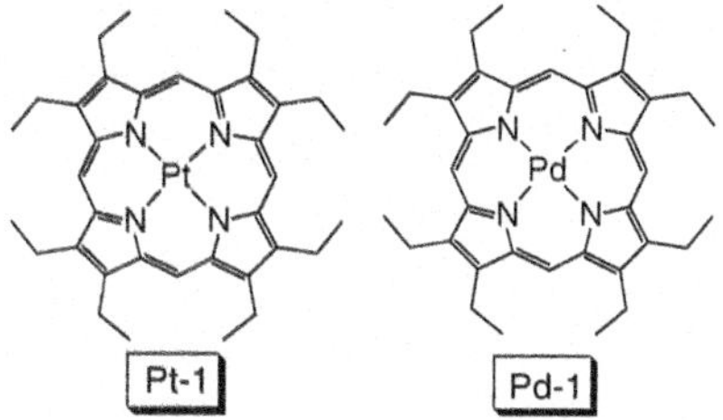

Fig. 7.10 Molecular structures of platinum and palladium octoethylporphyrin complex **Pt-1** and **Pd-1**

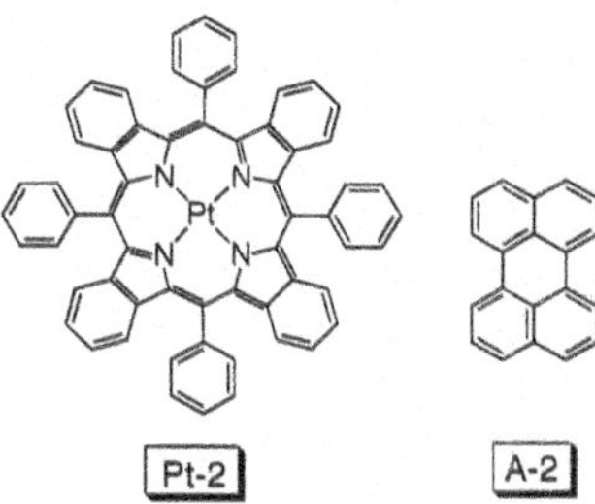

Fig. 7.11 Molecular structures of triplet photosensitizer platinum(II) tetraphenyltetrabenzoporphyrin (**Pt-2**) and the triplet acceptor (Perylene, **A-2**) [57]

the complex **Pt-1**, or other Pt porphyrin complexes, usually shows much longer triplet excited state lifetime than the Ru(II) polyimine complexes. The long-lived triplet excited state of the photosensitizer is beneficial for TTET process and the TTA upconversion.

Baluschev carried out TTA upconversion with focused solar light as the excitation source, and the external efficiency was 1 % (the excitation power density is 10 W cm^{-2}) [16]. Pd-1 (PdOEP) was used as triplet photosensitizer and DPA was used as triplet acceptor (Fig. 7.10).

Castellano shows that the **Pt-1**/DPA upconversion is effective even in polymer films with low glass transition temperature. Excitation is at 544 nm [56]. It is interesting that the upconversion works in the solid matrix in aerobic atmosphere and with excitation at low excitation power density of 6–27 mW cm^{-2}. This result paved the way for practical application of the TTA upconversion.

In order to use red light to perform the upconversion, a triplet photosensitizer with red-light absorption has to be used. Red-absorbing photosensitizer platinum (II) tetraphenyltetrabenzoporphyrin (**Pt-2**) (Fig. 7.11) and palladium porphyrin complex **Pd-1** were used as triplet photosensitizers (Fig. 7.8) [57–60].

The platinum (II) tetraphenyltetrabenzoporphyrin complex **Pt-2** (Fig. 7.11) shows strong absorption at 430 nm (Soret band) and an absorption at 611 nm (Q-band). The complex shows phosphorescence at 770 nm (1.61 eV, $\tau = 41.5$ μs). DPA is an inappropriate triplet acceptor in this case, due to its unmatched triplet state (T_1) energy level (1.77 eV), which is higher than the photosensitizer, and the TTET from the photosensitizer to the acceptor will be frustrated. Perylene (**A-2**) was selected as the triplet acceptor, for which the T_1 state energy level is 1.53 eV

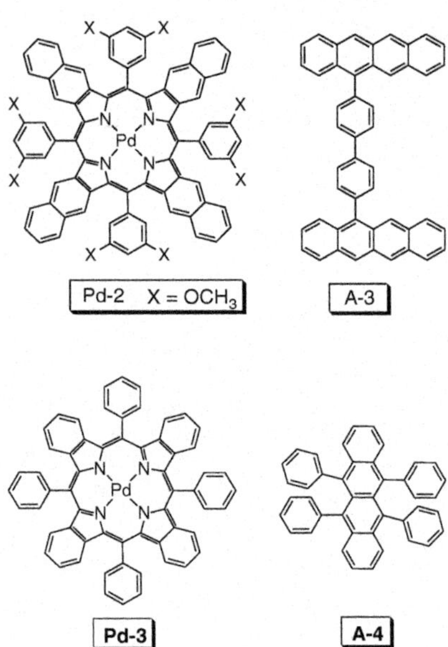

Fig. 7.12 Molecular structures of palladium complex **Pd-2** and the triplet acceptor **A-3** used for the upconversion. The compounds are from Ref. [58]

Fig. 7.13 Molecular structures of **Pd-3** (photosensitizer) and rubrene **A-4** (acceptor). The compounds are from Ref. [59]

(Fig. 7.11). The UV-vis absorption of perylene is in the region < 450 nm. With red excitation (635 nm laser), the blue/green fluorescence of perylene was observed. The upconversion quantum yield (Φ_{UC}) is 0.65 % [57]. Several factors may be responsible for the low upconversion quantum yield, such as the large size of the photosensitizer molecule, which may reduce the diffusion ability of the photosensitizer at triplet excited state [20, 21, 28].

For applications such as DSCs, the challenge is to effectively harvest the energy of the solar light in the red/near-IR region, where the normal organic dyes show poor absorption. Recently a palladium porphyrin complex that shows absorption in near-IR region was used for TTA-based upconversion (Fig. 7.12) [58]. The complex shows intense absorption at ca. 700 nm and the phosphorescence is at 916/942 nm. The near-IR (NIR) absorption of the Pd(II) complex is in particular significance since the low-energy NIR light of the solar radiance can be harvested with this photosensitizer. However, as the energy level of T_1 state of the complex is low, thus a triplet acceptor with low and matched T_1 energy level has to be used (**A-3**, Fig. 7.12), which shows green fluorescence emission in the region of 490–600 nm. The energy level of the triplet excited state of this compound is presumably lower than 1.32 eV (942 nm). With excitation at 700 nm, green emission of the acceptor/annihilator in the range of 490–600 nm was observed. The external upconversion quantum yield was determined as 4 %.

The authors also performed the upconversion with focused solar radiance. This is particularly interesting because the efficiency of the DSCs may be improved with the NIR absorbing upconversion schemes.

In order to harvest broadband excitation light, such as solar light, two photosensitizers were simultaneously used for TTA upconversion [59]. The two complexes used as triplet photosensitizers are **Pd-2** and **Pd-3**, respectively (Figs. 7.12, 7.13).

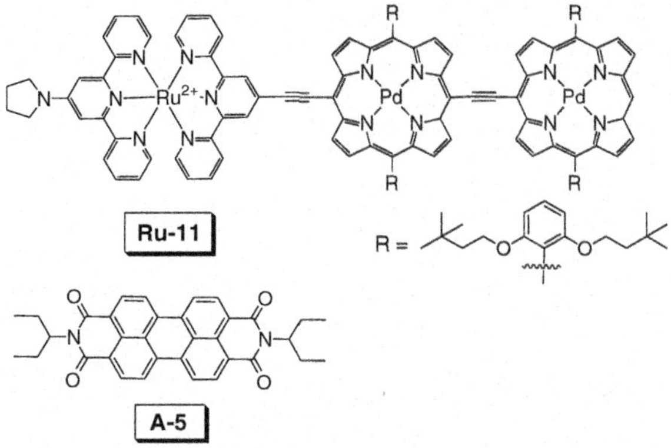

Fig. 7.14 Molecular structures of the near-IR absorbing photosensitizer **Ru-11** and the triplet acceptor **A-5**. The compounds are from Ref. [61]

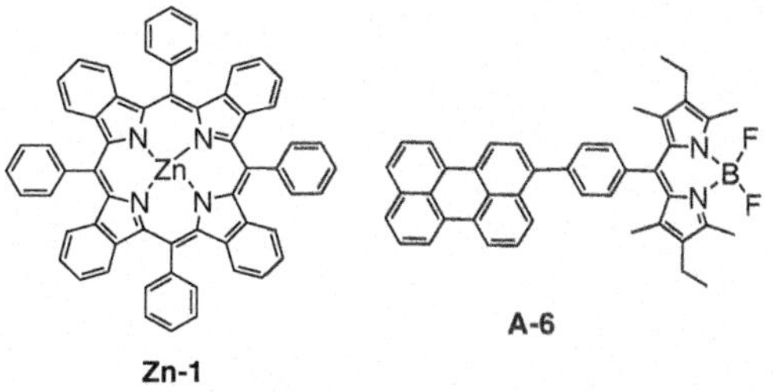

Fig. 7.15 ZnTPTBP **Zn-1** and **A-6** were used as triplet photosensitizer and triplet energy acceptor for TTA upconversion. The compounds Zn-1 and A-6 are from Refs. [63] and [17], respectively

The two complexes give absorption band at 630 nm and 700 nm, respectively. Thus, with the solar light as the excitation source, upconversion was observed with rubrene (**A-4**, Fig. 7.13) as the triplet acceptor.

In 2010, Castellano reported a TTA upconversion with NIR absorbing triplet photosensitizer **Ru-11** (Fig. 7.14) [61]. The excitation was carried out at 780 nm, and the upconverted emission of the perylenebisimide is at 541 nm. The upconversion quantum yield was determined as (0.75 ± 0.02) %.

Although the Pt(II)/Pd(II) porphyrin complexes have been successfully used as triplet photosensitizers for TTA upconversion, the limitation of these complexes is that the absorption/emission wavelength of the Pt(II)/Pd(II) complexes cannot be readily changed by molecular structure modification. Thus, it is desired that an alternative type of photosensitizer can be developed that shows tunable excitation/emission wavelength [3, 4, 20, 21].

Since some zinc porphyrin complexes show significant intersystem crossing, the application of ZnTPP (meso-tetraphenylporphine zinc) [62] and red-light excitable Zn(II) tetraphenyltetrabenzoporphyrin (TPTBP) **Zn-1** in TTA upconversion was studied (Fig.7.15) [63]. The triplet state lifetime of ZnTPTBP is 155.7 μs. TTA upconversion quantum yield of 0.76 % was observed with **A-6** as triplet acceptor. Note **A-6** is a perylene-BODIPY dyad, in which the perylene part is as the triplet acceptor, and this is a *hetero* annihilation between different chromophores. The S_1 state of the BODIPY part is with lower energy level than that of perylene; thus, the relation of $2E_{S1} > E_{T1}$ can be met [17].

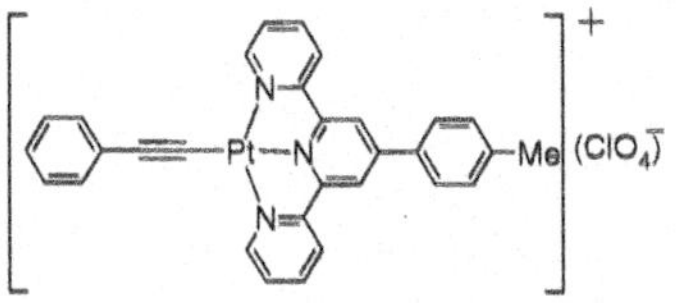

Fig. 7.16 The N^N^N Pt (II) acetylide complex **Pt-3** used for TTA upconversion. The upconversion with **Pt-3** was reported in Ref. [64]

7.2.2.1 Pt(II) Acetylide Complexes as the Triplet Photosensitizers: Tunable Photophysical Properties

N^N Pt(II) acetylide complexes are usually phosphorescent at room temperature, and the fluorescence of the ligands is completely quenched in the complexes, indicating efficient ISC process. The principal photophysical processes of the Pt(II) acetylide complexes are similar to that of the Ru(II) polyimine complexes [2–4]. Excitation into the ^{1}MLCT excited state is followed by an efficient ISC to the triplet excited state, which was identified as ^{3}MLCT/^{3}LLCT transition. Pt (II) acetylide complexes are featured with high phosphorescence quantum yield (up to 40 %) and readily tunable photophysical properties by simply changing the structure of the acetylide ligand [64–68]. In 2010, a trident Pt(II) acetylide complex, *N^N^N* Pt(II) phenylethynyl **Pt-3**, was used for TTA-based upconversion (Fig. 7.16) [64]. This complex shows moderate UV-vis absorption in the visible region (400 nm–550 nm) and a phosphorescence lifetime of 4.6 μs. The complex gives emission at 613 nm, as a broad emission band, due to the ^{3}MLCT nature of the excited state. Accordingly, DPA, a triplet acceptor with match energy level of T_1 excited state (1.77 eV, 700 nm), was selected as the triplet acceptor/annihilator/emitter. With selective excitation at either 500 nm or 514.5 nm (laser), the upconverted blue emission of DPA was observed in the range of 400 nm–500 nm. The upconversion quantum yield (Φ_{UC}) was determined as 0.2–1.1 %.

We propose that with N^N Pt(II) acetylides complexes that show strong absorption and longer triplet excited state lifetimes, the upconversion quantum yield (Φ_{UC}) may be greatly improved. Following this line, we prepared a fluorene, naphthalimide (NI), coumarin, and naphthaldiimide (NDI)acetylide Pt (II) complex, which shows intense absorption in visible region, as well as long-lived triplet excited states [3, 4].

7.2.2.2 Fluorene-Pt(II) Acetylide Complexes as the Triplet Photosensitizers

On attaching the fluorene chromophore to the Pt(II) center of the *N^N* Pt (II) bisacetylide complexes (**Pt-4–Pt-9**) (Fig. 7.17), the absorption in the visible region and long-lived triplet excited states was enhanced, especially for the complexes containing anthranyl (**Pt-6**, $\lambda_{abs} = 445$ nm, $\varepsilon = 8.54 \times 10^4$ M^{-1} cm^{-1}, $\tau_T = 66.7$ μs) and pyrenyl (**Pt-7**, ($\lambda_{abs} = 420$ nm, $\varepsilon = 9.70 \times 10^4$ M^{-1} cm^{-1},

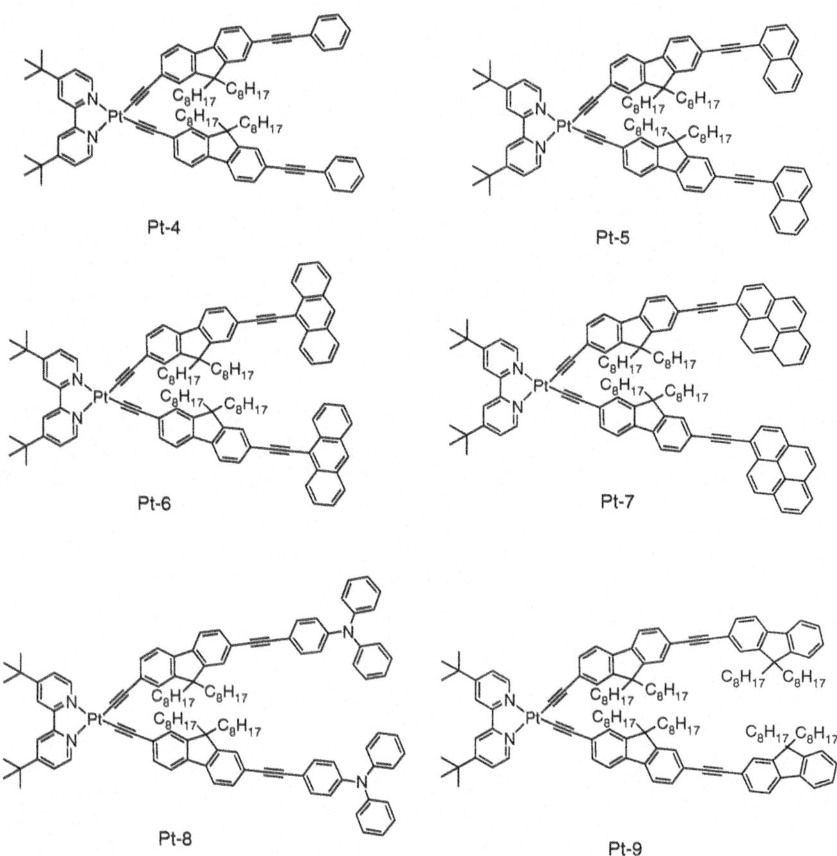

Fig. 7.17 The fluorene-*N^N* Pt(II) acetylide complexes **Pt-4–Pt-9** used for TTA upconversion [69]

$\tau_T = 54.7$ μs) fluorene ligands, compared to other complexes (τ_T is less than 2 μs). **Pt-4**, **Pt-5**, **Pt-8**, and **Pt-9** give strong emission at *ca.* 600 nm upon 445 nm laser excitation, whereas **Pt-7** exhibited weak luminescent and **Pt-6** is nonluminescent (Fig. 7.18a).

Strong upconverted blue fluorescence in the range of 400–550 nm with **Pt-7** as the triplet photosensitizer and triplet acceptor DPA **A-1** was observed along with minor upconversion with **Pt-5** as the triplet photosensitizer (Fig. 7.18b). For other complexes, however, the upconversion is negligible. Although the emissions of these complexes are much stronger than that of **Pt-7**, the lack of upconversion with **Pt-4**, **Pt-8**, and **Pt-9** is attributed to the short T_1 state lifetime and the weak absorption at the excitation wavelength; thus, the population of the triplet excited photosensitizers is under the threshold for TTA upconversion. Consequently, based on the result of **Pt-7** with a high upconversion quantum yield of 22.4 %, it is

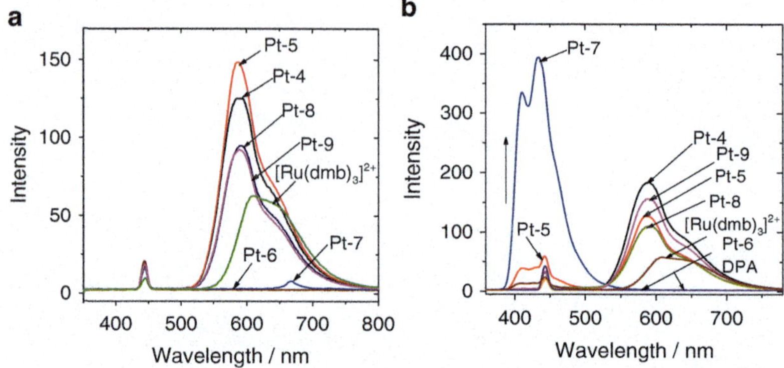

Fig. 7.18 Emission and upconversion with the complexes **Pt-4–Pt-9** as triplet photosensitizers upon 445 nm laser excitation. (**a**) Emission of the Pt(II) complexes without DPA. (**b**) Upconverted emission in the presence of DPA, $\lambda_{ex} = 445$ nm, 5 mW, *c*[photosensitizer] = 1.0×10^{-5} M, *c* [DPA] = 4.3×10^{-5} M in toluene, 25° C (Reprinted with permission from ref. [69]. Copyright © 2012, Royal Society of Chemistry). Later we also attached different aryl groups to the fluorene moiety, such as phenylacetylide **Pt-13**, naphthalimide-4-acetylide **Pt-10**, and in **Pt-11**, the fluorene moiety was altered to carbozale moiety (Fig. 7.19) [70]. With the fluorene linker between the arylacetylide and the Pt(II) center, the absorption of complexes in the visible range were intensified with longer triplet excited state lifetime and room temperature (RT) phosphorescence. For **Pt-10** ($\lambda_{abs} = 450$ nm, $\varepsilon = 6.21$ x10^4 M^{-1} cm^{-1}), **Pt-11** ($\lambda_{abs} = 458$ nm, $\varepsilon = 2.87 \times 104$ M^{-1} cm^{-1}), **Pt-12** ($\lambda_{abs} = 414$ nm, $\varepsilon = 3.91 \times 10^4$ M^{-1} cm^{-1}), **Pt-13** ($\lambda_{abs} = 424$ nm, $\varepsilon = 0.72 \times 10^4$ M^{-1} cm^{-1}), and **Pt-14** ($\lambda_{abs} = 373/431$ nm, $\varepsilon = 11.5/1.31 \times 10^4$ M^{-1} cm^{-1}), the triplet excited state lifetime was found to be 138.1 μs, 23.0 μs, 47.4 μs, 0.9 μs, and 0.7 μs, respectively. As a result, the significant upconversion for **Pt-10** ($\Phi_{UC} = 24.3$ %), **Pt-11** ($\Phi_{UC} = 14.7$ %), and **Pt-12**($\Phi_{UC} = 19.2$ %) as the triplet photosensitizer on addition of triplet acceptor DPA was observed as compared to **Pt-13** and **Pt-14** [71]. It should be pointed out that these high upconversion quantum yields are reasonable, although it was proposed that 11.1 % will be the maximal upconversion quantum yields [20, 21, 52]

proposed that weakly phosphorescent or non-phosphorescent transition metal complexes can be used as triplet photosensitizers for TTA upconversion, compared to the phosphorescent triplet photosensitizers [44, 69].

Herein, we propose to use a new parameter to evaluate the upconversion performance of the triplet photosensitizers (η, Eq. 5). Due to the prolonged T_1 excited state lifetime or the intensified absorption in the visible range, upconversion efficiency of **Pt-10** is 5.7-fold of **Pt-12** and **Pt-11** is 3.0-fold of **Pt-12**.

7.2.2.3 Coumarin-Pt(II) Acetylide Complexes as the Triplet Photosensitizers

Coumarin N^N Pt(II) acetylide complex has been reported, which shows intense absorption in visible spectral region (**Pt-15** in Fig. 7.20) [65]. DPA was used as triplet acceptor and upconversion quantum yield of 14.1 % was observed

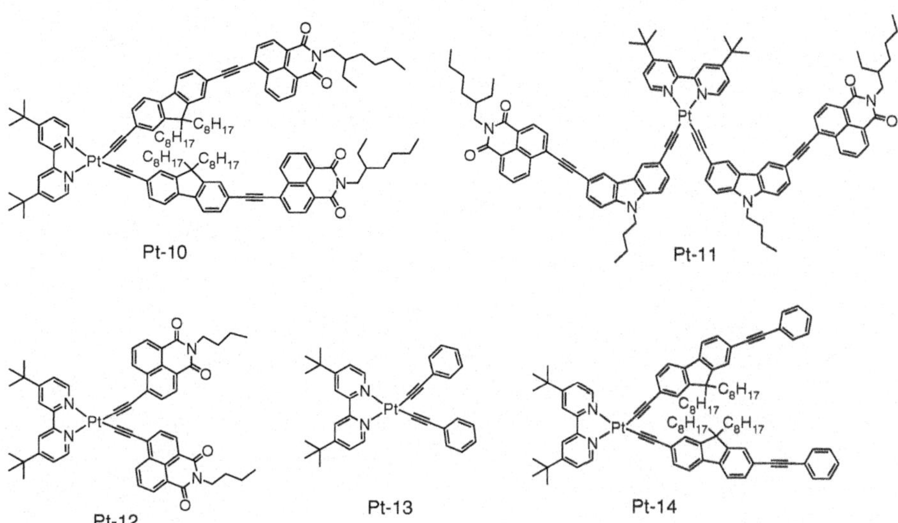

Fig. 7.19 Molecular structure of fluorine-N^N Pt(II) acetylide complexes **Pt-10–Pt-14** used for TTA upconversion[70]

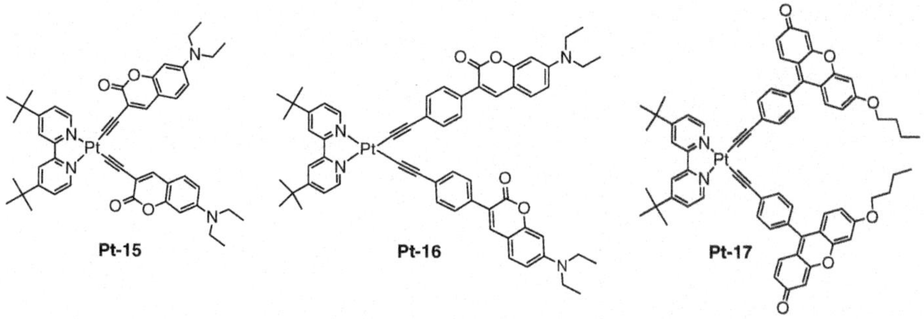

Fig. 7.20 N^N Pt(II) complexes **Pt-15**, **Pt-16**, and **Pt-17** that show intense absorption of visible light used as triplet photosensitizers for TTA upconversion. The compounds **Pt-15**, **Pt-16**, and **Pt-17** are from Refs. [65], [71], and [72], respectively

[65]. Under the same experimental conditions, the model complex *dbbpy* Pt (II) bisphenylacetylide gives upconversion quantum yield of 8.9 %. We compared the η value of the coumarin-containing complex **Pt-15** and the model complex *dbbpy* Pt(II) bisphenylacetylide **Pt-13**; the results show that the overall upconversion capability with **Pt-15** is improved by 7.3-fold over the model complex containing phenylacetylide. Recently, we reported Pt(*dbbpy*)(C^C-phenylene-coumarin) **Pt-16** with prolonged ^{3}IL lifetime of the T_1 state ($\tau_T = 20.15$ μs) due to the extension of the π-conjugation system inside the coumarin ligand, compared to that of the model complex *dbbpy* Pt(II) bisphenylacetylide ($\tau_T = 0.7$ μs) and **Pt-15** ($\tau_T = 2.44$ μs). As a result, the upconversion quantum yield of **Pt-16** was found to be increased up to 19.7 % [71].

On the other hand, if the coumarin moiety was replaced with the fluorescein chromophore in N^N Pt(II) bisacetylide complex **Pt-17** (Fig. 7.20), a decrease in upconversion quantum yield up to 10.7 % was observed due to the decrease in fluorescein localized intraligand triplet excited state lifetime ($\tau_T = 16.4$ μs) [72].

7.2.2.4 Rylene (NI or NDI)-Pt(II) Acetylide Complexes as the Triplet Photosensitizers

It should be noted that the absorption of **Pt-15** is still at short wavelength and the T_1 excited state lifetime is short (2.5 μs) [65]. Therefore, we prepared complex **Pt-18** (Fig. 7.21), in which the naphthalenediimide (NDI) was attached to the Pt(II) center via acetylide ligand [66]. Intense absorption in the visible region ($\lambda_{abs} = 583$ nm with $\varepsilon = 31{,}300$ M^{-1} cm^{-1}) and long-lived T_1 excited state was observed for the complex **Pt-18** ($\tau_T = 22.3$ μs); these photophysical properties are ideal for the complexes as triplet photosensitizers for TTA upconversion [20, 21]. The upconversion quantum yield (Φ_{UC}) of the complex was determined as 9.5 %. Under the same experimental conditions, no upconversion was observed for the model complex *dbbpy* Pt(II) bisphenylacetylide **Pt-13**.

A few years ago, some of us reported a naphthalimide (NI) acetylide-containing Pt(II) complex **Pt-12** (Fig. 7.19), which shows exceptionally long-lived ^{3}IL excited state ($\tau_T = 124$ μs) [37, 67]. Another Pt(II) bisacetylide complex **Pt-19** (Fig. 7.21) with long-lived T_1 excited state was reported by Castellano et al. [68]. Long-lived T_1 excited state of triplet photosensitizer is beneficial for the improvement of the efficiency of the TTET process, the critical process involved in the TTA upconversion. Exceptionally high upconversion quantum yields (Φ_{UC}) of 39.9 % were observed for **Pt-12** and 28.8 % for **Pt-19**.

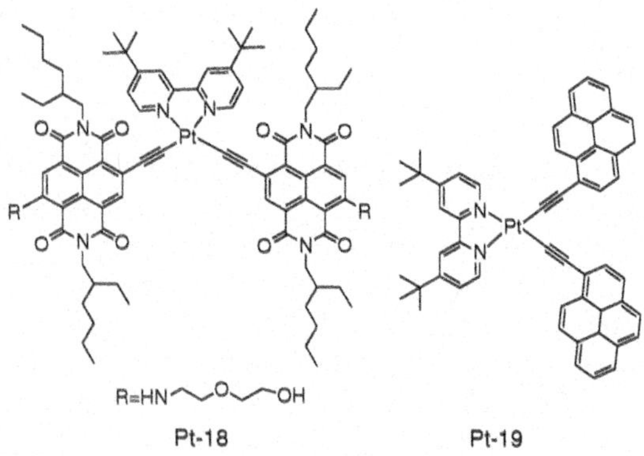

Fig. 7.21 N^N Pt (II) bisacetylide complexes that show prolonged T_1 excited state lifetimes ($\tau = 22.3$ μs for **Pt-18**, $\tau = 124.0$ μs for **Pt-12**, and $\tau = 73.6$ μs for **Pt-19**) used as triplet photosensitizers for TTA upconversion. The compounds are from Refs. [66], [37], and [68], respectively

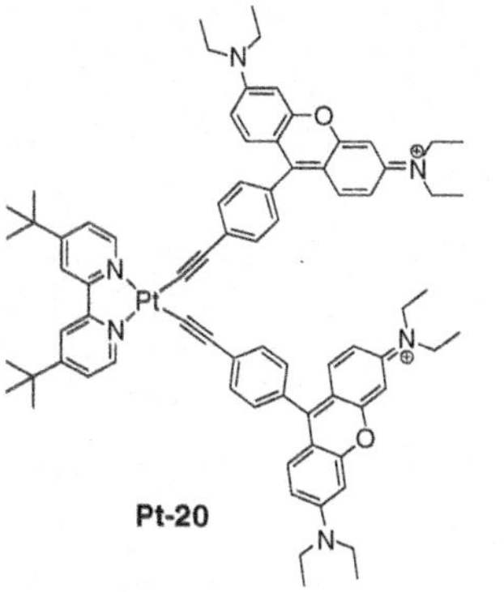

Fig. 7.22 *N^N* Pt (II) acetylide complex **Pt-20** containing rhodamine moiety. The complex shows strong absorption at 556 nm ($\varepsilon = 185{,}800\ M^{-1}\ cm^{-1}$) and long-lived non-emissive ^{3}IL excited state was observed ($\tau_T = 83.0\ \mu s$). The complex is from ref. [73]

Pt-20

7.2.2.5 Rhodamine-Pt(II) Acetylide Complexes as the Triplet Photosensitizers

A *N^N* Pt(II) acetylide complex with rhodamine moiety was reported by our group, in order to enhance the absorption in visible region and to access the long-lived ^{3}IL excited state localized on the rhodamine moiety (**Pt-20**. Fig. 7.22) [73]. The complex shows strong absorption at 556 nm ($\varepsilon = 185{,}800\ M^{-1}\ cm^{-1}$). By comparison, the intense absorption of **Pt-20** at 556 nm is due to the rhodamine acetylide ligand. Interestingly, only fluorescence (580 nm) and no phosphorescence was observed for **Pt-20** at either RT or 77 K. The assignment of fluorescence is based on the small Stokes shift of the emission band (24 nm), short luminescence lifetime (2.50 ns), and the O2-independent emission intensity [73].

Nanosecond time-resolved transient difference absorption spectra show that rhodamine-localized triplet excited state of **Pt-20** was populated upon excitation. The lifetime of the triplet excited state is 83.0 μs. This assignment of the triplet excited state as ^{3}IL state was supported by the position of the bleaching band and DFT calculations (spin density analysis of the triplet state of the complex).

Complex **Pt-20** was used as the triplet photosensitizer for TTA upconversion with perylene as the triplet acceptor, and upconversion quantum yield of 11.2 % was observed. Note that the overall upconversion capability of **Pt-20** is significant, due to its strong absorption at 556 nm ($\varepsilon = 185{,}800\ M^{-1}\ cm^{-1}$) [73].

7.2.2.6 BODIPY-Pt(II) Acetylide Complexes as the Triplet Photosensitizers

Recently, we reported the red-light excitable binuclear complexes **Pt-22–Pt-25** (Fig. 7.23), with two Pt(II) coordination centers connected to the π-core of the visible light-harvesting BODIPY ligands, showed red-shifted absorption (**Pt-23**, $\lambda_{abs} = 643$ nm, $\varepsilon = 42{,}300\ M^{-1}\ cm^{-1}$; **Pt-23**, $\lambda_{abs} = 643$ nm, $\varepsilon = 40{,}300\ M^{-1}\ cm^{-1}$) compared to the mononuclear BODIPY-Pt(II) complexes (**Pt-21** and **Pt-24**), (**Pt-21**, $\lambda_{abs} = 570$ nm, $\varepsilon = 38{,}300\ M^{-1}\ cm^{-1}$; **Pt-24**, $\lambda_{abs} = 602$ nm, $\varepsilon = 54{,}100\ M^{-1}\ cm^{-1}$) and exhibited long-lived BODIPY ligand-localized triplet excited states ($\tau_T = 57.9$–$72.4\ \mu s$) with nanosecond transient absorption spectra,

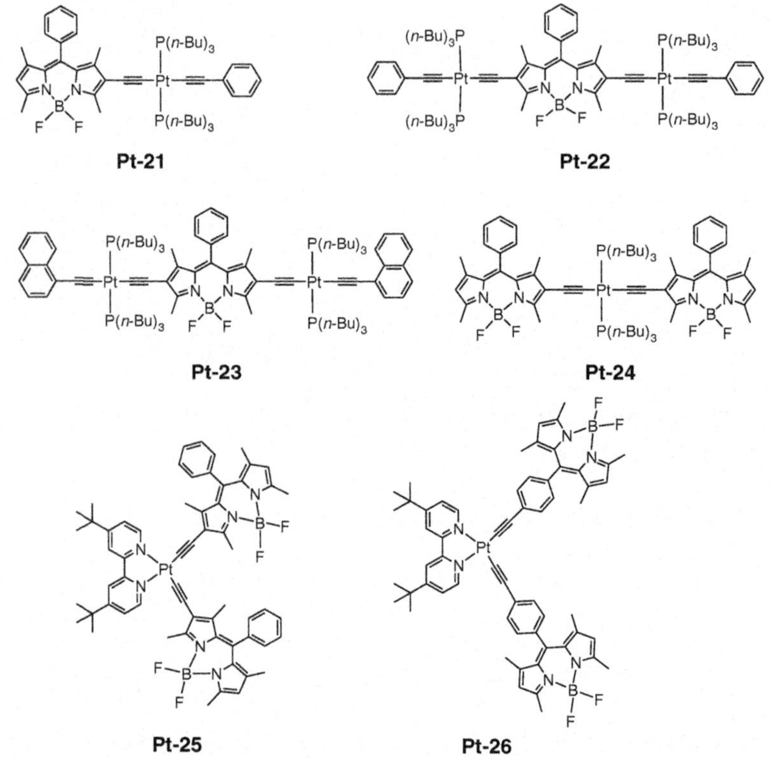

Fig. 7.23 BODIPY-Pt(II) acetylide complexes **Pt-21**, **Pt-22**, **Pt-23**, **Pt-24**, **Pt-25**, and **Pt-26**. The complex **Pt-21**–**Pt-24** and **Pt-25**–**Pt-26** is from Refs. [74] and [29], respectively

which is supported by spin density analysis. The platinum(II) bis(phosphine) bis (aryleneethynylene) complexes are used as triplet photosensitizers for the first time for red-light excited triplet–triplet annihilation (TTA)-based upconversion [74].

On addition of the triplet acceptor perylene to a solution of the complexes **Pt-21** and **Pt-24**, TTA upconverted blue emission in the range 400–550 nm was observed, upon excitation with a 589 nm laser. Interestingly, the upconverted emissions with **Pt-22** and **Pt-23** as the triplet photosensitizer were found to be much weaker than those with **Pt-21** and **Pt-24** as the triplet photosensitizer, and the TTA upconversion quantum yields of **Pt-21** ($\Phi_{UC} = 19.0$ %) and **Pt-24** ($\Phi_{UC} = 14.3$ %) are much higher than those of **Pt-22** ($\Phi_{UC} = 3.0$ %) and **Pt-23** ($\Phi_{UC} = 3.5$ %). **Pt-22** and **Pt-23** exhibited red-shifted fluorescence as compared to that of **Pt-21** and **Pt-24**, indicating that the energy levels of the T_1 states of **Pt-22** and **Pt-23** will probably be lower than those of **Pt-21** and **Pt-24**. With a view that a triplet acceptor perylenebisimide (PBI) with a lower T_1 state energy level (1.15–1.2 eV) than perylene (1.53 eV) may improve the TTA upconversion of the triplet photosensitizers **Pt-22** and **Pt-23**, a red laser (635 nm) was used for the excitation. Upconverted intense green emissions were observed for **Pt-22** (Fig. 7.24a) and **Pt-23** in the presence of

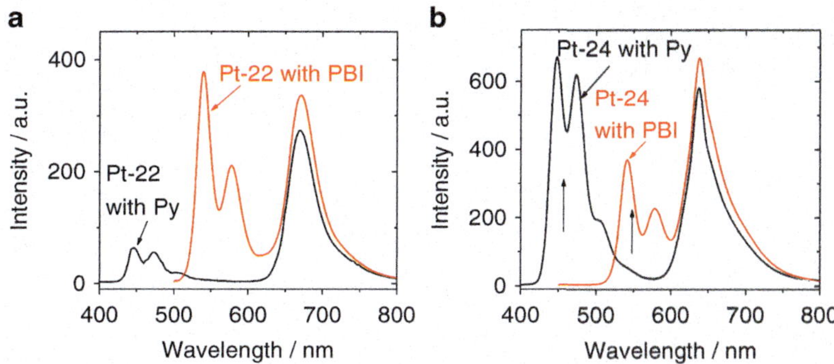

Fig. 7.24 Upconversions with (**a**) **Pt-22** and (**b**) **Pt-24** as the photosensitizers using perylene (Py) and PBI as the triplet acceptors. Excitation was with a 635 nm laser (10 mW, power density 70 mW cm^{-2}). c[photosensitizers] = 1.0×10^{-5} M. c[perylene] = 4.1×10^{-5} M. In deaerated toluene at 20 °C (Adapted from ref. [74] with permission)

PBI with increased TTA upconversion quantum yields 5.8 % (**Pt-22**) and 5.6 % (**Pt-23**) as compared to perylene triplet acceptor. The higher TTA upconversion quantum yields may be due to the larger driving force of the TTET between the triplet photosensitizer (**Pt-22** and **Pt-23**) and the triplet acceptor PBI [74].

On the other hand, opposite results were found for the photosensitizer **Pt-24** (Fig. 7.24b), using the triplet acceptor PBI which exhibits a lower T_1 state energy level than perylene. As a result, the energy gap between the triplet photosensitizer and the triplet acceptor became much larger. On the basis of this discussion, it was proposed that proper selection of the triplet acceptor is crucial for the TTA upconversion and an appropriate driving force for the TTET between the photosensitizer and the acceptor is needed for an upconversion system.

Two Pt(II) complexes in which the π-conjugation framework of BODIPY is either directly connected to the Pt(II) coordination center **Pt-25** or isolated from the Pt(II) center by the meso-phenyl moiety of the BODIPY chromophore were also reported (**Pt-26**, Fig. 7.23). The upconversion quantum yield was found to be 10.8 % and 7.4 % for **Pt-25** (Fig. 7.1) and **Pt-26**, respectively, in the presence of perylene triplet acceptor [29].

BF_2 or boron-containing fluorophores have been extensively used for molecular probes and molecular arrays due to the strong absorption in the visible region and high photostability [75–81]. Difluoroboron (BF_2) bound N^N Pt(II) bisacetylide complex (**Pt-27**, Fig. 7.25) was used as the triplet photosensitizer for triplet–triplet annihilation (TTA)-based upconversion, and an upconversion quantum yield of 8.9 % was observed with (DPA) as the triplet acceptor [82].

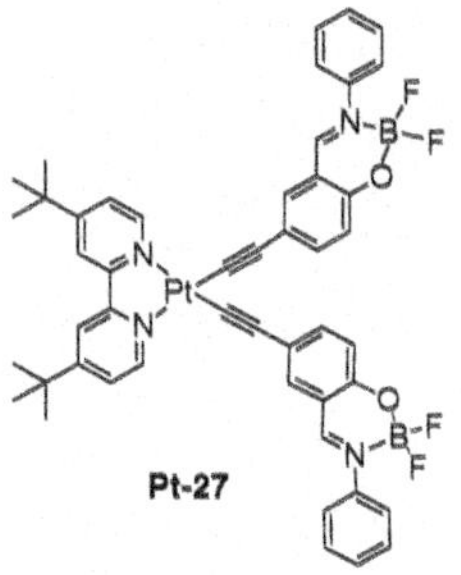

Fig. 7.25 BF_2 bound N^N Pt(II) acetylide complexes **Pt-27**. The complex is from ref. 82

7.2.2.7 Cyclometalated Pt(II)/Ir(III) Complexes as Triplet Photosensitizers

Most of cyclometalated Ir(III) or Pt(II) complexes show RT phosphorescence upon photoexcitation [1–4]. The photophysical properties of these complexes, such as the absorption wavelength, the emission wavelength, and the lifetime of the T_1 excited state, can be tuned by changing the C^N ligand. Cyclometalated Ir(III) complexes are normally used as the triplet emitters in organic light-emitting diode (OLED), due to their emissive triplet excited state. The emissive state of these complexes is generally assigned with ^{3}MLCT/^{3}IL mixed feature and the luminescence is characterized with long lifetime (in μs range). Therefore, these complexes can potentially be used as triplet photosensitizers for the TTA-based upconversion. However, it should be pointed out that the typical cyclometalated Ir(III) or Pt(II) complexes usually show weak absorption in the visible region and the lifetime of T_1 state is short (only a few μs) [1–4, 52].

Castellano et al. used complex **Ir-1** as the triplet photosensitizer for the TTA-based upconversion (Fig. 7.26) [83]. The lifetime of the triplet excited state of **Ir-1** is 1.55 μs and the energy level of the T_1 state is ca. 20,000 cm^{-1} (500 nm, 2.48 eV). However, the absorption of this complex is weak in visible spectral region. Pyrene and 3,8-di-(*tert*-butyl)pyrene were used as the triplet acceptor/annihilator, due to the appropriate energy level of the T_1 excited state (16,850 cm^{-1}, i.e., 593 nm, 2.09 eV). Upconverted blue fluorescence of pyrene was observed at 400 nm with selective excitation of **Ir-1** at 450 nm.

However, it should be pointed out that the UV-vis absorption of complex **Ir-1** is located in the UV and blue region and the molar absorption coefficient is only moderate. Furthermore, the lifetime of the T_1 excited state is short. Therefore, much room is left for the chemical modification of the molecular structure of the cyclometalated Ir(III) complexes to improve the UV-vis absorption property and thus to enhance the TTA upconversion with these complexes.

Inspired by Thompson's work [84], we prepared cyclometalated Ir(III) and Pt (II) complexes which contain coumarin ligand (**Ir-4** and **Ir-5**, Fig. 7.26, and **Pt-29–Pt-31**, Fig. 7.27) [85, 86]. The molecular design rationales are to prepare the Ir(III) and Pt (II) complexes that show intense absorption in the visible region and to access the long- lived ^{3}IL excited state. With DFT calculations, we predicted that

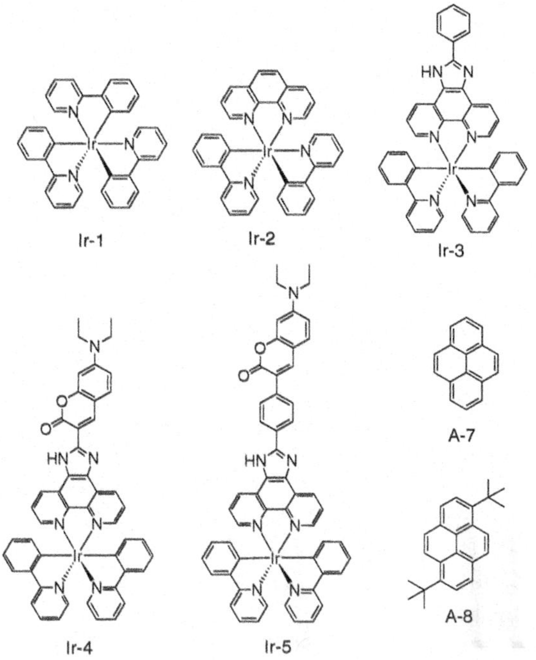

Fig. 7.26 Molecular structures of the triplet photosensitizers cyclometalated iridium complexes **Ir-1** (Ir(ppy)$_3$; ppy = 4-phenylpyridine), **Ir-2**, **Ir-3**, **Ir-4**, and **Ir-5** and the triplet acceptor pyrene and 3,8-di-*tert*-butylpyrene (**A-7** and **A-8**). Note the Ir(III) complexes are cations [83–85]

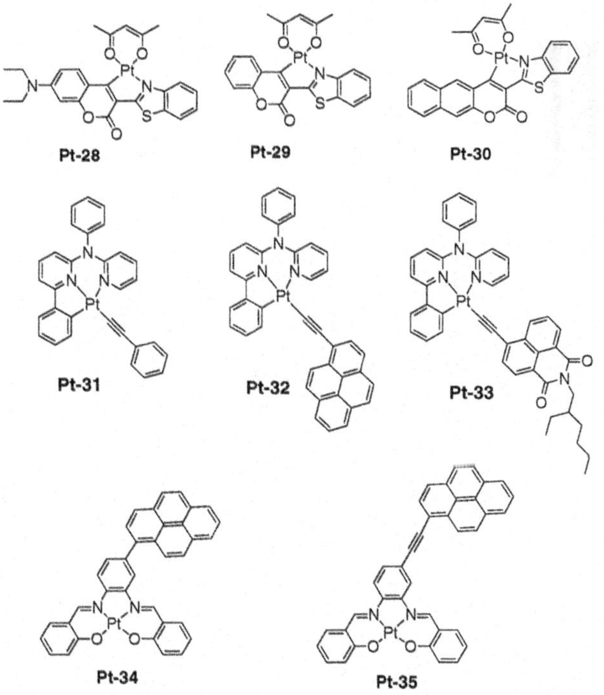

Fig. 7.27 Cyclometalated Pt (II) complexes **Pt-28–Pt-35** used for TTA upconversion. The complexes **Pt-28–Pt-30**, **Pt-31–Pt-33**, and **Pt-34–Pt-35** are from Refs. [86], [87], and [88], respectively

the T_1 energy level of the coumarin ligand will be close to the Ir(III) coordination center; thus, the intrinsic triplet excited state of the complex may be profoundly perturbed.

Upconversion with **Pt-29** ($\tau = 20.3$ μs, Fig. 7.25) as the photosensitizer and DPA as the acceptor gives upconversion quantum yield of 15.4 %. Upconversion with **Pt-30** and **Pt-31** (Fig. 7.27) gives much lower efficiency [86].

Recently, we reported tridentate cyclometalated platinum (II) acetylide complexes featuring a fused 5–6-membered metallacycle (C^N*N) Pt(II)L, where L = phenylacetylide (**Pt-32**), pyrenylacetylide (**Pt-33**), and naphthalimideacetylides (**Pt-34**. Fig. 7.27) with enhanced long-lived emissive triplet excited states for **Pt-33** ($\tau_T = 84$ μs) and **Pt-34** ($\tau_T = 136$ μs) due to ^{3}IL triplet excited states that localized on pyrene or NI acetylide ligands, respectively, when compared to the model complex **Pt-32** ($\tau_T = 9.2$ μs) in which the T_1 state is localized on the C^N*N ligand. Upconversion quantum yield of 19.5 % was observed with **Pt-34** with DPA as acceptor as the triplet photosensitizer was selectively photoexcited [87]. Pyrene based, the first Schiff base Pt(II) complexes **Pt-35** and **Pt-36** with long-lived ^{3}IL triplet excited state, 13.4 μs and 21.0 μs, exhibited upconversion quantum yield of 9.9 % and 17.7 %, respectively, in the presence of DPA triplet acceptor [88].

In contrast to the model complexes **Ir-2** and **Ir-3** (Fig. 7.26), both showing very weak absorption in visible region (e.g., $\varepsilon = 1{,}353\ M^{-1}\ cm^{-1}$ at 466 nm for **Ir-3**), the coumarin-containing **Ir-4** gives intense absorption ($\varepsilon = 70{,}920\ M^{-1}\ cm^{-1}$ at 466 nm) (Figs. 7.26 and 7.28). Furthermore, the T_1 excited state lifetimes of **Ir-2** and **Ir-3** are short (0.81 μs and 0.66 μs, respectively), and **Ir-4** shows profoundly prolonged T_1 excited state lifetime ($\tau_T = 75.5$ μs).

Interestingly, the coumarin-containing complexes **Ir-4** and **Ir-5** give much weaker emission than that of **Ir-2** and **Ir-3** (Fig. 7.28b) [85]. The emissive excited state of **Ir-4** and **Ir-5** was proposed to be ^{3}IL excited state by using nanosecond

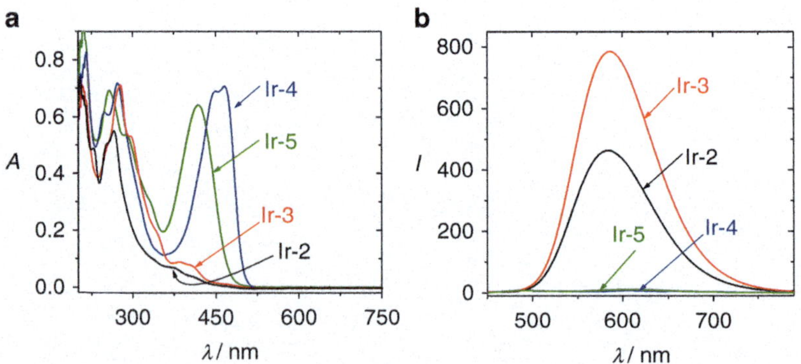

Fig. 7.28 (**a**) UV-vis absorption of **Ir-2**, **Ir-3**, **Ir-4**, and **Ir-5**. In CH_3CN (1.0×10^{-5} M; 20° C). (**b**) Emission spectra of the Ir^{III} complexes. **Ir-2**: $\lambda_{ex} = 386$ nm, **Ir-3**: $\lambda_{ex} = 407$ nm, **Ir-4**: $\lambda_{ex} = 462$ nm, **Ir-5**: $\lambda_{ex} = 421$ nm. In deaerated CH_3CN (1.0×10^{-5} M; 20 °C) (Reprinted with permission from ref. [85]. Copyright © 2011 WILEY-VCH Verlag Gmbh & Co. KGaA, Weinheim)

time-resolved transient absorption, 77 K emission spectra, and spin density analysis (DFT calculations). We propose that the weak emission of **Ir-4** and **Ir-5** does not necessarily deter the complexes from application for some photophysical process; thus, the complexes were used for TTA-based upconversion.

It is clear that the **Ir-4** and **Ir-5** are more efficient as triplet photosensitizers for TTA upconversion than the model complexes **Ir-2** and **Ir-3** (Fig. 7.29). For example, the upconversion quantum yields with **Ir-4** and **Ir-5** as triplet photosensitizers were determined as 21.3 % and 23.4 %, respectively. For **Ir-2** and **Ir-3**, however, no significant upconversion was observed (Fig. 7.29a) [85].

The upconversion are visible with unaided eyes. Herein, we noticed an interesting result that the quenched phosphorescence peak area of **Ir-4** and **Ir-5** is much smaller than that of the upconverted fluorescence peak area. The upconversion with **Ir-4** and **Ir-5** clearly shows that some photosensitizer molecules that are otherwise non-emissive were involved in the TTET process, that is, the dark excited states were effective as energy donor of the TTET process. Previously most transition

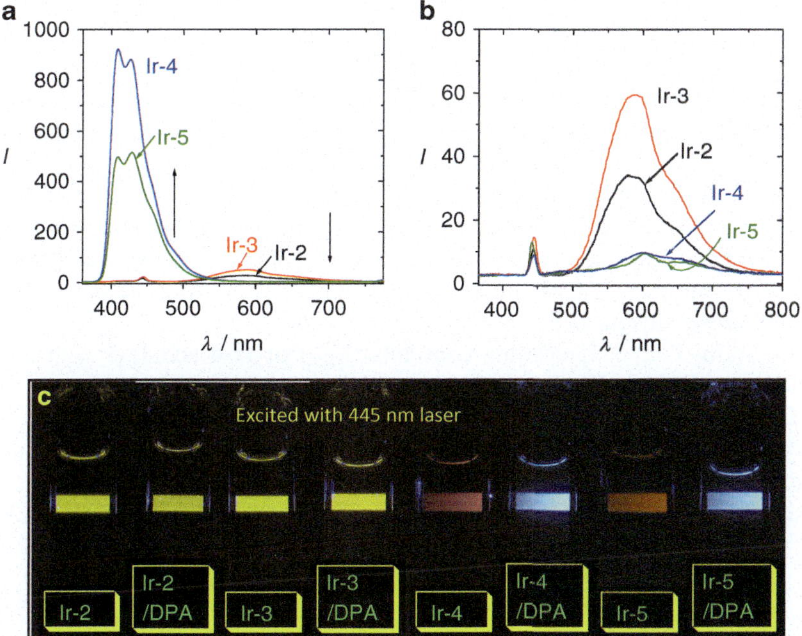

Fig. 7.29 Upconversion with Ir(III) complexes as the triplet photosensitizers and DPA as the triplet acceptor. (**a**) Upconversion emission spectra of the mixture of **Ir-2**, **Ir-3**, **Ir-4**, and **Ir-5** (1.0×10^{-5} M) with DPA (8.0×10^{-5} M). (**b**) Phosphorescence of photosensitizers alone ($\lambda_{ex} = 445$ nm, 5 mW). (c) Photographs of the upconversions. In deaerated CH_3CN. 20° C. (Reprinted with permission from ref. [85]. Copyright © 2011 WILEY-VCH Verlag Gmbh & Co. KGaA, Weinheim)

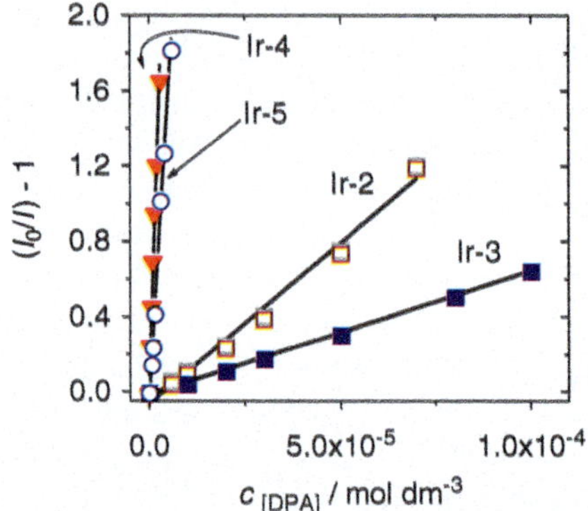

Fig. 7.30 Stern–Volmer plots generated from phosphorescence intensity quenching of complex **Ir-2** ($\lambda_{ex} = 386$ nm), **Ir-3** ($\lambda_{ex} = 410$ nm), **Ir-4** ($\lambda_{ex} = 475$ nm), and **Ir-5** ($\lambda_{ex} = 425$ nm). Phosphorescence was measured as a function of DPA concentration in CH_3CN. 1.0×10^{-5} mol dm^{-3}. 20° C (Reprinted with permission from ref. [85]. Copyright © 2011 WILEY-VCH Verlag Gmbh & Co. KGaA, Weinheim)

metal complexes used as triplet photosensitizers for TTA upconversion are phosphorescent. Our new concept to use dark triplet excited state to sensitize the TTET and the TTA upconversion will greatly increase the availability of the triplet photosensitizers for TTA upconversion [44, 84, 85].

The effect of the long-lived T_1 excited state on the efficiency of the TTET process is presented in Fig. 7.30. The slope of the quenching process with **Ir-4** and **Ir-5**, i.e., the quenching constants, is much larger than that with **Ir-2** and **Ir-3**. The quenching constants of **Ir-4** and **Ir-5** are 50-fold of that with **Ir-2** and **Ir-3** as the triplet photosensitizers.

The Stern–Volmer quenching constants (K_{SV}) of **Ir-4** and **Ir-5** with DPA as quencher were determined as 5.51×10^5 M^{-1} and 3.18×10^5 M^{-1}, respectively. For **Ir-2** and **Ir-3**, however, much smaller quenching constants of 1.71×10^4 M^{-1} and 6.57×10^3 M^{-1} were observed. Small Stern–Volmer quenching constants indicate relatively non-efficient TTET process.

Recently, we prepared a series of visible light-harvesting Ir(III) complex **Ir-6** ($\lambda_{abs} = 486$ nm, $\varepsilon = 1.09 \times 10^5$ M^{-1}cm^{-1}), **Ir-7** ($\lambda_{abs} = 484$ nm, $\varepsilon = 1.12 \times 10^5$ M^{-1} cm^{-1}) [89], **Ir-8** ($\lambda_{abs} = 473$ nm, $\varepsilon = 2.01 \times 10^4$ M^{-1} cm^{-1}), **Ir-9** ($\lambda_{abs} = 487$ nm, $\varepsilon = 4.20 \times 10^4$ M^{-1} cm^{-1}), **Ir-10** ($\lambda_{abs} = 437$ nm, $\varepsilon = 4.58 \times 10^4$ M^{-1}cm^{-1}) [90], **Ir-11** ($\lambda_{abs} = 403$ nm, $\varepsilon = 1.91 \times 10^4$ M^{-1} cm^{-1}), and **Ir-12** ($\lambda_{abs} = 466$ nm, $\varepsilon = 5.15 \times 10^4$ M^{-1} cm^{-1}) (Fig. 7.31) [89–91]. These complexes are with long-lived emissive T_1 excited state. The lifetimes of the triplet excited states of **Ir-6 - Ir-12** were determined as 7.6 μs, 54.5 μs, 65.9 μs, 34.8 μs, 2.7 μs, 56.1 μs, and 73.9 μs, respectively. These complexes were used as triplet photosensitizers for triplet–triplet annihilation upconversion, and upconversion quantum yields of 19.3 %, 12.7 %, 22.8 %, 9.8 %, 9.3 %, 0 %, and 23.7 % were observed for **Ir-6, Ir-7, Ir-8, Ir-9, Ir-10, Ir-11**, and **Ir-12**, respectively, upon 473 nm laser excitation on addition of DPA as triplet acceptor.

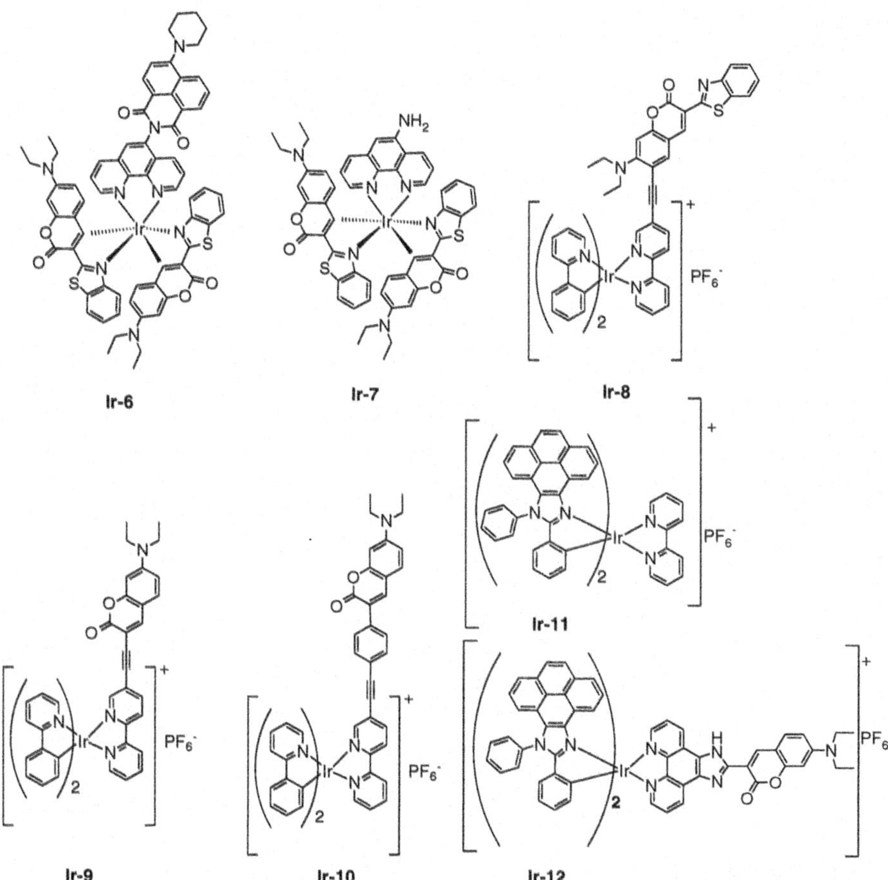

Fig. 7.31 Visible light-harvesting cyclometalated Ir(III) complexes **Ir-6–Ir-12** used for TTA upconversion. The complexes **Ir-6–Ir-7**, **Ir-8–Ir-10**, and **Ir-11–Ir-12** are from Refs [89], [90], and [91], respectively

Alongside, for the first time, we linked the π-conjugated core of naphthalenedimide (NDI) moiety to the Ir(III) center via C≡C triple bonds for the preparation of cyclometalated Ir(III) complex **Ir-13** (Fig. 7.32) which exhibited strong absorption of visible light ($\varepsilon = 11{,}000\ M^{-1}\ cm^{-1}$ at 542 nm) and a long-lived triplet excited state ($\tau_T = 130.0\ \mu s$) due to the NDI-localized intraligand triplet excited state (^3IL state). **Ir-13** was used as triplet–triplet annihilation upconversion in the presence of perylene acceptor upon 532 nm laser excitation. An upconversion quantum yield of 6.7 %, 238.6 μs of the upconverted luminescence lifetime of the blue emission band of the **Ir-13**/perylene mixture, and Stern–Volmer constant ($1.0 \times 10^6\ M^{-1}$) were observed, suggesting efficient TTET of **Ir-13** [92].

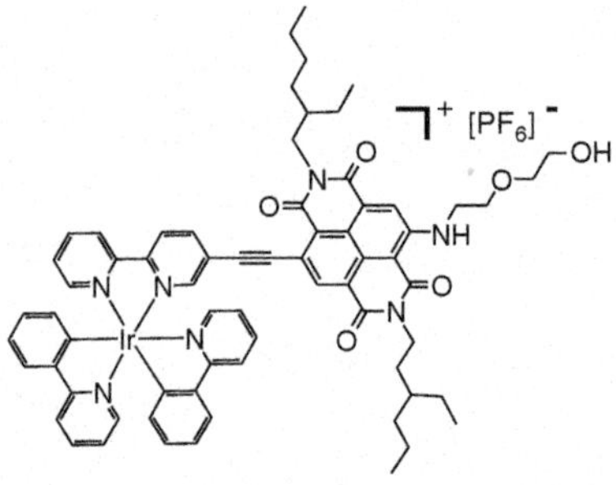

Fig. 7.32 NDI-cyclometalated Ir(III) complexes **Ir-13** used for TTA upconversion [92]

7.3 Re(I) Complexes as Triplet Photosensitizers

Re(I) complexes have attracted much attention due to the application as luminescent bioprobes and triplet photosensitizers for photocatalysis [93, 94]. As other transition metal complexes, most of these Re(I) complexes show weak absorption in the visible spectral region. For the first time, we linked BODIPY fluorophore to N^N Re(I) tricarbonyl chloride complexes (**Re-1** and **Re-2**) (Fig. 7.33) that give very strong absorption of visible light. **Re-1** (with BODIPY) and **Re-2** (with carbazole-ethynyl BODIPY) show exceptional strong absorption of visible light at 536 nm ($\varepsilon = 91{,}700\ M^{-1}\ cm^{-1}$) and 574 nm ($\varepsilon = 64{,}600\ M^{-1}\ cm^{-1}$) and long-lived triplet excited states ($\tau_T = 104.0$ μs for **Re-1**; $\tau_T = 127.2$ μs for **Re-2**), respectively [95].

Upconversion quantum yield of **Re-1** was ($\Phi_{UC} = 8.5$ %) 4-fold that of **Re-2** ($\Phi_{UC} = 2.0$ %) in the presence of perylene acceptor. The upconversion efficiency η value of **Re-1** was 15-fold that of **Re-2** with the 532 nm laser excitation. The Stern–Volmer quenching constants of **Re-1** ($K_{SV} = 6.75 \times 10^5\ M^{-1}$) and **Re-2** ($K_{SV} = 6.42 \times 10^5\ M^{-1}$) were found to be close to each other. Despite the similar quenching constants, **Re-2** shows much weaker TTA upconversion than **Re-1**, due to the weak absorption of **Re-2** at the excitation wavelength and the less efficient ISC of **Re-2**. It was observed that the TTA upconversion quantum yields of the Re (I) complexes **Re-1** and **Re-2** are 1.6-fold that of the efficient bis-iodo-BODIPY triplet photosensitizer, suggesting that the strong visible light absorption of the new Re(I) complex is effective, that is, the excitation energy harvested by the BODIPY ligand can be efficiently funneled to the triplet excited states of the complexes [95]. We also used coumarin ligands to prepare Re(I) complexes that show strong absorption of visible light and long-lived triplet excited states [96].

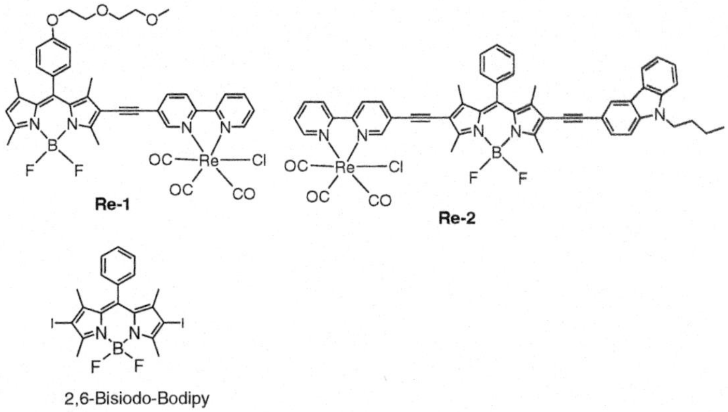

Fig. 7.33 Visible light-harvesting Re(I) complexes **Re-1** and **Re-2** used for TTA upconversion and bis-iodo-BODIPY [95]

7.4 Summary

In summary, visible light-harvesting transition metal complexes showing long-lived triplet excited states were developed and used as triplet photosensitizers for triplet–triplet annihilation (TTA) upconversion. Compared to the conventional transition metal complexes that have been used for TTA upconversion, the upconversion quantum yields were improved with the visible light-harvesting complexes. These complexes include the Ru(II), Ir(III), Pt(II), and Re(I). It should be pointed out that the application of the complexes is not limited to the TTA upconversion; these complexes may be also used for photocatalysis and photodynamic therapy, etc.

References

1. Flamigni L, Barbieri A, Sabatini C, Ventura B, Barigellet F (2007) Photochemistry and photophysics of coordination compounds: iridium. Top Curr Chem 281:143–203
2. Williams JAG (2007) Photochemistry and photophysics of coordination compounds: platinum. Top Curr Chem 281:205–268
3. Zhao J, Wu W, Sun J, Guo S (2013) Triplet photosensitizers: from molecular design to applications. Chem Soc Rev 42:5323–5351
4. Zhao J, Ji S, Wu W, Wu W, Guo H, Sun J, Sun H, Liu Y, Li Q, Huang L (2012) Transition metal complexes with strong absorption of visible light and long-lived triplet excited states: from molecular design to applications. RSC Adv 2:1712–1728
5. Kalyanasundaram K (1982) Photophysics, photochemistry and solar energy conversion with tris(bipyridyl)Ruthenium(II) and its analogues. Coord Chem Rev 46:159–244
6. Juris A, Balzani V, Barigelletti F, Campagna S, Belser P, Zelewsky AV (1988) Ru (II) polypyridine complexes: photophysics, photochemistry, electrochemistry, and chemiluminescence. Coord Chem Rev 84:85–277

7. Campagna S, Puntoriero F, Nastasi F, Bergamini G, Balzani V (2007) Photochemistry and photophysics of coordination compounds: ruthenium. Top Curr Chem 280:117–214
8. Wang X-Y, Guerzo AD, Schmehl RH (2004) Photophysical behavior of transition metal complexes having interacting ligand localized and metal-to-ligand charge transfer states. J Photochem Photobiol C 5:55–77
9. Medlycott EA, Hanan GS (2006) Synthesis and properties of mono- and oligo-nuclear Ru (II) complexes of tridentate ligands: the quest for long-lived excited states at room temperature. Coord Chem Rev 250:1763–1782
10. Armaroli N (2008) Electronic excited-state engineering. ChemPhysChem 9:371–373
11. Parker CA, Hatchard CG (1962) Sensitised anti-Stokes delayed fluorescence. In: An international symposium on carbohydrate chemistry. In: Proceedings of the Chemical Society, London, pp 386–387
12. Reinhard C, Valiente R, Güdel HU (2002) Exchange-induced upconversion in Rb2MnCl4:Yb3 +. J Phys Chem B 106:10051–10057
13. Haase M, Schafer H (2011) Upconverting nanoparticles. Angew Chem Int Ed 50:5808–5829
14. Etchart I, Berard M, Laroche M, Huignard A, Hernandez I, GillinWP CRJ, Cheetham AK (2011) Efficient white light emission by upconversion in Yb3+-, Er3+- and Tm3+-doped Y2BaZnO5. Chem Commun 47:6263–6265
15. Niu W, Wu S, Zhang S, Li L (2010) Synthesis of colour tunable lanthanide-ion doped NaYF4 upconversion nanoparticles by controlling temperature. Chem Commun 46:3908–3910
16. Baluschev S, Miteva T, Yakutkin V, Nelles G, Yasuda A, Wegner G (2006) Up-conversion fluorescence: noncoherent excitation by sunlight. Phys Rev Lett 97:143903
17. Turshatov A, Busko D, Avlasevich Y, Miteva T, Landfester K, Baluschev S (2012) Synergetic effect in triplet–triplet annihilation upconversion: highly efficient multi-chromophore emitter. ChemPhysChem 13:3112–3115
18. Baluschev S, Yakutkin V, Miteva T, Wegner G, Roberts T, Nelles G, Yasuda A, Chernov S, Aleshchenkov S, Cheprakov A (2008) A general approach for non-coherently excited annihilation up-conversion: transforming the solar-spectrum. New J Phys 10:013007
19. Monguzzi A, Mezyk J, Scotognella F, Tubino R, Meinardi F (2008) Upconversion-induced fluorescence in multicomponent systems: steady-state excitation power threshold. Phys Rev B 78:195112
20. Singh-Rachford TN, Castellano FN (2010) Photon upconversion based on sensitized triplet–triplet annihilation. Coord Chem Rev 254:2560–2573
21. Zhao J, Ji S, Guo H (2011) Triplet–triplet annihilation based upconversion: from triplet sensitizers and triplet acceptors to upconversion quantum yields. RSC Adv 1:937–950
22. Cheng YY, Khoury T, Clady RGCR, Tayebjee MJY, Ekins-Daukes NJ, Crossley MJ, Schmidt TW (2010) On the efficiency limit of triplet–triplet annihilation for photochemical upconversion. Phys Chem Chem Phys 12:66–71
23. Deng F, Blumhoff J, Castellano FN (2013) Annihilation limit of a visible-to-uv photon upconversion composition ascertained from transient absorption kinetics. J Phys Chem A 117:4412–4419
24. Singh-Rachford TN, Islangulov RR, Castellano FN (2008) Photochemical upconversion approach to broad-band visible light generation. J Phys Chem A 112:3906–3910
25. Simon YC, Weder C (2012) Low-power photon upconversion through triplet–triplet annihilation in Polymers. J Mater Chem 22:20817–20830
26. Castellano FN (2012) Transition metal complexes meet the rylenes. Dalton Trans 41:8493–8501
27. Huang X, Han S, Huang W, Liu X (2013) Enhancing solar cell efficiency: the search for luminescent materials as spectral converters. Chem Soc Rev 42:173–201
28. Monguzzi A, Tubino R, Meinardi F (2008) Upconversion-induced delayed fluorescence in multicomponent organic systems: role of Dexter energy transfer. Phys Rev B 77:155122

29. Wu W, Liu L, Cui X, Zhang C, Zhao J (2013) Red-light-absorbing diimine Pt(II) bisacetylide complexes showing near-IR phosphorescence and long-lived ^{3}IL excited state of Bodipy for application in triplet–triplet annihilation upconversion. Dalton Trans 42:14374–14379
30. Saltiel J, Marchand GR, Smothers WK, Stout SA, Charlton JL (1981) Concerning the spin-statistical factor in the triplet-triplet annihilation of anthracene triplets. J Am Chem Soc 103:7159–7164
31. Levin PP (2003) Kinetics of diffusion-controlled triplet–triplet annihilation of porphyrin in liquid and frozen thin layers of decanol. Dokl Phys Chem 388:10–12
32. Wu W, Sun J, Cui X, Zhao J (2013) Observation of the room temperature phosphorescence of Bodipy in visible light-harvesting Ru(II) polyimine complexes and application as triplet photosensitizers for triplet–triplet-annihilation upconversion and photocatalytic oxidation. J Mater Chem C 1:4577–4589
33. Islangulov RR, Kozlov DV, Castellano FN (2005) Low power upconversion using MLCT sensitizers. Chem Commun 30:3776–3778
34. Kozlov DV, Castellano FN (2004) Anti-Stokes delayed fluorescence from metal–organic bichromophores. Chem Commun 24:2860–2861
35. Ji S, Wu W, Wu W, Song P, Han K, Wang Z, Liu S, Guo H, Zhao J (2010) Tuning the luminescence lifetimes of ruthenium(II) polypyridine complexes and its application in luminescent oxygen sensing. J Mater Chem 20:1953–1963
36. Wu W, Wu W, Ji S, Guo H, Song P, Han K, Chi L, Shao J, Zhao J (2010) Tuning the emission properties of cyclometalated platinum(II) complexes by intramolecular electron-sink/arylethynylated ligands and its application for enhanced luminescent oxygen sensing. J Mater Chem 20:9775–9786
37. Guo H, Ji S, Wu W, Wu W, Shao J, Zhao J (2010) Long-lived emissive intra-ligand triplet excited states (^{3}IL): next generation luminescent oxygen sensing scheme and a case study with red phosphorescent diimine Pt(II) bis(acetylide) complexes containing ethynylated naphthalimide or pyrene subunits. Analyst 135:2832–2840
38. Winkler JR, Netzel TL, Creutz C, Sutin N (1987) Direct observation of metal-to-ligand charge-transfer (MLCT) excited states of pentaammineruthenium(ii) complexes. J Am Chem Soc 109:2381–2392
39. Medlycott EA, Hanan GS (2005) Designing tridentate ligands for ruthenium(II) complexes with prolonged room temperature luminescence lifetimes. Chem Soc Rev 34:133–142
40. Abrahamsson M, Jager M, Kumar RJ, Osterman T, Persson P, Becker H-C, Johansson O, Hammarstrom L (2008) Bistridentate Ruthenium(II)polypyridyl-type complexes with microsecond 3MLCT state lifetimes: sensitizers for rod-like molecular arrays. J Am Chem Soc 130:15533–15542
41. McClenaghan ND, Leydet Y, Maubert B, Indelli MT, Campagna S (2005) Excited-state equilibration: a process leading to long-lived metal-to-ligand charge transfer luminescence in supramolecular systems. Coord Chem Rev 249:1336–1350
42. Baba AI, Shaw JR, Simon JA, Thummel RP, Schmehl RH (1998) The photophysical behavior of d^6 complexes having nearly isoenergetic MLCT and ligand localized excited states. Coord Chem Rev 171:43–59
43. Ji S, Wu W, Wu W, Guo H, Zhao J (2011) Ruthenium(II) polyimine complexes with a long-lived ^{3}IL excited state or a ^{3}MLCT/^{3}IL equilibrium: efficient triplet sensitizers for low-power upconversion. Angew Chem Int Ed 50:1626–1629
44. Ji S, Guo H, Wu W, Wu W, Zhao J (2011) Ruthenium(II) polyimine–coumarin dyad with non emissive ^{3}IL excited state as sensitizer for triplet–triplet annihilation based upconversion. Angew Chem Int Ed 50:8283–8286
45. Loudet A, Burgess K (2007) BODIPY dyes and their derivatives: syntheses and spectroscopic properties. Chem Rev 107:4891–4932
46. Ulrich G, Ziessel R, Harriman A (2008) The chemistry of fluorescent Bodipy dyes: versatility unsurpassed. Angew Chem Int Ed 47:1184–1201

47. Benniston AC, Copley G (2009) Lighting the way ahead with boron dipyrromethene (Bodipy) dyes. Phys Chem Chem Phys 11:4124–4131
48. Bozdemir OA, Erbas-Cakmak S, Ekiz OO, Dana A, Akkaya EU (2011) Towards unimolecular luminescent solar concentrators: bodipy-based dendritic energy-transfer cascade with panchromatic absorption and monochromatized emission. Angew Chem Int Ed 50:10907–10912
49. Galletta M, Campagna S, Quesada M, Ulrich G, Ziessel R (2005) The elusive phosphorescence of pyrromethene–BF2 dyes revealed in new multicomponent species containing Ru(II)–terpyridine subunits. Chem Commun 33:4222–4224
50. Hanson K, Tamayo A, Diev VV, Whited MT, Djurovich PI, Thompson ME (2010) Efficient dipyrrin-centered phosphorescence at room temperature from bis-cyclometalated iridium(III) dipyrrinato complexes. Inorg Chem 49:6077–6084
51. Rachford AA, Ziessel R, Bura T, Retailleau P, Castellano FN (2010) Boron Dipyrromethene (Bodipy) phosphorescence revealedin $[Ir(ppy)_2(bpy\text{-}CtC{\equiv}Bodipy)]^+$. Inorg Chem 49:3730–3736
52. Ceroni P (2011) Energy up-conversion by low-power excitation: new applications of an old concept. Chem Eur J 17:9560–9564
53. Whited MT, Djurovich PI, Roberts ST, Durrell AC, Schlenker CW, Bradforth SE, Thompson ME (2011) Singlet and triplet excitation management in a bichromophoric near-infrared-phosphorescent BODIPY-benzoporphyrin platinum complex. J Am Chem Soc 133:88–96
54. Kalyanasundaram K (1992) Photochemistry of polypyridine and porphyrin complexes. Academic, New York, p 500
55. Papkovsky DB, O'Riordan TC (2005) Emerging applications of phosphorescent metalloporphyrins. J Fluoresc 15:569–584
56. Islangulov RR, Lott J, Weder C, Castellano FN (2007) Noncoherent low-power upconversion in solid polymer films. J Am Chem Soc 129:12652–12653
57. Singh-Rachford TN, Castellano FN (2010) Triplet sensitized red-to-blue photon upconversion. J Phys Chem Lett 1:195–200
58. Baluschev S, Yakutkin V, Miteva T, Avlasevich Y, Chernov S, Aleshchenkov S, Nelles G, Cheprakov A, Yasuda A, Müllen K, Wegner G (2007) Blue-green up-conversion: noncoherent excitation by NIR light. Angew Chem Int Ed 46:7693–7696
59. Baluschev S, Yakutkin V, Wegner G, Miteva T, Nelles G, Yasuda A, Chernov S, Aleshchenkov S, Cheprakov A (2007) Upconversion with ultrabroad excitation band: simultaneous use of two sensitizers. Appl Phys Lett 90:181103
60. Keivanidis PE, Baluschev S, Miteva T, Nelles G, Scherf U, Yasuda A, Wegner G (2003) Up-conversion photoluminescence in polyfluorene doped with metal(II)–octaethyl porphyrins. Adv Mater 15:2095–2098
61. Singh-Rachford TN, Nayak A, Muro-Small ML, Goeb S, Therien MJ, Castellano FN (2010) Supermolecular-chromophore-sensitized near-infrared-to-visible photon upconversion. J Am Chem Soc 132:14203–14211
62. Sugunan SK, Tripathy U, Brunet SMK, Paige MF, Steer Ronald P (2009) Mechanisms of low-power noncoherent photon upconversion in metalloporphyrin-organic blue emitter systems in solution. J Phys Chem A 113:8548–8556
63. Cui X, Zhao J, Yang P, Sun J (2013) Zinc(II) tetraphenyltetrabenzoporphyrin complex as triplet photosensitizer for triplet–triplet annihilation upconversion. Chem Commun 49:10221–10223
64. Du P, Eisenberg R (2010) Energy upconversion sensitized by a platinum(II) terpyridyl acetylide complex. Chem Sci 1:502–506
65. Sun H, Guo H, Wu W, Liu X, Zhao J (2011) Coumarin phosphorescence observed with *N^N* Pt (II) bisacetylide complex and its applications for luminescent oxygen sensing and triplet–triplet-annihilation based upconversion. Dalton Trans 40:7834–7841
66. Liu Y, Wu W, Zhao J, Zhang X, Guo H (2011) Accessing the long-lived near-IR-emissive triplet excited state in naphthalenediimide with light-harvesting diimine platinum (II) bisacetylide complex and its application for upconversion. Dalton Trans 40:9085–9089

67. Guo H, Muro-Small ML, Ji S, Zhao J, Castellano FN (2010) Naphthalimide phosphorescence finally exposed in a platinum(II) diimine complex. Inorg Chem 49:6802–6804
68. Pomestchenko IE, Luman CR, Hissler M, Ziessel R, Castellano FN (2003) Room temperature phosphorescence from a Platinum(II) diimine bis(pyrenylacetylide) complex. Inorg Chem 42:1394–1396
69. Li Q, Guo H, Ma L, Wu W, Liu Y, Zhao J (2012) Tuning the photophysical properties of N^N Pt(II) bisacetylide complexes with fluorene moiety and its applications for triplet–triplet-annihilation based upconversion. J Mater Chem 22:5319–5329
70. Guo H, Li Q, Ma L, Zhao J (2012) Fluorene as π-conjugation linker in N^N Pt(II) bisacetylide complexes and their applications for triplet–triplet annihilation based upconversion. J Mater Chem 22:15757–15768
71. Guo H, Qin H, Chen H, Sun H, Zhao J (2013) Phenylacetylide ligand mediated tuning of visible-light absorption, room temperature phosphorescence lifetime and triplet-triplet annihilation based up-conversion of a diimine Pt(II) bisacetylide complex. Dyes Pigments 99:908–915
72. Wu W, Zhao J, Wu W, Chen Y (2012) Room temperature long-lived triplet excited state of fluorescein in N^N Pt(II) bisacetylide complex and its applications for triplet-triplet annihilation based upconversions. J Organomet Chem 713:189–196
73. Huang L, Zeng L, Guo H, Wu W, Wu W, Ji S, Zhao J (2011) Room-temperature long-lived 3il excited state of rhodamine in an N^N Pt^{II} bis(acetylide) complex with intense visible-light absorption. Eur J Inorg Chem 29:4527–4533
74. Wu W, Zhao J, Sun J, Huang L, Yi X (2013) Red-light excitable fluorescent platinum(II) bis (aryleneethynylene) bis(trialkylphosphine) complexes showing long-lived triplet excited states as triplet photosensitizers for triplet–triplet annihilation upconversion. J Mater Chem C 1:705–716
75. Zhang X, Xiao Y, Qian X (2008) Highly efficient energy transfer in the light harvesting system composed of three kinds of boron–dipyrromethene derivatives. Org Lett 10:29–32
76. Coskun A, Akkaya EU (2006) Signal ratio amplification via modulation of resonance energy transfer:proof of principle in an emission ratiometric Hg(II) sensor. J Am Chem Soc 128:14474–14475
77. Sunahara H, Urano Y, Kojima H, Nagano T (2007) Design and synthesis of a library of BODIPY-based environmental polarity sensors utilizing photoinduced electron-transfer-controlled fluorescence ON/OFF switching. J Am Chem Soc 129:5597–5604
78. Wu Y, Peng X, Guo B, Fan J, Zhang Z, Wang J, Cui A, Gao Y (2005) Boron dipyrromethene fluorophore based fluorescence sensor for the selective imaging of Zn(II) in living cells. Org Biomol Chem 3:1387–1392
79. Cheng T, Xu Y, Zhang S, Zhu W, Qian X, Duan L (2008) A highly sensitive and selective off-on fluorescent sensor for cadmium in aqueous solution and living cell. J Am Chem Soc 130:16160–16161
80. Guo H, Jing Y, Yuan X, Ji S, Zhao J, Li X, Kan Y (2011) Highly selective fluorescent OFF–ON thiol probes based on dyads of BODIPY and potent intramolecular electron sink 2,4-dinitrobenzenesulfonyl subunits. Org Biomol Chem 9:3844–3853
81. Shao J, Guoa H, Ji S, Zhao J (2011) Styryl-BODIPY based red-emitting fluorescent OFF–ON molecular probe for specific detection of cysteine. Biosens Bioelectron 26:3012–3017
82. Liu Y, Li Q, Zhao J, Guo H (2012) BF_2-bound chromophore-containing N^N Pt (II) bisacetylide complex and its application as sensitizer for triplet–triplet annihilation based upconversion. RSC Adv 2:1061–1067
83. Zhao W, Castellano FN (2006) Upconverted emission from pyrene and di-tert-butylpyrene using $Ir(ppy)_3$ as triplet sensitizer. J Phys Chem A 110:11440–11445
84. Brooks J, Babayan Y, Lamansky S, Djurovich PI, Tsyba I, Bau R, Thompson ME (2002) Synthesis and characterization of phosphorescent cyclometalated platinum complexes. Inorg Chem 41:3055–3066

85. Sun J, Wu W, Guo H, Zhao J (2011) Visible-light harvesting with cyclometalated Iridium(III) complexes having long-lived ^{3}IL excited states and their application in triplet–triplet-annihilation based upconversion. Eur J Inorg Chem (21):3165–3173
86. Wu W, Wu W, Ji S, Guo H, Zhao J (2011) Accessing the long-lived emissive 3IL triplet excited states of coumarin fluorophores by direct cyclometallation and its application for oxygen sensing and upconversion. Dalton Trans 40:5953–5963
87. Wu W, Huang D, Yi X, Zhao J (2013) Tridentate cyclometalated platinum(II) complexes with strong absorption of visible light and long-lived triplet excited states as photosensitizers for tripletetriplet annihilation upconversion. Dyes Pigments 96:220–231
88. Wu W, Sun J, Ji S, Wu W, Zhao J, Guo H (2011) Tuning the emissive triplet excited states of platinum(II) Schiff base complexes with pyrene, and application for luminescent oxygen sensing and triplet–triplet-annihilation based upconversions. Dalton Trans 40:11550–11561
89. Ma L, Guo H, Li Q, Guo S, Zhao J (2012) Visible light-harvesting cyclometalated Ir(III) complexes as triplet photosensitizers for triplet–triplet annihilation based upconversion. Dalton Trans 41:10680–10689
90. Yi X, Zhang C, Guo S, Ma J, Zhao J (2014) Strongly emissive long-lived ^{3}IL excited state of coumarins in cyclometalated Ir(III) complexes used as triplet photosensitizers and application in triplet–triplet annihilation upconversion. Dalton Trans 43:1672–1683
91. Yi X, Yang P, Huang D, Zhao J (2013) Visible light-harvesting cyclometalated Ir(III) complexes with pyreno[4,5-d]imidazole C^N ligands as triplet photosensitizers for triplet-triplet annihilation upconversion. Dyes Pigments 96:104–115
92. Ma L, Guo S, Sun J, Zhang C, Zhao J, Guo H (2013) Green light-excitable naphthalenediimide acetylide-containing cyclometalated Ir(III) complex with long-lived triplet excited states as triplet photosensitizers for triplet–triplet annihilation upconversion. Dalton Trans 42:6478–6488
93. Fernández-Moreira V, Thorp-Greenwood FL, Coogan MP (2010) Application of d6 transition metal complexes in fluorescence cell imaging. Chem Commun 46:186–202
94. Choi AW-T, Louie M-W, Li SP-Y, Liu H-W, Chan BT-N, Lam TC-Y, Lin AC-C, Cheng S-H, Lo KK-W (2012) Emissive behavior, cytotoxic activity, cellular uptake, and PEGylation properties of new luminescent rhenium(I) polypyridine poly(ethylene glycol) complexes. Inorg Chem 51:13289–13302
95. Yi X, Zhao J, Sun J, Guo S, Zhang H (2013) Visible light-absorbing rhenium(I) tricarbonyl complexes as triplet photosensitizers in photooxidation and triplet–triplet annihilation upconversion. Dalton Trans 42:2062–2074
96. Yi X, Zhao J, Wu W, Huang D, Ji S, Sun J (2012) Rhenium(I) tricarbonyl polypyridine complexes showing strong absorption of visible light and long-lived triplet excited states as a triplet photosensitizer for triplet–triplet annihilation upconversion. Dalton Trans 41:8931–8940

Chapter 8
Triarylboron-Functionalized Metal Complexes for OLEDs

Zachary M. Hudson, Xiang Wang, and Suning Wang

Abstract Triarylboron functionalization was found recently to be a highly effective strategy in greatly enhancing the phosphorescence efficiency of Pt (II) compounds and the performance of Pt(II) compounds in electrophosphorescent OLEDs. This chapter examines the role of a dimesitylboron ($BMes_2$) group in the phosphorescence and electrophosphorescence of N^C-, C^C, and N^C^N-chelate Pt(II) compounds. The influence of the location of the $BMes_2$ group, substituent groups, ancillary ligands, and intramolecular hydrogen bonds on phosphorescent color/efficiency and the stability of the Pt(II) compounds are discussed in detail. The key focus is on the strategies of achieving efficient blue phosphorescent Pt (II) compounds. Preliminary electrophosphorescent data for $BMes_2$-functionalized Pt(II) compounds are also presented.

Keywords Triarylboron functional group • Platinum complexes in electroluminescence • Phosphorescent color tuning • Chelate ligand's impact on phosphorescent quantum efficiency

8.1 Introduction

Since the breakthrough discovery by Tang and coworkers of organic electroluminescent (EL) devices with an operating voltage below 10 V [1], tremendous progress on organic light-emitting diode (OLED) technologies has been achieved. Commercial goods such as televisions, cameras, and cellular phones based on OLED displays are now commercially available, and the market continues to rapidly expand. A key development that has enabled the production of energy-efficient OLEDs has been the implementation of phosphorescent molecules as the emitters in these devices, making it possible to achieve OLEDs with external quantum efficiencies (EQEs) of 20 % or higher [2–11].

Although a variety of phosphorescent materials incorporating different metal ions such as Ir(III) [12–16], Pt(II) [17–24], Cu(I) [25, 26], Os(II) [27–30], Ru

Z.M. Hudson • X. Wang • S. Wang (✉)
Department of Chemistry, Queen's University, Kingston, ON K7L 3N6, Canada
e-mail: wangs@chem.queensu.ca

W.-Y. Wong (ed.), *Organometallics and Related Molecules for Energy Conversion*, Green Chemistry and Sustainable Technology, DOI 10.1007/978-3-662-46054-2_8

(II) [31–36], Au(I) and Au(III) [37–40] have been used as dopants in OLEDs, Ir (III)-based compounds are the most commonly used because of their high emission quantum efficiencies as well as excellent chemical and thermal stabilities. However, despite intense global research efforts and the abundance of phosphorescent Ir (III) compounds reported in the literature [41–45], no stable blue OLEDs based on Ir(III) compounds have been achieved to date. This has motivated many teams to investigate alternative classes of compounds, as the development of efficient and stable blue phosphors for display and lighting applications remains a key challenge in OLED research today [41–45].

Our group has been interested in the development of Pt(II)-based phosphorescent emitters, particularly those with emission energies in the blue region of the visible spectrum. Compared to Ir(III) compounds, Pt(II) compounds are less well investigated due to their planar structures, which can lead to many problems when using these materials in electroluminescent devices. Specifically, a square-planar structure often leaves these materials highly prone to intermolecular quenching, the formation of excimers, or structural distortion in the excited state [6, 17–24, 46–48], all of which can lead to diminished phosphorescent quantum efficiencies (Φ_p). Conversely, the ligand-metal interactions in homoleptic square-planar Pt (II) compounds tend to be stronger than their Ir(III) counterparts due to reduced congestion around the metal center, and Pt(II) compounds in general emit at higher energies than the corresponding Ir(III) compounds as a result. For example, Ir (ppy)$_2$(acac) [14] emits at 516 nm while Pt(ppy)(acac) emits at 486 nm [49]. Furthermore, Pt(II) compounds are well known for their pronounced metal-to-ligand charge-transfer transitions, which can greatly facilitate the tuning of both the emission color and the emission quantum yields of these complexes [17–24, 46–48, 50–54].

Following the pioneering work of Forrest and coworkers on electrophosphorescent devices based on a Pt(II) porphyrin complex [17], PtOEP, many electrophosphorescent Pt(II) compounds have been developed for OLED applications [17–24, 46–48, 50–54]. Most of these have either the tridentate chelate structure or the bidentate chelate structure, as shown in Fig. 8.1. The conjugated chelate ligands are usually the dominant chromophore from which phosphorescence originates, while the X or X^Y chelate ligands usually act as ancillary ligands and play a smaller role in the emission of the complex. When designing a phosphorescent metal complex for OLEDs, the chromophoric ligand should in general

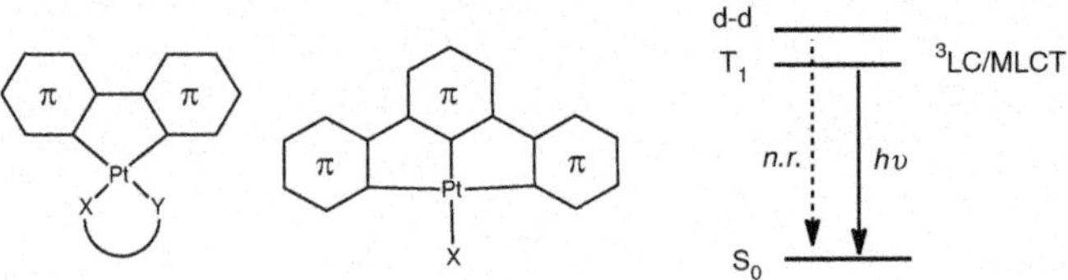

Fig. 8.1 *Left*: The structures of the most common phosphorescent Pt(II) compounds used in OLEDs. *Right*: A schematic presentation showing the relative energy levels of the ground state, the triplet excited state T_1, and the d-d states for efficient phosphorescent Pt(II) emitters

have the following features: (a) a triplet energy (T_1) appropriate for the desired emission color, (b) the ability to facilitate MLCT transitions and the mixing of MLCT excited states with the ligand-centered triplet state (^{3}LC) such that intersystem crossing is enhanced, (c) rigidity to reduce vibrational and rotational energy loss, and (d) sufficient bulk to minimize intermolecular interactions, unless such interactions are specifically desired. Ancillary ligands, on the other hand, should have a triplet energy significantly above that of the chromophoric ligand to minimize energy-transfer quenching and should be sufficiently rigid and bulky as well.

Because of the high triplet energy of blue phosphorescent compounds, the use of strong-field ligands as both the chromophore and ancillary ligands is critical to reducing thermal deactivation by metal-centered d-d excited states. In order to achieve high emission quantum yields, care must be taken to ensure that these d-d states lie sufficiently above the triplet state of the chromophore in energy (Fig. 8.1). To date, many different ligand systems have been designed with the aim of achieving bright blue phosphorescence in Pt(II) compounds, and several of these have been found to be highly effective in producing high-performance OLEDs [46–48, 50–54].

In this chapter, we examine the use of triarylboron-containing ligand systems used in highly efficient phosphorescent Pt(II) compounds and their OLED devices. We first introduce the key features of triarylboron compounds and explain why they are effective in enhancing the phosphorescent quantum efficiency of Pt (II) compounds. Next we discuss the various factors that have been found to have a significant impact on the phosphorescence of these materials using examples of complexes with different chromophoric and ancillary ligands.

8.2 Triarylboron Compounds and Their Role in the Phosphorescence of Metal Complexes

Triarylboron compounds have a trigonal planar geometry around the boron center and are the electron-transporting counterpart of triarylamines, the most commonly used class of hole transport materials in OLEDs [55]. Their use in OLEDs was first demonstrated by Shirota and coworkers in 1998, who showed that triarylboron compounds could be used as effective electron transport materials [56]. This feature has been further developed since, with several new classes of triarylboron-based electron transport materials having been reported in recent years [56–61]. For applications in OLEDs, it is necessary to protect the boron center from nucleophilic attack by electrophiles such as water, which cause irreversible decomposition of these materials via hydrolysis. This is typically achieved by introducing bulky *ortho*-substituents on the pendant aryl rings or by using bulky aryl groups such as anthryl [62, 63] as shown in Fig. 8.2. The presence of an unoccupied *p* orbital in triarylboron compounds provides Lewis acidity and electron-accepting ability,

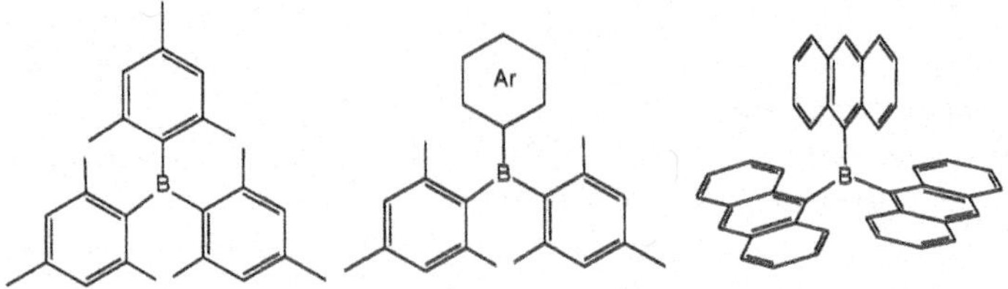

Fig. 8.2 Examples of sterically protected triarylboron compounds

Fig. 8.3 The charge-transfer (CT) process in a donor (N)-acceptor (B) system and the metal-to-ligand charge-transfer process in the $BMes_2$-metal system

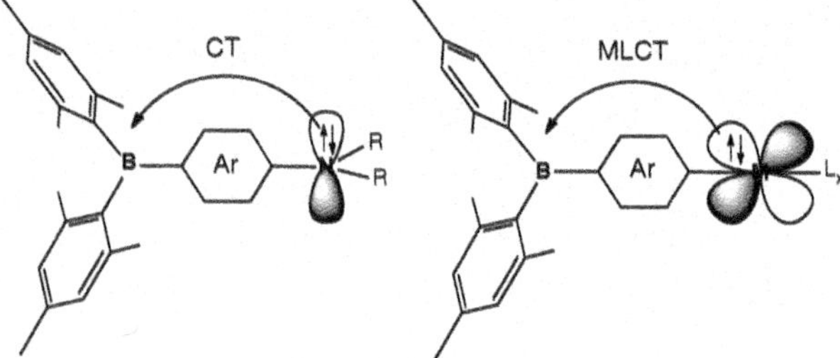

making them ideal for use in a wide variety of applications including anion sensing [64, 65], nonlinear optics [66–68], and optoelectronics [56–61, 69–72]. With appropriate steric protection, triarylboron compounds are in general stable under ambient conditions.

The most commonly used triarylboron unit in OLEDs materials is the -$BMes_2$ group (Mes = mesityl), as materials containing this moiety can be readily synthesized from commercially available $FBMes_2$. The aryl groups bound to the boron center are critical to stabilize the radical anion generated in electron injection/transport by delocalizing the charge through π-conjugation with the boron center. When these materials are further functionalized by electron donor groups such as amines, highly fluorescent molecules with quantum yields approaching 100 % can be achieved [69–72]. The highly efficient fluorescence in such donor-acceptor compounds originates from the highly polarized $N \rightarrow B$ ($\pi \rightarrow \pi^*$) charge-transfer transition mediated by the aryl linker with a high oscillator strength (Fig. 8.3).

The electron-accepting nature of the triarylboron unit makes it an attractive functional group for use in chromophoric ligands in Pt(II) compounds because it can enhance the MLCT transition and lower the energy of MLCT, as shown in Fig. 8.3. The impact of a $BMes_2$ group on the MLCT transitions in Pt(II) compounds may be illustrated using $Pt(B2bpy)Ph_2$ (**1**) as an example, reported by our group in 2007 [73]. (Fig. 8.4) Unlike the parent molecule $Pt(bpy)Ph_2$, which has a light yellow color, compound **1** has a deep red color caused by a distinct MLCT absorption band at 542 nm. Addition of fluoride ions changes the trigonal planar geometry of the three-coordinate boron centers to a four-coordinate tetrahedral geometry, leading to stepwise quenching of the low-energy MLCT peak as the 1:1 and 2:1 products are formed (Fig. 8.4), verifying that this transition was attributable

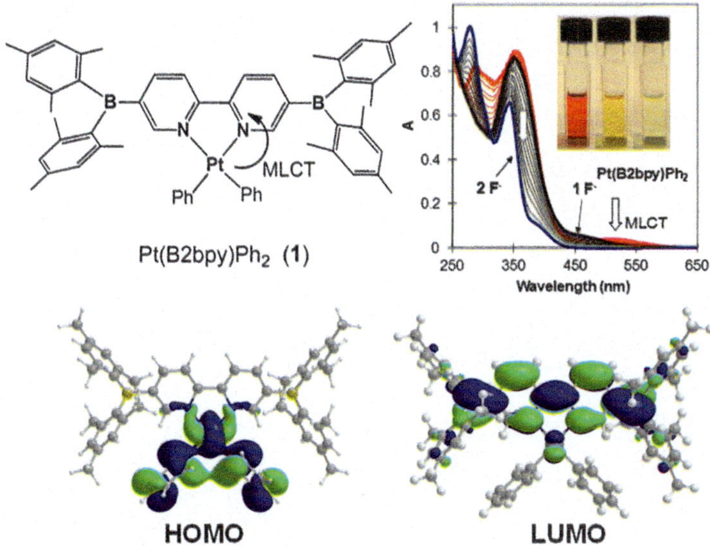

Fig. 8.4 The structure of compound **1**, the absorption spectral change upon the addition of NBu_4F in CH_2Cl_2, and the HOMO and LUMO diagrams of **1**

to the presence of the boron center. The low-energy MLCT and the clear separation of the MLCT state from the ^3LC state are believed to be responsible for the lack of phosphorescence in **1**. When one of the $BMes_2$ groups in **1** is replaced by an NPh_2 group, a low-lying N $\rightarrow$ B charge-transfer (CT) state is introduced on the chelate ligand that is close in energy to the MLCT state, giving the compound an orange phosphorescence, albeit at very low quantum efficiency (~0.1 %) [74].

In order to achieve bright phosphorescence, substantial mixing of both the ^3LC and MLCT states is thus essential. The first example of the use of a $BMes_2$ group to enhance the phosphorescent quantum efficiency of Pt(II) compounds was reported by Kitamura and coworkers, who found that the $BMes_2$-functionalized compound **2** has a much greater Φ_P (0.011) than the parent molecule **3** (0.002) [75]. This enhancement in Φ_P was attributed to the synergistic interactions of the MLCT and the ligand-based CT states in **2** (Fig. 8.5). Following this report, our group reported substantial boron-induced phosphorescent enhancement in compound **4**, which displays a green phosphorescence with $\lambda_{max} = 540$ nm and $\Phi_P = 0.088$ while the non-borylated parent molecule is non-emissive [76].

The use of the $BMes_2$ unit in other phosphorescent metal complexes [77] such as Ru(II) [78, 79], Re(I) [80], and Ir(III) [81–83] is also known, although the impact of the boron center on the phosphorescence efficiency of these metal complexes either appears to be negative or is not significant. For example, $Ir(ppy)_2(acac)$ emits at 516 nm with $\Phi_P = 0.53$ in 2-Me-THF [84] while the $BMes_2$-functionalized compounds **5** and **6** (Fig. 8.6) have either a lower Φ_P or a slightly improved efficiency ($\lambda_{em} = 605$ nm, $\Phi = 0.18$ in CH_2Cl_2 for **5** [81] and $\lambda_{em} = 512$ nm, $\Phi = 0.57$ in CH_3CN for **6** [82]).

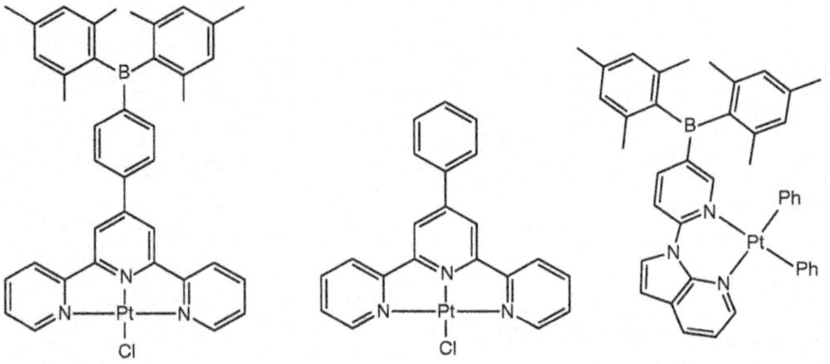

Fig. 8.5 The structures of compounds **2–4**

Fig. 8.6 The structures of compounds **5** and **6**

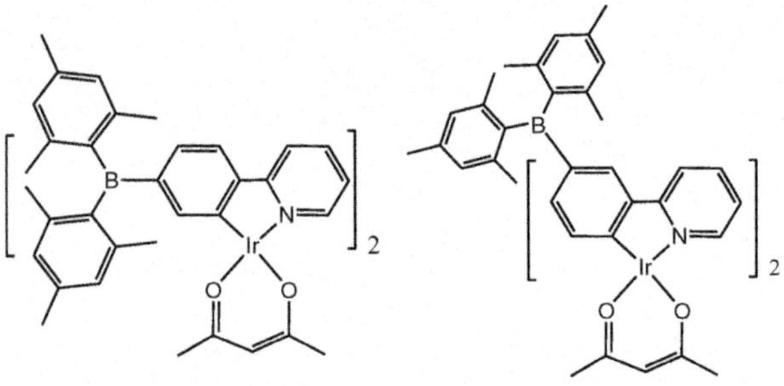

Although the $BMes_2$ unit can be effective in promoting phosphorescence in some of the N^N- or N^N^N-chelate Pt(II) compounds as shown above, Pt (II) compounds based on these types of chelate ligands are not suitable for use as emitters in OLEDs because of the low emission quantum efficiency. The most effective ligand systems for achieving bright phosphorescent Pt(II) compounds are based on N^C-, N^C^N-, or C^C-chelate ligands, which are examined in detail in the following sections.

8.3 $BMes_2$-Functionalized Pt(II) Compounds for OLEDs

8.3.1 N^C-Chelate Systems

8.3.1.1 Phenylpyridine (ppy) and Related Ligands

The Impact of $BMes_2$ Functionalization on Phosphorescence

The use of Pt(II)-phenylpyridines and related N^C-chelate Pt(II) compounds in OLEDs has been extensively examined by Thompson and others [17–22].

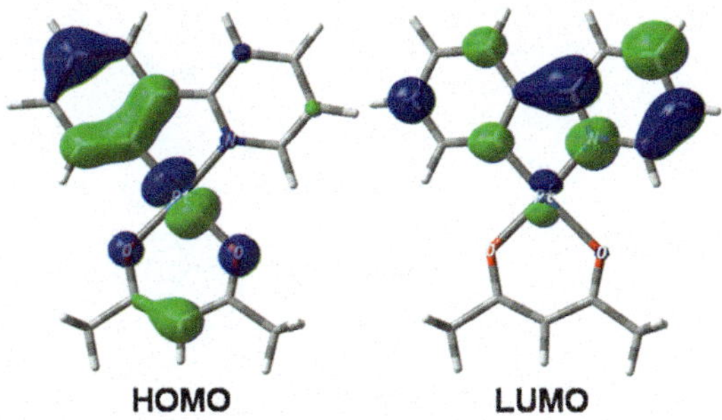

Fig. 8.7 The HOMO and LUMO diagrams of Pt(ppy)(acac)

Compared to N^N-chelate Pt(II) compounds, this class of materials produces much brighter phosphorescence in solution and in the solid state at ambient temperature. For example, Pt(ppy)(acac) and Pt(ppy)(dpm) (dpm = dipivaloylmethanoate) emit at 485 nm with $\Phi_P = 0.30$ and 0.35, respectively, in CH_2Cl_2 [85, 90]. The high emission quantum efficiency in these compounds can be explained by the efficient mixing of the emissive ^{3}LC state of the N^C-chelate ligand and the MLCT state, which enhances spin-orbit coupling and intersystem crossing from the S_1 to the T_1 state. The HOMO and LUMO diagrams of Pt(ppy)(acac) shown in Fig. 8.7 illustrate the LC-dominated first excited state (S_1 and T_1) and the involvement of the Pt-*d* orbital.

Nonetheless, these molecules often do not produce high efficiency OLEDs due to relatively low quantum efficiencies and their tendency to produce excimer emission. We therefore chose this class of compounds as the initial targets to examine the impact of $BMes_2$ functionalization on phosphorescence of Pt(II) compounds. The first group of compounds we examined are two isomers of **7** and **8** [86]. For isomer **8**, the emission wavelength is about 6–10 nm blueshifted from that of **7**, illustrating the influence of the location of the $BMes_2$ group on emission energy. Changing the ligand L from DMSO to pyridine in **7** and **8** led to a redshift of ~30 nm of the emission energy, with increased MLCT character and substantial increase in Φ_P. Nonetheless, **7** and **8** are very weak emitters ($\Phi = 0.004$–0.030) and not suitable for use in OLEDs. The low Φ_P values for **7** and **8** can be attributed to the presence of nonrigid ancillary ligands.

Indeed, replacing the two monodentate ancillary ligands in **7** and **8** with a chelating acetylacetonate (acac) ligand led to brightly phosphorescent compounds **9** and **10**, which emit at 538 nm and 527 nm, respectively [87], and are about 40–50 nm redshifted compared to Pt(ppy)(acac). However, both compounds have very impressive quantum yields of 0.77 ($\tau = 10.2$ μs) and 0.98 ($\tau = 9.5$ μs), respectively, in CH_2Cl_2 at ambient temperature, a great improvement over Pt(ppy)(acac) that has a Φ_P of 0.30 [85, 90]. The HOMO and LUMO diagrams for **9** and **10** resemble those of Pt(ppy)(acac), with the exception of the large contribution that the *p* orbital on boron makes to the LUMO of both molecules, accounting for the redshift in emission energy (Fig. 8.8). Furthermore, electrochemical data indicate that **9** and **10** undergo reversible reduction localized on the boron center. Also, the crystal structure shown in Fig. 8.8 illustrates the bulk of the $BMes_2$ unit, believed to greatly

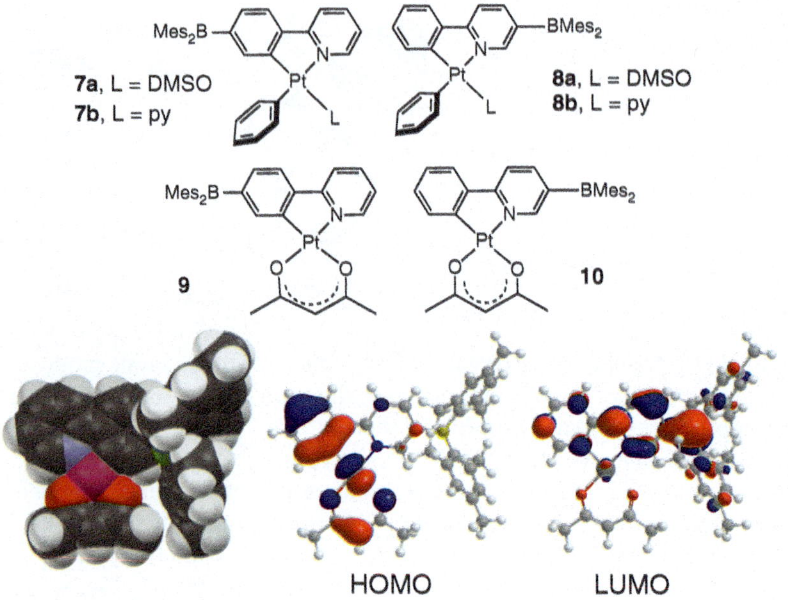

Fig. 8.8 The structures of compounds **7–10**, the crystal structure of **9**, and the HOMO-LUMO diagrams of **10**

inhibit the distortion of the molecule in the excited state, reducing nonradiative decay. Indeed, the rate of nonradiative decay, k_{nr}, for **10** was found to be $2.20 \times 10^3\ s^{-1}$, two orders of magnitude smaller than that of Pt(ppy)(acac). Further, the bulky $BMes_2$ group is also very effective in blocking intermolecular interactions, as no excimer emission was observed for either compound in solution or the solid state. These, along with the electron-accepting nature of the $BMes_2$ group, are responsible for the exceptionally high phosphorescent efficiency of compounds **9** and **10**.

To evaluate the impact of $BMes_2$ functionalization on the performance of Pt(II)-based OLEDs, we fabricated identical EL devices with the structure **I** (ITO/NPB (45 nm)/CBP (5 nm)/15 % Pt compound: CBP (15 nm)/TPBI (30 nm)/LiF (1 nm)/Al (100 nm)) for Pt(ppy)(acac) and **10**, respectively (Fig. 8.9). As shown by the EL data in Fig. 8.10, the performance of devices using **10** (Pt(BppyA)(acac)) as the dopant is far superior to that of Pt(ppy)(acac), showing no excimer emission. The maximum EQE of devices based on **10** was found to be 8.9 % at 100 cd/m^2, while that of the Pt(ppy)(acac) was found to be 6.9 % at a brightness of only 1 cd/m^2. To examine the electron-transporting properties of the $BMes_2$ group, electron-only devices with the structure of ITO/Cs_2CO_3 (1 nm)/Pt: CBP (10 wt%, 138 nm) or Pt (ppy)(acac):CBP (10 wt%, 111 nm)/LiF (1 nm)/Al (100 nm) for Pt(ppy)(acac) and **10** were fabricated. The device based on **10** exhibited a current density 3–4 orders of magnitude higher than that based on Pt(ppy)(acac), establishing unambiguously

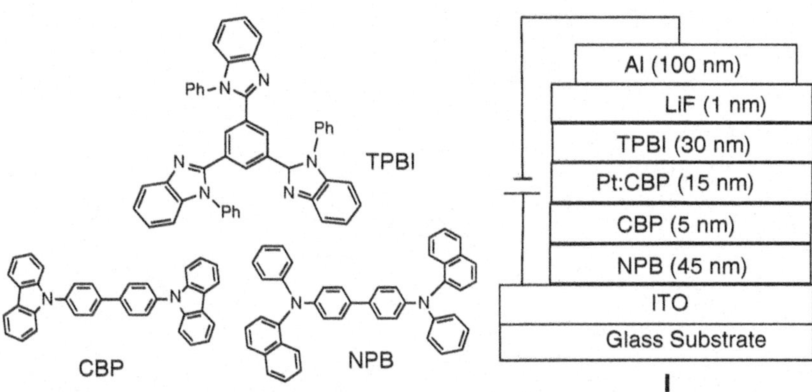

Fig. 8.9 The EL device structure **I** and the host and charge transport materials used for the evaluation of compound **10** and Pt(ppy)(acac)

that the triarylboron unit can greatly improve the electron mobility of phosphorescent materials in electroluminescent devices.

Further, a device work established that for devices using compound **10** as the dopant, the excitons are formed in the host and transferred to the dopant [88]. To facilitate exciton transfer to the dopant, a new device structure **II** for **10** was designed, in which the dopant **10** is deposited in both HTL and ETL layers. This device has a maximum EQE of 21 % and remains as high as 17 % at 1,000 cd/m^2 as shown in Fig. 8.11. This work firmly established that the incorporation of a $BMes_2$ group to N^C-chelate Pt(II) compounds can indeed greatly enhance their performance in electroluminescent devices.

Tuning the Emission Energy

The emission energy of **10** can be shifted to a longer wavelength by introducing a low-energy charge-transfer (CT) state on the chelate ligand. This is illustrated by compounds **11** and **12** shown in Fig. 8.12, which have an intense N $\rightarrow$ B CT transition localized on the N^C-chelate ligands, giving the free ligands fluorescent quantum efficiencies near unity. These two compounds emit at 596 and 590 nm with high Φ_P values of 0.79 and 0.91, respectively, in CH_2Cl_2 at ambient temperature [87, 89]. The values of k_{nr} for these two compounds ($3.08 \times 10^3\ s^{-1}$, $2.20 \times 10^3\ s^{-1}$) were found to be similar to that of **10**, while the k_r values ($1.17 \times 10^4\ s^{-1}$, $2.27 \times 10^4\ s^{-1}$) are about five to ten times smaller than those of **9** and **10**. This is caused by the much longer T_1 decay lifetime of the two compounds (67 μs for **11** and 40 μs for **12**). Examination of the absorption spectra and TD-DFT computational data indicates that the S_1 and the T_1 states in these compounds are almost entirely localized on the N^C-chelate ligand, with little contribution from the Pt(II) center. Further, the HOMO and LUMO in both molecules are localized on the amino and the boryl unit, respectively. These charge-separated excited states,

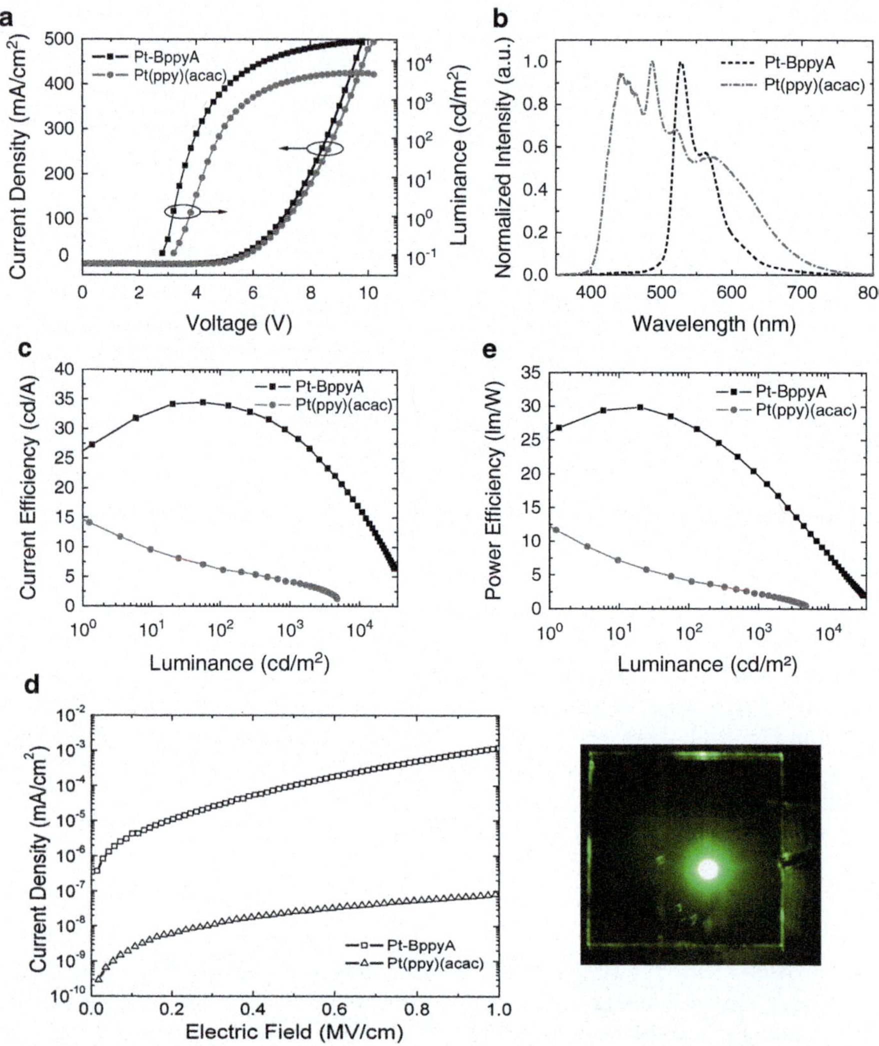

Fig. 8.10 (**a**)–(**e**): The electroluminescent data of Pt(BppyA)(acac) (**10**) and Pt(ppy)(acac) using device structure **I**. (**d**): the current density vs. electric field plot for the electron-only devices of **10** and Pt(ppy)(acac)

combined with a relatively small contribution from the metal center, are believed to be responsible for the long decay lifetime.

Orange EL devices with $\lambda_{em} = 582$ nm using **12** as the dopant (10 wt%) and a device structure similar to **II** were fabricated (Fig. 8.12). Although the EL spectra of the devices of **12** match the photoluminescence with a reasonable maximum EQE of 10.6 %, the performance remained substantially lower than that of devices employing **10** as the emissive material, also showed a greater efficiency roll-off.

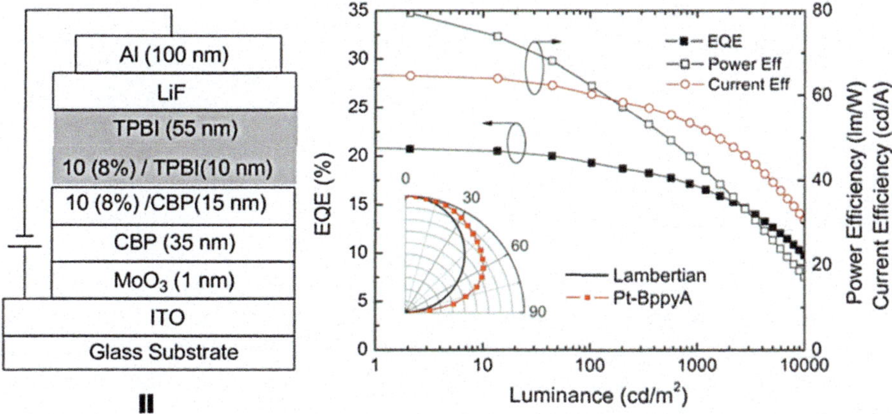

Fig. 8.11 The device structure **II** for **10** and the EQE, power, and current efficiency vs. luminance plot of the device

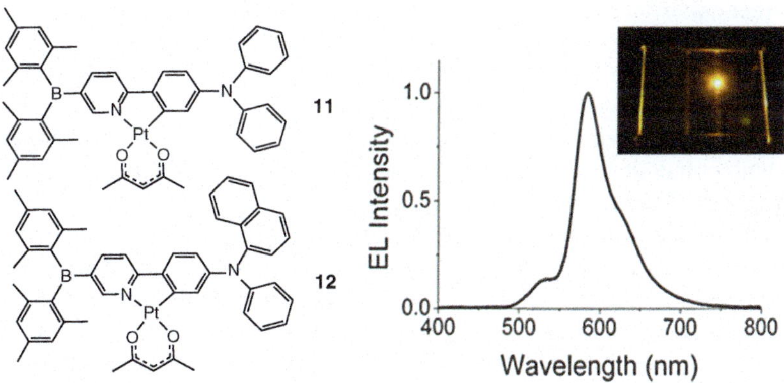

Fig. 8.12 The structures of compounds **11** and **12**, the EL spectrum, and a photograph of the EL device fabricated using **12**

This can be attributed to the long decay lifetime of the molecule, making the material more prone to the various exciton-quenching processes occurring in an EL device.

The emission color of the $BMes_2$-functionalized Pt-phenylpyridine compounds can also be tuned by introducing electron-donating or electron-withdrawing groups to the ppy chelate, as illustrated by compounds **13** and **14** (Fig. 8.13) [87]. The introduction of two fluorine atoms on the phenyl ring in compound **13** does not lead to a significant change in the emission energy ($\lambda_{max} = 530$ nm), relative to that of **10**, as the inductive electron-withdrawing effect of the F atom is offset by π donation from its lone pair electrons. The electron-donating methoxy group at the C5 position in **14**, however, leads to a large redshift of about 60 nm in the emission peak ($\lambda_{max} = 598$ nm) relative to that of **10**, caused mainly by a destabilization of

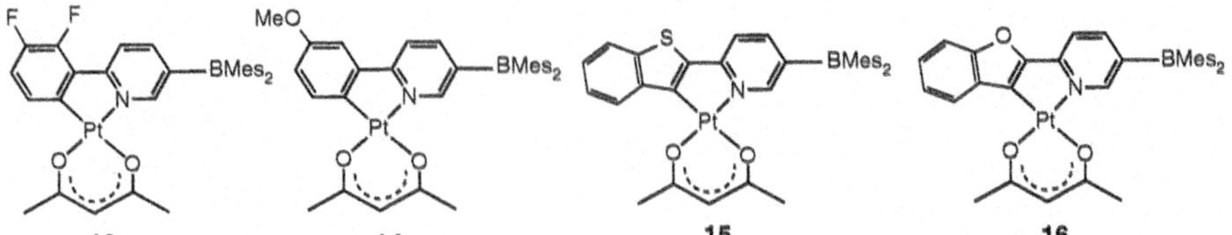

Fig. 8.13 The structures of compounds **13** and **14**

the HOMO level, which has a large contribution at C5. The Φ_P values of **13** and **14** (0.54 and 0.45, respectively) are greater than those of the corresponding non-borylated parent molecules (0.36 and 0.30) [90] but are much lower than those of **9** and **10**, as evidenced by considerably higher values of k_{nr} for **13** and **14** (5.57×10^4 s^{-1} and 3.26×10^4 s^{-1}, respectively). Compared to the parent molecules that emit at 489 nm and 541 nm, respectively, the emission energy of **13** and **14** is about 40–50 nm redshifted, following the same trend as that observed for **10** and Pt(ppy)(acac).

Further emission energy tuning of $BMes_2$-functionalized Pt-ppy compounds can be achieved by replacing the phenyl ring in ppy with a heterocycle, such as a benzothienyl or benzofuryl group [87] as illustrated by compounds **15** and **16** shown in Fig. 8.13. The phosphorescent peak of **15** and **16** (~650 nm) is more than 100 nm redshifted from that of **10**. However, in addition to the phosphorescent peak, both compounds also display fluorescent emission at 525 nm in CH_2Cl_2 and the solid state. This distinct singlet and triplet dual emission phenomenon was not observed for the non-borylated parent molecules [90]. Again, compared to the parent molecules, the phosphorescent peak of **15** and **16** is about 40 nm redshifted, while the Φ_P values (0.12 for **15** and 0.06 for **16**) are only slightly increased (0.04 and 0.02 for the parent molecules [90]). In the absorption spectra, the first absorption band of **15** and **16** is an intense ligand-centered band. It is very likely that the ^{1}MLCT state lies above the ^{1}LC state in these two compounds, which led to inefficient singlet to triplet intersystem crossing and low values of Φ_P. The non-conjugated lone pair electrons of the S and O atoms may also play a role in the nonradiative decay of the T_1 state in these compounds. Replacing the phenyl ring in ppy with a thienyl or furyl ring is thus not an effective approach in achieving bright Pt(II) phosphorescent systems in this case. The phosphorescent colors of the various $BMes_2$-functionalized Pt(II) acetylacetonates discussed above are shown in Fig. 8.14.

Impact of the Location of the $BMes_2$ Group on Phosphorescence

As shown by the MO diagrams of Pt(ppy)(acac) in Fig. 8.7, the placement of the electron-accepting $BMes_2$ group *meta* to the binding site of the Pt(II) atom on either the phenyl or the pyridyl ring can lead to the stabilization of the LUMO level due to the large orbital contributions at these locations. For this reason, these *meta*-isomers

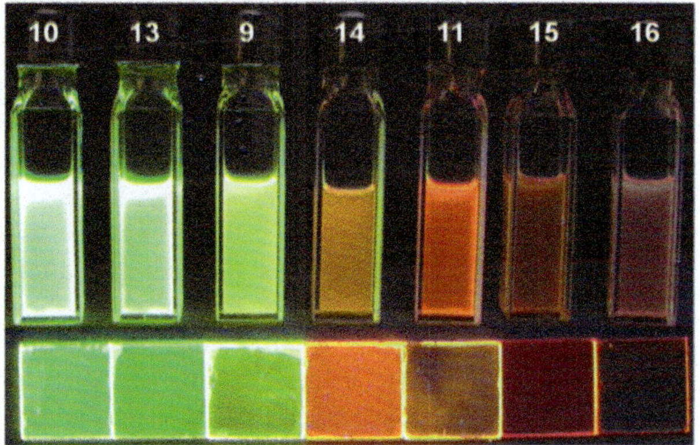

Fig. 8.14 Photographs showing the emission colors of representative examples of the $BMes_2$-functionalized Pt (II) compounds in CH_2Cl_2 (top) and in 10 wt% PMMA films (bottom) discussed in this section

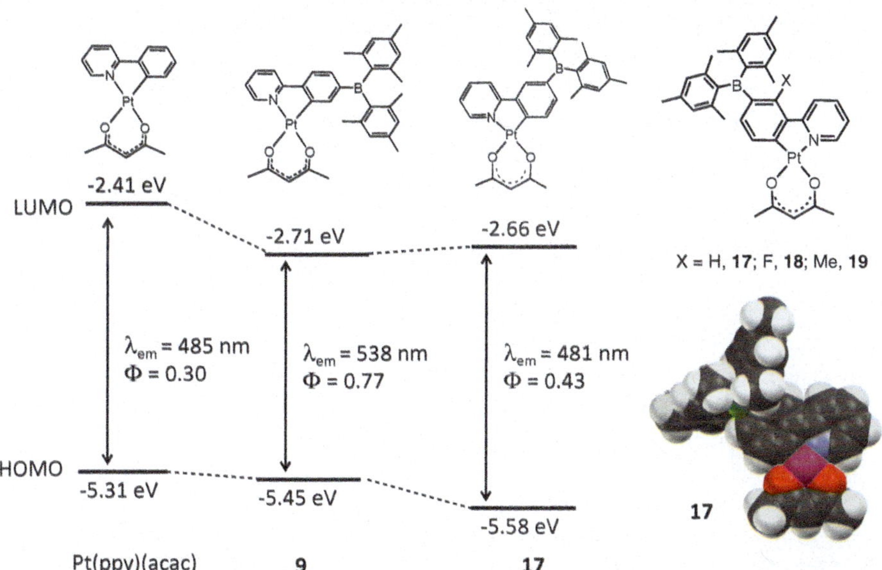

Fig. 8.15 A comparison of the HOMO and LUMO energy levels of Pt(ppy)(acac), compounds **9** and **17**, with the structures of compounds **17–19**

exhibit a decrease in emission energy relative to Pt(ppy)(acac). In contrast, the attachment of a $BMes_2$ group to the phenyl ring at the position *para* to the Pt binding site (the *para*-isomer) would be expected to stabilize the HOMO level with a relatively small impact on the LUMO, giving an overall blueshift. This phenomenon can also be observed in Ir(III) compounds **5** and **6**, in which the *para*-isomer (**6**) exhibits a blueshift of nearly 90 nm relative to the *meta*-isomer **5** (Fig. 8.6). The impact of the location of the $BMes_2$ group on the emission energies of Pt (II) compounds is illustrated by the three *para*-$BMes_2$-functionalized Pt-ppy compounds (**17–19**) in Fig. 8.15 [91].

In contrast to the green phosphorescent compound **9**, these three compounds emit blue-green light with λ_{max} at 481 nm, 475 nm, and 492 nm, respectively, and Φ_P values of 0.43, 0.25, and 0.26 in CH_2Cl_2. Compared to the *meta*-isomer **9**, the emission energy of **17** is blueshifted by 60 nm, consistent with the molecular orbital prediction. The impact of the $BMes_2$ location on the HOMO and LUMO energies and emission wavelengths is depicted in Fig. 8.15. The k_r value ($4.90 \times 10^4\ s^{-1}$) of **17** is smaller than that of **9** ($7.55 \times 10^4\ s^{-1}$), while its k_{nr} ($6.50 \times 10^4\ s^{-1}$) is much greater than that of **9** ($2.25 \times 10^4\ s^{-1}$), leading to a lower Φ_P, relative to **9**. Several factors may be responsible for this. First of all, as shown by the crystal structure of **17** in Fig. 8.15, the $BMes_2$ group at the *para*-position is much less effective in shielding the Pt(II) core and preventing the distortion of the Pt(II) core from distortion in the excited state, compared to that at the *meta*-position in **9**. Furthermore, because the Pt(II) core is more exposed in the *para*-isomer, the molecule is likely more prone to intermolecular quenching than the *meta*-isomer. In addition, the higher T_1 energy of **17** makes its T_1 state closer to the deactivating d-d excited states than that of **9**, increasing the possibility of thermal quenching. Substitution of this molecule with a fluorine atom in **18** resulted in a 6 nm blueshift and a substantial decrease in Φ_P, while methyl substitution (**19**) led to a 17 nm redshift and a similar decrease in Φ_P. The low Φ_P of **19** is likely caused by the steric crowding between the $BMes_2$ unit and the methyl group, leading to distortion of the chelate backbone, while the low Φ_P in **18** is most likely caused by deactivation of the high-energy T_1 excited state by nearby d-d states.

The pyridine ring in the phenylpyridine group of **17** may also be replaced by other *N*-heterocycles such as imidazolyl or benzimidazolyl without a negative impact on the emission efficiency, as illustrated by compound **17a** (Fig. 8.16) [92], which emits a blue-green color with $\lambda_{max} = 486$ nm that is ~5 nm redshifted compared to **17**, with a Φ_P of 0.50 in CH_2Cl_2.

One way to increase the emission quantum efficiency of the *para*-isomer **17** is to replace the acac ancillary ligand with a stronger chelate ligand, capable of raising the energy of the deactivating d-d excited states. One such example is picolinate (pic), as illustrated by compounds **20** and **21** in Fig. 8.16 [93]. Both compounds form the *N,N-trans*-isomer exclusively. Unlike compound **17** which does not show

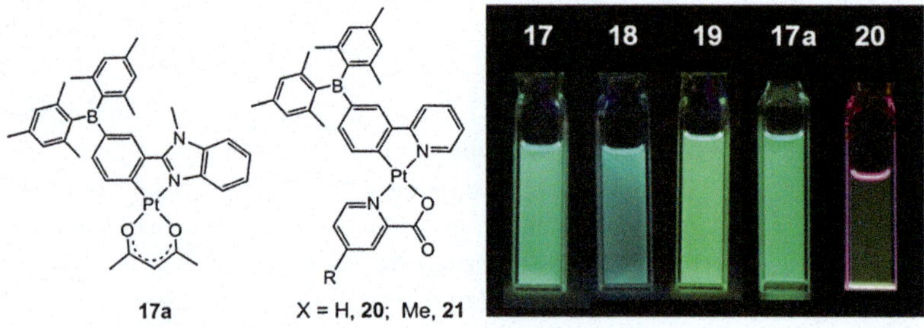

Fig. 8.16 The structures of **17a**, **20**, and **21**, with photos showing the emission color of **17a**, **17–20** in CH_2Cl_2 at a concentration of 1.0×10^{-5} M upon irradiation by a handheld UV lamp ($\lambda = 365$ nm)

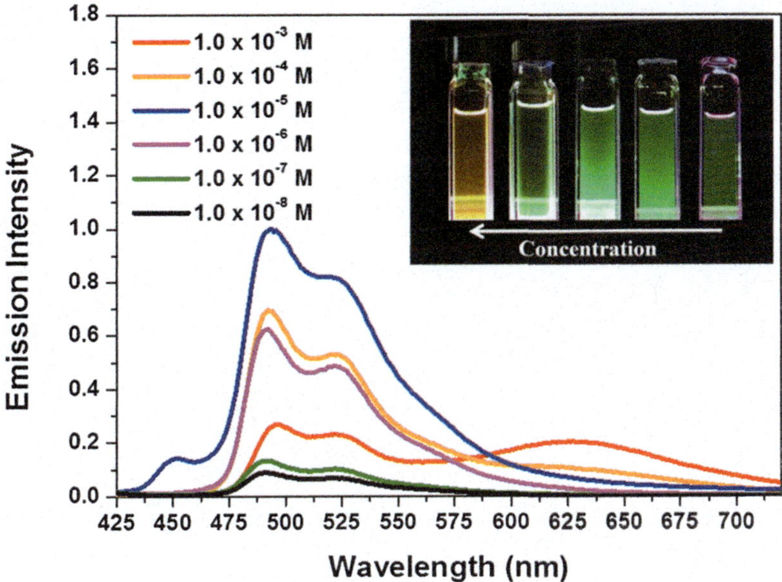

Fig. 8.17 The emission spectra of **20** in CH_2Cl_2 showing the formation of excimer emission at concentrations $> 1.0 \times 10^{-4}$ M

any evidence of excimer emission, these two compounds show distinct excimer emission in solution (Fig. 8.17). Crystal structures showed extended intermolecular π stacking interactions with a Pt···Pt separation distance of 3.71 Å in **20** and 4.12 Å in **21**. The methyl group in the pic ligand does somewhat reduce the intermolecular interactions. Nonetheless, both compounds were found to have a very low Φ_P (0.013 and 0.032, respectively) in CH_2Cl_2 with $\lambda_{em} = 493$ nm and 491 nm, respectively. TD-DFT data established that the two lowest singlet and triplet states in both compounds are inter-ligand CT transitions ($BMes_2$-ppy to pic) that effectively quench the emission from the higher-energy $BMes_2$-ppy ^3LC state. Consequently, picolinates were not found to be suitable ancillary ligands for high-performance $BMes_2$-ppy-based Pt(II) compounds.

The performance of the blue-green phosphorescent **17** and **18** in EL devices was evaluated using the device structure ITO/MoO_3 (1 nm)/CBP (35 nm)/Pt-dopant (12 wt%): CBP (15 nm)/TPBI (65 nm)/LiF (1 nm)/Al. The electroluminescence (EL) spectra of these devices match the photoluminescence (PL) spectra and no excimer emission was observed. The EQE for **17** is 4.7 % at 100 cd m^{-2} and 3.5 % at 1,000 cd m^{-2} while that of **18** is 6.5 % at 100 cd m^{-2} and 4.5 % at 1,000 cd m^{-2}. Although the EQE of both EL devices are not as high as that based on compound **10**, limited by their intrinsically low PL efficiency, these devices show very little efficiency roll-off in a wide range of doping levels (2 % - 12 %) as illustrated by the data in Fig. 8.18.

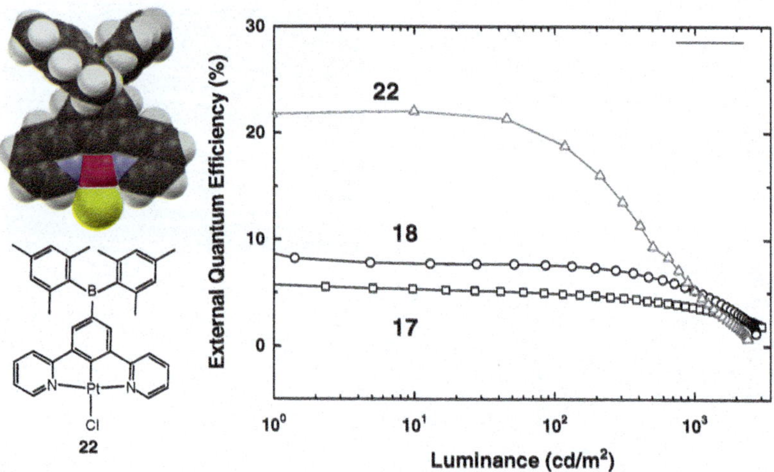

Fig. 8.18 The structure of compound **22** and the EQE vs. luminance plots of EL devices based on **17**, **18**, and **22** at a 12 wt% doping level

N^C vs. N^C^N Ligands

It has been shown by Williams and others that N^C^N-Pt(II) compounds are often brighter emitters and more effective as phosphorescent dopants in OLEDs compared to the corresponding N^C-chelate compounds [12]. This can be attributed to the greater rigidity imposed on the molecule by the tridentate chelate ligand, making it more difficult for the Pt(II) central core to undergo tetrahedral distortion in the excited state. Nonetheless, N^C^N-Pt(II) compounds based on ligands such as 1,3-dipyridylbenzene (dpb) are highly prone to excimer emission [50–54]. The introduction of the bulky $BMes_2$ to the N^C^N-Pt(II) compounds may therefore improve the performance of the Pt(II) compounds in EL devices. This is illustrated by the $BMes_2$-functionalized N^C^N-Pt(II) compound **22** (Fig. 8.18) [91]. Molecules of **22** form π-stacked dimers in the crystal lattice with a long Pt···Pt separation distance (4.91 Å). Nonetheless, compound **22** shows no evidence of excimer emission in solution or in the solid state, in sharp contrast to the non-borylated parent Pt(dpb)Cl, which is well known to produce excimer emission. Interestingly, however, the phosphorescent spectrum of **22** is very similar to that [50–54, 94] of Pt (dpb)Cl at a concentration of 10^{-6} M in CH_2Cl_2 with $\lambda_{em} = 485$ nm, indicating that these two molecules share the same origin of emission. Electrochemical and TD-DFT data indicated that the $BMes_2$ group stabilizes both HOMO and LUMO levels, both localized mostly on the N^C^N-chelate ligand. These stabilizations appear to counteract each other, and the emission energy of **22** appears similar to that of Pt(dpb)Cl as a result. It is also interesting to note that the emission energy of **22** is similar to that of both **17** and Pt(ppy)(acac). Significantly, **22** has a very impressive emission quantum yield (0.70 in CH_2Cl_2 and 0.76 in the solid state), higher than that of Pt(dpb)Cl (0.60 in CH_2Cl_2) and **17** (0.43 in CH_2Cl_2).

Furthermore, unlike the parent molecule, which has a very low solubility in organic solvents, **22** shows excellent solubility in common organic solvents facilitated by the bulky boron group. Unlike the N^C-chelate compound **17** in which the k_r value ($4. \times 10^4$ s^{-1}) is smaller than the k_{nr} (6.5×10^4 s^{-1}), the k_r (7.9×10^4 s^{-1}) of **22** is much greater than the k_{nr} (3.4×10^4 s^{-1}), in agreement with the higher rigidity and the higher Φ_P value of **22**. Blue-green EL devices using the same structure as those for **17** and **18** were also fabricated for **22**, with EL spectra again matching the PL very well for doping levels of 2–12 %. These devices have a much higher EQE than those of **17** and **18**, albeit much greater roll-off at high luminance (e.g., EQE = 15.3 % at 10 cdm^{-2}, 13.4 % at 100 cd m^{-2}, and 3.6 % at 1,000 cd m^{-2} at a doping level of 12 wt %). This work illustrates that $BMes_2$ functionalization of N^C^N-chelate Pt(II) compounds can also be highly effective in enhancing the phosphorescent efficiency of Pt(II)-based OLEDs.

8.3.2 *Phenyl-1,2,3-Triazolyl Chelate Ligands – The Quest for Blue Phosphorescence*

Although decorating ppy or dipyridylbenzene ligands with a $BMes_2$ group is very effective in enhancing the photoluminescent and electroluminescent quantum efficiencies of Pt(II) compounds, it would be difficult to achieve blue or deep-blue phosphorescence using these ligands because of the intrinsic T_1 energy associated with ppy or dpb. The addition of electron-withdrawing groups such as fluorine atoms to the chelate unit may be able to shift the emission energy somewhat toward blue, yet this often comes at the cost of reduced emission efficiency or increased excimer emission. Thus, to achieve blue phosphorescent Pt(II) compounds, it is necessary to use chelate ligands that have an intrinsically high T_1 energy. To shift the emission energy toward blue, one approach is to decrease the size of the aryl rings in the chelate ligand. Two classes of such chelate ligands that may be suitable for achieving blue phosphorescence are phenyl-R-1,2,3-triazolyl (Ph-R-trz), a N^C-chelate ligand, and phenyl-NHC (Ph-NHC, NHC = *N*-R-imidazolyl), a C^C-chelate ligand. In fact, TD-DFT computational data (Gaussian 09, B3LYP LanL2DZ for the Pt atom and 6–31 g (d) for non-Pt atoms) show that the T_1 energies of Pt(ph-Me-trz)(acac) and Pt(ph-NHC)(acac) (R = Me, Fig. 8.19) are 3.01 eV and 3.03 eV, respectively, much higher than that of Pt(ppy)(acac) (2.65 eV). In addition, these two types of ligands have been used in several deep-blue phosphorescent Ir(III) compounds [41 45] and blue Pt(II) compounds [47, 95],

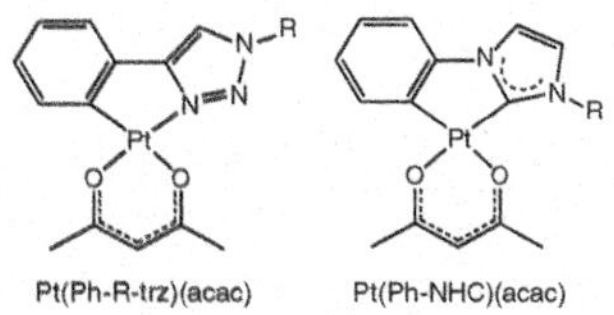

Pt(Ph-R-trz)(acac) Pt(Ph-NHC)(acac)

Fig. 8.19 The structures of Pt(Ph-R-trz)(acac) and Pt (Ph-NHC)(acac)

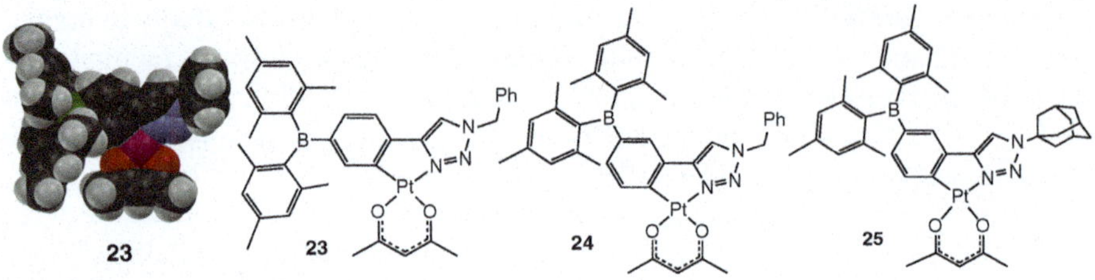

Fig. 8.20 The structures of compounds **23–25**

further supporting their potential in achieving efficient blue phosphorescence. We focus our discussion on the Ph-R-trz-based system first with the emphasis on the influence of different ancillary ligands.

8.3.2.1 *β*-Diketonato as the Ancillary Ligand

To illustrate the difference between the Ph-R-trz and ppy chelate chromophores, we take a look first at three Pt(II) compounds **23–25** in which the ancillary ligand is acac (Fig. 8.20) [96]. As observed in the $BMes_2$-ppy Pt(II) compounds, the *meta*-Pt compound **23** emits at a much longer wavelength (500 nm at 298 K in 2-Me-THF) than the *para*-isomer **24** (453 nm at 298 K in 2-Me-THF) at ambient temperature. DFT computational data showed that the HOMO and LUMO of **23** and **24** have a similar atomic orbital distribution as the ppy analogues of **9** and **17**. Thus, the 50 nm blueshift observed upon switching the $BMes_2$ group from *meta* to *para* to the Pt binding site is not surprising. Most significant is that compared to **9** and **17**, the emission energies of **23** and **24** are 30 nm blueshifted, demonstrating the effectiveness of replacing the pyridyl ring with triazole in achieving blue phosphorescence. However, **23** and **24** are much weaker emitters in solution at ambient temperature than **9** and **17**, with Φ_P values of 0.17 and <0.001, respectively, in Me-THF. In PMMA films, the Φ_P value of these two compounds becomes much higher (0.63 and 0.10, respectively). The low emission quantum efficiency of **23** and **24** in solution is caused mainly by thermal quenching due to solvent molecules, because these two compounds have a much longer decay lifetime (36 μs and 28 μs, respectively, at 77 K) than those of **9** and **17** (<12 μs at 77 K). Replacing the benzyl group in **24** by an adamantyl or methyl group did not change the quantum efficiency significantly. Compound **25** has the same emission energy and quantum efficiency as that of **24** with a decay lifetime of 30 μs at 77 K, an indication that the low Φ_P and the long decay lifetime of these compounds is not dependent on the R group of the trz unit. The drastically lower Φ_P of **24** and **25** compared to that of **23** is caused by the higher energy of the T_1 state in **24** and **25**, which is quenched more effectively by the deactivating d-d state than the lower T_1 state of **23**. Although *para*-$BMes_2$-ph-R-trz has a suitable T_1 energy for the production of blue phosphorescence as shown by **24** and **25**, because the trz unit is a much weaker donor than pyridine, significant d-d quenching is observed. Thus, in order to achieve bright blue phosphorescent Pt(II) compounds based on the *para*-$BMes_2$-ph-R-trz ligands, the ligand field strength around the Pt(II) center must be increased. This can be achieved by replacing the acac ancillary ligand with chelate ligands that are stronger donors, providing a stronger ligand field and high T_1 energy. Ligands that could potentially serve this role are picolinate and py-1,2,4-triazole.

8.3.2.2 Picolinate as the Ancillary Ligand

The impact of pic ancillary ligands on phosphorescence of $BMes_2$-ph-R-trz Pt (II) compounds is illustrated by compounds **26–29**. One interesting observation is that these compounds all adopt the *N,N-cis* structures shown in Fig. 8.21, in contrast to the ppy analogues of **20** and **21** that adopt the *N,N-trans* structure only. The key factor that favors the *N,N-cis* structure for **26–29** is the intramolecular N···H hydrogen bond between the trz unit and the py ring, which is not possible in **20** and **21**. As observed for the acac compounds **23–25**, the *meta*-isomer **26** is a blue-green emitter with $\lambda_{max} = 489$ nm in 2-Me-THF at ambient temperature and 483 nm at 77 K, while compounds **27–29** are blue emitters with $\lambda_{max} = 456$ nm at ambient temperature and at 452 nm at 77 K. Compared to **24** and **25**, the emission quantum efficiency of **27–29** (0.18, 0.24, 0.14, respectively, in 10 wt% PMMA) has a moderate improvement, but is still low. The decay lifetime of **27–29** (16–17 μs) is also much shorter than those of **24** and **25**. Nonetheless, the *para*-isomers of the Pt(II) compounds **27–29** have the tendency to exhibit excimer emission at a high doping level, while the same was not observed for the acac compounds **24** and **25** (Fig. 8.22). Crystal structural analyses also confirmed the presence of strong intermolecular stacking interactions in the pic Pt(II) compounds. The introduction

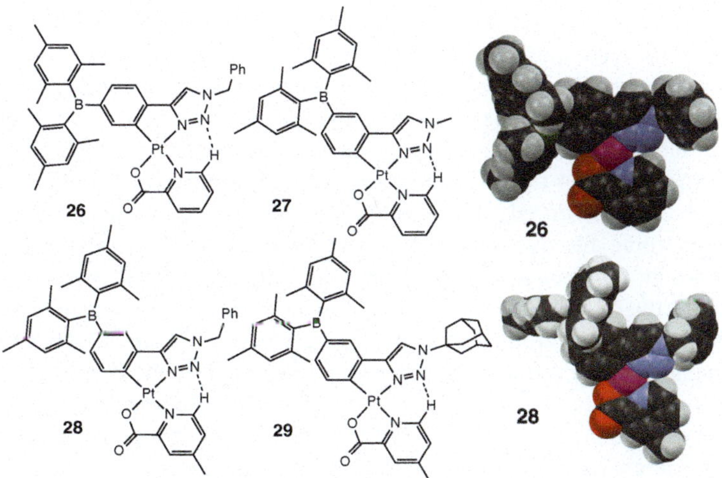

Fig. 8.21 The structures of compounds **26–29**

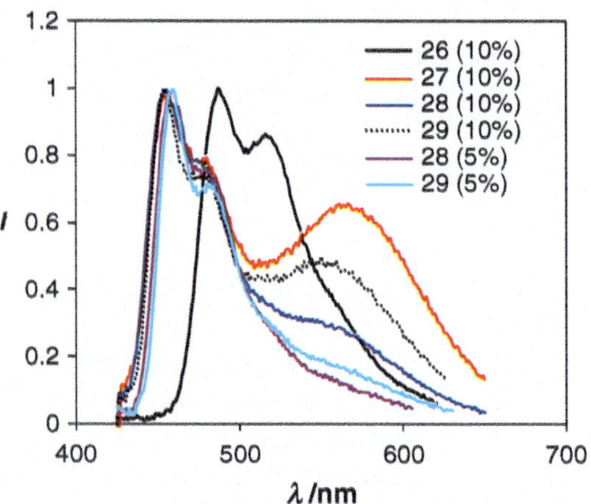

Fig. 8.22 The emission spectra of compounds **26**–**29** in PMMA at various doping levels

of the *p*-methyl group on the pic ligand in **28** only decreased the excimer emission slightly (comparing the emission spectra of **27** and **28** in 10 wt% PMMA in Fig. 8.22). The absence of excimer emission for the *meta*-isomer **26** illustrates that a bulky group at the position *meta* to the Pt binding site is more effective in blocking excimer emission than that at the *para*-position.

8.3.2.3 Pyridyl-1,2,4-trz as the Ancillary Ligand

Py-1,2,4-trz is an attractive ancillary ligand for the *para*-$BMes_2$-Ph-R-trz Pt (II) compounds. It has a high T_1 energy and is a stronger chelate ligand with a higher ligand field than the pic ligand. The impact of py-1,2,4-trz and the derivative ancillary ligands on the blue phosphorescence of $BMes_2$-Ph-R-trz Pt(II) compounds is illustrated with compounds **30**–**33** shown in Fig. 8.23. This class of compounds displays an interesting isomerism phenomenon. NMR and crystal structural analyses established that they exist in two isomeric forms, the N^1-*trans*-isomer (the major isomer) and the N^4-*trans*-isomer (the minor isomer), in both of which the 1,2,3-trz and the 1,2,4-trz rings are *trans* to each other. The corresponding *cis*-isomers were not observed. Again, intramolecular N···H hydrogen bonds are the key driving force for the preferential formation of the N^1-*trans*-isomer, which contains two intramolecular H bonds that provide the extra stability to the molecule, compared to the N^4-*trans* or *cis*-isomers in which only one or zero intramolecular N···H hydrogen bonds are possible. Both N^1-*trans* and N^4-*trans*-isomers display a blue phosphorescence peak at 460 nm that originates from the monomer emission. However, the emission spectra of the N^4-*trans*-isomer is dominated by an excimer emission peak at 544 nm (Fig. 8.24). It is not understood why the N^4-*trans*-isomer is more prone to excimer emission. As the N^1-*trans*-isomer is the major product, the following discussion concerns the N^1-*trans*-isomer only.

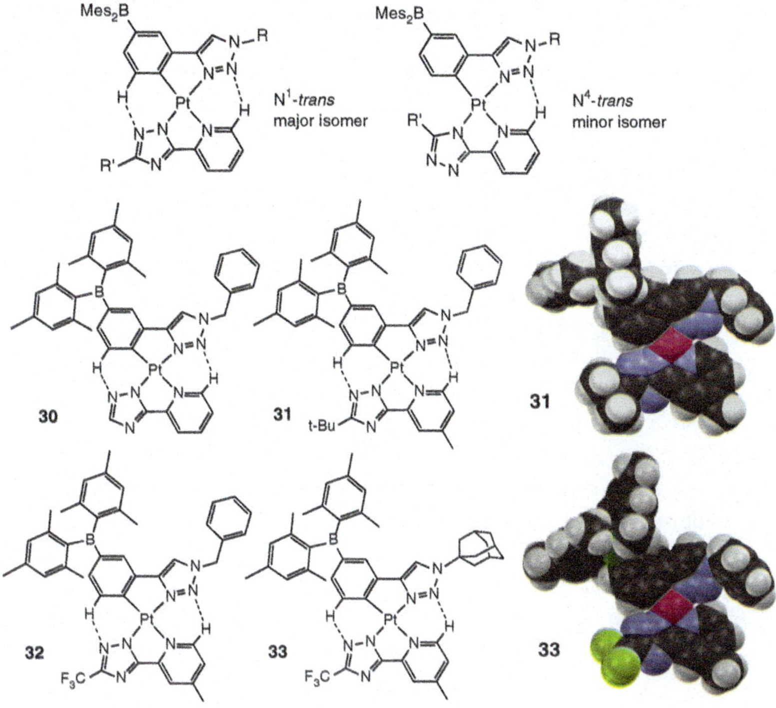

Fig. 8.23 The structures and isomers of compounds **30**–**33**

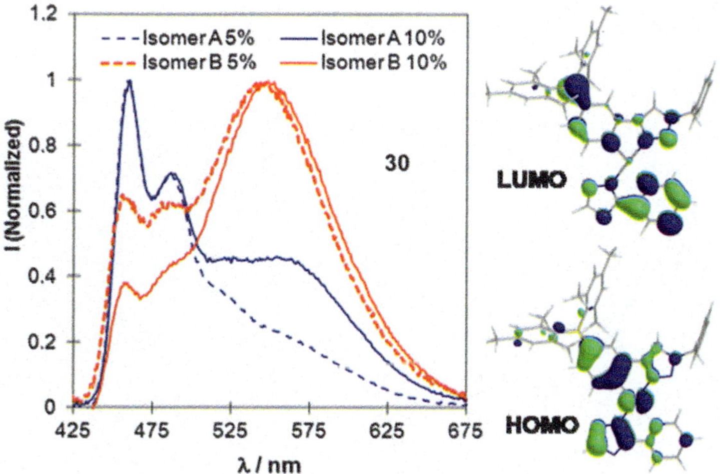

Fig. 8.24 The emission spectra of the *N*[1]-*trans*-isomer (A) and the *N*[4]-*trans*-isomer (B) of compound **30** in PMMA at 5 wt% and 10 wt% doping levels and the HOMO and LUMO diagrams of **30**

Compared to Pt(II) compounds with acac or pic as the ancillary ligand, compounds **30–32** are much brighter emitters with high emission quantum efficiencies. For example, compound **30** has Φ_P values of 0.82 and 0.59, respectively, in 5 and 10 wt% PMMA. Furthermore, the decay lifetimes (11–15 μs at 77 K) of this class of compounds are much shorter than those of the acac- or pic-based systems. These data indicate that the py-1,2,4-trz ligand is indeed very effective as an ancillary ligand in achieving bright blue phosphorescent Pt(II) compounds. This can be attributed to (a) the greater ligand field strength of the py-1,2,4-trz ligand, relative to that of pic and acac, that decreases quenching via the d-d state, and (b) the double intramolecular H bonds of the molecule that increase the rigidity of the Pt(II) core geometry, reducing distortion in the excited state. The introduction of a *t*-butyl group on the 1,2,4-trz ring in **31** led to a small redshift of the emission energy (λ_{em} = 474 nm in 2-Me-THF at 298 K, 457 nm in 2-Me-THF at 77 K, 465 nm in PMMA at 298 K) but a substantial increase in Φ_P to 0.97 and 0.65, respectively, at a 5 wt% and 10 wt% doping level, respectively, in PMMA. Furthermore, **31** has a much lesser tendency to give excimer emission than **30** does, although its emission peak is broader at ambient temperature due to the vibrational states introduced by the t-butyl group (Fig. 8.25). The introduction of a CF_3 group on the 1,2,4-trz ring in **32** led to a slight blueshift of the emission energy (460 nm at 298 K, 454 nm at 77 K in 2-Me-THF). However, despite the CF_3 group, compound **32** is highly prone to form excimers at doping concentrations > 2 % in PMMA. Nonetheless, it retains an impressive Φ_P value of 0.71 in 5 wt% PMMA with a white emission color and 0.47 in 10 wt% in PMMA with a light yellow emission color. The adamantyl-substituted molecule **33** displays similar monomer and excimer emission peaks as those of **32** but with a much lower emission quantum efficiency (0.27 at 5 wt%, 0.20 at 10 wt% in PMMA) due to its tendency to produce excimer emission at a very low doping level of 2 wt%. Thus, the benzyl group is more effective than the adamantyl group as a substituent for the 1,2,3-trz ring.

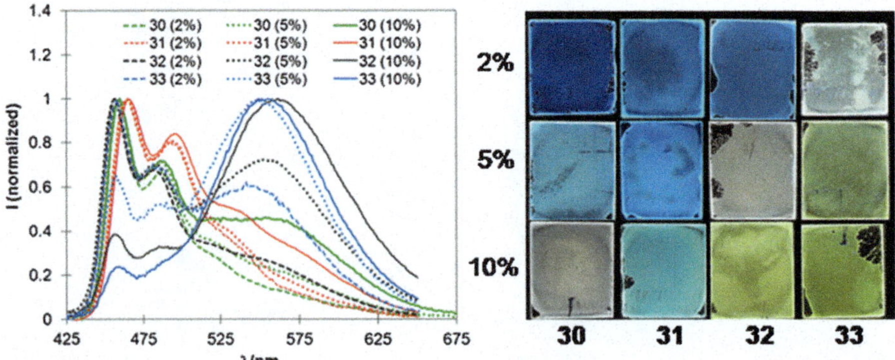

Fig. 8.25 *Left*: the emission spectra of compounds **30–33** (the N^1-*trans*-isomers) in PMMA at 2 wt %, 5 wt%, and 10 wt% doping levels. *Right*: photographs showing the emission colors of the same films

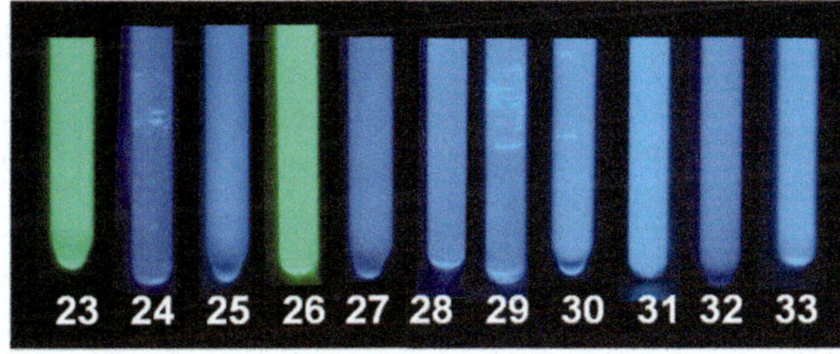

Fig. 8.26 Photographs showing the emission colors of compounds **23–33** in Me-THF at ~1.0×10^{-5} M at 77 K

TD-DFT computational studies and the phosphorescence spectra at 77 K indicate that the blue phosphorescence produced by compounds **30–33** shares the same origin, namely, a π-π^* transition that spreads over both the chromophoric ligand and the ancillary ligand with considerable MLCT character, as shown by the HOMO-LUMO diagrams of compound **30** in Fig. 8.24. Because of the intimate involvement of the py-1,2,4-trz ligand in the emissive state of the molecule, it is not surprising that replacement of a hydrogen atom in the triazole ring with either a *t*-butyl or –CF_3 group leads to shifts in the emission energy.

The above examples illustrate that for *para*-$BMes_2$-Ph-R-trz Pt(II) compounds, the monomer emission color is consistently blue (see the photographs in Fig. 8.26), although its emission efficiency varies drastically with the ancillary ligand. The py-1,2,4-trz ancillary ligand is most effective in achieving bright blue phosphorescent Pt(II) compounds. However, this ligand has a high tendency to produce bright excimer emission, which when combined with the appropriate contribution from the monomer emission can produce bright white phosphorescence (Fig. 8.25). For this reason, these molecules could be candidates for use as single dopants for white phosphorescent OLEDs.

The performance of this class of compounds in OLEDs is illustrated by EL devices using compound **31** or **32** as the dopant. Various host and charge transport materials were used to match the energy levels of these two compounds. The best results were observed using the device structures shown in Fig. 8.27. TmPyPB and CDBP were found to be good electron and hole transport materials, respectively, for these two emitters, because of their high triplet energy and the reasonable match of their energy levels with those of the host and the emitter. A blue EL device with 5 wt% of **31** was achieved with $\lambda_{em} = 467$ nm and CIE coordinates (x, y) of (0.19, 0.34). This device has a moderate EQE (8.3 % at 100 cd · m^{-2}). At a 10 wt% doping level of **31**, an intense excimer emission peak that is more prominent than that in the PL spectrum at the same doping level was observed in the EL spectrum, leading to a whitish emission color with CIE coordinates of (0.31, 0.44) and a much higher EQE of 15.6 % at 100 cd · m^{-2} (Fig. 8.28). These data indicate that the confinement and generation of the monomer triplet excitons of compound **31** in the EL device are much less efficient than the excitons formed from excimers. Therefore, better hosting materials would be the key to improve the performance of this compound

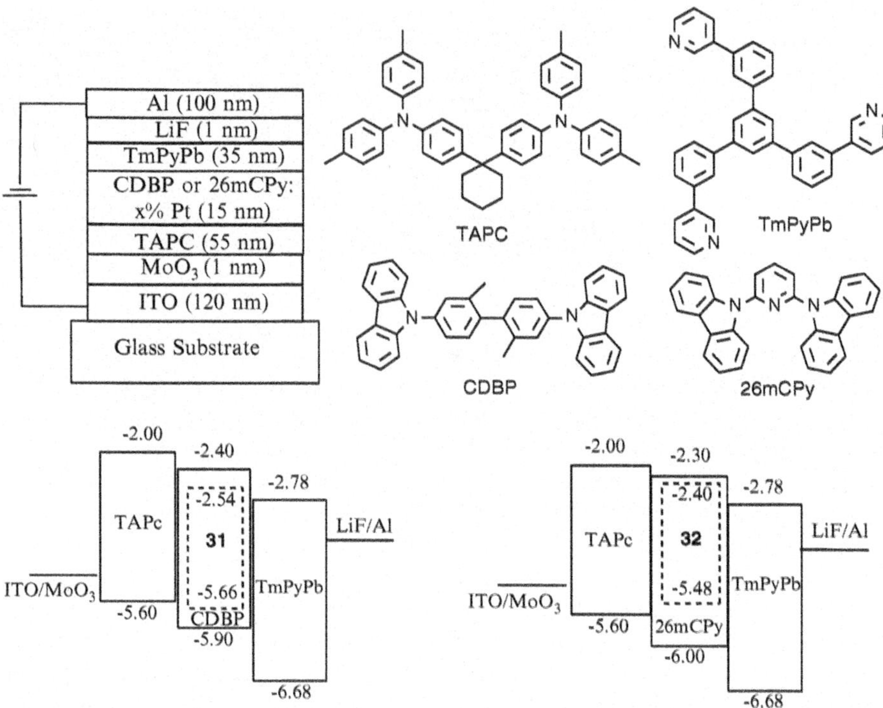

Fig. 8.27 The materials and the device structures used for evaluating the EL performance of compounds **31** and **32**

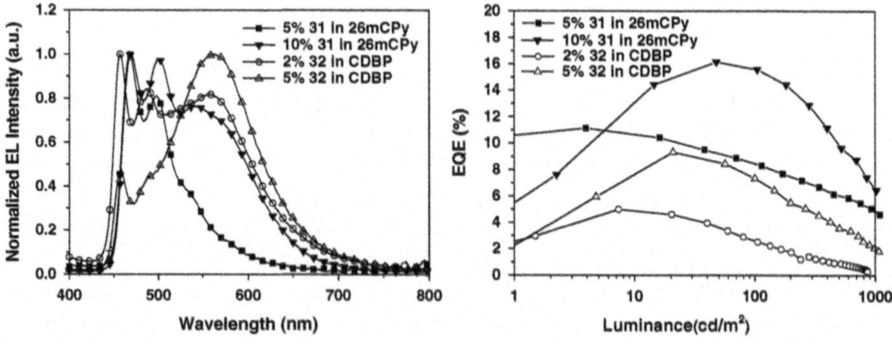

Fig. 8.28 The EL spectra and the EQE vs. luminance plots of EL devices based on **31** and **32**

in EL devices as a blue emitter. The EL devices of compound **32** consistently show a strong excimer emission peak even at a 2 wt% doping level, leading to white (CIE = (0.32, 0.42) at 2 wt%) or a whitish yellow emission color (CIE = (0.38, 0.48) at 5 wt%).

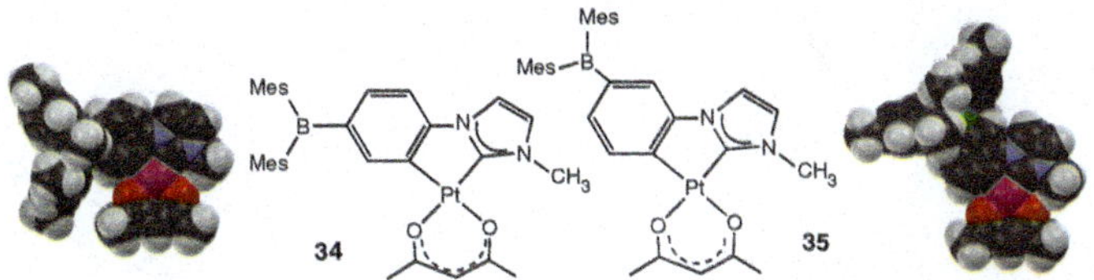

Fig. 8.29 The structures of compounds **34** and **35**

8.3.3 C^C-Chelate Ligands

Although the *para*-$BMes_2$-functionalized Ph-R-trz ligand is very effective in producing blue phosphorescent Pt(II) compounds, it is necessary to use strong chelate ligands such as the py-1,2,4-trz ligand as the ancillary ligand in order to achieve high Φ_P because of the weakly binding nature of the trz unit. The py-1,2,4-trz ancillary ligand, however, has the tendency to generate excimer emission, unless a bulky substituent group is present. If the 1,2,3-trz unit in the N^C-chelate ligand were replaced by a stronger donor such as an *N*-heterocyclic carbene (NHC), it may not be necessary to use the py-1,2,4-trz as the ancillary ligand. To demonstrate this, $BMes_2$-functionalized Ph-NHC ligands and their Pt(II) compounds **34** and **35** were synthesized and investigated [97], in which the ancillary ligand is acac (Fig. 8.29).

TD-DFT data indicated that the T_1 state for both compounds is dominated by the HOMO → LUMO transition, which is mainly localized on the C^C-chelate ligand with considerable contributions from the Pt *d* orbital. Consistent with the trend observed for the ppy and Ph-R-1,2,3-trz ligand systems, the *para*-isomer **35** has a higher T_1 energy than the *meta*-isomer **34**. Compound **34** emits at ~480 nm with a blue-green color while **35** emits at ~460 nm with a blue color in CH_2Cl_2 and in PMMA films (Fig. 8.29). Both compounds have impressive emission quantum efficiencies, 0.87/0.90 in CH_2Cl_2/10 wt% PMMA film for **34** and 0.41/0.86 for **35**. The decay lifetimes of both compounds (6.9 μs and 3.4 μs for **34** and **35**, respectively, in CH_2Cl_2 at 298 K) are much shorter than the related ppy and Ph-R-trz Pt(II) compounds. The high Φ_P of **34** and **35** can be mainly attributed to the greater ligand field strength of the C^C-chelate ligand, which raises the energy of d-d excited states, thus minimizing thermal quenching. Compared to the non-borylated C^C-chelate Pt(II) parent molecule, which emits at $\lambda_{em} = \sim 450$ nm with $\Phi_P = 0.07$, the $BMes_2$-functionalized Pt(II) compounds exhibit a more than tenfold increase in phosphorescent quantum efficiency. This can again be explained by the increased MLCT contribution to the emissive state due to the electron-withdrawing nature of the $BMes_2$ group, as well as reduced intermolecular interactions due to the presence of a bulky substituent. Significantly, no excimer emission was observed for either **34** or **35** in solution or the solid state (Fig. 8.30).

Preliminary evaluation of the performance of **34** and **35** in EL devices was carried out using the device structure of ITO/MoO_3 (1 nm)/CBP (35 nm)/mCP (5 nm)/mCP:Pt-emitter (12 %, 15 nm)/TPBi (65 nm)/LiF (1 nm)/Al. At 100 $cd \cdot m^{-2}$, current and power efficiencies of 50 cd/A and 34 lm/W were observed

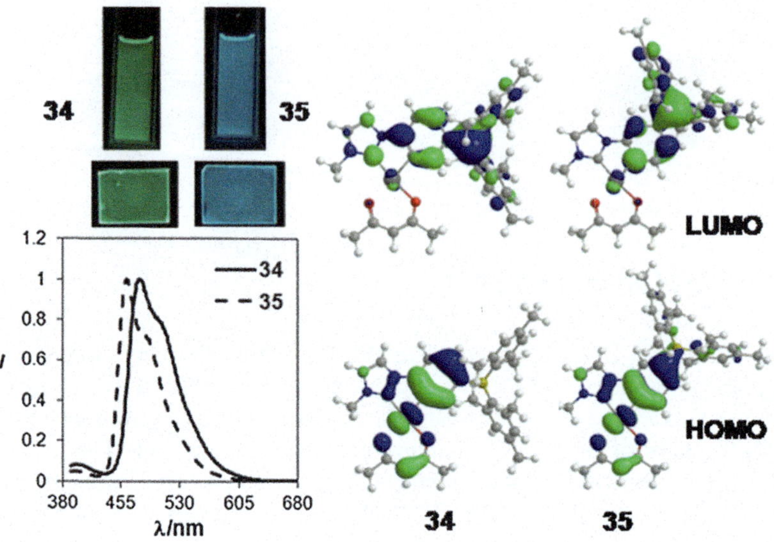

Fig. 8.30 The emission spectra of **34** and **35** in CH_2Cl_2 at 298 K with photographs showing the emission color of **34** and **35** in CH_2Cl_2 and in PMMA films at 10 wt%, alongside HOMO and LUMO diagrams of **34** and **35**

for the device incorporating **34** and 19 cd/A and 14 lm/W in the case of **35**. Maximum EQEs of 18 % and 9.8 % were observed for the devices of **34** and **35**, respectively. However, the EL spectra of both devices are substantially broadened relative to the PL spectra with CIE coordinates of (0.34, 0.53) and (0.27, 0.50), respectively, an indication that the excitons are not fully confined to the dopant in the EL devices. This again illustrates the importance and the challenge of developing suitable host and charge transport materials for blue phosphorescent dopants, which are critical to fully assessing the performance of these and related materials in OLEDs.

8.4 Summary

The key phosphorescent data for $BMes_2$-functionalized Pt(II) compounds presented in this chapter that are relevant for OLED applications are summarized in Table 8.1. The key strategies in achieving bright blue phosphorescent $BMes_2$-functionalized Pt(II) compounds discussed in this chapter are illustrated in Fig. 8.31. The examples shown in this chapter illustrate that functionalization of chelate chromophores with a $BMes_2$ group is a highly effective approach to achieve bright phosphorescent Pt (II) compounds. The $BMes_2$ group plays two roles: (a) as a bulky and rigid substituent that minimizes the distortion of the Pt(II) square-planar geometry in

Table 8.1 Phosphorescence data of compounds **9–35**

	In solution[a]				In PMMA (10 wt%) at 298 K		
Compd	λ_{em} (nm)	Φ_P	τ_P (μs) at 298 K	τ_P (μs) at 77 K	λ_{em} (nm)	Φ_P	τ_P (μs)
9	538	0.77	10.2	12.7	536	0.38	11.0
10	527	0.98	9.5	10.2	526	0.57	8.6
11	596	0.79	67.4	105	588	0.42	30.1
12	590	0.91	40.1				
13	530	0.56	7.9	16.3	527	0.32	8.5
14	598	0.46	14.1	18.6	590	0.27	13.2
15	525,648	0.12	14.5	13.6	535,648	0.03	13.0
16	525,654	0.06	9.5	8.9	519,658	0.02	10.9
17	481	0.43	8.7	9.4	476	0.35	9.5
18	475	0.25	9.6	10.2	472	0.21	7.2
19	492	0.26	8.7	12.2	487	0.30	8.7
20	493	0.013	8.0	16.2			
21	491	0.032	8.2	15.6			
22	485	0.70	8.9	9.5	489	0.76	6.6
23	500	0.17		36.1	493	0.63	
24	453	<0.001		27.8	471	0.10	
25	451	<0.001		30.2	455	0.09	
26	489	0.05		28.2	487	0.54	
27	456	<0.001		16.8	455,567	0.18	
28	457	<0.001		16.3	456	0.24	
29	456	0.005		15.7	454,559	0.14	
30	464	0.01		11.3	460	0.59	
31	474	0.10		9.6	466	0.65	
32	460	0.005		14.3	562	0.47	
33	460	0.005		14.6	556	0.20	
34	478	0.87	6.9		482	0.90	
35	462	0.41	3.4		464	0.86	

[a]For compounds **9–22**, **34**, and **35**, the spectra were recorded in CH_2Cl_2. For compounds **23–33**, spectra were recorded in 2-Me-THF

the excited state and reduces intermolecular interactions and (b) as an electron-accepting group that facilitates MLCT by the π-conjugation of the empty p orbital of the boron center. There are certainly other rigid and bulky groups that can replace $BMes_2$ in role (a). However, no substituents currently exist that can perform both roles as well as bulky boryl substituents. Zhou and coworkers reported the use of -P(O)Ph_2 and -S(O)$_2$Ph as the bulky and electron-withdrawing group in functionalizing ppy and used these ligands to form Pt(acac) compounds with bright phosphorescence. However, when incorporated into EL devices, these compounds gave performances lower than equivalent structures containing $BMes_2$-functionalized dopants [98], though future studies may lead to further

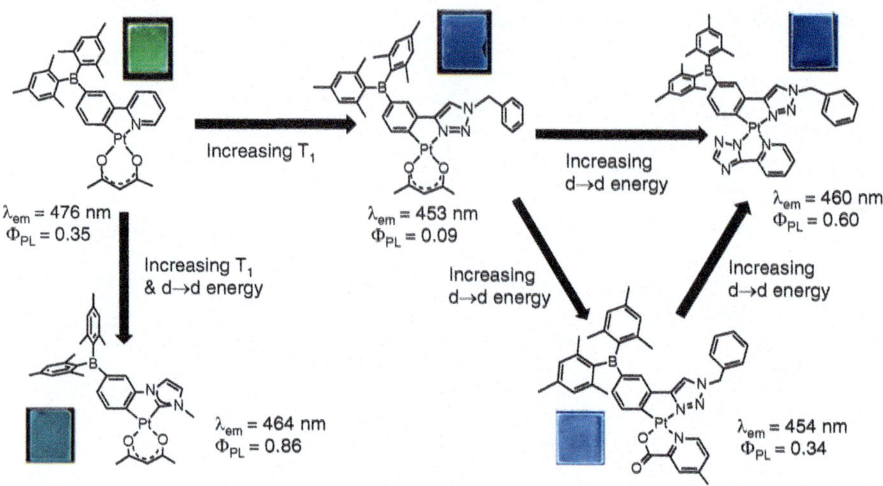

Fig. 8.31 The impact of the chelate chromophore and the ancillary ligand on the phosphorescence of *p*-BMes$_2$-functionalized Pt(II) compounds

improvements. The most important advantage of BMes$_2$ functionalization is that it allows us to access bright blue phosphorescent Pt(II) compounds that remain otherwise rare and underexplored. We have illustrated that the location of the BMes$_2$ group and the nature of the chelate chromophore and ancillary ligand (s) are the key factors for consideration in designing blue phosphorescent Pt (II) compounds. For the relatively weak chelate ligands such as Ph-R-1,2,3-trz, it is essential to employ ancillary ligands that have a much greater ligand field strength when bound to the Pt(II) center than acac and related ligands. Alternatively, one can employ strong Ph-NHC carbene chelate ligands, which can tolerate the use of simple ancillary ligands such as acac and retain a high efficiency of blue phosphorescence. Importantly, the EL data presented in this chapter for BMes$_2$-functionalized compounds are very preliminary. There remains much room for improvement, especially in the choice/development of the appropriate host and charge transport materials, which continues to present a challenge in the development of blue phosphorescent OLEDs.

References

1. Tang CW, VanSlyke SA (1987) Organic electroluminescent diodes. Appl Phys Lett 51:913–915
2. Thompson M (2007) The evolution of organometallic complexes in organic light-emitting devices. MRS Bull 32:694–701 (and references therein)
3. Baldo MA, Forrest SR, Thompson ME (2005) Ch. 6. In: Kafafi ZH (ed) Organic electroluminescence. Taylor & Francis, New York, pp 267–305

4. Evans RC, Douglas P, Winscom CJ (2006) Coordination complexes exhibiting room-temperature phosphorescence: Evaluation of their suitability as triplet emitters in organic light emitting diodes. Coord Chem Rev 250:2093–2126
5. Chou P-T, Chi Y (2007) Phosphorescent dyes for organic light-emitting diodes. Chem Eur J 13:380–395
6. Gareth Williams JA, Develay S, Rochester DL, Murphy L (2008) Optimising the luminescence of platinum(II) complexes and their application in organic light emitting devices (OLEDs). Coord Chem Rev 252:2596–2611
7. Yersin H (ed) (2008) Highly efficiency OLEDs with phosphorescent materials. Wiley-VCH, Weinheim
8. Wong W-Y, Ho C-L (2009) Heavy metal organometallic electrophosphors derived from multi-component chromophores. Coord Chem Rev 253:1709–1758
9. You Y, Park SY (2009) Phosphorescent iridium(III) complexes: toward high phosphorescence quantum efficiency through ligand control. Dalton Trans:1267
10. Chi Y, Chou P-T (2010) Transition-metal phosphors with cyclometalating ligands: fundamentals and applications. Chem Soc Rev 39:638–655
11. Xiao L, Chen Z, Qu B, Luo J, Kong S, Gong Q, Kido J (2011) Recent Progresses on Materials for Electrophosphorescent Organic Light-Emitting Devices. Adv Mater 23:926–952
12. Baldo MA, Lamansky S, Burrows PE, Thompson ME, Forrest SR (1999) Very high-efficiency green organic light-emitting devices based on electrophosphorescence. Appl Phys Lett 75:4–6
13. Adachi C, Baldo MA, Forrest SR, Thompson ME (2000) High-efficiency organic electrophosphorescent devices with tris(2-phenylpyridine)iridium doped into electron-transporting materials. Appl Phys Lett 77:904–906
14. Lamansky S, Djurovich P, Murphy D, Abdel-Razzaq F, Lee H-E, Adachi C, Burrows PE, Forrest SR, Thompson ME (2001) Highly phosphorescent bis-cyclometalated iridium complexes: synthesis, photophysical characterization, and use in organic light emitting diodes. J Am Chem Soc 123:4304–4312
15. Kang DM, Kang J-W, Park JW, Jung SO, Lee S-H, Park H-D, Kim Y-H, Shin SC, Kim J-J, Kwon S-K (2008) Iridium complexes with cyclometalated 2-cycloalkenyl-pyridine ligands as highly efficient emitters for organic light-emitting diodes. Adv Mater 20:2003–2007
16. Chen K, Yang C-H, Chi Y, Liu C-S, Chang C-H, Chen C-C, Wu C-C, Chung M-W, Cheng Y-M, Lee G-H, Chou P-T (2010) Homoleptic tris(pyridyl pyrazolate) IrIII complexes: en route to highly efficient phosphorescent OLEDs. Chem Eur J 16:4315–4327
17. Baldo MA, O'Brien DF, You Y, Shoustikov A, Sibley S, Thompson ME, Forrest SR (1998) Highly efficient phosphorescent emission from organic electroluminescent devices. Nature 395:151–154
18. Kwong RC, Sibley S, Dubovoy T, Baldo M, Forrest SR, Thompson ME (1999) Efficient, saturated red organic light emitting devices based on phosphorescent platinum(II) porphyrins. Chem Mater 11:3709–3713
19. D'Andrade BW, Brooks J, Adamovich V, Thompson ME, Forrest SR (2002) White light emission using triplet excimers in electrophosphorescent organic light-emitting devices. Adv Mater 14:1032–1036
20. Ma B, Djurovich PI, Garon S, Alleyne B, Thompson ME (2006) Platinum binuclear complexes as phosphorescent dopants for monochromatic and white organic light-emitting diodes. Adv Funct Mater 16:2438–2446
21. Kavitha J, Chang SY, Chi Y, Yu JK, Hu YH, Chou PT, Peng SM, Lee GH, Tao YT, Chien CH, Carty AJ (2005) In search of high-performance platinum (II) phosphorescent materials for the fabrication of red electroluminescent devices. Adv Funct Mater 15:223–229
22. Chang S-Y, Kavitha J, Hung J-Y, Chi Y, Cheng Y-M, Li EY, Chou P-T, Lee G-H, Carty AJ (2007) Luminescent platinum(II) complexes containing isoquinolinyl indazolate ligands: synthetic reaction pathway and photophysical properties. Inorg Chem 46:7064–7074
23. Chang S-Y, Cheng Y-M, Chi Y, Lin Y-C, Jiang C-M, Lee G-H, Chou P-T (2008) Emissive Pt (II) complexes bearing both cyclometalated ligand and 2-pyridyl hexafluoropropoxide ancillary chelate. Dalton Trans:6901–6911

24. He Z, Wong W-Y, Yu X, Kwok H-S, Lin Z (2006) Phosphorescent platinum (II) complexes derived from multifunctional chromophores: synthesis, structures, photophysics, and electroluminescence. Inorg Chem 45:10922–10937
25. Zhang Q, Zhou Q, Cheng Y, Wang L, Ma D, Jing X, Wang F (2004) Highly efficient green phosphorescent organic light-emitting diodes based on CuI complexes. Adv Mater 16:432–436
26. Jia WL, McCormick T, Tao Y, Lu J-P, Wang S (2005) New phosphorescent polynuclear Cu (I) compounds based on linear and star-shaped 2-(2′-pyridyl) benzimidazolyl derivatives: syntheses, structures, luminescence, and electroluminescence. Inorg Chem 44:5706–5712
27. Carlson B, Phelan GD, Kaminsky W, Dalton L, Jiang X, Liu S, Jen AKY (2002) Divalent osmium complexes: synthesis, characterization, strong red phosphorescence, and electrophosphorescence. J Am Chem Soc 124:14162–14172
28. Chien C-H, Liao S-F, Wu C-H, Shu C-F, Chang S-Y, Chi Y, Chou P-T, Lai C-H (2008) Electrophosphorescent polyfluorenes containing osmium complexes in the conjugated backbone. Adv Funct Mater 18:1430–1439
29. Chang S-H, Chang C-F, Liao J-L, Chi Y, Zhou D-Y, Liao L-S, Jiang T-Y, Chou T-P, Li EY, Lee G-H, Kuo T-Y, Chou P-T (2013) Emissive osmium(II) complexes with tetradentate bis (pyridylpyrazolate) chelates. Inorg Chem 52:5867–5875
30. Chi Y, Chou P-T (2007) Contemporary progresses on neutral, highly emissive Os(II) and Ru (II) complexes. Chem Soc Rev 36:1421–1431
31. Lyons CH, Abbas ED, Lee JK, Rubner MF (1998) Solid-state light-emitting devices based on the trischelated ruthenium(II) complex. 1. Thin film blends with poly(ethylene oxide). J Am Chem Soc 120:12100–12107
32. Handy ES, Pal AJ, Rubner MF (1999) Solid-state light-emitting devices based on the trischelated ruthenium(II) complex. 2. Tris(bipyridyl)ruthenium (II) as a high-brightness emitter. J Am Chem Soc 121:3525–3528
33. Bernhard S, Barron JA, Houston PL, Abruña HD, Ruglovksy JL, Gao X, Malliaras GG (2002) Electroluminescence in ruthenium(II) complexes. J Am Chem Soc 124:13624–13628
34. Buda M, Kalyuzhny G, Bard AJ (2002) Thin-film solid-state electroluminescent devices based on tris(2,2′-bipyridine)ruthenium(II) complexes. J Am Chem Soc 124:6090–6098
35. Gao FG, Bard AJ (2000) Solid-state organic light-emitting diodes based on Tris(2,2-′-bipyridine)ruthenium(II) complexes. J Am Chem Soc 122:7426–7427
36. Chou P-T, Chi Y (2006) Osmium- and ruthenium-based phosphorescent materials: design, photophysics, and utilization in OLED fabrication. Eur J Inorg Chem 2006:3319–3332
37. Ma Y, Zhou X, Shen J, Chao H-Y, Che C-M (1999) Triplet luminescent dinuclear-gold(I) complex-based light-emitting diodes with low turn-on voltage. Appl Phys Lett 74:1361–1363
38. Ma Y, Che C-M, Chao H-Y, Zhou X, Chan W-H, Shen J (1999) High luminescence gold(I) and copper(I) complexes with a triplet excited state for use in light-emitting diodes. Adv Mater 11:852–857
39. Wong KM-C, Zhu X, Hung L-L, Zhu N, Yam VW-W, Kwok H-S (2005) A novel class of phosphorescent gold(III) alkynyl-based organic light-emitting devices with tunable colour. Chem Commun (23):2906–2908
40. Yam VW-W, Tao C-H (2010) Chapter 5, 69–88. In: Corti C, Holiday R (ed) Gold: science and applications. Taylor & Francis Group, Boca Raton
41. So F, Kido J, Burrows P (2008) Organic light-emitting devices for solid-state lighting. MRS Bull 33:663–669
42. Cai X, Padmaperuma AB, Sapochak LS, Vecchi PA, Burrows PE (2008) Electron and hole transport in a wide bandgap organic phosphine oxide for blue electrophosphorescence. Appl Phys Lett 92:083308
43. Chiu Y-C, Hung J-Y, Chi Y, Chen C-C, Chang C-H, Wu C-C, Cheng Y-M, Yu Y-C, Lee G-H, Chou P-T (2009) En route to high external quantum efficiency (~12%), organic true-blue-light-emitting diodes employing novel design of iridium (III) phosphors. Adv Mater 21:2221–2225

44. Fu H, Cheng Y-M, Chou P-T, Chi Y (2011) Feeling blue? Blue phosphors for OLEDs. Mater Today 14:472–479
45. Lin C-H, Chang Y-Y, Hung J-Y, Lin C-Y, Chi Y, Chung M-W, Lin C-L, Chou P-T, Lee G-H, Chang C-H, Lin W-C (2011) Iridium(III) complexes of a dicyclometalated phosphite tripod ligand: strategy to achieve blue phosphorescence without fluorine substituents and fabrication of OLEDs. Angew Chem 123:3240–3244
46. Li K, Guan X, Ma C-W, Lu W, Chen Y, Che C-M (2011) Blue electrophosphorescent organoplatinum(III) complexes with dianionic tetradentate bis(carbene) ligands. Chem Commun 47:9075–9077
47. Unger Y, Meyer D, Molt O, Schildknecht C, Münster I, Wagenblast G, Strassner T (2010) Green–blue emitters: NHC-based cyclometalated [Pt(C^C*)-(acac)] complexes. Angew Chem Int Ed 49:10214–10216
48. Williams EL, Haavisto K, Li J, Jabbour GE (2007) Excimer-based white phosphorescent organic light-emitting diodes with nearly 100 % internal quantum efficiency. Adv Mater 19:197–202
49. Brooks J, Babayan Y, Lamansky S, Djurovich PI, Tsyba I, Bau R, Thompson ME (2002) Synthesis and characterization of phosphorescent cyclometalated platinum complexes. Inorg Chem 41:3055–3066
50. Cocchi M, Kalinowski J, Fattori V, Williams JAG, Murphy L (2009) Color-variable highly efficient organic electrophosphorescent diodes manipulating molecular exciton and excimer emissions. Appl Phys Lett 94:073309
51. Cocchi M, Virgili D, Fattori V, Rochester DL, Williams JAG (2007) N^C^N-coordinated platinum(II) complexes as phosphorescent emitters in high-performance organic light-emitting devices. Adv Funct Mater 17:285–289
52. Yang X, Wang Z, Madakuni S, Li J, Jabbour GE (2008) Efficient blue- and white-emitting electrophosphorescent devices based on platinum(II) [1,3-difluoro-4,6-di(2-pyridinyl)benzene] chloride. Adv Mater 20:2405–2409
53. Chang S-Y, Chen J-L, Chi Y, Cheng Y-M, Lee G-H, Jiang C-M, Chou P-T (2007) Blue-emitting platinum(II) complexes bearing both pyridylpyrazolate chelate and bridging pyrazolate ligands: synthesis, structures, and photophysical properties. Inorg Chem 46:11202–11212
54. Murphy L, Brulatti P, Fattori V, Cocchi M, Williams JAG (2012) Blue-shifting the monomer and excimer phosphorescence of tridentate cyclometalated platinum(II) complexes for optimal white-light OLEDs. Chem Commun 48:5817–5819
55. Shirota Y, Kageyama H (2007) Charge carrier transporting molecular materials and their applications in devices. Chem Rev 107:953–1010
56. Noda T, Shirota Y (1998) 5,5′-Bis(dimesitylboryl)-2,2′-bithiophene and 5,5″-Bis (dimesitylboryl)-2,2′:5′,2″-terthiophene as a novel family of electron-transporting amorphous molecular materials. J Am Chem Soc 120:9714–9715
57. Noda T, Ogawa H, Shirota Y (1999) A blue-emitting organic electroluminescent device using a novel emitting amorphous molecular material, 5,5′-Bis(dimesitylboryl)-2,2′-bithiophene. Adv Mater 11:283–285
58. Ravindranathan S, Feng X, Karlsson T, Widmalm G, Levitt MH (2000) Investigation of carbohydrate conformation in solution and in powders by double-quantum NMR. J Am Chem Soc 122:1102–1115
59. Tanaka D, Takeda T, Chiba T, Watanabe S, Kido J (2007) Novel electron-transport material containing boron atom with a high triplet excited energy level. Chem Lett 36:262–263
60. Tanaka D, Agata Y, Takeda T, Watanabe S, Kido J (2007) High luminous efficiency blue organic light-emitting devices using high triplet excited energy materials. Jpn J Appl Phys 46: L117–L119
61. Sun C, Hudson ZM, Helander MG, Lu Z-H, Wang S (2011) A polyboryl-functionalized triazine as an electron transport material for OLEDs. Organometallics 30:5552–5555

62. Yamaguchi S, Akiyama S, Tamao K (2001) Colorimetric fluoride ion sensing by boron-containing π-electron systems. J Am Chem Soc 123:11372–11375
63. Yamaguchi S, Shirasaka T, Akiyama S, Tamao K (2002) Dibenzoborole-containing π-electron systems: remarkable fluorescence change based on the "on/off" control of the $p\pi - \pi^*$ conjugation. J Am Chem Soc 124:8816–8817
64. Wade CR, Broomsgrove AEJ, Aldridge S, Gabbaï FP (2010) Fluoride ion complexation and sensing using organoboron compounds. Chem Rev 110:3958–3984 (and references therein)
65. Jäkle F (2010) Advances in the synthesis of organoborane polymers for optical, electronic, and sensory applications. Chem Rev 110:3985–4022
66. Collings JC, Poon S-Y, Le Droumaguet C, Charlot M, Katan C, Pålsson L-O, Beeby A, Mosely JA, Kaiser HM, Kaufmann D, Wong W-Y, Blanchard-Desce M, Marder TB (2009) The synthesis and one- and two-photon optical properties of dipolar, quadrupolar and octupolar donor–acceptor molecules containing dimesitylboryl groups. Chem Eur J 15:198–208
67. Yuan Z, Taylor NJ, Ramachandran R, Marder TB (1996) Third-order nonlinear optical properties of organoboron compounds: molecular structures and second hyperpolarizabilities. Appl Organomet Chem 10:305–316
68. Yuan Z, Entwistle CD, Collings JC, Albesa-Jové D, Batsanov AS, Howard JAK, Taylor NJ, Kaiser HM, Kaufmann DE, Poon S-Y, Wong W-Y, Jardin C, Fathallah S, Boucekkine A, Halet J-F, Marder TB (2006) Synthesis, crystal structures, linear and nonlinear optical properties, and theoretical studies of (p-R-phenyl)-, (p-R-phenylethynyl)-, and (E)-[2-(p-R-phenyl) ethenyl] dimes tylboranes and related compounds. Chem Eur J 12:2758–2771
69. Jia W-L, Bai D-R, McCormick T, Liu Q-D, Motala M, Wang R-Y, Seward C, Tao Y, Wang S (2004) Three-coordinate organoboron compounds BAr_2R (Ar = Mesityl, R = 7-Azaindolyl- or 2,2′-dipyridylamino-functionalized aryl or thienyl) for electroluminescent devices and supramolecular assembly. Chem Eur J 10:994–1006
70. Jia WL, Moran MJ, Yuan Y-Y, Lu ZH, Wang S (2005) (1-Naphthyl)phenylamino functionalized three-coordinate organoboron compounds: syntheses, structures, and applications in OLEDs. J Mater Chem 15:3326–3333
71. Jia WL, Feng XD, Bai DR, Lu ZH, Wang S, Vamvounis G (2004) Mes2B(p-4,4′-biphenyl-NPh(1-naphthyl)): a multifunctional molecule for electroluminescent devices. Chem Mater 17:164–170
72. Hudson ZM, Wang S (2009) Impact of donor–acceptor geometry and metal chelation on photophysical properties and applications of triarylboranes. Acc Chem Res 42:1584–1596
73. Sun Y, Ross N, Zhao S-B, Huszarik K, Jia W-L, Wang R-Y, Macartney D, Wang S (2007) Enhancing electron accepting ability of triarylboron via π-conjugation with 2,2′-Bipy and metal chelation: 5,5′-Bis(BMes2)-2,2′-bipy and its metal complexes. J Am Chem Soc 129:7510–7511
74. Sun Y, Wang S (2009) Conjugated triarylboryl donor–acceptor systems supported by 2,2-′-bipyridine: metal chelation impact on intraligand charger transfer emission, electron accepting ability, and "turn-on" fluoride sensing. Inorg Chem 48:3755–3767
75. Sakuda E, Funahashi A, Kitamura N (2006) Synthesis and spectroscopic properties of platinum (II) terpyridine complexes having an arylborane charge transfer unit. Inorg Chem 45:10670–10677
76. Zhao S-B, McCormick T, Wang S (2007) Ambient-temperature metal-to-ligand charge-transfer phosphorescence facilitated by triarylboron: Bnpa and its metal complexes. Inorg Chem 46:10965–10967
77. Hudson ZM, Wang S (2011) Metal-containing triarylboron compounds for optoelectronic applications. Dalton Trans 40:7805–7816
78. Sun Y, Hudson ZM, Rao Y, Wang S (2011) Tuning and switching MLCT phosphorescence of [Ru(bpy)3]2+ complexes with triarylboranes and anions. Inorg Chem 50:3373–3378
79. Wade CR, Gabbaï FP (2009) Cyanide anion binding by a triarylborane at the outer rim of a cyclometalated ruthenium(II) cationic complex. Inorg Chem 49:714–720

80. Lam S-T, Zhu N, Yam VW-W (2009) Synthesis and characterization of luminescent rhenium (I) tricarbonyl diimine complexes with a triarylboron moiety and the study of their fluoride ion-binding properties. Inorg Chem 48:9664–9670
81. Zhou G, Ho C-L, Wong W-Y, Wang Q, Ma D, Wang L, Lin Z, Marder TB, Beeby A (2008) Manipulating charge-transfer character with electron-withdrawing main-group moieties for the color tuning of iridium electro phosphors. Adv Funct Mater 18:499–511
82. You Y, Park SY (2008) A phosphorescent Ir(III) complex for selective fluoride ion sensing with a high signal-to-noise ratio. Adv Mater 20:3820–3826
83. Zhao Q, Li F, Liu S, Yu M, Liu Z, Yi T, Huang C (2008) Highly selective phosphorescent chemosensor for fluoride based on an iridium(III) complex containing arylborane units. Inorg Chem 47:9256–9264
84. Beyer B, Ulbricht C, Escudero D, Friebe C, Winter A, González L, Schubert US (2009) Phenyl-1H-[1,2,3]triazoles as new cyclometalating ligands for iridium(iii) complexes. Organometallics 28:5478–5488
85. Bossi A, Rausch AF, Leitl MJ, Czerwieniec R, Whited MT, Djurovich PI, Yersin H, Thompson ME (2013) Photophysical properties of cyclometalated pt(II) complexes: counterintuitive blue shift in emission with an expanded ligand π system. Inorg Chem 52:12403–12415
86. Rao Y-L, Wang S (2009) Impact of constitutional isomers of ($BMes_2$)phenyl-pyridine on structure, stability, phosphorescence, and Lewis acidity of mononuclear and dinuclear pt (II) complexes. Inorg Chem 48:7698–7713
87. Hudson ZM, Sun C, Helander MG, Amarne H, Lu Z-H, Wang S (2010) Enhancing phosphorescence and electrophosphorescence efficiency of cyclome-talated Pt(II) compounds with triarylboron. Adv Funct Mater 20:3426–3439
88. Wang ZB, Helander MG, Hudson ZM, Qiu J, Wang S, Lu ZH (2011) Pt(II) complex based phosphorescent organic light emitting diodes with external quantum efficiencies above 20 %. Appl Phys Lett 98:213301
89. Hudson ZM, Helander MG, Lu Z-H, Wang S (2011) Highly efficient orange electrophosphorescence from a trifunctional organoboron-Pt(II) complex. Chem Commun 47:755–757
90. Hudson ZM, Blight BA, Wang S (2012) Efficient and high yield one-pot synthesis of cyclometalated platinum(II) β-diketonates at ambient temperature. Org Lett 14:1700–1703
91. Rao Y-L, Schoenmakers D, Chang Y-L, Lu J-S, Lu Z-H, Kang Y, Wang S (2012) Bluish-green BMes2-functionalized PtII complexes for high efficiency PhOLEDs: impact of the BMes2 location on emission color. Chem Eur J 18:11306–11316
92. Rao Y-L, Goldbach V, Wang S. Unpublished work
93. Ko S-B, Lu J-S, Kang Y, Wang S (2013) Impact of a picolinate ancillary ligand on phosphorescence and fluoride sensing properties of $BMes_2$-functionalized platinum(II) compounds. Organometallics 32:599–608
94. Williams JAG, Beeby A, Davies ES, Weinstein JA, Wilson C (2003) An alternative route to highly luminescent platinum(II) complexes: cyclometalation with n∧c∧n-coordinating dipyridylbenzene ligands. Inorg Chem 42:8609–8611
95. Hang X-C, Fleetham T, Turner E, Brooks J, Li J (2013) Highly efficient blue-emitting cyclometalated platinum(II) complexes by judicious molecular design. Angew Chem Int Ed 52:6753–6756
96. Wang X, Chang Y-L, Lu J-S, Zhang T, Lu Z-H, Wang S (2014) Bright blue and white electrophosphorescent triarylboryl-functionalized C^N-chelate Pt(II) compounds: impact of intramolecular hydrogen bonds and ancillary ligands. Adv Funct Mater 24:1911–1927
97. Hudson ZM, Sun C, Helander MG, Chang Y-L, Lu Z-H, Wang S (2012) Highly efficient blue phosphorescence from triarylboron-functionalized platinum (II) complexes of N-heterocyclic carbenes. J Am Chem Soc 134:13930–13933
98. Zhou G, Wang Q, Wang X, Ho C-L, Wong W-Y, Ma D, Wang L, Lin Z (2010) Metallophosphors of platinum with distinct main-group elements: a versatile approach towards color tuning and white-light emission with superior efficiency/color quality/brightness trade-offs. J Mater Chem 20:7472–7484

Chapter 9
Organometallic Phosphors for OLEDs Lighting

Di Liu

Abstract White organic light-emitting diodes (OLEDs) are regarded as one of the most ideal semiconducting solid-state lighting sources to replace the conventional incandescent bulbs and fluorescent lamps. Phosphorescent OLEDs are advantageous over fluorescent ones in terms of higher efficiency by harvesting both singlet and triplet excitons. Transition-metal complexes represent the most successful phosphorescent materials for use in OLEDs. This chapter highlights the basic knowledge and application of the transition-metal complex phosphors that are frequently used in monochromic and white OLEDs, mainly including complexes of iridium(III), platinum(II), and copper(I). The elaboration generally covers the synthesis, excited states, photophysical properties, OLEDs performance, etc. Special emphasis is placed on structure-property-performance relationship of each organometallic phosphor. The advantages and challenges faced by both noble metals Ir and Pt and the normal metal Cu are discussed. How to achieve high-performance and low-cost white OLEDs based on organometallic phosphors is suggested.

Keywords Semiconducting solid-state lighting • White organic light-emitting diodes • Electrophosphorescence • Organometallic phosphors • Transition-metal complexes of Ir^{III}, Pt^{II}, and Cu^{I} • Metal-to-ligand charge-transfer transition

9.1 Introduction

Lighting consumes up to 20 % of the electricity in the world [1]. The incandescent bulbs and the fluorescent lamps are still the two most prevailing lighting sources currently. For the incandescent bulbs that are characterized by candoluminescence, the electricity-to-light conversion efficiency is only limited to 10 %, with the major electrical energy wasted through thermal irradiation. In the case of compact fluorescent lamps (CFLs), the energy conversion ratio is much increased in comparison

D. Liu (✉)
State Key Laboratory of Fine Chemicals, School of Chemistry, Dalian University of Technology, 2 Linggong Road, Dalian 116024, China
e-mail: liudi@dlut.edu.cn

W.-Y. Wong (ed.), *Organometallics and Related Molecules for Energy Conversion*, Green Chemistry and Sustainable Technology, DOI 10.1007/978-3-662-46054-2_9

with the incandescent bulbs. However, the usage of mercury in the fluorescent tubes inevitably causes environmental contamination, although it is only used in a trace amount. In the new century, people all over the world are faced with the same glorious but challenging issues, to reduce energy consumption and to protect the environment. Under this circumstance, the semiconducting solid-state lighting (SSL) products based on the white light-emitting diodes (LED) and the organic light-emitting diodes (OLEDs) are predicted as the next generation of general illumination systems that have great potentials to replace the traditional incandescent lamps and fluorescent tubes.

LEDs and OLEDs are such kinds of devices that contain inorganic/organic semiconducting material layers sandwiched between two electrodes and are capable of converting the injected carrier pairs into photons when driven by DC power. As the light-emitting products, LEDs and OLEDs are developed mainly for displays and lighting. In terms of energy saving, these SSLs are more advantageous over those traditional lighting tools. A representative estimate indicates that replacement of current white lighting technologies with 50 % efficient SSL would reduce the electricity used for lighting by 62 % and the total electrical energy consumption by 13 % in the USA [1]. In comparison with LEDs whose dimensions are usually small dots, the OLED-related products can be regarded as flat-panel lighting sources since OLEDs can be fabricated onto large area substrates, which is really a valuable merit. In addition, the OLEDs can also be made on flexible substrates, which bring about more utilization opportunities especially in some special situations. Accordingly, white OLEDs are attracting more and more research and industrial interests nowadays as one of the most promising solid-state lighting sources, although white OLEDs technique is temporarily not as mature as that of white LEDs.

Among methods for achieving practical white OLEDs, electrophosphorescence is the most effective due to its demonstrated potential for achieving 100 % internal quantum efficiency [2]. Electrophosphorescence, typically achieved by doping an organometallic phosphor into a conductive host, has been successfully used to generate the primary colors necessary for display applications. And efficient generation of broad spectral emission required of a white light source with a power efficiency (90 lm W^{-1}) exceeding that of incandescent bulbs (15 lm W^{-1}) and comparable with those of the fluorescent tube (65–100 lm W^{-1}) has been reported as well [3]. The organometallic phosphors, especially the second- and third-row transition-metal complexes possessing the d^6 or d^8 electron configuration (Ru^{II}, Os^{II}, Ir^{III}, Pt^{II}), are found particularly attractive because of their strong metal-ligand interaction and high luminescence efficiencies at room temperature [4–7]. These second- and third-row transition-metal phosphors have kept as the leading role of electrophosphorescence since its early observation in 1998 [2]. In addition, the first-row Cu^{I} (d^{10}) complexes have also attracted more and more attentions considering their high relative abundance, low cost, and nontoxic properties compared to the aforementioned noble metal complexes [8]. Most of these transition-metal phosphors have been successfully used in high-performance monochromic and white OLEDs so far.

The goal of this chapter is to report the recent advances of white OLEDs for lighting application and to describe the overall progress made in the last 20 years in phosphorescent transition-metal complexes for usage in both monochromic and white OLEDs. This report mainly focuses on the theoretical background of phosphorescence emission and material design concept, from the view of a chemist. First we will briefly introduce the general information of lighting. Subsequently we will describe the basic knowledge and the state of the art of white OLEDs. Then comes the elaboration of the organometallic phosphors for use in OLEDs. The organometallic phosphors discussed here are only limited to the transition-metal complexes of Ir^{III}, Pt^{II}, and Cu^{I} that are most commonly used in OLEDs, although many other analogues have also been developed for the same usage. It should be noted that this chapter is not meant to be an exhaustive literature overview, but rather an introduction to the field intended to present the basic concepts and to stimulate the interest of readers in this intriguing field.

9.2 Basic of Lighting

9.2.1 Evolution of Lighting Sources

Incandescent bulb had represented the most common lighting sources for a long period of time. It indeed lighted up the world. However, the energy conversion efficiency for incandescent bulbs is very low, with only less than 10 % of electrical energy converted into light, while the main part of energy is wasted by heat irradiation. The total power efficiency of a typical incandescent bulb is only 12–17 lm W^{-1}. Nowadays, incandescent bulbs have already been forbidden for use in some countries to reduce the energy consumption. In comparison, compact fluorescent lamps (CFLs), which are rapidly replacing incandescent bulbs, are more efficient with energy conversion efficiency of about 20 %. The power efficiency of fluorescent lamps is generally in the range of 65–100 lm W^{-1}, which has been set as the international efficiency standard for general illumination systems. However, the usage of highly toxic mercury in the fluorescent tubes inevitably causes environmental pollution, which conflicts the international policy to protect environments. Therefore, novel energy-saving, environment-friendly lighting sources are desired to act as the alternatives of these traditional lighting tools. Under this circumstance, the solid-state semiconducting lighting sources based on light-emitting diodes (LEDs) and organic light-emitting diodes (OLEDs) have emerged as new-generation lighting sources and attracted more and more research and industrial attentions to date.

LEDs and OLEDs generate light through the mechanism of electroluminescence from inorganic and organic semiconducting materials. Inorganic white LEDs are capable of fulfilling some solid-state lighting needs given that such devices have achieved power efficiencies of >44 lm W^{-1} and that their operational lifetime is

>9,000 h [1]. Products such as white LED flashlights are commercially available. The commercial white LED packages emitting cool white light, mainly based on blue-emitting GaN/InGaN LEDs covered with a yellow phosphor, are presently widely diffused and have reached efficacies of 132 lm W^{-1} with 50,000 h lifetime. However, some of the major challenges facing this technology for general illumination purpose are reducing the cost per lumen, obtaining an adequate green LED quantum efficiency, and finding efficient, long-life phosphorescent materials for wavelength down conversion [1]. Furthermore, the commercial available LEDs in current market are almost all small dots in dimension, which are mainly because it would be difficult to release the generated heat if it is made into large size. From this point of view, inorganic LEDs can only be said as dot lighting source.

Similarly, OLEDs laboratory demonstrations already have achieved the power efficiency of 90 lm W^{-1} [3]. And this efficiency is quite possible to be increased to over 120 lm W^{-1} by improving the light outcoupling, that is the photon extraction from the device [3]. This efficiency of OLEDs is much higher than that of incandescent bulbs and already comparable with those of fluorescent lamps, although a little lower than those of inorganic LEDs. Furthermore, in comparison with LEDs dot lighting source, the white OLEDs are regarded as large-area plane lighting sources. This suggests that there are considerable advantages to be gained by using OLEDs as solid-state lighting sources. Of course, the real commercialization of OLED lighting products depends mainly on further increasing the power efficiency and reducing the cost. Nevertheless, the worldwide countries are anticipating that solid-state lighting in the form of white OLEDs and white inorganic LEDs will decrease the energy consumption by a large extent in the near future.

9.2.2 Characterization of Lighting Sources

For comprehensive characterization and comparison of various approaches to white light sources, the operation lifetime, the color quality, the emission efficiencies, the cost, etc. are all important parameters.

There are two critical parameters that define the color quality of a white light source: the color rendering index (CRI) and the Commission Internationale de L'Eclairage (CIE) chromaticity coordinates. The CIE coordinates (x, y) locate the emission color in the chromaticity diagram shown in Fig. 9.1. The standard white light has CIE coordinates of (0.33, 0.33). However, there is a quite broad region of the diagram around this point that can be considered white light. The CRI is a number ranging from 0 to 100 that measures the ability of a source lighting an object needs to reproduce the true color of the object. CRI values less than 70 are unacceptable for indoor lighting applications. CRI values of white OLEDs are excellent being similar to those of incandescent bulbs (>90) and can be higher than those of fluorescent tubes and most common inorganic LEDs [9]. For blackbody and non-blackbody radiators, the chromaticity can also be specified by their color temperature and correlated color temperature (CCT), respectively [1]. The

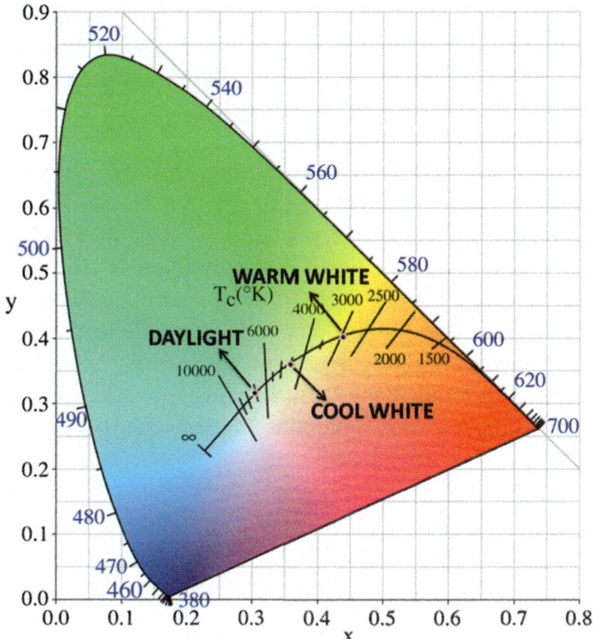

Fig. 9.1 The CIE chromaticity diagram

CCT is the temperature of a blackbody radiator that has a color that most closely matches the emission from a non-blackbody radiator. In general CCT values in the range of 2,500–6,500 K are required for lighting: for example, incandescent lamps have CCT $\approx$ 2,700 K (warn white), and fluorescent lamps can range from about 3,000 K to more than 4,000 K (cool white) [1]. Cool white light is less desirable by consumers, particularly for in-house lighting.

Light-emitting efficiency is another most important factor to evaluate the white light source. There are various efficiencies employed in evaluating the fundamental emission properties of many white light sources like OLEDs, such as quantum efficiency (including internal and external quantum efficiency), power efficiency, and luminance efficiency. For white OLED lighting sources, Forrest and coworkers have given almost authorized definition and measuring methods of these different emission efficiencies [10]. The internal quantum efficiency η_{int} (also termed as IQE) is the ratio of the total number of photons generated within the structure to the number of electrons injected, while the external quantum efficiency η_{ext} (also termed as EQE) is the ratio of the number of photons emitted by the OLEDs into the viewing direction to the number of electrons injected. Both η_{int} and η_{ext} can be expressed in terms of percentage. Generally η_{ext} can be measured experimentally by means of certain instrument and method, while η_{int} is obtained via theoretical calculation by taking into account the η_{ext} value and the fraction of light coupled out of the structure into the viewing direction. The luminous power efficiency η_{P} in lumen per watt (lm W^{-1}), also termed as power efficiency, represents the output light power from a device (measured in lumens, that is the unit of light intensity perceived by the human eye) per electrical power input (measured in watts). It is important to note that the efficacy of a light source takes into account the sensitivity of the human vision to the different wavelengths of the visible spectrum. The luminous efficiency η_L in candelas per ampere (cd A^{-1}), also frequently termed as luminance efficiency or current efficiency, is the ratio of the luminous intensity in forward direction and the current through the device. In many respects, η_L is equivalent to η_{ext}, with the expectation that η_L weights all the incident photons according to the photopic response of the human eye, while η_{ext} weights all photons equally. It should be noted that only photons emitted in the viewing direction are relevant for measuring the forward-viewing power efficiency η_P of white OLEDs for display applications. In contrast, standard techniques for measuring illumination quality white light sources must account for all emitted photons. Therefore, it is standard practice in the lighting industry to state the total power efficiency η_T based on the total number of photons emitted, requiring measurement with an integrating sphere.

9.3 White OLEDs

9.3.1 Basics of OLEDs

9.3.1.1 Mechanism

While common light-emitting diodes (LEDs) produce light by electroluminescence of III–V group mixed crystal inorganic semiconductors, organic light-emitting diodes (OLED) generate light from organic or organometallic molecules. The OLEDs generally have a multilayer architecture where the organic semiconducting material layers are sandwiched between the anode and cathode, one of which must be transparent to light (Fig. 9.2). Upon application of a DC bias voltage to the

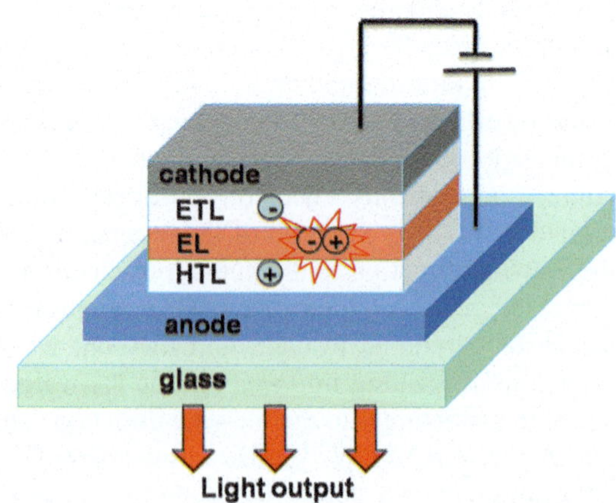

Fig. 9.2 Typical architecture of multilayer OLEDs

electrodes, electrons and holes are injected into the organic stacks, and they migrate through the respective charge-transporting layers and recombine on the emitting molecules, forming singlet and triplet excitons (excited states of molecules) that can generate light by radiative decay. The anode is required to have high work function for easy hole injection to organic layers, which is usually the indium tin oxide (ITO) thin film on glass substrate. The ITO film should have appropriate thickness to guarantee high transparence to visible light so that the light emitted from the device can be output from this side. The cathode used in OLEDs is low-work-function metal, such as Ca, Ba, Al, Mg-Ag alloy, which are usually thick enough to avoid light transmission. According to their function, the active organic materials used in OLEDs can be classified into the following types: hole-injecting layer (HIL) and electron-injecting layer (EIL), hole-transporting layer (HTL) and electron-transporting layer (ETL), and light-emitting layer (LE).

The hole-injecting layer (HI) usually has a rather shallow highest occupied molecular orbital (HOMO) or as low ionization potential (IP), so that it can promote the hole injection from the anode to organic layers due to small hole barrier at the anode interface. In a similar way, the electron-injecting layer (EI) is required to have suitably LUMO level so that electron injection from cathode is efficient due to small electron barrier. The hole-transporting layer (HT) and electron-transporting layer (ET) are characterized by significant high hole drift mobility and electron-transporting mobility, respectively. The HT materials are usually arylamine or carbazole derivatives, which have strong electron-donating properties, while the ET materials are those molecules containing strong electron-deficient or electron-withdrawing moieties. In principle, it is better for the HT and ET to have as high charge mobilities as possible to guarantee the good conductivity and low driving voltage of the whole device. It has been observed that the charge mobilities in a certain material are not only related to its chemical structure and electrochemical redox potentials but also determined by the molecular stacking order in the bulky phase. It has been well established that the good balance of positive and negative charges in the emitting layer is necessary if ideal emission performance of the OLEDs is desired. However, the electron-transporting mobilities of the current organic n-type materials are generally lower than the hole mobilities of organic p-type materials by one to three orders of magnitude, which keeps as the bottleneck problem of OLED field. In many cases some multifunctional material is utilized to simplify the device structure. For example, the HT layer usually simultaneously functions as the HI role if it possesses a rather shallow HOMO. Sometimes, an additional charge blocking layer, either hole-blocking layer (HB) that usually possesses a rather deep HOMO level or electron blocking layer (EB) that has a significantly high LUMO level is inserted next to the emitting layer to prevent the hole or electron from further migrating into the adjacent layer and to confine the excitons in the desired emitting layer. The emitting layer of an OLED is either a neat film of the emitting material or the doped film of the emitter material in a certain charge-transporting host matrix. According to the nature of singlet or triplet excitons that are responsible for the final electroluminescence from the device, the electroluminescence from the OLEDs can be divided into electrofluorescence and electrophosphorescence (vide infra), whose intrinsic efficiencies are greatly different from each other. The electroluminescence performance of multilayer OLEDs is possible to be good enough since the charge injection, charge transportation, and the luminous quantum yield of emitter can be optimized independently without sacrificing others.

It should be noted that the multilayer architectures are generally valid for those vacuum-deposited OLEDs, in which the active organic materials are low-molecular-weight and suitable for thermal sublimation. While for the OLEDs containing polymers or dendrimers that can only be processed by solution methods like spin coating or ink-jet printing, the device structure is usually much simple, e.g., single-layer device, since the deposition of the next layer by spin coating inevitably damages the underlayer film. In this case, some auxiliary materials such as HT and ET are usually blended into the emitting layer to improve the device conductivity and overall performance.

9.3.1.2 Electrofluorescence and Electrophosphorescence

It has been predicted by quantum spin statistics that recombination of free electrons and holes into the light-emitting layer in OLEDs generates about 25 % of singlet and 75 % of triplet excitons, as shown in Fig. 9.3 [11]. In electrofluorescent device where the light emitter is fluorophore, only singlet excitons can decay radiatively, while the radiative relaxation of triplet excitons is forbidden, resulting in the maximum internal quantum efficiency $\eta_{\text{int}}^{\text{max}}$ of 25 %. On the contrary, in the electrophosphorescent devices where the light emitter is phosphorescent material, all the electrical generated triplet excitons are capable of emitting phosphorescence at room temperature, at the same time the rest singlet excitons can be converted into triplet states through efficient intersystem crossing, which finally also decay radiatively by emitting phosphorescence. Therefore, all the electrical generated excitons can be harvested for light emission in electrophosphorescent devices,

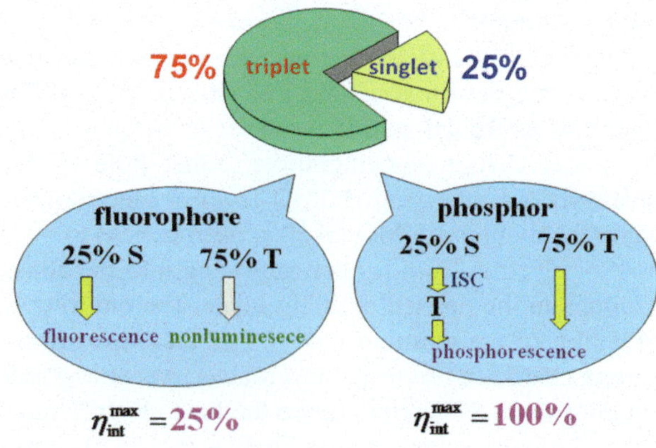

Fig. 9.3 Schematic showing the efficiency difference between electrofluorescence and electrophosphorescence

leading to the maximum theoretical internal quantum efficiency η_{int}^{max} of 100 % [2]. Evidently, the electrophosphorescent materials are superior to the fluorescent ones in terms of emission efficiency when used as emitters in OLEDs.

Since the milestone reports of utilization of phosphorescent materials in OLEDs in 1998, a large amount of phosphorescent compounds have been developed for application in both monochromic and white OLEDs. Different from the typical fluorescent materials that generally are either organic molecules or the complexes of some metals such as Al^{III} and Zn^{II}, the phosphorescent materials are typically organometallic complexes of transition metals, in which the metal ions usually have the d^n electron configuration. The strong spin-orbit coupling expected for these heavy-metal ions would lead to an efficient intersystem crossing from the singlet excited state to the triplet manifold. Furthermore, mixing singlet and triplet excited states through spin-orbit coupling, to a large extent, would partially remove the spin-forbidden nature of the $T_1 \rightarrow S_0$ radiative relaxation, resulting in the highly intense phosphorescent emission with internal quantum efficiency of unity by harvesting both singlet and triplet excitons in OLEDs (Fig. 9.3). At the same time, these emissive metal complexes usually have relatively short radiative decay times [12]. It was found that the transition-metal complexes possessing d^6 (Ru^{II}, Os^{II}, Ir^{III}), d^8 (Pt^{II}), and d^{10} electron configurations (Cu^I) are particularly attractive when used as light emitters in OLEDs, among so large amount of the phosphorescent transition-metal complexes reported so far.

9.3.2 *White OLEDs*

9.3.2.1 Architectures of White OLEDs

For lighting application, the white light emitted from OLEDs should be similar to the natural sunlight covering the full visible light range as much as possible. In general, the white light is composed either by three primary colors (red, green, and blue, termed as RGB) of light or two complementary colors (blue and yellow/orange, termed as BY) of light. However, most of the light-emitting compounds are monochromic luminescent with emission spectra covering only a limited wavelength region. Therefore, white OLEDs are usually made by integrating multiple light emitters within a certain device configuration, each of which emits a certain color of light. That is the multicomponent white OLEDs. On the other hand, it is also possible to gain white light from only one compound in OLEDs that is the single-component white OLEDs.

There are two different cases for single-component white OLEDs. One case is that the light-emitting molecule contains the RGB light-emitting moieties which are linked by covalent bonding. The typical examples are the single-molecule white-emitting polymers, in which the individual green- and red-emitting moieties are usually linked to the blue-emitting backbone, as shown by the structure in Fig. 9.4 [13]. Since the lower-energy components including green and red ones can get

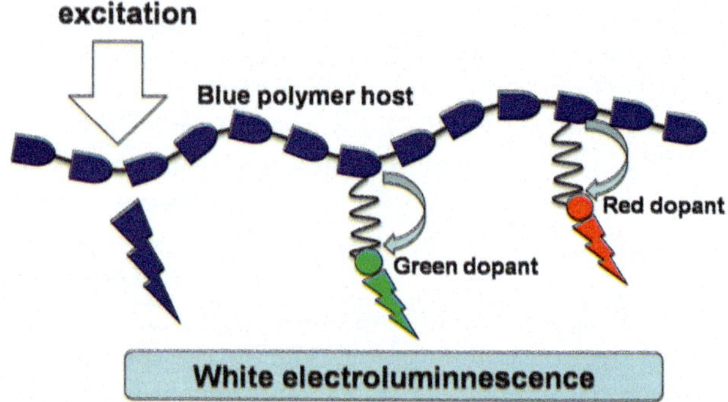

Fig. 9.4 Single-molecule white-emitting polymer (Reproduced with permission from Ref. [13]. Copyright 2007 Wiley-VCH Verlag GmbH & Co. KGaA)

excited by both direct charge recombination and energy transfer from higher-energy components, the lower-energy components are chemically doped at rather low concentration in order to balance the RGB light intensities to achieve the white emission. Another case for the single-component white OLEDs is that the light-emitting molecule can emit light from its different states, for example, from the single molecule and its excimer or electromer. Li [14] have discovered great discrepancy in the photoluminescence (PL) and electroluminescence (EL) of the carbazole-based molecule TECEB (Fig. 9.5a). The single molecule of TECEB is blue fluorescent in dilute solution (Fig. 9.5b), while white EL (Fig. 9.5c) was obtained from the OLEDs in which TECEB was controlled as the exclusive light emitter. It was observed that the white EL contains the blue fluorescence from the single molecule, and the green and red emissions were ascribed to the fluorescence and phosphorescence from the electromer of TECEB (electronic state shown in Fig. 9.5d) that was generated in electrical field of OLEDs.

The single-EML white OLEDs are fabricated by blending all the emitters of various colors together, frequently in a common host, to form the singe emission layer, as shown in Fig. 9.6a. This is the most straightforward strategy to make OLEDs so far. The first white OLED was fabricated in this way in 1994 by doping the orange-, green-, and blue-emitting dyes in the hole-transporting PVK host [15]. Based on the simple device structure, the single-EML white OLEDs can be fabricated by either solution processing technique or vacuum deposition method. When polymeric material is involved in white OLEDs, for example, used as the host, the emitting layer is easily obtained by solution processing the mixed solution of the polymer host doped with different emitting dyes. If small molecular material is utilized as the host in white OLEDs, the emission layer is obtained by vacuum co-evaporation of the host and the RGB dyes. In this case, several small molecular materials are evaporated together in the same vacuum chamber, implying that it will be quite difficult to accurately control the evaporation rate and the final doping concentration of each dye. For all single-EML white OLEDs based on both polymeric and small molecular hosts, since several emitting dyes are doped in the emission layer, the complicated energy transfer from higher-energy components to lower-energy ones must be taken into account in order to finely tune the emission

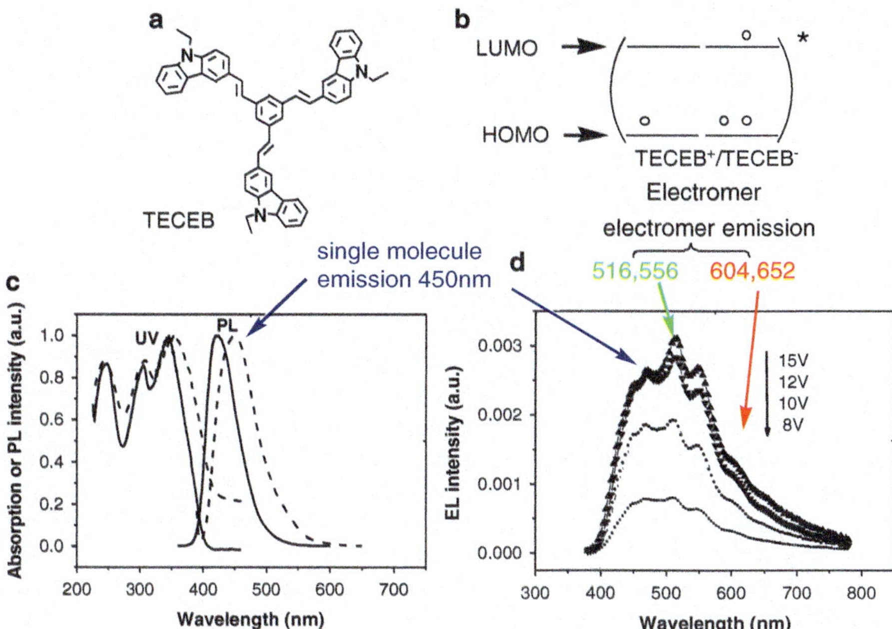

Fig. 9.5 (**a**) The chemical structure of the single-component white-emitting compound TECEB. (**b**) The electronic states of electromer. (**c**) The PL spectra (blue fluorescence) and (**d**) EL spectra (white emission) of TECEB (Figure 5c, d are reproduced with permission from Ref. [14]. Copyright 2004 Wiley-VCH Verlag GmbH & Co. KGaA)

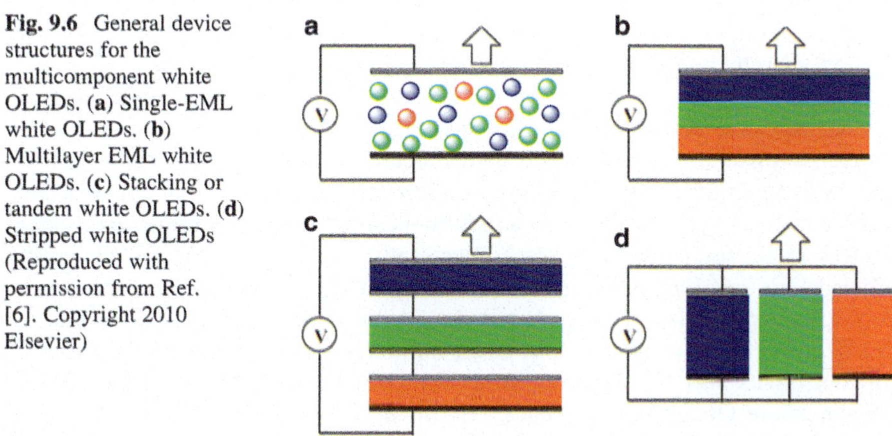

Fig. 9.6 General device structures for the multicomponent white OLEDs. (**a**) Single-EML white OLEDs. (**b**) Multilayer EML white OLEDs. (**c**) Stacking or tandem white OLEDs. (**d**) Stripped white OLEDs (Reproduced with permission from Ref. [6]. Copyright 2010 Elsevier)

spectra and color. In addition, phase separation is quite possible to occur in the multiple-doped emission layer, which must affect the performance and stability of the white OLEDs. In should be summarized that the single-EML device configuration is most frequently applied for polymeric white OLEDs up to now.

In multilayer EML white OLEDs, each RGB-emitting dye is set in individual emitting layer to produce the white light, as shown in Fig. 9.6b. Since multiple organic layers are integrated in the same device, the multilayer device configuration is not suitable for polymer-related solution method. The multilayer white OLEDs can only be fabricated by vacuum deposition, and all the emitting dyes and their host materials should be small molecules that are suitable for thermal sublimation. In multilayer white OLEDs, each color of light is emitted from an individual emitting layer. This makes it possible to optimize the emission color and performance of a certain emitting layer without sacrificing those of the other layers. Therefore, much better device performance can be expected for the multilayer white OLEDs in comparison with the single-EML white devices. However, the unwanted energy transfer between the adjacent layers is also possible in this kind of white devices, which is not good for the ideal white light color and even the emission efficiency, especially at high driving voltage region. Accordingly, a thin layer of certain charge-transporting material is inserted in between adjacent emitting layers to avoid energy transfer. In this way, Sun et al. [16] has fabricated a highly efficient multilayer white OLEDs by doping the red phosphor POIr, the green phosphor $Ir(ppy)_3$, and the blue fluorophore BCzVBI in common host CBP with CBP spacer layer at blue/red and blue/green interfaces that challenged the incandescent source by exhibiting peak external quantum efficiency of 18.7 % and power efficiency of 37.6 lm W^{-1}. The quality of white light strongly depends on the content of different colored light emitted from each emission layer. In order to balance the light intensity from each emitting layer, it is effective to adjust the film thickness of each emitting layer. In comparison with the single-EML white OLEDs, it is evident that the multilayer white OLEDs have the advantage of superior device performance, despite the relatively complicated fabrication.

As shown in Fig. 9.6c, the white light emission is also possibly achieved by the tandem (also termed as stacking) device configuration, in which the RGB monochromic sub-devices are connected in sequence to produce the white light. In tandem devices, the current density through each sub-device is the same. In comparison with the normal device, under a certain current density, higher driving voltage is required and the total light brightness from the tandem devices is also much increased by collecting emission from all sub-devices. Therefore, the tandem devices are characterized by greatly increased current efficiency, but the power efficiency is not changed. Apparently, if a certain brightness is required, the corresponding driving voltage is needed, and the tandem device will work under much lower current density relative to the normal device, which is finally favorable for good device stability. In comparison with the multilayer EML white OLEDs, the tandem white OLEDs are advantageous in terms of higher current density and longer device lifetime. However, in addition to the concurrent V_2O_5 and MoO_3, the successful search for suitable transparent internal connector between sub-stacks still remains a big challenge for tandem white OLEDs so far, which makes the tandem white OLEDs still less common as the other types of white OLEDs. White OLEDs can also be fabricated with the striped configuration in Fig. 9.6d, by laying out the RGB sub-devices in close proximity. The merits of this white OLEDs configuration include the ease of addressing the sub-devices and tuning the RGB light ratio in the resultant white light. However, this device strategy is still not as common as the aforementioned single-EML and multilayer white OLEDs.

9.3.2.2 Efficiency State of the Art

As stated before, the general lighting sources are required to reach a power efficiency of at least 60–70 lm W^{-1} of fluorescent tubes, which is the current benchmark for novel lighting sources. In order to fulfill the lighting application, great efforts have been made in the past decades in terms of both materials development and device fabrication technique to improve the overall efficiency of white OLEDs. Fortunately, as reported in scientific publications, with development of highly efficient phosphorescent materials, the efficiency of white OLEDs fabricated in research labs has gained great progress and many examples have already exceeded the fluorescent tube efficiency. As a whole, the vacuum-deposited white OLEDs based on small molecular materials usually realize higher efficiency than those solution-processed devices. This is mainly because the vacuum-deposited films have better quality and stability than those obtained from solution methods. However, the solution processing seems to be the final and the most ideal device fabrication technique suitable for practical application due to its simplicity and low cost.

Wu and Wong [17] demonstrated highly efficient white OLEDs with a peak forward-viewing power efficiency (PE) close to 40 lm W^{-1}, corresponding to a total PE of 50 lm W^{-1} if all photons emitted to the outside environment are measured, with an external quantum efficiency of 28.8 % and a peak current efficiency of 60 cd A^{-1}. Different from the traditional RGB three-color strategy, four-color system (RYGB) was applied to optimize the white light quality in this work. The blue-emitting FIrpic, green-emitting $Ir(mppy)_3$, yellow-emitting iridium complex 1, and red-emitting dendritic iridium complex Ir-G2 (structures shown in Fig. 9.7) were doped at suitable ratio in polymeric host PVK with the presence of electron-transporting OXD-7 to form the single emitting layer, which was obtained by spin coating their mixed solution. The usage of RYGB four-color system effectively guaranteed the electroluminescence spectra covering the entire visible light range and a good white light color quality. In addition, the famous hole-injecting material PEDOT:PSS (P8000) was introduced in this device to adjust the charge injection and balance. Finally the white OLEDs have the structure of ITO/PEDOT: PSS/PVK:OXD-7:Blue FIrpic:Green $Ir(mppy)_3$:Yellow Ir (1 or 2):Red Ir-G2/Ba/Al. The EL spectra under different blending ratio of the four emitters are shown in Fig. 9.7. The high efficiency was obtained by combining a carefully designed spectrum of the radiation to achieve a very high current efficacy with a modified hole-injecting layer that ensures reduced charge leakage, reduced power consumption, and remarkably improved charge-carrier balance. As claimed by the authors, that was the first report that the maximal forward-viewing power efficiency approached 40 lm W^{-1} for solution-processed white OLEDs.

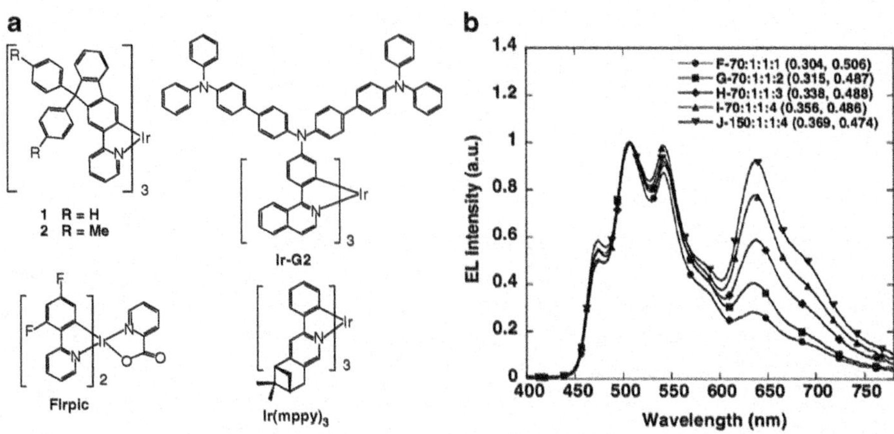

Fig. 9.7 Chemical structures of the RYGB iridium phosphors (**a**) and EL spectra of white OLEDs (**b**) (Adapted with permission from Ref. [17]. Copyright 2011 Wiley-VCH Verlag GmbH & Co. KGaA)

In 2012, Xie and Wong [18] reported a partially solution-processed complementary color white OLED with efficiency further improved in comparison with the above mentioned one. In this two-color white OLED system, the emitting layer was comprised of a blue-emitting FIrpic and an orange-emitting $(fbi)_2Ir(acac)$ or Ir $(Flpy\text{-}CF_3)_3$ doped in a dendritic host H2, which was obtained by spin coating. However, on top of this emitting layer, an electron-transporting SPPO13 layer was vacuum deposited to balance the charge injection and transportation. In addition, the low-conductivity PEDOT:PSS (P 8000) was also introduced as hole-injecting layer in between the anode and emitting layer. The white OLEDs have the structure of ITO/PEDOT:PSS/H2:Blue FIrpic:Orange $Ir(Flpy\text{-}CF_3)_3$/SPPO13/LiF/Al. In this way, a maximum forward-viewing power efficiency close to 50 lm W^{-1}, an external quantum efficiency of 26.6 %, and a peak current efficiency of 70.6 cd A^{-1} were obtained for that white OLEDs. Since illumination sources are typically characterized by their total power, a forward-viewing power efficiency of 50 lm W^{-1} corresponds to a total power efficiency of 85 lm W^{-1}. That represented a new world record for solution-processed white OLEDs and even comparable to the reported highest efficiencies of the white OLEDs made by thermal evaporation without outcoupling structures.

In comparison with the solution-processed white OLEDs, the vacuum-deposited white OLEDs usually have more complicated configurations. Li et al. [19] designed and synthesized novel iridium complexes containing CF_3- or F-substituted 2-phenylbenzothiazole as cyclometalating ligands, $(CF_3\text{-bt})_2Ir(acac)$ and $(F\text{-bt})_2Ir$ (acac), which exhibited quite efficient orange emission. In combination with the traditional blue iridium phosphor FIrpic, these orange iridium complexes were used to fabricate two-component multilayer white OLEDs with the blue and orange phosphors separately dispersed in CBP host to form two individual emitting layers. In order to optimize the charge injection and balance, PEDOT:PSS was introduced into this small molecular materials-based device. The white OLEDs have the structures of ITO/PEDOT:PSS/CBP:$(CF_3\text{-bt})_2Ir(acac)$ or $(F\text{-bt})_2Ir(acac)$/CBP: FIrpic/TPBI/LiF/Al. A maximum current efficiency of 68.6 cd A^{-1}, corresponding to forward-viewing external quantum efficiency of 26.2 % and a peak power efficiency of 34 lm W^{-1} were obtained for the white OLED containing $(F\text{-bt})_2Ir$ (acac). With multilayer EML structure, Ma and coworkers [20] also realized a high power efficiency of 41 lm W^{-1} in their white OLEDs.

All the above mentioned OLEDs are built in a standard substrate emitting architecture, where the outcoupling efficiency is approximately 20 %. The remaining 80 % of the photons are trapped by organic and substrate modes in equal amounts [3]. Hence, the greatest potential for a substantial increase in external quantum efficiency and power efficiency is to enhance the light outcoupling. In 2009, Leo and coworkers [3] demonstrated highly efficient white OLEDs with fluorescent tube efficiency (90 lm W^{-1} at a brightness of 1,000 cd m^{-2}) by combining a carefully chosen emitting layer with high-refractive-index substrates and using a periodic outcoupling structures. As shown in Fig. 9.8, the optimized white OLEDs have a detailed configuration of high-refractive-index substrate/MeO-TPD:NDP-2 (4 mol.%, 60 nm, hole-transport layer)/NPB (10 nm, electron-blocker layer)/TCTA:$Ir(MDQ)_2(acac)$ (6 nm, orange-red)/TCTA (2 nm)/ TPBI:FIrpic (4 nm, blue)/TPBI (2 nm)/TPBI:$Ir(ppy)_3$ (6 nm, green)/TPBi (10 nm, hole-blocking layer)/Cs:Bphen (40 nm, electron-transport layer)/Ag (100 nm, cathode). It is well known that the use of high-refractive-index glass substrates can substantially increase the amount of light coupled from the organic layers to the glass substrate (up to 80 %). As shown in Fig. 9.9b, increasing the refractive index of glass substrate from $n_{\text{low}} = 1.5$ to $n_{\text{high}} = 1.78$ causes the index mismatch between organic materials and substrate to vanish, enhancing light outcoupling into the high-refractive-index glass, so that all photons guided to organic modes by total internal reflection at the organic/glass interface in the low-refractive-index case are entering the glass substrate. The efficiency of these white OLEDs has the potential to be raised to 124 lm W^{-1} if the light outcoupling is further improved. Once again, the high efficiency could make white light OLEDs, with their soft area light and high color rendering qualities, the lighting sources of choice in the near future.

9.4 Organometallic Phosphors for OLEDs

Transition-metal complexes have emerged as the major phosphorescent materials predominantly used in OLEDs and other optoelectronic devices nowadays. Transition-metal complexes have unique advantages in electronic and photonic applications, since the electronic states can be changed in a controlled fashion within easily accessible ranges. Spin states may also be controlled by the strength and symmetry of the ligand field and the redox states of metal ions. A large amount of second- and third-, and even first-row transition-metal complexes have been

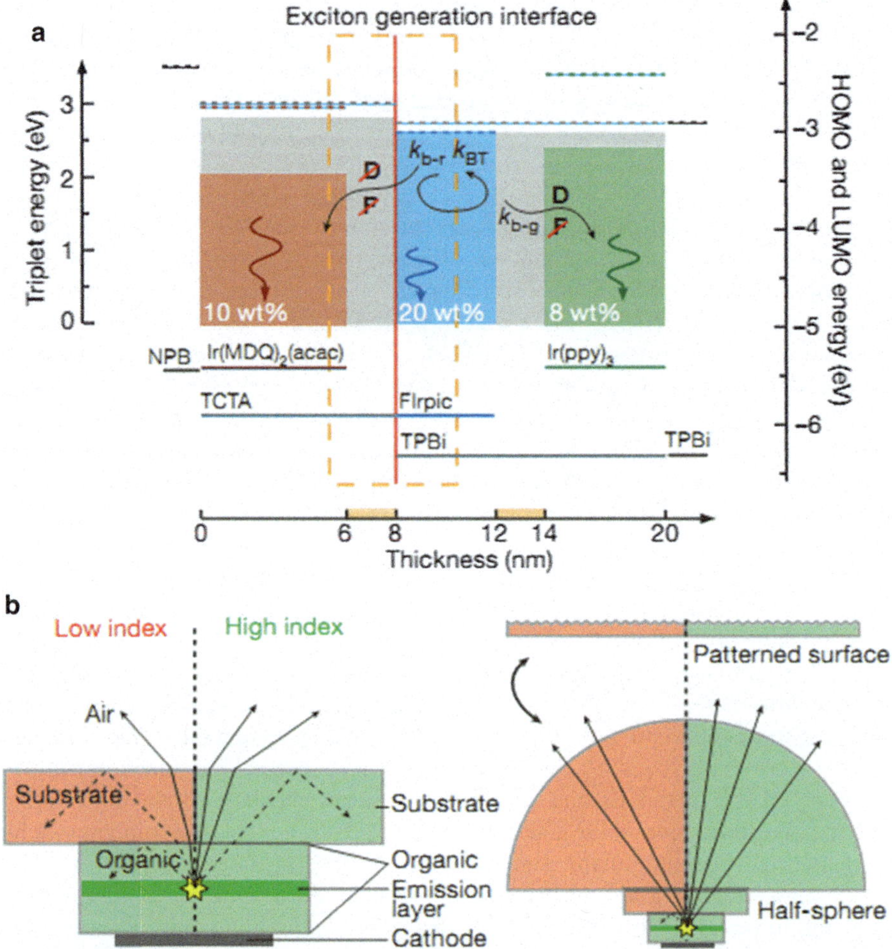

Fig. 9.8 Energy level diagram and light modes in an OLED. (**a**), the white OLED configuration and energy level diagram of the involved materials. (**b**), the *left* panel illustrates increasing the substrate refractive index could enhance light outcoupling into the high-refractive-index glass (Reproduced with permission from Ref. [3]. Copyright © 2009, right managed by Nature Publishing Group)

developed for application as light-emitting materials in various fields. However, to warrant the OLED applications, neutral as opposed to the ionic complexes are the main focus here, simply due to the higher volatility and the avoidance of internal defects that were produced by the current-induced migration of ionic species. Among all the transition metals, since the Ir^{III}, Pt^{II} complexes represent the most widely used group and possess the best performance, the Cu^{I} complexes are the cheapest and most suitable for large-scale practical application, and this report focuses on only these three transition metals and related materials.

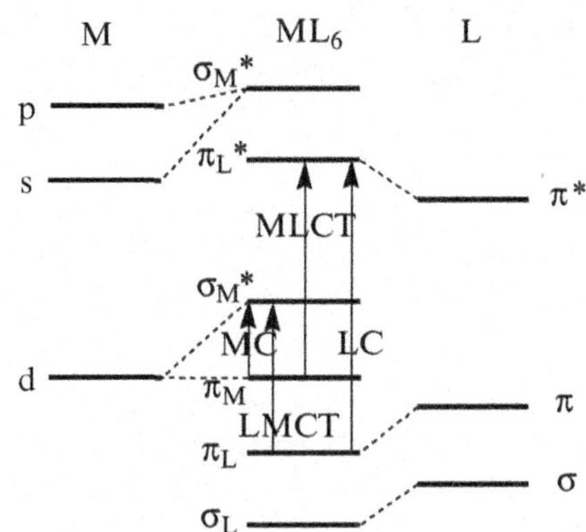

Fig. 9.9 Molecular orbitals diagram for an octahedral complex of a transition metal. The *arrows* indicate the four types of transitions based on localized MO configurations (Adapted with permission from Ref. [21]. Copyright © 2007, Springer-Verlag Berlin Heidelberg)

9.4.1 Electronic Transitions and Excited States

In order for a better understanding of the photophysical behaviors of transition-metal complexes including the light absorption and emission procedures, we had better start from learning the molecular orbital theory. It states that the electron density in each of the frontier molecular orbitals (MO) are not equally delocalized between metal and ancillary ligands, but preferentially located at the metal or the ligands. The electronic transition can be considered as the one-electron excitation that occurs among the associated frontier molecular orbitals [12].

An octahedral transition-metal complex (ML_6) that usually contains a d^6 metal center is taken as an example to elaborate the frontier MOs and related electronic transitions. As shown in Fig. 9.9, the various frontier MOs can be conveniently classified according to their predominant atomic orbital contributions as [21]: (1) strongly bonding, predominantly ligand-centered σ_L orbitals; (2) bonding, predominantly ligand-centered π_L orbitals; (3) essentially nonbonding, metal-centered π_M or d_π orbitals of t_{2g} symmetry; (4) antibonding, predominantly metal-centered $\sigma_M{}^*$ or $d\sigma^*$ orbital of e_g symmetry; (5) antibonding, predominantly ligand-centered $\pi_L{}^*$ orbitals; and (6) strongly antibonding, predominantly metal-centered $\sigma_M{}^*$ orbitals. In the ground electronic configuration of an octahedral complex of a d^n metal ion, orbitals of types 1 and 2 are completely filled, while n electrons reside in the orbitals of types 3 and 4. When light absorption and emission are concerned for these transition-metal complexes, four types of electronic transition and related excited states are frequently involved as follows.

Metal-Centered (MC) Excited States Typical heavy-metal complexes with a partially filled d shell at the metal center are characterized by low-energy MC excitation states, which arise from electron hopping between the nonbonding (d_π, orbital of type 3 in Fig. 9.9) and antibonding ($d_{\sigma*}$, orbital of type 4 in Fig. 9.9) orbitals. These d-d transitions are Laporte-forbidden, showing exceedingly low transition probability for the occurrence of both MC-based absorption and emission signals. Low-energy MC transitions are usually expected for metals of the first transition row like copper (Cu). For the second- and third-row transition-metal complexes with strong-field ligands in the spectrochemical series, the anticipated MC transitions are destabilized to the higher-energy region, making the metal-to-ligand charge-transfer state (MLCT) or ligand-centered excited state (LC) (vide infra) as the lowest-energy excited state. The singlet-to-triplet intersystem crossing is facilitated by the heavy-atom enhanced spin-orbit coupling. As a consequence, many of them have exhibited highly efficient, room-temperature phosphorescence in both fluid and solid states [12].

Metal-to-Ligand Charge-Transfer (MLCT) Transitions and States Metal-to-ligand charge-transfer (MLCT) states involve electronic transitions from a metal-based d_π orbital (type 3 in Fig. 9.9) to a ligand-based delocalized antibonding π_L^* orbital (type 5 in Fig. 9.9). Low-energy MLCT transitions are expected when the metal is easy to oxidize and the ligands are easy to reduce. These transitions are commonly observed in the middle- and late-transition-metal complexes possessing relatively low oxidation potentials. The transition process can be understood in this way that the metal center renders the easily accessed d_π electron, while the acceptor ligands accommodate this ejected electron in their unoccupied π_L^* orbital. Since the π_L^* orbital is usually delocalized over the ligand, the population to MLCT states may only cause minimum structural distortion, facilitating its radiative recombination, decay process with remarkable efficiency.

For the first-row transition-metal complexes, the MLCT states are generally quite reactive. With shifting the metal to the second- and third-row transition metals, the MLCT states of their complexes become stable because of the increased metal-ligand bonding strengths accompanied by the destabilized MC excited states and the consequent reduction of the radiationless deactivation. Moreover, MLCT transitions are strongly allowed processes, manifested by intense absorption bands in the visible or near-UV spectral regions. It should be pointed out that the excitation may generate both singlet and triplet MLCT states simultaneously. However, due to the strong spin-orbit coupling effect caused by these heavy-metal atoms, those singlet MLCT states will rapidly transfer into the triplet states through intersystem crossing. Again, the strong spin-orbit coupling removes the spin-forbidden nature of the electronic transition from the triplet excited state to the singlet ground state, resulting in the highly phosphorescent nature of these triplet MLCT states. The luminescent properties of MLCT states are particularly distinct for transition-metal complexes with d^6 and d^8 electron configurations.

Ligand-Centered (LC) $\pi\pi^*$ Transitions and Excited States Ligand-centered (LC) $\pi\pi^*$ excited states originate from electronic transitions between π orbitals that are mainly localized on the ligand chromophore. Low-energy LC $\pi\pi^*$ transitions are expected for the aromatic ligand with extended π and π^* orbitals when the metal center has minimized perturbation upon coordination. For the complexes of some main-group metal elements such as Zn^{II} and Al^{III} and their closed shell analogues that do not participate in the $\pi\pi^*$ transitions of the ligands, LC $\pi\pi^*$ transitions should be the predominate processes responsible for the light absorption.

Apparently, the spectroscopy behaviors of these main-group complexes are mainly determined by the ligand properties. While for transition-metal complexes, LC $\pi\pi^*$ transitions are frequently observed in their electronic absorption spectra, which correspond to the electronic transitions from the singlet ground state to the first singlet excited state of the ligands (^{1}LC). The excitation into ^{1}LC excited states usually corresponds to relative strong absorption intensity in the UV-Vis absorption spectra of the complexes. Subsequently a slow intersystem crossing may occur and generate the triplet excited state of the ligands (^{3}LC), which then decay by radiative and/or non-radiative ways. It should be noted that, due to the lack of direct incorporation of metal-center electron density, the intersystem crossing from singlet to triplet is not as fast as in the case of MLCT. This means that much less singlet-triplet mixing can be expected in the LC $\pi\pi^*$ states. As a result, the associated phosphorescence related to ^{3}LC excited states is subject to much longer radiative lifetime and relatively low efficiency due to possible quenching by any radiationless deactivation.

Ligand-to-Metal Charge-Transfer (LMCT) Transitions and States Low-energy ligand-to-metal charge-transfer (LMCT) transitions are expected for those metal complexes in which at least one of the ligands is easy to oxidize and the metal is easy to reduce (in high oxidation states). These complexes typically constitute early transition-metal complexes with cyclopentadienyl or with simple σ-bonded anionic ligands. For example, a class of d^{10} complexes involving Cu^I, Ag^I, and Au^I metals are good candidates that exhibit LMCT emission. In particular, the tetrametallic Cu^I clusters have been used in OLEDs and exhibited bright phosphorescence from the LMCT excited states. However, none of the complexes showing LMCT emission have shown performance data comparable to those fabricated using previously mentioned MLCT and LC $\pi\pi^*$ emitting materials.

In summary, the low-energy MC transitions are most applicable for the first-row transition-metal complexes in which the metal-ligand bond strength is severely weakened, and they are characterized by strong non-radiative decay feature and thus weak emission. The LMCT transitions are suitable for those complexes whose metal center is in high oxidation states and the ligands are strong electron-donating. And the LMCT-based emission is usually not strong too. The LC $\pi\pi^*$ transitions are distinct when the ligands are characterized by extended π structure and the metal center is main-group metal. In this case, the LC $\pi\pi^*$ transitions are responsible for both the light absorption and emission of these main-group metal complexes. The LC $\pi\pi^*$ transitions are also frequently involved together with MLCT for transition-metal complexes. In many cases, the excitation of transition-metal complexes generates MLCT or LC $\pi\pi^*$ excited states simultaneously. The energetic closeness and efficient overlap between ^{3}MLCT and ^{3}LC states may result in the mixed lowest-energy excited state (T_1) that is responsible for the phosphorescence.

9.4.2 Iridium(III) Phosphors

It is well known that the iridium complexes are the most successful phosphorescent materials used for OLEDs application. In comparison with the earlier reported platinum complexes such as PtOEP, the unique advantage of iridium complexes is their much shorter triplet lifetime that is typically in the order of several microsecond. This is indeed a precious merit since short triplet lifetime is required for the triplet excitons to prevent triplet-triplet annihilation and to guarantee high emission efficiency in OLEDs. In addition, iridium complexes are characterized by flexible color tenability and high phosphorescent quantum yields, which are all necessary valuable merits when OLEDs are involved in practical applications in lighting and display fields. It would be safe to say that, up to date, iridium complexes are the most successful series of organometallic phosphors among all the transition-metal complexes.

9.4.2.1 Structures and Syntheses

In general, the transition-metal complex phosphors used as phosphorescent materials in OLEDs field have the cyclometalated configuration. The Ir^{III} ion has a d^6 electron configuration, and the Ir^{III} complexes have the typical octahedral conformation with three bidentate ligands coordinated to the metal center. The Ir^{III} complexes used as phosphorescent materials can be classified into bis- and tris-cyclometalated complexes according to the number of the cyclometalating ligands around the iridium center. For clear understanding and identification, it is necessary to define cyclometalation and related concepts. As stated by Albrecht [22], cyclometalation refers to the transition metal-mediated activation of a C–R bond to form a metallacycle comprising a new metal-carbon σ bond. Typically, the reaction consists of two consecutive steps: initial coordination of the metal center via a donor group and subsequent intramolecular activation of the C–R bond, which closes the metallacycle. As a special type of chelation, cyclometalation is characterized by the formation of metal-carbon (M–C) σ bond. Therefore, the cyclometalating ligands refer to those only containing carbon atoms that will form C–M bond upon coordination, typically written as H-C^X, for example, the widely used bidentate cyclometalating H-C^N ligand. Those ligands that do not form C–M bond with metal center are usually called ancillary ligands.

According to whether all the ligands of an Ir^{III} complex are identical to each other, the cyclometalated iridium complexes can also be classified as homoleptic and heteroleptic. When different ligands are involved in an Ir^{III} complex, it is heteroleptic; otherwise, it is homoleptic when three bidentate ligands of same structure are incorporated to the iridium ion center. Accordingly, the bis-cyclometalated iridium complexes, frequently possessing structure of $(C\text{^}N)_2Ir(L\text{^}X)$, must be heteroleptic, and tris-cyclometalated $iridium^{III}$ complex can be either homoleptic (e.g., $Ir(C\text{^}N)_3$) or heteroleptic (e.g., $(C\text{^}N)_2Ir(C\text{^}N)'$).

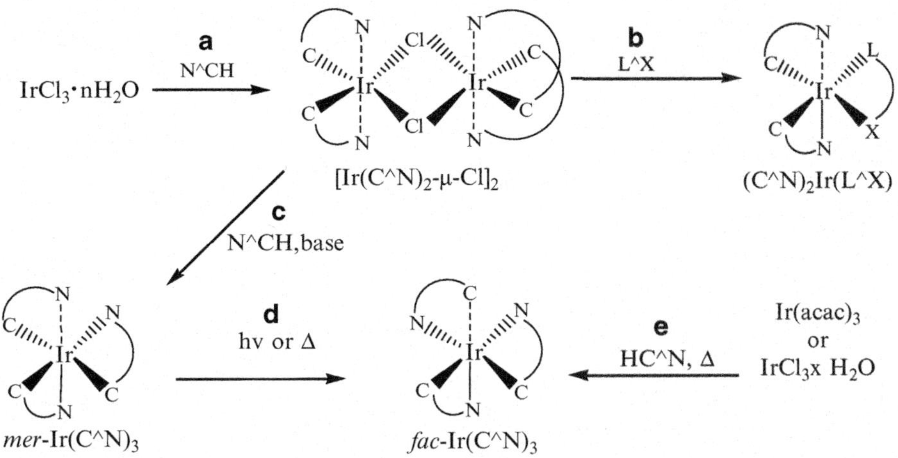

Fig. 9.10 Schematic representation of the synthetic strategies utilized for the synthesis of cyclometalated iridium(III) complexes. H-C^N: cyclometalating ligand

Figure 9.10 illustrates the most common synthetic strategies toward both homoleptic and heteroleptic and even the charged phosphorescent Ir^{III} complexes. The standard synthesis of an Ir^{III} complex is accomplished through a two-step process [23, 24] in which the first step is known as the Momoyama reaction (path a) that yields a chloride-bridged dinuclear Ir^{III} dimer $[Ir(C\text{^}N)_2\text{-}\mu\text{-}Cl]_2$. In the second step, the chloro-bridge in the dimer $[Ir(C\text{^}N)_2\text{-}\mu\text{-}Cl]_2$ can be split by the ancillary ligands (path b) leading to neutral (L^X = β-diketonates, picolinates, etc.) or charged bis-cyclometalated complexes (N^N = 2,2′-bipyridines, 1,10-phenanthrolines, etc.) with preferred trans-N,N configuration of the C^N ligands. Alternatively, the substitution of the chlorides in the dimer by a third cyclometalating ligand results in the tris-cyclometalated Ir^{III} complex $Ir(C\text{^}N)_3$ (path c). It should be noted that the tris-cyclometalated $Ir(C\text{^}N)_3$ have two isomers, the kinetically preferred meridional (*mer*) and the thermodynamically favored facial (*fac*) isomers. With cautious control of the reaction conditions, both isomers are possible to generate directly by path c. In has been proved that in solutions, applying thermal or photochemical energy, *mer*-isomers can be converted into the *fac*-form (path d) [25, 26]. The lower thermodynamic stability of the kinetically favored meridional formation is primarily due to the strongly trans-influencing aryl groups opposite to each other (in the *fac* isomer all three aryl groups are opposite to pyridyl or other neutral donor groups). The *fac*- and *mer*-isomers of $Ir(C\text{^}N)_3$ have different photophysical properties, the former of which are usually more efficient phosphorescent. Alternatively, the *fac*-$Ir(C\text{^}N)_3$ can also be directly prepared starting from the $Ir(acac)_3$ precursor (acac = acetoacetonate) or $IrCl_3 \cdot xH_2O$ (path e) [27].

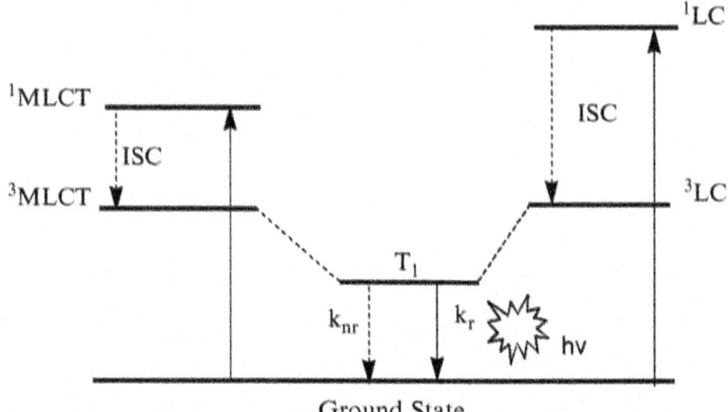

Fig. 9.11 Energetic closeness and overlap between ^{3}MLCT and ^{3}LC states result in the mixed lowest excited state (T_1) for transition-metal complexes, especially for Ir(III) complexes (Reproduced with permission from Ref. [28]. Copyright 2013 Elsevier)

9.4.2.2 Excited States and Color Tuning of Ir^{III} Phosphors

Excited states. In general, there are two basic types of electronic states in iridium (III) complexes that would compete for the LUMO, which is usually the only emitting state in a condensed matter. One of these is a ligand-centered^{3}LC($^3\pi$-π^*) triplet state, while the other is a triplet ^{3}MLCT state, as shown in Fig. 9.11 [28]. The d-orbital involvement in the bonding is expected to be higher for the MLCT state, which is reflected by the lower intensities of the corresponding absorption bands relative to those involving LC states. However, the energetic closeness and degree of overlap between ^{3}MLCT and ^{3}LC($^3\pi$-π^*) states result in the formation of a mixed lowest excited state (T_1). The excited molecule relaxes to the ground state through radiative (k_r) and non-radiative (k_{nr}) pathways. The accidental overlap between the two states mixes their photophysical properties. Depending on the chemical environment around the Ir(III) ion, i.e., the nature of cyclometalating ligands chelated to the metal, the origin of the emissive state is different. The energies of the lowest excited states play an important role as they respond to stimuli by adjusting the metal and ligand orbitals through substituent effects or tailoring the ligand structures, and this characteristic explains why Ir(III) complexes are the most outstanding emitters to date for OLED application [29]. Their metal-ligand-based luminescence provides the opportunity to change the emission energy and tune the color in the visible and near-infrared (NIR) range in the electrophosphorescent devices.

Color tuning strategies for iridium(III) complexes. Strong spin-orbit coupling leads to mixed singlet and triplet metal-to-ligand charge-transfer (MLCT) states as well as to mixed ligand-based emitting states. The metal-ligand-based emission enables an efficient tuning of the emission color by varying the ligands, and, thus, full-color applications based on phosphorescent iridium complexes can be realized. In general, the color tuning of iridium complexes can be achieved by adjusting the following aspects [19]: (1) cyclometalating ligands framework, (2) substituent effect in the ligands, (3) field strength of the ancillary ligand, and so on. This means that the cyclometalating ligand frameworks and the substituents on it and even the ancillary ligands combine to determine the optical and electronic parameters and the EL performance of the resultant iridium complexes. In other word, for

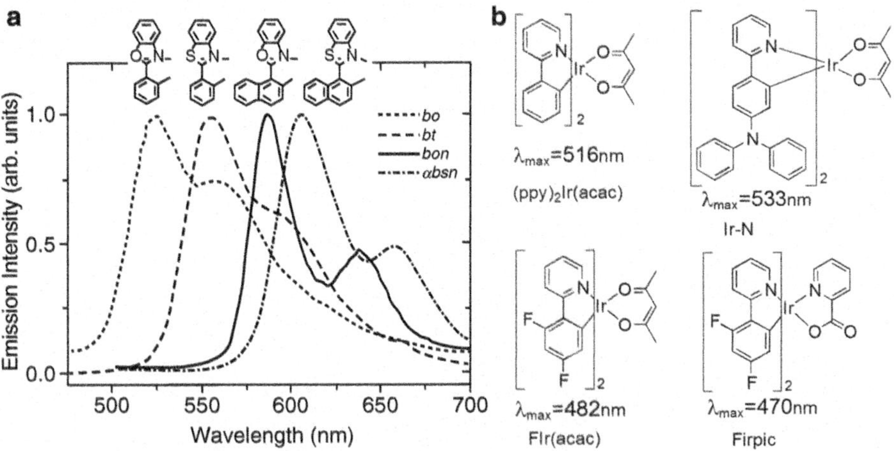

Fig. 9.12 (**a**) Solution photoluminescence spectra of *bo*$_2$Ir(acac), *bt*$_2$Ir(acac), *bon*$_2$Ir(acac), and α*bsn*$_2$Ir(acac). The structures of the individual C^N ligands are shown above the corresponding spectrum. (**b**) Influence of substituents and ancillary ligands on the emission wavelength of a group of blue- to green-emitting iridium complexes (Figure 9.12a is reproduced with permission from Ref. [23]. Copyright 2001 American Chemistry Society)

both homoleptic and heteroleptic iridium complexes, the chemical structure of the cyclometalating ligand is usually the fundamental aspect to determine the emission energy and quantum efficiency of the iridium complexes. In most cases the cyclometalating ligands are involved in the lowest-energy triplet excited state, either the ligand-centered state ^{3}LC or the ^{3}MLCT triplet state or their mixture, which is responsible for the phosphorescence. For example, the regular color tuning was achieved in a series of heteroleptic bis-cyclometalated iridium complexes (C^N)$_2$Ir(acac) (acac = acetylacetonate) with varying the C^N ligand structures [23], as shown in Fig. 9.12a. Substitution of S for O in a chromophore (*bo* → *bt*) leads to a 30-nm red shift, due to the higher polarizability and basicity of sulfur relative to oxygen, in this ligand-based excited state. Increasing the size of the ligand π system is expected to bathochromically shift electronic transitions, as is observed in converting a phenyl group to a naphthyl group (*bo* → *bon*), which leads to a 60-nm red shift. The effects of the naphthyl and sulfur substitutions are nearly additive, leading to an 80-nm red shift when comparing *bo* to α*bsn* complexes.

At the same time, the structure and even the position of the substituents on the ligand frameworks play an important role as well in tuning these physical parameters and EL behavior. As shown in Fig. 9.12b, (ppy)$_2$Ir(acac) (ppy = 2-phenylpyridine) is one of the prototype green-emitting iridium phosphor, which is known to exhibit room-temperature phosphorescence similar to its tris-cyclometalated complex Ir(ppy)$_3$, leading to the conclusion that the luminescence is mainly controlled by the common [Ir(C^N)2] fragment, while the emission wavelength is strongly dependent on the choice of C^N cyclometalating ligands. Incorporation of electron-donating or electron-withdrawing substituents on either pyridyl or phenyl moiety of the ppy ligand will lead to obvious change in emission wavelength. For example, introduction of two electron-withdrawing F (FIr(acac)) [30, 31] or electron-donating diphenylamine (Ir-N) [32] in 4 site of phenyl ring leads to blue and red shift of the phosphorescence of the corresponding iridium complexes, respectively. This is because these substituents have changed the frontier molecular orbitals levels and thus the energy band gaps of the iridium complexes. For most of iridium complexes, the HOMO consists of arene π and Ir d orbitals, while the LUMO is located largely on the N-related heterocycle that is usually electron-deficient. In general, the phosphorescence of the iridium complex red shifts when the electron-donating substituent is introduced into the HOMO-lying arene part and/or the electron-deficient group is incorporated in the LUMO-lying heterocycle moiety of C^N ligands. Hypsochromic shift will occur when the electron-deficient substituent is introduced into the HOMO-lying arene and/or the electron-rich group is incorporated into the LUMO-lying heterocycle [19, 32].

Similar to that of the C^N cyclometalates, the ancillary L^X chelates are also capable to modulate the photophysical properties, especially the emitting color, of heteroleptic complexes as a consequence of their intrinsic ligand field strength and electronic influences toward the central metal atoms. This can be illustrated by comparing the emission properties of FIr(acac) and FIrpic [30] (Fig. 9.12b). With changing the ancillary ligand from acac in FIr(acac) to pic in FIrpic, a hypsochromic shift by 12 nm is observed in their phosphorescence spectra. This is exclusively caused by the ancillary ligand picolinate possessing higher field strength, i.e., the stronger field strength of picolinate ligand has weaker donor strength, leading to the blue shift of iridium complex emission. By varying the ancillary ligands, a lot of deep-blue-emitting iridium complexes have been successfully invented for OLEDs application.

9.4.2.3 Ir^{III} Phosphors for OLEDs

Up to date, a large amount of Ir(III) complexes have been developed for application as phosphorescent emitters in OLEDs, with bright luminescence ranging from blue to red. It would be safe to say almost all the best performance OLEDs of each primary color reported so far are made using corresponding Ir(III) complexes as emitters. Based on the excellent performance in monochromatic OLEDs, these iridium phosphors are also used to fabricate white OLEDs, which exhibit impressive efficiency, indicating promising future of iridium phosphors in practical applications of OLEDs lighting and displays. In the following section, some typical examples of iridium complexes with various emitting colors are listed. It should be kept in mind that the iridium phosphors reported so far for OLEDs applications are not limited to only these examples.

Green-Emitting Ir^{III} Phosphors The earliest and best known example of Ir^{III} phosphors is the green-emitting, tris-cyclometalated complex *fac*-[Ir(ppy)$_3$] (structure shown in Fig. 9.13), which was originally synthesized and reported by Watts

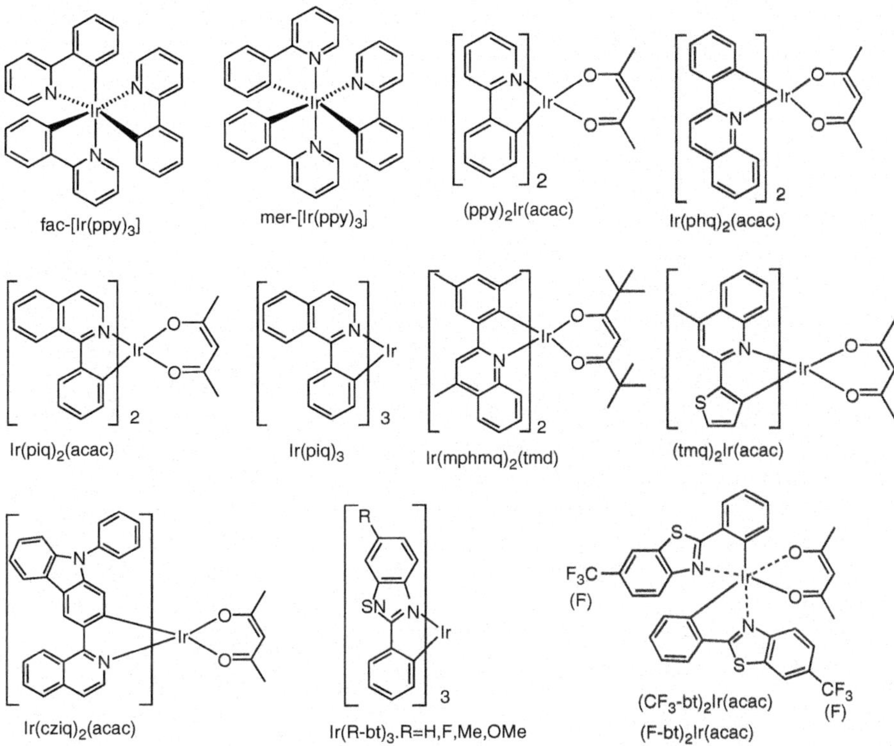

Fig. 9.13 Chemical structures of some typical red-, green-, yellow-, and orange-emitting iridium complexes

and coworkers in 1991 [27]. Many following cyclometalated iridium complexes of various emitting colors were developed from this prototype molecule by modifying the conjugation of cyclometalating ligands, introducing substituents, altering ancillary ligands, etc. *fac*-[Ir(ppy)$_3$] is strong green phosphorescent at room temperature with emission peak at 494 nm and a shoulder at longer wavelength, a high phosphorescence quantum yield Φ_{Ph} of 0.4 in solution, and a short triplet lifetime of 1.90 μs [27]. Quite a number of perfect green and white PhOLEDs have been made using *fac*-[Ir(ppy)$_3$] as doped emitter. In 2007, Kido and coworkers [33] developed a highly efficient green PhOLEDs with double emitting layer structure. In this device, wide-energy-gap CBP and TCTA were selected as hosts for *fac*-[Ir (ppy)$_3$] to form the two emitting layers. Hole injection and electron injection from the electrodes were balanced by placing chemically doped layers at the interface between the electrodes and the organic layers. In addition, a highly reflective Ag cathode was employed as an anode, instead of a conventional Al cathode to enhance the reflectivity of the cathode metal. In this way, an optimized device exhibited an external quantum efficiency (EQE) of 27 % (95 cd A^{-1}) and a high power efficiency of 97 lm W^{-1} at the brightness of 100 cd m^{-2}. Shortly afterwards, the same authors [34] further improved the efficiency of *fac*-[Ir(ppy)$_3$] based green PhOLEDs to

29 % at 100 cd m^{-2} and 26 % at 1,000 cd m^{-2} (corresponding to the ultra high power efficiencies of 133 lm W^{-1} at 100 cd m^{-2} and 107 lm W^{-1} at 1,000 cd m^{-2}), by using a newly synthesized electron-transporting material B3PYMPM.

The tris-cyclometalated iridium complex *fac*-[$Ir(ppy)_3$] has a configurational isomer, *mer*-[$Ir(ppy)_3$], as shown in Fig. 9.13. The *mer*-isomers (such as *mer*-[Ir $(ppy)_3$]) can be obtained as kinetic products at lower temperature and can be subsequently converted, either thermally or photochemically, into the more stable *fac* isomer [25, 26]. The two configurational isomers of tris-cyclometalated iridium complexes notably show differences in their photophysical properties, i.e., the *fac* isomers feature over an order of magnitude longer lifetimes and higher quantum efficiencies than their meridional counterparts [25]. As mentioned before, the bis-cyclometalated counterpart of the tris-cyclometalated $Ir(ppy)_3$ containing acac ancillary ligand, i.e., $(ppy)_2Ir(acac)$, is also a widely used green phosphor with good performance in its green PhOLEDs [35]. Furthermore, large amount of green-emitting iridium complexes based on ppy ligand framework but with various substituents to tune the electronic and optical parameters have been developed [32]. It should be mentioned that among all the iridium complexes of the three primary emitting colors, the overall performance of the green-emitting iridium complexes and their PhOLEDs are the best, and some of the green iridium complexes have been utilized to fabricate large-scale OLEDs products in industry.

Red-Emitting Ir^{III} Phosphors As shown in Fig. 9.13, the tris- and bis-cyclometalated iridium complexes based on the 2-phenylquinoline and 2-phenylisoquinoline cyclometalating ligands, e.g., $(Phq)_2Ir(acac)$ [36], $Ir(piq)_3$ and $(Piq)_2Ir(acac)$ [37], are among the most famous red-emitting iridium phosphors. In comparison with the 2-phenylpyridine-based green iridium complexes, the 2-phenylquinoline and 2-phenylisoquinoline based iridium complexes exhibit red-shifted phosphorescence by enlarging the π-conjugation in the major cyclometalating ligands. It is observed that the iridium complexes based on 2-phenylisoquinoline ligands usually show a red shift in phosphorescence relative to their isomers containing 2-phenylquinoline ligands. Both $(Phq)_2Ir(acac)$ and $(Piq)_2Ir(acac)$ and their tris-cyclometalated counterparts are widely used to fabricate efficient red PhOLEDs. In addition, many structure modifications on these red iridium complexes were performed, either by introducing substituents on the phenylquinoline and phenylisoquinoline ligands or replacing the phenyl ring with other aryl groups. For example, Kwon and coworkers [36] incorporated sterically crowded alkyl moieties into both main ligands as well as the ancillary ligand to form $(mphmq)_2Ir(tmd)$. Both the phenyl and the quinoline rings are fully methylated, and the diketone ancillary ligand is also decorated by bulky *tert*-butyl groups, with the purpose to get rid of the strong interactions between triplet excitons in OLEDs. The vacuum-deposited PhOLEDs with $(mphmq)_2Ir(tmd)$ as doped emitter exhibited saturated red emission with high efficiencies of 30.1 cd A^{-1} and 32.0 lm W^{-1}, which are dramatically improved in comparison with those data obtained from the control device with parent compound $(Phq)_2Ir(acac)$ as emitter. On the other hand, by replacing the phenyl ring with thiophene group, Cheng et al. [38] developed a novel cyclometalating ligand framework and corresponding iridium complex $(tmq)_2Ir(acac)$ (Fig. 9.13). By combination with a newly synthesized electron-transporting host material BIQS, $(tmq)_2Ir(acac)$ exhibited saturated red electroluminescence with excellent performance including an ideal CIE coordinates of (0.67, 0.33), a maximum brightness of 58,688 cd m^{-2} and the maximum external quantum efficiency, current efficiency, and power efficiency of 25.9 %, 37.3 cd A^{-1}, and 32.9 lm W^{-1}, respectively, which seems to be the highest values for deep-red PhOLEDs reported to date. In a similar way, Wong and coworkers [39] developed red-emitting iridium complex $Ir(Cziq)_2(acac)$ (i.e., compounds 3 and 4 in original literature) by using hole-transporting carbazole in combination with isoquinoline to design and prepare the cyclometalating ligand and corresponding cyclometalated iridium complexes, which also delivered good performance in its red PhOLEDs.

Yellow- and Orange-Emitting Ir^{III} Phosphors In addition to the three primary colors (RGB), yellow- and orange-emitting iridium complexes are also very important phosphorescent materials utilized in OLEDs field nowadays. The yellow- and orange-emitting materials are especially significant when they are used to make white OLEDs in combination with blue emitters. In comparison with the RGB color combination method, this kind of two-complementary-color strategy to generate white light is advantageous in terms of easy device fabrication and even high efficiency. In principle, the yellow- and orange-emitting color can be achieved by introducing certain substituents in the green and red-emitting iridium molecules. Besides, there are also some C^N cyclometalating ligand frameworks which are predominantly used to design yellow and orange iridium phosphors. 2-phenylbenzothiazole is just such an example, and a large group of yellow- and orange-emitting iridium complexes have been developed based on this ligand framework [40–43]. Li et al. [42] reported the synthesis and OLED applications of a group of tris-cyclometalated iridium complexes $Ir(r\text{-}bt)_3$ (i.e., compounds 1–4 in original paper, Fig. 9.13) containing pristine and substituted 2-phenylbenzothiazole ligands. By using the important catalyst of silver trifluoromethanesulfonate, this series of tris-cyclometalated iridium complexes based on this ligand framework was successfully prepared for the first time in 2011. Orange-emitting PhOLEDs were fabricated using these iridium complexes as doped emitters. An extremely high brightness of 95,800 cd m^{-2} and a maximum luminance efficiency of 87.9 cd A^{-1} (46.0 lm W^{-1}) were achieved for the pristine 2-phenylbenzothiazole ligand-based complex $Ir(bt)_3$. These performances represented a significant improvement for vacuum-deposited orange OLEDs and the new record of the efficiencies for orange OLEDs reported to that time. As mentioned above [19], the CF_3 and F decorated 2-phenylbenzothiazole were used as cyclometalating ligands to prepare heteroleptic bis-cyclometalated orange-emitting iridium complexes $(CF_3\text{-}bt)_2Ir(acac)$ and $(F\text{-}bt)_2Ir(acac)$, which then led to efficient white OLEDs with high efficiencies of 68.6 cd A^{-1} and 26.2 % in combination with the sky-blue FIrpic.

Sky-Blue and Deep-Blue-Emitting Ir^{III} Phosphors Blue-emitting iridium complexes were originally obtained by adjusting the ancillary ligands of those iridium complexes containing phenylpyridine derivatives as the cyclometalating ligands. The first blue phosphorescent dopant was FIrpic (Fig. 9.12b) with two fluorines in the phenyl unit to shift the HOMO level downward for high-triplet energy [30, 44]. The triplet energy was increased by the electron-withdrawing fluorine and a bulky picolinic acid ancillary ligand. FIrpic is currently the most widely used sky-blue phosphor for fabrication of both blue and white OLEDs, and excellent EL efficiencies were obtained for it. For example, Wong et al. [45] demonstrated a highly efficient blue PhOLED with FIrpic as doped emitter in a CN-containing host material mCPCN that exhibited a peak current efficiency of 58.7 cd A^{-1}. As far as we know, that was the highest value ever reported for FIrpic-based blue PhOLEDs so far. Li and coworkers [46] also reported a series of efficient sky-blue PhOLEDs by doping FIrpic into a group of novel CN-substituted bipolar host materials, which are characterized by high efficiencies up to 46 cd A^{-1} and slow efficiency roll-off. As stated in previous Sect. 9.3.2.2, many high-efficiency white OLEDs reported so far [3, 16–20] were fabricated with FIrpic as the blue-emitting component. Although the emission spectra of FIrpic were blueshifted, it exhibited only sky blue color with CIE coordinates of (0.17, 0.34), which is still far from the National Television Standards Committee (NTSC) standard blue values of (0.14, 0.08). The color coordinate was further blueshifted by changing the ancillary ligand from pic to bulky tetrakis(1-pyrazolyl)borate ligand (FIr6) [47]. The structures of some deep-blue iridium complexes are shown in Fig. 9.14. FIr6 has bulky tetrakis (1-pyrazolyl)borate ligand to reduce the conjugation of the main ligand through steric hindrance. The peak wavelength of FIr6 was 458 nm with CIE color coordinates of (0.16, 0.26), still being sky blue. Similarly, triazole (FIrtaz) and tetrazole (FIrN4) derivatives were adopted as the ancillary ligands to realize blue-emitting iridium complexes. The emission wavelengths of FIrtaz and FIrN4 were 459 and 460 nm, respectively [48, 49]. Apparently it is difficult to realize real blue color

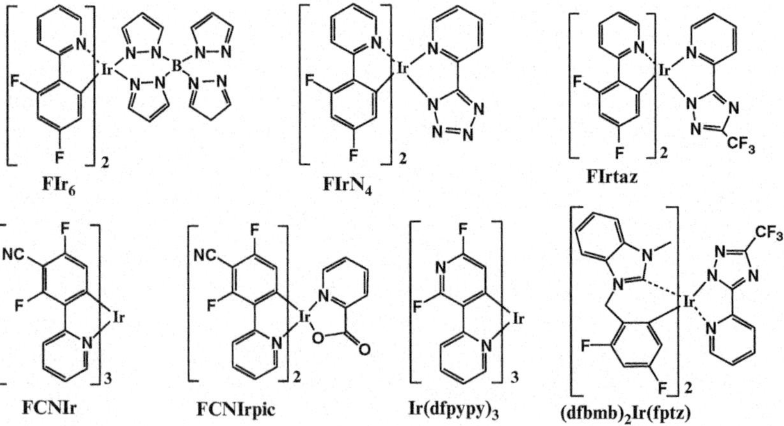

Fig. 9.14 Chemical structures of typical deep-blue iridium complexes

only by modifying the ancillary ligands, although the modification of the ancillary ligand indeed blueshifted the emission of these iridium phosphors.

Preparation of true blue, highly efficient phosphorescent iridium complexes has been long considered as a formidable challenge [50]. This task is far more difficult than those for preparing longer wavelength emission deriving from green, orange, and red phosphors. The key lies in the selection of suitable chromophores that possess both large intra-ligand $^3\pi$-π^* and ^{3}MLCT transition energies. In addition, all ancillary ligands must also have strong metal-ligand bonding interaction so that the d-d excited states or other unspecified quenching states are strongly destabilized to prevent thermal population to these higher-lying deactivating states [51, 52].

Incorporation of multiple strong electron-withdrawing groups is an effective strategy to enlarge the band gap of iridium complexes by reducing the HOMO levels. Phenylpyridine derivatives with cyano group (FCNIr or FCNIrpic) [53, 54] along with two fluorine units further shifted the color coordinate of blue PHOLEDs in this way. The best efficiency value of FCNIr-based blue PHOLED was 19.2 % with CIE coordinates of (0.15, 0.16) when it was doped in the diphenylphosphoryl-containing host materials [53]. A heteroleptic FCNIrpic with picolinic acid ancillary ligand was also effective as a deep-blue dopant, and the color coordinates of the blue PHOLED were (0.14, 0.17) [54]. The quantum efficiency value of the FCNIrpic-doped PHOLED was as high as 25.1 % and the quantum efficiency at 1,000 cd m^{-2} was 22.3 %. This is the best efficiency value reported in the deep-blue PHOLED up to now [54]. In a similar way, fluorinated dipyridine-type ligand was also used as the main ligand to construct deep-blue-emitting phosphorescent dopant $Ir(dfpypy)_3$ [55]. The use of difluoropyridine unit instead of common difluorophenyl further shifted the emission wavelength to blue region because of electron deficiency of pyridine ring. The emission wavelength was 438 nm with a shoulder at 463 nm and the color coordinates were (0.14, 0.12).

Strong σ-donating ligands such as carbenes were also used to realize true blue and even near-UV emitting iridium complexes [51]. Wu et al. [56] developed a weak conjugated benzyl carbene ligand to synthesize a complex of $(dfbmb)_2Ir(fptz)$ (dfbmb = 1-(2,4-difluorobenzyl)-3-methyl-benzimidazolium) with a high quantum yield ($\Phi = 0.73$). The $(dfbmb)_2Ir(fptz)$-based PhOLED with structure of ITO/NPB/TCTA/CzSi/CzSi:$(dfbmb)_2Ir(fptz)$/UGH2:$(dfbmb)_2Ir(fptz)$/UGH2/BCP/Cs_2CO_3/Al, exhibited a true blue light emission with CIE coordinates of (0.16, 0.13) with peak efficiencies of 6.0 %, 6.3 cd A^{-1}, and 4.0 lm W^{-1}.

9.4.3 Platinum(II) Phosphors

9.4.3.1 Structures and Excited States

The first phosphorescent material used in PhOLEDs to increase the IQE of OLEDs was a deep-red Pt(II) octaethylporphine (PtOEP) (structure shown in Fig. 9.15) [2]. In 1998, Forrest and coworkers demonstrated the first electrophosphorescent

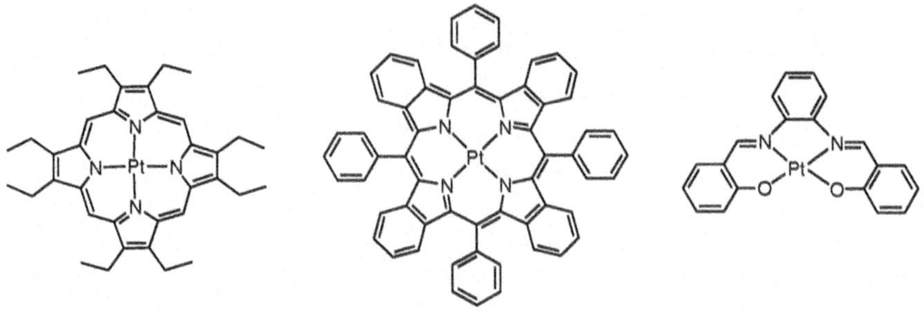

PtOEP Pt-TPTBP PtSB-2

Fig. 9.15 Chemical structures of some Pt-porphyrins and Pt-Schiff base phosphors

device by using PtOEP as doped emitter in a hole-transporting host material. The significance of this milestone work is to reveal the possibility to increase the internal quantum efficiency of OLEDs by harvesting both singlet and triplet excitons using phosphorescent emitters as opposite to the conventional fluorescent devices. From then on, large amount of transition-metal complexes including those of Pt^{II} have been developed for OLEDs application.

Pt^{II} ions have $5d^8$ electron configuration, and the 4-coordinate Pt^{II} complexes have square-planar geometries. It is notable that the square-planar geometry of Pt^{II} complexes allows formation of dimer and excimer and even aggregates through axial coordination, resulting in distinctively different photophysical properties when compared to the d^6 octahedral metal complexes [57]. Thus, in addition to the typical MLCT and ligand-centered π-π^* transition that occurred in the d^6 complexes, a new type of electronic transition, denoted as metal-metal-to-ligand charge transfer (MMLCT), became possible. This type of transition involves a charge transfer between the filled Pt-Pt bonding orbital and a vacant, ligand-based π^* molecular orbital [58].

9.4.3.2 Pt^{II} Phosphors for OLEDs

Similar to Ir^{III} complexes, Pt^{II} complexes are generally characterized by relatively short triplet lifetimes, high phosphorescent quantum yields, and tunable emission color over the entire visible light region, all of which are definitely favorable for ideal device performance when they are used as emitters in OLEDs. In addition, most of the Pt^{II} complexes have both good thermal stability and good solubilities in common organic solvents as well, making them capable of both vacuum sublimating and solution processing [59]. Based on their structural features, these Pt^{II} phosphors can be classified into three main categories for discussion according to the dentition number of the main ligands chelated to the Pt^{II} centers: (a) tetradentate ligands, including porphyrin and Schiff base; (b) tridentate ligands, including N^N^C-, N^C^N-, C^N^C-, and N^N^N-coordinating ligands; and (c) bidentate ligands, including N^C-, N^N-, and C^C-coordinating ligands. Obviously, the

N^N^C-, N^C^N-, C^N^C-type tridentate ligands and N^C- and C^C-type bidentate ligands are cyclometalating ligands since C–Pt bonds are formed in their complexes.

Pt^{II} Phosphors with Tetradentate Ligands This type of triplet emitters mainly includes Pt(II) porphyrin and Schiff base complexes. The pioneering phosphor of this type was PtOEP (Fig. 9.15) that originally started the electrophosphorescence research [2]. Pure red electroluminescence at 650 nm was achieved for PtOEP device with an IQE of 23 % and an EQE of 4 %, which shows great improvement in comparison with those with fluorescent molecules and typical rare earth metal complexes as emitters. However, due to the intrinsically long triplet lifetime of Pt phosphors with tetradentate ligands (e.g., phosphorescence lifetime of PtOEP is 91 μs), bulky substituents could be introduced to prevent intermolecular interactions between the triplet excitons in OLEDs. For example, Pt tetraphenyltetrabenzoporphyrin (Pt-TPTBP, in Fig. 9.15) [60] was designed in this way and exhibited an improved EQE of 8.0 % due to reduced triplet-triplet (T-T) annihilation especially at higher current densities. Che and coworkers have prepared many tetradentate Pt(II) Schiff base phosphors and tested their EL potentials in OLEDs as well as investigating their structure-property relationships [61, 62]. For example, OLEDs based on PtSB-2 (Fig. 9.15) showed a maximum EQE of 9.4 % and a lifetime of more than 20,000 h at a brightness of 100 cd m^{-2} [62].

Pt^{II} Phosphors with Tridentate Ligands There are four types of tridentate ligands to construct Pt(II) phosphors, including N^N^C-, N^C^N-, C^N^C-, and N^N^N-coordinating ligands. Among them, the former three are cyclometalating ligands since they contain C atoms and can form the C–Pt bond in the complexes. These ligands are thought to improve the rigidity of the corresponding Pt (II) complexes by suppressing the D_{2d} distortion to induce higher phosphorescent quantum yields, hence promoting the performance of electrophosphorescent devices.

Figure 9.16 illustrates some typical Pt(II) phosphors with tridentate N^N^C-, N^C^N-, C^N^C-, and N^N^N-coordinating ligands. Much work on Pt (II) complexes based on N^N^C-coordinating ligands and their emitting characteristics have been reported by Che et al. [63]. In solution, tridentate Pt(II) complexes bearing σ-alkynyl auxiliary ligands (PtNNC in Fig. 9.16) can emit a variety of colors from 550 to 630 nm, induced by ligands with different steric and electronic properties. However, most of these Pt complexes did not show acceptable performance when they are used as emitters in OLEDs. Fortunately, the luminescent performance of the tridentate Pt complexes could be improved by replacing the N^N^C-ligands with N^C^N ligands. It was proved that higher rigidity of the molecular skeleton generally leads to higher Φ_P. Williams et al. reported that the Pt–C bond lengths in Pt(N^C^N) complexes are around 1.90 Å, about 0.14 Å shorter than those in typical Pt(N^N^C) complexes [64]. Furthermore, the shorter Pt–C bond length is expected to deactivate the metal-centered d-d states by raising their energy, and hence lead to superior performance. This can be seen in a new

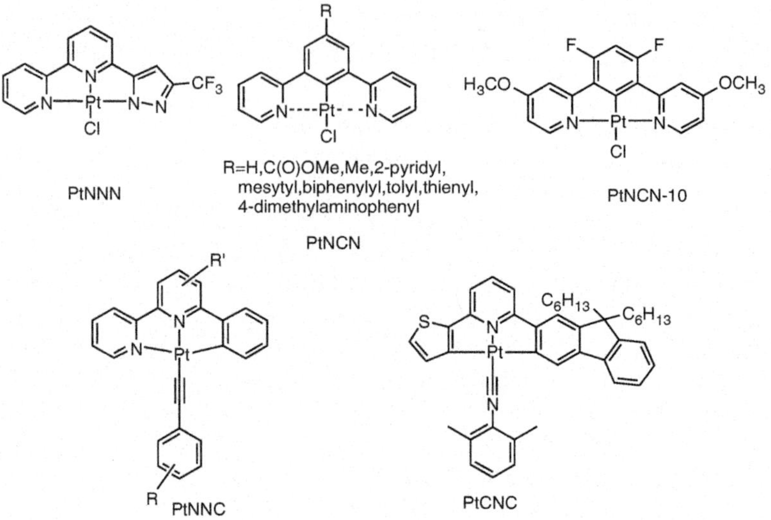

PtNNN

R=H,C(O)OMe,Me,2-pyridyl,
mesytyl,biphenylyl,tolyl,thienyl,
4-dimethylaminophenyl

PtNCN

PtNCN-10

PtNNC

PtCNC

Fig. 9.16 Chemical structures of some Pt(II) complexes based on tridentate ligands

series of Pt(N^C^N) complexes bearing various aryl substituents (PtNCN in Fig. 9.16) [65]. All of the new phosphors display Φ_P values from 0.46 to 0.65 with emission maxima from 481 to 588 nm. OLEDs employing this group of complexes PtNCN as emitters [66] can achieve maximum EQE values from 4 to 16 % and current efficiency values from 15 to 40 cd A^{-1}. It is worth mentioning that the electron-donating groups attached to the lateral pyridyl rings of the Pt complexes resulted in a blueshift in the monomer emission as well as excimer emission resulting from the interactions between emitter molecules. Accordingly, PtNCN-10 [67] showed a blueshift emission compared with the similar molecule containing methyl groups due to the stronger electron-donating ability of the methoxy group ($-OCH_3$). By tuning the doping concentration from 5 to 35 wt%, OLEDs based on PtNCN-10 could emit nearly any color from blue to yellowish red. The dependence of emission color on doping concentration offers a very simple and practical way to regulate OLED emission color by adjusting the contributions of monomer and excimer emission. The Pt(II) complexes containing the tridentate N^N^N- and C^N^C-coordinating ligands, typically possessing the molecular skeletons like PtNNN [68] and PtCNC [69] in Fig. 9.16, are also a group of green- to red-emitting phosphors. However, they are not as popular as the former two types of tridentate ligands-based Pt(II) due to poor emission.

Pt^{II} Phosphors with Bidentate Ligands Three types of bidentate ligands have been used to prepare Pt(II) complexes, i.e., N^C-, N^N-, and C^C-coordinating ligands. Complexes with one N^C-ligand (2-phenylpyridine-type or ppy-type) and one ancillary β-diketonato ligand (acetyl acetone or its derivatives) (Ptppy-1 in Fig. 9.17) represent the first and most developed type of EL phosphors, with good emission and tunable color. Theoretical calculations indicate that the HOMOs of

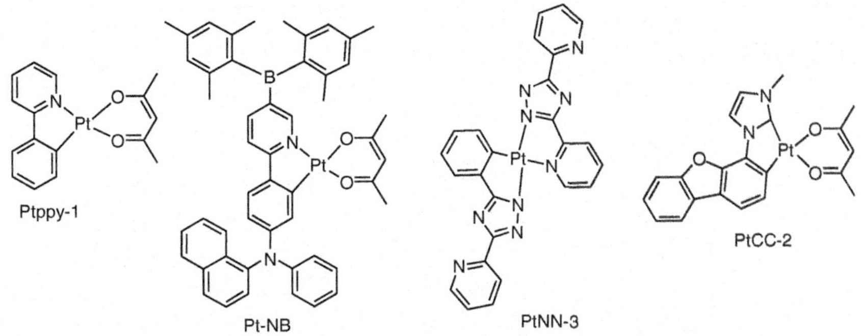

Fig. 9.17 Chemical structures of Pt(II) complexes containing bidentate ligands

this type of Pt(II) complexes are contributed mainly by the π orbitals of the phenyl ring and the dπ orbital of the Pt center, and LUMOs are located mainly on the orbitals of the pyridyl ring in the ligands. Therefore, it is possible to tune the emission color in a wide spectral range by incorporating electron-withdrawing or electron-donating substituents into both phenyl and pyridine rings of the ligands. In particular, some functional groups, such as the hole-transporting arylamine and the electron-transporting $Be(Mes)_2$, could be incorporated into the ppy ligand of Pt-NB (Fig. 9.17), making it multifunctional when used as emitter in OLEDs. Pt-NB exhibited outstanding EL performance with an EQE of 20.9 %, a CE of 64.8 cd A^{-1}, and a PE of 79.3 lm W^{-1} [70], which are among the best efficiencies achieved for Pt(II) phosphors so far and are quite close to those obtained from Ir(III) phosphors.

The N^N- and C^C-coordinating ligands-based Pt(II) complexes generally exhibited blueshifted phosphorescence compared to N^C-ligands-based ones. N^N ligands typically include N-heterocycle substituted pyrazole or triazole, like pyridyltriazole ligand for complex PtNN-3 in Fig. 9.17. This complex [71] features intense Pt-Pt intermolecular interactions due to the strong polarity induced by pyridyltriazolate as well as its square-planar geometry. Generally, interactions among the phosphor molecules will decrease EL efficiency. However, EL efficiencies could be enhanced by increasing the doping level of PtNN-3 and a maximum EQE of 19.7 %, a PE of 44.7 lm W^{-1}, and a CE of 62.5 cd A^{-1} were achieved at a doping level of 65 %. This interesting result may be attributed to energy-level matching among the functional layers within the devices to balance the hole/electron ratio and confine the recombination zone in the EML. Furthermore, the short lifetime of PtNN-3 in the solid film may play a critical role in reducing triplet-triplet annihilation and thus enhancing the device efficiencies. Carbene-based structures have been employed as C^C-coordinating ligands to prepare novel Pt phosphors. However, many C^C-ligands-based Pt(II) complexes are luminescent in ultraviolet (UV) light region, which makes them not suitable for fabricating OLEDs. Then modified carbenes, like the main ligand of complex PtCC-2 in Fig. 9.17, were developed, and the corresponding complex PtCC-2 exhibited moderate EL performance with peak EQE of 6.2 % in its OLEDs.

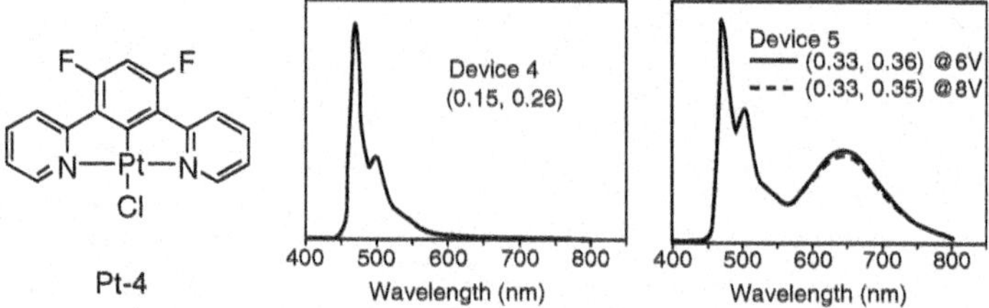

Fig. 9.18 Chemical structure of blue-emitting Pt complex Pt-4, and the EL spectra of its blue (*middle*) and white (*right*) OLEDs. Blue device (device 4): ITO/PEDOT:PSS/55 nm PVK/25 nm 26mCPy:OXD-7(49 %):Pt-4 (2 %)/40 nm BCP/CsF/Al; White device (device 5): ITO/PEDOT: PSS/30 nm TCTA/25 nm 26mCPy:Pt-4 (8 %)/40 nm BCP/CsF/Al (*Middle* and *right* figures reproduced with permission from Ref. [72]. Copyright 2008 Wiley-VCH Verlag GmbH & Co. KGaA)

Excimer of Pt^{II} Complexes and White OLEDs Pt(II) complexes are very different from those of 6-coordinate d^6 complexes that present an effectively spherical profile to their environment. Square-planar platinum(II) complexes with sterically undemanding ligands are essentially flat, and this allows close and intimate interactions, either with other identical molecules (e.g., intermolecular stacking or dimerization in the ground state or excimer formation in the excited state) or with other molecules (e.g., exciplex formation with Lewis bases). Therefore, the flat molecular conformation and thus excimer formation of these Pt(II) complexes could be used in fabrication of white OLEDs. For example, the tridentate ligand-based Pt(II) complex Pt-4 (Fig. 9.18) is intrinsically a highly efficient blue phosphor [72] with a maximum Φ_P of 0.8 in degassed dichloromethane, which was ascribed to both its rigid triplet state configuration and very high ligand field strength giving a strong destabilization to the metal-centered d-d excited states. A deep-blue device with Pt-4 doped in a cohost gave a maximum EQE of 16 % and a power efficiency of 20 lm W^{-1}. A WOLED containing 8 wt% of Pt-4 in 1,3-bis(N-carbazolyl) benzene (mCP) host exhibited a maximum EQE of 9.3 %, a PE of 8.2 lm W^{-1}, and CIE coordinates of (0.33, 0.35) at 1,300 cd m^{-2}, very close to the pure white point at (0.33, 0.33). Evidently, combination of the intrinsic blue phosphorescence of the monomer Pt(II) complex and the yellow emission from its excimer provides a practically facile way to fabricate white OLEDs.

9.4.4 Copper(I) Phosphors

As stated before, most organic electrophosphorescent materials, including the Ir^{III}, Pt^{II}, Ru^{II}, and Os^{II} complexes, are organometallic phosphors based on noble metals. Although excellent performances have been achieved for these noble metal complexes in PhOLEDs, the noble metal nature severely restricts their large-scale and long-term applications. Therefore, it is strongly desired to develop cheap phosphorescent materials for commercialization of PhOLEDs. Copper is such a type of transition metal characterized by high relative abundance, low cost, and nontoxic properties. Accordingly, the copper(I) complexes have been regarded as the most ideal phosphorescent materials for OLEDs application and have drawn more and more research attentions nowadays.

All the concurrent Cu(I) complexes can be classified into mononuclear and multinuclear complexes. Different from Ir^{III} and Pt^{II} complexes, almost all the Cu (I) complexes only contain Cu–N, Cu–P, and Cu–X(halogen) bonds between the ligands and metal center, some of which are chelate bonds and some others are σ bonds. There are few examples of Cu(I) complex containing Cu–C bond, except several Cu(I) complexes containing carbene ligands [73]. It seems that the ionic complexes of Cu(I) are more frequently observed as luminescent materials than other transition metals.

9.4.4.1 Distortion in Excited States

The MLCT excited state properties of Cu(I) complexes had been found for a long time [74, 75]. The application of Cu(I) complex as phosphorescent emitters in OLEDs was first reported in 1999 by Ma and coworkers [7]. However they did not become popular and were almost ignored for use in PhOLEDs for a long time due to a low photoluminescence (PL) and EL efficiency resulting from their typically excited-state distortion [76].

Cu(I) ions have d^{10} electron configuration and the four-coordinated Cu (I) complexes have tetrahedral geometry. The complete filling of d orbitals prevents MC d-d electronic transitions in Cu(I) complexes. Such transitions are available to d^9 Cu(II) complexes, however, and they are deactivated via ultrafast non-radiative processes. Accordingly, the most luminescent Cu complexes are not Cu(II) complexes but Cu(I) complexes [77]. The coordination behavior of Cu(I) complexes is strictly related to its electronic configuration. The complete filling of d orbitals (d^{10} configuration) leads to a symmetric localization of the electronic charge, which favors a tetrahedral disposition of the ligands around the metal center in order to locate the coordinative sites far from one another and minimize electrostatic repulsions, as shown in Fig. 9.19. However, these four-coordinated Cu(I) complexes readily suffer a Jahn-Teller based distortion (flattening) in the excited state and consequent formation of a five-coordinated exciplex, which promotes non-radiative decay and leads to weak emission efficiency (Fig. 9.19) [76, 78].

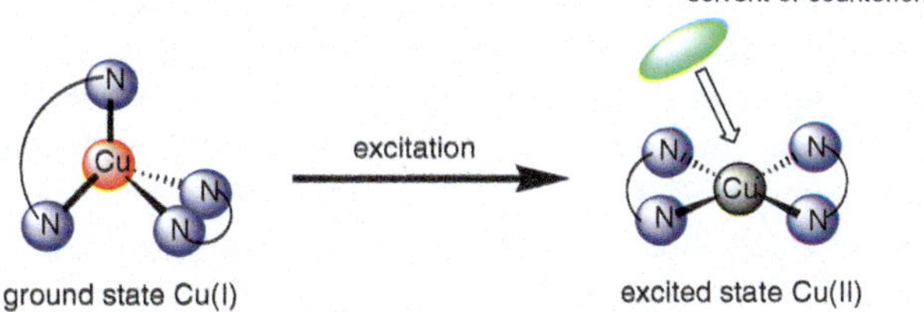

Fig. 9.19 Flattening distortion and subsequent nucleophilic attack by solvent, counterion, or other molecules following light excitation in Cu (I) complexes

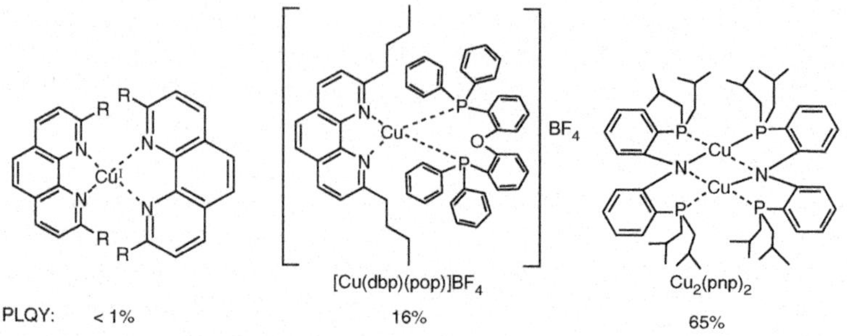

Fig. 9.20 Influence of sterically bulky substituents or PP ligands on PLQY of Cu(I) complexes

In order to improve the PL quantum yields (PLQY) of the Cu(I) complexes, the flattening distortion in excited states should be suppressed as much as possible. Incorporation of sterically bulky substituents on the ligands was proved to avoid this distortion. Introduction of bulky groups R instead of H atoms in phenanthroline ligands of complex $[Cu(NN)_2]^+$ (Fig. 9.20) was found to be favorable for high quantum yields. However, the PLQY of these modified $[Cu(NN)_2]^+$ complexes are still lower than 1 % [79]. On the other hand, the $[Cu(NN)(PP)]^+$ type complexes containing PP ligands have much higher PLQYs than the general $[Cu(NN)_2]^+$ complexes. For example, the PLQY of complexes $[Cu(dbp)(pop)]BF_4$ and $Cu_2(PNP)_2$ (Fig. 9.20) are increased to 16 % and 65 %, respectively [80]. The appearance of these high-PLQY $[Cu(NN)(PP)]^+$ complexes stimulated great research interests. Only from then on, the research on Cu(I) complexes as potential phosphorescent materials for OLED application was really started.

9.4.4.2 Cu^I Phosphors for PhOLEDs

Green Cu^I Phosphors In 2007, Tsuboyama and coworkers [81] prepared a series of mononuclear and binuclear Cu(I) complexes, among which the binuclear $Cu_2I_2(dppb)_2$ (Fig. 9.21) has the highest PLQY of 80 %. They fabricated a yellow PhOLED using $Cu_2I_2(dppb)_2$ as doped emitter and a current efficiency of 10.4 cd A^{-1} was obtained with emission peak at 562 nm. Recently, Adachi et al. [82] used this material as emitter to fabricate PhOLEDs. But the device structure was carefully controlled by selecting a high-triplet-energy host and an appropriate electron-transporting material. The EL emission was shifted to green region with peak at 517 nm. At the same time, device performance was greatly improved with a current efficiency of 30.6 cd A^{-1} and an EQE of 9.0 %. Osawa et al. [83] reported the properties of a three-coordinate Cu(I) complex (dtpb)CuBr (Fig. 9.21). The green-emitting PhOLED based on this complex exhibited an extremely high current efficiency of 65.3 cd A^{-1} and an EQE of 21.3 %. It is clear that the efficiencies of these green OLEDs based on Cu(I) complexes are already close to the values obtained from famous Ir^{III} complexes.

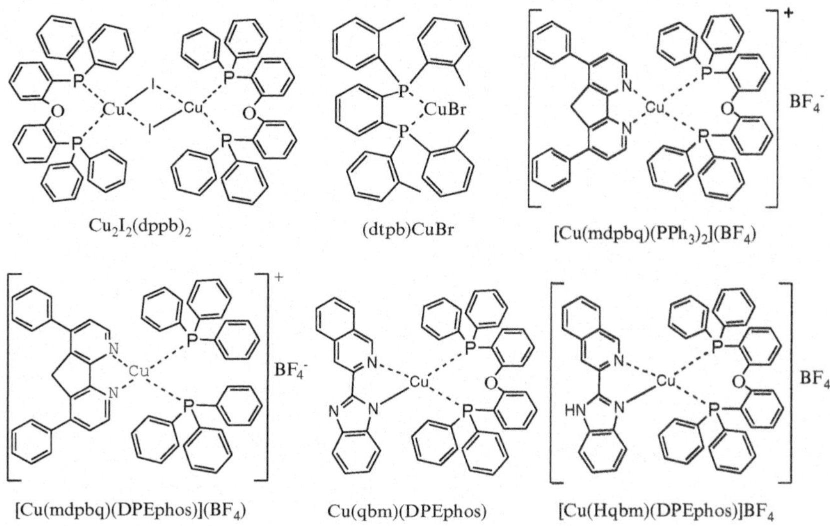

Fig. 9.21 Chemical structures of some Cu(I) complexes for use in PhOLEDs

Red Cu^{I} Phosphors A series of orange-red to red phosphorescent heteroleptic mononuclear Cu(I) complexes have been synthesized and fully characterized by Wang [84]. With highly rigid bulky biquinoline-type ligands, complexes [Cu (mdpbq)(PPh$_3$)$_2$](BF$_4$) and [Cu(mdpbq)(DPEphos)](BF$_4$) (Fig. 9.21) exhibit photoluminescence quantum yields of 0.56 and 0.43 and emission maximum of 606 nm and 617 nm in 20 wt% PMMA films, respectively. The complex [Cu (mdpbq)(DPEphos)](BF$_4$) exhibits the best device performance among this series of complexes. With the device structure of ITO/PEDOT/TCCz:[Cu(mdpbq) (DPEphos)](BF$_4$)(15 wt%)/TPBI/LiF/Al, saturated red electroluminescence was obtained with CIE coordinates of (0.61, 0.39) and with a current efficiency up to 6.4 cd A^{-1} and an EQE of 4.5 %. As claimed by the authors, that was the first report of the efficient mononuclear Cu(I) complexes with red emission.

Neutral Cu(I) Complexes Perform Better Than Charged Ones The ionic complexes that contain the main cationic complex part and the counterion seem to be more frequently observed for Cu(I) than other transition metals. Both neutral and charged Cu(I) complexes are possibly luminescent and have been used as phosphors in PhOLEDs. However, different behavior and properties have been observed for them. Wang and coworkers [85] prepared a group of mononuclear Cu (I) complexes, including Cu(qbm)(DPEphos) and [Cu(Hqbm)(DPEphos)](BF$_4$) in Fig. 9.21, based on the same ligands. These complexes were used as emitter to fabricate PhOLEDs and their device performance was compared. Multilayer organic light-emitting diodes (OLEDs) were fabricated with the device structure of ITO/PEDOT/TCCz:Cu complex (10 wt%)/BCP/Alq$_3$/LiF/Al, where TCCz is N-(4-(carbazol-9-yl)phenyl)-3,6-bis(carbazol-9-yl)carbazole and acts as the host for these Cu(I) complexes in the emitting layer. It was observed that the neutral

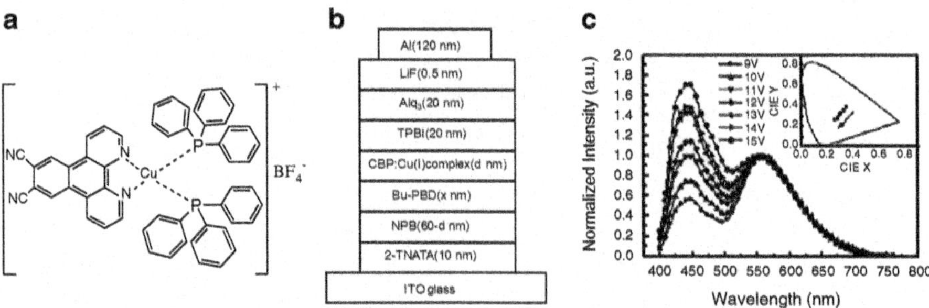

Fig. 9.22 White OLEDs configuration (**b**), Cu(I) phosphor structure (**a**), and the EL spectra (**c**) (Figure **c** reproduced with permission from Ref. [86]. Copyright 2006 American Institute of Physics)

mononuclear complex Cu(qbm)(DPEphos) exhibited a current efficiency of 8.87 cd A^{-1}, higher than that (5.58 cd A^{-1}) of its cationic counterpart [Cu(Hqbm)(DPEphos)](BF_4). In comparison with the charged counterparts, the neutral ones show blueshifted emissions and longer lifetimes. This indicates that the neutral Cu (I) complexes are more suitable for application as emitters in PhOLEDs than their charged counterparts. This is reasonable since the ionic species generally have lower volatility, and they may produce internal defects in OLEDs due to possible current-induced migration. This is consistent with the observation for the luminescent complexes of other transition metals.

White OLEDs Based on Cu(I) Phosphors Cu(I) phosphors were also used in combination with other emitting materials to make white OLEDs. Li and coworkers [86] fabricated a group of two-emitting-component white OLEDs by combining blue fluorescence and yellow phosphorescence. The Cu(I) complex (Fig. 9.22) was used as yellow emitter and *N,N′*-diphenyl-*N,N′*-bis(1-naphthyl)-(1,1′-benzidine)-4,4′-diamine (NPB) as blue fluorescent material. The white OLEDs have the structures of ITO/2-TNATA/NPB(60-*d*nm)/Bu-PBD/CBP:Cu(I) complex (*c* wt%, *d* nm)/TPBI/Alq_3/LiF/Al, in which 2-TNATA (4,4′,4″-tris[2-naphthyl(phenyl)amino]triphenylamine) acts as hole-injecting layer, NPB as hole-transporting blue-emitting layer, CBP:Cu(I) complex as yellow-emitting layer, and TPBI and Alq_3 as hole-blocking and electron-transporting layer, respectively. Interestingly, a thin layer of Bu-PBD layer sandwiched between two emission layers acts as a chromaticity-tuning layer. By carefully adjusting the film thickness of various functional layers and tuning the doping level of Cu(I) complex in CBP host, the optimized performance was obtained with CIE coordinates of (0.33, 0.36) and a maximum luminance of 2,466 cd m^{-2}, a peak current efficiency of 6.76 cd A^{-1}, and a power efficiency of 3.85 lm W^{-1}. Although these data were not comparable with those white OLEDs made from iridium complexes, this report is significant since it reveals the possibility to apply cheap Cu(I) phosphors to realize white light emission. It is quite reasonable to improve the overall performance of white OLEDs by designing novel highly efficient Cu(I) complexes. Actually, many newly developed Cu(I) complexes already show excellent performance of the intrinsic monochromatic emission [82, 83], which indicates the great potential of Cu(I) phosphors to fabricate high-performance and low-cost white OLEDs.

9.5 Conclusions and Perspectives

New generation of lighting sources are strongly desired to be energy-saving and environment-friendly. The semiconducting solid-state lighting sources including white LEDs and OLEDs have been established as the most ideal candidates as new-generation lighting sources. White OLEDs are intrinsically superior to LED products in terms of large area and flexible substrates and are more suitable for lighting application. Optimistic expectations predict that white OLEDs efficiencies will rapidly increase in the next 10 years, reaching values of 200 lm W^{-1} similar to those expected for inorganic LEDs. Although there have been reports of thermally activated delayed fluorescence (TADF) that also possibly reach an internal quantum efficiency of unity, all the high-performance white OLEDs are predominantly made from phosphorescent materials so far. Transition-metal complexes are intrinsically efficient phosphorescent at room temperature in comparison with organic molecules and are actually the dominant phosphorescent materials used in PhOLEDs now. It can be predicted that the transition-metal complex phosphors will retain as the high-efficiency materials for a long time in the future. While enjoying the excellent performance of these transition-metal complex phosphors, it should be kept in mind that their noble metal nature is not favorable for low cost. Fortunately, copper seems to be ideal alternatives for these noble metals due to high relative abundance and low-cost merits. The comparable efficiency of some PhOLEDs containing Cu(I) phosphors with those of famous iridium complexes lightened the future of copper products. However, the overall performance of the Cu(I) phosphors is still inferior to those for Ir and Pt complexes nowadays. Therefore, it is strongly desired to further develop novel chemical structures for highly efficient Cu (I) phosphors. At the same time, it is necessary to enhance light outcoupling by certain device techniques.

References

1. Anderade BWD, Forrest SR (2004) White organic light-emitting devices for solid-state lighting. Adv Mater 16:1585–1595
2. Baldo MA, Brien DFO, You Y, Shoustikov A, Sibley S, Thompson ME, Forrest SR (1998) Highly efficient phosphorescent emission from organic electroluminescent devices. Nature 395:151–154
3. Reineke S, Lindner F, Schwartz G, Seidler N, Walzer K, Lussem B, Leo K (2009) White organic light-emitting diodes with fluorescent tube efficiency. Nature 459:234–239
4. Hartmut Y (ed) (2008) Highly efficient OLEDs with phosphorescent materials. Wiley, Weinheim
5. Chi Y, Chou PT (2010) Transition-metal phosphors with cyclometalating ligands fundamentals and applications. Chem Soc Rev 39:638–655
6. Zhou GJ, Wong WY, Suo S (2010) Recent progress and current challenges in phosphorescent white organic light-emitting diodes (WOLEDs). J Photochem Photobio C Photochem Rev 11:133–156

7. Kalinowski J, Fattori V, Cocchi M, Williamsc JAG (2010) Light-emitting devices based on organometallic platinum complexes as emitters. Coord Chem Rev 255:2401–2425
8. Ma YG, Che CM, Chao HY, Zhou XM, Chan WH, Shen JC (1999) High luminescence gold (I) and copper(I) complexes with a triplet excited state for use in light-emitting diodes. Adv Mater 11:852–857
9. Farinola GM, Ragni R (2011) Electroluminescent materials for white organic light emitting diodes. Chem Soc Rev 40:3467–3482
10. Forrest SR, Bradley DDC, Thompson ME (2003) Measuring the efficiency of organic light-emitting devices. Adv Mater 15:1043–1048
11. Baldo MA, O'Brien DF, Thompson ME, Forrest SR (1999) Excitonic singlet-triplet ratio in a semiconducting organic thin film. Phys Rev B 60:14422–14428
12. Chou PT, Chi Y (2007) Phosphorescent dyes for organic light emitting diodes. Chem Eur J 13:380–395
13. Liu J, Xie ZY, Cheng YX, Geng YH, Wang LX, Jing XB, Wang FS (2007) Molecular design on highly efficient white electroluminescence from a single polymer system with simultaneous blue, green and red emission. Adv Mater 19:531–535
14. Li JY, Liu D, Ma CW, Lengyel O, Lee CS, Tung CH, Lee ST (2004) White-light emission from a single-emitting-component organic electroluminescent device. Adv Mater 16:1538–1541
15. Kido J, Hongawa K, Okuyama K, Nagai K (1994) White light-emitting organic electroluminescent devices using the poly (N-vinylcarbazole) emitter layer doped with three fluorescent dyes. Appl Phys Lett 64:815–817
16. Sun Y, Giebink NC, Kanno HS, Ma B, Thompson ME, Forrest SR (2006) Management of singlet and triplet excitons for efficient white organic light-emitting devices. Nature 440:908–912
17. Zou JH, Wu H, Lam CS, Wang CD, Zhu J, Zhong CM, Hu SJ, Ho CL, Zhou GJ, Wu HB (2011) Simultaneous optimization of charge-carrier balance and luminous efficacy in highly efficient white polymer light-emitting devices. Adv Mater 23:2976–2980
18. Zhang BH, Tan GP, Lam CS, Yao B, Ho CL, Liu LH, Xie ZY, Wong WY, Ding JQ, Wang LX (2012) High- efficiency single emissive layer white organic light emitting diodes based on solution-processed dendritic host and new orange-emitting iridium complex. Adv Mater 24:1873–1877
19. Wang RJ, Liu D, Ren HC, Zhang T, Yin HM, Liu GY, Li JY (2011) Highly efficient orange and white organic light-emitting diodes based on new orange iridium complexes. Adv Mater 23:2823–2827
20. Wang Q, Ding JQ, Ma DG, Cheng YX, Wang LX, Wang FS (2009) Manipulating charges and excitons within a single-host system to accomplish efficiency/CRI/color-stability trade-off for high-performance WOLEDs. Adv Mater 21:2397–2401
21. Balzani V, Bergamini G, Campagna S, Puntoriero F (2007) Photochemistry and photophysics of coordination compounds: overview and general concepts. In: Balzani V, Campagna S (eds) Photochemistry and photophysics of coordination compounds I. Springer, Berlin
22. Albrecht M (2010) Cyclometalation using d-block transition metals: fundamental aspects and recent trends. Chem Rev 110:576–623
23. Lamansky S, Djurovich P, Murphy D, Razzaq FA, Lee HE, Adachi C, Burrows PE, Forrest SR, Thompson ME (2001) Highly phosphorescent bis-cyclometalated iridium complexes: synthesis, photophysical characterization, and use in organic light emitting diodes. J Am Chem Soc 123:4304–4312
24. Lamansky S, Djurovich P, Murphy D, Razzaq FA, Kwong R, Tsyba I, Bortz M, Mui B, Bau R, Thompson ME (2001) Synthesis and characterization of phosphorescent cyclometalated iridium complexes. Inorg Chem 40:1704–1711
25. Tamayo AB, Alleyne BD, Djurovic PI, Lamansky S, Tysba I, Ho NN, Bau R, Thompson ME (2003) Phosphorescence quenching by conjugated polymers. J Am Chem Soc 125:7796–7797

26. McDonald AR, Lutz M, Chrzanowski LSV, Klink GPMV, Spek AL, Koten GV (2008) Probing the mer- to fac-isomerization of tris-cyclometallated homo- and heteroleptic (C, N)3 iridium (III) complexes. Inorg Chem 47:6681–6691
27. Dedeian K, Djurovic PI, Garces FO, Carlson G, Watts RJ (1991) A new synthetic route to the preparation of a series of strong photoreducing agents: fac-tris-ortho-metalated complexes of iridium(III) with substituted 2-phenylpyridines. Inorg Chem 30:1685–1687
28. Ho CL, Wong WY (2013) Charge and energy transfers in functional metallophosphors and metallopolyynes. Coord Chem Rev 257:1614–1649
29. Adachi C, Baldo MA, Forrest SR, Lamansky S, Thompson ME, Kwong RC (2001) High-efficiency red electrophosphorescence devices. Appl Phys Lett 78:1622–1624
30. Adachi C, Kwong RC, Djurovich P, Adamovich V, Baldo MA, Thompson ME, Forrest SR (2001) A mechanism for generating very efficient high-energy phosphorescent emission in organic materials. Appl Phys Lett 79:2082–2084
31. Ma YG, Gu X, Fei T, Zhang HY, Xu H, Yang B, Ma YG, Liu XD (2009) Tuning the emission color of iridium(III) complexes with ancillary ligands: a combined experimental and theoretical study. Eur J Inorg Chem 16:2407–2414
32. Zhou GJ, Ho CL, Wong WY, Wang Q, Ma DG, Wang LX, Lin ZY, Marder TB, Beeby A (2008) Manipulating charge-transfer character with electron-withdrawing main-group moieties for the color tuning of iridium electrophosphors. Adv Funct Mater 18:499–511
33. Watanabe S, Ide N, Kido J (2007) High-efficiency green phosphorescent organic light-emitting devices with chemically doped layers. Jpn J Appl Phys Part 1 46:1186–1188
34. Tanaka DK, Sasabe H, Li YJ, Su SJ, Takeda TS, Kido J (2007) Ultra high efficiency green organic light-emitting devices. Jpn J Appl Phys Part 2 Lett Express Lett 46(1–3):L10–L12
35. Adachi C, Baldo MA, Thompson Mark E, Forrest SR (2001) Nearly 100% internal phosphorescence efficiency in an organic light-emitting device. J Appl Phys 90:5048–5051
36. Han KD, Sung CN, Yun OH, Hwan YJ, Sik JW, Soo PJ, Chul SM, Hyuk KJ (2011) Highly efficient red phosphorescent dopants in organic light-emitting devices. Adv Mater 23:2721–2726
37. Su TH, Fan CH, Ou Yang, Yu H, Hsu LC, Cheng CH (2013) Highly efficient deep-red organic electrophosphorescent devices with excellent operational stability using bis(indoloquinoxalinyl) derivatives as the host materials. J Mater Chem C 1:5084–5092
38. Fan CH, Sun PP, Su TH, Cheng CH (2011) Host and dopant materials for idealized deep-red organic electrophosphorescence devices. Adv Mater 23:2981–2985
39. Ho CL, Wong WY, Gao ZQ, Chen CH, Cheah KW, Yao B, Xie ZY, Wang Q, Ma DG, Wang LX (2008) Red-light-emitting iridium complexes with hole-transporting 9-arylcarbazole moieties for electrophosphorescence efficiency/color purity trade-off optimization. Adv Funct Mater 18:319–331
40. Laskar IR, Chen TM (2004) Tuning of wavelengths: synthesis and photophysical studies of iridium complexes and their applications in organic light emitting devices. Chem Mater 16:111–117
41. Wang RJ, Deng LJ, Zhang T, Li JY (2012) Electrochemical properties and electroluminescence performance of orange-emitting iridium complexes Rhenium(i) tricarbonyl polypyridine complexes. Dalton Trans 41:6833–6841
42. Wang RJ, Liu D, Ren HC, Zhang T, Wang XZ, Li JY (2011) Homoleptic tris-cyclometalated iridium complexes with 2-phenylbenzothiazole ligands for highly efficient orange OLEDs. J Mater Chem 21:15494–15500
43. Wang RJ, Liu D, Zhang R, Deng LJ, Li JY (2012) Solution-processable iridium complexes for efficient orange-red and white organic light-emitting diodes. J Mater Chem 22:1411–1417
44. Holmes RJ, Forrest SR, Tung YJ, Kwong RC, Brown JJ, Garon S, Thompson ME (2003) Blue organic electrophosphorescence using exothermic host-guest energy transfer. Appl Phys Lett 82:2422–2424

45. Lin MS, Yang SJ, Chang HW, Huang YH, Tsai YT, Wu CC, Chou SH, Mondal E, Wong KT (2012) Incorporation of a CN group into mCP: a new bipolar host material for highly efficient blue and white electrophosphorescent devices. J Mater Chem 22:16114–16120
46. Deng LJ, Li JY, Wang GX, Wu LZ (2013) Simple bipolar host materials incorporating CN group for highly efficient blue electrophosphorescence with slow efficiency roll-off. J Mater Chem C 1:8140–8145
47. Tao Y, Wang Q, Yang C, Wang Q, Zhang Z, Zou T, Qin J, Ma D (2008) A simple carbazole/oxadiazole hybrid molecule: an excellent bipolar host for green and red phosphorescent OLEDs. Angew Chem Int Ed 47:8104–8107
48. Ye SJ, Wu MF, Chen CT, Song YH, Chi Y, Ho MH, Chen SF, Chen CH (2005) New dopant and host materials for blue-light-emitting phosphorescent organic electroluminescent devices. Adv Mater 17:285–289
49. Yang CH, Cheng YM, Chi Y, Hsu CJ, Fang FC, Wong KT, Chou PT, Chang CH, Tsai MH, Wu CC (2007) Blue-emitting heteroleptic iridium(III) complexes suitable for high-efficiency phosphorescent OLEDs. Angew Chem Int Ed 46:2418–2421
50. Ren X, Li J, Holmes RJ, Djurovich PI, Forrest SR, Thompson ME (2004) Ultrahigh energy gap hosts in deep blue organic electrophosphorescent devices. Chem Mater 16:4743–4747
51. Sajoto T, Djurovich PI, Tamayo A, Yousufuddin M, Bau R, Thompson ME, Holmes RJ, Forrest SR (2005) Blue and near-UV phosphorescence from Iridium complexes with cyclometalated pyrazolyl or N-heterocyclic carbene ligands. Inorg Chem 44:7992–8003
52. Lo SC, Shipley CP, Bera RN, Harding RE, Cowley AR, Burn PL, Samuel IDW (2006) Blue phosphorescence from iridium(III) complexes at room temperature. Chem Mater 18:5119–5129
53. Tanaka D, Takeda T, Chiba T, Watanabe S, Kido J (2007) Novel electron-transport material containing boron atom with a high triplet excited energy level. Chem Lett 36:262–263
54. Jeon SO, Jang SE, Son HS, Lee JY (2011) External quantum efficiency above 20 % in deep blue phosphorescent organic light-emitting diodes. Adv Mater 23:1436–1441
55. Yook KS, Jeon SO, Joo CW, Lee JY (2009) High efficiency deep blue phosphorescent organic light-emitting diodes. Org Electron 10:170–173
56. Chang CF, Cheng YM, Chi Y, Chiu YC, Lin CC, Lee GH, Chou PT, Chen CC, Chang CH, Wu CC (2008) Highly efficient blue-emitting iridium(III) carbene complexes and phosphorescent OLEDs. Angew Chem Int Ed 47:4542–4545
57. Ma B, Li J, Djurovich PI, Yousufuddin M, Bau R, Thompson ME (2005) Synthetic control of Pt···Pt separation and photophysics of binuclear platinum complexes. J Am Chem Soc 127:28–29
58. Lai SW, Che CM (2004) Luminescent cyclometalated diimine platinum(II) complexes: photophysical studies and applications. In: Balzani V, Campagna S (eds) Top Curr Chem 241:27–63. Springer
59. Yang XL, Yao CL, Zhou GJ (2013) Highly efficient phosphorescent materials based on platinum complexes and their application in organic light-emitting devices (OLEDs). Platin Metals Rev 57:2–16
60. Graham KR, Yang Y, Sommer JR, Shelton AH, Schanz KS, Xue J, Reynolds JR (2011) Extended conjugation platinum(II) porphyrins for use in near-infrared emitting organic light emitting diodes. Chem Mater 23:5305–5312
61. Xiang HF, Chan SC, Wu KKY, Che CM, Lai PT (2005) High-efficiency red electrophosphorescence based on neutral bis(pyrrole)-diimine platinum(II) complex. Chem Commun 11:1408–1410
62. Che CM, Kwok CC, Lai SW, Rausch AF, Finkenzeller WJ, Zhu N, Yersin H (2010) Photophysical properties and OLED applications of phosphorescent platinum(II) schiff base complexes. Chem Eur J 16:233–247
63. Lu W, Mi BX, Chan MCW, Hui Z, Che CM, Zhu N, Lee ST (2004) Light-emitting tridentate cyclometalated platinum(II) complexes containing σ-alkynyl auxiliaries: tuning of photo- and electrophosphorescence. J Am Chem Soc 126:4958–4971

64. Williams JAG, Beeby A, Davies ES, Weinstein JA, Wilson C (2003) An alternative route to highly luminescent platinum(II) complexes: cyclometalation with N∧C∧N-coordinating dipyridylbenzene ligands. Inorg Chem 42:8609–8611
65. Farley SJ, Rochester DL, Thompson AL, Howard JAK, Williams JAG (2005) Controlling emission energy, self-quenching, and excimer formation in highly luminescent N∧C∧N-coordinated platinum(II) complexes. Inorg Chem 44:9690–9703
66. Cocchi M, Virgili D, Fattori Rochester VDL, Williams JAG (2007) N∧C∧N-coordinated platinum(II) complexes as phosphorescent emitters in high-performance organic light-emitting devices. Adv Funct Mater 17:285–289
67. Cocchi M, Kalinowski J, Murphy L, Williams JAG, Fattori V (2010) Mixing of molecular exciton and excimer phosphorescence to tune color and efficiency of organic LEDs. Org Electron 11:388–396
68. Chen JL, Chang SY, Chi Y, Chen K, Cheng YM, Lin CW, Lee GH, Chou PT, Wu CH, Shih PI, Shu CF (2008) Pt II complexes with 6-(5-trifluoromethyl-pyrazol-3-yl)-2,2'-bipyridine terdentate chelating ligands: synthesis, characterization, and luminescent properties. Chem Asian J 3:2112–2123
69. Yam VWW, Tang RPL, Wong KMC, Lu XX, Cheung KK, Zhu N (2002) Syntheses, electronic absorption, emission, and ion-binding studies of platinum(II) C^N^C and terpyridyl complexes containing crown ether pendants. Chem Eur J 8:4066–4076
70. Wang ZB, Helander MG, Hudson ZM, Qiu J, Wang S, Lu ZH (2011) Pt(II) complex based phosphorescent organic light emitting diodes with external quantum efficiencies above 20 %. Appl Phys Lett 98:213301/1–213301/3
71. Bhansali US, Polikarpov E, Swensen JS, Chen WH, Jia H, Gaspar DJ, Gnade BE, Padmaperuma AB, Omary MA (2009) High efficiency organic photovoltaic cells based on a vapor deposited squaraine donor. Appl Phys Lett 95:233304/1–233304/3
72. Yang X, Wang Z, Madakuni S, Li J, Jabbour GE (2008) Efficient blue- and white-emitting electrophosphorescent devices based on platinum(II) [1,3-difluoro-4,6-di(2-pyridinyl)benzene] chloride. Adv Mater 20:2405–2409
73. Hu X, Rodriguez IC, Meyer K (2003) Copper complexes of nitrogen-anchored tripodal N-heterocyclic carbene. Ligands J Am Chem Soc 125:12237–12245
74. Irving H, Williams RJP (1953) The stability of transition-metal complexes. J Chem Soc 1953:3192–3210
75. Meyer TJ (1989) Chemical approaches to artificial photosynthesis. Acc Chem Res 22:163–170
76. McMillin DR, Mcnett KM (1998) Photoprocesses of copper complexes that bind to DNA. Chem Rev 98:1201–1219
77. Evans RC, Douglas P, Winscom CJ (2006) Coordination complexes exhibiting room-temperature phosphorescence: evaluation of their suitability as triplet emitters in organic light emitting diodes. Coord Chem Rev 250:2093
78. Xiang HF, Cheng JH, Ma XF, Zhou XG, Chruma JJ (2013) Near-infrared phosphorescence: materials and applications. Chem Soc Rev 42:6128–6185
79. Everly RM, McMillin DR (1989) Concentration-dependent lifetimes of cu(NN) + 2 systems: exciplex quenching from the ion pair state. Photochem Photobiol 50:711–716
80. Harkins SB, Peters JC (2005) A highly emissive Cu2N2 diamond core complex supported by a [PNP]- ligand. J Am Chem Soc 127:2030–2031
81. Tsuboyama A, Kuge K, Furugori M, Okada S, Hoshino M, Ueno K (2007) Photophysical properties of highly luminescent copper(I) halide complexes chelated with 1,2-bis(diphenylphosphino)benzene. Inorg Chem 46:1992–2001
82. Zhang QS, Komino T, Huang SP, Matsunami S, Goushi K, Adachi C (2012) Triplet exciton confinement in green organic light-emitting diodes containing luminescent charge-transfer Cu (I) complexes. Adv Funct Mater 22:2327–2336
83. Hashimoto M, Igawa S, Yashima M, Kawata I, Hoshino M, Osawa M (2011) Highly efficient green organic light-emitting diodes containing luminescent three-coordinate copper (I) complexes. J Am Chem Soc 133:10348–10351

84. Zhang QS, Ding JQ, Cheng YX, Wang LX, Xie ZY, Jing XB, Wang FS (2007) Novel heteroleptic cuI complexes with tunable emission color for efficient phosphorescent light-emitting diodes. Adv Funct Mater 17:2983–2990
85. Min JH, Zhang QS, Sun W, Cheng YX, Wang LX (2011) Neutral copper(I) phosphorescent complexes from their ionic counterparts with 2-(2′-quinolyl)benzimidazole and phosphine mixed ligands. Dalton T 40:686–693
86. Su ZS, Che GB, Li WL, Su WM, Li MT, Chu B, Li B, Zhang ZQ, Hu ZZ (2006) White-electrophosphorescent devices based on copper complexes using 2-(4-biphenylyl)-5-(4-tert-butyl-phenyl)-1,3,4-oxadiazole as chromaticity-tuning layer. Appl Phys Lett 88:213508/1–213508/3

Chapter 10
White Organic Light-Emitting Diodes Based on Organometallic Phosphors

Dongcheng Chen and Shi-Jian Su

Abstract Phosphorescent white organic light-emitting diodes (WOLEDs) employing organometallic phosphors as the emitters have attracted considerable attention in the past decade. Due to their capability of harvesting both singlet and triplet excitons to generate highly efficient devices, phosphorescent WOLEDs present potential applications in the next-generation solid-state lighting sources and in the flat-panel display with the assistance of the color filters. In this chapter, we attempt to give a brief overall introduction to the phosphorescent WOLEDs. The basic concepts, like electric-light conversion efficiency, parameters to assess the color quality of the emissive white light, device strategies to fabricate WOLEDs, and the common device fabrication procedures, are introduced firstly. These fundamental understandings are also favorable to comprehend the monochromatic OLEDs and the WOLEDs comprised of other emitters like fluorescent dyes. In particular, we focus on the discussion of phosphorescent WOLEDs with various device architectures and their corresponding device performances. We also note that further enhancement of the device lifetime with simultaneous realization of high efficiency and high quality of the emissive white light could make phosphorescent WOLEDs promising candidates for the alternative next-generation lighting sources.

Keywords White organic light-emitting diodes • Energy saving • Organometallic phosphors • Phosphorescence • Triplet excitons

10.1 Introduction

It is of a long history for mankind to utilize the artificial white light sources. The big breakthrough is Edison's invention, which for the first time transform the electricity energy into light radiation, resulting in the born of the incandescent bulb. The electric incandescent bulb is a heat light source, where the radiation light comes

D. Chen • S.-J. Su (✉)
State Key Laboratory of Luminescent Materials and Devices and Institute of Polymer Optoelectronic Materials and Devices, South China University of Technology, Guangzhou 510640, China
e-mail: mssjsu@scut.edu.cn

W.-Y. Wong (ed.), *Organometallics and Related Molecules for Energy Conversion*, Green Chemistry and Sustainable Technology, DOI 10.1007/978-3-662-46054-2_10

from the heat radiation of a metal with a high melting point, while it is cycled by the electricity. Quite a small ratio of the electric power is utilized to finally give light, because the power efficiency (η_p) of the bulb is quite low, typically with a value below 20 lm/W. The appearance of the fluorescent tube greatly promotes the utilization of the artificial white light sources. In contrast to the electric bulb, the fluorescent tube can be more efficient to transform the electricity into visible white light, yielding an enhanced η_p (20–80 lm/W). Besides, various emissive white lights with different quality could also be achieved from fluorescent tube, making its wider applications in different occasions, e.g., in the surgical operation room, which severely inquires the light source to vividly reproduce the color of the objects as illuminated in the sunshine. A fatal disadvantage of this kind of light-emitting fluorescent tubes is their containing of mercury, which could be harmful to the environment, and its disposal represents a great challenge. In the late 1990s, the achievement of wide bandgap inorganic semiconducting materials based on the V (III) and V(V) elements brings the inorganic blue-light-emitting diode (LED) to reality. As a result, white LEDs can be fabricated using a blue LED coated by the fluorescent powders with longer wavelength emissions or using red, green, and blue (RGB) sub-LEDs to give mixed white emission. Inorganic LEDs are solid-state lighting sources, which can get rid of the problems related to vacuum encapsulation required for incandescent bulbs and fluorescent tubes. Inorganic LEDs also possess many merits like low-driving voltage, energy saving, long lifetime, and fast response time. These merits promote white LEDs' wide application in solid-state lighting, and the process of replacement of traditional white sources like bulbs and even fluorescent tubes is undergoing. However, there are still some shortcomings for LEDs: (1) heat production in a LED during the operation is dramatic, leading to luminous decay when used for a long time; (2) fabrication of LEDs is quite complicated and costly; and (3) realization of white LEDs with high color-rendering index (CRI) above 80 is expensive due to the challenge of combining RGB sub-LEDs to fabricate a white LED.

Based on the pioneering work of Tang et al. in 1987 [1], Kido and coworkers firstly realized white organic light-emitting diodes (WOLEDs) via mixing three fluorescent dyes (blue, green, and orange) into one single emission layer (EML) [2]. Their work opened up a new approach to realize artificial white light sources. In contrast to the aforementioned inorganic white light sources, WOLEDs can be utilized as a lighting source with low cost, compatibility with flexible substrate, large size, and feasibility of realizing a high CRI. Besides, WOLEDs also combine most of the advantages of other cold light sources like LEDs while circumventing the heat dissipation problem. These potential advantages have attracted much research effort from both scientific field and industry.

According to the spin statistics, only 25 % of electric-generated excitons can be harvested by singlet energy levels in a fluorescent emitter, followed by a probable radiation process. Most of the electric-generated triplet excitons formed in the triplet energy levels cannot contribute to light emission in a fluorescent emitter due to the quantum spin forbidden, and this kind of excitons would finally decay in a non-radiation means. As thus, for a fluorescent OLED, in which a fluorescent dye plays a role of light-emitting material, internal quantum efficiency (η_{in}) could only be maximized to 25 %. When considering a light out-coupling factor of 0.2, a maximum of external quantum efficiency (η_{ext}) is limited to 5 %. This low efficiency has brought a bottleneck of the fluorescent OLEDs for application in artificial lighting sources. Fortunately, the discovery of the organometallic phosphors has rendered OLEDs to avoid this efficiency bottleneck since they can harness both the singlet and triplet excitons to give the probability of unit η_{in}. An organometallic phosphorescent emitter is generally comprised of a Ir(III), Pt(III), or other heavy metal central atoms and the peripheral ligands, in which the heavy metal atom effect leads to the mixing of singlet and triplet energy levels, thus enhancing the reverse intersystem crossing and giving probably 100 % exciton decay through radiation. Thereby, highly efficient phosphorescent WOLEDs can be obtained using the phosphorescent dyes to construct the light-emitting layers. We are focusing on the phosphorescent WOLEDs where organometallic materials are involved as emitters in the following discussion.

10.2 The Fundamental Concepts of Phosphorescent WOLEDs

10.2.1 Efficiency

η_{ext}, current efficiency (CE), and η_p are the mostly used parameters to characterize the device performances of a WOLED. η_{ext} is defined as the ratio of the total out-coupled photons divided by the injection of electron-hole pairs. It can be written as following:

$$\eta_{ext} = \eta_{in}\eta_{out} = \Upsilon_{e-h}\delta_s\eta_i\eta_{out} \tag{10.1}$$

where η_{in} is defined as the total number of generated photons within the devices per injected electron-hole pair; Υ_{e-h} is a balanced factor related to the ratio of minority carrier to the majority carrier to give $\Upsilon_{e-h} \leq 1$; δ_s is the branching ratio of the emissive excitons, which is 25 % for a fluorescent emitter and 100 % for a phosphorescent emitter; and η_i is the intrinsic quantum efficiency of radiation decays from both fluorescence and phosphorescence. η_{ext} is a parameter to assess the OLED performance independent of the perception of human's eyes.

CE is defined as following:

$$\mathrm{CE} = A\,L/I \tag{10.2}$$

where A is the effective light-emitting area of the device and L is the luminance under the corresponding current I. Noting that CE is a performance parameter related to naked eyes and has no connection with driving voltage, so it cannot give information about whether or not an OLED is energy saving.

Whereas, η_p is greatly affected by the driving voltage (V) and defined as the following:

$$\eta_p = \text{CE}\,\pi/V = A\,L\,\pi/(I\,V). \tag{10.3}$$

η_p is described as integral output light per consumed electric power, which is directly related to the consumption of electric energy, and tells whether a WOLED is energy saving or not. Accordingly, high η_p values are always the featured parameter for an efficient, energy-saving WOLED.

10.2.2 Quality of White Light

White light is generally characterized by three parameters: CIE (Commission Internationale d'Eclairage) coordinates, correlated color temperature (CCT), and color-rendering index (CRI).

A pair of two numbers [x, y] in 1931 chromaticity diagram is used to describe how the human eyes perceive the emission color of any light source. A pure ideal white light has a CIE coordinate of (0.333, 0.333), and the CIE coordinates of a perfect white lighting source should be close to the ideal white point. For illumination applications, CCT needs to be equivalent to that of a blackbody source between 3,000 and 7,500 K. Average daylight has a CCT of 6,500 K, and a warm white fluorescent lamp has a CCT of 3,000 K. CRI is used to assess the ability of a white lighting source to reproduce the true color of the illuminated objects compared with the color of the same objects illuminated by a reference source of comparable color temperature. CRI values range from 0 to 100, with 100 representing true color reproduction relative to the reference source. Typically, fluorescent lamps have CRI ratings between 60 and 99. CRI of the mercury lamp is near 50, and high-pressure sodium lamps can only have a CRI of 20. CRI above 80 is required for an indoor-lighting source.

10.2.3 How to Produce WOLEDs?

In principle, one can fabricate a WOLED with luminophors to emit RGB light. This RGB WOLED generally exhibits a high CRI value, due to the wide coverage of visible spectral region; however, it often bears a low η_p since highly efficient red and blue emitters are quite challenging to date. An effective strategy toward highly efficient WOLEDs is blending blue and yellow (BY) emission to yield white light. Due to the absence of green and deep-red emission, CRIs of these WOLEDs based

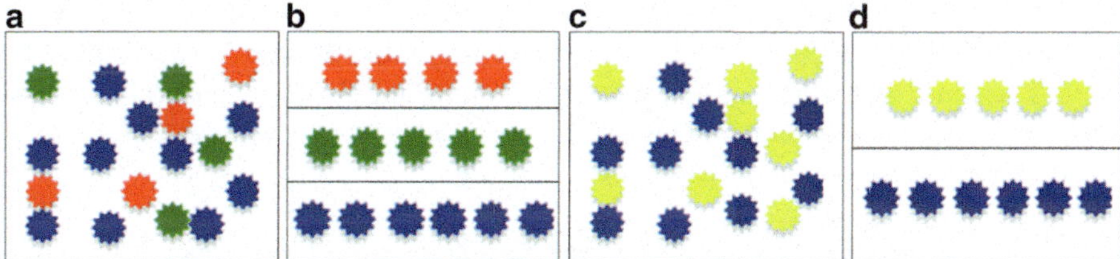

Fig. 10.1 Four basic device structures to fabricate the EML of the WOLEDs. (**a**) RGB emitters blended in one layer, (**b**) RGB emitters as separated deposited layers, (**c**) BY emitters blended in one layer, (**d**) BY emitters as separated deposited layers

on BY system are quite low, generally below 70, limiting their indoor-lighting applications.

From the device structure aspects, there are two commonly adopted structures to fabricate WOLEDs. One is blending the RGB or BY luminophors into one EML (Fig. 10.1a, c); another is depositing the RGB or BY luminophors as separated layers to realize multilayer white light emission (Fig. 10.1b, d). Noting that for a phosphor as a luminophor, host materials with a higher triplet state need to be incorporated to avoid the triplet exciton concentration-induced quenching and suppress the back energy transfer from the doped guest materials.

Another effective means to realize white light emission can be achieved from the rational design of the emissive molecules. Grafting all RGB or BY subunits into one polymer through covalent bonds can lead to white light electroluminescence from a single polymer. The single white light emission polymer is generally processed by simple solution spin-coating craft, which would lower the complexity and cost of device fabrication. Another approach utilizing the excimer/exciplex emission of the wide bandgap materials to cover the long-wavelength spectral region together with their intrinsic high-energy blue monomer emission to give white light is also widely investigated. This device configuration is suitable for a phosphor, like Pt(III) derivatives, having a planar structure to induce the excimer emission, to realize WOLEDs. Exciplex emissions are typically formed between an electron-rich donor and an electron-deficient acceptor material, and when combined with the intrinsic monomer emission from the donor or acceptor material, quite high quality of white light electroluminescence could be achieved. These approaches, to a certain extent, promote the selection of the white light emissive materials, since wide bandgap materials even some charge transport materials can be used as emissive materials to achieve high-quality white light emission.

10.2.4 Device Fabrication Procedures

In total, the current OLED fabrication procedures could be divided as two major categories: thermal vacuum evaporation and wet-coating process. The fabrication procedures of the WOLEDs have no obvious differences compared to those of monochromatic OLEDs. Thermal vacuum evaporation is widely applicable to deposit organic thin films relied on small molecular OLED materials in a high vacuum of 10^{-4} Pa or better. One of the most attractive advantages of thermal vacuum evaporation is that it enables the easy fabrication of multilayer devices, in which the thicknesses of its separated layers can be precisely controlled. The multilayer structures are favorable to the efficiency enhancement due to the feasibility of tailoring the balance of the electrons and holes and less limitation of selecting the surrounding functional materials. For wet-coating process, spin coating is one of the most widely adopted wet-processing approaches, because it can be quite simply conducted on a spin processor, which is of low cost and easy operation. Other wet-coating processes like roll-to-roll processing and ink-jet printing techniques are also widely investigated. Generally, polymer-based OLED materials should be solution processed, since polymer-based materials will decompose upon heating at a high temperature. Solution-processed small molecular materials especially dendrimers are also of interest, recently. These materials also present satisfied performance compared to their polymer counterparts. Solution processing offers a good alternative to fabricate OLEDs with low cost, compatibility with flexible substrates, a relatively small amount of wasted materials, and precise control of doping level. However, because orthogonal solvents are required to process the adjacent layers, this would limit the fabrication of multilayer solution-processed OLEDs. As a result, an OLED containing less functional layers is generally inefficient in relative to the thermal vacuum-fabricated multilayer OLEDs.

10.3 Phosphorescent WOLEDs Involving Organometallics

In this section, five parts are classified to introduce the phosphorescent materials used for phosphorescent WOLEDs. We firstly discuss the WOLEDs fabricated by all-phosphorescent dyes doped into a single layer or deposited as separated stacked multilayers. Then, the fluorescent/phosphorescent (F/P) hybrid WOLEDs typically employing the phosphors to achieve green, red, or yellow emission and fluorescent dyes to achieve blue emission are discussed. Also, WOLEDs utilizing a single polymer emission commonly grafted with phosphorescent dyes in the polymer chains were combined into this section. Introduction of WOLEDs achieved by excimer/exciplex emission is following. In addition, we also talk about tandem phosphorescent WOLEDs as an independent final part.

10.3.1 All-Phosphorescent WOLEDs

All emitters are comprised of phosphorescent dyes. In order to circumvent the concentration quenching, a rational choose of host materials is of importance to maximize the device performance. Several key factors should be considered to choose a rational host material: (1) the triplet energy level (E_T) of the host should be higher than that of the guest phosphor to suppress exothermic back energy transfer from the guest to the host; (2) suitable energy levels of the host materials approximately aligned with those of the neighboring transport materials can contribute to a low-driving voltage; (3) the host material should have bipolar-transport property to balance the electron and hole flow within the EML to achieve a high recombination efficiency and broaden the exciton recombination region to reduce the efficiency roll-off; (4) compatibility of the host material and the guest phosphor should be good to run around the potential phase separation, which is detrimental to the device lifetime; and (5) a high glass transition temperature (T_g) of the host material is also vital to withstand the heat generated during the device operation, and thus favorable to a longer device lifetime.

10.3.1.1 Single EML

D'Andrade et al. fabricated WOLEDs with a single EML employing all the RGB phosphorescent dyes, in which PQIr was used for red emission, $Ir(ppy)_3$ for green emission, and FIr6 for blue emission [3]. These phosphorescent dyes were co-deposited with UGH2, a host material with a high E_T above 3.0 eV, to serve as an electroluminescent layer. Chemical structures of the materials used and the device structures are shown in Fig. 10.2. It is shown that the LUMO level of TCTA is higher than the ones of all-phosphorescent dopants and the host UGH2, which could set up a barrier for the electrons traveling to the anode side. Similar conditions could be found for TPBI with a deep HOMO of 6.3 eV, confining the injected holes within the EML. Good confinement of the electrons and holes in the EML enhanced the efficiency of their developed WOLEDs. Moreover, the host UGH2 has a high E_T of 3.5 eV, which is the highest one among the host and the dopants. The high E_T of the host could suppress the back energy transfer from the dopant to the host, maintaining the efficient electrophosphorescence from the dopants. It was also demonstrated in their work that the direct triplet exciton formation on the blue phosphor FIr6 was favorable to the carrier harvesting and exciton recombination. Furthermore, a thin film craft was adopted in their devices; thus reduced driving voltage was required, and enhanced η_p was obtained. As a result, white devices having a η_p of (14 ± 1) lm/W at 10 mA/cm^2, a maximum η_p of (42 ± 4) lm/W and CIE coordinates that vary from (0.43, 0.45) at 0.1 mA/cm^2 to (0.38, 0.45) at 10 mA/cm^2, had been demonstrated. Besides, the RGB triple doped WOLEDs also had a high CRI of around 80. However, though inspiring η_p compared to that of an incandescent bulb had been achieved for their WOLEDs, it is still lagging behind in contrast to the high efficiency of a fluorescent tube, and the η_p values of their devices were not satisfied under a relatedly high luminance. This is probably attributed to the limitation of material selection when they fabricated the devices. Enhancement of device performances may be realized by replacement of an ETL

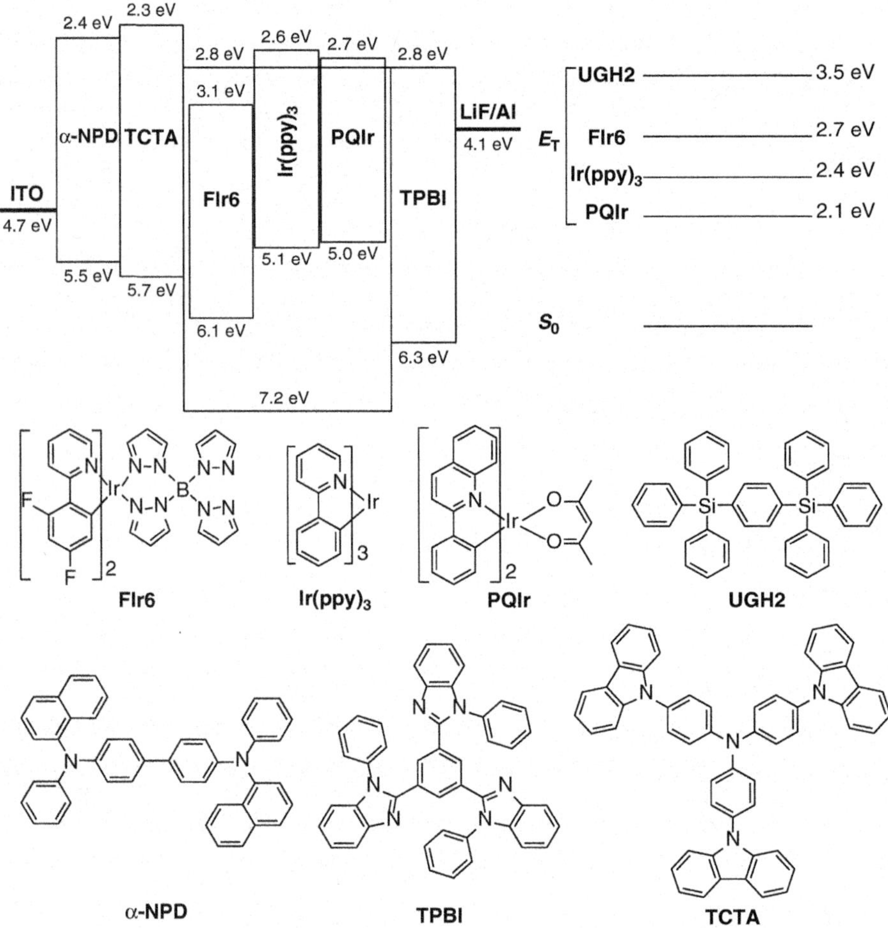

Fig. 10.2 Materials used and their corresponding energy levels (LUMOs, HOMOs, and E_Ts), as well as the adopted device structure

with a higher mobility and a higher triplet energy level than TPBI as well as a host material with better bipolar-transport ability than UGH2.

Yang and Ma et al. reported efficient WOLEDs by using several newly developed bipolar hosts, which are universal for blue, green, and orange phosphorescent dyes [4]. These bipolar hosts were developed based on UGH series hosts (see Fig. 10.3). All have a centered tetraarylsilane core, and asymmetric substitutions render these molecules with a hole-transport arylamine moiety and an electron-transport benzimidazole or oxadiazole unit. Compared to UGH2, these host materials have quite balanced electron- and hole-transport abilities, lower E_Ts near 2.7 eV but still high enough for a sky-blue phosphor FIrpic. A device configuration of ITO/MoO_3 (10 nm)/NPB (80 nm)/TCTA (5 nm)/p-BISiTPA or p-OXDSiTPA: 8 wt% FIrpic: 0.67 wt% $(fbi)_2Ir(acac)$ (20 nm)/TPBI (40 nm)/LiF (1 nm)/Al

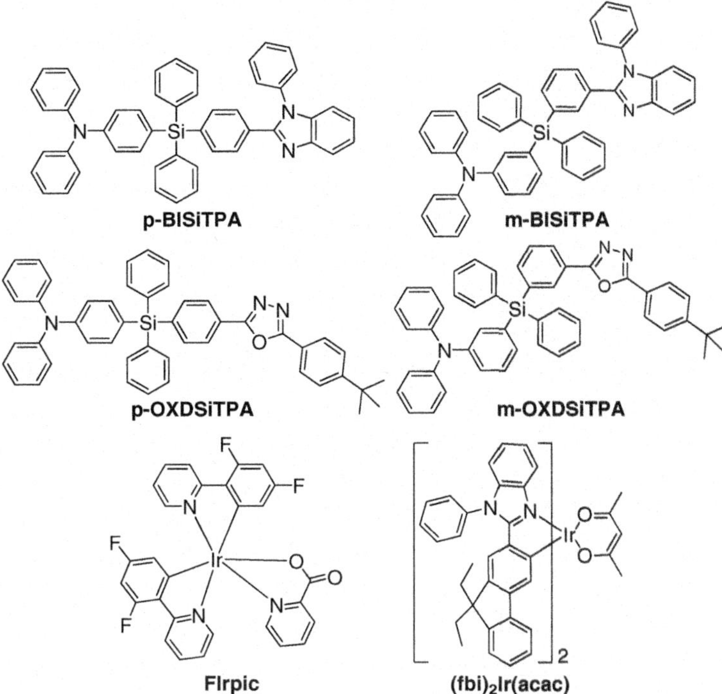

Fig. 10.3 Molecular structures of the developed host materials and the used blue and orange phosphors

(100 nm) was adopted for their WOLEDs. p-BISiTPA and p-OXDSiTPA were selected as typical hosts used for WOLEDs, and $(fbi)_2Ir(acac)$ was incorporated to give orange emission. Similar to the above devices reported by D'Andrade et al., all the phosphorescent dyes were doped into a single EML; also, TPBI was used as an ETL. OLEDs employing p-BISiTPA and p-OXDSiTPA as a host exhibited a maximal η_p of 42.7 and 51.9 lm/W and a low turn-on voltage of 3.1 and 2.9 V, respectively. These obtained efficiency values are much higher than that of the traditional incandescent bulb. Reduced efficiency roll-offs were also observed in their WOLEDs; however, due to the two-color system, these WOLEDs could only have a moderate CRI ranging from 60 to 67.

The extra work of the vacuum-evaporated phosphorescent WOLEDs in which the emission of white light comes from a single EML could be referred to the ref. [4–9]. These works were improved from the engineering of host materials, electron-transport materials, phosphorescent dyes, etc., typically with similar device architectures.

Different from the aforementioned work done by vacuum fabrication, we introduce here the fabrication of white phosphorescent OLEDs in which the organic layers were deposited by solution spin-coating processing. Wu and coworkers developed efficient phosphorescent OLEDs with a single active layer

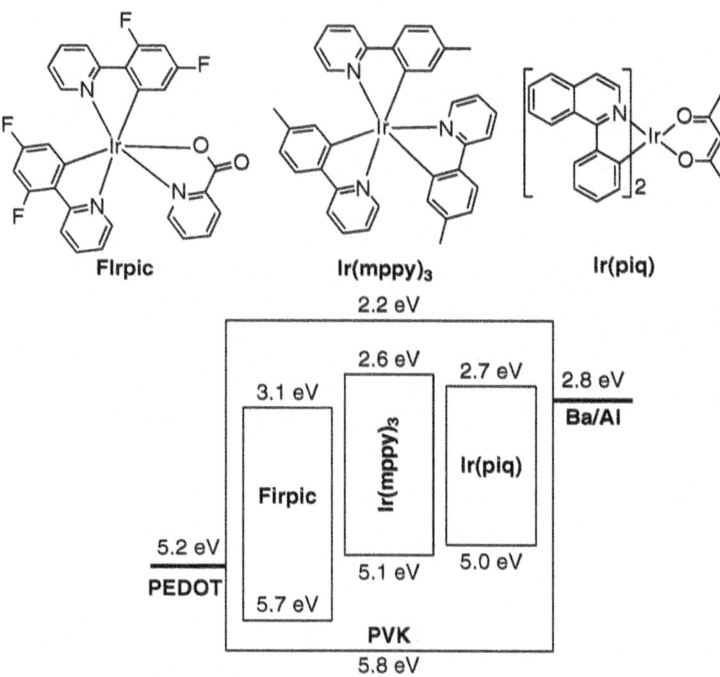

Fig. 10.4 Chemical structures of the doped phosphorescent dyes, the energy level schematic, and device structure of the investigated solution-processed WOLEDs

[10]. Figure 10.4 outlines the investigated device structures of the WOLEDs. RGB dyes, Ir(piq) for red emission, Ir(mppy)$_3$ for green emission, FIrpic for blue emission, and the host material PVK were dissolved in a chlorobenzene solution. PVK was selected as a host material due to its high-lying triplet state (3.0 eV), excellent film-forming property, high T_g, and hole-transport characteristics. OXD-7 was also used as a co-host to facilitate the electron transport, leading to the EML with good bipolar-transport property. A mixture was then obtained by mixing the host and the dopants together in the chlorobenzene solution through precisely controlling the doping ratios. These devices were fabricated with a single-active-layer architecture via solution processing, indicating a low-cost, straightforward approach toward efficient white phosphorescent OLEDs. An optimal device with triple doped phosphorescent dyes (FIpic:Ir(mppy)$_3$:Ir(piq) = 20:1:1) exhibited a maximal CE of 24.3 cd/A, a CCT of 5,010 K, and a CRI of 77. Due to the relatedly high driving voltage, a moderate η_p of 9.5 lm/W was obtained.

Though a simple device configuration was adopted to fabricate solution-processed polymer-based WOLEDs with moderate performances, the low η_p values limited its potential lighting application. Wu, Wong, and Choy et al. further promote the efficiency of solution-processed single-EML WOLEDs by simultaneous optimization of charge-carrier balance and luminous efficacy [11]. Two novel yellow phosphors were developed to be incorporated into a single EML to tune

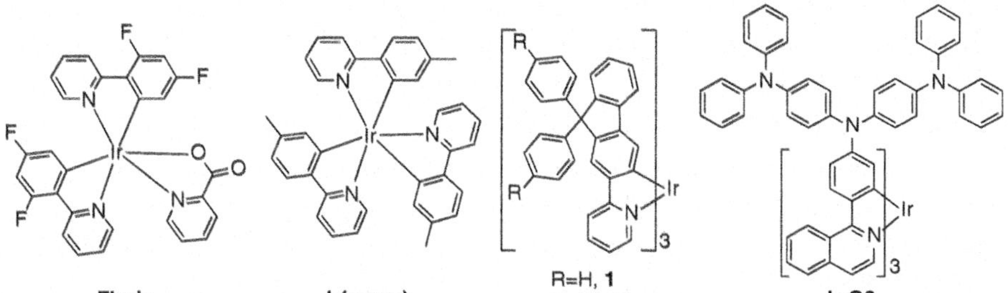

Fig. 10.5 Chemical structures of the blue, green, yellow, and red phosphorescent dyes

luminous efficacy of the WOLEDs (Fig. 10.5). FIrpic and yellow phosphors, *1* or *2*, were firstly doubly doped into a PVK:OXD-7 blending host matrix, to give an efficient single EML. WOLEDs employing this kind of EMLs with various doping ratios exhibited high CE values ca. 46.3 cd/A, corresponding to the η_p values ranging from 19 to 25.9 lm/W. However, the color quality of the above devices was moderate, with a CRI between 52 and 56. WOLEDs were further fabricated based on a blue, green, yellow, and red (BGYR) doping system to improve the color quality of the devices. An improved CRI of 76 could be achieved by using this doping system. A CE value as high as 45.5 cd/A, corresponding to a η_p of 20.2 lm/W, still can be achieved for a BGYR-based OLED. Moreover, if the anode buffer layer of PEDOT:PSS (4083) was in replacement of PEDOT:PSS P8000, a further enhanced peak η_p value of 37.4 lm/W in the forward-viewing direction was obtained for the above device. In order to obtain the total η_p, a factor of 1.34 (deduced by comparing the forward-viewing η_{ext} and the total η_{ext}) was applied to the forward-viewing efficiency, resulting in a total peak η_p of ~51 lm/W or a total η_p of 30 lm/W at a practically relevant brightness level. The obtained high efficiency in the solution-processed polymer-based WOLED is much higher than that of an incandescent bulb. These WOLEDs present a promising candidate for the white lighting sources given that the device lifetime is not a hindrance for the future material and device optimization.

10.3.1.2 Multi-EML

D'Andrade et al. also fabricated phosphorescent WOLEDs with multilayer EMLs, in which different layers emit different parts of the visible spectrum [12]. In fact, they had done this work report in 2002 prior to the aforementioned work with a single EML to give white phosphorescent OLEDs. Two typical device structures were reported in their work, and the corresponding device structures can be found in Table 10.1. Three phosphorescent dyes, FIrpic for blue emission, $Btp_2Ir(acac)$ for red emission, and $Bt_2Ir(acac)$ for yellow emission, were employed in a CBP host in a separated layer, respectively. Device 1 had a maximal η_p value of 6.4 ± 0.6 lm/W,

Table 10.1 Device structures of the phosphorescent WOLEDs

Device 1	Device 2
Cathode [LiF/Al]	Cathode [LiF/Al]
BCP (40 nm)	BCP (40 nm)
Yellow EML [8 wt% $Bt_2Ir(acac)$:CBP (2 nm)]	Red EML [8 wt% $Bt_2Ir(acac)$:CBP (2 nm)]
Red EML [8 wt% $Btp_2Ir(acac)$:CBP (2 nm)]	Hole/exciton blocker [BCP (3 nm)]
Blue EML [6 wt% FIrpic:CBP (20 nm)]	Blue EML [6 wt% FIrpic:CBP (20 nm)]
HTL [NPD (30 nm)]	HTL [NPD (30 nm)]
PEDOT:PSS (40 nm)	PEDOT:PSS (40 nm)
ITO/glass	ITO/glass

Table 10.2 Key chromaticity and efficiency values for WOLEDs

Device structure	Max η_{ext} [%]	Max CE [cd/A]	Max η_p [lm/W]	CRI	CIE(x, y) at 10 mA/cm^2
1	5.2 ± 0.5	11 ± 1	6.4 ± 0.6	83	(0.37, 0.40)
2	3.8 ± 0.4	6.1 ± 0.6	3.6 ± 0.4	50	(0.35, 0.36)

along with a maximal CE value of 11 ± 1 cd/A and a maximal η_{ext} of 5.2 ± 0.5 %. A high CRI of 83 was achieved for device 1, indicating the potential applications requiring a high-quality white light source. It is of interest that they placed a hole/exciton blocker between the red and blue EML; exciton formation and recombination regions could be tuned, resulting in most emission from the FIrpic-doped layer, thus leading to a matched white light spectrum mixed by the emissions from the blue and red phosphors. Device 2 performed a low CRI of 50, since green or yellow emission was absent; however, a CIE coordinate of (0.35, 0.36) was obtained, which is closer to the equi-energy ideal white light point of (0.333, 0.333) (see Table 10.2). Though yellow emission was absent, device 2 could find applications for flat-panel displays, because the human perception of the white light from the display panel will not be affected by the lack of yellow emission.

In order to further improve the efficiency of phosphorescent WOLEDs, Su and Kido et al. fabricated efficient blue and WOLEDs with a carrier- and exciton-confining structure [13]. A bipolar material DCzPPy was used as a host in their devices, and hole-transport host TCTA was utilized to assist the more balanced carrier recombination within the EML. The detailed information of the devices and materials are shown in Fig. 10.6. The blue device with a double emission layer has a η_p of 55 lm/W at the luminance of 100 cd/m^2. A very efficient WOLED was achieved with the insertion of ultrathin layers of orange-light-emitting PQ2Ir in the host material used for FIrpic. A record η_p of 53 lm/W at a display-relevant luminance of 100 cd/m^2 was obtained in the forward direction and rolls off slightly to 44 lm/W at an illumination-relevant luminance of 1,000 cd/m^2. Without the use of any out-coupling techniques, both are the highest values ever observed for WOLEDs. When a factor of 1.7 is applied to obtain the total efficiency from all

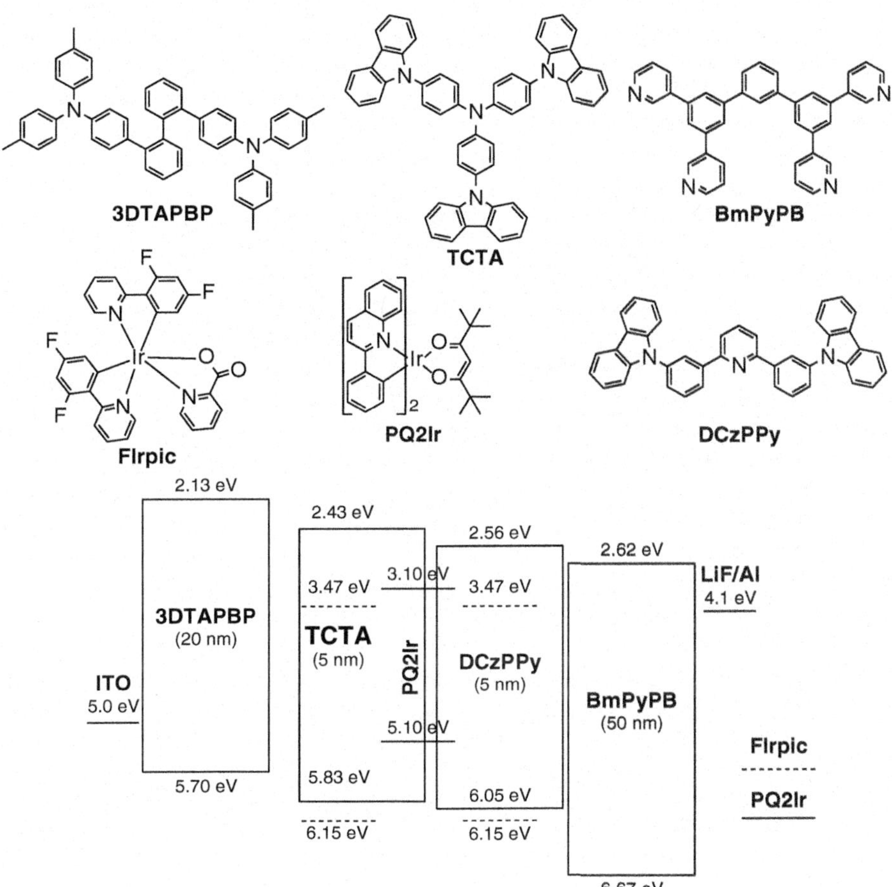

Fig. 10.6 Chemical structures of the materials, energy level diagrams, and device structures of the WOLEDs

directions, high efficiencies of 90.1 lm/W at 100 cd/m^2 and 74.8 lm/W at 1,000 cd/m^2 could be achieved for the developed WOLEDs. These values are comparable with the efficiency of a fluorescent tube, in spite of the absence of the light out-coupling technology.

Reineke et al. [14] reported a white phosphorescent OLED containing FIrpic, Ir(ppy)$_3$, and Ir(MDQ)$_2$(acac) as the emitting phosphors, where the blue emission layer was surrounded by red and green sublayers (2 nm of sublayer to suppress Förster-type energy transfer) of the emission layer to harvest unused excitons from the blue emission layer. Figure 10.7 outlines the energy level diagram of the devices in their study. They achieved a high record η_p of 90 lm/W at 1,000 cd/m^2 for a WOLED combining a high refractive index (1.78) glass substrate with a periodic light out-coupling structure. The obtained OLED has a CIE coordinate of (0.41,

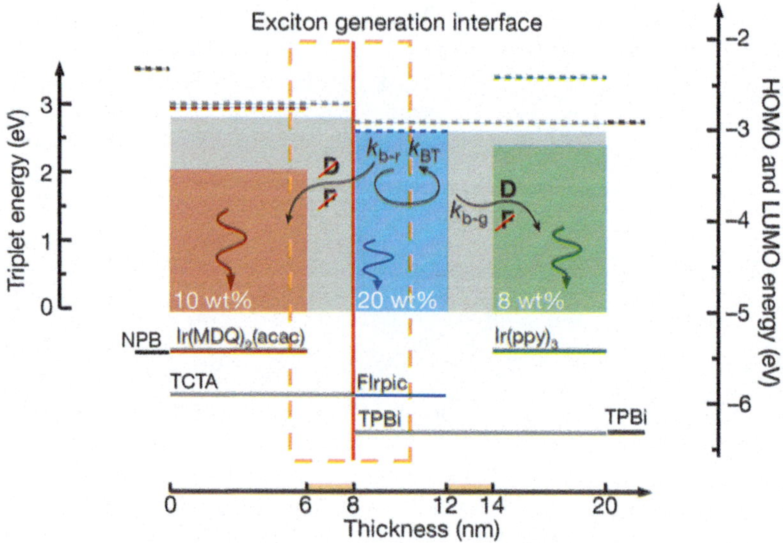

Fig. 10.7 Energy level diagrams of the highly efficient phosphorescent WOLEDs (Reprinted by permission from Macmillan Publishers Ltd: ref. [14], copyright 2009)

0.49) which indicates a yellow-like white light, owing to less contribution from blue emission. η_p of 124 lm/W could even be anticipated when using a 3D light extraction system with a more effective light out-coupling efficiency for their WOLEDs. This high η_p value will surpass the efficiency of a fluorescent tube, indicating a way for white organic light with efficiency beyond 100 lm/W.

10.3.2 Hybrid F/P WOLEDs

10.3.2.1 Small Molecule-Based F/P WOLEDs

Xu and Che et al. reported F/P WOLEDs using DNA as a blue-light source and (R-C^N^N)PtCl complexes as a yellow-green- or orange-light source (Fig. 10.8). For device I comprising complex 1 as an emitter, the peak η_{ext} and the peak η_p were determined to be 11 % and 12.6 lm/W, respectively. For device II containing complex 2 as an emitter, the peak η_{ext} and η_p were 11.8 % and 18.4 lm/W, respectively. The use of (RC^N^N)PtCl complexes provides an entry to a new family of electrophosphorescent platinum(II) emitters [15].

Novel bipolar molecule CPhBzIm (Fig. 10.9) was developed by Hung and Wong et al., and this material exhibits an excellent solid-state photoluminescence quantum yield of 69 %, triplet energy of 2.48 eV, and bipolar charge transport ability ($\mu_h \approx \mu_e \approx 10^{-6}$–$10^{-5}$ cm^2 V^{-1} s^{-1}). A non-doped deep-blue OLED was fabricated exhibiting promising performance with an η_{ext} of 3 % and a CIE coordinate of (0.16,

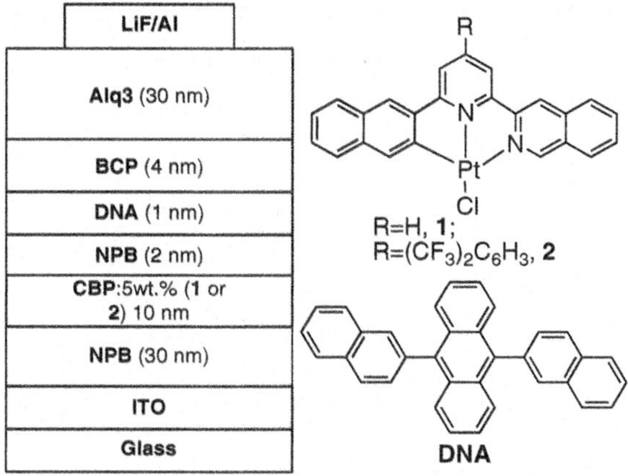

Fig. 10.8 Device structure of the F/P WOLEDs and chemical structures of blue emitter DNA and the orange phosphors

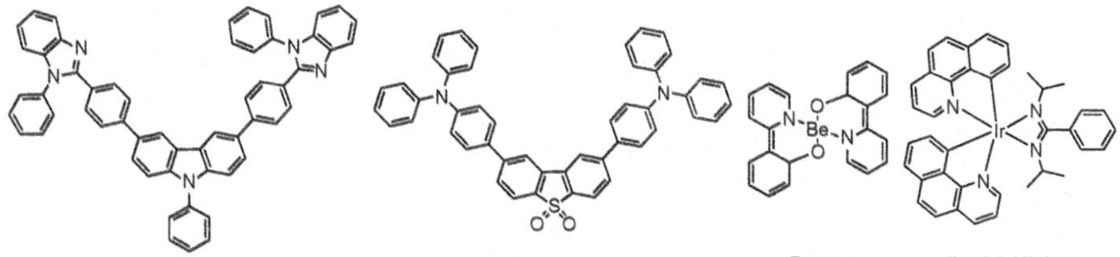

Fig. 10.9 Molecular structures of the materials used for F/P hybrid WOLEDs

0.05). When CPhBzIm served as a blue emitter and simultaneously a host material for a yellow-green phosphorescent dye $(pbi)_2Ir(acac)$, a simple single-doped, two-color-based WOLED was obtained with a peak η_p of 12.8 lm/W, corresponding to an η_{ext} of 7 %. The CIE coordinates of the WOLED are located at (0.31, 0.33), which is quite close to the ideal white point (0.333, 0.333) [16].

A high-efficiency and pure WOLED has been realized and reported by only doping one novel phosphorescent orange-light-emitting complex $(bzq)_2Ir(dipba)$ (see Fig. 10.9) into a deep-blue-emitting fluorescent complex $Bepp_2$ as an emissive layer [17]. A peak η_p of 48.8 lm/W and a peak η_{ext} of 27.8 % have been determined in a WOLED with a simple HTL-EML-ETL architecture. The η_p and η_{ext} at the applicable brightness of 1,000 cd/m^2 are 37.5 lm/W and 26.8 %, respectively.

F/P WOLEDs were fabricated combining an ideal sky-blue fluorophore, DADBT (Fig. 10.9), which has been proved as a highly efficient sky-blue fluorophore and an ideal host for orange phosphor by non-doped blue fluorescent device and an orange phosphor $Ir(2\text{-}phq)_3$. Tuning the doping concentrations could effectively separate and, respectively, utilize the singlet and triplet excitons in a single EML. The white device shows excellent electroluminescence performance with a low turn-on voltage of 2.4 V, a maximum total η_{ext} of 26.6 %, a CE of 53.5 cd/A, and a η_p of 67.2 lm/W, which further proves that high-efficiency F/P hybrid WOLEDs could be obtained by such a simple single-EML device structure using a doping concentration regulation strategy [18].

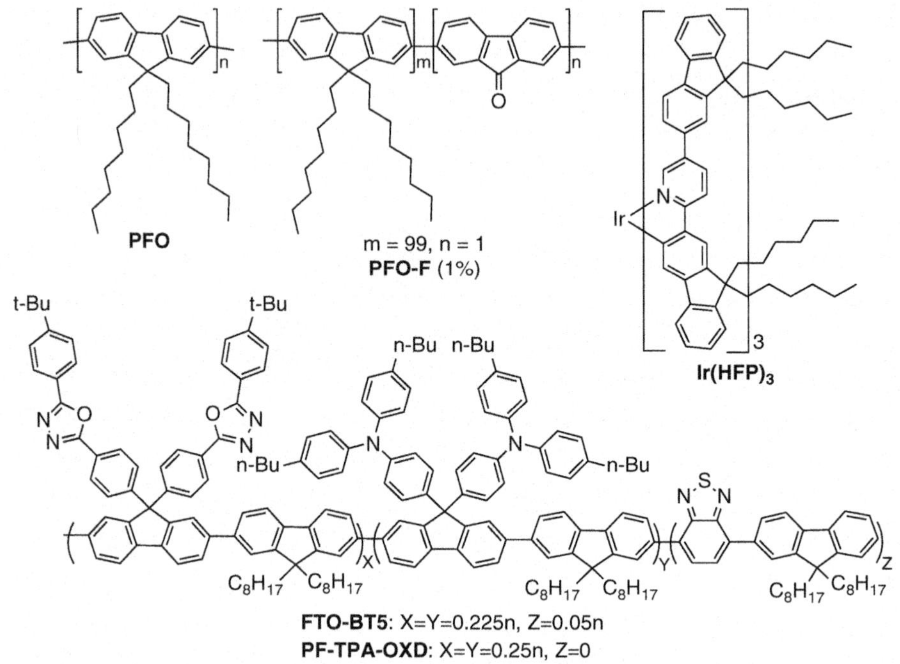

Fig. 10.10 Materials used for polymer-based F/P hybrid WOLEDs

10.3.2.2 F/P WOLEDs Using a F/P Polymeric Emitter

F/P WOLEDs were fabricated using the solution-processed method, by blending a fluorescent blue polymer emitter PFO, a yellow polymer emitter PFO-F, and a phosphorescent emitter $Ir(HFP)_3$ (Fig. 10.10) [19]. These devices have a simple active-layer configuration. Type I devices made from PFO co-spin coated with Ir $(HFP)_3$ have a CE of 4.3 cd/A. Type II devices made from blends of PFO, PFO-F, and $Ir(HFP)_3$ have a CE of 3 cd/A. The CIE coordinates, CCT, and CRI for type I and type II devices are (0.329, 0.321), 6,400 K, 92 and (0.352, 0.388), 4,600 K, 86, respectively. The EL spectra of these devices are quite stable under various current densities and the resulting luminance.

Efficient green-emitting polymer (FTO-BT5) was obtained through the incorporation of low-bandgap 2,1,3-benzothiadiazole (BT) moieties into the backbone of a blue-light-emitting polyfluorene copolymer (PF-TPA-OXD), which contains hole-transporting TPA and electron-transporting OXD pendent groups (Fig. 10.14) [20]. A highly efficient three-band F/P WOLED was prepared using PF-TPA-OXD as a blue-light-emitting host doped with green-emitting FTO-BT5 copolymer and a red phosphor Os(fppz). The WOLEDs reached a maximum η_{ext} of 4.1 % (8.3 cd/A) at a luminance of 402 cd/m^2 and a current density of 4.8 mA/cm^2. The resulting CIE coordinates located at around (0.33, 0.31) are quite stable under various current densities.

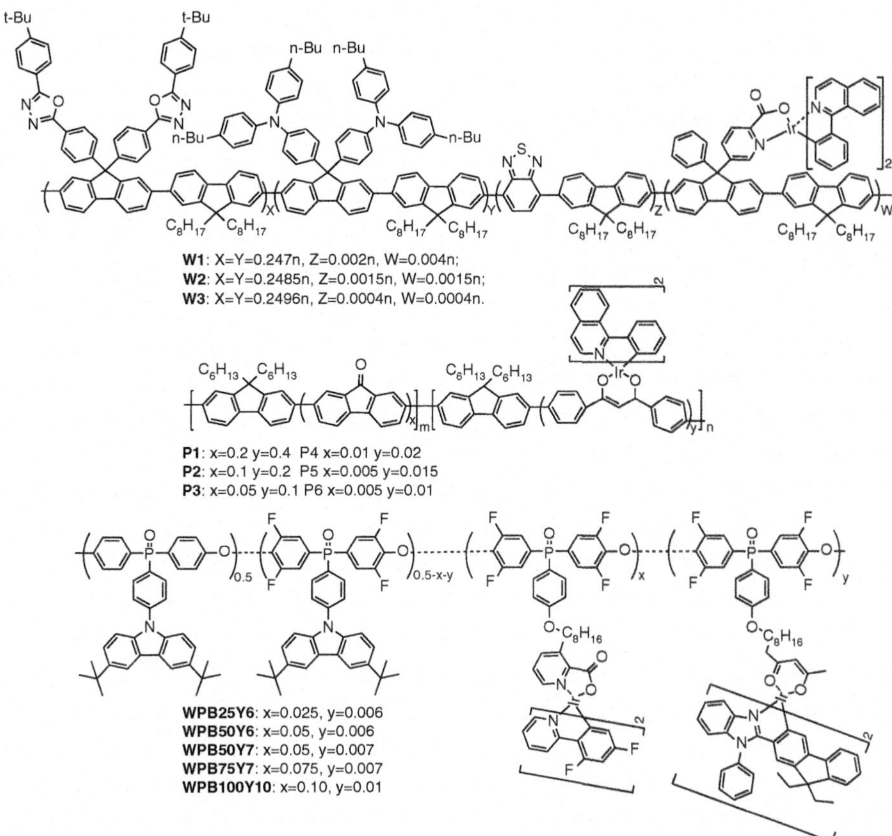

Fig. 10.11 White light-emitting polymers with covalently attached fluorescent and phosphorescent moieties

A green fluorophore and a red phosphor were covalently attached into the backbone and the side chains of the polyfluorene. Additionally, charge transport units like TPA and OXD pendants were also included to improve carrier injection and transport (Fig. 10.11) [21]. F/P hybrid WOLEDs were constructed with a configuration of ITO/PEDOT:PSS/W3/CsF/Al, in which the developed hybrid polymer W3 was used as a single emitter to give white emission. The WOLEDs exhibit a low turn-on voltage of 2.8 V and a luminance of ca. 103 cd/m^2 at below 6 V. The peak CE and η_p are 8.2 cd/A and 7.2 lm/W, respectively, with an almost constant CRI of 82 at various measured current densities.

Single fluorene-based copolymer to generate white light electroluminescence was also synthesized by Yang and Qin et al., where a green-emitting fluorenone and red-emitting iridium complex were incorporated into the copolymers (Fig. 10.11) [22]. Single-active-layer polymer light-emitting devices with a configuration of ITO/PEDOT/copolymer/CsF/Al have been fabricated. The P3-based device exhibits a maximum CE of 5.50 cd/A with a CIE coordinate of (0.32, 0.45) and a maximum luminance of 3,361 cd/m^2. The device from P5 shows a maximum CE of 3.25 cd/A with a CIE coordinate of (0.28, 0.32) and a maximum luminance of 1,015 cd/m^2.

Wang et al. reported efficient all-phosphorescent WOLEDs with a fluorinated poly(arylene ether phosphine oxide) backbone that has a high triplet energy and appropriate HOMO/LUMO levels, grafted with blue and yellow phosphors FIrpic and (fbi)Ir(acac) moieties (Fig. 10.11) [23]. Simultaneous blue and yellow triplet emissions were achieved to generate two-band white electroluminescence with a promising CE as high as 18.4 cd/A (8.5 lm/W, 7.1 %) and a CIE coordinate of (0.31, 0.43).

10.3.3 Excimer/Exciplex-Based Phosphorescent WOLEDs

The abovementioned methods are typically doping two or more phosphors into the host materials to generate two-band or three-band white light emission. In these systems, chromaticity stability is problematic due to the different efficiency of energy transfer from the host to each dopant in the mix, leading to an imbalanced white light emission or pronounced variations in color with luminance. One approach involving the utilization of the excimer or exciplex emission is attempted to produce WOLEDs with stable spectra under various electric fields. An excimer is identified as an excited state whose work function overlaps with two adjacent molecules of like composition, whereas an exciplex is noted as a state whose work function overlaps with neighboring, dissimilar molecules. Both excimer and exciplex lack a bound ground state; thus the color variation related to the energy transfer from the host or higher-energy dopants (blue emitter) to the lower-energy substances can be avoided. Due to the target topic of the utilization of organometallics in WOLEDs, we will only discuss the WOLEDs using the phosphorescent excimers or exciplexes as the emitters in the following. Unlike the pseudo-octahedral geometry associated with d^6 metal ions like Ir(III) and Pt (II) complexes associated with a d^8 metal ions are normally square planar. This square-planar structure leads to the strong face-to-face interaction between the neighboring molecules, easily enabling to cause excimer emission in the long-wavelength region. That is why most of the WOLEDs fabricated by organometallics are comprised of Pt(II) complexes. Through a rational design of molecular structures and/or a careful control of device conformation, high-quality white light electroluminescence can be achieved by simple doping and even single-doping device structures.

A novel approach to generate WOLEDs was demonstrated by Forrest et al. in 2002, with the use of excimer emission from the phosphor FPt1, coupled with FIrpic, or from FPt2 (Fig. 10.12) [24]. Square-planar Pt complexes easily form excimer in concentrated solutions and thin films. FPt1 and FPt2 have a broad emission spectrum spanning from 450 to 800 nm. The WOLED in a configuration of ITO/PEDOT:PSS/NPD/(CBP:FIrpic/FPt1)/BCP/LiF:Al has a maximum η_{ext} of

Fig. 10.12 Molecular structures of FIrpic, FPt1, and FPt2

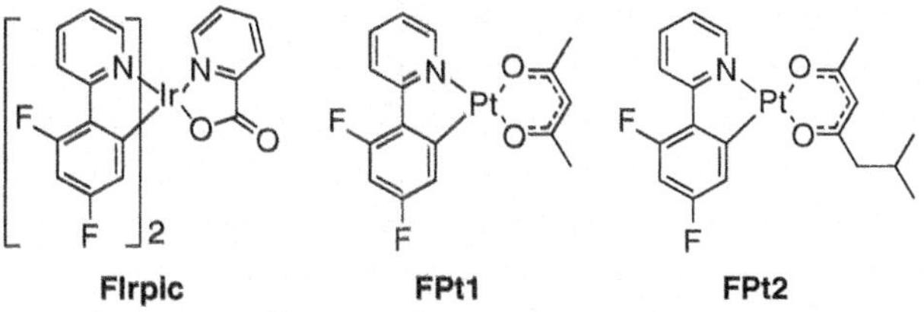

FIrpic FPt1 FPt2

4.0 ± 0.4 % corresponding to a CE of 9.2 ± 0.9 cd/A, a luminance of $31{,}000 \pm 3{,}000$ cd/m^2 at 16.6 V, a η_p of 4.4 ± 0.4 lm/W, a high CRI of 78, and a CIE coordinate of ca. (0.40, 0.44). FIrpic was doped to maintain the ratio of higher-energy emission to give a more balanced white color. When FIrpic was removed and FPt1 was replaced by FPt2 to form a single-doped EML, the emission from NPD was observed. As thus, the CIE coordinates shift to ca. (0.34, 0.35), giving relatively lower overall device efficiency.

Mixing excimer and exciplex emission was also demonstrated as an effective means to fabricate WOLEDs by Kalinowski, Cocchi, and Williams, et al. [25]. An emissive layer consisting of a Pt(II) phosphor named PtL^{2}Cl, co-doped with an electron-donating material m-MTDATA, was employed to yield white emission by mixing the intrinsic bluish-green emission of the mono-molecule of PtL^{2}Cl, the orange emission from the exciplex transition formed by PtL^{2}Cl and electron-donating m-MTDATA, and a more red-shifted emission related to the excimer transition formed between the similar adjacent PtL^{2}Cl molecules. CBP and TAPC were used to function as a spacer to tune the exciton recombination zones, respectively. The device diagram and adopted materials are shown in Fig. 10.13. Device I, having a CBP spacer, gives a more pronounced TPD emission peaking at 400 nm; this is attributed to the recombination zone locating at (TPD:PC)/CBP interface. This device has a CIE coordinate of (0.42, 0.47), a CRI of 84.02, and a CCT of 3,826 K at 22 V. In contrast, the TPD emission in device II with a TAPC spacer is not obvious, because the HOMO level of TAPC facilitates the hole penetration into the EML, enabling the exciton formation and recombination within the EML. As a result, device II has a CIE coordinate of (0.46, 0.45), a higher CRI of 90, and a CCT of 3,067 K. The TAPC spacer containing WOLED (device II) gives a maximum η_{ext} of (6.5 ± 0.5) % at low brightness levels ($L = 0.1$ cd/m^2) and (3.0 ± 0.5)% at a luminance $L = 500$ cd/m^2. The EL color stability under various voltages and luminances of device II is also better than device I.

Based on a blue emission cyclometalated Pt complex Pt-4 (Fig. 10.14), Li and Jabbour et al. reported an efficient single-doped WOLED in a configuration of ITO/PEDOT:PSS/TCTA/26mCPy:Pt-4(8 %)/BCP/CsF/Al [26]. The intrinsic blue emission from the isolated emission of Pt-4 and the red-shifted excimer emission from the interaction between the adjacent molecules render this WOLED with quite pure emission with a CIE coordinate of (0.33, 0.36). The η_p and η_{ext} of this device at 100 cd/m^2 are 7.3 lm/W and 9.3 %, respectively.

Cocchi and Williams also reported a phosphor PtL^{30}Cl (Fig. 10.14) with a tridentate N^C^N coordinating ligand similar to Pt-4 [27]. In order to achieve a

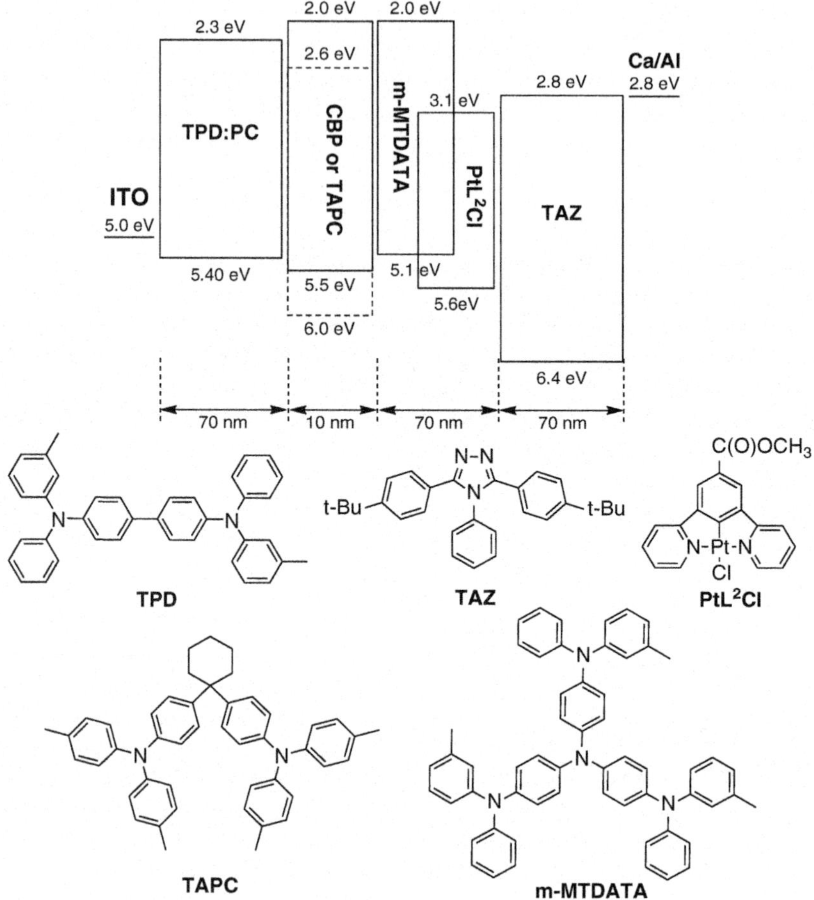

Fig. 10.13 Energy level diagrams and the adopted materials for the excimer/exciplex-based WOLEDs

Fig. 10.14 Molecular structures of Pt-4 and $PtL^{30}Cl$

more blue-shifted monomer emission, the electron-donating amino groups were introduced to the 4-position of the pyridyl rings to destabilize the LUMO without obviously affecting the HOMO. Thus, in relative to Pt-4, a wider bandgap for $PtL^{30}Cl$ was anticipated. The combination of monomer emission of the isolated $PtL^{30}Cl$ and the red-shifted emission emanating from the triplet excimer renders $PtL^{30}Cl$ of potential application for fabricating WOLEDs. A reasonable choice of

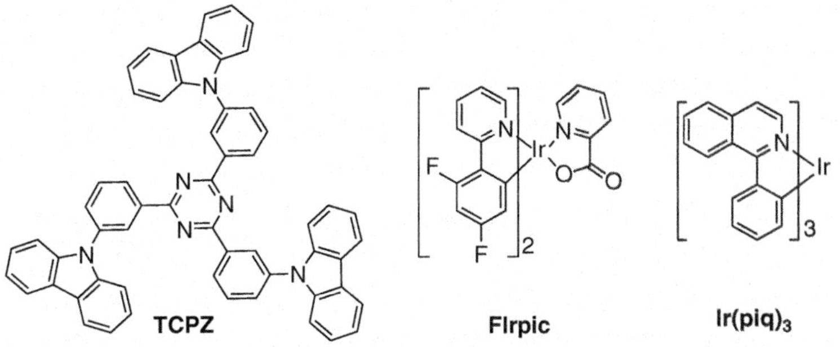

Fig. 10.15 Molecular structures of TCPZ and the adopted phosphorescent dopant

host materials (TCTA:TCP) and a careful adjustment of the doping ratio (20 wt.%) make the realization of PtL^{30}Cl-based efficient single-doped WOLEDs with a CRI of 87, a CIE coordinate of (0.36, 0.37), an η_{ext} of 3.7 %, and a CE of 7.4 cd/A.

It is of interest to demonstrate an effective means by Su et al. to obtain efficient WOLEDs with high CRIs by utilizing the exciplex emission emanating from the host material and the dopant [28]. Figure 10.15 depicts the structures of the EML materials used. TCPZ, a newly developed host material, is capable of small singlet-triplet exchange energy (ΔE_{ST}) and a low-lying LUMO level, which are favorable to lower the device driving voltage. When the conventional blue phosphor FIrpic was doped into this host material to serve as an EML in the OLEDs, a new band peaking at 535 nm emerges during the electroluminescent process, whereas this new band is not observed during their photoluminescence process. $Ir(piq)_3$ was inserted accompanied with FIrpic to fabricate WOLEDs in a configuration of ITO/TPDPES:TBPAH (20 nm)/TAPC (30 nm)/TCPZ:11 wt.% FIrpic (4.75 nm)/ TCPZ:4 wt.% $Ir(piq)_3$ (0.5 nm)/TCPZ:11 wt.% FIrpic (4.75 nm)/BmPyPB (40 nm)/ LiF (0.5 nm)/Al. This two-dopant three-band WOLED has a high CRI of 82 at the applicable brightness of 100 and 1,000 cd/m^2. An η_{ext} of 5.4 % and a η_p of 9.4 lm/W at 100 cd/m^2 were obtained for this novel OLED.

10.3.4 Tandem Phosphorescent WOLEDs

A tandem OLED is consisting of two or more vertically stacked sub-OLEDs linked by the charge generation layers (CGLs). The typical characteristic of a tandem OLED is that the current efficiency and luminance can scale linearly with the number of emitting units (sub-OLEDs). Generally, the driving voltage of a tandem OLED also increases linearly as a function of the number of the emitting units. Thereby, power conversion efficiency of a tandem OLED compared to a single OLED is not enhanced; in other words, a tandem OLED cannot save more electricity energy relative to a single one. However, tandem OLEDs are still of particular importance because of the potential long device lifetime which is associated with the suppression of thermal degradation due to the reduced excessive current.

The CGL is commonly comprised of a p-n heterojunction that is inversely linked between two emitting subunits to generate carriers (electrons and holes) under electricity excitation. The choice of CGLs plays an important role of achieving an excellent tandem OLED. Two basic requirements must be satisfied for a CGL: (1) electric potential drop across the CGL should be almost neglected to avoid of the extra increase of driving voltage of the whole tandem device and (2) the CGL should be homogeneous and transparent over the wavelength range of whole visible region in order not to block any emissive light. We are here only discussing the tandem WOLEDs containing organometallic phosphors.

Chen et al. reported two kinds of stacked WOLEDs employing Alq_3:20 wt. % Mg/MoO_3 as a charge generation layer (see Fig. 10.16) [29]. This CGL is transparent in a range between 400 and 800 nm. Tandem WOLEDs are fabricated by connecting the blue fluorescent unit and the orange phosphorescent emission

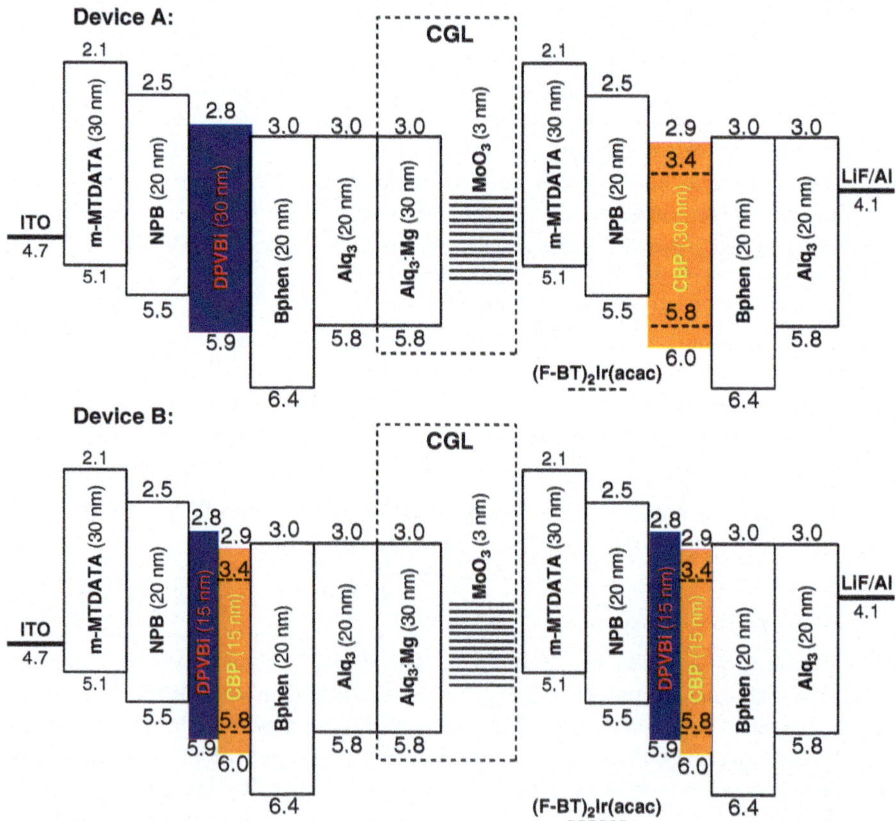

Fig. 10.16 The electronic level (eV) configurations of the developed stacked WOLEDs containing organometallic phosphors

unit separately or by connecting double-white light emission units in series. DPVBi and $(F\text{-}BT)_2Ir(acac)$ were used as a fluorescent blue emitter and a phosphorescent orange emitter, respectively. The former device with CGL-separated blue and orange emission units exhibited a better color stability under various voltages and brightness. Moreover, the efficiency of the former device is also higher than the latter. This difference of device performance is attributed to the avoidance of the movement of charge recombination zone and elimination of the Dexter energy transfer between blue and orange emission layers occurring in the latter. The former WOLED has a CE value of 35.9 cd/A and a CRI > 70 at 1,000 cd/m^2.

Based on the similar device architecture of the abovementioned device A using Alq_3:Mg/MoO_3 as a charge generation layer, Chen et al. further demonstrated charge-carrier separation takes place only in the MoO_3 layer [30]. Enhancement of device performance was obtained by simply adjusting the thickness of MoO_3. The stacked WOLED with a CE value of 39.2 cd/A has excellent color stability with the CIE coordinate only changing from (0.407, 0.405) to (0.398, 0.397) when luminance increases from 22 to 10,000 cd m^{-2}.

By utilizing 2,9-dimethyl-4,7-diphenyl-1,10-phenanthroline (BCP):Li/MoO_3 as an effective CGL, Ma et al. demonstrated that extremely high-efficiency tandem phosphorescent WOLEDs could be realized by using single-emitting-layer device configurations [31]. This stacked device achieved maximum forward-viewing CE of 110.9 cd A^{-1} and η_{ext} of 43.3 % at 1 $\mu A/cm^2$ with a stable CIE coordinate of (0.34, 0.41). The emitting phosphors in their study are FIrpic and $(fbi)_2Ir(acac)$ (Fig. 10.3).

Ma et al. adopted interface-modified C60/pentacene organic heterojunction accompanied with TPBI:Li_2CO_3 and TCTA:MoO_3 as charge generation layers to fabricate efficient phosphorescent WOLEDs [32]. Figure 10.17 shows the investigated device structures with or without C60/pentacene bilayer. It was demonstrated that C60/pentacene bilayer not only enhances the efficiency of the tandem devices but also lowers the device driving voltage, accordingly, unlike other tandem devices with an almost unchanged η_p relative to the single-unit device, and

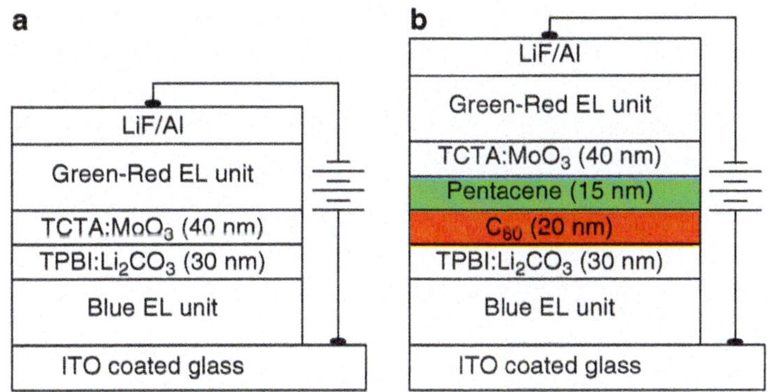

Fig. 10.17 Tandem device structures *w* (**a**) or *w/o* (**b**) C60/pentacene bilayers

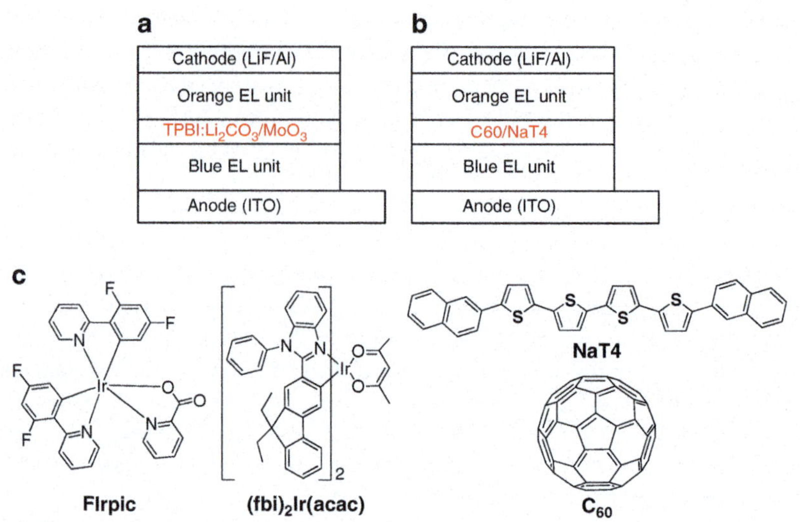

Fig. 10.18 Schematic diagrams of the tandem WOLEDs *w* (**a**) or *w/o* (**b**) C60/NaT4 bilayers; chemical structures (**c**) of key materials used are also shown

improved η_p was obtained for this tandem WOLEDs. The resulting tandem WOLED exhibits a maximum η_p of 53.8 lm/W and a maximum CE of 101.5 cd/A without any out-coupling techniques. More importantly, the efficiency roll-off is significantly suppressed, remaining η_p of 53 lm/W at 100 cd/m^2 and 45 lm/W at 1,000 cd/m^2. Meanwhile, they also demonstrated a high-efficiency white top-emitting OLEDs by introducing tandem structure using C60/pentacene as the CGL [33]. The resultant tandem top emissive WOLED possesses a CIE coordinate of (0.39, 0.43) and a CRI of 70, which retains high color stability toward driving voltage. The maximum forward-viewing η_{ext} and CE of the device are 16.9 % and 41.1 cd/A, respectively.

Furthermore, Ma et al. developed a new organic heterojunction comprised of an n-type C60 layer and a p-type 5,5‴-bis(naphth-2-yl)-2,2′:5′,2″:5″,2‴-quaterthiophene (NaT4) layer to replace the functionality of C60/pentacene bilayers. Figure 10.18 shows their investigated device structures. C60/NaT4 bilayer was also demonstrated with the ability to boost the η_p of tandem devices, to reduce efficiency roll-off, and to prolong the operational stability. The C60/NaT4-containing device has a maximum CE of 111.3 cd/A, a maximum η_p of 50.5 lm/W, and a maximum η_{ext} of 38.7 % [34].

10.4 Conclusions

In the past decade, WOLEDs, especially those containing organometallic phosphors, have seen a huge development under the continuous effort from the researchers in both academia and industry. Simultaneous improvement of device physics, device structures, and new excellent materials has put this area forward. Various device structures, e.g., EMLs with doping or nondoping, single layer or multilayer, and all phosphors or fluorescent/phosphorescent hybrid composition, are utilized to fabricate WOLEDs toward high-quality white emission with high efficiencies, ideal CIE coordinates, high CRIs, and suitable CCTs. The inner physical mechanism investigation of the WOLEDs is also favorable for us to rationally design and fabricate a white light-emitting device. It is worth noting that materials play the most important roles of further improving the device performance of WOLEDs. All functional materials like electrode materials, carrier injection materials, carrier transport materials, hosts, and the emitters in WOLEDs are required to be optimally chosen to maximize the performance. To date, phosphorescent WOLEDs capable of full utilization of the generated singlet and triplet excitons have already exhibited their potential application as a new class of artificial white light-emitting sources. Compared to the fluorescent tubes and the inorganic LEDs, one of the most challenging issues for the commercial application of the phosphorescent WOLEDs should be the device lifetime, which is of particular importance, however, is infrequently referred in the research papers. Future work on the achievement of phosphorescent WOLEDs with a long lifetime, high efficiency, and good color quality would really bring the realization of WOLEDs as the alternative next-generation lighting sources.

References

1. Tang CW, VanSlyke SA (1987) Organic electroluminescent diodes. Appl Phys Lett 51(12):913–915
2. Kido J, Hongawa K, Okuyama K, Nagai K (1994) White light-emitting organic electroluminescent devices using the poly(N-vinylcarbazole) emitter layer doped with three fluorescent dyes. Appl Phys Lett 64(7):815–817
3. D'Andrade BW, Holmes RJ, Forrest SR (2004) Efficient organic electrophosphorescent white-light-emitting device with a triple doped emissive layer. Adv Mater 16(7):624–628
4. Gong SL, Chen YH, Luo JJ, Yang CL, Zhong C, Qin JG, Ma DG (2011) Bipolar tetraarylsilanes as universal hosts for blue, green, orange, and white electrophosphorescence with high efficiency and low efficiency roll-off. Adv Funct Mater 21(6):1168–1178
5. Chien CH, Kung LR, Wu CH, Shu CF, Chang SY, Chi Y (2008) A solution-processable bipolar molecular glass as a host material for white electrophosphorescent devices. J Mater Chem 18(29):3461–3466
6. Yook KS, Lee JY (2012) Solution processed multilayer deep blue and white phosphorescent organic light-emitting diodes using an alcohol soluble bipolar host and phosphorescent dopant materials. J Mater Chem 22(29):14546–14550
7. Lin MS, Yang SJ, Chang HW, Huang YH, Tsai YT, Wu CC, Chou SH, Mondal E, Wong KT (2012) Incorporation of a CN group into mCP: a new bipolar host material for highly efficient blue and white electrophosphorescent devices. J Mater Chem 22(31):16114–16120
8. Hou LD, Duan LA, Qiao JA, Zhang DQ, Wang LD, Cao Y, Qiu Y (2011) Efficient solution-processed phosphor-sensitized single-emitting-layer white organic light-emitting devices: fabrication, characteristics, and transient analysis of energy transfer. J Mater Chem 21(14):5312–5318

9. Wang Q, Ding JQ, Ma DG, Cheng YX, Wang LX, Jing XB, Wang FS (2009) Harvesting excitons via two parallel channels for efficient white organic LEDs with nearly 100 % internal quantum efficiency: fabrication and emission-mechanism analysis. Adv Funct Mater 19(1):84–95
10. Wu HB, Zou JH, Liu F, Wang L, Mikhailovsky A, Bazan GC, Yang W, Cao Y (2008) Efficient single active layer electrophosphorescent white polymer light-emitting diodes. Adv Mater 20(4):696–702
11. Wu HB, Zou JH, Wu H, Lam CS, Wang CD, Zhu J, Zhong CM, Hu SJ, Ho CL, Zhou GJ, Choy WCH, Peng JB, Cao Y, Wong WY (2011) Simultaneous optimization of charge-carrier balance and luminous efficacy in highly efficient white polymer light-emitting devices. Adv Mater 23(26):2976–2980
12. D'Andrade BW, Thompson ME, Forrest SR (2002) Controlling exciton diffusion in multilayer white phosphorescent organic light emitting devices. Adv Mater 14(2):147–151
13. Su SJ, Gonmori E, Sasabe H, Kido J (2008) Highly efficient organic blue-and white-light-emitting devices having a carrier- and exciton-confining structure for reduced efficiency roll-off. Adv Mater 20(21):4189–4194
14. Reineke S, Lindner F, Schwartz G, Seidler N, Walzer K, Luessem B, Leo K (2009) White organic light-emitting diodes with fluorescent tube efficiency. Nature 459(7244):234–239
15. Yan BP, Cheung CCC, Kui SCF, Xiang HF, Roy VAL, Xu SJ, Che CM (2007) Efficient white organic light-emitting devices based on phosphorescent platinum (II)/fluorescent dual-emitting layers. Adv Mater 19(21):3599–3603
16. Hung WY, Chi LC, Chen WJ, Chen YM, Chou SH, Wong KT (2010) A new benzimidazole/carbazole hybrid bipolar material for highly efficient deep-blue electrofluorescence, yellow-green electrophosphorescence, and two-color-based white OLEDs. J Mater Chem 20(45):10113–10119
17. Peng T, Yang Y, Bi H, Liu Y, Hou ZM, Wang Y (2011) Highly efficient white organic electroluminescence device based on a phosphorescent orange material doped in a blue host emitter. J Mater Chem 21(11):3551–3553
18. Ye J, Zheng CJ, Ou XM, Zhang XH, Fung MK, Lee CS (2012) Management of singlet and triplet excitons in a single emission layer: a simple approach for a high-efficiency fluorescence/phosphorescence hybrid white organic light-emitting device. Adv Mater 24(25):3410–3414
19. Gong X, Ma WL, Ostrowski JC, Bazan GC, Moses D, Heeger AJ (2004) White electrophosphorescence from semiconducting polymer blends. Adv Mater 16(7):615–619
20. Wu FI, Shih PI, Tseng YH, Shu CF, Tung YL, Chi Y (2007) Highly efficient white-electrophosphorescent devices based on polyfluorene copolymers containing charge-transporting pendent units. J Mater Chem 17(2):167–173
21. Wu FI, Yang XH, Neher D, Dodda R, Tseng YH, Shu CF (2007) Efficient white-electrophosphorescent devices based on a single polyfluorene copolymer. Adv Funct Mater 17(7):1085–1092
22. Zhang K, Chen Z, Yang CL, Tao YT, Zou Y, Qin JG, Cao Y (2008) Stable white electroluminescence from single fluorene-based copolymers: using fluorenone as the green fluorophore and an iridium complex as the red phosphor on the main chain. J Mater Chem 18(3):291–298
23. Shao SY, Ding JQ, Wang LX, Jing XB, Wang FS (2012) White electroluminescence from all-phosphorescent single polymers on a fluorinated poly(arylene ether phosphine oxide) backbone simultaneously grafted with blue and yellow phosphors. J Am Chem Soc 134(50):20290–20293
24. D'Andrade BW, Brooks J, Adamovich V, Thompson ME, Forrest SR (2002) White light emission using triplet excimers in electrophosphorescent organic light-emitting devices. Adv Mater 14(15):1032–1036
25. Kalinowski J, Cocchi M, Virgili D, Tattori V, Williams JAG (2007) Mixing of excimer and exciplex emission: a new way to improve white light emitting organic electrophosphorescent diodes. Adv Mater 19(22):4000–4005

26. Yang XH, Wang ZX, Madakuni S, Li J, Jabbour GE (2008) Efficient blue- and white-emitting electrophosphorescent devices based on platinum(II) [1,3-difluoro-4,6-di(2-pyridinyl)benzene] chloride. Adv Mater 20(12):2405–2409
27. Murphy L, Brulatti P, Fattori V, Cocchi M, Williams JAG (2012) Blue-shifting the monomer and excimer phosphorescence of tridentate cyclometallated platinum(II) complexes for optimal white-light OLEDs. Chem Commun 48(47):5817–5819
28. Su SJ, Cai C, Takamatsu J, Kido J (2012) A host material with a small singlet-triplet exchange energy for phosphorescent organic light-emitting diodes: guest, host, and exciplex emission. Org Electron 13(10):1937–1947
29. Chen P, Xue Q, Xie W, Duan Y, Xie G, Zhao Y, Hou J, Liu S, Zhang L, Li B (2008) Color-stable and efficient stacked white organic light-emitting devices comprising blue fluorescent and orange phosphorescent emissive units. Appl Phys Lett 93(15):153508
30. Chen P, Xue Q, Xie WF, Xie GH, Duan Y, Zhao Y, Liu SY, Zhang LY, Li B (2009) Influence of interlayer on the performance of stacked white organic light-emitting devices. Appl Phys Lett 95(12):123307
31. Wang Q, Ding J, Zhang Z, Ma D, Cheng Y, Wang L, Wang F (2009) A high-performance tandem white organic light-emitting diode combining highly effective white-units and their interconnection layer. J Appl Phys 105(7):076101
32. Chen Y, Chen J, Ma D, Yan D, Wang L (2011) Tandem white phosphorescent organic light-emitting diodes based on interface-modified C-60/pentacene organic heterojunction as charge generation layer. Appl Phys Lett 99(10):103304
33. Wang Q, Chen Y, Chen J, Ma D (2012) White top-emitting organic light-emitting diodes employing tandem structure. Appl Phys Lett 101(13):133302
34. Chen Y, Tian H, Chen J, Geng Y, Yan D, Wang L, Ma D (2012) Highly efficient tandem white organic light-emitting diodes based upon C-60/NaT4 organic heterojunction as charge generation layer. J Mater Chem 22(17):8492–8498

Chapter 11
Kinetics and Mechanisms of Reduction of Protons and Carbon Dioxide Catalyzed by Metal Complexes and Nanoparticles

Shunichi Fukuzumi, Tomoyoshi Suenobu, and Yusuke Yamada

Abstract Kinetics and mechanisms of reduction of protons and CO_2 catalyzed by metal complexes and nanoparticles have been discussed in this chapter. Kinetic studies including deuterium kinetic isotope effects on heterogeneous catalysts for hydrogen evolution by proton reduction have been demonstrated to provide essential mechanistic information on bond cleavage and formation associated with electron transfer. The rate-determining steps in the catalytic cycles are clarified by kinetic studies, providing valuable information on observable intermediates. The most important intermediates in the catalytic reduction of protons and CO_2 are metal-hydride complexes, which can reduce protons and CO_2 to produce hydrogen and formic acid, respectively. The catalytic interconversion between hydrogen and a hydrogen storage compound has been made possible by changing pH, providing a convenient hydrogen-on-demand system in which hydrogen gas can be stored as a liquid (e.g., formic acid) or solid form (NADH) and hydrogen can be produced by the catalytic decomposition of the hydrogen storage compound.

Keywords Proton reduction • CO_2 reduction • Metal hydride • Kinetics • Nanoparticles

11.1 Introduction

The global annual energy consumption is increasing rapidly, whereas fossil fuels, which are currently the primary source of the energy, will be depleted eventually in the future. In addition, the burning of fossil fuels releases large amounts of carbon dioxide (CO_2) to the atmosphere, leading to global warming. Before the depletion of fossil fuels, which are the products of photosynthesis, artificial photosynthesis should be realized to produce solar fuels and to fix CO_2 using solar energy, which is

S. Fukuzumi (✉) • T. Suenobu • Y. Yamada
Department of Material and Life Science, Division of Advanced Science and Biotechnology, Graduate School of Engineering, Osaka University, ALCA, Japan Science and Technology Agency (JST), Suita, Osaka 565-0871, Japan
e-mail: fukuzumi@chem.eng.osaka-u.ac.jp

W.-Y. Wong (ed.), *Organometallics and Related Molecules for Energy Conversion*, Green Chemistry and Sustainable Technology, DOI 10.1007/978-3-662-46054-2_11

the most abundant on the earth [1–4]. In an ideal artificial photosynthesis, water is split using solar energy into hydrogen (H_2) and dioxygen (O_2), which in turn can be converted into water releasing its energy as electricity in H_2 fuel cells [5–10]. Once H_2 is formed from water, CO_2 can be reduced by H_2 to produce carbon monoxide or formate, which can be further reduced to methanol and methane [11–16]. Such reduction of CO_2 by H_2 provides the possibility of storing solar energy as these C1 compounds, contributing to reduced emission of CO_2. There have so far been many reviews on each step of artificial photosynthesis, i.e., light harvesting and charge separation [17–23], proton reduction [24–28], CO_2 reduction [29–33], and water oxidation [34–42]. However, the detailed catalytic mechanisms of reduction of protons for hydrogen evolution and reduction of CO_2 have yet to be fully clarified. Kinetic studies certainly help in understanding catalytic mechanisms of proton reduction and CO_2 reduction, which are catalyzed by metal complexes and nanoparticles. Thus, in this review, we have chosen to focus on kinetics and mechanisms of reduction of protons and CO_2 catalyzed by metal complexes and nanoparticles.

11.2 Mechanisms of Catalytic Hydrogen Evolution

11.2.1 Cobalt Hydride Complexes

Platinum is currently used as the most efficient catalyst for the reduction of protons to H_2 [43]. Because of the scarcity and high cost of platinum, replacement of platinum by more earth-abundant metals such as cobalt as proton reduction catalysts has attracted much attention [44–48]. In order to develop efficient proton reduction catalysts, it is quite important to elucidate mechanisms of the catalytic proton reduction. The mechanism of the proton reduction by Co complexes has been clarified using a dinuclear Co complex with bis(pyridyl)pyrazolato (bpp^-) and terpyridine (trpy) ligands, $[Co^{III}_2(trpy)_2(\mu\text{-bpp})(OH)(OH_2)](PF_6)_4$ (**1**$(PF_6)_4$, Fig. 11.1) [49].

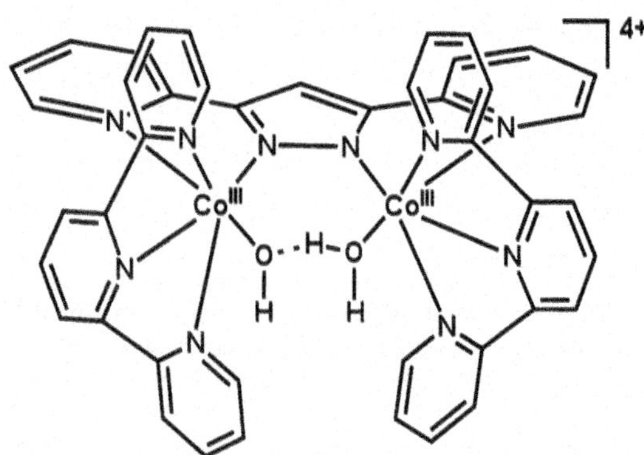

Fig. 11.1 Structure of $[Co^{III}_2(trpy)_2(\mu\text{-bpp})(OH)(OH_2)]^{4+}$ ($\mathbf{1}^{4+}$)

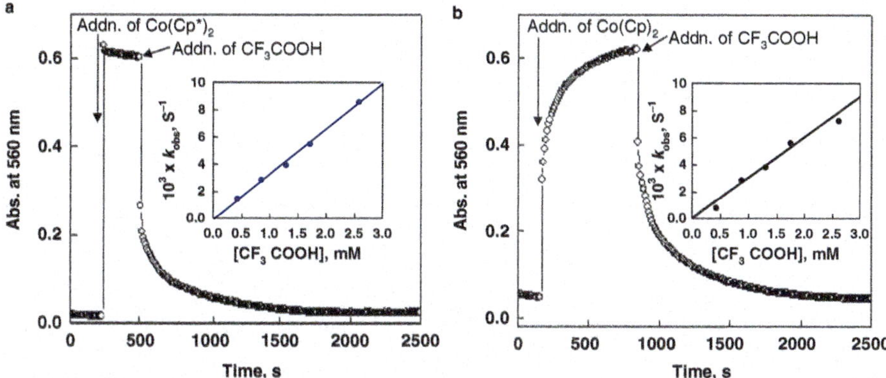

Fig. 11.2 Time profiles of absorbance at 560 nm due to (**a**) the Co^ICo^I complex (**1**, 0.087 mM) and (**b**) the $Co^{II}Co^I$ complex ($\mathbf{1}^+$, 0.087 mM) with CF_3COOH (0.87 mM) in deaerated MeCN at 298 K. *Insets*: Plots of k_{obs} vs. concentration of CF_3COOH for the second step reaction of (**a**) **1** and (**b**) $\mathbf{1}^+$ with CF_3COOH (Reprinted with the permission from Ref. [49]. Copyright 2011 American Chemical Society)

The dinuclear cobalt(III) complex ($\mathbf{1}^{4+}$) undergoes step-by-step reduction by decamethylcobaltocene ($Co(Cp^*)_2$, $Cp^* = \eta^5$-pentamethylcyclopentadienyl) to **1**, because the one-electron oxidation potential of $Co(Cp^*)_2$ ($E_{ox} = -1.53$ V vs. SCE) is lower than the one-electron reduction potential of $\mathbf{1}^+$ ($E_{red} = -1.09$ V vs. SCE) [49]. Addition of 4 equiv. of $Co(Cp^*)_2$ to a deaerated MeCN solution containing $\mathbf{1}^{4+}$ resulted in rapid formation of **1**, which exhibits visible and NIR absorption bands at 560 and 1,050 nm [49]. Addition of 10 equiv. of CF_3COOH to the MeCN solution of **1** resulted in the two-step decay of absorbance at 560 nm due to **1** (Fig. 11.2a) [49]. The two-step reaction of **1** with CF_3COOH suggests that the protonation of **1** affords a hydride complex, $[(Co^{III}\text{–}H)(Co^{III}\text{–}H)]^{2+}$, which is in equilibrium with **1**, followed by the reaction of the hydride complex with protons to produce H_2 [49]. The first-order dependence of k_{obs} with respect to the concentration of CF_3COOH in the second step indicates that the protonation of the Co(III)–H moiety is the rate-determining step to produce H_2 [49].

When cobaltocene ($Co(Cp)_2$, $Cp = \eta^5$-cyclopentadienyl) was used as a reductant, $\mathbf{1}^+$ was obtained by the three-electron reduction of $\mathbf{1}^{4+}$ with $Co(Cp)_2$, because the one-electron oxidation potential of $Co(Cp)_2$ ($E_{ox} = -0.9$ V vs. SCE) is more negative than the E_{red} value of $\mathbf{1}^{2+}$ (−0.78 V vs. SCE) but less negative than the E_{red} value of $\mathbf{1}^+$ (−1.09 V vs. SCE) [49]. The reaction of $\mathbf{1}^+$ with CF_3COOH also exhibited a two-step decay (Fig. 11.2b) [49].

Based on the two-step kinetics, the reaction mechanism of H_2 production from $\mathbf{1}^+$ is shown in Scheme 11.1 [49]. Three-electron reduction of $\mathbf{1}^{4+}$ by 3 equiv. of Co $(Cp)_2$ occurs to produce $\mathbf{1}^+$ (Eq. 11.1). $\mathbf{1}^+$ is protonated by CF_3COOH to produce the hydride complex ($[Co^{II}Co^{III}\text{–}H]^{2+}$), which is in equilibrium with $\mathbf{1}^+$ (Eq. 11.2, the first step in Fig. 11.2b). The formation of hydride complex was confirmed by the ^1H NMR spectra which exhibit a typical Co(III)–H peak at $\delta = -8.64$ ppm [49]. The

Scheme 11.1 Reaction mechanism of H_2 production with $[Co^{II}Co^{I}]^{+}$

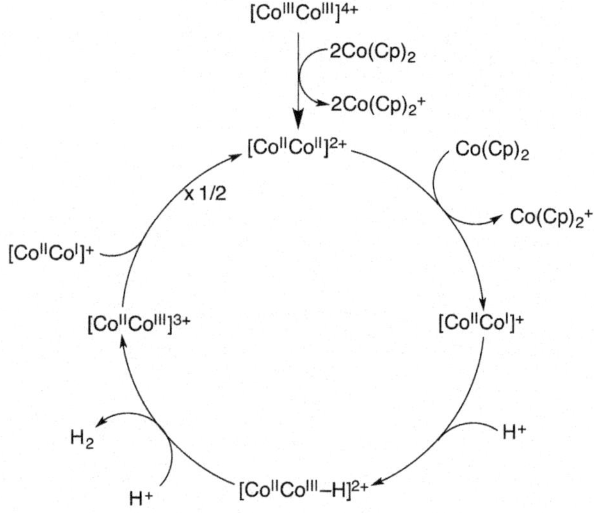

hydride complex reacts with protons to produce H_2 and $\mathbf{1}^{3+}$ (Eq. 11.3). This is the rate-determining step for the H_2 production, because the k_{obs} values for the second step are proportional to the proton concentration (inset of Fig. 11.2b) [49]. $\mathbf{1}^{3+}$ is reduced by $\mathbf{1}^{+}$ to produce two equivalent $\mathbf{1}^{2+}$ (Eq. 11.4) [49]:

$$\left[Co^{III}Co^{III}\right]^{4+} + 3Co(Cp)_2 \rightarrow \left[Co^{II}Co^{I}\right]^{+} + 3Co(Cp)_2^{+} \quad (11.1)$$

$$\left[Co^{II}Co^{I}\right]^{+} + H^{+} \rightleftharpoons \left[Co^{II}Co^{III} - H\right]^{2+} \quad (11.2)$$

$$\left[Co^{II}Co^{III} - H\right]^{2+} + H^{+} \rightarrow \left[Co^{II}Co^{III}\right]^{3+} + H_2 \quad (11.3)$$

$$\left[Co^{II}Co^{I}\right]^{+} + \left[Co^{II}Co^{III}\right]^{3+} \rightarrow 2\left[Co^{II}Co^{II}\right]^{2+} \quad (11.4)$$

$$-d\left[\left[Co^{II}Co^{I}\right]^{+}\right]/dt = k[H^{+}]\left[\left[Co^{II}Co^{I}\right]^{+}\right]/(1 + K[H^{+}]) \quad (11.5)$$

According to Scheme 11.1, the rate of decay of $\mathbf{1}^{+}$ is given by Eq. 11.5, where k is the rate constant of protonation of the hydride complex to produce H_2 (Eq. 11.3) and K is the protonation equilibrium constant of $\mathbf{1}^{+}$ to produce the hydride complex, $[Co^{II}Co^{III}–H]^{2+}$ (Eq. 11.2). Based on the kinetic results in Fig. 11.2, the k and K values of $\mathbf{1}^{+}$ at 298 K were determined to be 2.9 M^{-1} s^{-1} and 5.3×10^{2} M^{-1}, respectively [49]. Similarly the k and K values of **1** at 298 K were also determined to be 3.3 M^{-1} s^{-1} and 1.1×10^{3} M^{-1}, respectively [49]. The K value of **1** is twice larger than that of $\mathbf{1}^{+}$, because **1** has two Co^{I} sites as compared with $\mathbf{1}^{+}$ which has one Co^{I} site. The similar k values between $\mathbf{1}^{+}$ and **1** suggest that the two Co^{I} sites in **1** act rather independently in the reaction with proton.

A cobalt tetraaza-macrocyclic complex $[Co^{III}(CR)Cl_2]^{+}$ (CR = 2,12-dimethyl-3,7,11,17-tetraazabicyclo(11.3.1)-heptadeca-1(17),2,11,13,15-pentaene) has been

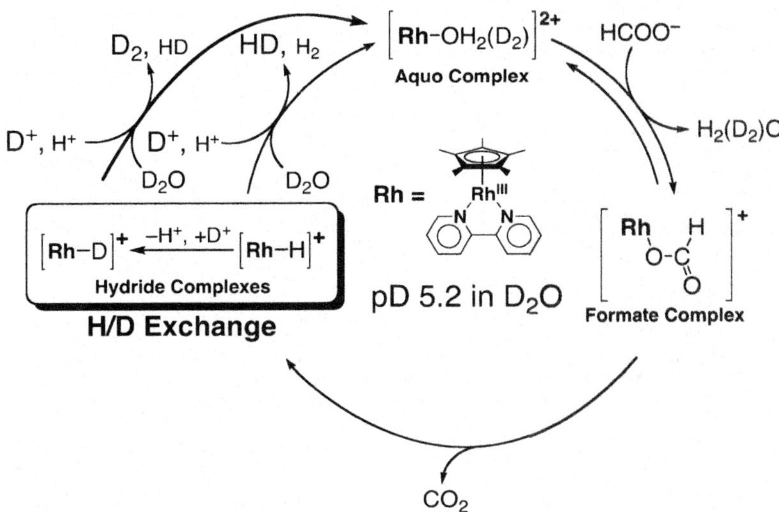

Scheme 11.2 Catalytic mechanism of decomposition of HCOOH by **2** in D_2O

reported to act as an efficient proton reduction catalyst in photocatalytic hydrogen evolution with ascorbate (HA^-) and ascorbic acid (H_2A) as an electron donor and a proton donor, respectively, and $[Ru(bpy)_3]^{2+}$ as a photocatalyst in water [50]. The catalytic activity and stability of $[Co^{III}(CR)Cl_2]^+$ were higher than that of other cobalt complexes such as cobaloxime derivatives [51–53] to afford a high turnover number (TON = 1,000) [50]. The Co(III) complex was reduced with HA^- to produce the Co(II) complex [50]. The Co(II) complex was further reduced by electron transfer from the excited state of $[Ru(bpy)_3]^{2+}$ ($[Ru(bpy)_3]^{2+*}$ where * denotes the excited state) to produce the Co(I) complex, which reacts with protons to yield H_2 and the Co(II) complex similar to $\mathbf{1}^+$ in Scheme 11.1 [50]. In this case, however, formation of the Co(III)-hydride complex has not been detected [50]. The detailed photocatalytic mechanism of hydrogen evolution is discussed in the next section.

11.2.2 Rhodium Hydride Complexes

A water-soluble rhodium-aqua complex, $[Rh^{III}(Cp^*)(bpy)(H_2O)](SO_4)$ (**2**(SO_4), bpy = 2,2′-bipyridine), acts as an efficient catalyst for H_2 evolution from HCOOH in an aqueous solution at 298 K [54]. The kinetic study revealed the catalytic mechanism of the catalytic decomposition of HCOOH to H_2 and CO_2 with **2** (SO_4) as shown in Scheme 11.2 [54]. The rate of H_2 evolution increased linearly with increasing concentrations of **2**(SO_4) as shown in Fig. 11.3a. On the other hand, the TOF value increased with increasing [HCOOH] to reach a limiting value as shown in Fig. 11.3b. Such a saturation behavior indicates that the formation of the formate complex is in equilibrium with $HCOO^-$, followed by β-hydrogen

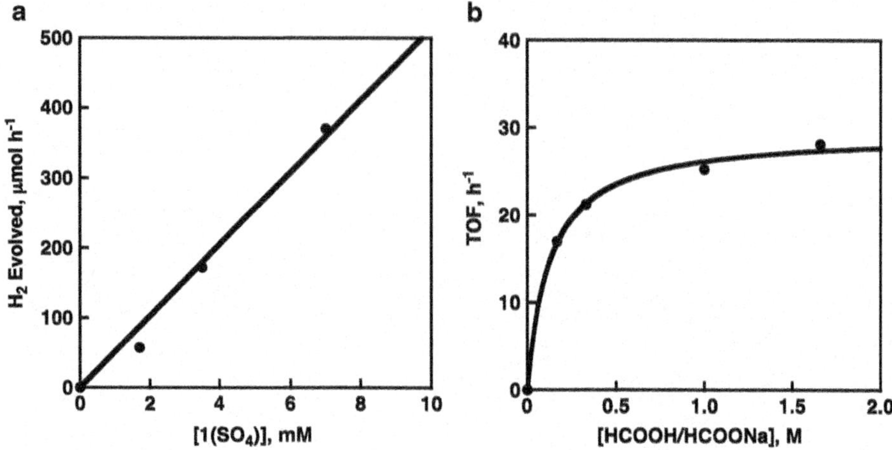

Fig. 11.3 (**a**) Plot of rate of H_2 evolution vs. the concentration of **2**(SO_4) in the decomposition of HCOOH/HCOONa (1.7 M) catalyzed by **2**(SO_4) in deaerated H_2O at pH 4.1 at 293 K. (**b**) Plot of TOFs vs. the concentration of HCOOH/HCOONa in the decomposition of HCOOH/HCOONa catalyzed by **2**(SO_4) (7.0 mM) in deaerated H_2O at pH 3.8 at 298 K (Reproduced from Ref. [54] by permission of John Wiley & Sons Ltd)

elimination from the formate complex to produce the Rh^{III}-hydride complex, which becomes the rate-determining step at large concentrations of $HCOO^-$. The formate complex ($[Rh^{III}(Cp^*)(OC(O)H)(bpy)]^+$) was detected by the electrospray ionization (ESI)-mass spectrometry at $m/z = 439.2$ [54]. When pH was changed, the maximum TOF value was obtained at pH 3.9, which corresponds to pK_a of HCOOH. No catalytic reactivity was observed at pH higher than pK_a of **2**(SO_4), indicating that the hydroxo complex $[Rh^{III}(Cp^*)(OH)(bpy)]^+$ has no catalytic reactivity [54].

When HCOOH was replaced by DCOOH, the catalytic decomposition of DCOOH in H_2O with **2**(SO_4) afforded not only HD but also H_2 [54]. A significant deuterium kinetic isotope effect was observed in the catalytic decomposition of DCOOH because the rate-determining step is the β-deuterium elimination from the formate complex to produce the Rh^{III}-D complex (vide supra) [54]. The formation of H_2 suggests that the deuteride species ($[Rh^{III}(Cp^*)(D)(bpy)]^+$), formed by deuteride transfer from $DCOO^-$ to $[Rh^{III}(Cp^*)(bpy)(H_2O)]^{2+}$, undergoes rapid H/D exchange with H_2O to afford $[Rh^{III}(Cp^*)(H)(bpy)]^+$ that reacts with H^+ to produce H_2. When the decomposition of HCOOH was performed with **2**(SO_4) in D_2O, D_2 was formed as a major product (73 %) together with HD (24 %) and H_2 (3 %) [54]. In this case, hydride transfer from $HCOO^-$ to $[Rh^{III}(Cp^*)(bpy)(H_2O)]^{2+}$ occurs to afford $[Rh^{III}(Cp^*)(H)(bpy)]^+$ that undergoes H/D exchange with D^+ to produce $[Rh^{III}(Cp^*)(D)(bpy)]^+$ [54]. Then, the reaction of $[Rh^{III}(Cp^*)(D)(bpy)]^+$ with D^+ yields D_2 as the main product [54]. The unexchanged hydride complex $[Rh^{III}(Cp^*)(H)(bpy)]^+$ reacts with D^+ and a small amount of H^+ derived from HCOOH to yield HD and a small amount of H_2, respectively [54]. Thus, rapid

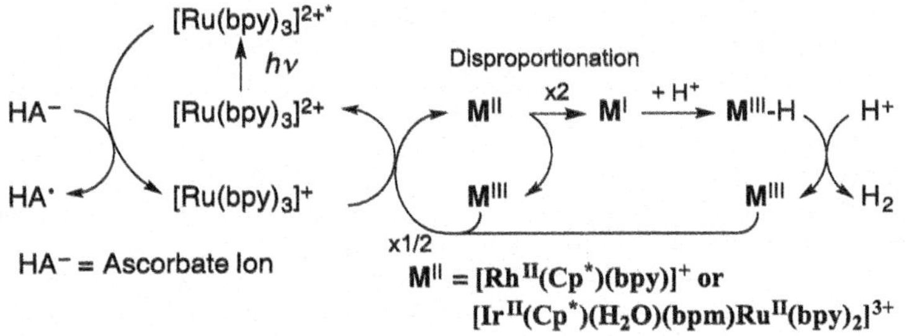

Scheme 11.3 Catalytic mechanism of hydrogen evolution with **2** or **3** as a proton reduction catalyst, ascorbate as an electron donor, and $[Ru(bpy)_3]^{2+}$ as a photocatalyst

H/D exchange between the hydride (or deuteride) species and proton (or deuteron) occurs as shown in Scheme 11.2, suggesting that the formal hydride species has a protic character.

The protic character of the hydride species was confirmed by formation of $[Rh^{I}(Cp^{*})(bpy)]$ by deprotonation from $[Rh^{III}(Cp^{*})(H)(bpy)]^{+}$ with a base (NaOH) [54]. Such a protic character of metal-hydride species was reported for the corresponding Ir complex with the same ligand as the Rh complex, i.e., $[Ir^{III}(Cp^{*})(H)(bpy)]^{+}$, which undergoes efficient H/D exchange with deuteron [55, 56]. DFT calculations showed that the positive charge of metal-hydride (M-H) (+0.571) was larger for $[Ir^{III}(Cp^{*})(H)(bpy)]^{+}$ as compared to the value for $[Rh^{III}(Cp^{*})(H)(bpy)]^{+}$ (+0.481) [54].

The Rh(III) complex (**2**(SO_4)) can also be used as a proton reduction catalyst in photocatalytic hydrogen evolution with ascorbate (HA^{-}) as an electron donor and $[Ru(bpy)_3]^{2+}$ as a photocatalyst [57]. The photocatalytic mechanism is shown in Scheme 11.3, where photoinduced electron transfer from HA^{-} to $[Ru(bpy)_3]^{2+*}$ (* denotes an excited state) occurs to produce $[Ru(bpy)_3]^{+}$, which reduces $[Rh^{III}(Cp^{*})(bpy)]^{2+}$ to $[Rh^{II}(Cp^{*})(bpy)]^{+}$, which was detected as a transient absorption band at 750 nm in Fig. 11.4a [57]. Disproportionation of $[Rh^{II}(Cp^{*})(bpy)]^{+}$ occurs to produce $Rh^{I}(Cp^{*})(bpy)$ and $[Rh^{II}(Cp^{*})(bpy)]^{+}$ as indicated by the second-order decay of absorbance at 750 nm due to $[Rh^{II}(Cp^{*})(bpy)]^{+}$ (see the second-order plot in inset of Fig. 11.4b) [57]. $Rh^{I}(Cp^{*})(bpy)$ is protonated to produce the hydride complex ($[Rh^{III}(Cp^{*})(H)(bpy)]^{+}$), which reacts with proton to produce H_2, accompanied by regeneration of $[Rh^{III}(Cp^{*})(bpy)]^{2+}$ [57]. In the same manner, when a heterodinuclear iridium–ruthenium complex $[Ir^{III}(Cp^{*})(H_2O)(bpm)Ru^{II}(bpy)_2](SO_4)_2$ (**3**$(SO_4)_2$, bpm = 2,2-bipyrimidine) was used in place of **2**(SO_4), photocatalytic H_2 evolution was confirmed to proceed via disproportionation of $[Ir^{II}(Cp^{*})(H_2O)(bpm)Ru^{II}(bpy)_2]^{3+}$. Thus, disproportionation of $[Rh^{II}(Cp^{*})(bpy)]^{+}$ or $[Ir^{II}(Cp^{*})(H_2O)(bpm)Ru^{II}(bpy)_2]^{3+}$ is the key step to convert one-electron process induced by one photon to the two-electron process for H_2 evolution (Scheme 11.3). This shows sharp contrast to the case of $[Co^{III}(CR)Cl_2]^{+}$, which is reduced by HA^{-} to produce the Co(II) complex, which is further reduced to the Co (I) complex via photoinduced electron transfer from $[Ru(bpy)_3]^{2+*}$ to the Co (II) complex (vide supra) [50].

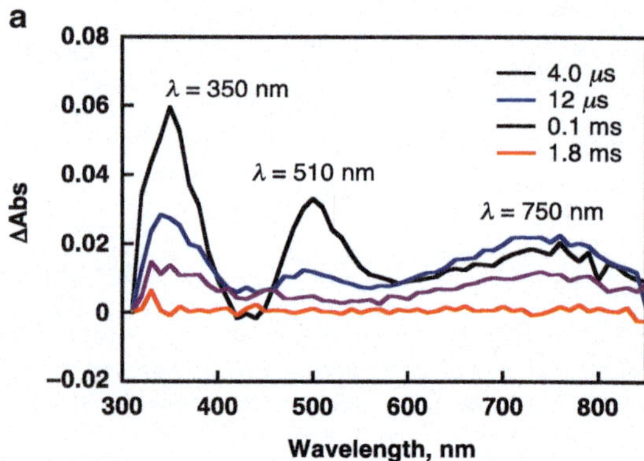

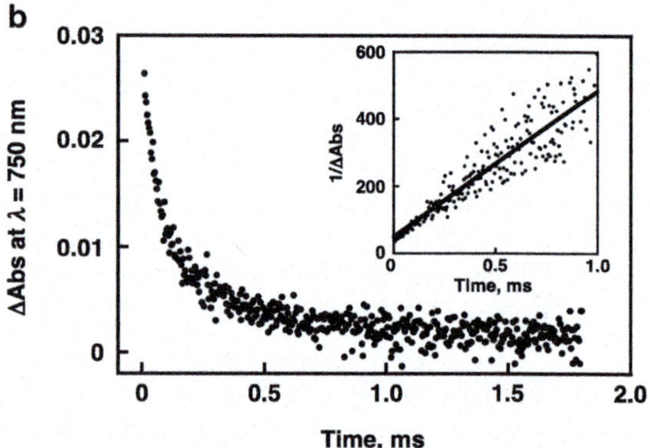

Fig. 11.4 (**a**) Transient absorption spectra of **2** (SO_4) $(1.6 \times 10^{-4}$ M) after laser excitation of $[Ru(bpy)_3]^{2+}$ $(8.0 \times 10^{-5}$ M) at $\lambda = 455$ nm in the presence of H_2A (0.8 M) and NaHA (0.3 M) in deaerated H_2O at pH 3.6 at 298 K. (**b**) Decay time profile of absorbance at $\lambda = 750$ nm due to $[Rh^{II}(Cp^*)(bpy)]^+$. *Inset*: Second-order plot (Reproduced from Ref. [57] by permission of John Wiley & Sons Ltd)

The same type of photocatalytic H_2 evolution with ascorbate and $[Ru(bpy)_3]^{2+}$ occurs using $[Rh^{III}(dmbpy)_2Cl_2]Cl$ (dmbpy = 4,4′-dimethyl-2,2′-bipyridine) as a proton reduction catalyst [58]. The catalytic reactivity of $[Rh^{III}(dmbpy)_2Cl_2]Cl$ was higher than that of $[Rh^{III}(Cp^*)(bpy)]^{2+}$ to afford high TON and TOF values (1,010 and 857 h^{-1}) at pH 4.0 [58]. The high catalytic activity may result from formation of colloidal rhodium nanoparticles during the photocatalytic reaction, which are known to promote the reduction of protons into H_2 [59]. This possibility was ruled out, because the addition of a large excess of mercury had no significant effect on the catalytic activity of $[Rh^{III}(dmbpy)_2Cl_2]Cl$. Mercury is known to form amalgam with colloidal metal or to adsorb to nanoparticular metal catalysts, and mercury poisoning has been reported for rhodium colloids [60].

11.2.2.1 Iridium Hydride Complexes

A heterodinuclear iridium–ruthenium complex $[Ir^{III}(Cp^*)(H_2O)(bpm)Ru^{II}(bpy)_2](SO_4)_2$ (**3**$(SO_4)_2$, bpm = 2,2-bipyrimidine) also acts as an efficient catalyst for H_2 evolution from HCOOH in an aqueous solution at 298 K [61]. The maximum TOF

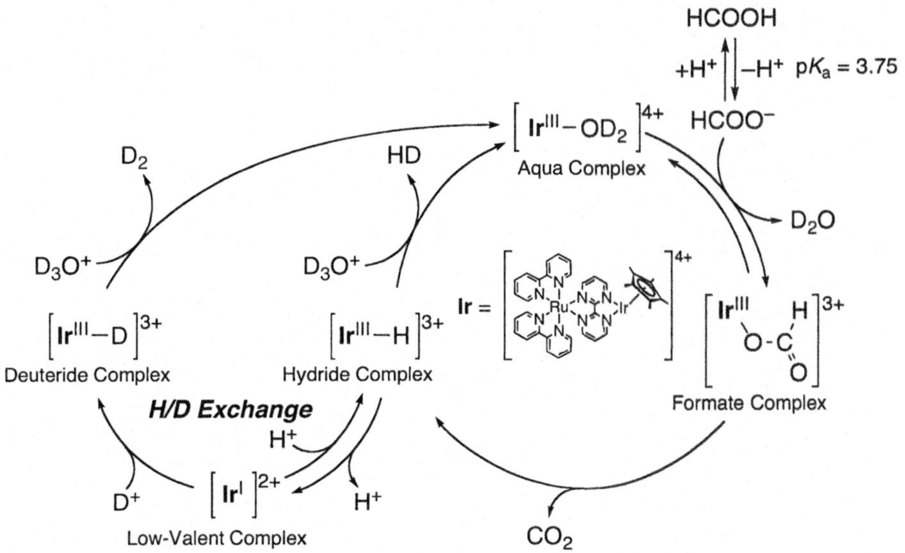

Scheme 11.4 Catalytic mechanism of decomposition of HCOOH by **3** in D_2O

value (426 h^{-1}) was obtained at pH 3.8 which agrees with the pK_a value of HCOOH [61]. The TOF value is much higher than that of $[Rh^{III}(Cp^*)(bpy)(H_2O)](SO_4)$ under the same experimental conditions (TOF = 27 h^{-1}) [61]. The catalytic mechanism is shown in Scheme 11.4, which is similar to the case of $[Rh^{III}(Cp^*)(bpy)(H_2O)](SO_4)$ in Scheme 11.2 [61]. The reaction of $\mathbf{3}^{4+}$ with $HCOO^-$ affords the formate complex ($[Ir^{III}(Cp^*)(O(CO)H)(bpm)Ru^{II}(bpy)_2]^{3+}$), followed by β-hydrogen elimination to give the Ir–hydride complex ($[Ir^{III}(Cp^*)(H)(bpm)Ru^{II}(bpy)_2]^{3+}$) which reacts with H^+ to produce H_2, accompanied by regeneration of $\mathbf{3}^{4+}$ [61]. Rapid H/D exchange between the hydride (or deuteride) species and proton (or deuteron) also occurs as the case of $\mathbf{2}^{2+}$ in Scheme 11.2 [61]. In Scheme 11.4, however, the rate-determining step in the overall hydrogen evolution reaction is not the β-hydrogen elimination step but the reaction of the hydride complex with H^+ to evolve H_2 at pH 3.8 [61]. The Arrhenius plots for TOF in D_2O (red circles) vs. H_2O (black circles) in Fig. 11.5 afforded $A_H/A_D = 3.1 \times 10^{-5}$, $E_a(D) - E_a(H) =$ 8.2 kcal mol^{-1}, and an unusually large KIE value at 298 K (KIE = 40) [61]. Such values for the Arrhenius parameters $A_H/A_D << 1$ and $E_a(D) - E_a(H) >$ 1.2 kcal mol^{-1} together with a large KIE value at 298 K (KIE > 9) are generally taken to unambiguously demonstrate the involvement of tunneling [62–65]. Because a protic character of metal-hydride species is more enhanced for the Ir–hydride complex as compared with the Rh-hydride complex, the reaction of the Ir–hydride complex with proton becomes the rate-determining step.

The catalytic activity for hydrogen evolution from formic acid was further enhanced by using a C^N cyclometalated organoiridium complex, $[Ir^{III}(Cp^*)\{4\text{-}(1H\text{-pyrazol-1-yl-}\kappa N^2)\text{benzoic acid-}\kappa C^3\}(H_2O)]_2SO_4$ ($[\mathbf{4}]_2 \cdot SO_4$, Fig. 11.6), as a

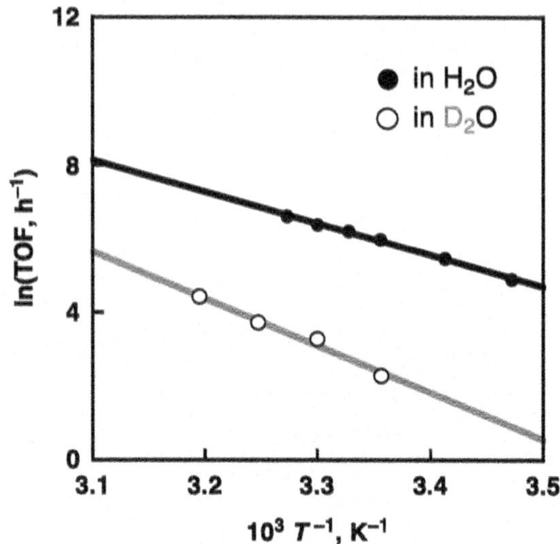

Fig. 11.5 Arrhenius plots of TOF for the decomposition of HCOOH/HCOONa (3.1 M) catalyzed by **3**$(SO_4)_2$ (0.3 mM) in deaerated H_2O (*closed circles*) or D_2O (*open circles*) at pH 3.8 or pD 3.8, respectively (Reprinted with the permission from Ref. [61]. Copyright 2010 American Chemical Society)

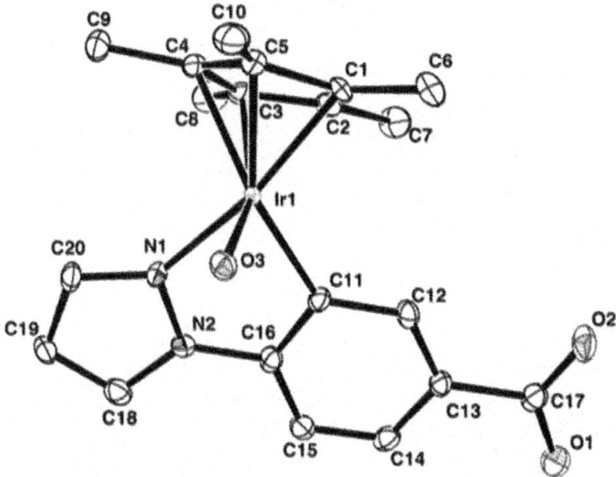

Fig. 11.6 ORTEP drawing of **4**. Hydrogen atoms are omitted for clarity (Reprinted with the permission from Ref. [68]. Copyright 2012 American Chemical Society)

catalyst with TOF value over 2,000 h^{-1} at 298 K (Fig. 11.7) [66]. The catalytic mechanism is shown in Scheme 11.5 [66].

As pH was increased, the Ir(III) complex $[\mathbf{Ir}^{\mathbf{A}}–H_2O]^+$ released protons from the carboxy group and the aqua ligand to form the corresponding benzoate complex $[\mathbf{Ir}^{\mathbf{B}}–H_2O]^0$ and the hydroxo complex $[\mathbf{Ir}^{\mathbf{B}}–OH]^-$, respectively (Scheme 11.6). The pK_a values of $[\mathbf{Ir}^{\mathbf{A}}–H_2O]^+$ and $[\mathbf{Ir}^{\mathbf{B}}–H_2O]^0$ were determined from the spectral titration to be $pK_{a1} = 4.0$ and $pK_{a2} = 9.5$, respectively [66]. The saturation behavior of TOF of hydrogen evolution with increasing concentration of [HCOOH] + [HCOOK] at pH 2.8 (Fig. 11.7) indicates that hydrogen is produced via the formate complex of $[\mathbf{Ir}^{\mathbf{A}}–H_2O]^+$, followed by β-elimination to produce the hydride

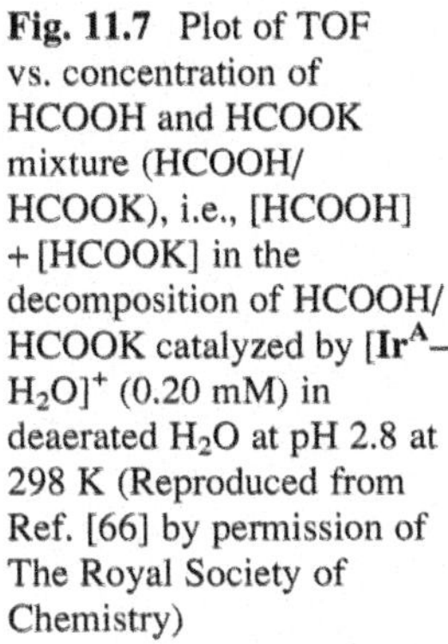

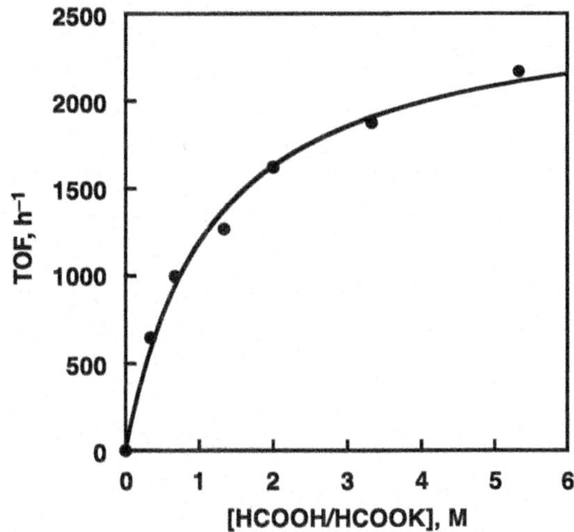

Fig. 11.7 Plot of TOF vs. concentration of HCOOH and HCOOK mixture (HCOOH/HCOOK), i.e., [HCOOH] + [HCOOK] in the decomposition of HCOOH/HCOOK catalyzed by [**Ir**^{A}–H_2O]$^+$ (0.20 mM) in deaerated H_2O at pH 2.8 at 298 K (Reproduced from Ref. [66] by permission of The Royal Society of Chemistry)

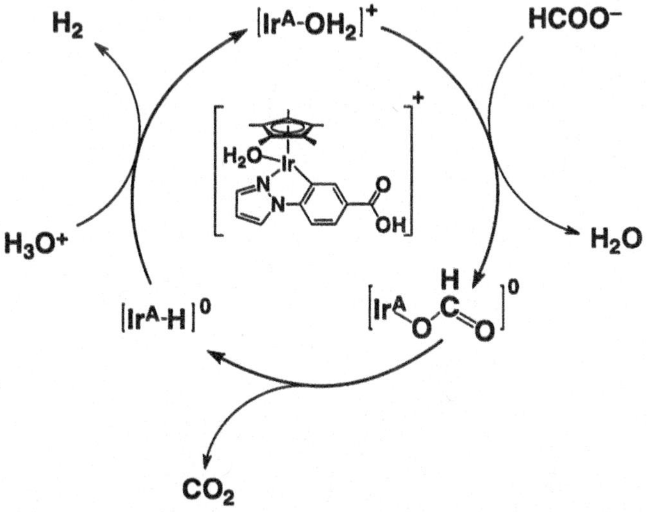

Scheme 11.5 Catalytic mechanism of H_2 evolution and decomposition of $HCOO^-$ with **4** in H_2O

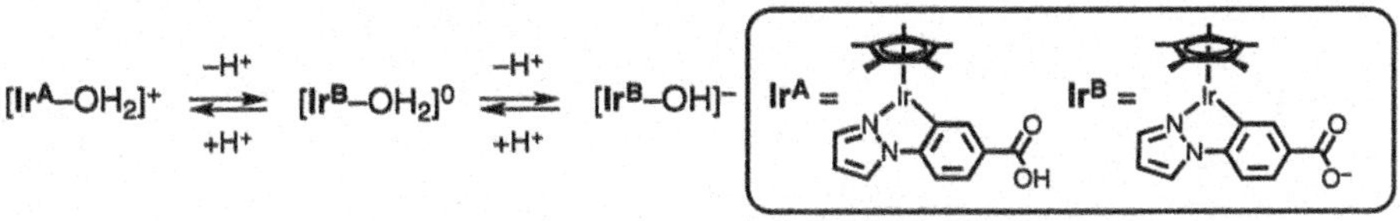

Scheme 11.6 Acid–base equilibria of iridium aqua complexes

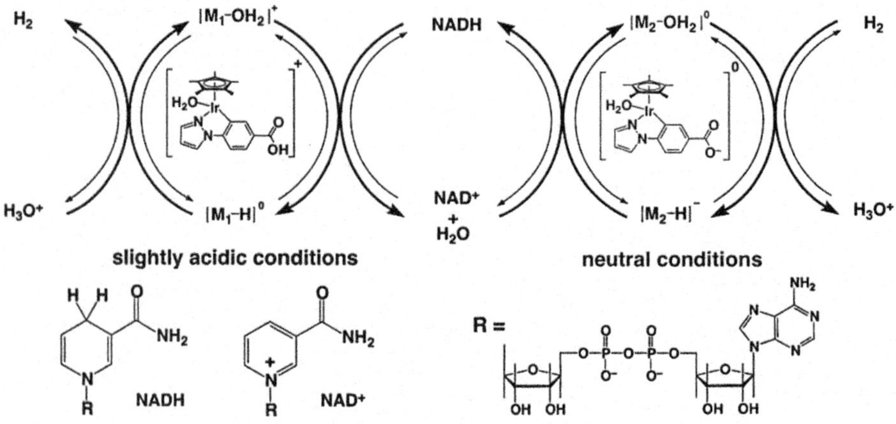

Scheme 11.7 Catalytic mechanism of interconversion between H_2 and NADH with **4** in H_2O

complex, which reacts with proton to produce H_2 (Scheme 11.6) [66]. The activation energy was determined to be 18.9 kcal mol^{-1}, which is much smaller than the activation energy of the decomposition of formic acid without catalysts (78 kcal mol^{-1}) [67].

The C^N cyclometalated organoiridium complex $[\mathbf{Ir}^{\mathbf{A}}–H_2O]^+$ can also act as an efficient catalyst for hydrogen evolution from NADH (dihydronicotinamide adenine dinucleotide), which is a natural electron and proton source in respiration and CO_2 fixation [68], in water at pH 4.1 [69]. NADH has been frequently used as an electron and proton source in photocatalytic hydrogen evolution with a photocatalyst and a proton reduction catalyst [70–73]. Under acidic conditions, NADH can reduce thermally proton to produce H_2 and NAD^+ by the catalysis of $[\mathbf{Ir}^{\mathbf{A}}–H_2O]^+$ (Eq. **11.6**). The catalytic cycle is shown in Scheme 11.7 [69]. Under acidic

$$\begin{array}{c}\mathrm{NADH} + \mathrm{H}^+ \rightarrow \mathrm{NAD}^+ + \mathrm{H}_2 \\ [\mathbf{Ir}^{\mathbf{A}} - \mathrm{H_2O}]^+\end{array} \tag{11.6}$$

conditions, hydride transfer from NADH to $[\mathbf{Ir}^{\mathbf{A}}–H_2O]^+$ occurs to produce NAD^+ and the Ir(III)-hydride complex $[\mathbf{Ir}^{\mathbf{A}}–H]^0$, which reacts with H_3O^+ to produce H_2, accompanied by regeneration of $[\mathbf{Ir}^{\mathbf{A}}–H_2O]^+$ [69]. Under basic conditions, however, the catalytic cycle was reversed, when H_2 can reduce the deprotonated carboxylate form $[\mathbf{Ir}^{\mathbf{B}}–H_2O]^0$ to produce the Ir(III)-hydride complex, which reduces NAD^+ to NADH, accompanied by regeneration of the deprotonated form of $[\mathbf{Ir}^{\mathbf{B}}–H_2O]^0$ [69]. Thus, interconversion between NADH and H_2 at ambient pressure and temperature can be efficiently catalyzed by $[\mathbf{Ir}^{\mathbf{A}}–H_2O]^+$ and $[\mathbf{Ir}^{\mathbf{B}}–H_2O]^0$ depending on pH.

According to Scheme 11.6, the Ir(III) complex $[\mathbf{Ir}^{\mathbf{A}}–H_2O]^+$ is converted to the hydroxo complex $[\mathbf{Ir}^{\mathbf{B}}–OH]^-$ at pH 13.6. When ethanol was added to an aqueous solution of $[\mathbf{Ir}^{\mathbf{B}}–OH]^-$ at pH 13.6, hydride transfer from ethanol to $[\mathbf{Ir}^{\mathbf{B}}–OH]^-$

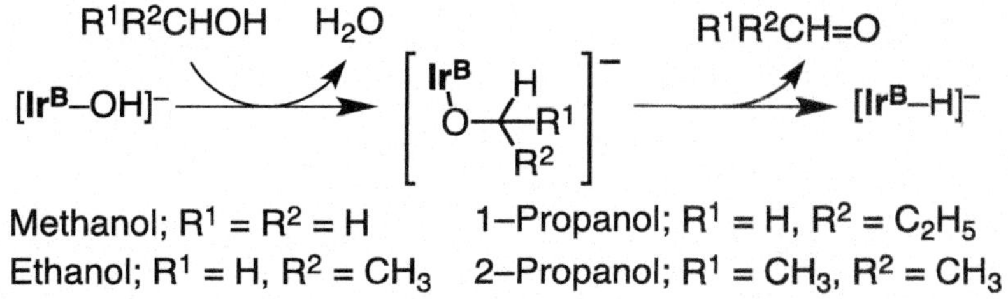

R^1R^2CHOH H_2O $R^1R^2CH{=}O$

$[Ir^B\text{–}OH]^- \rightarrow [Ir^B\text{–}O\text{–}CHR^1R^2]^- \rightarrow [Ir^B\text{–}H]^-$

Methanol; $R^1 = R^2 = H$ 1–Propanol; $R^1 = H$, $R^2 = C_2H_5$
Ethanol; $R^1 = H$, $R^2 = CH_3$ 2–Propanol; $R^1 = CH_3$, $R^2 = CH_3$

Scheme 11.8 Dehydrogenation reaction of alcohols by $[Ir^B\text{–}OH]^-$ in H_2O

occurred to produce acetaldehyde and the hydride complex $[\mathbf{Ir^B}\text{–}H]^-$ [74]. When CD_3CD_2OH in place of CH_3CH_2OH was added to an aqueous solution of $[\mathbf{Ir^B}\text{–}OH]^-$, a kinetic deuterium isotope effect (KIE) for the formation of $[\mathbf{Ir^B}\text{–}D]^-$ was observed to be $k_H/k_D = 2.1$ [74]. The observation of KIE indicates that the reaction of ethanol with $[\mathbf{Ir^B}\text{–}OH]^-$ involves the C–H bond cleavage. Thus, the β-hydrogen elimination of the ethoxy complex which is produced by the replacement of a hydroxy (OH) ligand of $[\mathbf{Ir^B}\text{–}OH]^-$ by a ethoxy (CH_3CH_2O) ligand, may be the rate-determining step for formation of the hydride complex $[\mathbf{Ir^B}\text{–}H]^-$ (Scheme 11.8). Other alcohols can also reduce $[\mathbf{Ir^B}\text{–}OH]^-$ to produce the hydride complex $[\mathbf{Ir^B}\text{–}H]^-$ [74].

The hydride complex $[\mathbf{Ir^B}\text{–}H]^-$ is stable at pH 14. When pH was decreased to 0.8 by adding H_2SO_4, however, the hydride complex $[\mathbf{Ir^B}\text{–}H]^-$ was converted to an aqua complex $[\mathbf{Ir^A}\text{–}H_2O]^+$ as shown by Fig. 11.8a, accompanied by evolution of hydrogen (H_2) [74]. The conversion between the hydride complex $[\mathbf{Ir^B}\text{–}H]^-$ and the aqua complex $[\mathbf{Ir^A}\text{–}H_2O]^+$ accompanied by H_2 evolution was repeated by alternate change in pH between 12 and 2 in the presence of excess amount of ethanol as shown in Fig. 11.8b (Scheme 11.9) [74]. Without changing pH, however, no catalytic H_2 evolution from ethanol occurred with $[\mathbf{Ir^A}\text{–}OH]^0$ [74].

Photoirradiation of the hydride complex $[\mathbf{Ir^B}\text{–}H]^-$ resulted in the conversion to the [C,C] cyclometalated complex $[\mathbf{Ir^C}\text{–}H]^-$ (Scheme 11.10) [74]. In contrast to the [C,N] cyclometalated Ir–hydride complex $[\mathbf{Ir^B}\text{–}H]^-$, the [C,C] cyclometalated Ir–hydride complex $[\mathbf{Ir^C}\text{–}H]^-$ can react with water to produce H_2 under basic conditions as shown in Fig. 11.9. The turnover number (TON) of H_2 evolution from isopropanol with $[\mathbf{Ir^C}\text{–}H]^-$ (Eq. 11.7) increases linearly with time to reach 3.3 (2.5 h), whereas $[\mathbf{Ir^B}\text{–}H]^-$ has no catalytic reactivity even at elevated temperature at 323 K (Fig. 11.9). TON for H_2 evolution from isopropanol with $[\mathbf{Ir^C}\text{–}H]^-$ increases with increasing temperature to be 26 (1.0 h) at 353 K (Fig. 11.9) [74]. The enhanced catalytic activity of $[\mathbf{Ir^C}\text{–}H]^-$ results from the electronic donating effect of phenylpyrazole ligand on the metal center with a [C,C] cyclometalated iridium as indicated by the upfield shift of a hydride signal bonded to Ir^{III} center ($\delta = -17.48$) as compared with that of $[\mathbf{Ir^B}\text{–}H]^-$ ($\delta = -14.34$) [74]:

$$(CH_3)_2CHOH \xrightarrow[{[\mathbf{Ir^C} - H]^-}]{} (CH_3)_2CO + H_2 \tag{11.7}$$

a
Absorbance
pH 12
pH 2
Wavelength, nm
b
Absorbance at λ = 350 nm
Cycles

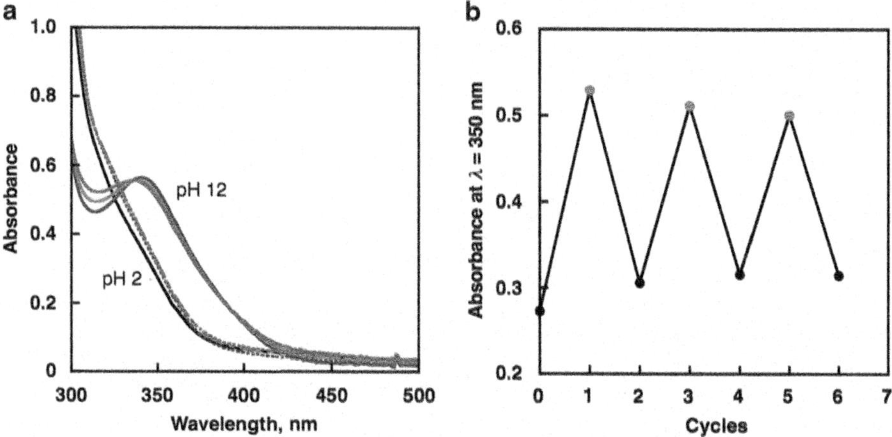

Fig. 11.8 (**a**) UV-vis absorption spectral change of an aqueous solution of [**Ir**$^{\mathbf{B}}$–OH]$^-$ (0.12 mM) and ethanol (82 mM) by alternate change in pH. (**b**) Changes of absorbance at $\lambda = 350$ nm due to the formation of a hydride complex [**Ir**$^{\mathbf{B}}$–H]$^-$ in the reaction of [**Ir**$^{\mathbf{B}}$–OH]$^-$ (0.12 mM) with ethanol (82 mM) in water (pH 11.8–12.2) and due to the hydrogen evolution in the reaction of the hydride complex [**Ir**$^{\mathbf{B}}$–H]$^-$ with proton in water at 298 K (pH 2.0–3.3) by adding an aqueous solution of H_2SO_4 (5.0 M) or NaOH (5.0 M) (Reprinted with the permission from Ref. [74]. Copyright 2012 American Chemical Society)

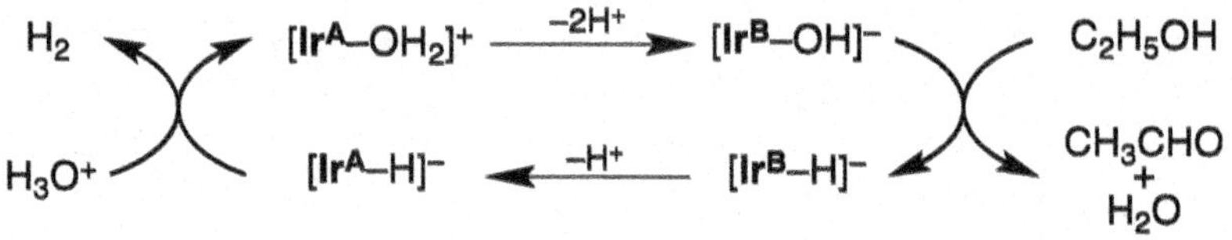

H2
H3O+
[IrA–OH2]+
−2H+
[IrB–OH]−
[IrA–H]−
−H+
[IrB–H]−
C2H5OH
CH3CHO
+
H2O

Scheme 11.9 Catalytic mechanism of H_2 evolution from ethanol with [Ir^A–H_2O]$^+$ in H_2O

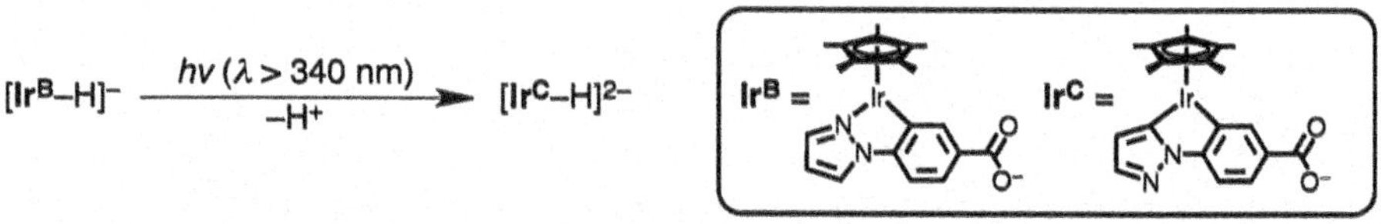

[IrB–H]−
hν (λ > 340 nm)
−H+
[IrC–H]2−
IrB =
IrC =

Scheme 11.10 Conversion from a [C,N] to [C,C] cyclometalated Ir complex under photoirradiation

11.2.2.2 Ruthenium Hydride Complexes

The catalytic activity of hydrogen evolution from alcohols has been reported to be remarkably enhanced by using ruthenium complexes containing pincer-type ligands [75]. Catalytic hydrogen evolution occurred in methanol containing KOH (8.0 M) with [RuHCl(CO)(HN($C_2H_4P^iPr_2$)$_2$)] (**5**), which exhibited high activities

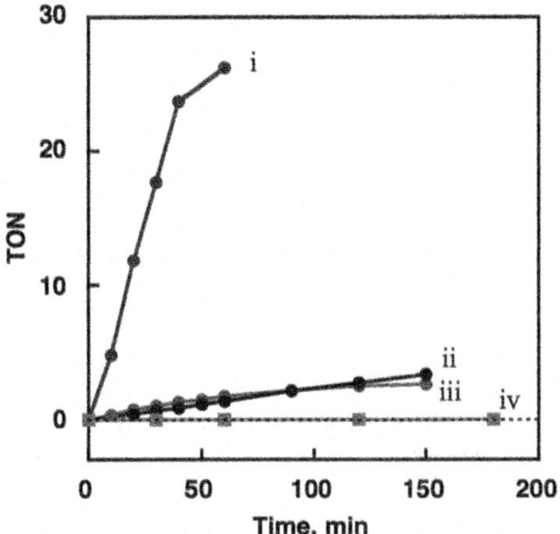

Fig. 11.9 Time course of H_2 evolution from 2-propanol (4.3 M) catalyzed by 5 (55 μM) in water (pH 11.9) at 353 K *i*) and 323 K (*ii*) and that from ethanol (5.7 M) catalyzed by $[\mathbf{Ir}^{C}\text{–H}]^-$ (55 μM) in water (pH 11.9) at 323 K (*iii*). Time course of H_2 evolution from 2-propanol (4.3 M) catalyzed by $[\mathbf{Ir}^{B}\text{–H}]^-$ (55 μM) in water (pH 11.9) at 323 K (*iv*) (Reprinted with the permission from Ref. [74]. Copyright 2012 American Chemical Society)

up to TOF = 4,700 h^{-1} and TON = 350,000 at 368 K [75]. Under catalytic conditions, both formate and carbonate ions were observed as traces of the reaction mixtures, indicating that the formate is an intermediate in this dehydrogenation sequence and that CO_2 is trapped as carbonate [75]. The overall stoichiometry of the hydrogen evolution from methanol with NaOH is given by Eq. 11.8. The Ru-hydride species were observed in solution under catalytic conditions [75]:

$$CH_3OH + 2NaOH \xrightarrow{\mathbf{5}} 3H_2 + Na_2CO_3 \quad (11.8)$$

Base-free hydrogen evolution from methanol without formation of CO has recently been achieved by using the ruthenium-based PNP pincer complex (**6**: Ru-MACHO-BH) in Eq. 11.9 [76]. The combination of Ru-MACHO-BH (**6**) with $Ru(H)_2(dppm)_2$ (**7**) further enhanced the catalytic activity for hydrogen evolution from neutral methanol [76]. A long-term experiment gave a 26 % yield of H_2 (relative to H_2O) and a TON > 4,200 [76]. In this case full conversion of all "available" hydrogen atoms in methanol to H_2 has been achieved by synergetic homogeneous catalysis of **6** and **7** [76]:

$$CH_3OH + H_2O \xrightarrow{\mathbf{6},\ \mathbf{7}} 3H_2 + CO_2 \quad (11.9)$$

Hydrogen is also produced by the electrocatalytic reduction of protons with a Ru (II) complex $[Ru^{II}(tpy)(bpy)(S)]^{2+}$ (tpy = 2,2′:6′,2″-terpyridine, bpy = 2,2-′-bipyridine, S = solvent) in acetonitrile (MeCN) [77]. The Ru(II)-hydride complex

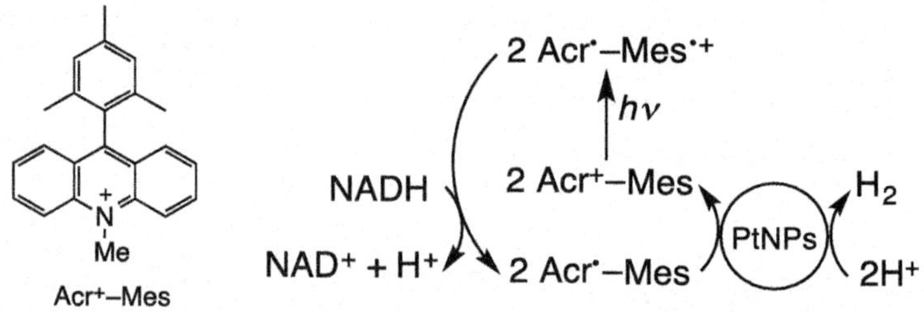

Scheme 11.11 Chemical structure of Acr^+–Mes and overall photocatalytic cycle for H_2 evolution

$[Ru^{II}(tpy)(bpy)(H)]^+$ is produced by the reaction of the ligand-based two-electron-reduced species $[Ru^{II}(tpy^{\bullet -})(bpy^{\bullet -})(MeCN)]^0$ with water (Eq. 11.10) [77]:

$$[Ru^{II}(tpy^-)(bpy^-)(MeCN)]^0 + H_2O \rightarrow [Ru^{II}(tpy)(bpy)(H)]^+ + OH^- \quad (11.10)$$

Further reduction of the hydride to $[Ru^{II}(tpy^{\bullet -})(bpy)(H)]^0$ at −1.41 V (vs. NHE) is proposed to trigger the catalytic water reduction via formation of the dihydrogen–dihydride complex $[Ru^{II}(tpy)(bpy)(H_2)]^+$ (Eqs. 11.11, 11.12, and 11.13) [77]. However, this intermediate has yet to be detected. In the presence of an acid,

$$[Ru^{II}(tpy^-)(bpy)(H)]^0 + H_2O \rightarrow [Ru^{II}(tpy^-)(bpy)(H_2)]^+ + OH^- \quad (11.11)$$

$$[Ru^{II}(tpy^-)(bpy)(H_2)]^+ + e^- \rightarrow [Ru^{II}(tpy^-)(bpy^-)(H_2)]^0 \quad (11.12)$$

$$[Ru^{II}(tpy^-)(bpy^-)(H_2)]^0 + MeCN \rightarrow H_2 + [Ru^{II}(tpy^-)(bpy^-)(MeCN)]^0 \quad (11.13)$$

$[Ru^{II}(tpy)(bpy)(H)]^+$ can react with H^+ to produce H_2 [77]. Many other metal hydrides are known to catalyze electrochemical reduction of protons to H_2 [78–83].

11.2.2.3 Proton-Coupled Electron Transfer to Metal Nanoparticles

Pt nanoparticles (PtNPs) act as the best catalyst for catalytic reduction of protons to H_2 [43]. Photocatalytic H_2 evolution occurred efficiently using NADH as a sacrificial electron donor, 9-mesityl-10-methylacridinium ion (Acr^+–Mes) [84] as an organic photocatalyst, and PtNPS as a proton reduction catalyst (Scheme 11.11) [85]. Photoexcitation of Acr^+–Mes resulted in intramolecular electron transfer from the Mes moiety to the singlet excited sate of the Acr^+ moiety to produce the electron-transfer state ($Acr^{\bullet}$–$Mes^{\bullet +}$) [84, 86–88]. NADH is oxidized by the $Mes^{\bullet +}$ moiety of $Acr^{\bullet}$–$Mes^{\bullet +}$ to produce two equivalents of $Acr^{\bullet}$–Mes. Electron transfer from $Acr^{\bullet}$–Mes to PtNPs with protons resulted in H_2 evolution [85].

The kinetics and mechanism of the PtNP-catalyzed hydrogen evolution by an $Acr^{\bullet}$–Mes were studied by simultaneous determination of the rate of hydrogen

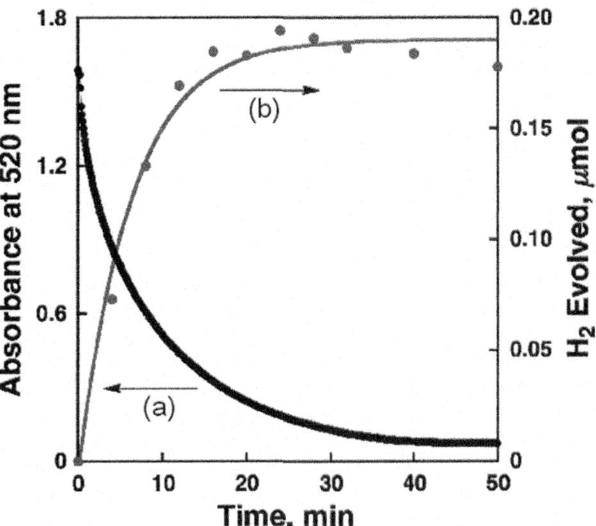

Fig. 11.10 (a) Time profile of electron transfer from Acr•–Mes to spherical PtNPs with the diameter of 4.5 nm (0.1 μg), monitored by decrease in absorbance at 520 nm due to Acr•–Mes in a (pH 5.0, 50 mM) CH_3COOH/CH_3COONa buffer and MeCN [1:1 (v/v)] mixed solution. (b) Time profile of H_2 evolution (Reproduced from Ref. [85] by permission of John Wiley & Sons Ltd)

evolution and the rate of electron transfer from Acr•–Mes to PtNPs [85]. The rate of H_2 evolution in a (pH 5.0, 50 mM) CH_3COOH/CH_3COONa buffer and MeCN [1:1 (v/v)] mixed solution is virtually the same as the rate of electron transfer from Acr•–Mes to PtNPs, which was monitored by decrease in absorbance at 520 nm due to Acr•–Mes as shown in Fig. 11.9 [85]. This indicates that electron transfer from Acr•–Mes to PtNPs is the rate-determining step for the catalytic H_2 evolution. The rate constant of electron transfer from Acr•–Mes to PtNPs (k_{et}) is proportional to proton concentration (Fig. 11.10) [85]. When CH_3COOH/CH_3COONa buffer (pH 4.5, 50 mM) in H_2O was replaced by CH_3COOD/CH_3COONa in D_2O, an inverse kinetic isotope effect (KIE = 0.68) was observed in electron transfer from Acr•–Mes to PtNPs [85]. Such an inverse kinetic isotope effect results from the difference in the zero-point energy for Pt–H (Pt–D) bond at the transition state as compared with that before electron transfer when the interaction between Pt and H^+ (or D^+) is much smaller as shown in Fig. 11.11 [85]. This indicates that proton-coupled electron transfer (PCET) from Acr•–Mes to PtNPs producing a Pt–H bond is the rate-determining step (r.d.s.) in the catalytic hydrogen evolution. The inverse KIE (0.68) in Fig. 11.12 shows sharp contrast to the large KIE (40) observed for the hydrogen evolution from formic acid, catalyzed by an Ir–hydride complex ($[Ir^{III}(Cp^*)(H)(bpm)Ru^{II}(bpy)_2]^{3+}$) when the heterolytic Ir–H bond cleavage by proton is the rate-determining step in Scheme 11.4 (vide supra) [61]. Based on the results in Figs. 11.10 and 11.12, the PtNP-catalyzed H_2 evolution mechanism was proposed as shown in Scheme 11.12 [85]. PCET from Acr•–Mes to PtNPs produces the Pt–H bond, followed by rapid elimination of H_2 from two Pt–H bonds.

When Acr^+–Mes was replaced by 2-phenyl-4-(1-naphthyl)quinolinium ion ($QuPh^+$–NA) [89], photocatalytic H_2 evolution also occurred efficiently with NADH and PtNPs (Scheme 11.12) [90]. However, the rate constant of electron transfer from QuPh•–NA to PtNPs was invariant with pH [90], in contrast to the

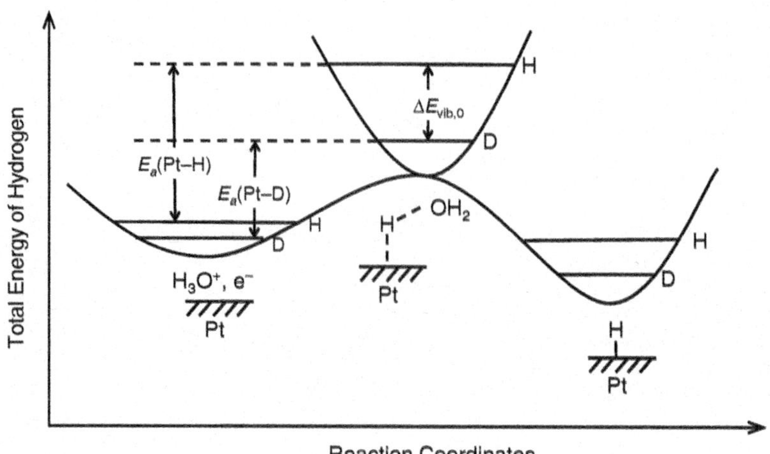

Fig. 11.11 Illustration of the PCET pathway to produce the Pt–H or Pt–D bond

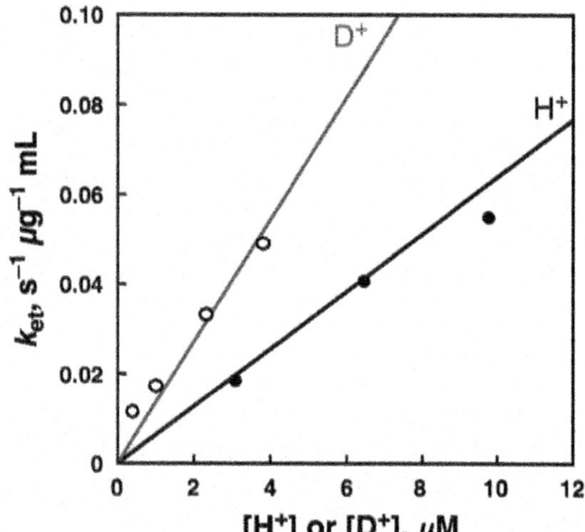

Fig. 11.12 Dependence of k_{et} on [H$^+$] or [D$^+$] for electron transfer from Acr$^\bullet$–Mes to spherical PtNPs with the diameter of 4.5 nm in H_2O/MeCN [1:1 (v/v)] containing CH_3COOH/CH_3COONa buffer (50 mM) or in D_2O/MeCN [1:1 (v/v)] containing CH_3COOD/CH_3COONa buffer (50 mM) at 298 K (Reproduced from Ref. [85] by permission of John Wiley & Sons Ltd)

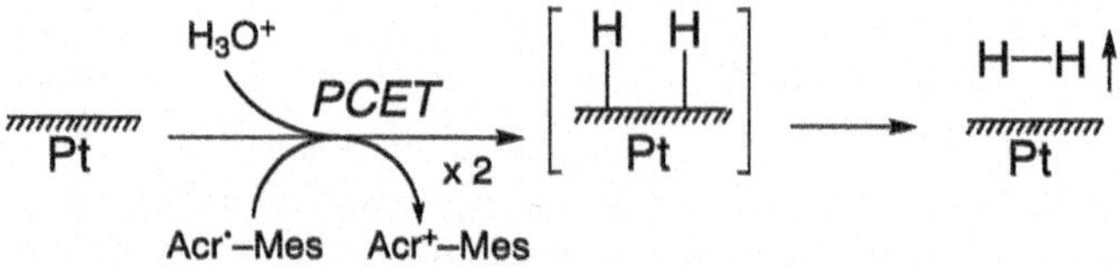

Scheme 11.12 Mechanism of H_2 evolution on Pt surfaces

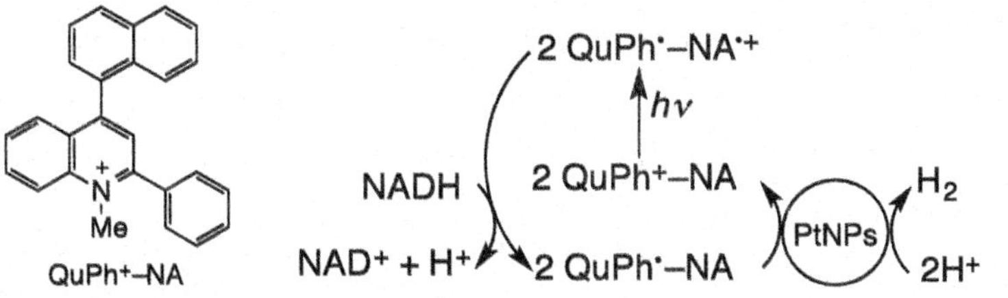

Scheme 11.13 Chemical structure of $QuPh^+$–NA and overall photocatalytic cycle for H_2 evolution

case of PCET from $Acr^•$–Mes to PtNPs in which the rate constant was proportional to proton concentration (Fig. 11.10) [85]. Thus, electron transfer from $QuPh^•$–NA to MNPs occurs without assistance of proton because of the much stronger reducing ability of $QuPh^•$–NA as compared with $Acr^•$–Mes judging from the significantly more negative oxidation potential of $QuPh^•$–NA ($E_{ox} = -0.90$ V vs. SCE) [89] than that of $Acr^•$–Mes ($E_{ox} = -0.57$ V vs. SCE) [84]. Because the rate of hydrogen evolution was much slower than the rate of electron transfer from $QuPh^•$–NA to MNPs and the hydrogen evolution was also pH independent at pH < 10, the rate-determining step of the catalytic H_2 evolution may be elimination of hydrogen from two Pt–H bonds [90]. Thus, the rate-determining step for the catalytic H_2 evolution is changed depending on the reducing ability of one-electron reductants (Scheme 11.13).

11.2.2.4 Kinetics and Mechanisms of Catalytic CO_2 Reduction

The catalytic reduction of CO_2 by H_2 has attracted significant interest because catalytic transformation of CO_2 would be promising for the production of fuels as liquid hydrogen sources and value-added chemicals [91–96]. However, the reactions involving CO_2 are commonly carried out at high pressure [97–106], which may not be economically suitable and also poses safety concerns. In order to improve the catalytic activity for the CO_2 reduction, it is of primary importance to elucidate the catalytic mechanism.

Kinetics and mechanism of the catalytic reduction of CO_2 by H_2 to produce formic acid (HCOOH) were reported by using $[Ir^{III}(Cp^*)(L)(H_2O)](SO_4)$ and $[Ru^{II}(\eta^6\text{-}C_6Me_6)(L)(H_2O)](SO_4)$ (L = bpy or 4,4′-OMe-bpy) as catalysts in water [107]. The rates of the catalytic reduction of CO_2 by H_2 with $[Ir^{III}(Cp^*)(L)(H_2O)](SO_4)$ under acidic conditions in H_2O are affected by the pressure of H_2 and CO_2. Turnover number (TON) of the catalytic reduction of CO_2 (2.5 MPa) by H_2 with $[Ir^{III}(Cp^*)(L)](SO_4)$ increased with increasing H_2 pressure at pH 3.0 at 40 °C to reach a constant value (Fig. 11.13a), whereas TON was proportional to CO_2 pressure at 5.5 MPa of H_2 (Fig. 11.13b). The reactions of $[Ir^{III}(Cp^*)(L)$

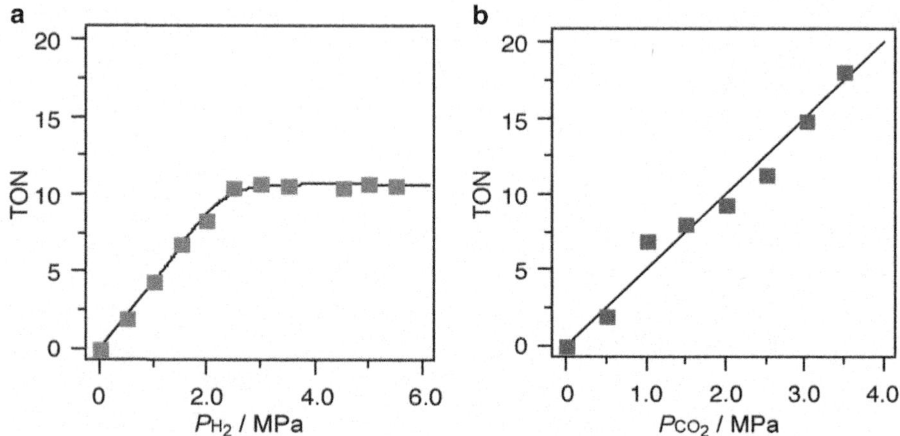

Fig. 11.13 (**a**) Dependence of TONs at 0.5 h on H_2 pressure for the reduction of CO_2 (2.5 MPa) by H_2 catalyzed by **6**(SO_4) (20 μmol) at pH 3.0 in a citrate buffer solution (20 cm^3) at 40 °C. (**b**) Dependence TONs at 0.5 h on CO_2 pressure for the reduction of CO_2 by H_2 (5.5 MPa) catalyzed by the **6**(SO_4) (20 μmol) at pH 3.0 in a citrate buffer solution (20 cm^3) at 40 °C (Reproduced from Ref. [107] by permission of The Royal Society of Chemistry)

(H_2O)](SO_4) with H_2 at pH 3.0 in a citrate buffer solution provide the Ir(III)–hydride complexes [Ir^{III}(Cp^*)(L)(H)]$_2$(SO_4), which were detected by the ESI-mass spectra and ^{1}H NMR spectra. Because TON of the catalytic reduction of CO_2 by H_2 was proportional to CO_2 pressure, the rate-determining step may be the reaction of the Ir(III)–hydride complex with CO_2 to produce the formate complex as shown in Scheme 11.14. In such a case, the rate of formation of HCOOH in the catalytic reduction of CO_2 by H_2 with [Ir^{III}(Cp^*)(L)(H_2O)](SO_4) is given by Eq. 11.14,

$$\mathrm{d[HCOOH]/d}t = k_1k_2[[\mathrm{Ir{-}OH_2}]^{2+}]P_{\mathrm{H2}}P_{\mathrm{CO2}}/(k_{-1} + k_1P_{\mathrm{H2}}) \tag{11.14}$$

where k_1 is the rate constant of the reaction of the aqua complexes [Ir^{III}(Cp^*)(L)(H_2O)]$^{2+}$ with H_2, k_{-1} is the rate constant of the back reaction, k_2 is the rate constant of the reaction of the hydride complex [Ir^{III}(Cp^*(L)(H))]$^+$ with CO_2, and [[Ir–OH_2]$^{2+}$]$_0$ is the initial concentration of [Ir^{III}(Cp^*)(L)(H_2O)]$^{2+}$ [107]. Under the conditions such that $k_1P_{H2} >> k_{-1}$, the rate of formation of HCOOH becomes constant at large H_2 pressure as observed in Fig. 11.13a [107].

When [Ir^{III}(Cp^*)(L)(H_2O)](SO_4) was replaced by [Ru^{II}(η^6-C_6Me_6)(L)(H_2O)](SO_4), TON was proportional to H_2 pressure at 40 °C (Fig. 11.14a), whereas TON exhibited a saturation behavior with increasing CO_2 pressure (Fig. 11.14b) [107]. In such a case, the rate-determining step was changed from the reaction of the Ir(III)–hydride complex with CO_2 to produce the formate complex to the reaction of the Ru(III)–aqua complex with H_2 to produce the Ru(III)–hydride complex (Scheme 11.15) [107]. The rate of formation of HCOOH is given by Eq. 11.15,

Scheme 11.14 Catalytic mechanism of CO_2 reduction by H_2 with $[Ir^{III}(Cp^*)(L)(H_2O)](SO_4)$ to form HCOOH in H_2O

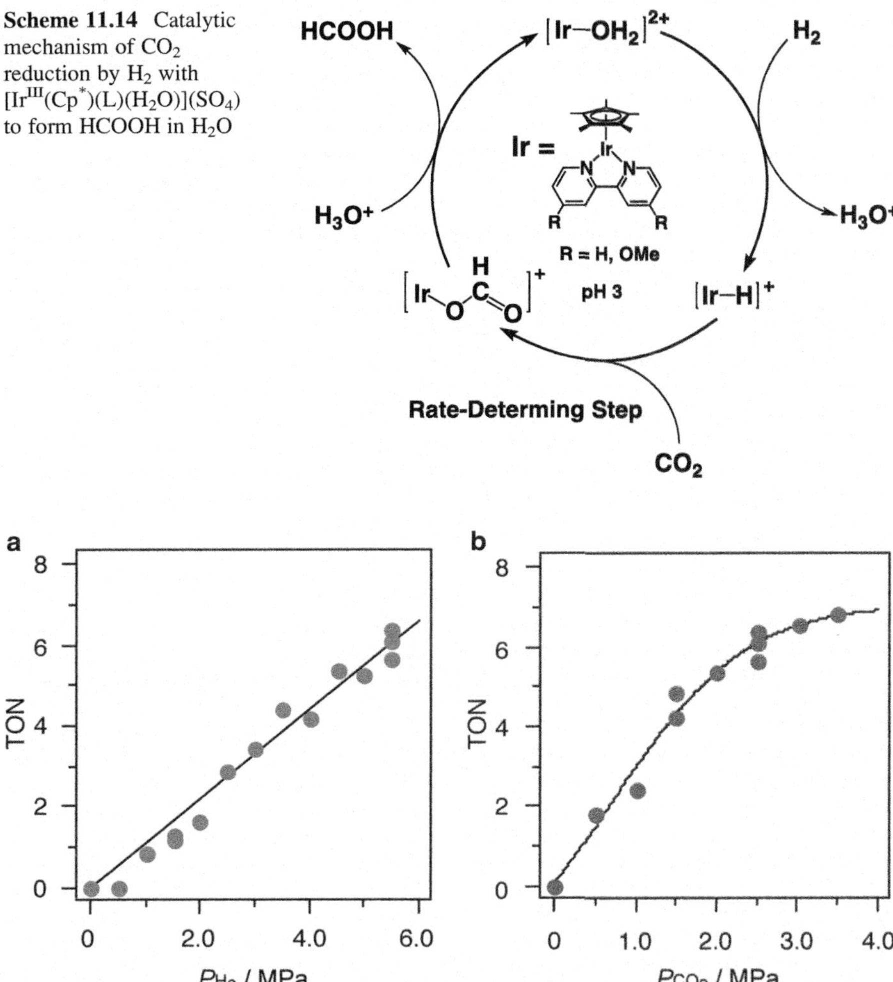

Fig. 11.14 (**a**) Dependence of TONs at 3 h on H_2 pressure for the reduction of CO_2 (2.5 MPa) by H_2 catalyzed by **2**(SO_4) (20 μmol) at pH 3.0 in a citrate buffer solution (20 cm^3) at 40 °C. (**b**) Dependence of TONs at 3 h on CO_2 pressure for the reduction of CO_2 by H_2 (5.5 MPa) catalyzed by **2**(SO_4) (20 μmol) at pH 3.0 in a citrate buffer solution (20 cm^3) at 40 °C (Reproduced from Ref. [107] by permission of The Royal Society of Chemistry)

which indicates that the rate becomes constant at large CO_2 pressure as observed in Fig. 11.14b [107]. The Ru(III)–hydride complex, which was prepared independently by the reaction of the Ru(III)–aqua complex with $NaBH_4$, reacted with CO_2 to produce the formate complex, which was detected by ESI-mass and ^{1}H NMR spectra [108, 109]:

$$\mathrm{d[HCOOH]}/\mathrm{d}t = k_1k_2[[\mathrm{Ru{-}OH_2}]^{2+}]P_{\mathrm{H2}}P_{\mathrm{CO2}}/(k_{-1} + k_2P_{\mathrm{CO2}}) \qquad (11.15)$$

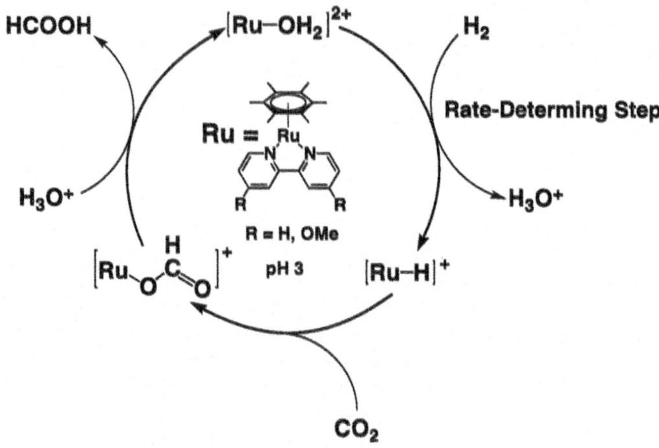

Scheme 11.15 Catalytic mechanism of CO_2 reduction by H_2 with $[Ru^{II}(\eta^6\text{-}C_6Me_6)(L)(H_2O)](SO_4)$ to form HCOOH in H_2O

The change in the rate-determining step in the catalytic reduction of CO_2 by H_2 between the Ir and Ru complexes results from the stronger Ir–H bond as compared with the Ru–H bond as indicated by the higher Ir–H stretching frequency (2,056 cm^{-1}) than the Ru–H stretching frequency (1,899 cm^{-1}) [107]. The stronger Ir–H bond facilitates the formation of the Ir–H bond, but decelerates the Ir–H bond cleavage by CO_2, which becomes the rate-determining step in the Ir complex-catalyzed CO_2 reduction by H_2. Conversely the weaker Ru–H bond facilitates the Ru–H bond cleavage by CO_2 but decelerates the formation of the Ru–H bond, which becomes the rate-determining step. The initial TOF for the catalytic reduction of CO_2 by H_2 with $[Ir^{III}(Cp^*)(L)(H_2O)](SO_4)$ was improved from 1 h^{-1} (L = bpy) to 27 h^{-1} (L = 4,4′-OMe-bpy) [107]. Thus, the more electron-rich Ir–H complex exhibits the higher catalytic reactivity. The X-ray crystal structures of the Ir–H and Ru–H complexes are shown in Fig. 11.15 [55, 109]. In both cases, the hydride complexes adopt a distorted octahedral coordination which has a terminal hydride ligand.

The catalytic activity for the reduction of CO_2 to HCOOH was enhanced by using a C^N cyclometalated organoiridium complex ($[Ir^{III}(Cp^*)\{4\text{-}(1H\text{-pyrazol-1-yl-}\kappa N^2)\text{benzoic acid-}\kappa C^3\}(H_2O)]_2SO_4$ ($[\mathbf{Ir^A}\text{–}H_2O]^+$)), which was employed for the catalytic decomposition of HCOOH to H_2 under acidic conditions in Fig. 11.7 (vide supra) [66]. At pH 7.5, the carboxylic acid is deprotonated to produce the more electron-rich Ir complex ($[\mathbf{Ir^B}\text{–}H_2O]^0$) when the direction of the reaction was reversed and the catalytic reduction of CO_2 by H_2 with $[\mathbf{Ir^B}\text{–}H_2O]^0$ occurred to produce formate at ambient pressure and temperature as shown in Fig. 11.16, where TON increased linearly with time to exceed over 100 [66]. Turnover frequency (TOF) increased with decrease in pH to afford the highest value at pH 8.8 and decreased with further increase in pH to reach zero at pH 10.4 [66]. The pH dependence of TOF is similar to pH dependence of the amount ratios of $[\mathbf{Ir^B}\text{–}H_2O]^0$

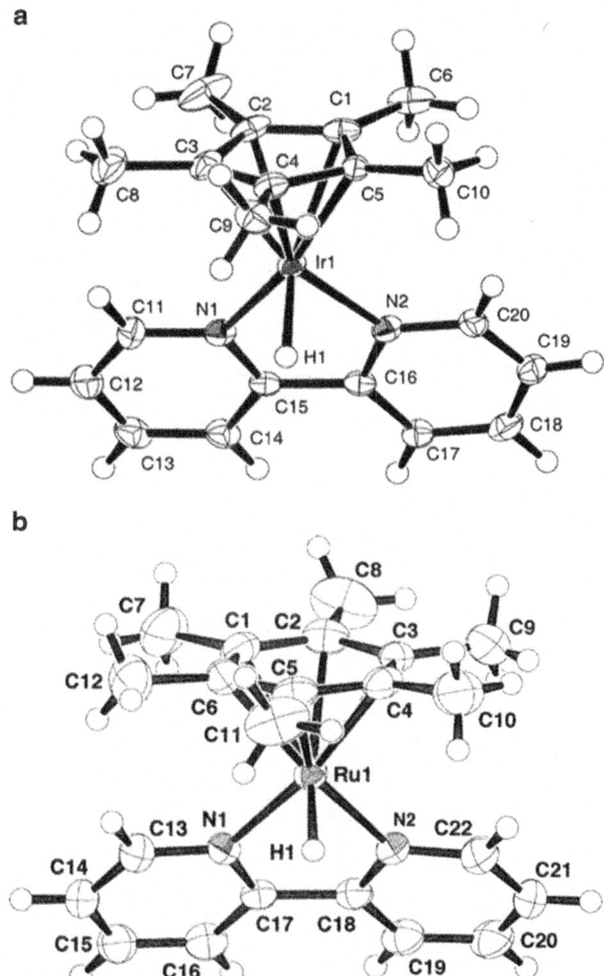

Fig. 11.15 ORTEP drawings of (**a**) $[Ir^{III}(Cp^*)(bpy)(H)](PF_6)$ [109] and (**b**) $[Ru^{II}(\eta^6\text{-}C_6Me_6)(H)](CF_3SO_3)$ [55]. The counter anions are omitted for clarity (Reprinted with the permission from Ref. [55, 109]. Copyright 2003 American Chemical Society)

over $[\mathbf{Ir^B}\text{–}OH]^-$ and HCO_3^- over CO_3^{2-} (red line and red dashed line in Fig. 11.17, respectively). Thus, the reduction of HCO_3^- by H_2 is catalyzed by $[\mathbf{Ir^B}\text{–}H_2O]^0$ at pH 8.8. The TOF value at pH 8.8 increased linearly with increasing concentration of CO_2, which is converted to mixture of HCO_3^- and CO_3^{2-} (Fig. 11.18) [66]. Thus, the rate-determining step in the catalytic reduction of CO_2 to formate by H_2 is the insertion of CO_2 to the Ir–H complex $[\mathbf{Ir^B}\text{–}H]^0$ to produce the formate complex in Scheme 11.16 [66].

A dinuclear Cp*Ir catalyst with 4,4,6,6-tetrahydroxy-2,2-bipyrimidine as a bridging ligand (see the crystal structure in Fig. 11.19) can also catalyze the reduction of CO_2 by H_2 at ambient pressure at pH 8.4 with TOF = 70 h^{-1} at 298 K [110]. Mononuclear Cp*Ir complexes with biazole ligands also act as

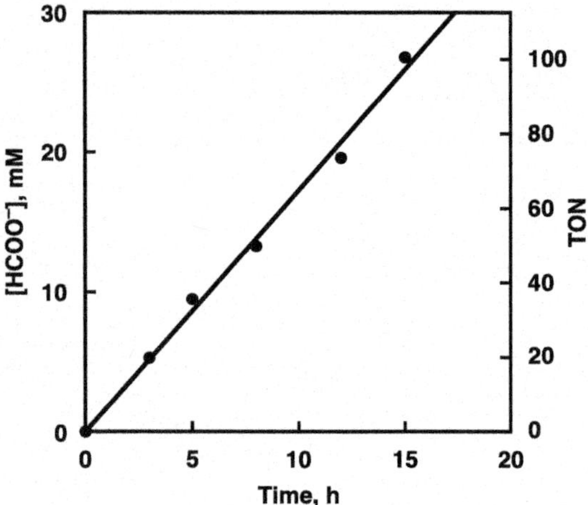

Fig. 11.16 Time course of the concentration of formate and TON for the formate formation in the reduction of CO_2 by H_2 catalyzed by $[\mathbf{Ir^B}\text{–}H_2O]^0$ (0.26 mM) under atmospheric pressure of H_2 (50 mL/min) and CO_2 (50 mL/min) in deaerated H_2O at 303 K at pH 7.5 (Reproduced from Ref. [66] by permission of The Royal Society of Chemistry)

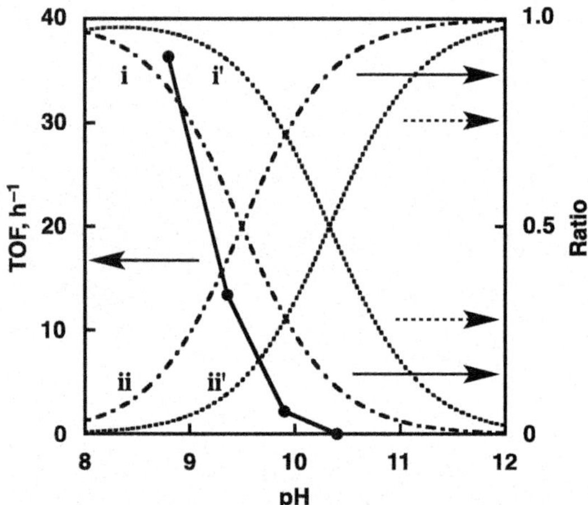

Fig. 11.17 pH dependence of the formation rate (TOF) of formate in the catalytic generation of formate from H_2, HCO_3^-, and CO_3^{2-} ($[HCO_3^-]+[CO_3^{2-}]=2.0$ M) catalyzed by $[\mathbf{Ir^B}\text{–}H_2O]^0$ (0.18 mM) in deaerated H_2O at 333 K (*solid line*). Alternate *long* and *short dashed lines*, (*i*) and (*ii*) show the amount ratios of complex $[\mathbf{Ir^B}\text{–}H_2O]^0$ and $[\mathbf{Ir^B}\text{–}OH]^-$, respectively, to the total amount of these complexes. *Dashed lines*, (*i'*) and (*ii'*) show the ratios of HCO_3^- and CO_3^{2-}, respectively (Reproduced from Ref. [66] by permission of The Royal Society of Chemistry)

efficient catalysts for reduction of CO_2 by H_2 to formate at ambient pressure and temperature [111].

Iridium complexes mentioned above act as efficient catalysts for the selective decomposition of formic acid to H_2 and CO_2 without formation of CO under acidic conditions at ambient temperature [55, 66, 110, 111]. Thus, the catalytic interconversion between hydrogen and formic acid has been made possible by changing pH

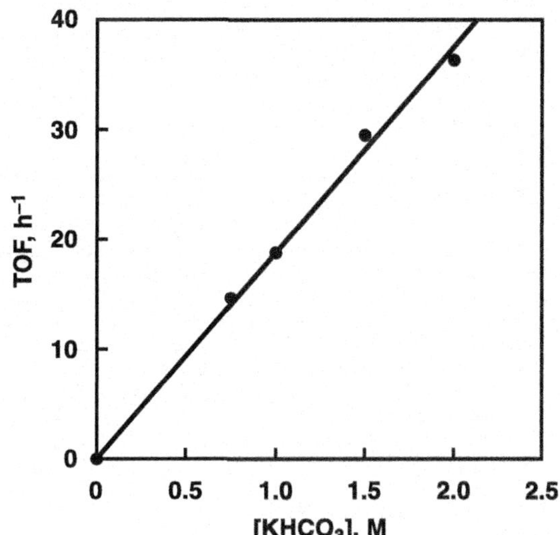

Fig. 11.18 Plot of TOF vs. the concentration of $KHCO_3$ and K_2CO_3 mixture ($KHCO_3/K_2CO_3$), i.e., $[KHCO_3] + [K_2CO_3]$ in the hydrogenation reaction of $KHCO_3/K_2CO_3$ with H_2 catalyzed by $[\mathbf{Ir^B}–H_2O]^0$ (0.18 mM) under atmospheric pressure of H_2 in deaerated H_2O at pH 8.8 at 333 K (Reproduced from Ref. [66] by permission of The Royal Society of Chemistry)

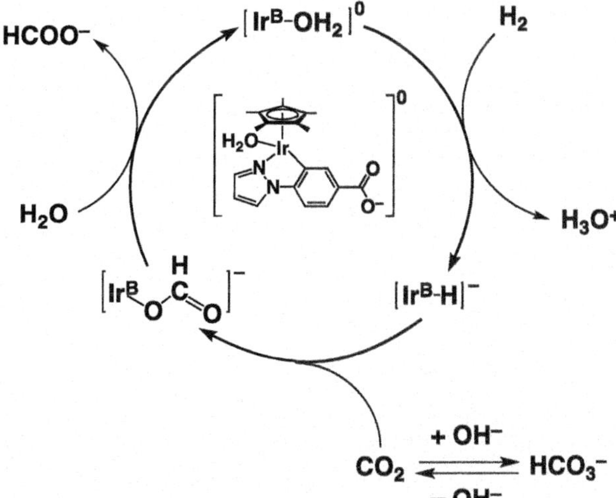

Scheme 11.16 Catalytic mechanism of CO_2 reduction by H_2 with $[\mathbf{Ir^A}–H_2O]^+$ to form $HCOO^-$ in H_2O

with the same catalyst, providing a convenient hydrogen-on-demand system in which hydrogen (gas) can be stored as formic acid (liquid) and whenever needed hydrogen is produced by the catalytic decomposition of formic acid [11, 66, 112]. Formic acid can also be directly used as a fuel in direct formic acid fuel cells, which have recently attracted much attention due to high electromotive force, limited fuel crossover, and high practical power densities at low temperatures as compared with direct methanol fuel cells [113–116].

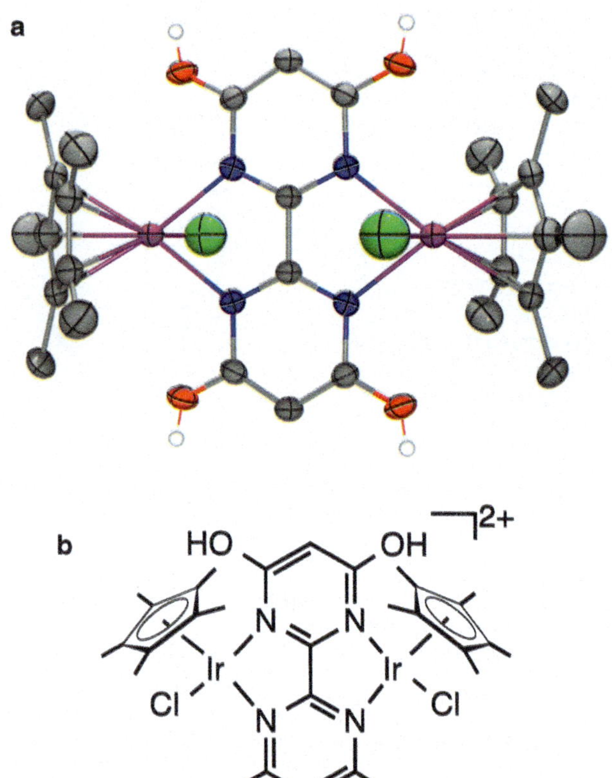

Fig. 11.19 (**a**) X-ray crystal structure and (**b**) the structural formula of a dinuclear Cp*Ir catalyst with 4,4,6,6-tetrahydroxy-2,2-bipyrimidine as a bridging ligand employed for efficient CO_2 reduction to formate by H_2 (Reprinted by permission from Macmillan Publishers Ltd: Ref. [110], copyright 2012)

11.3 Conclusions

We have overviewed kinetic studies on catalytic reduction of protons and CO_2 in mainly homogeneous phase, providing valuable mechanistic insights. Kinetic studies including deuterium kinetic isotope effects on heterogeneous catalysts for hydrogen evolution have also been demonstrated to provide essential mechanistic information on bond cleavage and formation associated with electron transfer. Kinetic studies have also enabled us to determine the rate-determining steps in the catalytic cycles, providing valuable information on observable intermediates, which can be detected by various methods. The most important intermediates in the catalytic reduction of protons and CO_2 are metal-hydride complexes, which can reduce protons and CO_2 to produce hydrogen and formic acid, respectively. Metal η^1-CO_2 complexes that are formed by a nucleophilic attack to low-valent metal complexes with the central carbon are responsible for the two-electron reduction of CO_2 to CO [33, 99, 117–119]. The key remaining challenge is not just two-electron reduction of CO_2 with two protons to formic acid or carbon monoxide (CO) but multiple proton-coupled electron transfers to produce further reduced products such as methanol and methane. Such multi-electron reduction beyond two-electron reduction of CO_2 has so far been achieved in the heterogeneous systems by photocatalysis and electrocatalysis [13, 16, 120–132]. A series of different homogeneous catalysts have also been employed to achieve the catalytic reduction of CO_2 by H_2 to methanol in a single vessel to promote the various steps of the CO_2 reduction sequence [133]. Recently the homogenously catalyzed reduction of CO_2 by H_2 to methanol has been achieved by a single ruthenium phosphine complex [134, 135]. Methanol can also be obtained by the disproportion of formic acid catalyzed by an Ir complex ($[Ir^{III}(Cp^*)(bpy)(H_2O)](OTf)_2$) [136]. Further kinetic studies on such homogeneously catalyzed multi-electron reduction of CO_2 by H_2 may elucidate the catalytic mechanisms, which will certainly help develop efficient catalysts for production of carbon-neutral alternatives to fossil fuels.

Acknowledgements The authors gratefully acknowledge the contributions of their collaborators and coworkers cited in the references and support by an ALCA (Advanced Low Carbon Technology Research and Development) program from the Japan Science and Technology Agency and funds from the Ministry of Education, Culture, Sports, Science, and Technology, Japan.

References

1. Lewis NS, Nocera DG (2006) Powering the planet: chemical challenges in solar energy utilization. Proc Natl Acad Sci U S A 103:15729–15735
2. Gray HB (2009) Powering the planet with solar fuel. Nat Chem 1:7–7
3. Thomas JM (2014) Reflections on the topic of solar fuels. Energy Environ Sci 7:19–20
4. Faunce TA, Lubitz W, Rutherford AW, MacFarlane D, Moore GF, Yang P, Nocera DG, Moore TA, Gregory DH, Fukuzumi S, Yoon KB, Armstrong FA, Wasielewski MR, Styring S (2013) Energy and environment policy case for a global project on artificial photosynthesis. Energy Environ Sci 6:695–698
5. Züttel A, Borgschulte A, Schlapbach L (eds) (2011) Hydrogen as a future energy carrier. Wiley-VCH, Weinheim
6. Nocera DG (2012) The artificial leaf. Acc Chem Res 45:767–776
7. Kärkäs MD, Johnston EV, Verho O, Åkermark B (2013) Artificial photosynthesis: from nanosecond electron transfer to catalytic water oxidation. Acc Chem Res 47:100–111
8. Wen F, Li C (2013) Hybrid artificial photosynthetic systems comprising semiconductors as light harvesters and biomimetic complexes as molecular cocatalysts. Acc Chem Res 46:2355–2364
9. Bensaid S, Centi G, Garrone E, Perathoner S, Saracco G (2012) Towards artificial leaves for solar hydrogen and fuels from carbon dioxide. ChemSusChem 5:500–521
10. Fukuzumi S, Hong D, Yamada Y (2013) Bioinspired photocatalytic water reduction and oxidation with earth-abundant metal catalysts. J Phys Chem Lett 4:3458–3467
11. Fukuzumi S (2008) Bioinspired energy conversion systems for hydrogen production and storage. Eur J Inorg Chem 2008:1351–1362
12. Balzani V, Credi A, Venturi M (2008) Photochemical conversion of solar energy. ChemSusChem 1:26–58
13. Habisreutinger SN, Schmidt-Mende L, Stolarczyk JK (2013) Photocatalytic reduction of CO_2 on TiO_2 and other semiconductors. Angew Chem Int Ed 52:7372–7408

14. Liao F, Zeng Z, Eley C, Lu Q, Hong X, Tsang SCE (2012) Electronic modulation of a copper/zinc oxide catalyst by a heterojunction for selective hydrogenation of carbon dioxide to methanol. Angew Chem Int Ed 51:5832–5836
15. Benson EE, Kubiak CP, Sathrum AJ, Smieja JM (2009) Electrocatalytic and homogeneous approaches to conversion of CO_2 to liquid fuels. Chem Soc Rev 38:89–99
16. Liu Q, Wu D, Zhou Y, Su H, Wang R, Zhang C, Yan S, Xiao M, Zou Z (2014) Single-crystalline, ultrathin $ZnGa_2O_4$ nanosheet scaffolds to promote photocatalytic activity in CO_2 reduction into methane. ACS Appl Mater Interfaces 6:2356–2361
17. Wasielewski MR (2009) Self-assembly strategies for integrating light harvesting and charge separation in artificial photosynthetic systems. Acc Chem Res 42:1910–1921
18. Gust D, Moore TA, Moore AL (2009) Solar fuels via artificial photosynthesis. Acc Chem Res 42:1890–1898
19. Guldi DM, Sgobba V (2011) Carbon nanostructures for solar energy conversion schemes. Chem Commun 47:606–610
20. Fukuzumi S (2008) Development of bioinspired artificial photosynthetic systems. PCCP 10:2283–2297
21. D'Souza F, Ito O (2012) Photosensitized electron transfer processes of nanocarbons applicable to solar cells. Chem Soc Rev 41:86–96
22. Fukuzumi S, Ohkubo K (2012) Assemblies of artificial photosynthetic reaction centres. J Mater Chem 22:4575–4587
23. Fukuzumi S, Ohkubo K, Suenobu T (2014) Long-lived charge separation and applications in artificial photosynthesis. Acc Chem Res 47:1455–1464
24. Fukuzumi S, Yamada Y, Suenobu T, Ohkubo K, Kotani H (2011) Catalytic mechanisms of hydrogen evolution with homogeneous and heterogeneous catalysts. Energy Environ Sci 4:2754–2766
25. Fukuzumi S, Yamada Y (2012) Catalytic activity of metal-based nanoparticles for photocatalytic water oxidation and reduction. J Mater Chem 22:24284–24296
26. Dempsey JL, Brunschwig BS, Winkler JR, Gray HB (2009) Hydrogen evolution catalyzed by cobaloximes. Acc Chem Res 42:1995–2004
27. Eckenhoff WT, Eisenberg R (2012) Molecular systems for light driven hydrogen production. Dalton Trans 41:13004–13021
28. Halpin Y, Pryce MT, Rau S, Dini D, Vos JG (2013) Recent progress in the development of bimetallic photocatalysts for hydrogen generation. Dalton Trans 42:16243–16254
29. Schneider J, Jia H, Muckerman JT, Fujita E (2012) Thermodynamics and kinetics of CO_2, CO, and H^+ binding to the metal centre of CO_2 reduction catalysts. Chem Soc Rev 41:2036–2051
30. Schulz M, Karnahl M, Schwalbe M, Vos JG (2012) The role of the bridging ligand in photocatalytic supramolecular assemblies for the reduction of protons and carbon dioxide. Coord Chem Rev 256:1682–1705
31. Fukuzumi S, Suenobu T (2013) Hydrogen storage and evolution catalysed by metal hydride complexes. Dalton Trans 42:18–28
32. Jessop PG, Joó F, Tai C-C (2004) Recent advances in the homogeneous hydrogenation of carbon dioxide. Coord Chem Rev 248:2425–2442
33. Kobayashi K, Tanaka K (2014) Approach to multi-electron reduction beyond two-electron reduction of CO_2. Phys Chem Chem Phys 16:2240–2250
34. Concepcion JJ, Jurss JW, Brennaman MK, Hoertz PG, Patrocinio AOT, Murakami Iha NY, Templeton JL, Meyer TJ (2009) Making oxygen with ruthenium complexes. Acc Chem Res 42:1954–1965
35. Sala X, Maji S, Bofill R, García-Antón J, Escriche L, Llobet A (2013) Molecular water oxidation mechanisms followed by transition metals: state of the art. Acc Chem Res 47:504–516
36. Meyer TJ, Huynh MHV (2003) The remarkable reactivity of high oxidation state ruthenium and osmium polypyridyl complexes. Inorg Chem 42:8140–8160

37. Lv H, Geletii YV, Zhao C, Vickers JW, Zhu G, Luo Z, Song J, Lian T, Musaev DG, Hill CL (2012) Polyoxometalate water oxidation catalysts and the production of green fuel. Chem Soc Rev 41:7572–7589
38. Liu X, Wang F (2012) Transition metal complexes that catalyze oxygen formation from water: 1979–2010. Coord Chem Rev 256:1115–1136
39. Cao R, Lai W, Du P (2012) Catalytic water oxidation at single metal sites. Energy Environ Sci 5:8134–8157
40. Singh A, Spiccia L (2013) Water oxidation catalysts based on abundant 1st row transition metals. Coord Chem Rev 257:2607–2622
41. Cady CW, Crabtree RH, Brudvig GW (2008) Functional models for the oxygen-evolving complex of photosystem II. Coord Chem Rev 252:444–455
42. Duan L, Tong L, Xu Y, Sun L (2011) Visible light-driven water oxidation-from molecular catalysts to photoelectrochemical cells. Energy Environ Sci 4:3296–3313
43. Le Goff A, Artero V, Jousselme B, Tran PD, Guillet N, Métayé R, Fihri A, Palacin S, Fontecave M (2009) From hydrogenases to noble metal–free catalytic nanomaterials for H_2 production and uptake. Science 326:1384–1387
44. Thoi VS, Sun Y, Long JR, Chang CJ (2013) Complexes of earth-abundant metals for catalytic electrochemical hydrogen generation under aqueous conditions. Chem Soc Rev 42:2388–2400
45. Wang M, Chen L, Sun L (2012) Recent progress in electrochemical hydrogen production with earth-abundant metal complexes as catalysts. Energy Environ Sci 5:6763–6778
46. Losse S, Vos JG, Rau S (2010) Catalytic hydrogen production at cobalt centres. Coord Chem Rev 254:2492–2504
47. Esswein AJ, Nocera DG (2007) Hydrogen production by molecular photocatalysis. Chem Rev 107:4022–4047
48. Wang M, Na Y, Gorlov M, Sun L (2009) Light-driven hydrogen production catalysed by transition metal complexes in homogeneous systems. Dalton Trans 2009:6458–6467
49. Mandal S, Shikano S, Yamada Y, Lee Y-M, Nam W, Llobet A, Fukuzumi S (2013) Protonation equilibrium and hydrogen production by a dinuclear cobalt–hydride complex reduced by cobaltocene with trifluoroacetic acid. J Am Chem Soc 135:15294–15297
50. Varma S, Castillo CE, Stoll T, Fortage J, Blackman AG, Molton F, Deronzier A, Collomb M-N (2013) Efficient photocatalytic hydrogen production in water using a cobalt(III) tetraaza-macrocyclic catalyst: electrochemical generation of the low-valent Co(I) species and its reactivity toward proton reduction. Phys Chem Chem Phys 15:17544–17552
51. Guttentag M, Rodenberg A, Kopelent R, Probst B, Buchwalder C, Brandstätter M, Hamm P, Alberto R (2012) Photocatalytic H_2 production with a rhenium/cobalt system in water under acidic conditions. Eur J Inorg Chem 2012:59–64
52. Krishnan CV, Brunschwig BS, Creutz C, Sutin N (1985) Homogeneous catalysis of the photoreduction of water. 6. Mediation by polypyridine complexes of ruthenium(II) and cobalt (II) in alkaline media. J Am Chem Soc 107:2005–2015
53. Krishnan CV, Sutin N (1981) Homogeneous catalysis of the photoreduction of water by visible light. 2. Mediation by a tris(2,2′-bipyridine)ruthenium(II)-cobalt(II) bipyridine system. J Am Chem Soc 103:2141–2142
54. Fukuzumi S, Kobayashi T, Suenobu T (2008) Efficient catalytic decomposition of formic acid for the selective generation of H_2 and H/D exchange with a water soluble rhodium complex in aqueous solution. ChemSusChem 1:827–834
55. Abura T, Ogo S, Watanabe Y, Fukuzumi S (2003) Isolation and crystal structure of a water-soluble iridium hydride: a robust and highly active catalyst for acid-catalyzed transfer hydrogenations of carbonyl compounds in acidic media. J Am Chem Soc 125:4149–4154
56. Suenobu T, Guldi DM, Ogo S, Fukuzumi S (2003) Excited-state deprotonation and H/D exchange of an iridium hydride complex. Angew Chem Int Ed 42:5492–5495

57. Fukuzumi S, Kobayashi T, Suenobu T (2011) Photocatalytic production of hydrogen by disproportionation of one-electron-reduced rhodium and iridium–ruthenium complexes in water. Angew Chem Int Ed 50:728–731
58. Stoll T, Gennari M, Serrano I, Fortage J, Chauvin J, Odobel F, Rebarz M, Poizat O, Sliwa M, Deronzier A, Collomb M-N (2013) $[Rh^{III}(dmbpy)_2Cl_2]^+$ as a highly efficient catalyst for visible-light-driven hydrogen production in pure water: comparison with other rhodium catalysts. Chem Eur J 19:782–792
59. Amouyal E, Koffi P (1985) Photochemical production of hydrogen from water. J Photochem 29:227–242
60. Weddle KS, Aiken JD, Finke RG (1998) Rh(0) nanoclusters in benzene hydrogenation catalysis: kinetic and mechanistic evidence that a putative $[(C_8H_{17})_3NCH_3]^+[RhCl_4]^-$-ion-pair catalyst is actually a distribution of Cl^- and $[(C_8H_{17})_3NCH_3]^+$ stabilized Rh (0) nanoclusters. J Am Chem Soc 120:5653–5666
61. Fukuzumi S, Kobayashi T, Suenobu T (2010) Unusually large tunneling effect on highly efficient generation of hydrogen and hydrogen isotopes in pH-selective decomposition of formic acid catalyzed by a heterodinuclear iridium-ruthenium complex in water. J Am Chem Soc 132:1496–1497
62. Kwart H (1982) Temperature dependence of the primary kinetic hydrogen isotope effect as a mechanistic criterion. Acc Chem Res 15:401–408
63. Bercaw JE, Chen GS, Labinger JA, Lin B-L (2008) Hydrogen tunneling in protonolysis of platinum(II) and palladium(II) methyl complexes: mechanistic implications. J Am Chem Soc 130:17654–17655
64. Pan Z, Horner JH, Newcomb M (2008) Tunneling in C – H oxidation reactions by an oxoiron (IV) porphyrin radical cation: direct measurements of very large H/D kinetic isotope effects. J Am Chem Soc 130:7776–7777
65. Kohen A, Klinman JP (1998) Enzyme catalysis: beyond classical paradigms. Acc Chem Res 31:397–404
66. Maenaka Y, Suenobu T, Fukuzumi S (2012) Catalytic interconversion between hydrogen and formic acid at ambient temperature and pressure. Energy Environ Sci 5:7360–7367
67. Ruelle P, Kesselring UW, Ho N-T (1986) Ab initio quantum-chemical study of the unimolecular pyrolysis mechanisms of formic acid. J Am Chem Soc 108:371–375
68. Dau H, Haumann M (2008) The manganese complex of photosystem II in its reaction cycle-basic framework and possible realization at the atomic level. Coord Chem Rev 252:273–295
69. Maenaka Y, Suenobu T, Fukuzumi S (2011) Efficient catalytic interconversion between NADH and NAD^+ accompanied by generation and consumption of hydrogen with a water-soluble iridium complex at ambient pressure and temperature. J Am Chem Soc 134:367–374
70. Kotani H, Ono T, Ohkubo K, Fukuzumi S (2007) Efficient photocatalytic hydrogen evolution without an electron mediator using a simple electron donor-acceptor dyad. Phys Chem Chem Phys 9:1487–1492
71. Hasobe T, Sakai H, Mase K, Ohkubo K, Fukuzumi S (2013) Remarkable enhancement of photocatalytic hydrogen evolution efficiency utilizing an internal cavity of supramolecular porphyrin hexagonal nanocylinders under visible-light irradiation. J Phys Chem C 117:4441–4449
72. Yamada Y, Miyahigashi T, Kotani H, Ohkubo K, Fukuzumi S (2012) Photocatalytic hydrogen evolution with Ni nanoparticles by using 2-phenyl-4-(1-naphthyl)quinolinium ion as a photocatalyst. Energy Environ Sci 5:6111–6118
73. Amao Y (2011) Solar fuel production based on the artificial photosynthesis system. ChemCatChem 3:458–474
74. Maenaka Y, Suenobu T, Fukuzumi S (2012) Hydrogen evolution from aliphatic alcohols and 1,4-selective hydrogenation of NAD^+ catalyzed by a C, N and a C, C cyclometalated organoiridium complex at room temperature in water. J Am Chem Soc 134:9417–9427
75. Nielsen M, Alberico E, Baumann W, Drexler H-J, Junge H, Gladiali S, Beller M (2013) Low-temperature aqueous-phase methanol dehydrogenation to hydrogen and carbon dioxide. Nature 495:85–89

76. Monney A, Barsch E, Sponholz P, Junge H, Ludwig R, Beller M (2014) Base-free hydrogen generation from methanol using a bi-catalytic system. Chem Commun 50:707–709
77. Chen Z, Glasson CRK, Holland PL, Meyer TJ (2013) Electrogenerated polypyridyl ruthenium hydride and ligand activation for water reduction to hydrogen and acetone to iso-propanol. Phys Chem Chem Phys 15:9503–9507
78. Bullock RM, Appel AM, Helm ML (2014) Production of hydrogen by electrocatalysis: making the H-H bond by combining protons and hydrides. Chem Commun 50:3125–3143
79. Chen S, Ho M-H, Bullock RM, DuBois DL, Dupuis M, Rousseau R, Raugei S (2014) Computing free energy landscapes: application to Ni-based electrocatalysts with pendant amines for H_2 production and oxidation. ACS Catal 4:229–242
80. Raugei S, Chen S, Ho M-H, Ginovska-Pangovska B, Rousseau RJ, Dupuis M, DuBois DL, Bullock RM (2012) The role of pendant amines in the breaking and forming of molecular hydrogen catalyzed by nickel complexes. Chem Eur J 18:6493–6506
81. Rose MJ, Gray HB, Winkler JR (2012) Hydrogen generation catalyzed by fluorinated diglyoxime-iron complexes at low overpotentials. J Am Chem Soc 134:8310–8313
82. Helm ML, Stewart MP, Bullock RM, DuBois MR, DuBois DL (2011) A synthetic nickel electrocatalyst with a turnover frequency above 100,000 s^{-1} for H_2 production. Science 333:863–866
83. DuBois DL (2014) Development of molecular electrocatalysts for energy storage. Inorg Chem 53:3935–3960
84. Fukuzumi S, Kotani H, Ohkubo K, Ogo S, Tkachenko NV, Lemmetyinen H (2004) Electron-transfer state of 9-mesityl-10-methylacridinium ion with a much longer lifetime and higher energy than that of the natural photosynthetic reaction center. J Am Chem Soc 126:1600–1601
85. Kotani H, Hanazaki R, Ohkubo K, Yamada Y, Fukuzumi S (2011) Size- and shape-dependent activity of metal nanoparticles as hydrogen-evolution catalysts: mechanistic insights into photocatalytic hydrogen evolution. Chem Eur J 17:2777–2785
86. Ohkubo K, Kotani H, Fukuzumi S (2005) Misleading effects of impurities derived from the extremely long-lived electron-transfer state of 9-mesityl-10-methylacridinium ion. Chem Commun 2005:4520–4522
87. Fukuzumi S, Kotani H, Ohkubo K (2008) Response: why had long-lived electron-transfer states of donor-substituted 10-methylacridinium ions been overlooked? Formation of the dimer radical cations detected in the near-IR region. Phys Chem Chem Phys 10:5159–5162
88. Hoshino M, Uekusa H, Tomita A, Koshihara S, Sato T, Nozawa S, Adachi S, Ohkubo K, Kotani H, Fukuzumi S (2012) Determination of the structural features of a long-lived-electron-transfer state of 9-mesityl-10-methylacridinium ion. J Am Chem Soc 134:4569–4572
89. Kotani H, Ohkubo K, Fukuzumi S (2012) Formation of a long-lived electron-transfer state of a naphthalene-quinolinium ion dyad and the π-dimer radical cation. Faraday Discuss 155:89–102
90. Yamada Y, Miyahigashi T, Kotani H, Ohkubo K, Fukuzumi S (2011) Photocatalytic hydrogen evolution under highly basic conditions by using Ru nanoparticles and 2-phenyl-4-(1-naphthyl)quinolinium ion. J Am Chem Soc 133:16136–16145
91. Aresta M, Dibenedetto A, Angelini A (2014) Catalysis for the valorization of exhaust carbon: from CO_2 to chemicals, materials, and fuels. Technological use of CO_2. Chem Rev 114:1709–1742
92. Saeidi S, Amin NAS, Rahimpour MR (2014) Hydrogenation of CO_2 to value-added-products—a review and potential future developments. J CO_2 Util 5:66–81
93. Wang W, Wang S, Ma X, Gong J (2011) Recent advances in catalytic hydrogenation of carbon dioxide. Chem Soc Rev 40:3703–3727
94. Kondratenko EV, Mul G, Baltrusaitis J, Larrazabal GO, Perez-Ramirez J (2013) Status and perspectives of CO_2 conversion into fuels and chemicals by catalytic, photocatalytic and electrocatalytic processes. Energy Environ Sci 6:3112–3135
95. Centi G, Quadrelli EA, Perathoner S (2013) Catalysis for CO_2 conversion: a key technology for rapid introduction of renewable energy in the value chain of chemical industries. Energy Environ Sci 6:1711–1731

96. Olah GA, Goeppert A, Prakash GKS (2009) Chemical recycling of carbon dioxide to methanol and dimethyl ether: from greenhouse gas to renewable, environmentally carbon neutral fuels and synthetic hydrocarbons. J Org Chem 74:487–498
97. Jessop PG, Ikariya T, Noyori R (1994) Homogeneous catalytic-hydrogenation of supercritical carbon-dioxide. Nature 368:231–233
98. Jessop PG, Ikariya T, Noyori R (1995) Homogeneous hydrogenation of carbon-dioxide. Chem Rev 95:259–272
99. Tanaka K, Ooyama D (2002) Multi-electron reduction of CO_2 via Ru-CO_2, -C(O)OH, -CO, -CHO, and -CH_2OH species. Coord Chem Rev 226:211–218
100. Wesselbaum S, Hintermair U, Leitner W (2012) Continuous-flow hydrogenation of carbon dioxide to pure formic acid using an integrated scCO_2 process with immobilized catalyst and base. Angew Chem Int Ed 51:8585–8588
101. Ziebart C, Federsel C, Anbarasan P, Jackstell R, Baumann W, Spannenberg A, Beller M (2012) Well-defined iron catalyst for improved hydrogenation of carbon dioxide and bicarbonate. J Am Chem Soc 134:20701–20704
102. Jeletic MS, Mock MT, Appel AM, Linehan JC (2013) A cobalt-based catalyst for the hydrogenation of CO_2 under ambient conditions. J Am Chem Soc 135:11533–11536
103. Huff CA, Sanford MS (2013) Catalytic CO_2 hydrogenation to formate by a ruthenium pincer complex. ACS Catal 3:2412–2416
104. Li Y-N, He L-N, Liu A-H, Lang X-D, Yang Z-Z, Yu B, Luan C-R (2013) In situ hydrogenation of captured CO_2 to formate with polyethyleneimine and Rh/monophosphine system. Green Chem 15:2825–2829
105. Drake JL, Manna CM, Byers JA (2013) Enhanced carbon dioxide hydrogenation facilitated by catalytic quantities of bicarbonate and other inorganic salts. Organometallics 32:6891–6894
106. Badiei YM, Wang W-H, Hull JF, Szalda DJ, Muckerman JT, Himeda Y, Fujita E (2013) Cp*Co(III) catalysts with proton-responsive ligands for carbon dioxide hydrogenation in aqueous media. Inorg Chem 52:12576–12586
107. Ogo S, Kabe R, Hayashi H, Harada R, Fukuzumi S (2006) Mechanistic investigation of CO_2 hydrogenation by Ru(II) and Ir(III) aqua complexes under acidic conditions: two catalytic systems differing in the nature of the rate determining step. Dalton Trans 4657–4663
108. Hayashi H, Ogo S, Fukuzumi S (2004) Aqueous hydrogenation of carbon dioxide catalysed by water-soluble ruthenium aqua complexes under acidic conditions. Chem Commun 2004:2714–2715
109. Hayashi H, Ogo S, Abura T, Fukuzumi S (2003) Accelerating effect of a proton on the reduction of CO_2 dissolved in water under acidic conditions. Isolation, crystal structure, and reducing ability of a water-soluble ruthenium hydride complex. J Am Chem Soc 125:14266–14267
110. Hull JF, Himeda Y, Wang W-H, Hashiguchi B, Periana R, Szalda DJ, Muckerman JT, Fujita E (2012) Reversible hydrogen storage using CO_2 and a proton-switchable iridium catalyst in aqueous media under mild temperatures and pressures. Nat Chem 4:383–388
111. Manaka Y, Wang W-H, Suna Y, Kambayashi H, Muckerman JT, Fujita E, Himeda Y (2014) Efficient H_2 generation from formic acid using azole complexes in water. Catal Sci Technol 4:34–37
112. Fujita E, Muckerman JT, Himeda Y (2013) Interconversion of CO_2 and formic acid by bio-inspired Ir complexes with pendent bases. Biochim Biophys Acta Bioenerg 1827:1031–1038
113. Ge J, Chen X, Liu C, Lu T, Liao J, Liang L, Xing W (2010) Promoting effect of vanadium ions on the anodic Pd/C catalyst for direct formic acid fuel cell application. Electrochim Acta 55:9132–9136
114. Wang R, Liu J, Liu P, Bi X, Yan X, Wang W, Ge X, Chen M, Ding Y (2014) Dispersing Pt atoms onto nanoporous gold for high performance direct formic acid fuel cells. Chem Sci 5:403–409
115. Cai W, Liang L, Zhang Y, Xing W, Liu C (2013) Real contribution of formic acid in direct formic acid fuel cell: investigation of origin and guiding for micro structure design. Int J Hydrog Energy 38:212–218

116. Ji X, Lee KT, Holden R, Zhang L, Zhang J, Botton GA, Couillard M, Nazar LF (2010) Nanocrystalline intermetallics on mesoporous carbon for direct formic acid fuel cell anodes. Nat Chem 2:286–293
117. Morris AJ, Meyer GJ, Fujita E (2009) Molecular approaches to the photocatalytic reduction of carbon dioxide for solar fuels. Acc Chem Res 42:1983–1994
118. Takeda H, Ishitani O (2010) Development of efficient photocatalytic systems for CO_2 reduction using mononuclear and multinuclear metal complexes based on mechanistic studies. Coord Chem Rev 254:346–354
119. Chen Z, Concepcion JJ, Brennaman MK, Kang P, Norris MR, Hoertz PG, Meyer TJ (2012) Splitting CO_2 into CO and O_2 by a single catalyst. Proc Natl Acad Sci U S A 109:15606–15611
120. Izumi Y (2013) Recent advances in the photocatalytic conversion of carbon dioxide to fuels with water and/or hydrogen using solar energy and beyond. Coord Chem Rev 257:171–186
121. Yan S, Wang J, Zou Z (2013) An anion-controlled crystal growth route to Zn_2GeO_4 nanorods for efficient photocatalytic conversion of CO_2 into CH_4. Dalton Trans 42:12975–12979
122. Roy SC, Varghese OK, Paulose M, Grimes CA (2010) Toward solar fuels: photocatalytic conversion of carbon dioxide to hydrocarbons. ACS Nano 4:1259–1278
123. Hamdy MS, Amrollahi R, Sinev I, Mei B, Mul G (2014) Strategies to design efficient silica-supported photocatalysts for reduction of CO_2. J Am Chem Soc 136:594–597
124. Yu J, Jin J, Cheng B, Jaroniec M (2014) A noble metal-free reduced graphene oxide-CdS nanorod composite for the enhanced visible-light photocatalytic reduction of CO_2 to solar fuel. J Mater Chem A 2:3407–3416
125. Andrews E, Ren M, Wang F, Zhang Z, Sprunger P, Kurtz R, Flake J (2013) Electrochemical reduction of CO_2 at Cu nanocluster/($10\hat{1}0$) ZnO electrodes. J Electrochem Soc 160:H841–H846
126. Le M, Ren M, Zhang Z, Sprunger PT, Kurtz RL, Flake JC (2011) Electrochemical reduction of CO_2 to CH_3OH at copper oxide surfaces. J Electrochem Soc 158:E45–E49
127. Schouten KJP, Kwon Y, van der Ham CJM, Qin Z, Koper MTM (2011) A new mechanism for the selectivity to C-1 and C-2 species in the electrochemical reduction of carbon dioxide on copper electrodes. Chem Sci 2:1902–1909
128. Costentin C, Robert M, Savéant J-M (2013) Catalysis of the electrochemical reduction of carbon dioxide. Chem Soc Rev 42:2423–2436
129. Qiao J, Liu Y, Hong F, Zhang J (2014) A review of catalysts for the electroreduction of carbon dioxide to produce low-carbon fuels. Chem Soc Rev 43:631–675
130. Peterson AA, Nørskov JK (2012) Activity descriptors for CO_2 electroreduction to methane on transition-metal catalysts. J Phys Chem Lett 3:251–258
131. Yan Y, Zeitler EL, Gu J, Hu Y, Bocarsly AB (2013) Electrochemistry of aqueous pyridinium: exploration of a key aspect of electrocatalytic reduction of CO_2 to methanol. J Am Chem Soc 135:14020–14023
132. Cole EB, Lakkaraju PS, Rampulla DM, Morris AJ, Abelev E, Bocarsly AB (2010) Using a one-electron shuttle for the multielectron reduction of CO_2 to methanol: kinetic, mechanistic, and structural insights. J Am Chem Soc 132:11539–11551
133. Huff CA, Sanford MS (2011) Cascade catalysis for the homogeneous hydrogenation of CO_2 to methanol. J Am Chem Soc 133:18122–18125
134. Wesselbaum S, vom Stein T, Klankermayer J, Leitner W (2012) Hydrogenation of carbon dioxide to methanol by using a homogeneous ruthenium-phosphine catalyst. Angew Chem Int Ed 51:7499–7502
135. Li Y, Junge K, Beller M (2013) Improving the efficiency of the hydrogenation of carbonates and carbon dioxide to methanol. ChemCatChem 5:1072–1074
136. Miller AJM, Heinekey DM, Mayer JM, Goldberg KI (2013) Catalytic disproportionation of formic acid to generate methanol. Angew Chem Int Ed 52:3981–3984

Chapter 12
Molecular Catalysts and Organometallics for Water Oxidation

Khurram Saleem Joya

Abstract Water can be used as a cheap and renewable source of electrons and protons to make nonfossil fuel-based chemical energy carriers for a sustainable power supply. However, water oxidation is an intricate chemical process and an energy-intensive reaction involving the removal of four electrons with the release of four protons at the same time. Inside the thylakoid membrane in plant leaves is embedded a manganese-calcium molecular cluster in natural photosystem II (PS-II), which represents an excellent model for designing an artificial equivalent of the photosynthesis for light-to-fuel conversion via water splitting. Inspired by the natural PS-II, the scientific community has been striving hard during the last two decades to develop a bio-inspired catalytic system for water oxidation. However, a truly biomimetic catalytic system matching the performance of photosystem for efficient water splitting operating with four consecutive proton-coupled electron transfer (PCET) steps to generate oxygen and hydrogen for hundred thousands of cycles at high rate is yet to be demonstrated. In this chapter, we provide an insight regarding the biomimetic approaches to make molecular and organometallic water oxidation complexes that have been investigated recently in homogeneous solution catalysis using chemical oxidants or as surface-immobilized heterogeneous species for electro-assisted catalytic systems. After comparing their catalytic activities and stabilities, an overview of the mechanistic aspects is also discussed.

Keywords Water oxidation • Renewable energy • Molecular catalysts • Photosystem II • Organometallics • Sunlight conversion • Solar fuel

K.S. Joya (✉)
Leiden Institute of Chemistry, Gorlaeus Laboratory, Leiden University, 9502, 2300 RA Leiden, The Netherlands

Division of Physical Sciences and Engineering, KAUST Catalysis Center (KCC), King Abdullah University of Science and Technology (KAUST), 4700 KAUST, Thuwal 23955-6900, Saudi Arabia

Department of Chemistry, University of Engineering and Technology, GT Road, 54890 Lahore, Punjab, Pakistan
e-mail: khurramsj@chem.leidenuniv.nl

W.-Y. Wong (ed.), *Organometallics and Related Molecules for Energy Conversion*, Green Chemistry and Sustainable Technology, DOI 10.1007/978-3-662-46054-2_12

12.1 Introduction

The quest for renewable fuels and alternative energy carriers from earth-abundant materials is currently one of the greatest scientific and technological challenges [1]. It is expected that the future energy demand will get doubled by 2050 from the present consumption value of 15 TW [2]. For the production of renewable and clean fuel, easy and abundant supply of protons and electrons is required, and water is the most abundant chemical feedstock (covers about 71 % of the earth's surface) for obtaining the reducing equivalents that can be used to produce energy carriers [3, 4]. However, splitting water to release protons and electrons at tremendous rate is an energy-demanding chemical reaction and needs efficient catalytic system obtained from renewable and inexpensive sources [5]. There is enormous amount of sunlight that strikes the earth's exterior (at a rate of 1.2×105 TW), and its effective use offers a potential solution to make protons and electrons via water oxidation process [6]. So the generation of hydrogen or carbon-neutral nonfossil fuels using sunlight and water as the raw materials represents an attractive solution for environmentally clean and renewable energy sources providing a mean to capture CO_2 and its useful conversion into energy-bearing chemical feedstock [6, 7].

The most challenging task is to develop a stable, efficient, and easily accessible water oxidation catalyst (WOC) from earth-abundant materials that is capable of multi-electron oxidation of water with simultaneous O_2 release at a high rate and optimal activity. This parallels the development of electrochemically driven water oxidation systems with molecular catalysts and oxide-based systems [8]. Natural PS-II offers an excellent model for developing an artificial solar-to-fuel conversion device, where a manganese-calcium oxygen-evolving complex (OEC) is involved in the process of four-electron water oxidation to generate dioxygen and release of four protons in a four-step consecutive proton-coupled electron transfer cycle [8, 9]. It also shows how a self-repairing structural framework of water oxidation catalytic chemistry can be applied to couple efficient charge separation with a four-step proton-coupled electron transfer pathway [9]. Mechanistically, incident light activation of the P680 chlorophylls induces the generation of a $P680^{+\cdot}$ radical cation by charge separation [9, 10]. The $P680^{+\cdot}$ cation radical is then re-reduced by the oxidation of tyrosine Yz in the vicinity. This acquires the oxidizing power in the Mn-cluster towards water oxidation by extracting four electrons from two water molecules with simultaneous dioxygen release as depicted in Scheme 12.1.

Both the water oxidation and reduction reactions are pH dependent. In catalytic systems, the two half reactions (pH = 0) are given as:

Oxidation:

$$2H_2O \rightarrow 4H^+ + 4e^- + O_2 \quad E^\circ = 1.23\ V(\text{vs. NHE}) \tag{12.1}$$

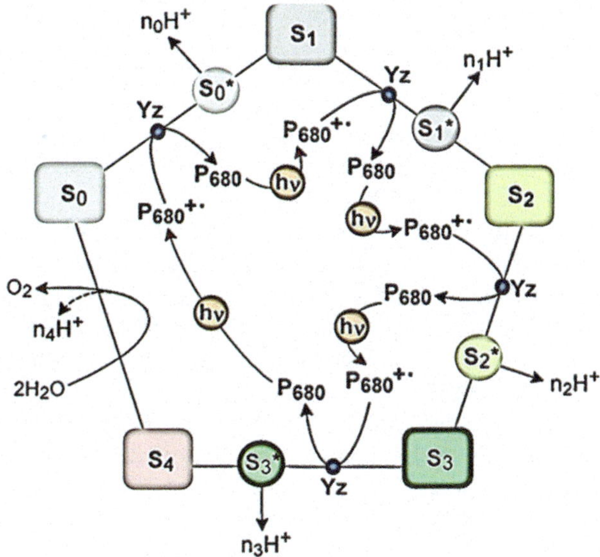

Scheme 12.1 Extended Kok cycle of light-induced water oxidation driven by $P680^{+\cdot}$ cation radical with tyrosine (Yz) acting as intermediate. The $S_{1–4}$ transient states are indicated by a* (the exact pathway of proton release is not known yet) (For further details, see Ref. [10])

Reduction:

$$4H^+ + 4e^- \rightarrow 2H_2 \quad E^\circ = 0.00\ \text{V}\ (\text{vs. NHE}) \tag{12.2}$$

and the overall four-electron water oxidation reaction is given by:

$$2H_2O_{(l)} \rightarrow 4e^- + 4H^+_{(aq)} + O_{2(g)} \quad E^\circ\text{cell} = 1.23\ \text{V};\ \Delta G = 475\ \text{kJ/mol} \tag{12.3}$$

Inspired by natural phenomenon, in molecular complexes, under acidic conditions in contact with the catalytic site [11], the water oxidation proceeds according to the following mechanisms:

$$2H_2O_{(l)} + [\text{Cat}-(OH_2)]^{2+} \rightarrow [\text{Cat}-(OH)]^{2+} + 2H_2O_{(l)} + H^+ + e^- \tag{12.3a}$$

$$\rightarrow [\text{Cat}(=O)]^{2+} + 2H_2O_{(l)} + 2(H^+ + e^-) \tag{12.3b}$$

$$\rightarrow [\text{Cat}-(OOH)]^{2+} + H_2O_{(l)} + 3(H^+ + e^-) \tag{12.3c}$$

$$\rightarrow [\text{Cat-}(OO)]^{2+} + H_2O_{(l)} + 4(H^+ + e^-)$$

$$\rightarrow [\text{Cat-}(OH_2)]^{2+} + O_{2(g)} + 4(H^+ + e^-) \tag{12.3d}$$

In the change in the chemical potential of the protons at the catalytic site as induced by an alkaline surrounding [11], the four-electron reaction proceeds according to following pathway:

$$4OH^- + [\text{Cat-}(OH_2)]^{2+} \rightarrow [\text{Cat-}(OH)]^{2+} + H_2O_{(l)} + 3OH^- + e^- \tag{12.3e}$$

$$\rightarrow [Cat(=OH)]^{2+} + 2H_2O_{(l)} + OH^- + 2e^- \quad (12.3f)$$

$$\rightarrow [Cat\text{-}(OOH)]^{2+} + 2H_2O_{(l)} + OH^- + 3e^- \quad (12.3g)$$

$$\begin{aligned} &\rightarrow [Cat\text{-}(OO)]^{2+} + 3H_2O_{(l)} + 4e^- \\ &\rightarrow [Cat\text{-}(OH_2)]^{2+} + O_{2(g)} + 2H_2O_{(l)} + 4e^- \end{aligned} \quad (12.3h)$$

In order to circumvent higher-energy intermediates in neutral pH condition, a good water oxidation catalyst must release four electrons and protons in four consecutive proton-coupled electron transfer steps [6, 11]. With four PCET steps (12.3a to 12.3d and 12.3e to 12.3h), a transformation of the reaction to a pH-independent reference frame provides a unified picture that is particularly useful when aiming for the design of complex device topologies [11].

The PS-II catalytic machinery has inspired many efforts in recent years to mimic the catalytic design and devised an artificial WOC using ruthenium and iridium metals in the molecular complexes and organometallics [10]. A major task is to establish an efficient and stable oxygen-evolving catalyst that displays multi-electron oxidation activity for hundred thousands of cycles [11]. There are many water-splitting models based on noble metal complexes, organometallics, and inorganic metal oxide catalysts, but none of them have proven good overall efficiency for water splitting that is considered as a bottleneck in constructing an artificial photosynthetic device [12]. In this chapter, we give an overview of the biomimetic approaches that were undertaken to make molecular and organometallic water oxidation complexes for homogeneous solution catalysis and surface-immobilized heterogeneous electrochemical assemblies for solar fuels.

12.2 Homogeneous Versus Heterogeneous Water Oxidation Catalysts

On the catalytic induction, the water oxidation reaction is induced by use of an external chemical oxidant, also known as a catalyst activator. One electron chemical oxidation is very useful as it gives more control over the reaction, kinetics, and mechanistic understanding [13]. The molecular catalytic species can be employed for water oxidation reaction, either by suspending in homogeneous solution or on the inert surfaces of nanoparticles, inorganic oxides, and conducting anodic materials [14]. The classical chemical oxidant that is used in labs for testing the molecular complexes for water oxidation in aqueous solution is cerium ammonium nitrate $(NH_4)_2[Ce(NO_3)_6]$, or CAN. Recently, some research groups also reported the use of $NaIO_4$ as primary oxidant in water oxidation. As long as the molecular complex stays intact, it is a homogeneous chemical species, but at times it tends to undergo self-oxidation or decomposition to generate metal nanoparticles or oxide materials and turns out as a heterogeneous entity. So, there is a lot of debate about

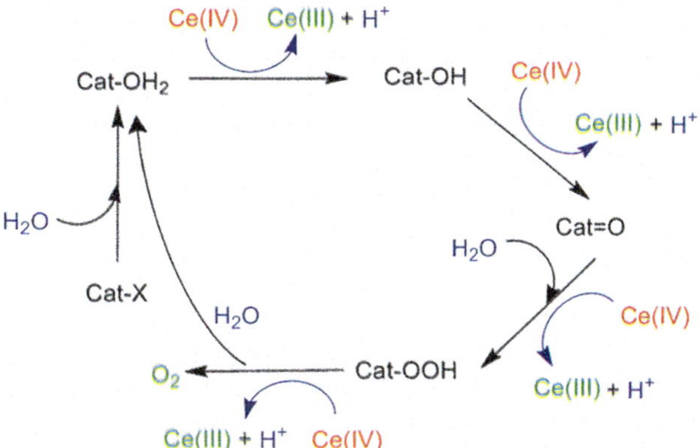

Scheme 12.2 Catalytic mechanism of water oxidation using CAN with molecular catalyst in aqueous phase

the integrity and homogeneity of the molecular water oxidation catalysts in harsh oxidizing conditions and their chemical fate during catalytic phase [15]. Metal oxides of transition metals (RuO_x, IrO_x, CoO_x, MnO_x, etc.), on the other hand, have also been studied extensively for water oxidation as pure heterogeneous phase catalysis; however, these inorganic materials are required in bulk quantities for application in catalytic systems [16].

In solution catalysis, the catalytic species are suspended in the homogeneous phase, preferably in aqueous acidic solutions (0.1 M HNO_3 or $HClO_4$), and Ce (IV) is added externally in excess to activate the metal complex (Eq. 12.2). Cerium ammonium nitrate with its tetravalent cerium is a monomeric species in aqueous solution and is thought to act as a single-electron oxidant [17]:

$$\mathrm{Cat} + 2\mathrm{OH}_{2(l)} + 4\mathrm{Ce}^{\mathrm{IV}}_{(\mathrm{aq})} \rightarrow 4\mathrm{H}^{+}_{(\mathrm{aq})} + \mathrm{O}_{2(\mathrm{q})} + 4\mathrm{Ce}^{\mathrm{III}}_{(\mathrm{aq})} \tag{12.4}$$

Here the four Ce(IV) species are changed to Ce(III) while abstracting four electrons during water oxidation to make one molecule of O_2 as shown in Scheme 12.2. However, EXAFS spectroscopy and density functional theory calculations have revealed that Ce(IV) generates oxo- and/or hydroxo-bridging binuclear complex, and a part of the oxygen is suggested to be generated from the NO_3^- ions in the CAN rather than from water [10]. This work has already put question marks on many studies using CAN in homogeneous phase. Moreover, the high concentration of the chemical oxidant probably promotes decomposition or decay of the catalyst. This suggests that one needs to be careful while using these strong oxidizing chemicals in water oxidation studies.

12.3 Molecular Approaches to Catalytic Water Oxidation

The molecular chemistry for water oxidation catalysis has started almost forty years ago aiming for biomimicking the Mn-Ca cluster in natural PS-II in design, PCET pathway, and stability to perform for hundred thousands of cycles at high rate

[18]. The first example of a molecular catalytic system for water oxidation came in the early 1970s with the introduction of a di-μ-oxo-bridged di-manganese (Mn^{III}–Mn^{IV}) 2,2′-bipyridine complex by M. Calvin [19]. This work was followed by several other studies during the next two decades of molecular water oxidation catalysts using essentially the same ligand architecture, with binuclear ruthenium and manganese motifs [9]. Only recently molecular catalysts with other metals like iridium, iron, cobalt, and copper in mononuclear or multisite regime have been introduced for water oxidation reaction in solution [20]. At present, the field is growing, and many research groups funded with huge budgets are focusing on the molecular science for water oxidation and solar fuels. Here we classified the water oxidation catalysis via molecular species into three groups—metal complexes, organometallics, and polynuclear polyoxometalate (POM)—and the description is given in the following sections.

12.3.1 Water Oxidation with Metal Complexes in Solution-Phase Catalysis

The development of bio-inspired water oxidation catalysts started with the advent of few Mn- and Ru-based complexes; however, the catalytic efficiency was very low [11]. In recent reports, transition metals (like Ru, Ir, Mn, Fe, Co) coordinated to nitrogen-based ligands are synthesized to develop homogeneous WOCs, and the catalysis was investigated using CAN as chemical catalyst activator. In mono-metal or multisite Mn or Ru-derived molecular complexes, bi- or tri-nitrogen ligands like bpy (bpy = 2,2′-bipyridine), phen (1,10-phenanthroline), or terpy (terpy = 2,2′:6′,2″-terpyridine) are employed to make catalysts for oxygen evolution from water oxidation [9].

12.3.1.1 Mn-Based Catalysts for Water Oxidation

During early development in manganese-based WOCs, a tetramanganese complex, $Mn_4O_4L_6$ (L = diphenylphosphinate), with a Mn_4O_4 ($2Mn^{III}$, $2Mn^{IV}$) cubane-like core mimicking the active site of the PS-II water oxidation cluster was introduced, but the catalytic activity and homogeneity are still debatable [21]. Two years later, a diaqua Mn-terpy dimer, $[(terpy)(H_2O)Mn(\mu\text{-}O)_2Mn(terpy)(H_2O)]^{3+}$, with di-μ-oxo motif, was reported giving maximum turnover number (TON) of 4, but the catalyst was found to undergo decomposition forming permanganate ions after 6 h [22]. For a di-Mn catalyst, $[Mn_2(mcbpen)_2(OH_2)_2]^{2+}$, with a tetra nitrogen carboxy ligand *N*-methyl-*N′*-carboxymethyl-*N*,*N′*-bis(2-pyridylmethyl)ethane-1,2-diamine (mcbpen), TON of 20 was reported, and membrane inlet mass spectrometry (MIMS) analysis showed that each of two oxygen atoms in the evolved O_2 came

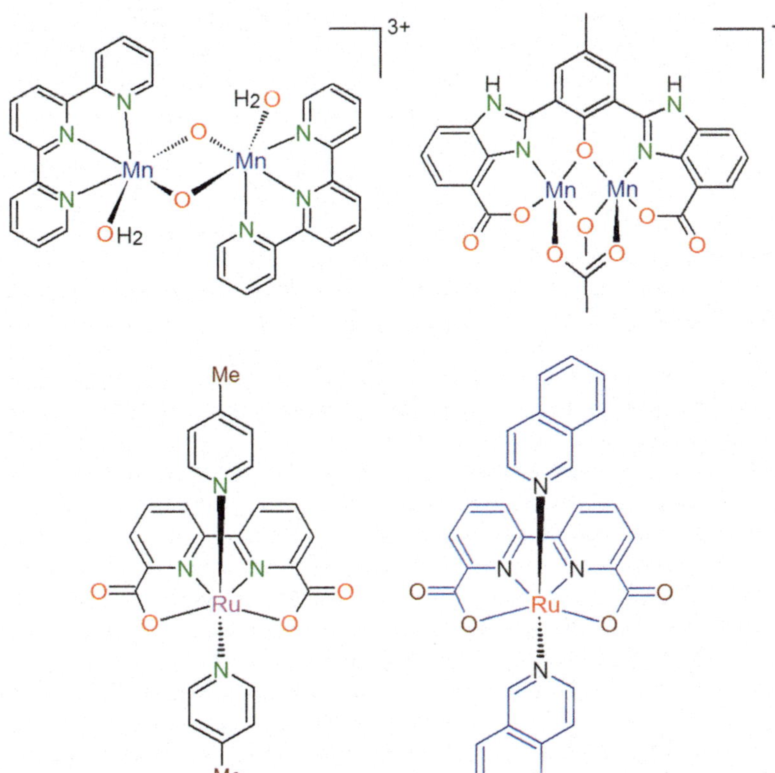

Fig. 12.1 (*Top*) Dinuclear Mn-derived complexes and (*bottom*) single-site [(dcabpy)-Ru-(isq)$_2$] complex (6,6′-dicarboxy-2,2′-bipyridine (dcabpy); isq = isoquinoline), for homogeneous water oxidation

both from water and TBHP (tert-butyl hydrogen peroxide) used as an oxidant [23]. Recently, a bio-inspired dinuclear Mn-complex was presented for water oxidation with central 4-methylphenol ligand and two benzimidazole units having carboxylate arms bridging two manganese sites (Fig. 12.1). With 480-fold excess of $[Ru(bpy)_3]^{2+}$ as single-electron oxidant, the catalyst operated with an initial turn-over rate (TOF) of 0.027 s^{-1} lasting for an hour with a TON of < 30. This is claimed to be the most stable molecular system based on two manganese sites [24].

12.3.1.2 Ruthenium and Iridium Complexes for Oxygen Generation

The bi-ruthenium tetra aqua tetrakis-bipyridine $[(bpy)_2(H_2O)Ru^{III}(\mu\text{-}O)Ru^{III}(H_2O)(bpy)_2]^{4+}$ with a $Ru^{III}(\mu\text{-}O)Ru^{III}$ core, also known as blue dimer, is considered the first working homogeneous water oxidation catalyst [25]. This study was followed

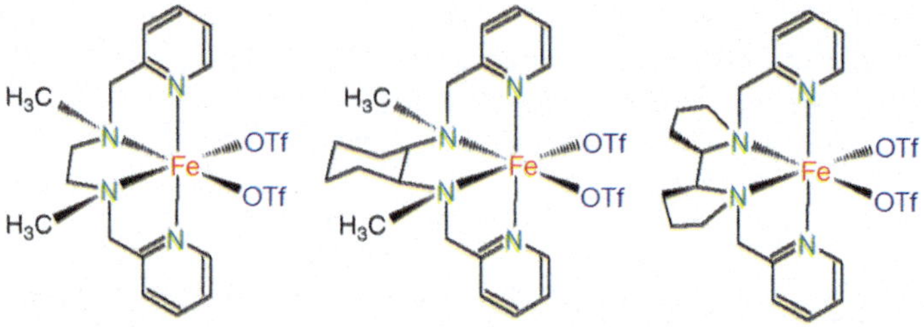

Fig. 12.2 Iron-based single-site coordination complexes with nitrogen surroundings for water oxidation

by a tetra aqua terpy-Ru dimer, $[(terpy)(H_2O)_2Ru(\mu\text{-}O)Ru(terpy)(H_2O)_2]^{4+}$ analogous to the Mn-dimer, but the catalyst deactivated just after 1 turnover. The major reason for the catalytic decomposition was ascribed to be the presence of the μ-oxo bridge between two adjacent metal centers [26]. This problem was addressed by developing a dinuclear Ru complex with a Hbpp-type bridging mode $[Ru_2^{II}(bpp)(trpy)_2(H_2O)_2]^{3+}$ [Hbpp = 2,2′-(1H-pyrazole-3,5-diyl)bis(pyridine)]. In this catalyst, two Ru metals have been deliberately placed in a close proximity using a dinucleating bridging ligand Hbpp instead of a μ-oxo bridge. This modification of the catalyst design avoids the decomposition by reductive cleavage and makes it more active than other Ru-based homogeneous WOCs [27].

Recently, mono-metal Ru-catalyst complexes are reported for water oxidation suggesting that one site can also perform the water catalysis using a chemical oxidant. The best example for a rapid WOC is the $[(dcabpy)\text{-}Ru\text{-}(isq)_2]$ (dcabpy = 6,6′-dicarboxylic acid-2,2′-bipyridine, isq = isoquinoline) that shows an unprecedented oxygen generation rate greater than 300 s^{-1} (Fig. 12.1). The complex also generates more than 8,000 TONs showing its good stability in aqueous acids [28]. Single-site half-sandwiched iridium complexes with di-nitrogen bpy, phen, and bpm (2,2′-bipyrimidine) ligands in combination with relatively strong donating Cp* moiety (Cp* is pentamethylcyclopentadiene) are also described to produce oxygen very rapidly using CAN in aqueous solutions. However, the water oxidation mechanism is still debatable [20].

12.3.1.3 Abundant Transition Metals Derived Molecular Complexes for Water Oxidation

Besides Mn-, Ru-, and Ir-based molecular complexes, other common transition metal-based catalysts have also been studied for oxygen evolution in solution. Fe-, Co-, and Cu-derived molecular complexes are the most prominent among them. Iron-based coordination complexes with nitrogen surroundings have shown high efficiency for long-term homogeneous water oxidation (Fig. 12.2). Turnover numbers in excess of 350 and 1,000 were obtained using cerium ammonium nitrate at pH = 1 and sodium periodate at pH = 2, respectively [29]. Recently, cobalt-derived molecular complexes are also shown to be active for water oxidation. However, the catalysis is undertaken in electrochemical conditions with the homogeneous suspension of the catalyst in solution.

12.3.2 Water Oxidation Catalysis with Organometallics in Homogeneous Solution Phase

Recently, iridium- and ruthenium-based organometallics are also described to be effective for homogeneous-phase water oxidation. The first working example was a group of iridium organometallic complexes with (2-pyridyl)phenylate anion (2-ph-py)]-type ligand motif. Up to 2,500 TONs have been reported in the presence of Ce(IV), though the rate was lower than for the ruthenium analogues [30]. Recently, Brudvig and Crabtree have prepared single-site half-sandwiched iridium complexes by combining relatively strong donating Cp* motif with a 2-(2-pyridyl)phenylate ligand (Fig. 12.3). With this new system, an oxygen generation rate of 0.9/s was achieved in the homogeneous phase [31]. Many other Ir-organometallics have also been studied for water oxidation; however, they were found to be transformed into heterogeneous nanoparticle species acting as the WOC for water oxidation by CAN (Fig. 12.3). Crabtree and coworkers recently suggested that the homogeneous Ir(IV) species are observed that relates organometallic iridium complexes to homogeneous water oxidation complex using $NaIO_4$ as chemical oxidant.

12.3.3 Polynuclear Polyoxometalate for Water Oxidation in Homogeneous Phase

Another class of molecular water oxidation catalysts comprises of polyoxometalate (POM) of Ru and Co. In ruthenium catalysts, the close placing of the Ru–Ru atoms at 0.318 nm is considered to be a key factor for O–O bond formation. Recently, a tetra-cobalt POM catalyst with Co_4O_4-core $[Co_4(H_2O)_2(PW_9O_{34})_2]^{10-}$ was investigated showing a high TOF ≥ 5 in pH = 8 solution. However, an electrochemical study suggests the formation of a heterogeneous CoO_x species for oxygen evolution [32]. A list of molecular complexes and organometallics for water oxidation in homogeneous solution using chemical oxidants is given in Table 12.1.

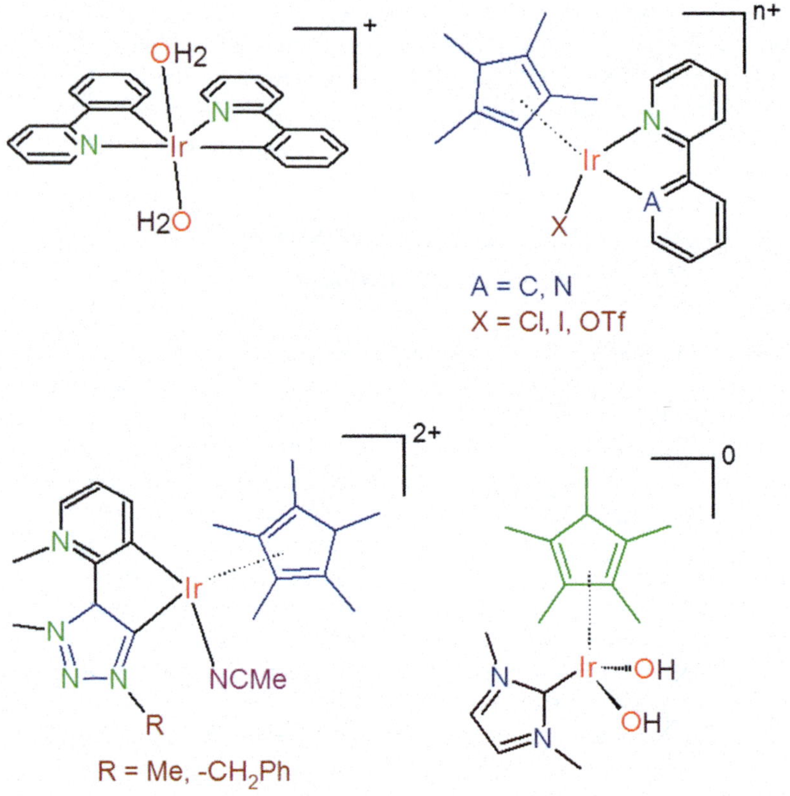

Fig. 12.3 Mono-site iridium-Cp*-based organometallics for homogeneous water oxidation

12.4 Surface Electrochemical Assemblies for Water Oxidation

In order to develop a solar fuel generation device, the molecular bio-inspired catalysts need to be transferred to electrode surface in an electrochemical setup. Many molecular complexes have been investigated in electrolytic cell, but the oxygen evolution current densities (J) were very low. Molecular electrocatalytic assemblies operate at a potential in excess to the thermodynamic water-oxidation potential, i.e., $E^{\circ} = 1.23$ V (vs. NHE, pH = 0). This excess potential or the overpotential is usually 500–700 mV that renders the large-scale application of molecular system for water oxidation [9]. Many Ru complexes along with few Ir and Co complexes were studied under electrochemical condition. The catalysts were either suspended in the solution in contact with the anode or were immobilized on electrodes with surface-anchoring groups. Catalytic modifications with carboxylic [–O–(C=O)–], phosphonate [–$(O)_2$–(P=O)–], or silyl [–(O_3)–Si–]

Table 12.1 Water oxidation catalytic systems derived from molecular complexes and organometallics for homogeneous oxygen evolution using catalyst activator or external chemical oxidant(s)

Catalysts	System	Conditions	TOF	TON[a]	Ref.
Blue dimer	Homogeneous, aqueous	Ce(IV)	0.0042 s^{-1}	13	[25]
Mn-terpy dimer	Homogeneous, aqueous	Ce(IV) in NaClO	–	4	[22]
Ru-terpy dimer	Homogeneous, aqueous	Ce(IV) in $HClO_4$	–	~1	[26]
Ru_2-Hbpp.terpy$_2$	Homogeneous, aqueous	Ce(IV) in triflic acid	–	~15	[27]
Terpy-Ru(N–N)[b]	Homogeneous, aqueous	Ce(IV) in 0.1 M acids	~0.02 s^{-1}	~600	[13, 20]
Cp*-Ir-(ph-py)Cl[c]	Homogeneous, aqueous	Ce(IV) in MeCN/H_2O	0.9 s^{-1}	1,500	[31]
Fe^{III}-TAML[d]	Homogeneous, pH = 0.7	Ce(IV) in H_2O	1.3 s^{-1}	~12	[9]
[Co_4O_4] cubane	Homogeneous, photochemical	$[Ru(bpy)_3]^{2+}$ in $Na_2S_2O_8$ solution	0.02 s^{-1}	40	[10]
Co_4O_4–POM[e]	Homogeneous, pH = 8	$[Ru(bpy)_3]^{2+}$	~5 s^{-1}	75	[32]
Fe-tetradentate N-ligand	Homogeneous, pH = 1,2	Ce(IV) in H_2O	150–850 h^{-1}	~1,000	[29]
[(bipyridine)-Ir-Cp*-(OH_2)]	Homogeneous, pH = 1–2	Ce(IV) in aq. HNO_3	150–850 h^{-1}	~1,000	[34]

[a]Total turnover produced at the end of catalysis operation
[b]N–N is di-nitrogen ligands
[c]Cp* is pentamethylcyclopentadiene and ph-py is 2-(2-pyridyl)phenylate
[d]TAML is tetraamido macrocyclic ligand
[e]Co-polyoxometalate $[Co_4(H_2O)_2(PW_9O_{34})_2]^{10-}$ complex

functionalities have been shown to make a covalent linkage with conducting TiO_2, SnO_2, or ITO (indium tin oxide) surfaces in water oxidation assemblies [1, 10].

12.4.1 Molecular Complexes for Electrochemical Water Oxidation Catalysis

In surface-linked ruthenium complexes, up to 30,000 turnover numbers were achieved at a potential of 1.80 V (vs. NHE, pH = 0) in an acidic medium, but at low current density, $J < 50\ \mu A\ cm^{-2}$ [33]. This low current density for O_2 evolution suggests a low performance for the electrochemical water oxidation. Co-porphyrin- and Co-corrole-type systems have also been proposed, but the catalytic activity remains low with low O_2 yield. We recently showed that the mono-site iridium-Cp* complexes can be efficiently anchored to an ITO surface via COOH or PO_3H_2

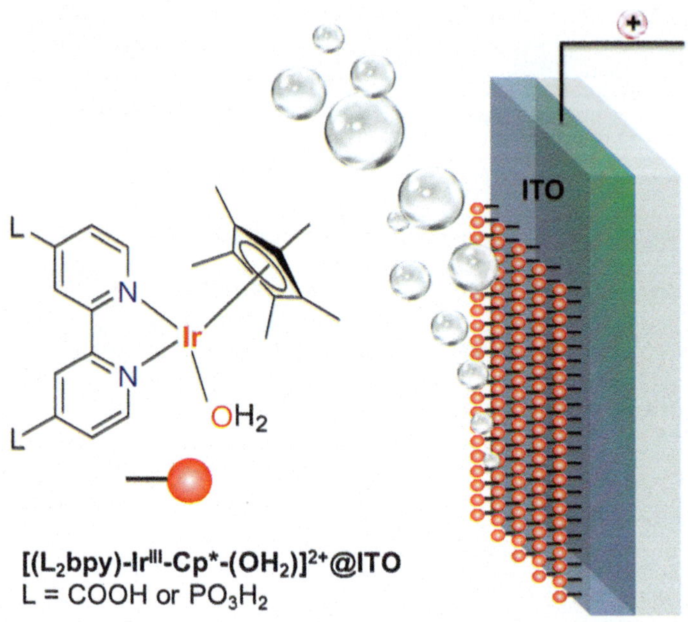

Fig. 12.4 A surface-immobilized electrochemical water oxidation assembly showing Cp*-iridium-derived complexes with L_2bpy (L = COOH, PO_3H_2) ligands for electro-driven oxygen evolution

linkers (Fig. 12.4). During long-term controlled potential electrolysis (CPE) of water (Fig. 12.5), the electrocatalytic system generates greater than 200,000 TONs for O_2 generation with a TOF of more than 6.7 s^{-1}. The oxygen evolution current density was >1.75 mA cm^{-2} in a pH = 4 solution [34]. Another interesting report describes a copper bipyridine (Cu-bpy) hydroxo-complex, $[(bpy)Cu(\mu\text{-}OH)]_2^{2+}$, for alkaline water electrolysis. The Cu hydroxo-complex self-assembles in a pH > 11 solution and shows a turnover frequency up to 100 s^{-1} [35]. However, the system operates at a high overpotential, 700–900 mV in alkaline electrolyte, which may limit its large-scale application for the H_2/O_2 evolution via water electrolysis, especially when used for liquid fuel production (in combination with a CO_2 reduction module).

12.4.2 *POM-Based Catalyst for Electrochemical Water Oxidation*

Another example for the electro-driven water oxidation was introduced in the form of a polyoxometalate diruthenium-substituted catalyst for oxygen evolution. A POM is an inorganic structure with two metal sites bridged by oxygen and hydroxo-units and is potentially more robust than the organic molecular catalysts

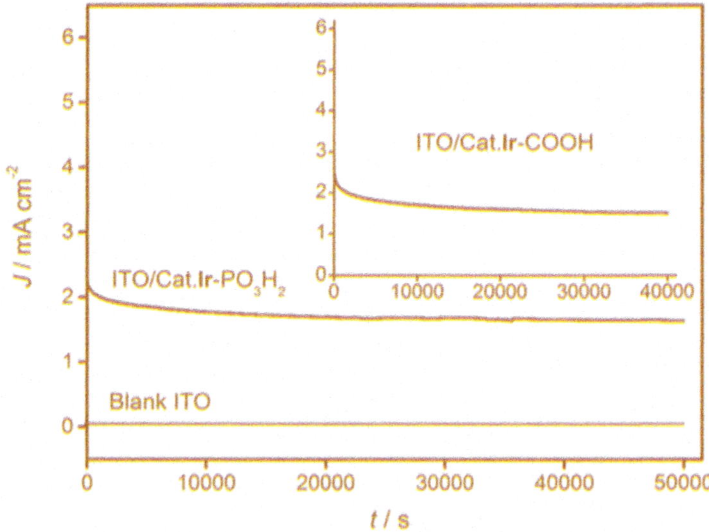

Fig. 12.5 Controlled potential water electrolysis with the ITO, ITO/Cat.**Ir**–PO_3H_2, and (*inset*) ITO/Cat.**Ir**–COOH in deoxygenated 0.1 M buffer solution (pH = 4) at 1.75 V (vs. NHE). Catalyst density is 1.55–1.75 × 10^{-10} mol cm^{-2}. Reproduced from ref. [34] by permission of John Wiley & Sons Ltd

for optimal catalytic performance. The water electrolysis was conducted in aqueous phosphate buffer (pH = 8), and a catalytic current is observed at 0.55 V that increases with the applied potential up to a maximum of 1.05 V. A Tafel slope of ~120 mV per decade was observed, which excludes the formation of Ru-oxide during catalysis [36]. Recently, a tetraruthenate core POM assembly was integrated with a multiwalled carbon nanotube (MWCNT)/ITO surface to form a nanostructured oxygen-evolving assembly for electrochemical water oxidation (Fig. 12.6). In a pH = 7 solution, TOFs from 36 to 306/s were obtained, depending on the applied potential, which are higher than the previous practices. The excellent performance of the system is attributed to the synergistic coupling of the Ru4-POM with MWCNT on ITO surface which provides good electrical plugging of the hybrid material to make the nanoscale structure [37]. A list of molecular complexes for electrochemical oxygen evolution is given in Table 12.2, which summarizes various concepts related to molecular complexes developed during water oxidation.

12.5 Mechanism of Water Oxidation

Formation of O_2 is realized by the extraction of four electrons and protons in four consecutive PCET steps from two water molecules in Mn_4CaO_5 cluster embedded in the photosystem II. The stepwise increase in the oxidation state of the Mn^{n+} gives

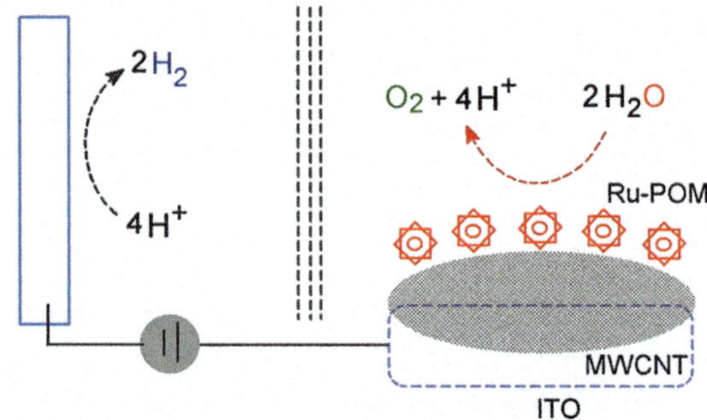

Fig. 12.6 Nanostructured oxygen-evolving material comprises of a tetraruthenate core POM integrated with MWCNT surface on ITO anode for electrochemical water oxidation

Table 12.2 Electrocatalytic water oxidation systems derived from molecular complexes

Catalysts	System	Conditions	TOF, *J*[a]	TON	Ref.
Mn_4O_4 cubane complex	Heterogeneous, photo- and photo-electrochemical	1.0 V (Ag/AgCl) Xe lamp	–	1,000	[10, 18]
PO_3H_2-terpy-Ru dimer	Heterogeneous, pH = 6	1.5 V (Ag/AgCl)	–	3	[12]
Ru_2-Hbpp. t-trpy	Heterogeneous, pH = 1	1.17 V (Ag/AgCl) in triflic acid	–	120	[9, 10]
Ru_2-Btpyan	Heterogeneous, pH = 4	ITO +1.7 V (Ag/AgCl)	0.12 mA/cm^2	33,500	[8]
Mebimpy-Ru (N–N)[b]	Heterogeneous, pH = 5	1.85 V (NHE) in 0.1 M acetate buffer	0.36 s^{-1}, 14.8 μA/cm^2	11,000	[33]
Terpy-Ru(N–N)Ru(bpy)$_3$L[c]	Heterogeneous, pH = 1	1.80 V (NHE) in 0.1 M $HClO_4$	0.3 s^{-1}, 6.7 μA/cm^2	8,900	[33, 40]
[$(PO_3H_2)_2$-bpy]-Ir-Cp*-(OH_2)[d]	Heterogeneous, pH = 4	1.75 V (NHE) in 0.1 M acetate buffer	~6.5 s^{-1}	2 × 10^5	[34]
[(bpy)Cu (μ-OH)]$_2^{2+}$	Heterogeneous, pH > 4	700–900 mV overpotential in alkaline electrolyte	~100 s^{-1}		[35]
[Ru(dcabpy) (py-pic)$_2$][e]	CNT/ITO, pH = 7[e]	1.4 V (vs. NHE)	0.22 mA/cm^2	11,000	[10]

[a]*J* = current density (in mA/cm^2)
[b]N–N is di-nitrogen-based ligands and Mebimpy is 2,6-bis(1-methylbenzimidazol-2-yl)-pyridine
[c]Tpy-Ru(N–N)Ru(bpy)$_3$ L is $[\{4,4'\text{-}(H_2O_3PCH_2)_2\text{-bpy}\}_2Ru^{II}(bpm)Ru^{II}(tpy)\text{-}(OH_2)]^{4+}$
[d]Cp* is pentamethylcyclopentadiene and bpy is 2,2′-bipyridine
[e]py is pyrene and pic = 4-picolin; CNT is carbon nanotube

rise to the accumulation of the four oxidizing equivalents, and consecutive PCET steps enable the accumulation of four redox equivalents while circumventing high-energy intermediates during the multi-electron water oxidation cycle [38]. In artificial biomimetic molecular catalytic systems, the maximum oxygen evolution activity is thought to be constrained by a minimum overpotential of ~0.4 V. In these systems the cycle proceeds through an HOO* intermediate, and there is a constant difference of 3.2 ± 0.1 eV in the affinity between the HOO* and the HO* intermediates [39]. In many WOCs, a detailed mechanistic insight for water oxidation and O–O bond formation is difficult to obtain [9, 11]. Both single-site and binuclear Ru complexes are unable to construct a four-step PCET mechanism for water oxidation. In mono-site ruthenium complexes, an HOO* intermediate is formed at high overpotential from a $[Ru^{V}=O]^{3+}$ species generated by an electron transfer pathway in a non-PCET rate-limiting step [13]. The formation of dioxygen in binuclear Ru systems is realized either by intramolecular O–O bond formation through oxo-oxo coupling at two catalytic sites without HOO* generation, which is prone to catalytic deactivation after few cycles, or by nucleophilic OH_2 insertion generating a higher-energy HOO* intermediate in a non-PCET step [40].

In iridium organometallics and metal complexes $[(Cp^*)Ir^{III}(N–C)OH_2]^+$ and $[(Cp^*)Ir^{III}(N–N)OH_2]^{2+}$, the Ir-oxo species $[Ir^{V}=O)]^{n+}$ ($n = 1$ for N–C and 2 for N–N) is suggested to be generated from $[Ir^{III}\text{-}(OH_2)]^{n+}$ oxidation with two protons released [31]. The O–O bond formation possibly proceeds by the nucleophilic attack of an OH_2 to a $[Ir^{V}=O)]^{n+}$ intermediate (Scheme 12.3). At the same time, a proton is thought to leave from the attacking OH_2 to make a hydroxide ion that enhances the nucleophilicity for Ir-oxo complex [31, 34]. It is already defined that on second aqua insertion, the next electron removal is not coupled with the proton transfer. Most likely the $[Ir^{III}–OOH)]^{(n-1)+}$ complex is generated by deprotonation accompanying the second OH_2 ligation to the $[Ir^{V}=O)]^{n+}$. This can eject an electron to form a $[Ir^{IV}–OOH)]^{n+}$-type intermediate in the next step (Scheme 12.3) and then loses the fourth electron and proton and generates dioxygen while re-coordinating to another OH_2 to regenerate the aqua complex for the next catalytic turnover [31]. As the catalyst shows a fast rate for oxygen evolution, the last OH_2 ligation is not likely to be the rate-limiting step, and the $[Ir^{V}–OO)]^{n+}$ may not be a reaction intermediate but a lowered transition state or transient species [Scheme 12.3 (II)]. Most probably the large bite angle of the bpy-Ir and Cp* provides a wide gap to facilitate rapid OH_2 insertion at the beginning of the second half of the water oxidation cycle. Moreover, the strong electron-donating character of the Cp* helps to stabilize the complex and intermediates during the catalytic cycle [34].

12.6 Conclusions and Outlook

Catalytic water oxidation to make hydrogen or other nonfossil carbon-neutral fuels using sunlight is an attractive and feasible approach for future renewable energy carriers. Establishing a stable and efficient light-induced electrochemical molecular water oxidation system is a key step towards the development of a bio-inspired solar-to-fuel conversion device [41]. With the advent of modern technology and chemical science developments, the picture is getting clear towards a stable and efficient WOC derived from molecular complexes and organometallics. Significant scientific steps have been made in the recent years related to water-splitting science and solar fuel research [42]. Thermodynamically speaking, water oxidation process

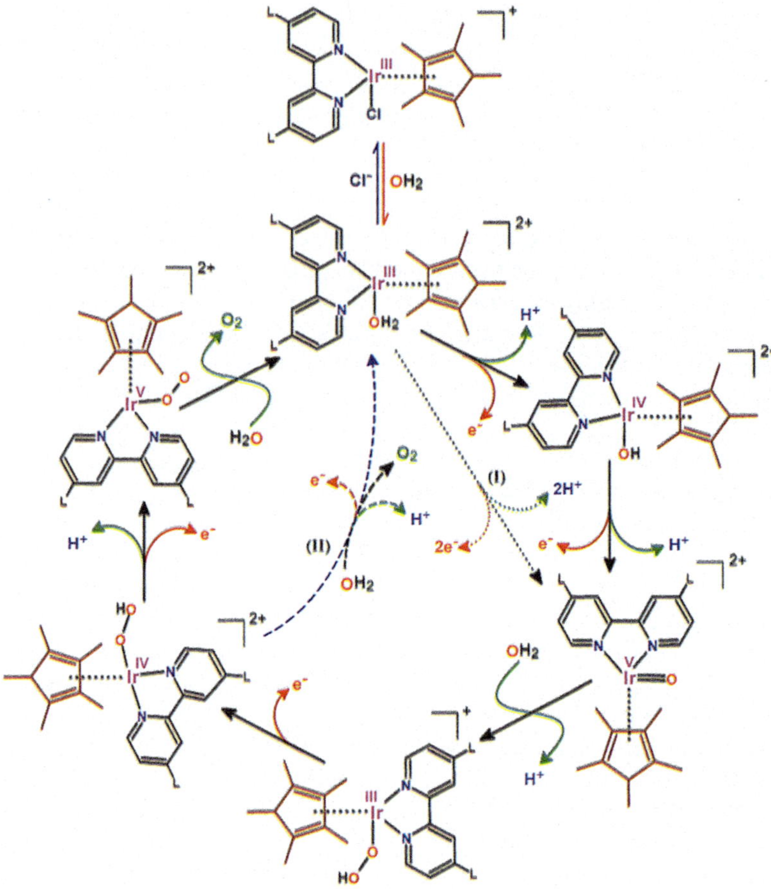

Scheme 12.3 A plausible water oxidation mechanism and O–O bond formation by the [(N–N)Ir^{III}(Cp*)–OH$_2$]$^{2+}$ complex (Reproduced from Ref. [34] by permission of John Wiley & Sons Ltd)

is an energy-intensive regime, but the smart design of the catalytic material biomimicking the Mn_4CaO_5 cluster in the photosynthetic system could make it possible to split water with ease, aiming to operate with small activation barriers and at moderate overpotential. From noble metal complexes to transition metal organometallics, multinuclear to mono-site systems, various water oxidation catalysts have been investigated both in a homogeneous solution and immobilized on surfaces in electrochemical or photo-electrochemical setups. In this chapter, we attempted to give an overview of the recent progress in the design and implementation of molecular catalysts and organometallics for water oxidation in homogeneous phase and surface electrochemical assemblies. The field of water oxidation catalysis is growing, and the next step is the synergistic combination of the various components to make the artificial device for conversion of water, sunlight, or carbon dioxide into oxygen, hydrogen, or low-carbon fuels for a sustainable and greener future.

References

1. Joya KS, Joya YF, Ocakoglu K et al (2013) Water splitting catalysis and solar fuel devices: artificial leaf on the move. Angew Chem Int Ed 52:10426–10437
2. McEvoy JP, Brudvig GW (2006) Water-splitting chemistry of photosystem II. Chem Rev 106:4455–4483
3. Joya KS, de Groot HJM (2014) Artificial leaf goes simpler and more efficient for solar fuel generation. ChemSusChem 7:73–76
4. Najafpour MM, Ehrenberg T, Wiechen M et al (2010) Calcium manganese(III) oxides ($CaMn_2O_4.xH_2O$) as biomimetic oxygen-evolving catalysts. Angew Chem Int Ed 49:2233–2237
5. Vallés-Pardo JL, Guijt MC, Iannuzzi M, Joya KS et al (2012) Ab-initio molecular dynamics study of water oxidation reaction pathways in mono-Ru catalysts. ChemPhysChem 13:140–146
6. Young KJ, Martini LA, Milot RL et al (2012) Light-driven water oxidation for solar fuels. Coord Chem Rev 256:2503–2520
7. Barton EE, Rampulla DM, Bocarsly AB (2008) Selective solar-driven reduction of CO_2 to methanol using a catalyzed p-GaP based photoelectrochemical cell. J Am Chem Soc 130:6342–6344
8. Wada T, Tsuge K, Tanaka K (2001) Syntheses and redox properties of bis(hydroxoruthenium) complexes with quinone and bipyridine ligands. Water-oxidation catalysis. Inorg Chem 40:329–337
9. Joya KS, de Groot HJM (2012) Biomimetic molecular water splitting catalysts for hydrogen generation. Int J Hydrog Energy 37:8787–8799
10. Joya KS, Vallés-Pardo JL, Joya YF et al (2013) Molecular catalytic assemblies for electrodriven water splitting. ChemPlusChem 78:35–47
11. Joya KS (2011) Molecular catalytic system for efficient water splitting. Wöhrmann Printing Press, Zutphen, The Netherlands
12. Cao R, Wenzhen Laia W, Du P (2012) Catalytic water oxidation at single metal sites. Energy Environ Sci 5:8134–8157
13. Wasylenko DJ, Ganesamoorthy C, Henderson MA et al (2010) Electronic modification of the $[RuII(tpy)(bpy)(OH_2)]^{2+}$ scaffold: effects on catalytic water oxidation. J Am Chem Soc 132:16094–16106
14. Deng Z, Tseng H-W, Zong R et al (2008) Preparation and study of a family of dinuclear Ru (II) complexes that catalyze the decomposition of water. Inorg Chem 47:1835–1848
15. Schley ND, Blakemore JD, Subbaiyan NK et al (2011) Distinguishing homogeneous from heterogeneous catalysis in electrode-driven water oxidation with molecular iridium complexes. J Am Chem Soc 133:10473–10481
16. Joya KS, Joya YF, de Groot HJM (2014) Ni-based electrocatalyst for water oxidation developed in-situ in a HCO_3^-/CO_2 system at near-neutral pH. Adv Energy Mater. doi:10.1002/aenm.201301929
17. Hurst JK (2005) Water oxidation catalyzed by dimeric μ-oxo bridged ruthenium diimine complexes. Coord Chem Rev 249:313–328
18. Yamazaki H, Shouji A, Kajita M et al (2010) Electrocatalytic and photocatalytic water oxidation to dioxygen based on metal complexes. Coord Chem Rev 254:2483–2491
19. Calvin M, Cooper SR (1974) Solar energy by photosynthesis: manganese complex photolysis. Science 185:376
20. Limburg B, Bouwman E, Bonnet S (2012) Molecular water oxidation catalysts based on transition metals and their decomposition pathways. Coord Chem Rev 256:1451–1467
21. Ruettinger WF, Campana C, Dismukes GC (1997) Synthesis and characterization of $Mn_4O_4L_6$ complexes with cubane-like core structure: A new class of models of the active site of the photosynthetic water oxidase. J Am Chem Soc 119:6670–6671

22. Limburg J, Vrettos JS, Liable-Sands LM et al (1999) A functional model for O–O bond formation by the O_2-evolving complex in photosystem II. Science 283:1524–1527
23. Poulsen AK, Rompel A, McKenzie CJ (2005) Water oxidation catalyzed by a dinuclear Mn complex: a functional model for the oxygen-evolving center of photosystem II. Angew Chem Int Ed 44:6916–6920
24. Karlsson EA, Lee B-L, Åkermark T et al (2011) Photosensitized water oxidation by use of a bioinspired manganese catalyst. Angew Chem Int Ed 50:11715–11718
25. Gersten SW, Samuels GJ, Meyer TJ (1982) Catalytic oxidation of water by an oxo-bridged ruthenium dimer. J Am Chem Soc 104:4029–4032
26. Lebeau EL, Adeyemi SA, Meyer TJ (1998) Water oxidation by $[(tpy)(H_2O)_2Ru^{III}OR-u^{III}(H_2O)_2(tpy)]^{4+}$. Inorg Chem 37:6476–6484
27. Sens C, Romero I, Rodrıguez M et al (2005) A new Ru complex capable of catalytically oxidizing water to molecular dioxygen. J Am Chem Soc 126:7798–7799
28. Duan L, Bozoglian F, Mandal S et al (2012) A molecular ruthenium catalyst with water-oxidation activity comparable to that of photosystem II. Nat Chem 4:418–423
29. Fillol JL, Codolà Z, Garcia-Bosch I et al (2011) Efficient water oxidation catalysts based on readily available iron coordination complexes. Nat Chem 3:807–813
30. McDaniel ND, Coughlin FJ, Tinker LL et al (2008) Cyclometalated iridium(III) aquo complexes: efficient and tunable catalysts for the homogeneous oxidation of water. J Am Chem Soc 130:210–217
31. Blakemore JD, Schley ND, Balcells D et al (2010) Half-Sandwich iridium complexes for homogeneous water-oxidation catalysis. J Am Chem Soc 132:16017–16029
32. Stracke JJ, Finke RG (2011) Electrocatalytic water oxidation beginning with the cobalt polyoxometalate $[Co_4(H_2O)_2(PW_9O_{34})_2]^{10-}$: Identification of heterogeneous CoOx as the dominant catalyst. J Am Chem Soc 133:14872–14875
33. Concepcion JJ, Jurss JW, Hoertz PG et al (2009) Catalytic and surface-electrocatalytic water oxidation by redox mediator–catalyst assemblies. Angew Chem Int Ed 48:9473–9476
34. Joya KS, Subbaiyan NK, D'Souza F et al (2012) Surface-immobilized single-site iridium complexes for electrocatalytic water splitting. Angew Chem Int Ed 51:9601–9605
35. Barnett SM, Goldberg KI, Mayer JM (2012) A soluble copper–bipyridine water-oxidation electrocatalyst. Nat Chem 4:498–502
36. Howells AR, Sankarraj A, Shannon C (2004) Diruthenium-substituted polyoxometalate as an electrocatalyst for oxygen generation. J Am Chem Soc 126:12258–12259
37. Toma FM, Sartorel A, Iurlo M et al (2010) Efficient water oxidation at carbon nanotube–polyoxometalate electrocatalytic interfaces. Nat Chem 2:826–831
38. Dau H, Limberg C, Reier T et al (2010) The mechanism of water oxidation: from electrolysis via homogeneous to biological catalysis. ChemCatChem 2:724–761
39. Rossmeisl J, Dimitrievski K, Siegbahn P et al (2007) Comparing electrochemical and biological water splitting. J Phys Chem C 111:18821–18823
40. Concepcion JJ, Tsai M-K, Muckerman JT et al (2010) Mechanism of water oxidation by single-site ruthenium complex catalysts. J Am Chem Soc 132:1545–1557
41. Herrero C, Lassalle-Kaiser B, Leibl W et al (2008) Artificial systems related to light driven electron transfer processes in PSII. Coord Chem Rev 252:456–468
42. Joya KS, de Groot HJM (2013) Electrochemical in situ surface enhanced Raman spectroscopic characterization of a trinuclear ruthenium complex, Ru-red. J Raman Spectrosc 44:1195–1199

Chapter 13
Recent Development in Water Oxidation Catalysts Based on Manganese and Cobalt Complexes

Lawrence Yoon Suk Lee and Kwok-Yin Wong

Abstract Energy directly harvested from sunlight offers an ultimate method of meeting the needs for clean energy with minimal impact on our environment. Intensive research efforts are currently being put on the development of efficient conversion system that can transform solar energy into fuel via light-driven water splitting to generate H_2 and O_2, learning from Nature's photosynthesis to collect and store solar energy in chemical bonds. Especially, the development of efficient water oxidation catalysts is one of the key issues for achieving artificial photosynthetic devices. From a practical point of view, it is highly desirable to replace noble metal catalysts, which have been quite successful so far, by earth-abundant metal catalysts. In recent years, there has been noticeable progress in the development of water oxidation catalysts (WOCs) based on earth-abundant metals. This review chapter covers the most significant achievements in WOCs based on manganese and cobalt complexes, with emphasis on recent developments in the last three years.

Keywords Water oxidation • Proton-coupled electron transfer • Photocatalysis • Electrocatalysis • Manganese complex • Cobalt complex • Homogeneous catalysis

13.1 Introduction

The civilization of mankind has accelerated with the discovery and use of energy in a wise and productive way. Our society now has become so dependent on the energy mostly generated by burning the fossil fuels that a single day without it is difficult to imagine. After the exponential growth through the industrial era, the realization that the use of traditional fossil fuels produces much carbon dioxide, creating a significant negative impact on the environment such as global warming, has sparked vast efforts in search for substitutes of fossil fuels from the academic circles to the industries. The fossil fuels on which we are heavily relying for our

---

L.Y.S. Lee • K.-Y. Wong (✉)
Department of Applied Biology and Chemical Technology, The Hong Kong Polytechnic University, Hung Hom, Kowloon, Hong Kong SAR
e-mail: kwok-yin.wong@polyu.edu.hk

W.-Y. Wong (ed.), *Organometallics and Related Molecules for Energy Conversion*, Green Chemistry and Sustainable Technology, DOI 10.1007/978-3-662-46054-2_13

daily activities currently provide more than 80 % of total energy consumption worldwide and have started to show the signs of depletion. Oils, for example, are predicted to deplete in next 50 years [1]. The current situation has put the search for clean and renewable energy sources as the highest priority and one of the most urgent matters for sustaining the progress and development of our society. Novel renewable carbon-free or carbon-neutral energy sources must be identified and generated in next 10-50 years. Among all the energy options available, solar energy is the most promising and the only source of truly renewable energy. Earth receives solar energy at the rate of approximately 120,000 TW (1 TW $=10^{12}$ W), which vastly exceeds the current annual worldwide energy consumption rate of approximately 17 TW [2]. Many strategies are under intensive development for a better utilization of this endless energy from the sun, including solar to electricity (photovoltaics) and solar to fuel (photosynthesis). Nonetheless, the efficiencies of such solar-energy conversion have not yet met the standards required by consumers market. Most techniques developed so far carry some weaknesses, either in low conversion efficiency or economic viability. There are also technical limitations in scaling up which set a high barrier to the market. Nature has shown us a hint: photosynthesis that occurs every day in the plants provides an excellent example of efficient utilization of solar energy on the large scale. Solar-energy conversion via water splitting occurs in the reaction center of photosystem II (PS-II) in cyanobacteria, algae, and green plants. The advance in understanding the natural photosynthetic water-splitting chemistry can offer a solid basis for unraveling the mechanisms of water splitting into oxygen and hydrogen gases. Hydrogen produced from the water splitting can serve as an ultimate clean energy and used in fuel cells. It can be also used in industries for chemical synthesis. In industry, a large amount of ammonia is currently being synthesized using hydrogen that is produced from fossil fuels, such as natural gas by steam reforming. Alternate production of hydrogen from water can avoid the massive coproduction of CO_2 in the conventional method and serve as the storage of solar energy in chemical bonds.

Extensive efforts have been devoted to the research of mimicking Nature's photosynthesis process, the artificial photosynthesis, to harness solar energy for clean and sustainable fuel production. Artificial photosynthesis is a collective system of multiple processes: light harvesting, charge separation, catalytic oxidation and reduction of water, and CO_2 fixation. There have been remarkable advances in photovoltaic researches that new materials efficient in harvesting the sunlight and separating charges are now available [3–7]. Many excellent water reduction catalysts for hydrogen evolution [8–11] and CO_2 fixation using homogeneous catalysts under mild conditions [12–14] have been developed. The most difficult step in artificial photosynthesis is water oxidation. Over the last three decades, many water oxidation catalysts were developed. However, the majority of these effective catalysts are based on rare and precious metals. Contrast to our achievements, Nature only uses a Mn_4Ca cluster as an active catalyst. It is made of abundant and thus cheap metals and still beats our catalysts with incomparable efficiency and stability. The manganese complexes in Nature are known for all oxidation states between Mn^{II} and Mn^{V}, thereby making multiple one-electron oxidation processes possible. Despite of less successful examples so far, all of these facts make manganese complexes a potential candidate for water oxidation catalyst.

In this review chapter, we will start from the understanding of how Nature efficiently handles the process of splitting water molecule and briefly summarize the early efforts to mimic this process using non-noble metal complexes. As inexpensive water oxidation catalysts are highly desirable for the development of photocatalytic solar cells and fuel production from water and sunlight, manganese and cobalt catalysts appear as very relevant choices. We mainly focus on recent achievements in the development of water oxidation catalysts based on earth-abundant manganese and cobalt with emphasis on homogeneous molecular complexes.

13.2 Water Oxidation in Nature

13.2.1 Oxidation of Water

In the process of splitting water, the oxidation part is considered as a bottleneck for the realization of artificial photosynthesis. The electrochemical oxidation of water or hydroxide ion involves the following reactions.

$$4e^- + 4H^+ + O_2 \rightleftharpoons 2H_2O \quad E^\circ = 1.23\ \text{V}\ \textit{vs.}\ \text{SHE}\ (\text{pH} = 0) \tag{13.1}$$

$$4e^- + 2H^+ + O_2 \rightleftharpoons 2OH^- \quad E^\circ = 0.41\ \text{V}\ \textit{vs.}\ \text{SHE}\ (\text{pH} = 14) \tag{13.3}$$

These four-electron oxidations of water afford a good measure of the thermodynamic requirements for the electrochemical processes. The electrons required for these redox processes are continuously supplied from an electrode, either by simultaneous or sequential transfer of multiple electrons. However, the photochemical processes of these reactions are largely limited by the transfer of necessary electrons which is induced through one-photon absorption by a sensitizer molecule under highly specific photon–flux–density conditions. The alternative one-electron oxidation of hydroxide ion (Eq. 13.3) by photochemical approach requires UV light irradiation, since the standard potential is rather high. The two-electron oxidation of water (Eq. 13.4) generally produces hydrogen peroxide. Water oxidation is not only a thermodynamically demanding reaction ($E^\circ = 1.23$ V *vs.* SHE at pH 0) but also represents tremendous molecular complexity from the mechanistic perspectives. This reaction involves the removal of four protons and four electrons from two water molecules, with the concurrent formation of bonding between two oxygen atoms. Thus, the water oxidation process is recognized as the bottleneck for the development of solar-energy conversion devices, such as artificial photosynthesis, and remains as one of the most important and urgent research fields. One way of facilitating this complicated task would be *via* a pathway known as proton-coupledelectron transfer (PCET). This coupling of electron transfer to proton transfer allows the total charge of a chemical species to remain unchanged and avoids the charge buildup and high-energy intermediates and thus is of fundamental importance in many biological and chemical processes, including the water oxidation in Nature.

$$e^- + OH^{\bullet} \rightleftharpoons OH^- \quad E^\circ = \tilde{}2\ V\ vs.\ SHE \tag{13.3}$$

$$2e^- + 2H^+ + H_2O_2 \rightleftharpoons 2H_2O \quad E^\circ = 1.77\ V\ vs.\ SHE(pH = 0) \tag{13.4}$$

13.2.2 Photosystem II (PS-II)

In Nature, light-driven oxygen generation occurs at the oxygen-evolving complex (OEC) of photosystem II (PS-II) present in photosynthetic organisms, such as cyanobacteria, algae, and green plants [15–20]. PS-II is a multi-subunit enzyme complex that features a chlorophyll-based antenna system and a special chlorophyll pair to absorb photons and separate charges. Upon the absorption of visible light (λ=~400–700 nm) by chlorophyll antenna, the immediate and efficient charge separation at the chlorophyll pair results in the reduction of quinones and the formation of a strong oxidant, $P_{680}{}^{+}$ (E°=1.2 V *vs.* NHE). Through a sequential PCET involving a chain of redox-active cofactors in a so-called Kok cycle [21], four electrons from two water molecules are subtracted, releasing dioxygen at the unique tetranuclear manganese cluster embedded in PS-II [22–24]. This cluster is called the oxygen-evolving complex (OEC) as two water molecules are bound to and oxidized to molecular oxygen.

It is striking that there are only minor differences on the level of this water-splitting complex found from a wide range of organisms, even after billions of years of evolution. The oxidation of water in the OEC has continued since the advent of cyanobacteria with little alteration. This uniqueness in structure and function as well as the complexity has made PS-II, especially OEC, the subject of intense research efforts in order to gain a detailed understanding of light-driven water-splitting process [15, 16, 18, 19, 25–29]. In PS-II, more than a thousand water molecules are found, providing extensive hydrogen-bonding networks that may serve as channels for protons from the manganese cluster to the reduction site. Recent studies with a resolution down to 1.9 Å have unveiled many structural details of OEC. Five metal atoms (four Mn and one Ca) connected by five oxygen atoms as oxo-bridges are suggested to form a cubane-like Mn_4CaO_5 cluster where four water molecules are bound [15]. In this suggested structure, each metal ion has three μ-oxo bridges in a Mn_3CaO_4 cubane structure which is connected to another Mn ion by a mono-μ-oxo bridge. The linking mode of metal ions is still a matter of debate and several other models were suggested, for example, one in which the fourth manganese is connected to an μ-oxo ligand of the Mn_3Ca cubane and one in which it is connected to a Mn ion of this cubane through carboxylate bridges. The fourth dangling Mn was proposed to undergo stepwise oxidation coupled with proton liberation. The formation of a Mn^V=O unit was also suggested for this dangling Mn atom, on which water attack is responsible for the O–O bond formation [15, 16].

13.2.3 Homogeneous and Heterogeneous Catalytic Water Oxidation

Homogeneous catalytic systems for water oxidation have the advantages in studying the detailed mechanisms and detecting the active intermediates, such as high-valent metal-oxo species [22, 30, 31]. The properties of catalysts can be readily tuned through careful ligand modification, which also allows a controlled coupling of the catalysts to photosensitizers. An important issue on homogeneous catalysts is their structural integrity during the water oxidation, especially those metal complexes with organic ligands. Organic ligands are prone to the oxidation by the oxidants used in water oxidation, resulting in the formation of insoluble metal oxides or hydroxide particles after the hydrolysis of released metal ions. The highly oxidizing conditions required for O_2 evolution can also cause the oxidation and decomposition of even durable inorganic ligands, such as polyoxometalate, in some cases. The resulting metal oxides or hydroxides are also known as robust and highly active catalysts for water oxidation. The question that whether catalysis is accomplished by the soluble molecular catalyst or the molecular catalyst is merely a precursor for an insoluble heterogeneous metal oxide, which is the true catalyst for water oxidation, has become the subject of intensive debate [32]. Thus, very careful examination would be required for homogeneous catalytic systems when analyzing the kinetics and detecting the intermediates in water oxidation.

13.3 Manganese Complex Catalysts

Since the recognition of its involvements in biological redox systems in the 1970s, manganese complexes have long been contributed to the studies related to the structure and function of the OEC in PS-II [22, 33–37]. In the effort of understanding the structural and mechanistic aspects of the OEC, a number of such oxomanganese complexes of varying nuclearity, ligands, and type and number of oxo-bridges had been synthesized and characterized in the 1980s. However, most of these complexes which were designed as a structural model for the OEC have not been successful in generating O_2 from water except for a handful of examples such as *trans*-$Mn^{IV}(ans)_2Cl_2$ (ans=*N*-alkyl-3-nitrosalicylimide) [38–40], $[(bpy)_2Mn^{III}(\mu\text{-}O)_2Mn^{IV}(bpy)_2)]^{3+}$ (bpy=2,2′-bipyridine, μ is a bridging mode) [41], and *o*-phenylene-linked dimanganese porphyrin complex [42].

13.3.1 Early Manganese Complex Catalysts: Dimer and Related Catalysts

Inspired by the manganese porphyrin dimer [42] and μ-oxo ruthenium dimer [43], both of which were thought to catalyze water oxidation through a terminal oxo-ligand of a high-valent metal ion M=O, another di-μ-oxo dinuclear complex

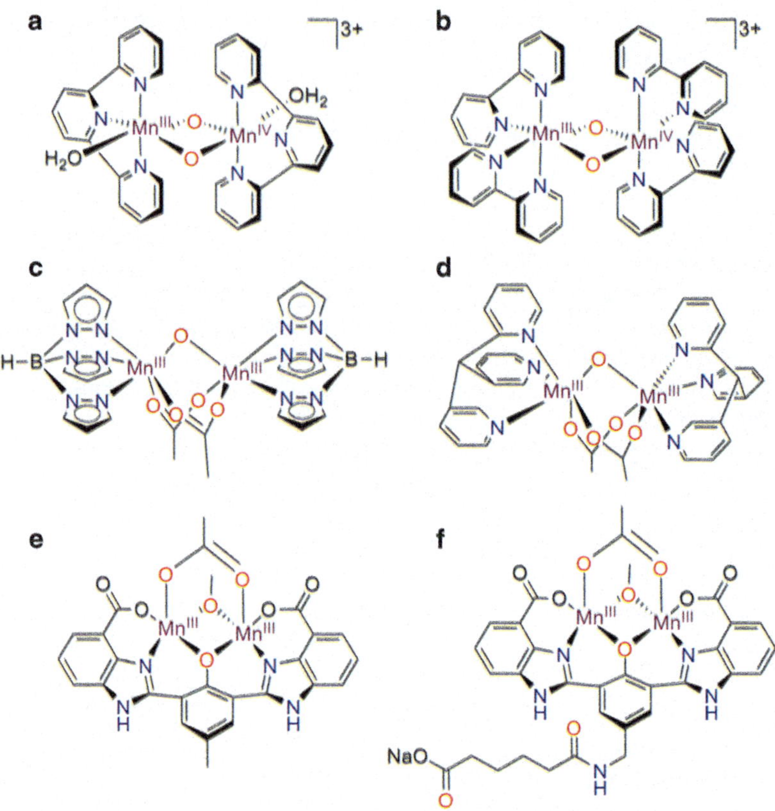

Fig. 13.1 Dimeric manganese complexes investigated as water oxidation catalysts (WOCs) [41, 44, 61, 62, 70, 73]

with terpyridine ligands $[(tpy)(H_2O)Mn^{III}(\mu\text{-}O)_2Mn^{IV}(tpy)(H_2O)]^{3+}$ (tpy = 2,2′:6′,2″-terpyridine; Mn dimer, Fig. 13.1a) was synthesized for catalytic O_2 evolution by Limburg's group [44, 45]. A homogeneous O_2 evolution from the reaction of Mn dimer and an oxygen donor (sodium hypochlorite (NaClO) or Oxone®) in an aqueous solution was reported with a maximum turnover number (TON) of 4 for 6 h. Several other reports followed to show the catalytic O_2 evolution by similar dinuclear manganese complexes and Oxone [46–50]. There are still controversies regarding the water oxidation catalyzed by the Mn dimer, because the evolved oxygen can also be transferred from Oxone (HSO_5^-) or the organic peroxide *t*-BuOOH, not water. Although they tried an ^{18}O-isotope-labeling experiment to demonstrate the O atom originated from water not the oxidant, the isotope exchange between ClO^- and $H_2{}^{18}O$ was completely ignored to puzzle the ^{18}O-isotope-labeling results. Their proposed catalytic cycle for O_2 evolution involved a hypothesized intermediate of di-μ-oxo Mn^V–Mn^V dimer with terminal manganese-oxo (Mn=O) group. This intermediate was proposed to react with OH^- to produce O_2 and di-μ-oxo diaquo-Mn^{III}–Mn^{III} dimer, yet it had not been detected during the reaction. Later, in another reaction of Mn dimer with a one-electron Ce^{IV} oxidant, a smaller amount of O_2 was detected with a lower TON of less than unity (0.54), suggesting non-catalytic O_2 evolution by Mn dimer which eventually decomposed to permanganate ions by ligand oxidation after 6 h [44]. Other groups also investigated the catalytic water oxidation by Mn dimer in an aqueous phase using a non-oxygen-containing oxidant Ce^{IV} or $[Ru(bpy)_3]^{3+}$, but no O_2 evolution was observed [51–53]. Baffert and coworkers concluded that Mn dimer is not a suitable homogeneous catalyst for water oxidation due to the formation of an inactive tetranuclear linear complex with a μ-oxo bridge from Mn^{IV}–Mn^{IV} state of Mn dimer [52, 54]. Later, studies using resonance Raman spectroscopy by Limburg and coworkers revealed that the exchange of oxygen atoms between OCl^- and water is rather fast ($t_{1/2} < 10$ s) and thus may account for another pathway toward O_2 evolution [45]. More recent results of ^{18}O-isotope-labeling experiments gave deeper insights on possible explanations for the water oxidation by μ-oxo Mn dimers [55, 56]. Based on the mass spectrometric analysis of O_2 evolution experiments in $H_2{}^{18}O$-enriched aqueous solutions, either no (H_2O_2, *t*-BuOOH) or only 50 % ($HSO_5{}^-$) of ^{18}O-isotope label was found to be incorporated into the O_2 produced. The reactions with peroxides, therefore, seem to follow a different pathway from the water oxidation reaction, which involves the transfer of at least one oxygen atom from the oxidant to a manganese center. Very recently, a new mechanism of the O_2 evolution catalyzed by Mn dimer was proposed by Hatakeyama and coworkers based on density functional theory (DFT) calculations [57]. In order to explain the results of the isotope study without including the Mn^V=O intermediate, they considered the dinuclear manganese complex in the presence of excess oxidants (OCl^- or $HSO_5{}^-$) to be neutralized by three counterions that enables HO^- to form from Mn-bound water (OCl^- acting as an activator) and an O–O bond to form by accepting an electron transferred from HO^-.

In 1986, Kaneko and coworkers [41, 58] first reported the O_2 evolution catalyzed by $[(bpy)_2Mn^{III}(\mu\text{-}O)_2Mn^{IV}(bpy)_2)]^{3+}$ (Fig. 13.1b) using Ce^{IV} as oxidant in heterogeneous phase, with the manganese dimer adsorbed on the surface of clay. Several attempts have followed to directly attach the Mn dimer and similar dinuclear manganese complexes onto supporting layers of clay minerals, such as kaolin, montmorillonite, and mica [51, 59–64]. The adsorption of dimanganese complexes on the supporting layers induced the catalytic evolution of O_2 from the reaction of water with Ce^{IV}, with maximum TONs ranging from 15 to 17. Yagi and coworkers observed that the supporting layers are essential for preventing the decomposition of Mn dimer to MnO_4^- by disproportionation [51]. The use of kaolin or mica as a solid support not only suppressed the formation of MnO_4^- but also provided a surface of highly concentrated Mn dimers, which was favorable for the cooperative catalysis. These aluminum silicates serve as very hydrophilic and layered surfaces for the metal complexes to strongly bind to, and they are also stable to oxidation and hydrolysis, which make them suitable for water oxidation under highly acidic conditions. The adsorption-induced water oxidation with Ce^{IV} was not limited to Mn dimers. The manganese complexes $[(HB(pz)_3)Mn^{III}(\mu\text{-}O)(OAc)_2Mn^{III}((pz)_3BH)]$ (pz=pyrazole, Fig. 13.1c) and $[(tpdm)Mn^{III}(\mu\text{-}O)(OAc)_2Mn^{III}(tpdm)]^{2+}$ (tpdm=*tris* (2-pyridyl)methane, Fig. 13.1d) which are inactive in homogeneous phase were reported to form active heterogeneous water oxidation catalysts when immobilized on kaolin or montmorillonite [61, 62]. Electron-withdrawing or electron-donating groups introduced to the tpy ligands of Mn dimer/mica system strongly affected the rate and extent of water oxidation using Ce^{IV} as oxidant, suggesting that the Mn-dimer derivatives were also active catalysts for water oxidation [64].

UV/Vis spectroscopy revealed that the immobilization of Mn dimer greatly altered its electronic properties, indicating strong interactions of the Mn dimer with the clay surfaces [61, 63]. More importantly, kinetic studies and electron paramagnetic resonance (EPR) spectroscopy suggested that the catalysis involves the cooperation of two or more manganese centers, which was facilitated by the adsorption on interlayers of clay minerals [60, 62, 65]. It was proposed that these manganese aggregates are built up from oxo-linked manganese centers of an oxidation state $\geq$+3, as the adsorption of Mn^{II} ions on montmorillonite did not lead to an active catalyst [62]. Although the dimerization of two Mn dimers has been demonstrated to occur *via*oxo-bridging of manganese centers, and the precise structures of resulting Mn_4 species were obtained using X-ray crystallography [52, 66], the exact nuclearity, structure, and geometry of them on the clay support are not yet confirmed. In fact, there is still a question on what species is truly responsible for the catalytic water oxidation. Recently, Najafpour and coworkers reported detailed studies on the surface characterizations of clay or mica that has adsorbed Mn-dimer molecules. Using various techniques including XPS, FT-IR, XRD, SEM, and TEM, they showed that it takes only a few minutes of water oxidation to cover the surface with nanosized manganese oxides which is proposed to be the active catalyst [67, 68]. Similar conversion of chloro-bridged iridium dimer complex to the iridium oxide nanoparticles was reported during the water oxidation with Ce^{IV} [69].

Another interesting homogeneous dinuclear manganese complex based on a ligand containing two imidazole groups was reported by Karlsson and coworkers (Fig. 13.1e) [70]. They improved previously reported ligand [46, 71, 72] by substituting benzylic amines with more oxidatively stable imidazole groups and introducing negatively charged carboxylate groups which dramatically reduced the redox potentials of the metal center. In the presence of a single-electron oxidant $[Ru(bpy)_3]^{3+}$, this complex showed catalytic oxidation of water to molecular oxygen giving a TON of *ca.* 25 after 1 h. ^{18}O-labeling experiments confirmed that both oxygen atoms of dioxygen produced are derived from water. Moreover, it was also capable of catalyzing photochemical water oxidation using either $[Ru(bpy)_3]^{2+}$ or $[Ru(bpy)_2(deeb)]^{2+}$ (deeb=4,4′-bis(ethoxycarbonyl)-2,2′-bipyridine). The oxidation-resistant ligand environment was claimed to allow homogeneous catalytic oxidation of water by use of the mild single-electron oxidant, although the involvement of heterogeneous manganese oxide as an active catalyst is still in question.

A series of related dinuclear manganese complexes were reported by Arafa and coworkers while we were preparing this review chapter [73]. They introduced a variety of substituents as non-innocent distal groups in order to investigate their impacts on promoting proton transfer processes during water oxidation catalysis. In the presence of $[Ru(bpy)_3]^{3+}$, the complex possessing distal carboxylate group was reported to evolve O_2 most efficiently among all the derivatives with a TON of 12 (Fig. 13.1f). This was ascribed to the decreased redox potentials in the manganese complex, which would improve the O–O bond formation. In addition, DFT calculations supported that the distal carboxylates in the ligand framework allow the accommodation of a bridging hydroxide/aqua molecule to yield a hydrogen-bonded scaffold, which might facilitate the oxidation of water.

13.3.2 Dye-Sensitized Manganese Catalytic Systems

In order to utilize sunlight as a driving force for water oxidation, oxygen-evolving electrocatalysts were coupled to light-absorbing dyes and sacrificial electron acceptors in attempts to engineer homogenous photocatalysts. Dyes with an oxidation potential higher than +0.82 V (*vs.* NHE) are required to drive water oxidation in a solution of pH 7. Conventional $[Ru(bpy)_3]^{2+}$ and $[Ir(bpy)_3]^{2+}$ dyes are routinely used to harvest visible light and to generate the sufficient oxidizing driving force for the activation of catalyst. The activated catalyst then extracts electrons from water molecule to complete the redox cycle.

Sun and coworkers reported a Mn^{II} complex covalently linked to $[Ru(bpy)_3]^{2+}$ as a first step of mimicking PS-II and showed that after initial electron transfer from the photoexcited Ru^{II} to an external electron acceptor, methyl viologen (MV^{2+}), intramolecular electron transfer takes place from the Mn^{II} to the photogenerated Ru^{III} in acetonitrile solution (Fig. 13.2) [74, 75]. They also showed that external dinuclear manganese complex could be oxidized by the tyrosyl radical photogenerated by intramolecular electron transfer in aqueous solution [76], and Mn dimer covalently linked to Ru photosensitizer *via* a tyrosine unit could be oxidized by photogenerated Ru^{III} [77]. Burdinski and coworkers used a similar approach with mono-, di-, and trinuclear manganese complexes with phenolate ligands covalently linked to $[Ru(bpy)_3]^{2+}$ [78]. They showed that intramolecular electron transfer takes place from a Mn^{II}-trimer moiety to the photochemically generated Ru^{III}, as well as from a phenolate ligand to the Ru^{III} in a Mn^{IV}-monomer derivative [79]. Although catalytic O_2 evolution was not demonstrated, these works by Sun and Burdinski have offered mechanistic insight into the electron transfer coupled with proton transfer which very likely plays an important role in photosynthetic water oxidation.

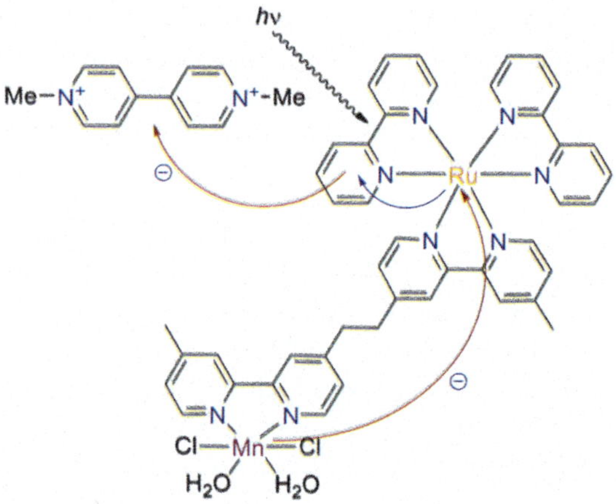

Fig. 13.2 Photoinduced electron transfer pathway of a Mn^{II} complex linked to $[Ru(bpy)_3]^{2+}$ dye in the presence of MV^{2+}. The *curved arrows* show the transfer of an electron from excited Ru^{II} complex to MV^{2+} and intramolecular electron transfer from coordinated Mn^{II} to photogenerated Ru^{III} [74]

A few studies on the photochemical and catalytic properties of oxomanganese complexes immobilized on solid supports have followed. Yagi and coworkers incorporated $[Ru(bpy)_3]^{2+}$ and $[(H_2O](tpy)Mn^{III}(\mu\text{-}O)_2Mn^{IV}(tpy)(H_2O)]^{3+}$ within a mica support *via* cation exchange [80]. This configuration could bring the Ru dye and Mn dimer close enough for reductive quenching of excited dye by the Mn dimer. However, the isolation of these reactive centers within a mica support limited the oxidative quenching by the dissolved $S_2O_8^{2-}$. Consequently, the efficiency of the mica/Mn-dimer/Ru-dye photocatalytic system was not very high that it yielded a TON of 3.4 after 17 h of visible light irradiation.

Co-attachment of both dye and manganese catalyst onto the surface of a support, such as TiO_2 nanoparticle, was expected to help resolving this problem. This would ensure the oxidative quenching by the conduction band of TiO_2 as well as the proximity-enhanced reductive quenching of excited dye. Similar attempts have been made with Pt nanoparticles attached onto TiO_2 which had been shown to split water molecule by UV light irradiation in the gas phase but not in the liquid phase [81]. Brudvig and coworkers reported the direct deposition of a Mn dimer $[H_2O(tpy)Mn^{III}(\mu\text{-}O)_2Mn^{IV}(tpy)H_2O](NO_3)_3$ onto TiO_2 nanoparticles, either by substituting one of its water ligands with TiO_2 nanoparticle [65] or using a robust chromophoric covalent linker [82, 83]. Although these two systems successfully demonstrated the visible light-induced formation of Mn^{IV} tetramers [65] and a higher oxidation state of Mn^{IV}–Mn^{IV} [83], photocatalytic O_2 evolution was not observed. Both configurations required the activation with Ce^{4+} as the primary one-electron oxidant for the splitting of water.

13.3.3 Manganese–Corrole Systems

Corroles are naturally occurring macrocyclic ligands and known to stabilize the coordinated metal centers in high-valent oxidation states. In 2007, Gao and

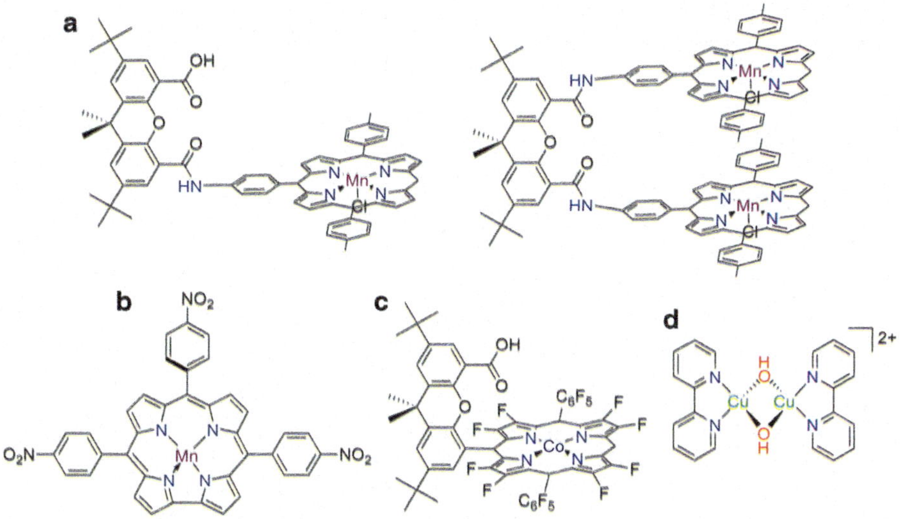

Fig. 13.3 Manganese and other WOCs based on corrole system [84, 86–88]

coworkers showed water oxidation electrochemically catalyzed by a manganese complex where a xanthene backbone functionalized with mono- and bis-corrole units was used to bind the metal (Fig. 13.3a) [84]. In the latter complex, two corrole units are configured in a face-to-face manner, keeping the two manganese centers in a close and rigid environment that is beneficial for stabilizing them in high-valent states. From a potential-controlled electrolysis in a 3:2 CH_3CN:CH_2Cl_2 solution containing $n$$Bu_4NOH$, O_2 evolution was detected with the potential scanning from 1.0 to −1.5 V. In the absence of mechanistic details, electrocatalysis at 0.85 V for 15 min at 20 °C showed that the binuclear manganese complex was able to generate O_2 more efficiently (46 μM) than the mononuclear complex (16 μM). The moderate difference in their catalytic efficiency suggested that both catalysts operated through a common single-site mechanism which was further verified by DFT calculations indicating that both manganese–corrole catalysts react *via* nucleophilic addition of water or hydroxide to a Mn^{IV}=O intermediate [85]. Bis-corrole manganese catalyst may react *via* O–O coupling, but this route was energetically less favorable than the nucleophilic attack. The authors prepared another manganese (III) corrole complex in a subsequent report in order to get a better understanding of manganese–corrole system and investigate the possibility of O–O bond formation by the attack of water (Fig. 13.3b) [86]. The nucleophilic attack of hydroxide on a high-valent Mn^{V}=O species that was generated by adding the oxidant triggered rapid oxygen evolution, as verified by real-time MS measurements, and it was suggested as a potential mechanism for oxidation of water to molecular oxygen. Mn^{V}=O and Mn^{IV}-OO^{-} species were identified as catalytic intermediates by means of electronic absorption measurements, high-resolution mass spectroscopy, and ^{18}O-isotope-labeling experiments. Although only very small amounts of O_2

were detected with a TON of 0.01, it is a valuable finding that water oxidation can proceed *via* nucleophilic addition of hydroxide to a mononuclear $Mn^{V}{=}O$.

This corrole system has been extended to other metal catalysts. Nocera and coworkers reported a series of cobalt–corrole water oxidation catalysts, an octafluorinated "hangman" complex being the best one, which will be described more in detail later (Fig. 13.3c) [87]. Mayer and Goldberg also showed one of the highest rate catalytic water oxidations using a copper–bipyridine complex formed by a self-assembly reaction (Fig. 13.3d) [88].

13.3.4 Manganese Cubane Systems

Recent high-resolution X-ray crystallographic studies on single crystals of PS-II elucidated the three-dimensional structure of its many subunits and the environments around the OEC [15, 16, 19, 20]. Inspired by the manganese cluster in natural PS-II, many multinuclear catalysts have been developed to mimic the structural and functional properties of OEC. Several multinuclear manganese complexes of various configurations have been reported, and most of them were thoroughly reviewed in recent years [89, 90]. It is noteworthy that a plentiful number of tetranuclear manganese complexes with different motif of the Mn_4O_5 core have been prepared as more relevant catalysts for the oxygen evolution from water. Among such a variety of tetramanganese complexes, however, only a few were investigated for water oxidation and showed only mild catalytic activities.

Dismukes and coworkers developed a tetramanganese complex $[Mn_4O_4L_6]$ (L= diphenylphosphinate, $(MeOPh)_2PO_2^-$) with a ($2Mn^{III}$–$2Mn^{IV}$)-cubane core [35, 91], where the manganese centers are bridged by diphenylphosphinate ligand, and successfully isolated the one-electron oxidized $[Mn_4O_4L_6]^+$ (Mn^{III}–$3Mn^{IV}$) cubane complex (Fig. 13.4) [92]. UV light-induced O_2 release from $[Mn_4O_4L_6]$ was observed by laser–desorption–ionization time-of-flight mass spectrometry (LDI-TOF-MS) by vaporizing and ionizing the species with a pulsed N_2 laser operating at 337 nm [93–95]. With ^{18}O-isotope-labeled experiment, the release of O_2 and a L^- ligand from $[Mn_4O_4L_6]$ was confirmed, and a so-called butterfly structure $[Mn_4O_2L_5]^+$ was proposed. When $[Mn_4O_4L_6]$ complex reacted with the hydrogen atom donor, phenothiazine, in CH_2Cl_2 solution, two of the $[Mn_4O_4L_6]$'s corner oxygens were released as two labile water molecules, resulting in the formation of $[Mn_4O_2L_6]$ and $[Mn_4O_2L_5]^+$, "incomplete cubane" and "butterfly" structures (Fig. 13.4) [96]. Thus, they demonstrated that the $[Mn_4O_4L_6]$ cubane structure has an intrinsically reactive topology that facilitates both chemical reduction to water and photo-rearrangement to an O_2 molecule.

The illumination with visible light and small external potential of 1.2 V (*vs.* NHE) triggered sustained photocurrents and oxygen evolution from the $[Mn_4O_2L_6]$ complex suspended in a proton-conducting Nafion membrane coated onto a conducting electrode in aqueous electrolyte at pH=6.5 with >1,000 catalytic cycles and its maximum TOF of 0.075 s^{-1} [97]. This was further developed to a

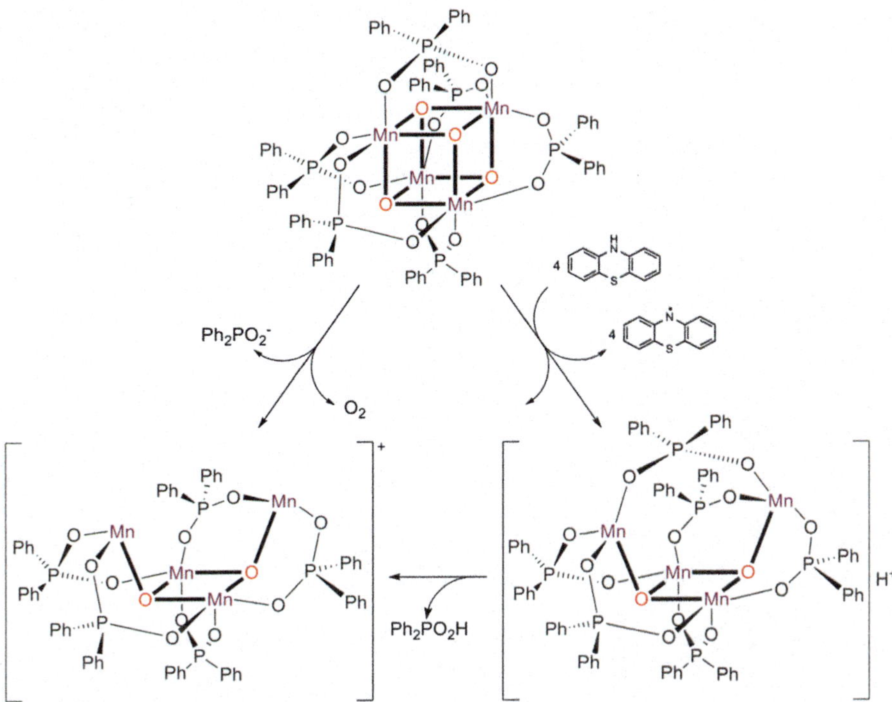

Fig. 13.4 The structure of manganese cubane complex and its reaction pathways to the incomplete cubane and butterfly structures [92]

composite photoanode by combining Mn-cubane catalyst/Nafion membrane hybrid with a $[Ru^{II}(bpy)_2(COO)_2bpy]$ light sensitizer supported onto a TiO_2 film/FTO electrode (FTO = fluorine-doped tin oxide) [98]. Upon the irradiation of visible light, the Ru^{II} dye injects an electron into the conduction band of TiO_2 from which it flows into an external circuit, generating a potent one-electron oxidant, $[Ru^{II}(bpy)_2(COO)_2bpy]^+$. By coupling this charge separation with the $[Mn_4O_4L_6]$ catalyst, a photoelectrochemical multilayer device system has been constructed to oxidize water using only visible light, without any external potential. The average peak photocurrent density was *ca.* 31 $\mu A \cdot cm^{-2}$ (faradic yield of O_2 evolution = 10 %, TOF = 47 ± 10 h^{-1}); however, the instability of the photocurrent increased with irradiation time. A self-repair mechanism was suggested to be involved, in which a "butterfly" structure was induced by the photolytic disruption of a diphenylphosphinate ligand and the evolution of O_2 followed by reassembly with a water molecule regained the cubane structure.

Recently, Spiccia and coworkers performed a detailed study on the catalytic mechanism of the $[Mn_4O_4L_6]^+$ in a Nafion membrane by using *in situ* Mn K-edge X-ray absorption spectroscopy (XAS) and TEM [99]. A decomposition product of $[Mn_4O_4L_6]$ within the Nafion polymer matrix was revealed by spectroscopic analyses which was suggested to be Mn^{II}-type compounds ($Mn(H_2O)_6^{2+}$). Subsequent electrochemical oxidation of these compounds was proposed to form $Mn^{III/IV}$ oxides that are believed to be catalytically active species responsible for water oxidation.

A variety of manganese catalysts were further confirmed to lose their integrity on loading into acidic Nafion films and form Mn^{II}-type species, which were then converted into MnO_x nanoparticles on electrooxidation [100, 101]. Prolonged water oxidation catalysis was observed by these 6–20 nm-sized MnO_x nanoparticles that were generated from precursors, $[Mn^{IV}(Me_3TACN)(OCH_3)_3]^+$ and $[Me_3TACN]_2Mn^{III}_2(\mu\text{-}O)(\mu\text{-}CH_3COO)_2]^{2+}$ (Me_3TACN=N,N',N''-trimethyl-1,4,7-triazacyclononane) [101].

13.3.5 *Other Manganese Systems*

Anxolabéhère-Mallart and coworkers studied the sequential electrochemical oxidation of a mononuclear Mn^{II}–OH_2 complex with a non-porphyrinic six-coordinate ligand to Mn^{III}–OH and Mn^{IV}=O species (Fig. 13.5) [102]. The Mn^{IV} complex was generated in cyclic voltammetric scans through either sequential or direct electrochemical oxidation, which was further characterized by UV/Vis and EPR spectra. The analysis of X-ray absorption spectroscopy (XAS) data revealed a gradual shortening of the Mn-O bond as the complex was oxidized to a higher oxidation state. DFT calculations suggested that the bonding in Mn-O species is best described as a Mn^{III}–$O^{\cdot}$ complex and it might be involved in the crucial role in natural and artificial water-splitting reactions.

The encaging of a catalyst in pores of metal-organic frameworks (MOF) was reported by Hansen and Das to improve the activity of water oxidation catalysts [103]. They demonstrated that Mn dimers ($[(tpy)Mn^{III}(\mu\text{-}O)_2Mn^{IV}(tpy)]^{3+}$) encaged in a Cr-based MOF catalyzed water oxidation in the presence of Ce^{IV} for more than 7 days with an initial TOF of 40 h^{-1}, and the isolation of single molecules of catalyst in small pores (one catalyst per pore) can sustain its activity.

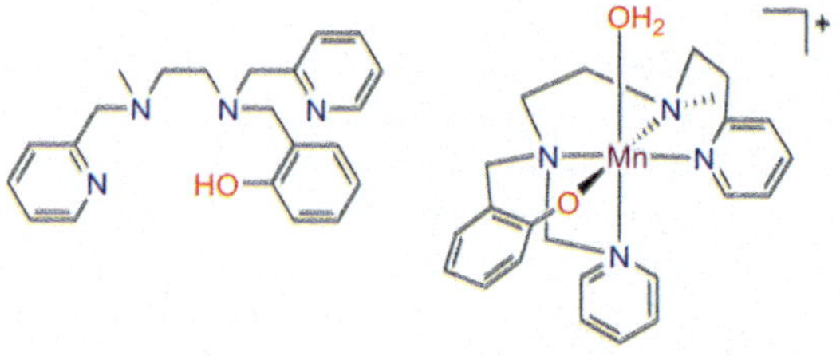

Fig. 13.5 Structures of a $[N_4O^-]$ ethane-bridged ligand and a Mn^{II}–OH_2 complex with the ligand developed by Anxolabéhère-Mallart and coworkers [102]

13.4 Cobalt Complex Catalysts

Cobalt ions (Co^{2+}) have been known to catalyze the oxidation of water since the early 1980s [104–106]. In the presence of $[Ru(bpy)_3]^{3+}$ and persulfates ($S_2O_8^{2-}$), aqueous solutions of Co^{II} salts or cobalt oxides were reported to evolve oxygen. However, the interests in these catalysts diminished as they tend to aggregate quickly to form insoluble and catalytically inactive precipitates. In 2008, Nocera and Kanan revitalized cobalt complexes as an interesting water oxidation catalyst by reporting that the potential-controlled electrolysis of Co^{II} salts in pH 7 phosphate buffer at 1.3 V (*vs.* NHE) resulted in the deposition of a dark precipitate on the electrode from which oxygen was evolved [107]. This solid material, **CoPi**, was found to be active catalyst as catalytic water oxidation was observed on it even in the absence of Co^{2+} ions. **CoPi** has attracted considerable interests as it operated at low overpotentials and self-repairs under catalytic conditions in the presence of protic buffers [108]. The phosphate ions have been suggested to play a major role both as proton acceptors and self-repairing agents for their ability to continuously re-integrate solubilized Co^{2+} ions into the film during catalysis through precipitation of Co^{3+} ions. Characterization of deposited material with various techniques including X-ray diffraction, extended X-ray absorption fine structure (EXAFS), TEM, and X-ray photoelectron spectroscopy (XPS) identified it as an amorphous cobalt oxide/hydroxide with its structural unit of active phase being a cluster of edge-sharing molecular cobaltates where interconnected complete or incomplete Co^{III}-oxo cubanes are condensed in the extended phase [2, 109–113].

In general, two possible active intermediates in transition metal complex-catalyzed water oxidation can be considered (Fig. 13.6). Upon oxidation in water, the catalyst is first converted to a molecular metal-oxo species (LM=O) that may either oxidize the water or get decomposed to a metal oxide (MO) [114]. In the latter case, the catalyst is only a precursor for MO which is the real active species for water oxidation. The relative stability and reactivity of LM=O and MO would determine the reaction pathway. It will depend on the nature of metal, redox potential, chelating ability, and the robustness of the ligands with respect to oxidation, as well as the oxidation method. With a strongly binding ligand to the metal, k_2 exceeds k_3 making LM=O the active catalyst, whereas the metal catalyst with a less stable ligand will have a faster conversion to another form of active catalyst MO ($k_3 >> k_2$).

Nocera's work has demonstrated that Co^{II} or Co^{III} complexes are converted to their oxide forms following the second pathway described above, and they act as precursors for active water oxidation catalyst, CoO_x. A large number of reports on successful heterogeneous CoO_x catalysts for oxygen evolution from water have

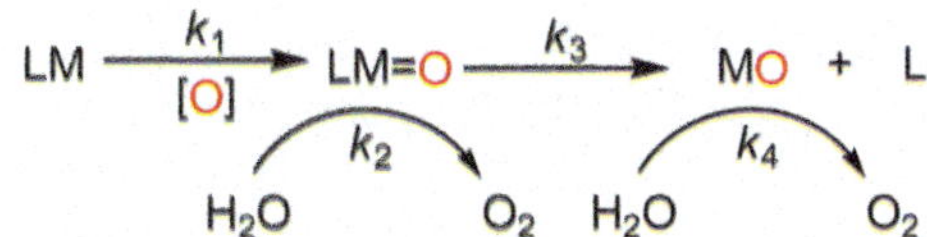

Fig. 13.6 Reaction pathways of molecular metal complex in the catalytic water oxidation [114]

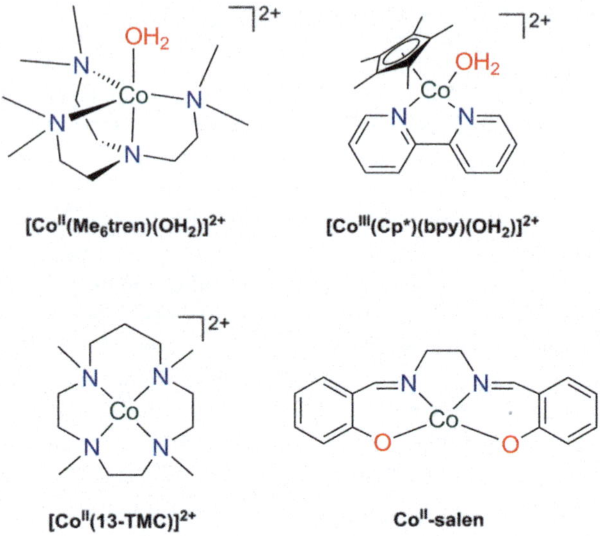

Fig. 13.7 Water-soluble cobalt complexes that act as precursors to other active WOCs [118, 119]

$[Co^{II}(Me_6tren)(OH_2)]^{2+}$ $[Co^{III}(Cp^*)(bpy)(OH_2)]^{2+}$

$[Co^{II}(13\text{-}TMC)]^{2+}$ Co^{II}-salen

followed [115] as well as controversial cobalt polyoxometalates [116, 117]. Here, we will only briefly discuss a few examples of Co complexes as precursor for active water oxidation catalyst and focus more on recently developed molecular cobalt complexes which do not decompose to heterogeneous CoO_x phase.

13.4.1 Cobalt Complexes as a Precursor for Active Catalysts

Hong and coworkers reported photocatalytic oxygen evolution from water-soluble mononuclear cobalt complexes $[Co^{II}(Me_6tren)(OH_2)]^{2+}$ (Me_6tren=tris(N,N′,N″-dimethylaminoethyl)amine), $[Co^{III}(Cp^*)(bpy)(OH_2)]^{2+}$ ($Cp^*=\eta^5$-pentamethylcyclopentadienyl), and $[Co^{II}(13\text{-}TMC)]^{2+}$ (13-TMC=1,4,7,10-tetramethyl-1,4,7,10-tetraazacyclotridecane) in combination with $[Ru(bpy)_3]^{2+}$ and $S_2O_8{}^{2-}$ (Fig. 13.7) [118]. ^{1}H NMR measurements suggested that these complexes were converted to different species after photocatalytic water oxidation at catalyst concentration >2.5 mM. X-ray photoelectron spectroscopy (XPS) revealed that the surface of the formed nanoparticles is mainly covered with $Co(OH)_x$, which can act as the actual catalyst.

More recently, Fu and coworkers reported a mononuclear cobalt(II) complex based on N,N′-bis(salicylidene)ethylenediamine ligand (Co^{II}-salen, Fig. 13.7) [119]. This cobalt salen complex is hydrolytically stable, but formation of nanoparticles during the photochemical reaction was observed from dynamic light scattering (DLS) measurements. Electrospray ionization mass spectrometry (ESI-MS), ^{1}H NMR, and XPS analysis indicated that the nanoparticles have a mixed Co^{III} species composed of oxides and/or hydroxides. In this case, Co^{II}-salen acted

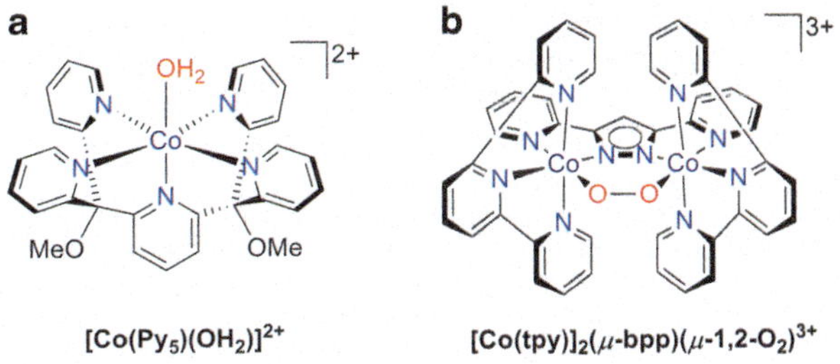

Fig. 13.8 Homogeneous cobalt WOCs with rigid ligands that stabilize the complex [120, 122]

a $[Co(Py_5)(OH_2)]^{2+}$

b $[Co(tpy)]_2(\mu\text{-bpp})(\mu\text{-}1,2\text{-}O_2)^{3+}$

as a precursor to an active heterogeneous cobalt catalyst which showed one of the highest TONs of 854.

13.4.2 Cobalt Catalyst with Rigid Ligands

Homogeneous mononuclear cobalt catalysts have hardly been reported so far, partially due to facile catalyst degradation as the ligand opposite to Co=O bond is labile during catalysis. Berlinguette's group demonstrated the prevention of ligand dissociation with ($[Co^{II}(PY5)(OH_2)]^{2+}$ (PY5 = 2,6-(bis(bis-2-pyridyl)-methoxymethane)pyridine, Fig. 13.8a) by using a pentadentate pyridine ligand [120]. The rigidity of the PY5 ligand and coordination of the other four pyridines to cobalt prevent the pyridine *trans* to the aqua ligand to easily dissociate from the complex. From the cyclic voltammetric (CV) experiments at various pH, they showed that the complex is stable to undergo a well-defined PCET process to a higher oxidation state $[Co^{III}–OH]^{2+}$, which was oxidized to furnish $[Co^{IV}–OH]^{3+}$ required for water oxidation in the presence of base. The complex $[Co^{II}(PY5)(OH_2)]^{2+}$ was claimed to be oxidatively stable from neutral to mildly basic (pH 7.6–10.3), and the catalytic oxygen evolution was observed with a high TOF of 79 s^{-1}. Very similar CVs were also observed from $[Co(OH_2)_6]^{2+}$ which readily decomposed to cobalt nanoparticles upon oxidation as in the cases of several other cobalt complexes [121]. Further kinetic and electrochemical studies concluded that $[Co^{II}(PY5)(OH_2)]^{2+}$ acts as a molecular catalyst, but it still remains unknown whether the nanoparticles formed during electrocatalytic water oxidation contribute to the catalytic activities and, if then, to what extent.

Rigsby and coworkers reported a binuclear cobalt(III) complex that was supported by a bridging bispyridylpyrazolate (bpp) ligand ($[Co(tpy)]_2(\mu\text{-bpp})(\mu\text{-}1,2\text{-}O_2)^{3+}$, Fig. 13.8b) [122]. They ascribed the catalyst's stability during electrocatalytic water oxidation to the non-labile and kinetically inert bpp bridging ligand framework which is believed to minimize ligand exchange and decomposition of catalyst under the forcing conditions. A catalytic wave responsible for water oxidation was observed at 1.91 V (*vs.* NHE) with a quasi-reversible one-electron oxidation peak at 1.56 V, suggesting that $Co^{III}–Co^{III}$ dimer is oxidized by two electrons before catalytic turnover is initiated. Controlled-potential electrolysis at

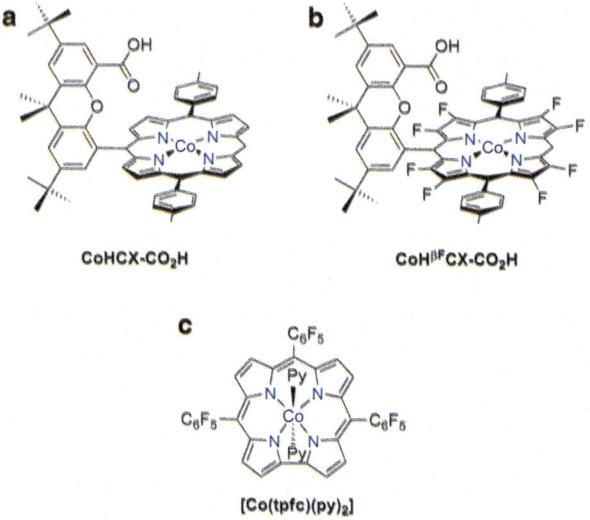

Fig. 13.9 Cobalt WOCs based on corrole system [87, 126]

2.0 V exhibited stable steady-state currents for 5 h, which could be modulated by changing the ancillary ligand from tpy to a more electron-donating bis-(*N*-methylimidazolyl)-pyridine ligand. However, the authors could not completely exclude the possibility that O_2 evolution arises from the decomposition of the Co complexes.

13.4.3 Cobalt–Corrole Systems

Nocera's group subsequently reported another series of interesting water oxidation catalysts based on cobalt–corroles, bolstered by the studies on $[Co^{II}(PY5)(OH_2)]^{2+}$ [120] and Mn corroles bearing nitroaromatic *meso* groups [86]. The cobalt (II) hangman porphyrin complex was previously shown to promote the $4e^-/4H^+$ reduction of oxygen to water [123, 124], which is the reverse reaction of water oxidation. Using fluorinated phenyl groups, two Co^{III} hangman corrole complexes, CoHCX-CO_2H and CoH$^{\beta F}$CX-CO_2H, were prepared (Fig. 13.9a and 9b) [87]. For the latter complex, the β-pyrrolic positions of the macrocycle were also fluorinated in addition to the fluorinated phenyl groups, aiming at increasing the oxidizing power of the Co complex. Electrochemical studies revealed a catalytic water oxidation step at +1.25 V (*vs.* Ag/AgCl) at neutral pH with a TOF of 0.81 s^{-1}. On the basis of the pH dependence of catalytic oxygen evolution, a PCET mechanism was suggested, and a Co^{IV} species was postulated as the active catalytic state for water oxidation. The exact electronic structure of the metal center and the corrole ligand, however, remained unclear, as either Co^{IV}–corrole or Co^{III}–corrole$^{\bullet+}$ is possible. The fluorination also resulted in boosting the stability of catalyst during the electrochemical water oxidation based on the results from UV/Vis, LD-MS MALDI-TOF, and high-resolution ESI-MS measurements. The hanging group attached to corrole unit was suggested to play an important role in pre-organizing water molecule within the cleft for the critical O–O bond formation step.

Several other transition metal complexes containing β-octafluoro hangman corrole ligand were reported to catalyze the O–O bond formation, for example, Fe, Ru, Ir, and Mn. Lai and coworkers compared these metals for their catalytic abilities using density functional theory (DFT) calculations [125]. They showed that the Co^{V} catalyst in its Co^{IV}–corrole$^{\bullet+}$ $S=1$ state is the most efficient water oxidant, determined by the ease of two-electron reduction and the $OH^{\bullet+}$ affinity of the M^{IV}=O of the corresponding M^{V}=O·H_2O complex.

Very recently, a simpler form of cobalt–corrole catalyst with bifunctionality of oxygen evolution and hydrogen production was reported by Lei and coworkers [126]. They prepared a cobalt–corrole catalyst without a hanging unit, $[Co(tpfc)(py)_2]$ (tpfc=5,10,15-tris(pentafluorophenyl)corrole, Fig. 13.9c), and coated it onto ITO electrode for electrochemical water oxidation and proton reduction. A TOF of $0.20\ s^{-1}$ at 1.4 V (*vs.* Ag/AgCl) was observed for catalytic water oxidation, while the stability of catalyst was confirmed with UV/Vis, scanning electron microscope (SEM) and energy disperse X-ray spectroscopy (EDX). Based on the pH-dependent water oxidation and theoretical studies, the dissociation of a Co-bound pyridine group and a PCET process were proposed.

13.4.4 Cobalt–Cubane Systems

Following Kanan and Nocera's investigations on **CoPi** [107], photochemical water oxidation by tetracobalt(III)-oxo cubane complex $[Co_4O_4(OAc)_4(py)_4]$ was reported by McCool and coworkers in 2011 (Fig. 13.10a) [127]. The synthesis and properties of this cobalt cubane was previously described and reported as an active catalyst for the oxidation of benzyl alcohols [128, 129]. Using a standard photoexcitation system of $[Ru(bpy)_3]^{2+}$ as a photosensitizer and $S_2O_8^{2-}$ as an electron acceptor, they reported a TON>40. Some decomposition with release of Co^{2+} ions (<5 %) occurred during the catalysis, but control experiments ruled out the possibility of observed catalytic activities from species other than $[Co_4O_4(OAc)_4(py)_4]$. Furthermore, Co^{II}- and $[Co_4O_4]$-dependent activities differed markedly in terms of kinetics and lag phases. Sartorel and coworkers further examined the effect of subtle changes in the pyridine through the introduction of various substituents and reported that *p*-methoxy derivative yields the highest quantum efficiency of 80 % and TON of 140 [130, 131]. The photochemical quantum yield was reported to depend on pH and catalyst concentration. The oxidation potential of the one-electron oxidation process, corresponding to the formation of formal $[3Co^{III}, Co^{IV}]^+$ species, was suggested to follow the expected Hammet linear free energy relationship, moving to less positive values with the

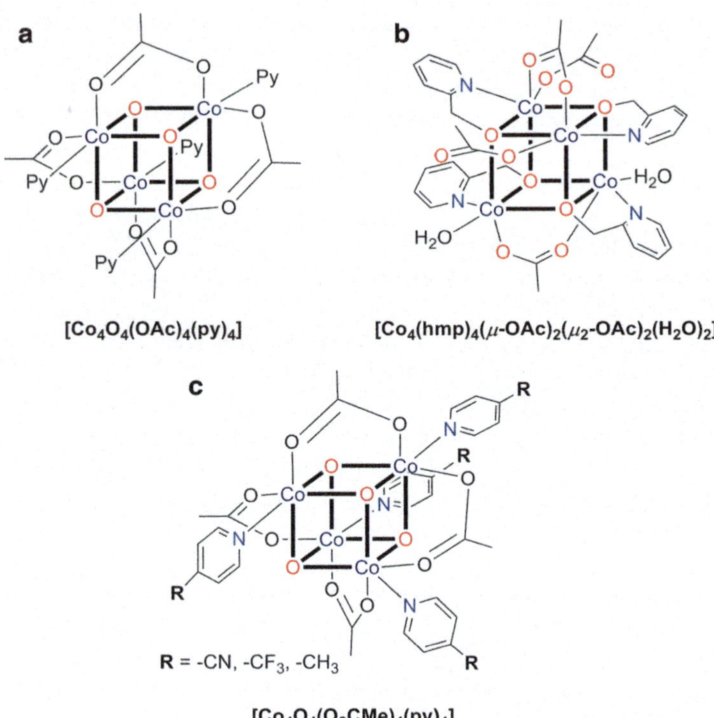

Fig. 13.10 Cobalt WOCs based on cubane structure [127, 135, 136]

more electron-donating groups on pyridine. It is noteworthy that $[Co_4O_4(OAc)_4(py)_4]$ can be considered as a simplified model for **CoPi**, so some properties of this cubane complex have been investigated to understand PCET properties of **CoPi** [132, 133].

Symes and coworkers reported catalytic water oxidation by similar cobalt–cubane complexes $[Co_4O_4(CO_2Me)_2(bpy)_4](ClO_4)_2$ and $[Co_4O_4(CO_2Py)_2(bpy)_4](ClO_4)_2$ (CO_2Py=4-carboxypyridine) which were appended to a Re^I photosensitizer ($[Re(phen)(CO)_3(CH_3CN)]PF_6$, phen=1,10-phenanthroline), and their light-driven oxygen evolution was studied as model systems of PS-II [134]. Different from PS-II, the Co^{III} centers in these cubanes were mainly stabilized by rather rigid bidentate ligands.

Recently, a Co^{II}-based cubane water oxidation catalyst was investigated by Evangelisti and coworkers. The catalyst was designed to closely mimic the PS-II by having a high-spin state of Co^{II} and introducing monodentate acetate and aqua ligands to $[Co^{II}_4(hmp)_4(\mu\text{-}OAc)_2(\mu_2\text{-}OAc)_2(H_2O)_2]$ (hmp=2-(hydroxymethyl) pyridine, Fig. 13.10b), which provided a flexible environment of the catalyst without compromising its stability. It is noteworthy that aqua ligands were implemented to cobalt–cubane structure for the first time in an attempt to take the same O–O formation pathways *via* water attack and exchange processes as in PS-II. The photocatalytic activity of $[Co^{II}_4(hmp)_4(\mu\text{-}OAc)_2(\mu_2\text{-}OAc)_2(H_2O)_2]$, measured in the presence of $[Ru(bpy)_3]^{2+}$ and $S_2O_8^{2-}$, exhibited pH-dependent O_2 generation with a TOF of 7.0 s^{-1} and TON of 40 at pH 9.

In 2014, Sun's group reported the immobilization of cobalt–cubane system ($[Co^{III}_4O_4(O_2CMe)_4(py)_4]$, Fig. 13.10c) onto a Nafion film-coated fluorine-doped tin oxide (FTO) electrode and an $\alpha\text{-}Fe_2O_3$ photoanode [136]. A sharply increased current wave attributed to oxygen evolution was observed at 1.0 V (*vs.* Ag/AgCl) from the cyclic voltammograms of cobalt–cubane-doped Nafion film, revealing that cobalt–cubane complex is catalytically active. The catalyst was stable in a prolonged application of applied potential of 1.2 V in neutral phosphate buffer. In contrast to Mn_4O_4 cubane doped in Nafion that was recently evidenced to be the precursor of manganese oxide [99], no such decomposition or nanoparticle formation was determined from CV, DLS, and SEM results. They further integrated cobalt–cubane complex with $\alpha\text{-}Fe_2O_3$ as a composite photoanode in photoelectrochemical experiments, where a significant shift of the onset potential by 400 mV as well as improved photocurrents was observed.

13.4.5 Other Mononuclear Cobalt Systems

Leung and coworkers investigated a homogeneous water oxidation catalysis by $[Co^{II}(qpy)(OH_2)_2]^{2+}$ (qpy=2,2′:6′,2″:6″,2‴-quaterpyridine, Fig. 13.11a) [114]. They observed photochemical oxygen evolution from water at pH 8.0 in the presence of $[Ru^{II}(bpy)_3]^{2+}$ and $S_2O_8^{2-}$ with a maximum TON of 335. The results from ESI-MS and DLS measurement after the photocatalysis revealed that the complex was stable under photooxidation conditions with less than 1 % decomposition, and no decomposed ligand nor colloidal particles were detected.

Cobalt porphyrin complexes were investigated by Nakazono and coworkers as single-site catalysts for water oxidation [137]. Among the three water-soluble cobalt porphyrins, **CoTPPS** (Co^{III}-5,10,15,20-tetrakis(4-sulfonatophenyl)porphyrin, Fig. 13.11b) showed the most active photochemical O_2 evolution activity with a maximum TON of 122. The reaction of **CoTPPS** was pH dependent and the maximum activity was reported at pH 11. Although DLS data confirmed that there is no nanoparticle formation (CoO_x), the catalytic activity of **CoTPPS** dramatically decreased with time to 15 % of the initial value, due to the decomposition of **CoTPPS** in the presence of $S_2O_8^{2-}$.

Wang and Groves recently reported another homogeneous cobalt–porphyrin catalyst, **Co^{II}-TDMImP** (Co^{II}–5,10,15,20-tetrakis-(1,3-dimethylimidazolium-2-yl)porphyrin, Fig. 13.11c), with a highly electron-deficient*meso*-dimethylimidazolium porphyrin. [138]. A strong catalytic current with an onset potential of 1.2 V (*vs.* Ag/AgCl) was observed in the cyclic voltammogram of **Co^{II}-TDMImP** without significant release of Co^{2+} ions. Mechanistic investigation indicated the

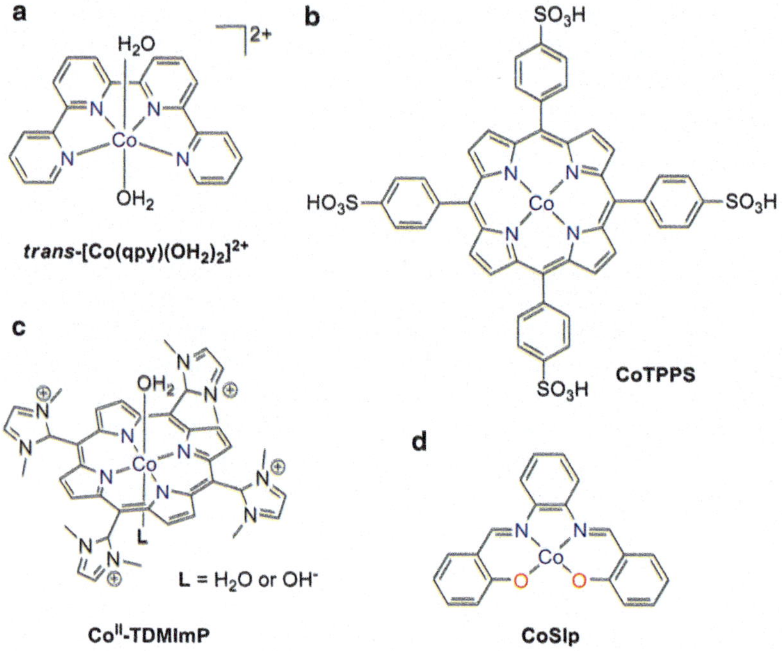

Fig. 13.11 Recently developed mononuclear cobalt WOCs [114, 137–139]

generation of a Co^{IV}–O porphyrin cation radical as the reactive oxidant, which has accumulated two oxidizing equivalents above Co^{III}-resting state of the catalyst. Interestingly, the buffer base in solution was shown to play several critical roles during the catalysis by facilitating both redox-coupled proton transfer processes leading to the reactive oxidant and subsequent O–O bond formation.

Pizzolato and coworkers reported a cobalt(II) complex with a salophen ligand (**CoSlp**, salophen = (*N,N'*-bis(salicylaldehyde)-1,2-phenylenediamine, Fig. 13.11d) as a mononuclear water oxidation catalyst [139]. Both electrochemical and visible light-driven photochemical water oxidation were demonstrated at neutral pH. **CoSlp** was stable enough not to undergo any major changes even after prolonged reactions under photochemical conditions, where $[Ru^{II}(bpy)_3]^{2+}$ and $S_2O_8^{2-}$ were engaged as photosensitizer and sacrificial electron acceptor, respectively. However, its reactivity (TON = 17) was not very impressive. The results of laser flash photolysis suggested the fast formation of a formal Co^{IV} derivatives by two consecutive one-electron oxidation of **CoSlp** by $[Ru^{II}(bpy)_3]^{3+}$ under irradiation conditions. A possible mechanism involving a photogenerated Co^{IV}-oxo intermediate that undergoes a rate determining nucleophilic attack by water molecule, followed by the O-O bond generation within the structural motif of Co^{II}-hydroperoxide, was suggested.

13.5 Conclusion and Outlook

The conversion of water into hydrogen and oxygen holds enormous potential for the future of our society. Nature has given us an excellent example of how to accomplish this goal of clean and renewable energy system. Artificial photosynthesis, a mimicking of Nature's photosynthetic engine, marks up a central and urgent scientific challenge of our days. It would require a close interplay of many different disciplines to realize this, in particular the oxidation of water that is currently considered as the bottleneck.

In this review chapter, recent developments in water oxidation catalysts based on manganese and cobalt were discussed. These two first-row transition metals share common advantages for developing artificial water oxidation system. First of all, both are relatively abundant on Earth, and therefore systems based on them will have strong economic viability and compatibility. Various high-oxidation states accessible for manganese and cobalt with potentials close to the thermodynamic potential required for water oxidation render them good candidates for artificial OECs. Lastly, both can form robust and stable metal-oxo frameworks under harsh oxidizing environments. Great progress has been achieved in the past decade using these elements for water splitting to hydrogen and oxygen. It is still a crucial challenge for chemists to design and identify more efficient and stable catalysts based on earth-abundant elements in order to develop a practical artificial photosynthetic system. A number of issues still remain for most of molecular catalysts of manganese and cobalt. Stability is one of several key concerns. The lifetimes for most WOCs range from a few minutes to several days during catalysis. Practically, much longer lifetime with millions to billions of TON is needed. Another concern to keep in mind when designing WOCs is the working potential which should be close to the thermodynamic requirement for water oxidation. This will minimize the overpotential required and thus the energy loss in electrocatalytic water oxidation. This is also important in designing the photocatalysts for water oxidation, as most commonly used photosensitizers have a low oxidation potential.

References

1. BP Statistical review of world energy 2013. http://www.bp.com/statisticalreview. Accessed 25 Apr 2014
2. Nocera DG (2012) The artificial leaf. Acc Chem Res 45(5):767–776
3. Kim H-S, Lee J-W, Yantara N et al (2013) High efficiency solid-state sensitized solar cell-based on submicrometer rutile TiO_2 nanorod and $CH_3NH_3PbI_3$ perovskite sensitizer. Nano Lett 13(6):2412–2417
4. Mathew S, Yella A, Gao P et al (2014) Dye-sensitized solar cells with 13 % efficiency achieved through the molecular engineering of porphyrin sensitizers. Nat Chem 6(3):242–247
5. Haid S, Marszalek M, Mishra A et al (2012) Significant improvement of dye-sensitized solar cell performance by small structural modification in π-conjugated donor–acceptor dyes. Adv Funct Mater 22(6):1291–1302

6. Chirilă A, Reinhard P, Pianezzi F et al (2013) Potassium-induced surface modification of Cu (In, Ga)Se2 thin films for high-efficiency solar cells. Nat Mater 12(12):1107–1111
7. Yang W, Duan H-S, Bob B et al (2012) Novel solution processing of high-efficiency earth-abundant Cu2ZnSn(S, Se)4 solar cells. Adv Mater 24(47):6323–6329
8. Karunadasa HI, Montalvo E, Sun Y et al (2012) A molecular MoS_2 edge site mimic for catalytic hydrogen generation. Science 335(6069):698–702
9. Voiry D, Yamaguchi H, Li J et al (2013) Enhanced catalytic activity in strained chemically exfoliated WS_2 nanosheets for hydrogen evolution. Nat Mater 12(9):850–855
10. Du P, Eisenberg R (2012) Catalysts made of earth-abundant elements (Co, Ni, Fe) for water splitting: recent progress and future challenges. Energy Environ Sci 5(3):6012–6021
11. Yang S, Gong Y, Zhang J et al (2013) Exfoliated graphitic carbon nitride nanosheets as efficient catalysts for hydrogen evolution under visible light. Adv Mater 25(17):2452–2456
12. Hull JF, Himeda Y, Wang W-H et al (2012) Reversible hydrogen storage using CO_2 and a proton-switchable iridium catalyst in aqueous media under mild temperatures and pressures. Nat Chem 4(5):383–388
13. Chun J, Kang S, Kang N et al (2013) Microporous organic networks bearing metal-salen species for mild CO2 fixation to cyclic carbonates. J Mater Chem A 1(18):5517–5523
14. Wong W-L, Lee LYS, Ho K-P et al (2014) A green catalysis of CO_2 fixation to aliphatic cyclic carbonates by a new ionic liquid system. Appl Catal A 472:160–166
15. Umena Y, Kawakami K, Shen J-R et al (2011) Crystal structure of oxygen-evolving photosystem II at a resolution of 1.9 Å. Nature 473(7345):55–60
16. Ferreira KN, Iverson TM, Maghlaoui K et al (2004) Architecture of the photosynthetic oxygen-evolving center. Science 303(5665):1831–1838
17. Rutherford AW, Faller P (2003) Photosystem II: evolutionary perspectives. Philos Trans R Soc Lond B Biol Sci 358(1429):245–253
18. Loll B, Kern J, Saenger W et al (2005) Towards complete cofactor arrangement in the 3.0 Å resolution structure of photosystem II. Nature 438(7070):1040–1044
19. Sproviero EM, Gascón JA, McEvoy JP et al (2008) Computational studies of the O_2-evolving complex of photosystem II and biomimetic oxomanganese complexes. Coord Chem Rev 252 (3–4):395–415
20. Guskov A, Kern J, Gabdulkhakov A et al (2009) Cyanobacterial photosystem II at 2.9 Å resolution and the role of quinones, lipids, channels and chloride. Nat Struct Mol Biol 16 (3):334–342
21. Kok B, Forbush B, McGloin M (1970) Cooperation of charges in photosynthetic O_2 evolution–I. A linear four step mechanism. Photochem Photobiol 11(6):457–475
22. Yagi M, Kaneko M (2001) Molecular catalysts for water oxidation. Chem Rev 101(1):21–36
23. Mayer JM (2004) Proton-coupled electron transfer: a reaction chemist's view. Annu Rev Phys Chem 55:363–390
24. Jablonsky J, Lazar D (2008) Evidence for intermediate S-states as initial phase in the process of oxygen-evolving complex oxidation. Biophys J 94(7):2725–2736
25. Rutherford AW, Boussac A (2004) Water photolysis in biology. Science 303(5665):1782–1784
26. Yano J, Kern J, Sauer K et al (2006) Where water is oxidized to dioxygen: structure of the photosynthetic Mn_4Ca cluster. Science 314(5800):821–825
27. Petrie S, Stranger R, Gatt P et al (2007) Bridge over troubled water: resolving the competing photosystem II crystal structures. Chem Eur J 13(18):5082–5089
28. Meyer TJ, Huynh MHV, Thorp HH (2007) The possible role of proton-coupled electron transfer (PCET) in water oxidation by photosystem II. Angew Chem Int Ed 46(28):5284–5304
29. Dau H, Zaharieva I (2009) Principles, efficiency, and blueprint character of solar-energy conversion in photosynthetic water oxidation. Acc Chem Res 42(12):1861–1870
30. Sartorel A, Bonchio M, Campagna S et al (2013) Tetrametallic molecular catalysts for photochemical water oxidation. Chem Soc Rev 42(6):2262–2280

31. Joya KS, Vallés-Pardo JL, Joya YF et al (2013) Molecular catalytic assemblies for electrodriven water splitting. ChemPlusChem 78(1):35–47
32. Artero V, Fontecave M (2013) Solar fuels generation and molecular systems: is it homogeneous or heterogeneous catalysis? Chem Soc Rev 42(6):2338–2356
33. Pecoraro VL, Baldwin MJ, Gelasco A (1994) Interaction of manganese with dioxygen and its reduced derivatives. Chem Rev 94(3):807–826
34. Manchanda R, Brudvig GW, Crabtree RH (1995) High-valent oxomanganese clusters: structural and mechanistic work relevant to the oxygen-evolving center in photosystem II. Coord Chem Rev 144:1–38
35. Rüttinger W, Dismukes GC (1997) Synthetic water-oxidation catalysts for artificial photosynthetic water oxidation†. Chem Rev 97(1):1–24
36. Mukhopadhyay S, Mandal SK, Bhaduri S et al (2004) Manganese clusters with relevance to Photosystem II. Chem Rev 104(9):3981–4026
37. Young KJ, Martini LA, Milot RL et al (2012) Light-driven water oxidation for solar fuels. Coord Chem Rev 256(21–22):2503–2520
38. Matsushita T, Fujiwara M, Shono T (1981) Reactions of dichloromanganese(IV) Schiff base complexes with water as model for oxidation in photosystem II. Chem Lett 10(5):631–634
39. Fujiwara M, Matsushita T, Shono T (1985) Reaction of dichloromanganese(IV) Schiff-base complexes with water as a model for water oxidation in photosystem II. Polyhedron 4(11):1895–1900
40. Matsushita T, Spencer L, Sawyer DT (1988) Synthesis and characterization of binuclear manganese complexes: redox models for the water oxidation cofactor of photosystem II. Inorg Chem 27(7):1167–1173
41. Ramaraj R, Kira A, Kaneko M (1986) Oxygen evolution by water oxidation mediated by heterogeneous manganese complexes. Angew Chem Int Ed 25(9):825–827
42. Naruta Y, M-a S, Sasaki T (1994) Oxygen evolution by oxidation of water with manganese porphyrin dimers. Angew Chem Int Ed 33(18):1839–1841
43. Geselowitz D, Meyer TJ (1990) Water oxidation by μ-oxobis[bis(bipyridine)oxoruthenium (V)]$^{4+}$. An oxygen-labeling study. Inorg Chem 29(19):3894–3896
44. Limburg J, Vrettos JS, Liable-Sands LM et al (1999) A functional model for O-O bond formation by the O_2-evolving complex in Photosystem II. Science 283(5407):1524–1527
45. Limburg J, Vrettos JS, Chen H et al (2001) Characterization of the O_2-evolving reaction catalyzed by [(terpy)(H_2O)$Mn^{III}(O)_2Mn^{IV}$(OH_2)(terpy)]$(NO_3)_3$ (terpy=2,2′:6,2″-Terpyridine). J Am Chem Soc 123(3):423–430
46. Poulsen AK, Rompel A, McKenzie CJ (2005) Water oxidation catalyzed by a dinuclear Mn complex: a functional model for the oxygen-evolving center of Photosystem II. Angew Chem Int Ed 44(42):6916–6920
47. Chen TR, Olack G et al (2006) Speciation of the catalytic oxygen evolution system: [MnIII/IV2(μ-O)2(terpy)2(H2O)2](NO3)3+HSO5. Inorg Chem 46(1):34–43
48. Tagore R, Crabtree RH, Brudvig GW (2008) Oxygen evolution catalysis by a dimanganese complex and its relation to photosynthetic water oxidation. Inorg Chem 47(6):1815–1823
49. Nayak S, Nayek HP, Dehnen S et al (2011) Trigonal propeller-shaped [$Mn^{III}{}_3M^{II}$Na] complexes (M=Mn, Ca): structural and functional models for the dioxygen evolving centre of PSII. Dalton Trans 40(12):2699–2702
50. Seidler-Egdal RK, Nielsen A, Bond AD et al (2011) High turnover catalysis of water oxidation by Mn(II) complexes of monoanionic pentadentate ligands. Dalton Trans 40 (15):3849–3858
51. Yagi M, Narita K (2004) Catalytic O_2 evolution from water induced by adsorption of [(OH_2)(Terpy)Mn(μ-O)$_2$Mn(Terpy)(OH_2)]$^{3+}$ complex onto clay compounds. J Am Chem Soc 126 (26):8084–8085
52. Baffert C, Romain S, Richardot A et al (2005) Electrochemical and chemical formation of [Mn4IVO5(terpy)4(H2O)2]6+, in relation with the Photosystem II oxygen-evolving center model [Mn2III, IVO2(terpy)2(H2O)2]3+. J Am Chem Soc 127(39):13694–13704

53. Kurz P, Berggren G, Anderlund MF et al (2007) Oxygen evolving reactions catalysed by synthetic manganese complexes: a systematic screening. Dalton Trans 38:4258–4261
54. Kanady JS, Tsui EY, Day MW et al (2011) A synthetic model of the Mn3Ca subsite of the oxygen-evolving complex in Photosystem II. Science 333(6043):733–736
55. Beckmann K, Uchtenhagen H, Berggren G et al (2008) Formation of stoichiometrically ^{18}O-labelled oxygen from the oxidation of ^{18}O-enriched water mediated by a dinuclear manganese complex-a mass spectrometry and EPR study. Energy Environ Sci 1(6):668–676
56. Shevela D, Koroidov S, Najafpour MM et al (2011) Calcium manganese oxides as oxygen evolution catalysts: O_2 Formation Pathways Indicated by ^{18}O-Labelling Studies. Chem Eur J 17(19):5415–5423
57. Hatakeyama M, Nakata H, Wakabayashi M et al (2012) New reaction model for O–O bond formation and O_2 evolution catalyzed by dinuclear manganese complex. J Phys Chem A 116 (26):7089–7097
58. Ramaraj R, Kira A, Kaneko M (1987) Heterogeneous water oxidation by a dinuclear manganese complex. Chem Lett 16(2):261–264
59. Narita K, Kuwabara T, Sone K et al (2006) Characterization and activity analysis of catalytic water oxidation induced by hybridization of $[(OH_2)(terpy)Mn(\mu\text{-}O)_2Mn(terpy)(OH_2)]^{3+}$ and clay compounds. J Phys Chem B 110(46):23107–23114
60. Yagi M, Narita K, Maruyama S et al (2007) Artificial model of photosynthetic oxygen evolving complex: catalytic O_2 production from water by di-μ-oxo manganese dimers supported by clay compounds. Biochim Biophys Acta Bioenerg 1767(6):660–665
61. Kurz P (2009) Oxygen evolving reactions catalysed by manganese-oxo-complexes adsorbed on clays. Dalton Trans 31:6103–6108
62. Berends H-M, Homburg T, Kunz I et al (2011) K10 montmorillonite supported manganese catalysts for the oxidation of water to dioxygen. Appl Clay Sci 53(2):174–180
63. Gao Y, Crabtree RH, Brudvig GW (2012) Water oxidation catalyzed by the tetranuclear Mn complex $[Mn^{IV}{}_4O_5(terpy)_4(H_2O)_2](ClO_4)_6$. Inorg Chem 51(7):4043–4050
64. Yamazaki H, Igarashi S, Nagata T et al (2012) Substituent effects on core structures and heterogeneous catalytic activities of $Mn^{III}(\mu\text{-}O)_2Mn^{IV}$ dimers with 2,2′:6′,2″-terpyridine derivative ligands for water oxidation. Inorg Chem 51(3):1530–1539
65. Li G, Sproviero EM, Snoeberger Iii RC et al (2009) Deposition of an oxomanganese water oxidation catalyst on TiO_2 nanoparticles: computational modeling, assembly and characterization. Energy Environ Sci 2(2):230–238
66. Chen CM-N, Duboc C et al (2005) New linear high-valent tetranuclear manganese-oxo cluster relevant to the oxygen-evolving complex of photosystem II with oxo, hydroxo, and aqua coordinated to a single Mn(IV). Inorg Chem 44(25):9567–9573
67. Najafpour M, Boghaei D (2009) Heterogeneous water oxidation by bidentate Schiff base manganese complexes in the presence of cerium(IV) ammonium nitrate. Transit Met Chem 34(4):367–372
68. Najafpour MM, Isaloo MA (2014) Mechanism of water oxidation by nanolayered manganese oxide: a step forward. RSC Adv 4(13):6375–6378
69. Junge H, Marquet N, Kammer A et al (2012) Water oxidation with molecularly defined iridium complexes: insights into homogeneous versus heterogeneous catalysis. Chem Eur J 18(40):12749–12758
70. Karlsson EA, Lee B-L, Åkermark T et al (2011) Photosensitized water oxidation by use of a bioinspired manganese catalyst. Angew Chem Int Ed 50(49):11715–11718
71. Huang P, Magnuson A, Lomoth R et al (2002) Photo-induced oxidation of a dinuclear Mn2II, II complex to the Mn2III, IV state by inter- and intramolecular electron transfer to RuIIItris-bipyridine. J Inorg Biochem 91(1):159–172
72. Lee B-L, Kärkäs MD, Johnston EV et al (2010) Synthesis and characterization of oligonuclear Ru, Co and Cu oxidation catalysts. Eur J Inorg Chem 2010(34):5462–5470
73. Arafa WAA, Karkas MD, Lee B-L et al (2014) Dinuclear manganese complexes for water oxidation: evaluation of electronic effects and catalytic activity. Phys Chem Chem Phys. doi:10.1039/C3CP54800G

74. Sun L, Berglund H, Davydov R et al (1997) Binuclear ruthenium–manganese complexes as simple artificial models for Photosystem II in green plants. J Am Chem Soc 119(30):6996–7004
75. Sun L, Hammarstrom L, Norrby T et al. (1997) Intramolecular electron transfer from coordinated manganese(II) to photogenerated ruthenium(III). Chem Commun (6):607–608
76. Magnuson A, Frapart Y, Abrahamsson M et al (1998) A biomimetic model system for the water oxidizing triad in Photosystem II. J Am Chem Soc 121(1):89–96
77. Sun L, Raymond MK, Magnuson A et al (2000) Towards an artificial model for Photosystem II: a manganese(II, II) dimer covalently linked to ruthenium(II) tris-bipyridine via a tyrosine derivative. J Inorg Biochem 78(1):15–22
78. Burdinski D, Bothe E, Wieghardt K (1999) Synthesis and characterization of *tris*(bipyridyl) ruthenium(II)-modified mono-, di-, and trinuclear manganese complexes as electron-transfer models for Photosystem II. Inorg Chem 39(1):105–116
79. Burdinski D, Wieghardt K, Steenken S (1999) Intramolecular electron transfer from Mn or ligand phenolate to photochemically generated Ru^{III} in multinuclear Ru/Mn complexes. Laser flash photolysis and EPR studies on Photosystem II models. J Am Chem Soc 121 (46):10781–10787
80. Yagi M, Toda M, Yamada S et al (2010) An artificial model of photosynthetic photosystem II: visible-light-derived O_2 production from water by a di-μ-oxo-bridged manganese dimer as an oxygen evolving center. Chem Commun 46(45):8594–8596
81. Sato S, White JM (1980) Photodecomposition of water over Pt/TiO_2 catalysts. Chem Phys Lett 72(1):83–86
82. McNamara WR, Snoeberger RC, Li G et al (2008) Acetylacetonate anchors for robust functionalization of TiO_2 nanoparticles with Mn(II)–terpyridine complexes. J Am Chem Soc 130(43):14329–14338
83. Li G, Sproviero EM, McNamara WR et al (2009) Reversible visible-light photooxidation of an oxomanganese water-oxidation catalyst covalently anchored to TiO_2 nanoparticles. J Phys Chem B 114(45):14214–14222
84. Gao Y, Liu J, Wang M et al (2007) Synthesis and characterization of manganese and copper corrole xanthene complexes as catalysts for water oxidation. Tetrahedron 63(9):1987–1994
85. Privalov T, Sun L, Åkermark B et al (2007) A computational study of O–O bond formation catalyzed by mono- and bis-Mn^{IV}–corrole complexes. Inorg Chem 46(17):7075–7086
86. Gao Y, Tr Å, Liu J et al (2009) Nucleophilic attack of hydroxide on a Mn^{V} oxo Complex: a model of the O–O bond formation in the oxygen evolving complex of Photosystem II. J Am Chem Soc 131(25):8726–8727
87. Dogutan DK, McGuire R, Nocera DG (2011) Electocatalytic water oxidation by cobalt(III) hangman β-octafluoro corroles. J Am Chem Soc 133(24):9178–9180
88. Barnett SM, Goldberg KI, Mayer JM (2012) A soluble copper–bipyridine water-oxidation electrocatalyst. Nat Chem 4(6):498–502
89. Mullins CS, Pecoraro VL (2008) Reflections on small molecule manganese models that seek to mimic photosynthetic water oxidation chemistry. Coord Chem Rev 252(3–4):416–443
90. Cady CW, Crabtree RH, Brudvig GW (2008) Functional models for the oxygen-evolving complex of photosystem II. Coord Chem Rev 252(3–4):444–455
91. Ruettinger WF, Campana C, Dismukes GC (1997) Synthesis and characterization of Mn4O4L6 complexes with cubane-like core structure: a new class of models of the active site of the photosynthetic water oxidase. J Am Chem Soc 119(28):6670–6671
92. Ruettinger WF, Ho DM, Dismukes GC (1999) Protonation and dehydration reactions of the $Mn_4O_4L_6$ cubane and synthesis and crystal structure of the oxidized cubane $[Mn_4O_4L_6]^+$: a model for the photosynthetic water oxidizing complex. Inorg Chem 38(6):1036–1037
93. Ruettinger W, Yagi M, Wolf K et al (2000) O2 evolution from the manganese–oxo cubane core Mn4O46+: a molecular mimic of the photosynthetic water oxidation enzyme? J Am Chem Soc 122(42):10353–10357
94. Yagi M, Wolf KV, Baesjou PJ et al (2001) Selective photoproduction of O_2 from the Mn_4O_4 cubane core: a structural and functional model for the photosynthetic water-oxidizing complex. Angew Chem Int Ed 40(15):2925–2928

95. Wu J-Z, De Angelis F, Carrell TG et al (2005) Tuning the photoinduced O_2-evolving reactivity of $Mn_4O_4^{7+}$, $Mn_4O_4^{6+}$, and $Mn_4O_3(OH)^{6+}$ manganese–oxo cubane complexes. Inorg Chem 45(1):189–195
96. Ruettinger WF, Dismukes GC (2000) Conversion of core oxos to water molecules by 4e−/4H + reductive dehydration of the Mn4O26+ core in the manganese–oxo cubane complex Mn4O4(Ph2PO2)6: a partial model for photosynthetic water binding and activation. Inorg Chem 39(5):1021–1027
97. Brimblecombe R, Swiegers GF, Dismukes GC et al (2008) Sustained water oxidation photocatalysis by a bioinspired manganese cluster. Angew Chem Int Ed 47(38):7335–7338
98. Brimblecombe R, Koo A, Dismukes GC et al (2010) Solar driven water oxidation by a bioinspired manganese molecular catalyst. J Am Chem Soc 132(9):2892–2894
99. Hocking RK, Brimblecombe R, Chang L-Y et al (2011) Water-oxidation catalysis by manganese in a geochemical-like cycle. Nat Chem 3(6):461–466
100. Jiao F, Frei H (2010) Nanostructured manganese oxide clusters supported on mesoporous silica as efficient oxygen-evolving catalysts. Chem Commun 46(17):2920–2922
101. Singh A, Hocking RK, Chang SLY et al (2013) Water oxidation catalysis by nanoparticulate manganese oxide thin films: probing the effect of the manganese precursors. Chem Mater 25(7):1098–1108
102. Lassalle-Kaiser B, Hureau C, Pantazis DA et al (2010) Activation of a water molecule using a mononuclear Mn complex: from Mn-aquo, to Mn-hydroxo, to Mn-oxyl via charge compensation. Energy Environ Sci 3(7):924–938
103. Hansen RE, Das S (2014) Biomimetic di-manganese catalyst cage-isolated in a MOF: robust catalyst for water oxidation with Ce(IV), a non-O-donating oxidant. Energy Environ Sci 7(1):317–322
104. Shafirovich VY, Khannanov NK, Strelets VV (1980) Chemical and light-induced catalytic water oxidation. Nouv J Chim 4(2):81–84
105. Brunschwig BS, Chou MH, Creutz C et al (1983) Mechanisms of water oxidation to oxygen: cobalt(IV) as an intermediate in the aquocobalt(II)-catalyzed reaction. J Am Chem Soc 105(14):4832–4833
106. Ghosh PK, Brunschwig BS, Chou M et al (1984) Thermal and light-induced reduction of the ruthenium complex cation $Ru(bpy)_3^{3+}$ in aqueous solution. J Am Chem Soc 106 (17):4772–4783
107. Kanan MW, Nocera DG (2008) *In-situ* formation of an oxygen-evolving catalyst in neutral water containing phosphate and Co^{2+}. Science 321(5892):1072–1075
108. Lutterman DA, Surendranath Y, Nocera DG (2009) A self-healing oxygen-evolving catalyst. J Am Chem Soc 131(11):3838–3839
109. Dau H, Limberg C, Reier T et al (2010) The mechanism of water oxidation: from electrolysis *via* homogeneous to biological catalysis. ChemCatChem 2(7):724–761
110. McAlpin JG, Surendranath Y, Dincǎ M et al (2010) EPR evidence for Co(IV) species produced during water oxidation at neutral pH. J Am Chem Soc 132(20):6882–6883
111. Kanan MW, Surendranath Y, Nocera DG (2009) Cobalt-phosphate oxygen-evolving compound. Chem Soc Rev 38(1):109–114
112. Risch M, Khare V, Zaharieva I et al (2009) Cobalt–oxo core of a water-oxidizing catalyst film. J Am Chem Soc 131(20):6936–6937
113. Kanan MW, Yano J, Surendranath Y et al (2010) Structure and valency of a cobalt–phosphate water oxidation catalyst determined by *in Situ* X-ray spectroscopy. J Am Chem Soc 132(39):13692–13701
114. Leung C-F, Ng S-M, Ko C-C et al (2012) A cobalt(II) quaterpyridine complex as a visible light-driven catalyst for both water oxidation and reduction. Energy Environ Sci 5(7):7903–7907
115. Jiao F, Frei H (2010) Nanostructured cobalt and manganese oxide clusters as efficient water oxidation catalysts. Energy Environ Sci 3(8):1018–1027

116. Stracke JJ, Finke RG (2011) Electrocatalytic water oxidation beginning with the cobalt polyoxometalate $[Co_4(H_2O)_2(PW_9O_{34})_2]^{10-}$: identification of heterogeneous CoO_x as the dominant catalyst. J Am Chem Soc 133(38):14872–14875
117. Yin Q, Tan JM, Besson C et al (2010) A fast soluble carbon-free molecular water oxidation catalyst based on abundant metals. Science 328(5976):342–345
118. Hong D, Jung J, Park J et al (2012) Water-soluble mononuclear cobalt complexes with organic ligands acting as precatalysts for efficient photocatalytic water oxidation. Energy Environ Sci 5(6):7606–7616
119. Fu S, Liu Y, Ding Y et al (2014) A mononuclear cobalt complex with an organic ligand acting as a precatalyst for efficient visible light-driven water oxidation. Chem Commun 50(17):2167–2169
120. Wasylenko DJ, Ganesamoorthy C, Borau-Garcia J et al (2011) Electrochemical evidence for catalytic water oxidation mediated by a high-valent cobalt complex. Chem Commun 47(14):4249–4251
121. Wasylenko DJ, Palmer RD, Schott E et al (2012) Interrogation of electrocatalytic water oxidation mediated by a cobalt complex. Chem Commun 48(15):2107–2109
122. Rigsby ML, Mandal S, Nam W et al (2012) Cobalt analogs of Ru-based water oxidation catalysts: overcoming thermodynamic instability and kinetic lability to achieve electrocatalytic O_2 evolution. Chem Sci 3(10):3058–3062
123. McGuire R Jr, Dogutan DK, Teets TS et al (2010) Oxygen reduction reactivity of cobalt (II) hangman porphyrins. Chem Sci 1(3):411–414
124. Dogutan DK, Stoian SA, McGuire R et al (2010) Hangman corroles: efficient synthesis and oxygen reaction chemistry. J Am Chem Soc 133(1):131–140
125. Lai W, Cao R, Dong G et al (2012) Why is cobalt the best transition metal in transition-metal hangman corroles for O-O bond formation during water oxidation? J Phys Chem Lett 3(17):2315–2319
126. Lei H, Han A, Li F et al (2014) Electrochemical, spectroscopic and theoretical studies of a simple bifunctional cobalt corrole catalyst for oxygen evolution and hydrogen production. Phys Chem Chem Phys 16(5):1883–1893
127. McCool NS, Robinson DM, Sheats JE et al (2011) A Co_4O_4 "cubane" water oxidation catalyst inspired by photosynthesis. J Am Chem Soc 133(30):11446–11449
128. Beattie JK, Hambley TW, Klepetko JA et al (1998) The chemistry of cobalt acetate—IV. The isolation and crystal structure of the symmetric cubane, tetrakis[(μ-acetato)(μ_3-oxo) (pyridine)cobalt(III)]·chloroform solvate, $[Co_4(\mu_3\text{-}O)_4(\mu\text{-}CH_3CO_2)_4(C_5H_5N)in4]\cdot 5CHCl_3$ and of the dicationic partial cubane, trimeric, [(μ-acetato)(acetato)tris(μ-hydroxy(μ_3-oxo) hexakispyridinetricobalt(III)]hexafluorophosphate·water solvate, $[Co_3(\mu_3\text{-}O)(\mu\text{-}OH)_3(\mu\text{-}CH_3CO_2(CH_3CO_2(C_5H_5N)_6[PF_6]_2\cdot 2H_2O$. Polyhedron 17(8):1343–1354
129. Chakrabarty R, Bora SJ, Das BK (2007) Synthesis, structure, spectral and electrochemical properties, and catalytic use of cobalt(III)–oxo cubane clusters. Inorg Chem 46(22):9450–9462
130. Berardi S, La Ganga G, Natali M et al (2012) Photocatalytic water oxidation: tuning light-induced electron transfer by molecular Co_4O_4 cores. J Am Chem Soc 134(27):11104–11107
131. La Ganga G, Puntoriero F, Campagna S et al (2012) Light-driven water oxidation with a molecular tetra-cobalt(III) cubane cluster. Faraday Discuss 155:177–190
132. McAlpin JG, Stich TA, Ohlin CA et al (2011) Electronic structure description of a $[Co(III)_3Co(IV)O_4]$ cluster: a model for the paramagnetic intermediate in cobalt-catalyzed water oxidation. J Am Chem Soc 133(39):15444–15452
133. Symes MD, Surendranath Y, Lutterman DA et al (2011) Bidirectional and unidirectional PCET in a molecular model of a cobalt-based oxygen-evolving catalyst. J Am Chem Soc 133(14):5174–5177
134. Symes MD, Lutterman DA, Teets TS et al (2013) Photo-active cobalt cubane model of an oxygen-evolving catalyst. ChemSusChem 6(1):65–69
135. Evangelisti F, Güttinger R, Moré R et al (2013) Closer to Photosystem II: a Co_4O_4 cubane catalyst with flexible ligand architecture. J Am Chem Soc 135(50):18734–18737

136. Zhang B, Li F, Yu F et al (2014) Electrochemical and photoelectrochemical water oxidation by supported cobalt–oxo cubanes. ACS Catal 4(3):804–809
137. Nakazono T, Parent AR, Sakai K (2013) Cobalt porphyrins as homogeneous catalysts for water oxidation. Chem Commun 49(56):6325–6327
138. Wang D, Groves JT (2013) Efficient water oxidation catalyzed by homogeneous cationic cobalt porphyrins with critical roles for the buffer base. Proc Natl Acad Sci U S A 110 (39):15579–15584
139. Pizzolato E, Natali M, Posocco B et al (2013) Light driven water oxidation by a single site cobalt salophen catalyst. Chem Commun 49(85):9941–9943

Chapter 14
Hydrogen Activation in Water by Organometallic Complexes

Luca Gonsalvi, Federica Bertini, Antonella Guerriero, and Irene Mellone

Abstract Hydrogen has found important applications as reducing agent for chemical transformations and is nowadays considered as one of the most promising energy vectors able to fuel devices to produce electricity on demand (direct hydrogen fuel cells). Crucial to its application is the understanding at the molecular level of how hydrogen interacts with (transition) metals which are commonly used as catalysts to lower the energy barrier to split the H_2 molecule into its components and allow transfer and reactivity. In this chapter, selected examples of hydrogen activation by water-soluble organometallic complexes are summarized.

Keywords Homogenous phase hydrogen activation • Ruthenium, rhodium, iron organometallic complexes • Water-soluble sulfonated and amphiphilic phosphines • Hydrogen storage by organometallics

14.1 Introduction

Hydrogen is one of the most widely used chemicals and has found various applications, the most common for the reduction of unsaturated bonds in molecules (hydrogenation) and as energy vector, a field which has recently received large attention from academia and industry in the quest for novel processes and technologies towards the development of a hydrogen economy [1].

In order to react with other molecules, hydrogen needs to be activated, i.e. the H-H covalent bond must be cleaved efficiently under mild conditions, either in a homolytic or heterolytic way. Over the last 80 years, many different ways to achieve this target have been discovered, often based on catalyzed processes in solid–gas phases (heterogeneous) or in liquid–gas phase (homogeneous). The presence of metals, often including noble and non-noble transition ones, was shown to be crucial to lower the activation barriers associated with hydrogenation reactions. It was also discovered that some natural enzymes such as hydrogenases,

L. Gonsalvi (✉) • F. Bertini • A. Guerriero • I. Mellone
Consiglio Nazionale delle Ricerche, Istituto di Chimica dei Composti Organometallici (CNR-ICCOM), Via Madonna del Piano 10, 50019 Sesto Fiorentino (Firenze), Italy
e-mail: l.gonsalvi@iccom.cnr.it

W.-Y. Wong (ed.), *Organometallics and Related Molecules for Energy Conversion*, Green Chemistry and Sustainable Technology, DOI 10.1007/978-3-662-46054-2_14

containing Fe only or Fe-Ni moieties, are nature's reply to hydrogen activation, and many research groups around the world have been inspired by these observations trying to mimic the active sites of such enzymes to achieve efficient catalysts for such a process.

Already in 1973, James' monography *Homogeneous Hydrogenation* [2] cited ca. 2,000 references related to research of the past 20 years. Among these, only a few recognized the role of metals as active sites for hydrogen splitting, for example, Calvin's report (1938) on quinone hydrogenation based on Cu(I) acetate [3], although it was already known at that time that hydrogen can bind to transition metals such as Fe to give $H_2Fe(CO)_4$ [4]. However, the first fully characterized transition metal complex showing the presence of M-H bonds, namely, $[Cp_2Re(H)_2](BF_4)$, was reported later on 1955 by Wilkinson et al. [5].

Setting the ground for metal hydride chemistry, many authors contributed to the development of this research area, which grew to a mature field when it was discovered that transition metal hydrides were pivotal for homogeneously catalyzed hydrogenations and olefin hydroformylation [6]. Another important step in the understanding of the nature of hydrogen activation by metals was to establish that this process can occur both in homolytic and heterolytic ways. The next section of this chapter will summarize the fundamental aspects related to these two mechanisms.

Only after some decades, when the use of water as a benign, cheap, abundant and harmless solvent for chemical reactions was proposed to tackle the problems related to the use of toxic organic solvents, chemists considered of interest to run homogenous catalytic processes such as hydrogenation and hydroformylation in this medium. It was soon discovered that the general understanding that water and transition metal hydrides were poorly compatible had significant exceptions and in fact water-soluble complexes could be synthesized, isolated and characterized or clearly identified in water-phase catalytic reactions. The role of water in hydrogen activation mechanism was later identified as non-innocent, e.g. in some cases this solvent can take part in the splitting of dihydrogen by assisting its heterolytic cleavage in the pathway to metal coordination. The understanding of this role has in turn deep consequences in the design of catalysts and evaluation of catalytic processes run in water. In this chapter, selected examples of water-soluble hydrides will be discussed in dedicated sections.

14.2 Hydrogen Coordination and Activation by Metals: The Principles

Already in the 1950s, a considerable debate existed on how hydrogen molecules interact with metal in order to cleave their covalent bond giving initially a $M(\eta^2\text{-}H_2)$ *nonclassical* dihydrogen complex, where H_2 behaves as a ligand and weak Lewis base [7, 8]. The first identified example of this class of complexes is Kubas'

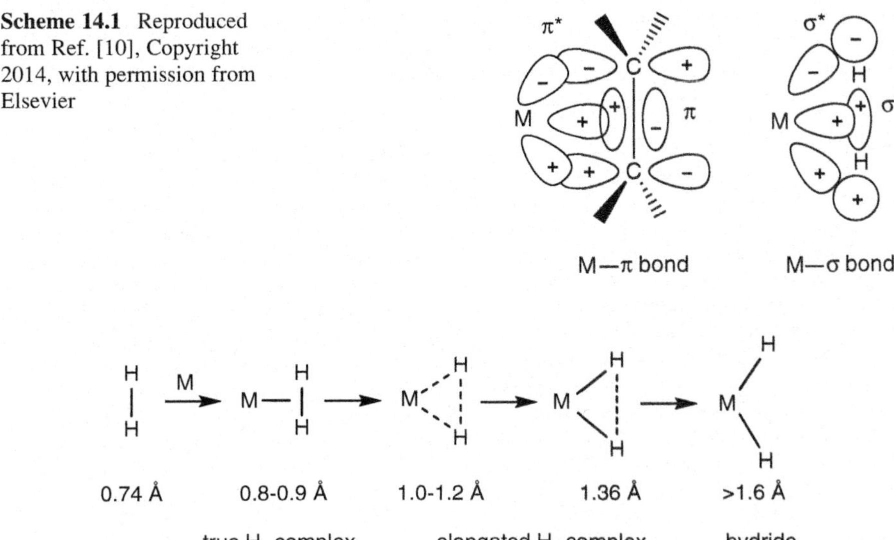

Scheme 14.1 Reproduced from Ref. [10], Copyright 2014, with permission from Elsevier

Scheme 14.2 Reproduced from Ref. [11], Copyright 2014, with permission from Elsevier

tungsten carbonyl [W(CO$_3$)(P^{i}Pr$_3$)$_2$(η^2-H$_2$)] complex [9]. The commonly accepted orbital interaction considers the Dewar-Chatt-Duncanson model, similar to olefin π-coordination, with electron donation from the σ-bond of the ligand to an empty metal orbital accompanied by backdonation from a filled d-orbital into the σ^*-orbital of the ligand (Scheme 14.1). Theoretical studies clarified further important details of these interactions, namely, the possibility of a nonclassical, 3-centre bonding, the role of backdonation and the effect of *trans* ligands on binding and activation, the rotation of H_2 ligand. A recent review article by Kubas has recently appeared in the literature covering these and other issues related to dihydrogen coordination to metals [10], following on considerations by the same author on the role of such a class of compounds for hydrogen storage and production [11].

In the 1990s, further studies based on X-ray crystallography and NMR data highlighted that intermediate bonding situations between nonclassical dihydrogen and dihydride complexes could be possible, the so-called *elongated* H_2 complexes. Scheme 14.2 summarizes the commonly accepted range of H-H bond lengths corresponding to the various cases, from intact H_2 molecule to fully activated (dihydride situation). It was observed that in passing from M(η^2-H$_2$) to M(H)$_2$, d_π basicity increased with the H-H bond distance [10, 12].

In general, first-row transition metals, electron-withdrawing ancillary ligands and cationic complexes favour binding of H_2 to the metal centre and shorten H-H distance. A strong influence on $d_{H\text{-}H}$, hence on the nature of the resulting complex upon H_2 activation, is also dictated by the nature of the ligand trans to H_2. Strong π-acceptors such as CO reduce backdonation giving rather short $d_{H\text{-}H}$ values, at ca. 0.9 Å.

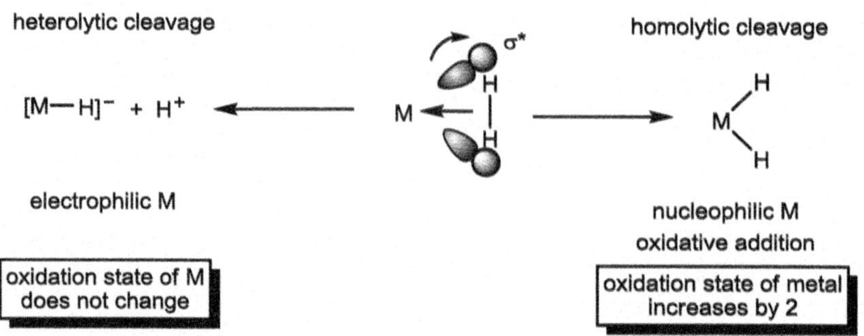

Scheme 14.3 Reproduced from Ref. [10], Copyright 2014, with permission from Elsevier

14.2.1 Homolytic Versus Heterolytic Activation

It is worth here summarizing the two main mechanisms underlying hydrogen activation by metals, i.e. homolytic versus heterolytic cleavage of H_2. As a general observation, while homolytic pathway is favoured by highly basic transition metal centres, the heterolytic counterpart is preferred in the case of electrophilic metal complexes or complexes with highly π-acidic *trans* ligands. A simplified view of the dual pathways for H_2 bond cleavage is shown in Scheme 14.3 [10].

In the gas phase, homolytic cleavage of molecular hydrogen is thermodynamically more favoured than the heterolytic one ($\Delta H° = 436$ vs. 1,675 kJ mol^{-1}, respectively), but this can be compensated in water phase by the high proton hydration energy ($\Delta H° = -1{,}084$ kJ mol^{-1}) together with the binding energy of H^- to the metal centres.

The homolytic mechanism is ubiquitous in transition metal-catalyzed processes such as hydrogenation and hydroformylation [13–15], as early demonstrated by Vaska and Wilkinson for Ir(I) and Rh(I) complexes, respectively [16, 17]. After an initial transient state that was later understood to be an σ-complex of H_2, both H atoms are transferred to the metal centre which increases its oxidation state by two (oxidative addition), yielding a *cis*-dihydride metal complex.

In the case of heterolytic splitting, polarization of the coordinated H-H molecule occurs giving a hydride (H^-) and proton (H^+) which is captured either intermolecularly by an external base (B^-, intermolecular process) or by a basic site on the ancillary ligands bound to the metal centre or counter-anion (intramolecular process), the latter exploited, for example, in Noyori-type asymmetric ketone hydrogenation [18, 19]. A general scheme for heterolytic H-H activation is shown in Scheme 14.4 [10]. The relevance to catalytic hydrogenation has been reviewed [20]. The heterolytic pathway is generally preferred in polar media, including water, and is considered as energetically more facile, as all reactions are in principle reversible, no change in the oxidation state of the metal is required, and generally little or no ligand rearrangement must occur to give the final metal

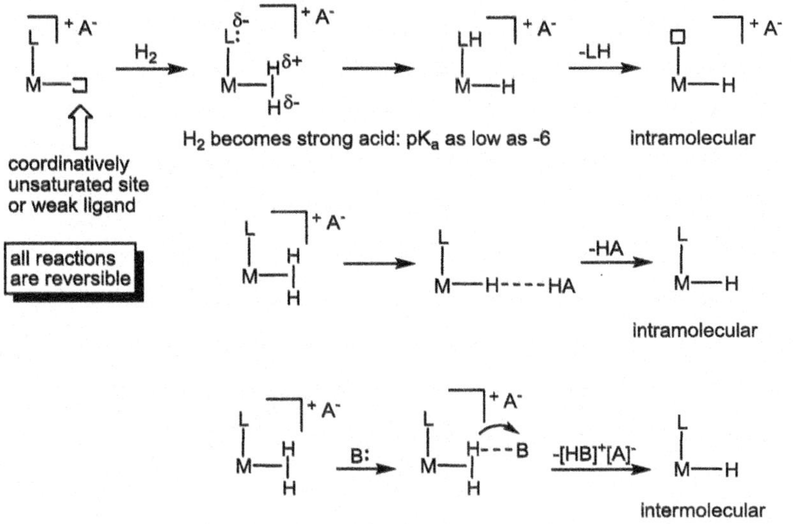

coordinatively
unsaturated site
or weak ligand

all reactions
are reversible

H_2 becomes strong acid: pK_a as low as -6

intramolecular

intramolecular

intermolecular

Scheme 14.4 Reproduced from Ref. [10], Copyright 2014, with permission from Elsevier

hydride product [21–23]. One of the most widely found cases of heterolytic hydrogen activation is H/D exchange which is a powerful tool for mechanistic investigations. It is of great importance also to other fields of research other than catalysis as it is related to living organisms and enzymatic activation of H_2 in dehydrogenases.

Using these two activation pathways, hydrogen can behave as an amphoteric ligand, both as a Lewis base by σ-donation and as Lewis acid by electron donation to σ*-orbitals. In turn, this involves that coordinated dihydrogen can span a rather broad range of pK_a values, usually in the range 5–16, but including examples of superacidic behaviour (−6 in pseudo-aqueous scale) as for complex $[Os(dppe)_2(CO)(\eta^2\text{-}H_2)]^{2+}$ [24].

14.3 Hydrogen Activation by Ru Water-Soluble Complexes

In this chapter, an overview of selected examples of Ru complexes and organometallic compounds which were shown to activate hydrogen in water will be given. The first paragraph will include the reactivity of a narrow class of complexes bearing water as ligands (aquo complexes). In the following two paragraphs, emphasis will be put on phosphine-stabilized complexes, with particular attention to anionic sulfonated phosphines (the most widely used) and a class of cagelike neutral aminophosphines based on 1,3,5-triaza-7-phosphaadamantane (PTA).

$$2\,[\mathrm{Ru}(\eta^6\text{-}\mathrm{C_6H_6})\mathrm{Cl_2}]_2 + 5\,\mathrm{H_2} \rightarrow [\mathrm{H_4Ru_4}(\eta^6\text{-}\mathrm{C_6H_6})_4]^{2+} + 8\mathrm{Cl}^- + 6\mathrm{H}^+$$

Scheme 14.5 Synthesis of **4** by reaction of **1** with H_2

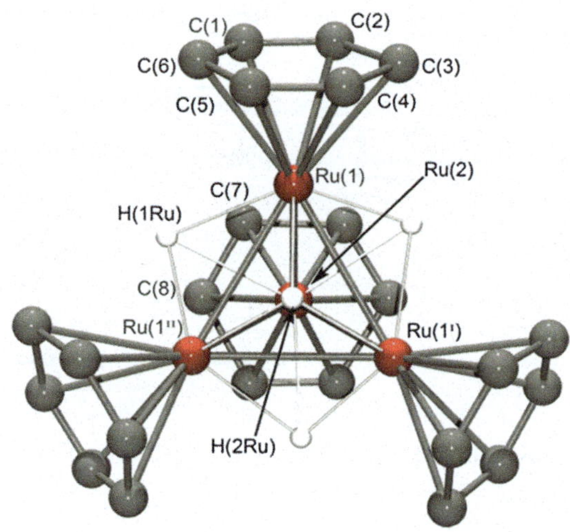

Fig. 14.1 X-ray crystal structure of cluster cation $[H_4Ru_4(\eta^6\text{-}C_6H_6)_4]^{2+}$ (**4**)

14.3.1 Aquo Complexes

The old prejudice that haunted early organometallic chemists against the compatibility of water phase and this class of complexes was ruled out as early as 1972, when Zelonka and Baird [25] published on the reactivity of $[Ru(\eta^6\text{-}C_6H_6)Cl_2]_2$ (**1**) with D_2O to give a mixture of hydrolysis products which was later confirmed to be $[Ru(\eta^6\text{-}C_6H_6)Cl_2(D_2O)]$ and $[Ru(\eta^6\text{-}C_6H_6)Cl(D_2O)_2]^+$.

The tris-aquo analogue $[Ru(\eta^6\text{-}C_6H_6)(H_2O)_3](SO_4)_2$ (**2**) was later isolated and fully characterized by Merbach and colleagues [26], by reaction of solid $[Ru(H_2O)_6](tos)_2$ (**3**; tos = *p*-toluenesulfonate) with either 1,3- or 1,4-cyclohexadiene in EtOH followed by addition of $Ag(SO_4)_2$. The silver salt was needed to remove any trace of chloride anions and shift the aquation equilibrium towards the trisubstituted derivative.

All these complexes were shown to be able to activate H_2 in water. Süss-Fink and coworkers studied the reactivity of water solutions of **1** (given as described mixtures of variously substituted aquo derivatives) with hydrogen under different conditions [27]. At 20 °C under 1.5 bar of hydrogen pressure, the reaction gave the electron-deficient (58e) brown tetrametallic tetrahydrido cluster $[H_4Ru_4(\eta^6\text{-}C_6H_6)_4]^{2+}$ (**4**), according to the formal stoichiometry given in Scheme 14.5, which was fully characterized both in solution by NMR and in the solid state by single crystal X-ray diffraction (Fig. 14.1) [28].

By increasing the temperature to 55 °C and the hydrogen pressure to 60 bar, the 60-e hexahydrido analogue of **4**, namely, $[H_6Ru_4(\eta^6\text{-}C_6H_6)_4]^{2+}$ (**5**), was obtained

$$2\,[Ru(\eta^6\text{-}C_6H_6)Cl_2]_2 + 6H_2 \rightarrow [H_6Ru_4(\eta^6\text{-}C_6H_6)_4]^{2+} + 8Cl^- + 6H^+$$

Scheme 14.6 Synthesis of **5** by reaction of **1** with H_2

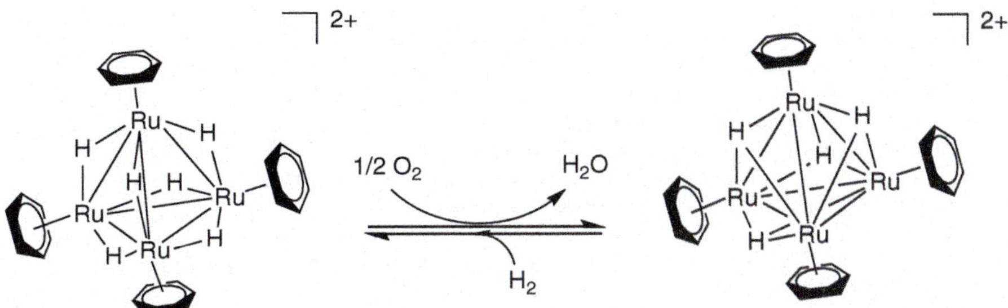

Scheme 14.7 Reproduced from Ref. [27], Copyright 2014, with permission from Elsevier

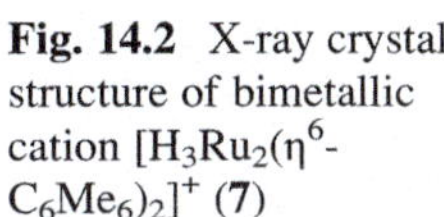
Fig. 14.2 X-ray crystal structure of bimetallic cation $[H_3Ru_2(\eta^6\text{-}C_6Me_6)_2]^+$ (**7**)

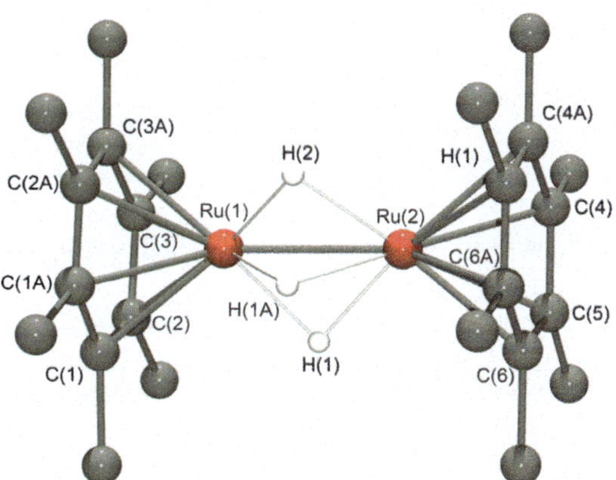

(Scheme 14.6). This violet complex is rather unstable towards oxygen, reverting back to **4** with elimination of water, according to Scheme 14.7.

^{1}H NMR spectra of complexes **4** and **5** show a single hydride resonance at room temperature, suggesting four equivalent μ^3-H hydrides for **4** and six equivalent μ^2-H hydrides for **6**, respectively. However, ^{1}H NMR spectra and T_1 measurements of **5** at −120 °C in THF-*d*$_4$/MeOH-*d*$_4$ (1:1) established that this complex should exist in solution at low temperature and in the solid state as the tetrahydro-dihydrogen cluster $[H_4Ru_4(\eta^6\text{-}C_6H_6)_4(H_2)]^{2+}$, a proposal which was further supported by DFT calculations [29].

The same authors explored the reactivity of other water-soluble Ru-arene dimers with hydrogen, in particular using the more bulky analogue of **1**, i.e. $[Ru(\eta^6\text{-}C_6Me_6)Cl_2]_2$ (**6**). Interestingly, even under forcing conditions ($pH_2 = 60$ bar, 55 °C), a lower nuclearity cluster was obtained, e.g. the 30-e bimetallic complex $[H_3Ru_2(\eta^6\text{-}C_6Me_6)_2]^+$ (**7**) [30], whose X-ray crystal structure is shown in Fig. 14.2.

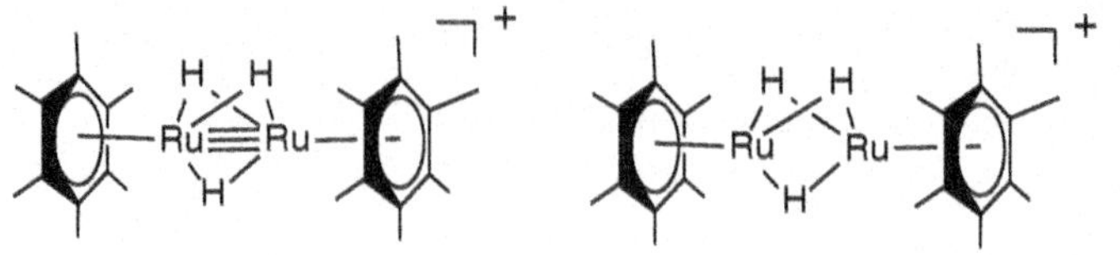

Scheme 14.8 Reproduced from Ref. [27], Copyright 2014, with permission from Elsevier

$$1.5\ [Ru(\eta^6\text{-}C_6Me_4H_2)Cl_2]_2 + 3H_2 \rightarrow [H_3Ru_3(\eta^6\text{-}C_6Me_4H_2)_3Cl]^{2+} + 5Cl^- + 3H^+$$

Scheme 14.9 Synthesis of 9 by reaction of 8 with H_2

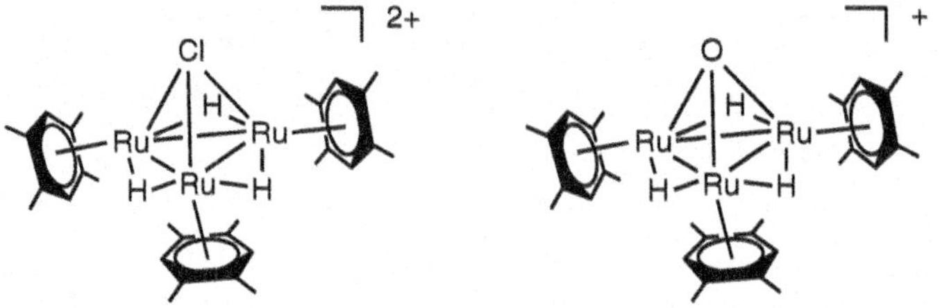

$$[H_3Ru_3(\eta^6\text{-}C_6Me_4H_2)_3Cl]^{2+} + H_2O \longrightarrow [H_3Ru_3(\eta^6\text{-}C_6Me_4H_2)_3(H_3O)]^+ + HCl + H^+$$

Scheme 14.10 Reproduced from Ref. [27], Copyright 2014, with permission from Elsevier

Although the Ru-Ru bond distance obtained from the solid-state structure is as short as 2.468 Å, compatible with a (formally expected) metal-metal triple bond (Scheme 14.8, left), DFT calculations showed that the bonding situation is better described as two RuHRu three-centre, two-electron interaction (Scheme 14.8, right).

For intermediate steric bulkiness of the arene ring, as for $[Ru(\eta^6\text{-}C_6Me_4H_2)Cl_2]_2$ (**8**), reaction with hydrogen in water gave trinuclear tris-hydrido clusters as in Scheme 14.9. Interestingly, the intermediate chloro-capped hydrogen activation product $[H_3Ru_3(\eta^6\text{-}C_6Me_4H_2)_3Cl]^{2+}$ (**9**), which was isolated as chloride salt, hydrolyzes to the oxo-capped monocation analogue $[H_3Ru_3(\eta^6\text{-}C_6Me_4H_2)_3(H_3O)]^+$ (**10**) in water in the presence of $NaBF_4$ (Scheme 14.10) [31, 32].

It is worth mentioning here the reactivity with hydrogen of the simplest Ru-aquo complex, namely, $[Ru(H_2O)_6](tos)_2$ (**3**), as it paved the way for a rich chemistry and mechanistic observations and is considered as a milestone achievement for researchers working in the field of aqueous-phase Ru chemistry and catalysis.

In 1998, Merbach and colleagues described the reactivity of **3** with H_2 gas under pressure [33]. In detail, a 0.1 M solution of **3** in D_2O was pressurized to 40 bar in a HPNMR tube, and the corresponding ^1H NMR spectrum displayed a signal at −7.68 ppm, in the chemical shift range expected upon formation of hydride species. As Ru(0) was observed to form under these conditions, the authors repeated the

$$[Ru(H_2O)_5(\eta^2\text{-}H_2)]^{2+} - H^+ = [Ru(H_2O)_5(H)]^+ + D^+ = [Ru(H_2O)_5(\eta^2\text{-}HD)]^{2+}$$

Scheme 14.11 H/D exchange pathways from **11** to **13** via intermediate **12**

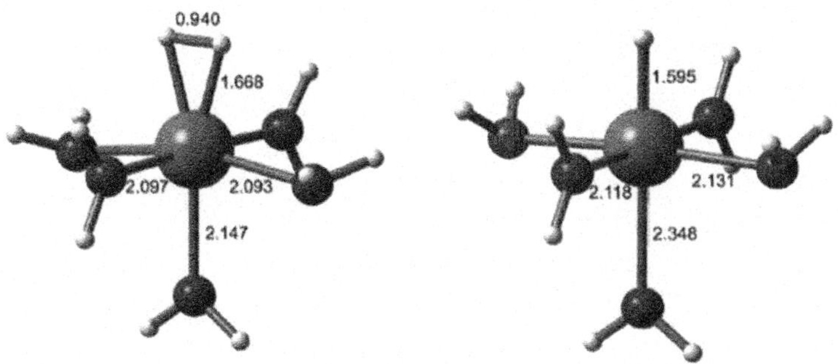

Fig. 14.3 Calculated structures of $[Ru(H_2O)_5(\eta^2\text{-}H_2)]^{2+}$ (**11**) and $[Ru(H_2O)_5H]^+$ (**13**) (Reprinted with permission from Ref. [36]. Copyright 1993 American Chemical Society)

experiment in the presence of *p*-toluenesulfonic acid (0.13 M). After 19 h, the spectrum showed singlets at −7.65 ppm, attributed to $[Ru(H_2O)_5(\eta^2\text{-}H_2)]^{2+}$ (**11**), 4.62 ppm (free H_2), and triplets at 4.59 (free HD, $^1J_{HD} = 42.8$ Hz) and −7.68 ppm due to the H/D exchange product $[Ru(H_2O)_5(\eta^2\text{-}HD)]^{2+}$ (**12**, $^1J_{HD} = 31.2$ Hz). From $^1J_{HD}$ values, it was possible to calculate [34] the Ru-bound H-H distance as 0.889 Å, only slightly longer than the bond length of free hydrogen (0.740 Å), indicative of little backdonation from the metal *d*-orbitals to the σ*-antibonding orbital of the (η^2-H_2) ligand, which should then have a Lewis acidity character. This also explains why the presence of an acid stabilized solutions of **11** for more than three days under H_2 pressure. The authors proposed that a pH-dependent equilibrium was active in the H/D exchange for this system in D_2O, involving as intermediate the monohydrido complex $[Ru(H_2O)_5H]^+$ (**13**) (Scheme 14.11).

Further studies were carried out on this system [35], including ^{1}H, ^{2}H and ^{17}O NMR spectroscopies and DFT calculations (Fig. 14.3). Kinetic studies gave for **11** a formation rate and equilibrium constants of $k_f = (1.7 \pm 0.2) \times 10^{-3}$ kg mol^{-1} s^{-1} and $K_{eq} = 4.0 \pm 0.5$ mol kg^{-1}, suggesting that the reaction of **3** with H_2 to give **11** follows an I_d mechanism as for related small ligands [36].

14.3.2 Hydroxylated and Methoxylated Phosphines

The use of water-soluble ancillary ligands is a common strategy to convey the properties of transition metal complexes in water. Among the most widely used ligands, water-soluble phosphines have received particular attention, and especially sulfonated phosphines (i.e. bearing a SO_3^- substituent, generally on a phenyl group linked to the P donor atom) are a well-developed class of such ligands. Examples of

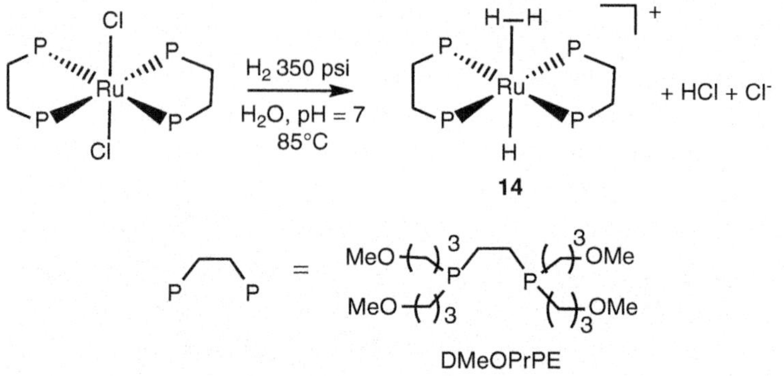

Scheme 14.12 Hydrogen activation by *trans*-[Ru(DMeOPrPE)$_2$Cl$_2$] yielding **14**

hydrogen activation in water brought about by such complexes will be given in the next paragraph.

An interesting although less represented class of water-soluble diphosphines bearing either hydroxy- or methoxy-ending groups was developed by Tyler and coworkers. Ligand 1,2-bis(bis(methoxypropylphosphino)ethane) (DMeOPrPE) was reacted with [Ru(cod)Cl$_2$]$_n$ to yield *trans*-[Ru(DMeOPrPE)$_2$Cl$_2$] [37]. This species reacted smoothly (Scheme 14.12) under a pressure of H_2 (25 bar) in water (buffered, pH = 7) at 85 °C in ca. 2 h to give *trans*-[Ru(DMeOPrPE)$_2$H(η^2-H_2)]$^+$ (**14**).

^{31}P{^{1}H}NMR gave a single resonance at 63.4 ppm in MeOH-d_4, whereas the corresponding ^{1}H NMR resonances were observed as a broad singlet at −6.6 and a quintet at −11.4 ppm. T_1 measurements (500 MHz, −20 °C) gave a value of 21.1 ms corresponding to an H-H distance of 0.86 and 1.06 Å for slow and fast rotation, respectively. This complex was used to probe the existence of DHHB (dihydrogen hydrogen bonding) due to the inertness of the η^2-H_2 to water substitution, using a combination of NMR-based texts including rotational dynamics.

The study was extended to other *trans*-[Ru(diphosphine)$_2$Cl$_2$] complexes obtained with DMeOPrPE analogues (Scheme 14.13), differing for the length of alkyl chain bound to P and the kind of terminal group (OH vs. OMe). Their behaviour towards hydrogen activation and the stability of the product *trans*-[Ru (diphosphine)$_2$H(η^2-H_2)]$^+$ obtained by stepwise H_2 addition/heterolysis pathway were compared to water-insoluble analogues [38]. In fact, water solubility was facilitated in all cases by the cationic nature of the complexes. The general conclusion was that with less electron-donating ligands (DHMPE, DPPE), substitution reactions exchanging η^2-H_2 with H_2O occurred, whereas with more electron-donating ligands (DHPrPE, DMeOPrPE, DEPE, DMPE), no substitution was observed even at concentrations of 55 M after 1 week at 75 °C. This is in line with the higher d_π-σ^* backbonding interaction ruling the strength of the M-(η^2-H_2) bond.

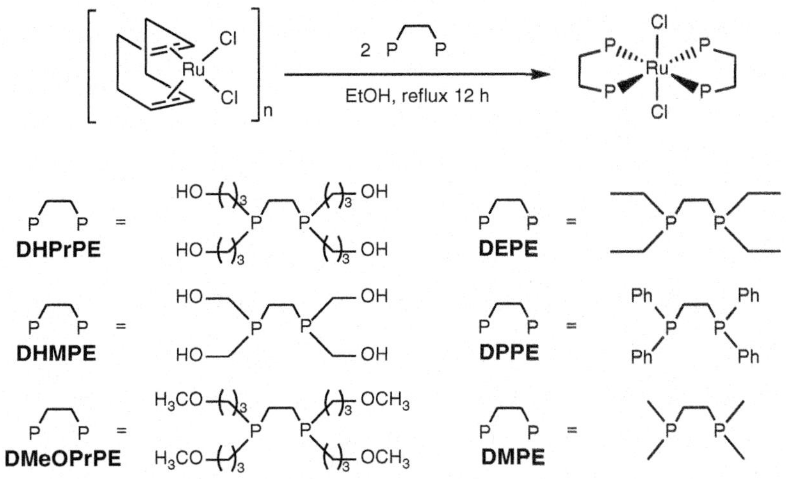

Scheme 14.13 Syntheses of water-soluble *trans*-[Ru(WSDP)$_2$Cl$_2$] (WSDP = water-soluble diphosphines)

14.3.3 Sulfonated Phosphines

Sulfonated phosphines, commonly obtained by reaction of the water-insoluble parent compounds with oleum/sulfuric acid under controlled conditions, constitute without doubts the most represented class of water-soluble phosphines which have received attention from academia and industry for use in aqueous-phase catalyzed processes. The well-known Ruhrchemie/Rhône-Poulenc rhodium-catalyzed olefin hydroformylation process, which makes use of TPPTS [TPPTS = tris(*m*-sulfonatophenyl)phosphine trisodium salt], is considered as a paradigm in this field and has inspired further research efforts in ligand design and optimization by many authors over the years [39].

Although it is not the purpose of this chapter to describe the use of transition metal complexes bearing sulfonated ligands in catalysis, as this topic has already been thoroughly reviewed in the past [40–43], it is worth mentioning a few examples where clearly identified Ru-hydrido complexes found specific applications.

The water-soluble complex ruthenium complex [H_2Ru(CO)(TPPMS)$_3$] [**15**, TPPMS = (*m*-sulfonatophenyl)phosphine sodium salt] was shown to be an efficient catalyst precursor for the aqueous-biphasic hydroformylation of terminal, substituted and cyclic alkenes [44]. Ionic strength of the reaction media and temperature were often observed to determine the selectivity towards the final product. For example, hydroformylation of 4-penten-1-ol in water using [HRu(CO)(TPPTS)$_3$] (**16**) gave preferentially the linear product 6-hydroxy-hexan-1-al, whereas 2-hydroxy-3-methyltetrahydropyran derived from the corresponding branched aldehyde with an increase of the ionic strength [45].

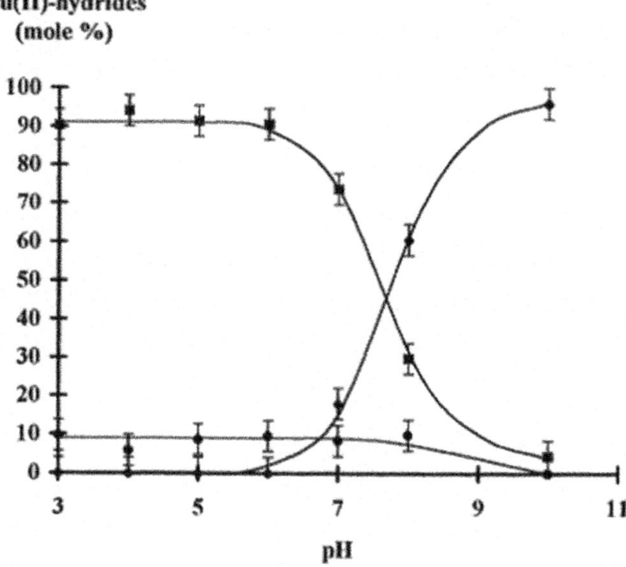

Fig. 14.4 Distribution of water-soluble ruthenium(II)-hydrides as a function of pH, based on the average of ^1H and ^{31}P NMR integrated intensities. ■ [HRuCl(TPPMS)$_3$], ♦ [H$_2$Ru(TPPMS)$_4$], ● [HRuCl(TPPMS)$_2$]$_2$. [Ru] = 2.4×10^{-2} M, [TPPMS] = 7.2×10^{-2} M, [KCl] = 0.2 M, 50 °C, H_2, P_{total} = 1 bar (Reprinted from Ref. [48], Copyright 1988, with permission from Elsevier)

Complexes [HRu(CO)(CH$_3$CN)(L)$_3$][BF$_4$] (**17**, L = TPPTS; TPPMS) were used as catalyst precursors for the hydroformylation of eugenol, estragole, safrole and trans-anethole under moderate conditions in biphasic media and their activities. Higher activities were observed using TPPTS, due to the higher water solubility of such ligand compared to TPPMS [46].

Joó and colleagues established some of the key effects related to the formation and molecular distribution of water-soluble Ru(II) hydrido complexes stabilized by TPPMS and derived from hydrogen activation in water. By a combination of pH-potentiometric and NMR studies, it was observed that the conversion of dimeric [RuCl$_2$(TPPMS)$_2$]$_2$ to [HRuCl(TPPMS)$_3$] (**18**) was favoured at pH $\leq$ 3.3, whereas at pH $\geq$ 7, the dihydride *cis*-[H$_2$Ru(TPPMS)$_4$] (**19**), characterized by a ^1H NMR signal at $\delta_H = -10.4$ ppm, formed preferentially [47, 48]. Figure 14.4 shows the distribution curves of such species at various pH values. The observed pH-dependent hydrogen activation has a profound influence of the catalytic properties of [RuCl$_2$(TPPMS)$_2$]$_2$, for example, in the chemoselective hydrogenation of cinnamaldehyde. Complex **18** catalyzed the selective hydrogenation of the C=C bond, while the C=O group could be reduced efficiently in the presence of **19** giving cinnamol as main product. Further studies showed that high selectivity to C=O reduction could also be achieved at constant (buffered) pH = 3.04 by increasing the hydrogen pressure to 8 bar. Higher pressures would promote the conversion of **18** to **19** even at acidic pH values (Scheme 14.14) [49].

0.5 $[RuCl_2(TPPMS)_2]_2$ + TPPMS + H_2 ⇌ $[HRuCl(TPPMS)_3]$ + H^+ + Cl^-

18

+ TPPMS + H_2

TPPMS =

P

SO_3Na

$[H_2Ru(TPPMS)_4]$ + H^+ + Cl^-

19

Scheme 14.14 Reproduced from Ref. [49] with permission of John Wiley & Sons Ltd

The active role of water in the selective C=O bond hydrogenation by **19** was studied in detail by DFT calculations with the inclusion of a $(H_2O)_3$ cluster in addition to a continuum model [50]. The catalytic cycle, modelled using PH_3 in place of the full phosphine ligand and acrolein instead of cinnamaldehyde, involves at first dissociation of a TPPMS ligand from **19** to form the active penta-coordinate species $[H_2Ru(TPPTS)_3]$ (**20**), followed by substrate coordination and hydrogen transfer from the Ru centre. The second hydrogenation step comes from metal-coordinated H_2O. H_2 coordination regenerates the initial active species **20** by eliminating the unsaturated alcohol. The authors concluded that in the proposed mechanism the role of the protonating agent (H_2O) is adequate in basic media to promote C=O reduction, whereas the stronger H_3O^+ is needed (acidic conditions) to bring about C=C bond hydrogenation, justifying the experimentally observed selectivity at different pH values.

The formation of different classical and nonclassical Ru hydrides derived from water-phase hydrogen activation starting from $[RuCl_2(TPPMS)_2]_2$ was recently reinvestigated by Laurenczy and Joó [51]. In detail, under 1 bar H_2 at 333 K in the absence of added phosphine or halide, the monohydride dimer $[RuHCl(TPPMS)_2]_2$ (**21**) is formed as stable compound, without any sign of Cl bridge cleavage up to 100 bar and further heating. In the presence of NaCl (0.102 M), complex **18** was observed to form. Under acidic conditions (pH = 3.01), raising the H_2 pressure to > 5 bar, and in the presence of added TPPMS, the *trans*-isomer of complex **19**, namely, *trans*-$[H_2Ru(TPPMS)_4]$ (**22**), was formed. At variance with **19**, this complex is characterized by a ^1H NMR quintet signal at $\delta_H = -7.7$ ppm and a $^{31}P\{^1H\}$NMR singlet at $\delta_P = -57.2$ ppm. Under basic or neutral conditions, even in the presence of a Ru-TPPMS ratio of 1:4 or 1:5, the Ru-containing species was *cis-fac*-$[H_2Ru(H_2O)(TPPMS)_3]$ (**23**). Interestingly, under moderate H_2 pressure (ca. 5 bar), at pH = 10.0, solutions containing $[RuCl_2(TPPMS)_2]_2$ and TPPMS (total Ru-P = 1:4) gave a broad singlet at $\delta_H = -7.2$ ppm, which remained unchanged up to 100 bar pressure. T_1(min) measurements gave a value of 18 ms, indicative of a coordinated dihydrogen ligand, and four hydrogen atoms in fast exchange on NMR timescale. The unprecedented $[RuH_2(\eta^2\text{-}H_2)(TPPMS)_3]$ (**24**) was proposed for this species, formally arising from **23** by replacement of H_2O with H_2 at elevated pressures. The study was complemented by reactivity studies of $[RuCl_2(TPPMS)_2]_2$ with sodium formate (transfer hydrogenation conditions).

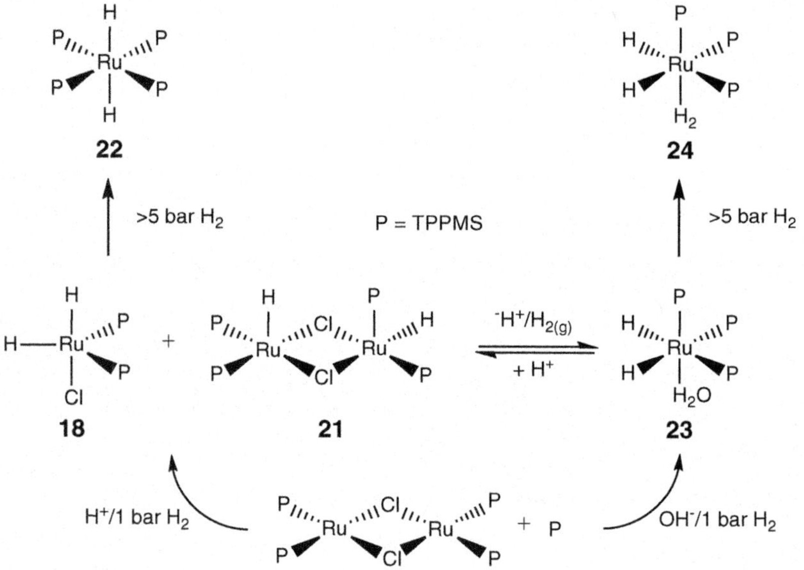

Scheme 14.15 Reproduced from Ref. [51] with permission from the Royal Society of Chemistry

A summary of the various situations observed under different conditions is shown in Scheme 14.15.

The combination of Ru and sulfonated phosphines such as TPPMS and TPPTS have found important applications in the field of hydrogen storage and production from organic compounds in water phase, namely, the reversible activation of formic acid/formate to hydrogen and carbon dioxide mixtures [52, 53]. This system allows for efficient storage of hydrogen using CO_2 as feedstock for hydrogenation to formate, while the reverse reaction, formic acid dehydrogenation, gives fuel cell-grade hydrogen generation on demand. The process runs efficiently at moderate pressures and temperature and constitutes a zero carbon footprint cycle for energy storage. Laurenczy et al. reported [54, 55] that catalytic systems generated either in situ from $[Ru(H_2O)_6](tos)_2$ (tos = toluene-4-sulfonate) or commercial hydrated $RuCl_3$ in the presence of *m*-TPPTS catalyzed the dehydrogenation reaction already at 25 °C, reaching a TOF of 460 h^{-1} at 120 °C. Noteworthy, this system showed to be active in a wide temperature range between 25 and 170 °C, giving in all cases conversions of 90–95 %. Furthermore, the hydrogen produced was of high purity and no CO formation was observed by FTIR (detection limit 2 ppm) even at high temperatures, making this catalytic system suitable for fuel cell applications. The influence of different hydrophilic ligands on the FA dehydrogenation in the presence of $RuCl_3 \times 3H_2O$ has also been investigated in detail. These studies showed that ligand basicity, its hydrophilic properties as well as steric effects were the main parameters which influenced the catalytic activity [56].

Joó and Laurenczy [57] showed that stirring aqueous solutions of HCO_2Na with $[RuCl_2(TPPMS)_2]_2$ in an atmospheric gas burette at 40–80 °C yielded substantial amounts of virtually CO-free gas (<10 ppm). The turnover number [TON = mol

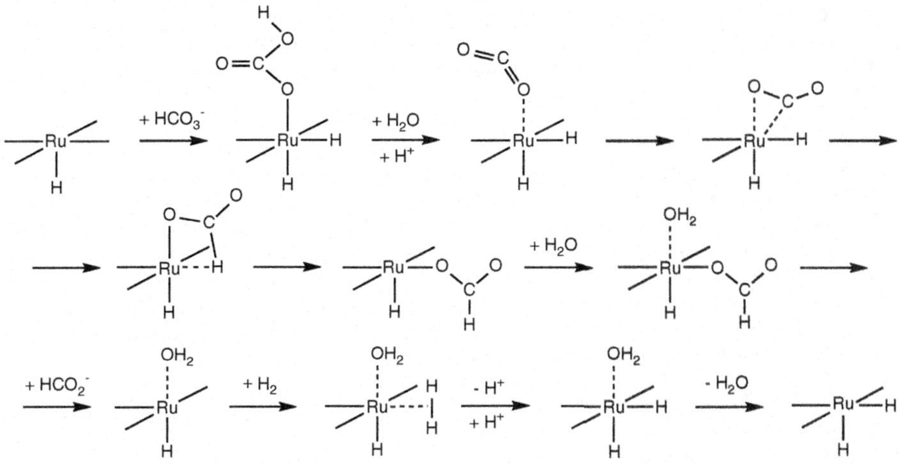

Scheme 14.16 Reprinted from Ref. [59], Copyright 2006, with permission from Elsevier

reacted substrate (mol catalyst)$^{-1}$] achieved at 80 °C in 1 h was 120, and the maximum amount of the gas evolved corresponded to 47 % of the theoretical yield. In closed pressure tubes, at 80 °C the final pressure was measured as 6.2 bar, equivalent to a 37 % yield of bicarbonate. The formate/bicarbonate equilibrium could be shifted to the desired direction (i.e. hydrogen release or storage) by simple pressure change. This was demonstrated by experiments using a medium pressure sapphire NMR tube, where an aqueous solution of $H^{13}CO_3Na$ was pressurized with 100 bar H_2 at 83 °C in the presence of $[RuCl_2(TPPMS)_2]_2$ and TPPMS, and the reaction was monitored by recording ^{13}C NMR spectra of the solution. In 200 min, 90 % of substrate was hydrogenated to $H^{13}CO_2Na$. At this point the pressure was released, and after closing the tube, the reaction mixture was left to equilibrate at 83 °C, leading to decomposition of formate. The hydrogenation/decomposition cycle was repeated for three consecutive cycles showing good and steady efficiency.

These studies follow on earlier reports by the same authors on homogeneous hydrogenation of aqueous hydrogen carbonate to formate under mild conditions [58], where a series of Ru, Rh, Ir and Pd complexes bearing water-soluble phosphines such as TPPMS and 1,3,5-triaza-7-phosphaadamantane (PTA, see next chapter) were tested. Density functional theory (DFT) studies later allowed to propose a reaction mechanism for catalytic bicarbonate hydrogenation in aqueous phase (Scheme 14.16), highlighting the role of Ru-hydride, Ru-dihydrogen and Ru-aquo complexes, respectively [59].

14.3.4 Amphiphilic Neutral Phosphines

Among neutral water-soluble phosphines, the cagelike aminophosphine 1,3,5-triaza-7-phosphaadamantane (PTA) has received special attention for synthetic

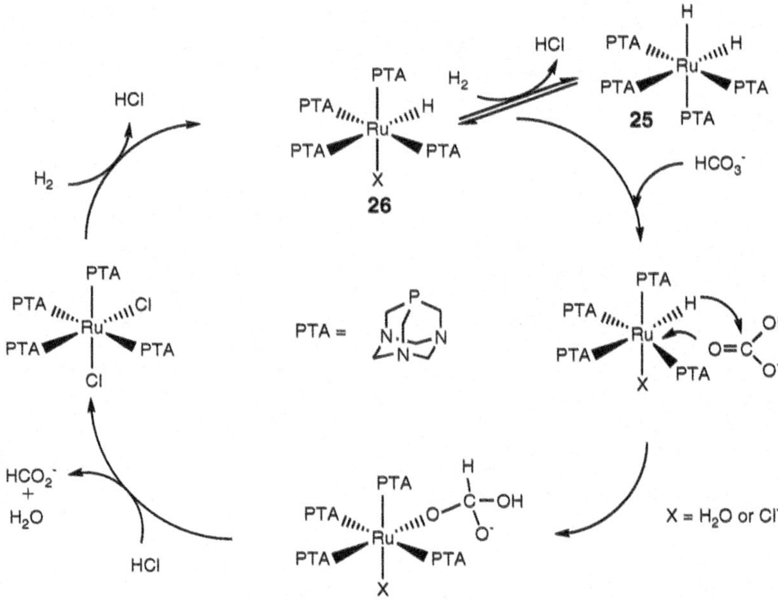

Scheme 14.17 Water-phase sodium bicarbonate hydrogenation mechanism catalysed by **26**

coordination chemistry and catalytic applications, as it combines high Lewis basicity and small Tolman cone angle (103°) with easier functionalization compared to sulfonated phosphines [60–62]. A large library of PTA derivatives is nowadays available, due to the contributions of many research groups worldwide. Due to the interest in medicinal and catalytic applications, many Ru complexes bearing PTA and derivatives were obtained during the years, some of which were used for hydrogenation reactions and hydrogen activation in water.

Complex *cis*-$[RuCl_2(PTA)_4]$, prepared by reduction of $RuCl_3$ in ethanol in the presence of PTA, was used as efficient catalyst for the regioselective C=O bond hydrogenation of unsaturated aldehydes to the corresponding alcohols under transfer hydrogenation conditions and to convert CO_2 and bicarbonate to formate in the absence of amine or other additives under mild conditions and pressure of H_2 [63]. Also in this case, the pH of the aqueous media was important to determine the nature of Ru-hydrido species formed. At pH = 12.0, complex *cis*-$[H_2Ru(PTA)_4]$ (**25**) was observed to form, whereas complex *cis*-$[HRu(PTA)_4X]^n$ (**26**, $X = Cl^-$ or H_2O; n = 0, +1) formed at pH = 2.0. In the presence of an excess of PTA, complex $[HRu(PTA)_5]^+$ (**27**) instead appeared as determined by NMR measurements. The kind of complex formed at different pH values had strong influence on the rate of CO_2 hydrogenation, which gave the highest initial reaction rate (TOF = 807.3 h^{-1}) at pH = 5.86 and decreased at lower or higher pH. On the base of these results, a mechanism based on **26** as catalytically active species was proposed (Scheme 14.17).

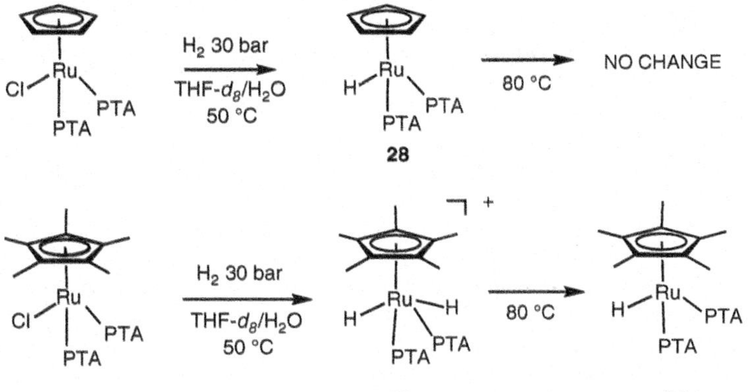

Scheme 14.18 Hydrogen activation in water by piano-stool complexes [Cp^RRu(PTA)$_2$Cl] (Cp^R = C_5H_5, C_5Me_5)

Piano-stool Ru(II) complexes bearing PTA, namely, [CpRuCl(PTA)$_2$] and [Cp*RuCl(PTA)$_2$] (Cp = η^5-cyclopentadienyl; Cp* = η^5-pentamethyl cyclopentadienyl), were prepared in our laboratories [64] and showed good activity in the regioselective C = C bond reduction of α, β-unsaturated ketones (benzylidene acetone) under water/octane biphasic conditions under a pressure of hydrogen. Using high-pressure ^{31}P NMR, [CpRuCl(PTA)$_2$] was observed to activate hydrogen at 50 °C under 30 bar pressure in H_2O/THF-d_8 to form the monohydride species [CpRuH(PTA)$_2$] (**28**) which remained stable upon heating to 80 °C. In contrast, [Cp*RuCl(PTA)$_2$] gave the dihydride complex [Cp*RuH$_2$(PTA)$_2$]Cl (**29**) which converted to the monohydride [Cp*RuH(PTA)$_2$] (**30**) after heating to 80 °C (Scheme 14.18). Complex **29** was also identified as the active species generated in water during $NaHCO_3$ hydrogenation under 100 bar H_2 pressure at 80 bar [65].

Frost et al. reported that **28** is also active for the same reaction at room temperature and low pressures of H_2 (10–150 psi), with modest TOFs [66]. The study includes the evaluation of pH effects. At pH > 7 and pH < 3.6, poor catalytic performances were observed, with the highest activity at pH = 4.7. At pH 2.1, the selectivity to products changed with the type of buffer used (HBF_4/NaH_2PO_4, 99 % of 4-phenylbutan-2-one; HCl/NaH_2PO_4, 77.5 % of 4-phenylbut-3-en-2-ol). By NMR methods, the authors proposed that the active catalytic species was [CpRuII(PTA)(PTAH)]$^+$ (**31**) resulting from protonation of PTA.

A more detailed view of the mechanism of hydrogen activation by these complexes was obtained by theoretical DFT calculations [67, 68], highlighting the non-innocent role of water in facilitating heterolytic H-H bond cleavage in solution by extensive hydrogen bonding networking, which helps in lowering the energy barriers of the overall process. PTA is also involved in H-H activation and in particular through quaternization of one of the basic N atoms, giving **31** as the

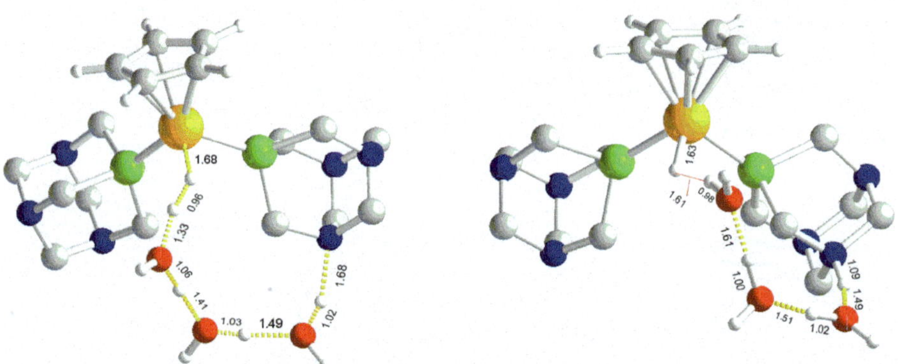

Fig. 14.5 Optimized structures of the transition state for heterolytic hydrogen splitting and the water-stabilized monohydride $[CpRuH(PTA)(PTAH)]^+.(H_2O)_3$ [**31.(H$_2$O)$_3$**] (Reprinted with permission from Ref. [67]. Copyright 2007 American Chemical Society)

most stable product (deepest energy minimum). In Fig. 14.5 are shown the optimized structures of the transition state for heterolytic hydrogen splitting and the water-stabilized monohydride $[CpRuH(PTA)(PTAH)]^+.(H_2O)_3$ [**31.(H$_2$O)$_3$**]. Calculated pK_a values established the highly acidic nature of Ru-dihydrogen complexes in water, while the lower acidity for **30** (4.7) compared to **29** (0.3) agrees with the experimental observation of the former as a stable intermediate at 50 °C.

14.4 Hydrogen Activation by Other Transition Metals in Water

Rhodium water-soluble complexes bearing either sulfonated phosphines such as TPPMS or PTA were studied for their catalytic properties and found to be able to bring about H_2 activation. Besides catalyzing the amine-free CO_2 hydrogenation in sodium formate solutions [69], complex $[RhCl(TPPMS)_3]$ was studied together with its TPPTS and PTA analogues for isotope exchange reactions in H_2-D_2O and D_2-H_2O systems, and the results were compared with the activities of their Ru counterparts in the temperature range 20–70 °C [70]. The specific rates of H-D exchange were observed to be pH dependent. For $[RhCl(PTA)_3]$, $[RhCl(TPPMS)_3]$ and $[RhCl(TPPTS)_3]$, the obtained values of TOF (h^{-1}) were 908, 806 and 989, respectively. Theoretical DFT calculations showed that the protonation of hydride ligands in $[Rh(H)_2Cl(PR_3)_3]^+$ (PR_3 = TPPMS, TPPTS, PTA) by H_3O^+ occurs via dihydrogen-bonded adducts, giving the cationic hydrido-dihydrogen complexes $[RhH(\eta^2\text{-}H_2)Cl(PR_3)_3]^+$. Two possible mechanisms were proposed [71].

The water-soluble analogue of Wilkinson's catalyst $[RhCl(TPPMS)_3]$ by reaction with H_2 under atmospheric pressure gave the monohydride product [HRh

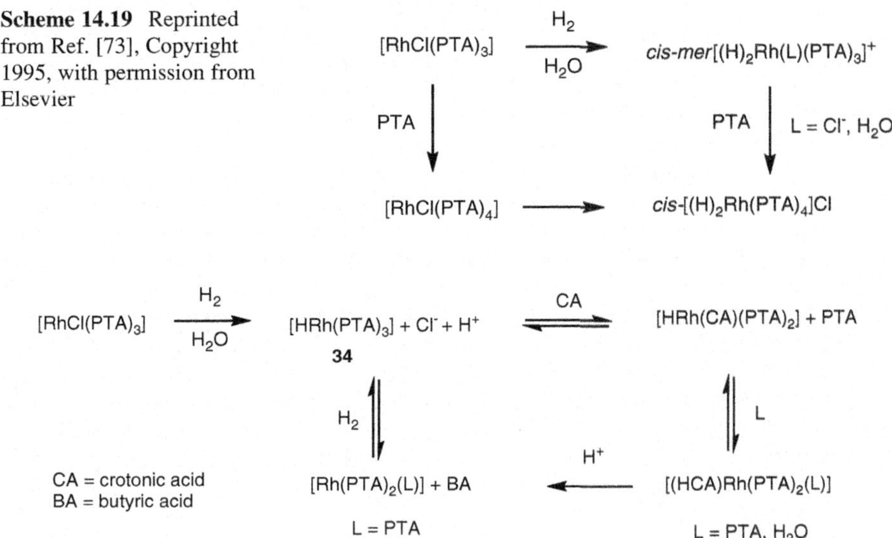

Scheme 14.19 Reprinted from Ref. [73], Copyright 1995, with permission from Elsevier

Scheme 14.20 Reprinted from Ref. [74], Copyright 1996, with permission from Elsevier

(TPPMS)$_3$] (**32**) instead of the oxidative addition product [Rh(H)$_2$Cl(TPPMS)$_3$] (**33**) as for the original water-insoluble analogue, accompanied by a pH drop close to 2.0 [72]. The authors proposed a monohydride-based mechanism to explain the change in selectivity for maleic and fumaric acid hydrogenation using this system, although they could not exclude hydrodechlorination from **33** (however not detected) to **32** in water phase.

In contrast to the Rh/TPPMS system, hydrogenation reactions in the presence of [RhCl(PTA)$_3$] in water gave a more complicated situation, and PTA oxide was formed extensively, accompanied by formation of colloidal metal. In H_2O/EtOH, hydrogenation of cinnamaldehyde gave TOF = 250 h^{-1} at 78 °C. In the presence of an excess of free ligand, metal dihydrides were proposed to form, as shown in Scheme 14.19, based on NMR analysis [73].

A more detailed study involving the role of pH effects on the nature and distribution of complexes derived from hydrogen activation by [RhCl(PTA)$_3$] was then carried out by the same authors [74]. Starting from catalytic tests on the reduction of olefins and oxo acids, a detailed kinetic study was carried out with crotonic acid as substrate. It was observed that the highest rate of the reactions was obtained at pH = 4.7 and that proton liberation was associated with the reaction of the catalyst with H_2 in aqueous solutions. These results, confirmed by H/D exchange experiments in D_2O, suggested that, similarly to [RhCl(TPPMS)$_3$], a reductive dehydrochlorination of an intermediate Rh(III)-dihydride must take place leading to a mechanism involving [HRh(PTA)$_3$] (**34**) as the active species. From kinetic measurements, the hydrogenation of crotonic acid to butyric acid was proposed to follow the mechanism shown in Scheme 14.20.

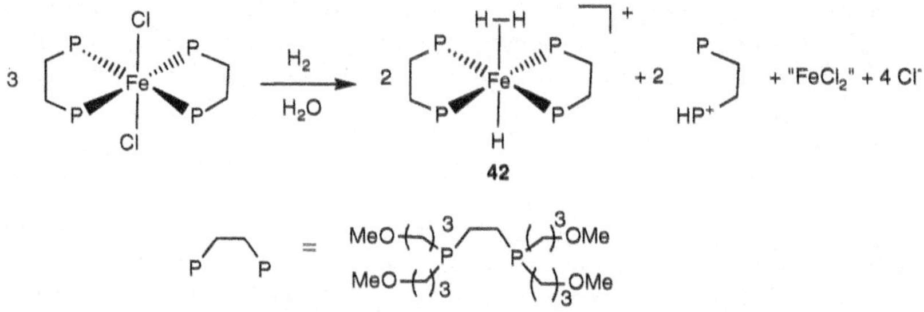

Scheme 14.21 Hydrogen activation by *trans*-[Fe(DMeOPrPE)$_2$Cl$_2$] yielding **42**

The water-soluble analogues of Vaska's complex *trans*-[IrCl(CO)(PPh$_3$)$_2$] were obtained using TPPMS and PTA instead of PPh$_3$. Hydrogen activation by *trans*-[IrCl(CO)(TPPMS)$_2$] gave different products, again depending on pH of the aqueous phase. The total amount of protons produced per moles of Ir is ca. zero at pH <3 and is ca. 1 at pH $=11$. Under acidic conditions, oxidative addition products [(H)$_2$IrCl(CO)(TPPMS)$_2$] (**35a**) or [(H)$_2$Ir(CO)(TPPMS)$_2$]$^+$ (**35b**), generated upon chloride dissociation, may form. Conversely, in less acidic or basic solutions, heterolytic hydrogen splitting occurs, leading to [(H)$_3$Ir(CO)(TPPMS)$_2$] (**36**) or [HIr(CO)(TPPMS)$_3$] (**37**) in the presence of excess of TPPMS [75].

The iridium(III) complexes [Cp*IrCl$_2$(PTA)] (**38**) and [Cp*IrCl(PTA)$_2$]Cl (**39**) were used as catalysts for the hydrogenation of hydrogen carbonate in water. By reaction with H$_2$ (100 bar), the former gave quantitatively the dihydride derivative [Cp*Ir(H)$_2$(PTA)] (**40**) as shown by multinuclear NMR spectroscopy [76]. The presence of the two hydride ligands was indicated by the ^{1}H and ^{31}P{^{1}H} NMR analysis showing a doublet at $\delta_H = -18.4$ ppm ($^1J_{PH} = 30$ Hz) in the hydride region of the corresponding ^{1}H NMR spectrum, which was simplified into a singlet in the ^{1}H{^{31}P} NMR spectrum and a triplet in the ^{31}P NMR spectrum at $\delta_P = -68.5$ ppm ($^2J_{PH} = 30$ Hz). Under the same conditions, the latter gave the cationic monohydride [Cp*IrH(PTA)$_2$]$^+$ (**41**), characterized by a triplet at $\delta_H = -17.9$ ppm ($^2J_{PH} = 30$ Hz) in the ^{1}H NMR spectrum and by a doublet at $\delta_P = -76.3$ ppm ($^2J_{PH} = 30$ Hz) in the ^{31}P{^{1}H} NMR spectrum.

The Fe(II) analogue of *trans*-[Ru(DMeOPrPE)$_2$Cl$_2$] reacted with H$_2$ in water to give *trans*-[Fe(DMeOPrPE)$_2$H(η^2-H$_2$)]$^+$ (**42**), characterized by a doublet in the ^{31}P NMR spectrum at $\delta_P = 88.9$ ppm ($^2J_{PH} = 45$ Hz) and a quintet at $\delta_H = -15.1$ ppm ($^2J_{PH} = 45$ Hz) and a broad singlet at $\delta_H = -10.9$ ppm in the ^{1}H NMR spectrum measured at 233 K. T_1(min) measurements (500 MHz) gave a value of 19.5 ms corresponding to an H-H distance of 0.85 Å (slow rotation) and of 1.07 Å (fast rotation) [77]. Proton release was observed to be associated with hydrogen activation, giving solutions with pH $= 2.6$, also due to the contribution of the buffering capacity of the free ligand. In absence of proton sponge (PS), which was used to obtain quantitative formation of **42**, the overall stoichiometry implies ligand decoordination and protonation, as shown in Scheme 14.21.

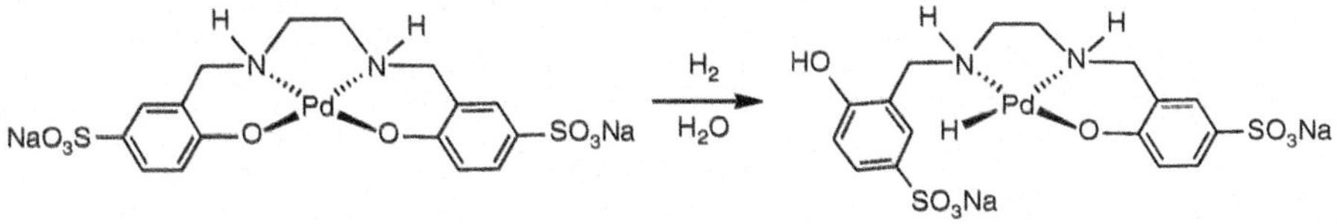

Scheme 14.22 Synthesis of complex **44** involving heterolytic hydrogen splitting

More detailed mechanistic experiments were then carried out, showing that in water, *trans*-[Fe(DMeOPrPE)$_2$Cl$_2$] reacts instantaneously to give *trans*-[Fe(DMeOPrPE)$_2$(H$_2$O)Cl]$^+$, which then activates hydrogen to yield complex **42**, with the intermediate formation of *trans*-[Fe(DMeOPrPE)(H)Cl] (**43**) resulting from heterolytic H-H splitting. In contrast, hydrogen activation by the same Fe (II) initial complex proceeds in nonaqueous solvents such as toluene-d_8 with initial formation of *trans*-[Fe(DMeOPrPE)(η^2-H$_2$)Cl]$^+$ (**43**), requiring a chloride scavenger (TlPF$_6$) and a base (PS) to reach complete conversion to **42** [78].

Pd(II) complexes of sulfonated tetrahydrosalen (sulfosalan, HSS) were obtained by Joó et al. and applied in hydrogenation and redox isomerization of allylic alcohols in neat water or water/organic solvent biphasic systems. DFT calculations allowed to establish that hydrogen is activated heterolytically, giving the Pd(II)-hydride complex [HPd(HSS-Hphen)] (**44**), where one of the phenolate oxygens acts as a base for the proton resulting from H-H splitting (Scheme 14.22) [79].

Acknowledgements Financial contributions by CNR and ECRF projects EFOR and Firenze Hydrolab^2 are gratefully acknowledged. Italian Ministry for Education and Research MIUR is also thanked for supporting this research through projects PRIN 2009 and Premiale 2011.

References

1. Zuttel A, Borgschulte A, Schlapbach L (2008) Hydrogen as a future energy carrier. Wiley, Weinheim
2. James BR (1973) Homogeneous hydrogenation. Wiley, New York
3. Calvin M (1938) Homogeneous catalytic hydrogenation. Trans Faraday Soc 34:1181–1191
4. Hieber W, Leutert F (1931) Zur Kenntnis des koordinativ gebundenen Kohlenoxyds: Bildung von Eisencarbonylwasserstoff. Naturwissenschaften 19:360–361
5. Green MLH, Pratt L, Wilkinson G (1958) Biscyclopentadienylrhenium hydride. J Chem Soc 3916–3922
6. Cotton FA, Wilkinson G, Murillo CA, Bochmann M (1999) Advanced inorganic chemistry, 6th edn. Wiley, New York
7. Halpern J (1959) Homogeneous catalytic activation of molecular hydrogen by metal ions and complexes. J Phys Chem 63:398–403
8. Halpern J (1959) The catalytic activation of hydrogen in homogeneous, heterogeneous, and biological systems. Adv Catal 11:301–370
9. Kubas GJ, Ryan RR, Swanson BJ, Vergamini PJ, Wasserman HJ (1984) Molecular hydrogen complexes of the transition metals. 4. Preparation and characterization of M(CO)$_3$(PR$_3$)$_2$(η^2-H$_2$) (M = molybdenum, tungsten) and evidence for equilibrium dissociation of the H-H bond to give MH$_2$(CO)$_3$(PR$_3$)$_2$. J Am Chem Soc 108:7000–7009

10. Kubas GJ (2014) Activation of dihydrogen and coordination of molecular H_2 on transition metals. J Organomet Chem 751:33–49
11. Kubas GJ (2009) Hydrogen activation on organometallic complexes and H_2 production, utilization, and storage for future energy. J Organomet Chem 694:2648–2653
12. Szymczak NK, Tyler DR (2008) Aspects of dihydrogen coordination chemistry relevant to reactivity in aqueous solution. Coord Chem Rev 252:212–230
13. Jessop PG, Morris RH (1992) Reactions of transition metal dihydrogen complexes. Coord Chem Rev 121:155–284
14. Kubas GJ (2005) Catalytic processes involving dihydrogen complexes and other sigma-bond complexes. Catal Lett 104:79–101
15. Bianchini C, Peruzzini M (2001) Dihydrogen metal complexes in catalysis. In: Peruzzini M, Poli R (eds) Recent advances in hydride chemistry. Elsevier, Amsterdam, pp 271–297
16. Vaska L, DiLuzio JW (1962) Activation of hydrogen by a transition metal complex at normal conditions leading to a stable molecular dihydride. J Am Chem Soc 84:679–680
17. Osborn JA, Jardine FH, Wilkinson GJ (1966) The preparation and properties of tris(triphenylphosphine)halogenorhodium(I) and some reactions thereof including catalytic homogeneous hydrogenation of olefins and acetylenes and their derivatives. J Chem Soc A 1711–1732
18. Noyori R (2002) Asymmetric catalysis: science and opportunities (nobel lecture). Angew Chem Int Ed 41:2008–2022 and references therein
19. Noyori R, Koizumi M, Ishii D, Okhuma T (2001) Asymmetric hydrogenation via architectural and functional molecular engineering. Pure Appl Chem 73:227–232
20. See for example: Clapham SE, Hadzovic A, Morris RH (2004) Mechanisms of the H_2-hydrogenation and transfer hydrogenation of polar bonds catalyzed by ruthenium hydride complexes. Coord Chem Rev 248:2201–2237
21. Brothers PJ (1981) Heterolytic activation of hydrogen by transition metal complexes. Prog Inorg Chem 28:1–61
22. Morris RH (2001) Non-classical hydrogen bonding along the pathway to the heterolytic splitting of dihydrogen. In: Peruzzini M, Poli R (eds) Recent advances in hydride chemistry. Elsevier, Amsterdam, pp 1–38
23. Kubas GJ (2004) Heterolytic splitting of H-H, Si-H, and other σ bonds on electrophilic metal centres. Adv Inorg Chem 56:127–178
24. Rocchini E, Mezzetti A, Ruegger H, Burckhardt U, Gramlich V, Del Zotto A, Martinuzzi P, Rigo P (1997) Heterolytic H_2 activation by dihydrogen complexes. Effects of the ligand X in $[M(X)H_2\{Ph_2P(CH_2)_3PPh_2\}_2]^{n+}$ (M = Ru, Os; X = CO, Cl, H). Inorg Chem 36:711–720
25. Zelonka RA, Baird MC (1972) Reactions of benzene complexes of ruthenium(II). J Organomet Chem 35:C43–C46
26. Stebler-Röthlisberger M, Hummel W, Pittet PA, Bürgi HB, Ludi A, Merbach AE (1988) Triaqua(benzene)ruthenium(II) and triaqua(benzene)osmium(II): synthesis, molecular structure, and water-exchange kinetics. Inorg Chem 27:1358–1363
27. Süss-Fink G (2014) Water-soluble arene ruthenium complexes: from serendipity to catalysis and drug design. J Organomet Chem 751:2–19
28. Meister G, Rheinwald G, Stoeckli-Evans H, Süss-Fink G (1994) Hydrogen activation by arene ruthenium complexes in aqueous solution. Part 2. Build-up of cationic tri- and tetra-nuclear ruthenium clusters with hydrido ligands. J Chem Soc Dalton Trans 3215–3223
29. Süss-Fink G, Plasseraud L, Maisse-François A, Stoeckli-Evans H, Berke H, Fox T, Gautier R, Saillard J-Y (2000) The cluster dication $[H_6Ru_4(C_6H_6)_4]^{2+}$ revisited: the first cluster complex containing an intact dihydrogen ligand? J Organomet Chem 609:196–203
30. Jahncke M, Neels A, Stoeckli-Evans H, Süss-Fink G (1998) Reactions of the cationic complex $[(\eta^6\text{-}C_6Me_6)_2Ru_2(\mu^2\text{-}H)_3]^+$ with nitrogen-containing heterocycles in aqueous solution. J Organomet Chem 561:227–235
31. Süss-Fink G, Meister G, Haak S, Rheinwald G, Stoeckli-Evans H (1997) Organometallic clusters and water: theme and variations. New J Chem 21:785–790
32. Süss-Fink G, Meister A, Meister G (1995) Clusters and water: build-up of multinuclear organometallic compounds in aqueous solution. Coord Chem Rev 143:97–111

33. Aebischer N, Frey U, Merbach AE (1998) Formation and in situ characterization of the first dihydrogen aqua complex: $[Ru(H_2O)_5(H_2)]^{2+}$. Chem Commun 2303–2304
34. Maltby PA, Schlaf M, Steinback M, Lough AJ, Morris RH, Klooster WT, Kloetze TF, Srivastava RC (1996) Dihydrogen with frequency of motion near the ^1H larmor frequency. Solid-state structures and solution NMR spectroscopy of osmium complexes trans-[Os(H · ·H) $X(PPh_2CH_2CH_2PPh_2)_2]^+$ (X = Cl, Br). J Am Chem Soc 118:5396–5407
35. Grundler PV, Yazyev OV, Aebischer N, Helm L, Laurenczy G, Merbach AE (2006) Kinetic studies on the first dihydrogen aquacomplex, $[Ru(H_2)(H_2O)_5]^{2+}$: formation under H_2 pressure and catalytic H/D isotope exchange in water. Inorg Chim Acta 359:1795–1806
36. Aebischer N, Laurenczy G, Ludi A, Merbach AE (1993) Monocomplex formation reactions of hexaaquaruthenium(II): a mechanistic study. Inorg Chem 32:2810–2814
37. Szymczak NK, Zakharov LN, Tyler DR (2006) Solution chemistry of a water-soluble η^2-H_2 ruthenium complex: evidence for coordinated H_2 acting as a hydrogen bond donor. J Am Chem Soc 128:15830–15835
38. Szymczak NK, Braden DA, Crossland JC, Turov Y, Zakharov LN, Tyler DR (2009) Aqueous coordination chemistry of H_2: why is coordinated H_2 inert to substitution by water in trans-Ru $(P_2)_2(H_2)H^+$-type complexes (P_2 = a chelating phosphine)? Inorg Chem 48:2976–2984
39. Bahrmann H, Bogdanovic S, van Leeuwen PWNM (2004) Higher alkenes. In: Cornils B, Herrmann WA (eds) Aqueous-phase organometallic catalysis. Wiley, Weinheim, pp 391–409
40. Liu S, Xiao J (2007) Toward green catalytic synthesis – transition metal-catalyzed reactions in non-conventional media. J Mol Catal A Chem 270:1–43
41. Kalck P, Dessoudeix M (1999) Inter-facial catalysis using various water-compatible ligands in supramolecular systems. Coord Chem Rev 192:1185–1198
42. Gimenez-Pedros M, Aghmiz A, Claver C, Masdeu-Bultò AM, Sinou D (2003) Micellar effect in hydroformylation of high olefin catalysed by water-soluble rhodium complexes associated with sulfonated diphosphines. J Mol Catal A Chem 200:157–163
43. Kohlpaintner CW, Fischer RW, Cornils B (2001) Aqueous biphasic catalysis: Ruhrchemie/ Rhône-Poulenc oxo process. Appl Catal A Gen 221:219–225
44. Baricelli PJ, Lujano E, Rodriguez M, Fuentes A, Sanchez-Delgado RA (2004) Synthesis and characterization of $Ru(H)_2(CO)(TPPMS)_3$ and catalytic properties in the aqueous-biphasic hydroformylation of olefins. Appl Catal A Gen 263:187–191
45. Sullivan JT, Sadula J, Hanson BE, Rosso RJ (2004) The hydroformylation of 4-penten-1-ol and 3-buten-1-ol in water with $HRh(CO)(TPPTS)_3$ and the effects of solution ionic strength. J Mol Catal A Chem 214:213–218
46. Melean LG, Rodriguez M, Romero M, Alvarado ML, Rosales M, Baricelli PJ (2011) Biphasic hydroformylation of substituted allylbenzenes with water-soluble rhodium or ruthenium complexes. Appl Catal A Gen 394:117–123
47. Joó F, Kovacs J, Benyei AC, Kathó A (1998) Solution pH: a selectivity switch in aqueous organometallic catalysis – hydrogenation of unsaturated aldehydes catalyzed by sulfonatophenylphosphane-Ru complexes. Angew Chem Int Ed 37:969–970
48. Joó F, Kovacs J, Benyei AC, Kathó A (1998) The effects of pH on the molecular distribution of water soluble ruthenium(II) hydrides and its consequences on the selectivity of the catalytic hydrogenation of unsaturated aldehydes. Catal Today 42:441–448
49. Papp G, Elek J, Nadasdi L, Laurenczy G, Joó F (2003) Dramatic pressure effects on the selectivity of the aqueous/organic biphasic hydrogenation of trans-cinnamaldehyde catalyzed by water-soluble Ru(II)-tertiary phosphane complexes. Adv Synth Catal 345:172–174
50. Rossin A, Kovacs G, Ujaque G, Lledos A, Joó F (2006) The active role of the water solvent in the regioselective C = O hydrogenation of unsaturated aldehydes by $[RuH_2(mtppms)_x]$ in basic media. Organometallics 25:5010–5023
51. Papp G, Horvath H, Laurenczy G, Szatmari I, Katho A, Joó F (2013) Classical and non-classical phosphine-Ru(II)-hydrides in aqueous solutions: many, various, and useful. Dalton Trans 42:521–529

52. Joó F (2008) Breakthroughs in hydrogen storage – formic acid as a sustainable storage material for hydrogen. ChemSusChem 1:805–808
53. Grasemann M, Laurenczy G (2012) Formic acid as a hydrogen source – recent developments and future trends. Energy Environ Sci 5:8171–8181
54. Fellay C, Dyson PJ, Laurenczy G (2008) A viable hydrogen-storage system based on selective formic acid decomposition with a ruthenium catalyst. Angew Chem Int Ed 47:3966–3968
55. Fellay C, Yan N, Dyson PJ, Laurenczy G (2009) Selective formic acid decomposition for high-pressure hydrogen generation: a mechanistic study. Chem Eur J 15:3752–3760
56. Gan W, Fellay C, Dyson PJ, Laurenczy G (2010) Influence of water-soluble sulfonated phosphine ligands on ruthenium catalyzed generation of hydrogen from formic acid. J Coord Chem 63:2685–2694
57. Papp G, Csorba J, Laurenczy G, Joó F (2011) A charge/discharge device for chemical hydrogen storage and generation. Angew Chem Int Ed 50:10433–10435
58. Joó F, Laurenczy G, Nadasdi L, Elek J (1999) Homogeneous hydrogenation of aqueous hydrogen carbonate to formate under exceedingly mild conditions – a novel possibility of carbon dioxide activation. Chem Commun 971–972 and references therein
59. Kovacs G, Schubert G, Joó F, Papai I (2006) Theoretical investigation of catalytic HCO_3^- hydrogenation in aqueous solutions. Catal Today 115:53–60
60. Phillips AD, Gonsalvi L, Romerosa A, Vizza F, Peruzzini M (2004) Coordination chemistry of 1,3,5-Triaza-7-phosphaadamantane (PTA). Transition metal complexes and related catalytic, medicinal and photo-luminescent applications. Coord Chem Rev 248:955–993
61. Bravo J, Bolaño S, Gonsalvi L, Peruzzini M (2010) Coordination chemistry of 1,3,5-Triaza-7-phosphaadamantane (PTA) and derivatives. Part II. The quest for tailored ligands, complexes and related applications. Coord Chem Rev 248:555–607
62. Gonsalvi L, Guerriero A, Hapiot F, Krogstad DA, Monflier E, Reginato G, Peruzzini M (2013) Lower and upper rim-modified PTA derivatives: coordination chemistry and applications in catalytic reactions in water. Pure Appl Chem 85:385–396
63. Laurenczy G, Joó F, Nadasdi L (2000) Formation and characterization of water-soluble hydrido-ruthenium(II) complexes of 1,3,5-triaza-7-phosphaadamantane and their catalytic activity in hydrogenation of CO_2 and HCO_3^- in aqueous solution. Inorg Chem 39:5083–5088
64. Akbayeva DN, Gonsalvi L, Oberhauser W, Peruzzini M, Vizza F, Brüggeller P, Romerosa A, Sava G, Bergamo A (2003) Synthesis, catalytic properties and biological activity of new water soluble ruthenium cyclopentadienyl PTA complexes $[(C_5R_5)Ru(PTA)_2Cl]$ (R = H, Me; PTA = 1,3,5-triaza-7-phosphaadamantane). Chem Commun 264–265
65. Bosquain SS, Dorcier A, Dyson PJ, Erlandsson M, Gonsalvi L, Peruzzini M, Laurenczy G (2007) Aqueous phase carbon dioxide and bicarbonate hydrogenation catalysed by cyclopentadienyl ruthenium complexes. Appl Organomet Chem 21:947–951
66. Frost BJ, Mebi CA (2004) Aqueous organometallic chemistry: synthesis, structure, and reactivity of the water-soluble metal hydride, $CpRu(PTA)_2H$. Organometallics 23:5317–5323
67. Rossin A, Gonsalvi L, Phillips AD, Maresca O, Lledos A, Peruzzini M (2007) Water-assisted H-H bond splitting mediated by $[CpRu(PTA)_2Cl]$ (PTA = 1,3,5-triaza-7-phosphaadamantane). A DFT Anal Organomet 26:3289–3296
68. Kovacs G, Rossin A, Gonsalvi L, Lledos A, Peruzzini M (2010) Comparative DFT analysis of ligand and solvent effects on the mechanism of H_2 activation in water mediated by half-sandwich complexes $[Cp'Ru(PTA)_2Cl]$ (Cp' = C_5H_5, C_5Me_5; PTA = 1,3,5-triaza-7-phosphaadamantane). Organometallics 29:5121–5131
69. Zhao G, Joó F (2011) Free formic acid by hydrogenation of carbon dioxide in sodium formate solutions. Catal Commun 14:74–76
70. Kovacs G, Nadasdi L, Laurenczy G, Joó F (2003) Aqueous organometallic catalysis. Isotope exchange reactions in H_2–D_2O and D_2–H_2O systems catalyzed by water-soluble Rh- and Ru-phosphine complexes. Green Chem 5:213–217

71. Kovacs G, Schubert G, Joó F, Papai I (2005) Theoretical mechanistic study of rhodium (I) phosphine-catalyzed H/D exchange processes in aqueous solutions. Organometallics 24:3059–3065 and references therein
72. Joó F, Csiba P, Benyei A (1993) Effect of water on the mechanism of hydrogenations catalysed by rhodium phosphine complexes. J Chem Soc Chem Commun 1602–1604
73. Darensbourg DJ, Stafford NW, Joó F, Reibenspies JH (1995) Water-soluble organometallic compounds. 5. The regio-selective catalytic hydrogenation of unsaturated aldehydes to saturated aldehydes in an aqueous two-phase solvent system using 1,3,5-triaza-7-phosphaadamantane complexes of rhodium. J Organomet Chem 488:99–108
74. Joó F, Nadasdi L, Benyei AC, Darensbourg DJ (1996) Aqueous organometallic chemistry: the mechanism of catalytic hydrogenations with chlorotris(1,3,5-triaza-7-phosphaadamantane) rhodium(I). J Organomet Chem 512:45–50
75. Kovacs J, Todd TD, Reibenspies JH, Joó F, Darensbourg DJ (2000) Water-soluble organometallic compounds. 9. Catalytic hydrogenation and selective isomerization of olefins by water-soluble analogues of Vaska's complex. Organometallics 19:3963–3969
76. Erlandsson M, Landaeta VR, Gonsalvi L, Peruzzini M, Phillips AD, Dyson PJ, Laurenczy G (2008) Methylcyclopentadienyl iridium PTA complexes and their application in catalytic water phase carbon dioxide hydrogenation (PTA = 1,3,5-triaza-7-phosphaadamantane). Eur J Inorg Chem 620–627
77. Gilbertson JD, Szymczak NK, Tyler DR (2004) H_2 activation in aqueous solution: formation of trans-$[Fe(DMeOPrPE)_2H(H_2)]^+$ via the heterolysis of H_2 in water. Inorg Chem 43:3341–3343
78. Gilbertson JD, Szymczak NK, Crossland JL, Miller WK, Lyon DK, Foxman BM, Davis J, Tyler DR (2007) Coordination chemistry of H_2 and N_2 in aqueous solution. Reactivity and mechanistic studies using trans-$Fe^{II}(P_2)_2X_2$-type complexes (P_2 = a chelating, water-solubilizing phosphine). Inorg Chem 46:1205–1214
79. Voronova K, Purgel M, Udvardy A, Benyei AC, Kathó A, Joó F (2013) Hydrogenation and redox isomerization of allylic alcohols catalyzed by a new water-soluble Pd-tetrahydrosalen complex. Organometallics 32:4391–4401

Chapter 15
Metal-Organic Frameworks as Platforms for Hydrogen Generation from Chemical Hydrides

Yanying Zhao and Qiang Xu

Abstract Metal-organic frameworks (MOFs), a new class of emerging materials with porosity, crystalline, high interior surface area, controllable structures, high thermal stability, and high yield with low cost, are showing the potential applications for hydrogen storage/release. With respect to physical hydrogen storage (compression, liquefaction, and physisorption), the chemical hydrogen storage is free from extreme processing conditions and safety risk. In this chapter, we select recent and significant advances in the development of MOFs as platforms for hydrogen generation from chemical hydrides and highlight special emphasis on enhanced kinetics and thermodynamics for (1) hydrogen generation from chemical hydrides confined in MOFs, (2) MOF-supported metal nanoparticle-catalyzed hydrogen generation from chemical hydrides, and (3) hydrogen generation from chemical hydrides catalyzed by catalysts formed using MOFs as precursors.

Keywords Metal-organic frameworks • Hydrogen storage • Hydrogen generation • Chemical hydrides • Nanoconfined hydrides • Complex metal hydrides • MOF-supported metal nanoparticle

15.1 Introduction

Metal-organic frameworks (MOFs) are generally constructed by metal-oxygen or metal-nitrogen coordination bonds between inorganic vertices (metal ions or clusters) and organic linkers to form infinite polymer systems, which have been regarded as one of the most rapidly development fields in materials and chemistry

Y. Zhao
National Institute of Advanced Industrial Science and Technology (AIST),
1-8-31 Midorigaoka, Ikeda, Osaka 563-8577, Japan

Department of Chemistry, Zhejiang Sci-Tech University, Hangzhou 310018, China

Q. Xu (✉)
National Institute of Advanced Industrial Science and Technology (AIST),
1-8-31 Midorigaoka, Ikeda, Osaka 563-8577, Japan
e-mail: q.xu@aist.go.jp

W.-Y. Wong (ed.), *Organometallics and Related Molecules for Energy Conversion*,
Green Chemistry and Sustainable Technology, DOI 10.1007/978-3-662-46054-2_15

science. Combined with a low framework density and high thermal stability, along with high yield by using cheap starting materials, MOFs are showing the research competition and exponentially explosive growth in the structure-dependent applications [1], such as molecular adsorption/storage and separation [2–13], energy storage and conversion [14–16], sensing [17, 18], hazardous materials removal [19–23], and catalysis[24–30], as well as various proof-of-concept demonstrations [31–35]. As a kind of multifunctional materials by tuning the metal-containing units [secondary building units (SUBs)] along with organic ligand linkers, the most attractive features of MOFs are not only the intriguing structural topologies including their crystalline nature, high and permanent porosity [36–39], various and controllable structures [40–46], and uniform pore sizes in the nanoscale range (from several angstroms up to 10 nm) [47–51] but also their high interior surface areas (typically, BET surface area of 1,000–3,000 m^2/g; the Langmuir surface area record is more than 10,000 m^2/g) [5, 52–55]. Among unique properties of interest, MOFs and MOF-derived materials are demonstrating their potential advantages on clean and renewable energy technologies, such as hydrogen [56, 57], Li-ion batteries [58–62], fuel cells [15, 63–67], and supercapacitors [68–72], which make them objects of extensive study, further industrial-scale production and application.

Among the most important renewable clean energy sources [73–75], hydrogen, one of the lightest and most potential energy carrier because of its higher energy content by weight compared with gasoline (120 $MJ \cdot kg^{-1}$ compared to 44 $MJ \cdot kg^{-1}$ for gasoline), is widely considered as an optimum energy source as a result of the serious issues arising from the use of large amounts of fossil fuels and consequent nonrenewable energy depletion and environmental pollution. Hydrogen is a globally accepted most promising and clean fuel, and when reacted with oxygen in a proton exchange membrane (PEM) fuel cell [76], it can generate electric power with sole water as a by-product and hence environmentally friendly relative to the dwindling fossil fuels. Due to the low boiling point (−252.87 °C) and low volume density (0.089888 g/L) in the gaseous state at 1 atm, however, there are some technological barriers for hydrogen energy application, such as compact, handling pressure and temperature, recycling of by-products, and cost-effective and convenient plus safe transport of hydrogen. These shortcomings become urgent to be improved for portable electronic devices or mobile vehicles applications. The 2017 DOE (Department of Energy) targets are a gravimetric capacity of 5.5 wt. % H_2 and a volumetric capacity of 40 g/L H_2 that should be achieved at absolute pressures below 12 bar and temperatures between −40 and +85 °C for hydrogen storage or onboard generation [77].

Hydrogen storage is clearly one of the key challenges in developing hydrogen economy [78]. Current approaches/technologies for hydrogen storage include compressed gas, cryogenic liquid, and solid fuel as physical and chemical combination with materials. The traditional hydrogen storage, by modifying its physical state in gaseous or liquid form in pressurized and cryogenic tanks, provides viable means for stationary hydrogen storage, but challenges remain in their applications for onboard vehicles because the liquid hydrogen requires the addition of a refrigeration unit to maintain a cryogenic state, thus adding weight and energy costs.

High-pressure storage of hydrogen gas is consequently limited by the weight of the storage canisters and potential for developing leaks. Moreover, storage of hydrogen in liquid or gaseous form poses important safety problems for onboard transport applications. As a result, pressurized hydrogen systems are eliminated from consideration due to the constraints of the large physical volume required and the energy loss when compressing the gas to high pressures (typically 5,000–10,000 psi), which guide researchers toward the development of solid storage materials [79]. Recently, a wide range of porous materials such as zeolites, silica, and carbon and metal-organic frameworks, usually correlated with specific surface area, are important solid-state materials that have been adopted to the potential hydrogen storage media [34, 44, 80–82]. However, high storage capacities are typically achieved usually at a low temperature (normally −223 to −196 °C) when using these porous physisorption materials. Despite their ready reversibility, it is far difficult to maintain in a vehicle application due to the weak interactions between molecular hydrogen and adsorbents and the extreme working conditions.

Compared with the physical hydrogen storage abovementioned, chemical storage media containing hydrogen in chemically bonded or complexed forms, or incorporated into small organic molecules, such as metal [83–87], boron/nitrogen/aluminum based [88–92] and complex [93–95] hydrides, clathrates [96–98], formic acid [99–102], and other liquid organic hydrogen storage materials (LOHs) [103–106], usually provide excellent hydrogen gravimetric storage capacities and the volumetric densities, and some of them can even surpass that of liquid hydrogen. Over the past decades, chemical hydrides have received considerable attention as promising hydrogen storage materials due to their attractive features, including safe storability, relative stability, and high hydrogen density [107–110]. However, they often suffer from high thermal stability and poor reversibility, as well as sensitivity to air, like the high temperature required to reform or desorb hydrogen, slow hydrogen release kinetics or deterioration with successive cycling, and different loading/unloading logistics and heat dissipation issues. Thus, further research is required to develop systems with improved performance in order to justify their implementation in industrial applications. In particular, the development of new catalysts should become a primary focus of the scientific community. The chemical hydrogen storage materials, including sodium borohydrides, sodium alanates, ammonia borane, hydrazine, hydrazine borane, and formic acid, are easy to fuel under friendly environments and work effectively under ambient temperature and pressure with suitable catalysts [111, 112]. Recently porous MOFs are demonstrating their superiority on improving thermodynamics and reaction kinetics of hydrogen production from chemical hydrides. Consequently, selecting efficient MOF materials to conveniently release hydrogen from chemical hydrides is thereupon becoming desired urgently under friendly conditions, including life cycles, safety, and environmental impact.

In this chapter, the topic is focused on the state of the art of representative MOFs for the improvement of thermodynamics and kinetics on hydrogen generation from chemical hydrides. We summarize the exploration and application of MOFs for hydrogen generation from chemical hydrides based on their three structural elements such as metal vertex, organic ligands, and/or pore system. Examples are followed on hydrogen generation from chemical hydrides in three aspects: (1) MOFs act as host materials for nanoconfining chemical hydrides, (2) the frameworks act as host materials for supporting metal nanoparticle catalysts, and (3) MOFs act as precursors for forming catalysts by decomposition of frameworks. At last, the problems need to be focused, and prospective directions are discussed on MOF-based hydrogen generation from hydrides.

15.2 Hydrogen Generation from Nanoconfined Hydrides in MOFs

Metal-organic frameworks have been proven versatile and accessible by a variety of methods and approaches, including postsynthetic modification (PSM), deprotection (PSD), and exchange (PSE) [113–118]. Early stages of exploration was first to develop a substantial foundation of MOF synthetic chemistry and molecular adsorption/separation. In recent years, the focus of scientific interest has shifted toward the development of multifunctional applications of MOFs. Though MOF-based catalysis was proposed more than 20 years ago [119] and 5 years later demonstrated experimentally [120], only recently has there been extensive experimental and theoretical exploration in catalysis [1, 24–30, 121]. To date, a number of porous MOF materials have been investigated to obtain chemical hydrides at the nanoscale. It is now well established that reducing the hydride particle sizes to the nanoscale is an effective and interesting strategy for enhancing both the kinetics and the thermodynamic properties.

Due to the permanent nanoscaled cavities and open windows offering congenital condition for small molecules to access, porous materials have exhibited a potential for use as a hydrogen sources vessel to encapsulate hydride materials, such as ammonia borane, dimethyl borane (DMAB), borohydrides, and alanates [122–124]. In particular, poor cycling stability, particle agglomeration, and coarsening can be averted if the hydride particles are nanoconfined within porous materials. Despite considerable achievements, new approaches that are easier to operate and more practicable remain to be developed. Recent studies indicate that nanoconfinement of hydride materials in MOFs not only improves the dehydrogenation kinetics and thermodynamics of the process but also prevents unwanted by-product generation. Herein, we review the recent examples of the use of MOFs as templates for nanoconfining hydrides on the improvement of dehydrogenation kinetics and thermodynamics.

15.2.1 Ammonia Borane and Its Derivatives

Ammonia borane (NH_3BH_3, AB) has low molecular weight of 30.7 g mol^{-1} and high stoichiometric gravimetric capacity storage of 19.6 wt. %, exceeding the 2015 targets (9 wt. % and 81 g/L between −15 and 85 °C) set by the US department, which makes it an attractive candidate for chemical hydrogen storage. To date, considerable works involving the release of hydrogen from the thermal dehydrogenation of AB have been reported [109, 125, 126]. Thermolysis of pristine AB is a three-step process (Eqs. 15.1, 15.2 and 15.3).

$$NH_3BH_3 \rightarrow 1/n[NH_2 - BH_2] + H_2 \quad (15.1)$$

$$1/n[NH_2 - BH_2] \rightarrow 1/n[NH - BH]_n + H_2 \quad (15.2)$$

$$1/n[NH - BH]_n \rightarrow BN + H_2 \quad (15.3)$$

Since the third thermal decomposition step requires high temperature (>1,200 °C) [127], AB may practically provide only ~2 equivalents of H_2, corresponding to 13 wt. %, for solid-state hydrogen storage according to the first two steps. In the first decomposition step, AB can rapidly decompose at approximately its melting point (~112 °C) and releases 1 equiv. of H_2 (6.5 wt. %), but its solid-state decomposition at moderate temperature range is very slow and typically involves a long induction period. A similar kinetics problem is that the second decomposition step occurs at a broad temperature range centered at ~150 °C. Another major problem of the thermolysis approach is the impurity contamination of hydrogen output. The thermally activated H_2 generation from AB is typically accompanied with the evolution of diverse unexpected gaseous by-products, such as ammonia, aminoborane [BH_2NH_2], diborane (B_2H_6), borazine $(HNBH)_3$, and aminodiborane [$BH_2NH_2BH_3$] [125]. As a result, the practical application of AB is still handicapped by its slow thermal dehydrogenation kinetics below 100 °C and H_2 release rate, and the formation of detrimental volatile by-products, and severe material foaming during the desorbed hydrogen and irreversibility. Most of all, the elimination of ammonia is the most critical issue because a very small amount of ammonia (ppm level) poisons the catalysts of proton exchange membrane (PEM) fuel cells. It was first reported by Autrey and coworkers that confining AB in mesoporous silica scaffold can improve the kinetics of hydrogen generation and depress the volatile by-products in the AB thermal decomposition [122]. Here we only sum up the metal-organic frameworks as host matrix to confine AB on enhancing thermal decomposition thermodynamics and kinetics and elimination of undesirable by-products.

The successful synthesis of JUC-32-Y [128] (Y(BTC)(H_2O) · DMF (H_3BTC = 1, 3, 5-benzene-tricarbocylic acid, DMF = N, N′-dimethylformamide))-confined AB is denoted as AB@JUC-32-Y and shown in Fig. 15.1, which is the first example of the use of an MOF as the host material to confine AB system [129]. 8 wt. % AB was introduced by the infusion method in anhydrous methanol solvent at room

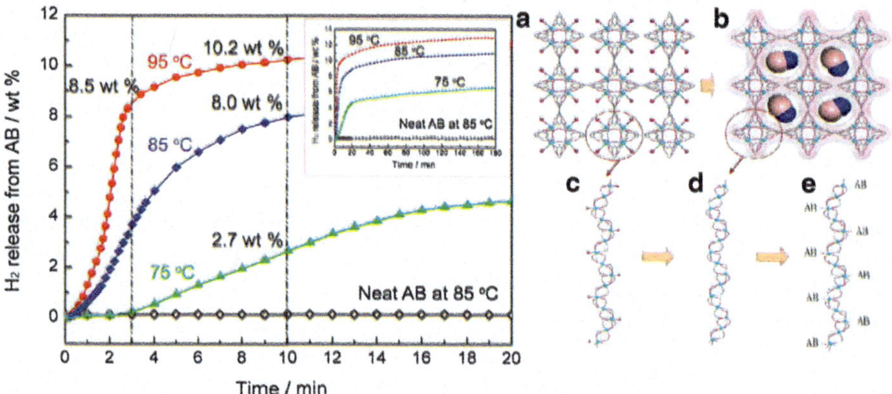

Fig. 15.1 *Left*: Time dependences of hydrogen release from AB@JCU-32-Y at different temperature and neat AB at 85 °C. *Right*: 3D structure image views of (*a*) JUC-32-Y and (*b*) AB@JCU-32-Y and views of 1D chains of JUC-32-Y (*c*) before and (*d*) after removal of terminal H_2O and (*e*) after interaction with AB (Reprinted with the permission from Ref. [129]. Copyright 2010 American Chemical Society)

temperature in a glove box. The confinement AB@JCU-32-Y, with an AB:JUC-32-Y molar ratio of 1:1, led to a big decrease (~30 °C) in the decomposition temperature and remarkable increase in the hydrogen release rate because the coordinated metal Y^{3+} sites of JUC-32-Y which interacted strongly with AB greatly enhance the hydrogen release kinetics and completely prevent the formation of undesired products of ammonia and borazine. At reduced temperature of 95 °C, AB inside JUC-32-Y could release 8.5 wt. % hydrogen within only 3 min and reached 10.2 wt. % hydrogen in only 10 min. At even a low temperature of 85 °C, AB could release 8.0 wt. % hydrogen within 10 min, while neat AB does not release any hydrogen at this temperature. In another report, Yao and coworkers reviewed the nanoconfinement effect on improving the kinetics at a relatively low temperature and the prevention/reduction of undesirable gas formation [130]. However, AB@JUC-32-Y is largely added weight and low AB loading due to the heavy Y metal. Thereafter it is highly desirable to have a lightweight MOF with stable and suitable nanopore channels that can hold more than AB molecules.

Subsequently, Srinivas, Yildirim, and coworkers used a lightweight MOF, Mg-MOF-74 (Mg_2(DOBDC), DOBDC = 2, 5-dioxido-1, 4-benzenedicarboxylate) [131, 132] to accommodate AB with a large mass fraction up to approximately 26 wt. % AB [129], which is regarded as a promising candidate for nanoconfinement and catalytic decomposition of AB for the improvements on the hydrogen release temperature and absence of unwanted by-products [133]. In rigid one-dimensional Mg-MOF-74 framework, after removal of the terminal water molecules upon heating under vacuum, the coordinately unsaturated open Mg^{2+} metal sites come into being. All different AB loadings confined in Mg-MOF-74 demonstrated that H_2 generation started immediately at low temperatures without an induction period. The systematic investigations concluded that the

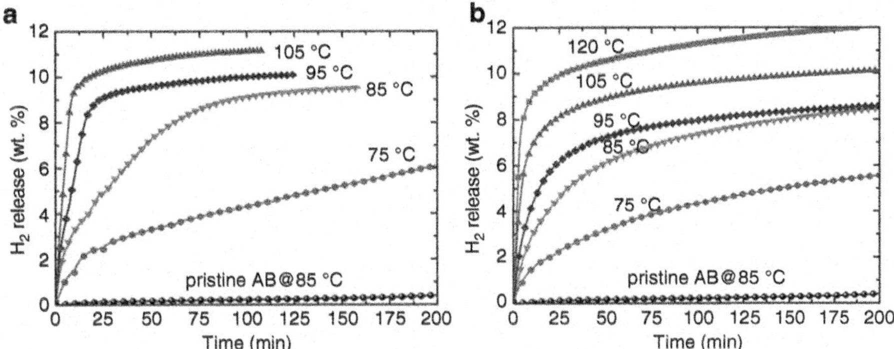

Fig. 15.2 The isothermal H_2 desorption kinetics of AB-Mg-MOF-74 at different temperatures for AB/Mg mole ratios of (**a**) 0.5 and (**b**) 1 (Reproduced from Ref. [133] by permission of John Wiley & Sons Ltd)

dehydrogenation kinetics of the AB@Mg-MOF-74 system depends on the level of AB loading, thus indicating the significant catalytic role that Mg metal centers play. For the heavy AB loading systems, the intermolecular interactions of AB which cause the kinetics are quite similar to the pristine AB due to the formation of bulk phase AB from combining nanoconfined AB with excess crystalline AB aggregated, which exactly support the importance of nanoconfinement and AB interaction with the active metal centers for promoting efficient and clean H_2 release. Furthermore, the nanoconfinement of AB molecules within the one-dimensional pores of Mg-MOF-74 decreased the temperature of dehydrogenation to below 100 °C, as shown in Fig. 15.2. The dehydrogenation of AB@Mg-MOF-74 is a single-step process without endothermic AB melting peaks which implies the direct solid-state decomposition of the nanoconfined AB. Most importantly, the AB@Mg-MOF-74 system could offer clean hydrogen delivery by suppressing the detrimental by-products of ammonia, borazine, and diborane. Other heavy metal centers, such as Ni-MOF-74 and HKUST-1 ($Cu_3(TMA)_2(H_2O)_3$, TMA = benzene-1, 3, 5-tricarboxylate) [134], are not suitable for the nanoconfinement of AB molecules perhaps because of inappropriate pore size/shape, metal types, and coordination.

In another report, Srinivas et al. showed a promising hydrogen storage capability of AB confined in nanoporous Zn-MOF-74 [135–137] (Zn_2(DHBDC), DHBDC = 2,5-dihydroxy-1,4-benzenedicarboxylate). The 1AB@Zn-MOF-74 and 2AB@Zn-MOF-74, with the AB:Zn ratios of 1:1 and 2:1, respectively, exhibited controllable and enhanced kinetics at reduced temperature. The 1AB@Zn-MOF-74 with ~20 wt. % AB loading showed no trace of any volatile by-products during the dehydrogenation by pyrolysis, while amounts of ammonia, boranzine, and diborane were observed for 2AB@Zn-MOF-74 due to overfilling of MOF with excess AB depositing outside the pores [138]. The 1AB@Zn-MOF-74 showed a considerably lowered onset desorption temperature of 60 °C and released about 10 wt. % H_2 at above 100 °C within a few minutes and about 9 wt. % H_2 at temperature between

85 and 75 °C within 2 h. 1AB@Zn-MOF-74 could release about 8 wt. % H_2 even at temperature as low as 65 °C. More importantly, the dehydrogenation mechanism of nanoconfined AB is concluded that the hydrogen is released from the weakened B-H bonding and subsequent B-O bond is formed during the thermal decomposition of AB due to the interaction of electropositive B in $-BH_3$ group and electronegative N in $-NH_3$ with the carboxylate ligands and the metal centers in the MOF. As a result, the metal center has prominent effect on the dehydrogenation temperature as well as kinetics for limited nanoconfinement within the pore channels.

Recently, Srinivas, Yildirim, and coworkers loaded AB to another flexible Fe-MIL-53 framework (MIL: Material of the Institut Lavoisier, $[Fe^{III}(OH)(BDC)]$, BDC = 1,4-benzenedicarboxylate) [139, 140] with coordinatively saturated Fe sites with AB:Fe molar ratios of 0.5:1, 1:1, and 1.5:1 [141]. The encapsulation of AB in Fe-MIL-53 exhibits fast hydrogen release of 1.38 equiv. of H_2 around 100 °C within 30 min, whereas an ultimate release of 0.84 equiv. of H_2 was observed from pristine AB during a longer time. Along with the increasing of AB loading, the activation energy barriers for H_2 release of AB@Fe-MIL-53 increases, but they are still lower than that of pristine AB [122, 142, 143]. Furthermore, AB@Fe-MIL-53 can release 1.22 equiv. of H_2 at 80 °C. In particular, AB@Fe-MIL-53 shows instant H_2 release similar to the AB@MOF-74(Mg and Zn). The research results indicate the flexible pores also trapped the B- and N-containing residues to depress the release of by-products. More importantly, when compared to the unsaturated metal MOFs, the flexible pores in Fe-MIL-53 did not exhibit much improvement in dehydrogenation kinetics and thermodynamics, suggesting that the dehydrogenation property of confined AB is largely governed by coordinatively unsaturated metal sites in MOF pores.

In 2011, Si et al. selected chromium (III) terephthalate MIL-101 [144] with zeotype cubic structure with a giant cell volume and hierarchy of extra-large pore sizes as a host material to prepare and characterize a series of nanocomposites of MIL-101 and Ni-MIL-101 with different amounts of AB [145]. When dehydrogenation by pyrolysis, MIL-101- and Ni-MIL-101-encapsulated ABs start to evolve hydrogen at about 50 °C and give broad desorption peaks centered at 75 and 85 °C, respectively, which are lower than other nanocomposite systems and, more importantly, lower than the proton exchange membrane (PEM) fuel cell working temperature (80 °C), without undesirable ammonia, borazine, and diborane. Thus, the MIL-101 is effective catalyst to change the thermal decomposition mechanism of AB, and the introduction of Ni catalyst within frameworks is a feasible method for improving both thermodynamics and kinetics of dehydrogenation from AB.

Employing postsynthetic chemical modification of MIL-101(Cr) [146], Chan and coworkers prepared the $NHCOCH_3$-MIL-101, together with NO_2-MIL-101 and NH_2-MIL-101, to confine AB by solution impregnation method. The AB@NH_2-MIL-101 and AB@$NHCOCH_3$-MIL-101 nanocomposites can generate ~1.3 equiv. of hydrogen within 20 min at 85 °C [147]. The dehydrogenation improvement of AB@NH_2-MIL-101 and AB@$NHCOCH_3$-MIL-101 over AB@MIL-101 may be closely related to the interaction between AB and functional groups of MIL-101 s. Comparison of four nanocomposites of MIL-101 indicated that the modifications of

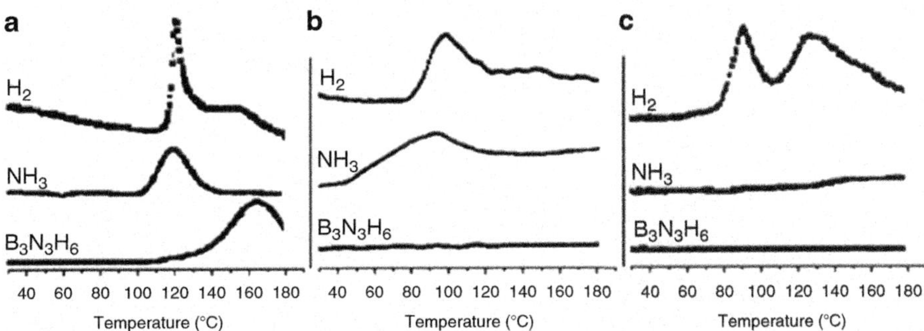

Fig. 15.3 TPD-MS spectra of (**a**) pristine AB, (**b**) AB@MIL-101 (AB/MIL-101 = 1:1 wt/wt), and (**c**) AB/Pt@MIL-101 (AB/1 % Pt@MIL-101 = 1:1 wt/wt) (Reprinted with the permission from Ref. [148]. Copyright 2012 American Chemical Society)

the functional groups in MOFs are effective alternative options in improving dehydrogenation thermodynamics and kinetics of AB. MOFs with effective functional groups can improve hydrogen release from AB at lower temperature and higher purity as compared to untreated MIL-101 s.

In 2012, our group also loaded AB into MIL-101 to form AB@MIL-101 [148]. TPS/MS (temperature-programmed desorption mass spectrometry) indicated that AB@MIL-101 started to evolve H_2 at 70 °C with a broad peak centered at 95 °C, which largely lowered the dehydrogenation temperature compared to the pristine AB. However, the evolution of ammonia was not completely suppressed, as shown in Fig. 15.3, while no borazine was detected. Interestingly, no noticeable peaks corresponding to NH_3 and borazine were detected for AB/Pt@MIL-101; the dehydrogenation temperature shifted to much lower temperature (see Sect. 15.3.1 in detail).

ZIFs (zeolite imidazolate frameworks), composed of imidazolate linkers and tetra-coordinated metal ions (Co^{2+}, Zn^{2+}), are usually exceptionally chemically robust, thermally stable, and commercially available. Large surface areas, large cavities, and small pore apertures make them particularly interesting as candidates for heterogeneous catalysis [149, 150]. Very recently, Zhong et al. employed ZIF-8 ([Zn(MeIm)$_2$], MeIm-2-methylimidazolate) [151, 152], a sodalite zeolite-type structure with nanopore structural feature, to catalyze the dehydrogenation of pyrolysis AB through solid-state mixing and compared the results to those obtained from the nanoconfined composite analogue prepared using solvent-based impregnation techniques [153]. It was deduced that Zn ions can promote hydrogen release from AB as a catalyst by comparing the same zinc concentration between the milled mixture of AB/$ZnCl_2$ and hand-milled 90 wt. % AB/ZIF. For comparison, the isothermal kinetics of AB/ZIF-8, AB/$ZnCl_2$, and AB@ZIF-8-MeOH (immersing AB into ZIF-8 in an anhydrous methanol solvent) have all better results on both the rate and volume of hydrogen released compared to neat AB, as shown in Fig. 15.4. In the first 10 min, approximately 0.91 and 0.656 equiv. of hydrogen were released from the 75 and 90 wt. % AB/ZIF-8 samples, respectively. After 2 h of hydrogen generation, 1.18 and 1 equiv. of hydrogen were evolved, respectively. The rate of

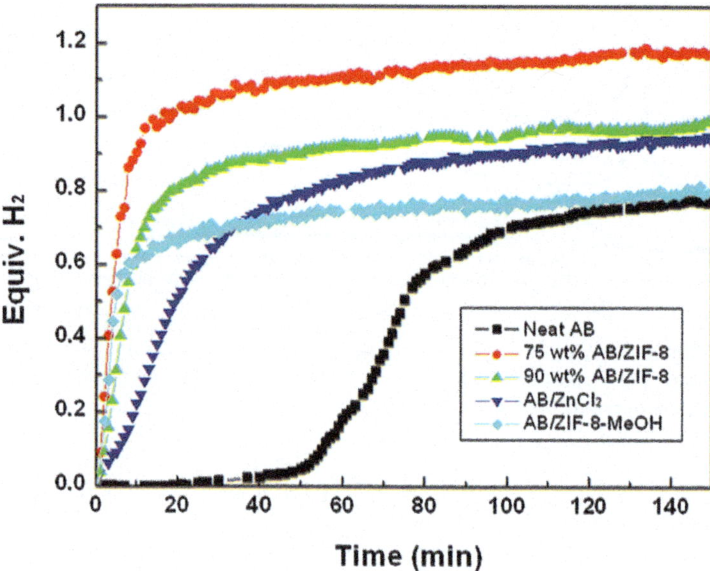

Fig. 15.4 Hydrogen evolved over time at 90 °C (Reprinted with the permission from Ref. [153]. Copyright 2012 American Chemical Society)

hydrogen release from ABZIF-8-MeOH is similar to that from 75 to 90 wt. % AB. ZIF-8 samples, but the amount of hydrogen released is much less than that of those two samples. Therefore, the nanoconfinement effect is not a key factor for the promotion of AB dehydrogenation in this case, and it is speculated that the advantage of ZIF-8 is due to the Zn ions being homogeneously disperse on the surface of ZIF-8. PXRD patterns of AB/ZIF-8 after pyrolysis show no major changes in the ZIF-8 framework, indicating that ZIF-8 is stable during dehydrogenation and potentially reusable as a catalyst.

Moreover, Fisher and coworkers employed ZIF-8 to load the derivative of AB, dimethylamine borane (DMAB = $H_3B \cdot NMe_2H$), inside its pores by a vapor-phase infiltration method [154]. Most interestingly, as shown in Fig. 15.5, by heterogeneous catalysis of ZIF-8, the dehydrocoupling of DMAB can yield $(H_2B \cdot NMe_2)_2$, together with hydrogen release at room temperature. Thus, the dehydrocoupling effect greatly improved the dehydrogenation thermodynamics of the DMAB@ZIF-8, because the dehydrocoupling of neat DMAB requires temperatures as high as 130 °C.

15.2.2 Complex Metal Hydrides

Complex metal hydrides (CMHs) [90, 93, 94, 155–158], due to their commercially availability and high gravimetric hydrogen densities, have been considered as potential candidates for solid hydrogen storage materials. The most important

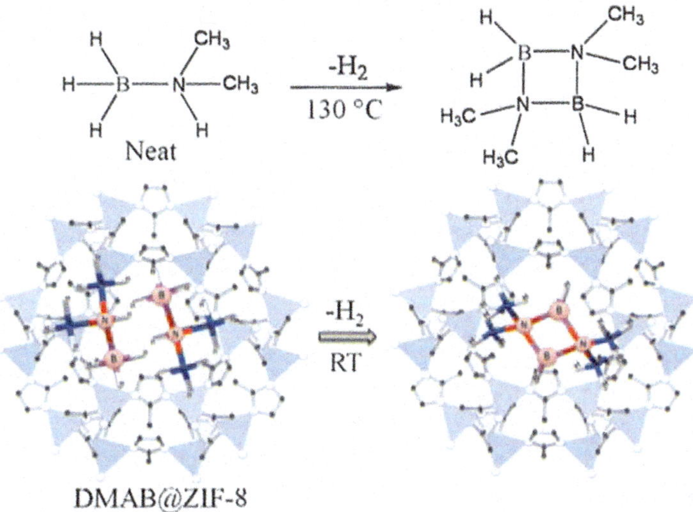

Fig. 15.5 Conceptual representation of the catalytic dehydrogenation and cyclization of DMAB to selectively yield inside ZIF-8 (Reproduced with permission from Ref. [154] by permission of Jon Wiley & Sons Ltd)

complex hydrides include borohydrides, alanates, and transition metal hydrides [159–164]. For fuel cell vehicles, the goal for DOE onboard storage systems is to achieve reversible storage at high density but moderate temperature and hydrogen pressure. Thus, sluggish hydrogen release kinetics and thermodynamically too high stability at elevated temperatures are two major obstacles that need to be overcome, owing to the strong covalent bonds between hydrogen atoms and the central atoms of the molecular anions. To this end, a large amount of efforts in recent research activity has been devoted to improvements in their thermodynamic and kinetic aspects with a focus on the fundamental dehydrogenation and rehydrogenation properties and on providing guidance for material design in terms of tailoring thermodynamics and promoting kinetics for hydrogen storage.

Up to now, nanosizing and scaffolding have emerged as important strategies to control the kinetics, reversibility, and equilibrium pressure for hydrogen storage in light metal hydride systems [165–174]. Reducing the size of metal hydrides to the nanometer range allows fast kinetics for both hydrogen release and subsequent uptake [165–170]. Reversibility of the hydrogen release is impressively facilitated by nanoconfining the materials in a carbon or metal-organic framework scaffold, in particular for reactions involving multiple solid phases [171–179]. It has become clear that nanoconfinement is a strong tool to change physicochemical properties of complex metal hydrides, which might not only be of relevance for promoting kinetics and tailoring thermodynamics of dehydrogenation and rehydrogenation at ambient environment but also for other applications such as rechargeable batteries.

15.2.2.1 Borohydrides

The study of complex metal hydrides as candidate hydrogen storage materials began with $LiBH_4$ usually synthesized by the reaction of ethyl lithium with diborane (B_2H_6), yielding a material containing a gravimetric hydrogen content of 18.4 % [180]. However, pure $LiBH_4$ is thermodynamically very stable, releasing hydrogen only at elevated 400 °C under 1 bar of H_2 and releases 13.5 wt.% of hydrogen by decomposition into LiH and B according to the reaction $LiBH_4 = LiH + B + 3/2H_2$ [181]. Dehydrogenation of $LiBH_4$ is reversible, while it suffers from poor hydrogen uptake kinetics with limited rehydrogenation shown at 600 °C and 150 bar. The end products lithium hydride and boron absorb hydrogen at 600 °C in a hydrogen pressure of 35 MPa for 12 h or at 727 °C under hydrogen pressure of 15 MPa for over 10 h to form $LiBH_4$ [182]. However, its use in hydrogen storage applications is problematic because of poor hydrogen desorption/absorption kinetics, the release of volatile gases (B_2H_6), and the formation of stable by-products ($Li_2B_{12}H_{12}$) [183–185]. Diborane emission and $Li_2B_{12}H_{12}$ production limit the amount of accessible hydrogen uptake in the material over multiple cycles. Like confinement of AB and its derivatives, there have been numerous reports demonstrating improvement of the hydrogenation/dehydrogenation kinetics of $LiBH_4$ through incorporation of $LiBH_4$ into porous SBA-15, carbon aerogels, ordered mesoporous carbon (CMK-3), activated carbon, and CNT materials, which lower the desorption temperature, improve the reversible formation of $LiBH_4$, and limit the amount of volatile by-products [172, 177–179, 185–199]. Up to date, only one literature came from Yu's group with the employment of metal-organic framework, HKUST-1, to confine borohydride for hydrogen storage/release if some significant contributions were not left out.

In 2011, Guo, Yu, and coworkers loaded 84 wt.% $LiBH_4$ into dehydrated HKUST-1 pores in ether solution to synthesize $LiBH_4$@HKUST-1 with an interaction between $LiBH_4$ and Cu^{2+} ions [200]. The dehydrogenation of $LiBH_4$@HKUST-1 started from around 60 °C, which is dramatically lower than that for the pristine $LiBH_4$ (380 °C), together with the release of diborane. With increasing temperature, the amount of released gas was significantly increased, and the dehydrogenation kinetics was accelerated. After heating up to 200 °C, a total gas release of 4.8 mmol g^{-1} was observed for the $LiBH_4$@HKUST-1 sample, indicating a partial decomposition of loaded $LiBH_4$ below this temperature (7 mmol g^{-1} for a complete decomposition of the confined $LiBH_4$ to H_2), because the coordinated water molecules in HKUST-1 can react with $LiBH_4$ to release H_2 during the process of loading $LiBH_4$ into hydrated HKUST-1 at room temperature. More importantly, the nanofinement by nanostructural materials and the consequent redox reaction between $LiBH_4$ and Cu–O units enabled dehydrogenation to occur in $LiBH_4$@Cu-MOFs at a much lower temperature.

15.2.2.2 Alanates

Though alanates show poor kinetics with dehydrogenation taking place at temperatures well above 200 °C, and reversibility is also achieved only under high temperature and pressure conditions, aluminum-based metal hydrides have offered high possibilities for meeting the requirements of onboard applications. The recent hydrogen storage research, particularly for automotive applications, has generated renewed interest in aluminum-based hydrides due to their capacity to store up to 11 wt. % hydrogen with volumetric capacities up to 150 g H_2/L (more than twice that of liquid hydrogen), since the breakthrough by Bogdanović and coworkers, who demonstrated that transition metal-doped $NaAlH_4$ could reach a reversible storage capacity of more than 5 wt. % [155, 201, 202] and reversibly release/absorb hydrogen under mild conditions [203, 204]. Hydrogen can be released from these materials by low-temperature thermolysis (<100 °C), making them well suitable for proton exchange membrane fuel cells and other low temperature applications. The decomposition reaction equations of bulk $NaAlH_4$ are usually described by the three steps as follows:

$$NaAlH_4 \leftrightarrow (1/3)Na_3AlH_6 + (2/3)Al + H_2 \quad (3.7\ \text{wt\%}\ H_2) \tag{15.4}$$

$$(1/3)Na_3AlH_6 \leftrightarrow NaH + (1/3)Al + (1/2)H_2 \quad (1.9\ \text{wt\%}\ H_2) \tag{15.5}$$

$$NaH \leftrightarrow Na + (1/2)H_2 \quad (1.9\ \text{wt\%}\ H_2) \tag{15.6}$$

Theoretically, the $NaAlH_4$ can release 3.7 wt. % of hydrogen for Eq. (15.4) and 1.9 wt. % (relative to the original $NaAlH_4$) for each of Eqs. (15.5) and (15.6). The final decomposition of NaH (step 3) is usually considered unreachable for mobile applications, requiring too high temperatures at useful pressures. Addition of a catalyst (Ti or other transition metal) is necessary for both Eqs. (15.4) and (15.5) to be accessible and reversible under ambient conditions. The kinetics of the above reversible reactions of $NaAlH_4$ can be improved markedly by the addition of suitable materials, such as Ti-based dopants [205–210]. Other transition metals containing additives have also proven to be effective catalysts for improving the hydrogenation and dehydrogenation kinetics of $NaAlH_4$ [211, 212]. Most of the transition metal-based catalysts improve both the hydrogenation and dehydrogenation kinetics by decreasing the activation energy of the respective reactions [168, 213–217]. However, the role of these additions has to be carefully understood since they not only function as catalysts but also facilitate the transport of the species to the surface before desorption. In contrast, nanoconfined $NaAlH_4$ decomposes in a single step without the need for a catalyst as reported by experimental and theoretical methods [171, 218–225]. In this regard, the studies focused on metal-organic frameworks proposed it as a template for formation of nanoscale $NaAlH_4$ on dehydrogenation and reversible hydrogen storage [226–228].

In 2009, Allendorf and coworkers [226] loaded $NaAlH_4$ to HKUST-1 in THF, and on average there are eight THF molecules and eight formula units of $NaAlH_4$ per large pore. And then the thermodynamics and kinetics of H_2 desorption were

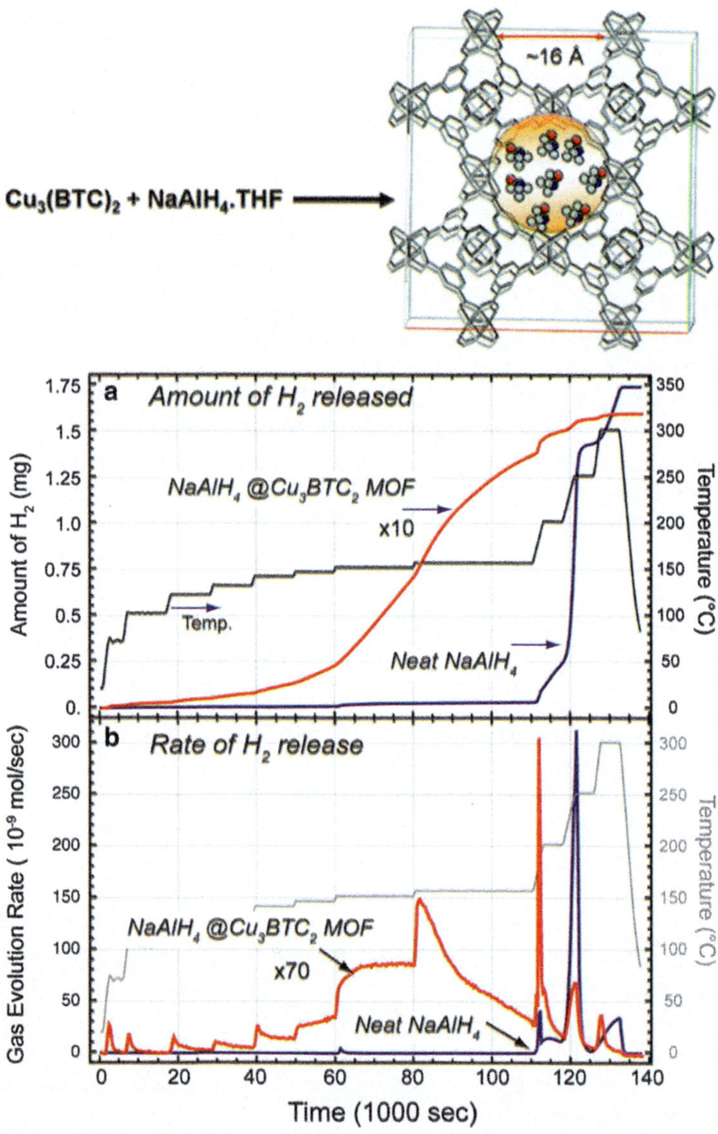

Fig. 15.6 (**a**) Amount of H_2 and (**b**) rate of H_2 released from $NaAlH_4$@MOF compared with neat $NaAlH_4$ (Reprinted with the permission from Ref. [226]. Copyright 2009 American Chemical Society)

investigated quantitatively. For the pyrolysis of nanoscale hydrides in MOF, it can start desorbing H_2 at 70 °C and desorb ~80 % of the total H_2 at 155 °C, during which MOF is not decomposed (Fig. 15.6). In contrast, bulk $NaAlH_4$ shows the onset dehydrogen temperature of 150 °C and 70 % of the total H_2 is desorbed at 250 °C. The results show that the size of $NaAlH_4$ clusters with 1 nm exerts a greater

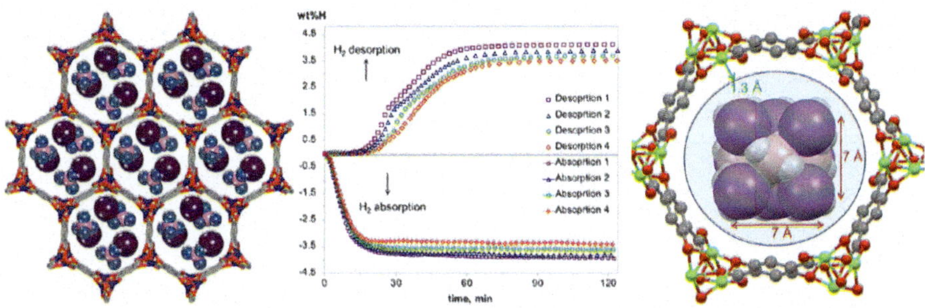

Fig. 15.7 Desorption and absorption of $NaAlH_4$@Ti-MOF-74(Mg) for four consecutive cycles (Reprinted with the permission from Ref. [228]. Copyright 2012 American Chemical Society)

influence on the thermodynamics and reaction rates than other factors, such as interactions between $NaAlH_4$ and pore walls and so on [227]. When confining both hydrogen storage materials and catalysts in pores of MOFs, the sample of Ti-doped nano-$NaAlH_4$@MOF-74 exhibits an onset temperature for hydrogen desorption of ~50 °C and releases 3.6 wt. % in 2.5 h at 150 °C, which is similar to that of undoped nano-$NaAlH_4$@MOF-74 (4.5 wt. %) at 200 °C, as shown in Fig. 15.7. Although the presence of titanium is not necessary for the increase in desorption kinetics, it enables rehydrating to be fully reversible, where it displays minimal capacity loss (from 4.1 to 3.6 wt. % for hydrogen generation) in four dehydrogenation/ rehydrogenation cycles under H_2 pressure [228].

15.2.3 Conclusions

To store/release hydrogen for onboard applications, hydrides are promising due to their high hydrogen contents and moderate hydrogen release temperature. A number of MOF materials have been explored to nanoconfine hydrides and further overcome the critical barriers of lowering the temperature and improving the kinetics for hydrogen release. As shown in Table 15.1, the abovementioned nanoconfinement or MOF-confined hydrides as the catalysts suggested that the synergistic effect of nanoconfinement and metal catalyst centers of MOFs contributes to the enhanced hydrides dehydrogenation kinetics. However, none of them can satisfy all the requirements for mobile applications according to the targets of DOE. The recent efforts on overcoming the critical barriers by nanoconfinement of hydrides in porous MOF materials show the satisfaction of hydrogen release from hydrides in operational temperature and kinetics for PEM fuel cells. As a result, some issues must be focused in the future research. (1) The nanoconfining MOF composite material of the hydrides plus framework system has the reduction of theoretical hydrogen capacity and density due to no contribution to hydrogen capacity from the framework itself. In this case, investigations on selecting suitable lighter metal and MOF materials with low density, suitable pore volume, and high

Table 15.1 Chemical hydrides nanoconfined in MOFs

MOF	CH[a]	T^{Onset} (°C)[b]	T^{Peak} (°C)[c]	Ref.
JUC-32-Y	AB	50	84	[129]
Mg-MOF-74	AB	65	100	[133]
Zn-MOF-74	AB	60	100	[138]
Ti-MOF-74(Mg)	$NaAlH_4$	50	200	[228]
Fe-MIL-53	AB	~60	102	[141]
Ni-MIL-101	AB	50	75	[145]
MIL-101	AB	50	85	[145]
MIL-101(Cr)	AB	~80	~90	[145]
NO_2-MIL-101	AB	~85	~105	[147]
NH_2-MIL-101	AB	68	~73	[147]
$NHCOCH_3$-MIL-101	AB	60	~87	[147]
MIL-101	AB	70	95	[148]
ZIF-8	AB	~70	85 ~ 88	[153]
ZIF-8	DMAB		25	[154]
			40	
HKUST-1	$LiBH_4$	60	110	[200]
HKUST-1	$NaAlH_4$	70	100	[226, 227]

[a]*CH* chemical hydride, *AB* ammonia borane(NH_3BH_3), *DMAB* dimethylamine borane (H_3BNMe_2H)
[b]Starting temperature of hydrogen generation
[c]Decomposition peak of dehydrogenation temperature (°C)

surface area will be sufficiently crucial. In addition, the increase of loading rate is also very important, and some loss might be a worthwhile sacrifice to improve the thermodynamics of the hydrogen release reaction or to improve the kinetics of the reverse reaction. (2) The chemistry reaction mechanism of nanoconfinement is not entirely clear. Systematic investigations on the thermodynamic change related to the framework characteristics such as pore size, organic linker of the framework, and the role of metallic ions should be done both theoretically and experimentally. The demonstrations of the intrinsic characteristics and interactions between the hydrides and MOF should be critical to understand the fundamentals of nanoconfinement effect. (3) The cheap and easy ways to realize the regeneration of hydrides off-board become urgent. The irreversibility of chemical hydrogen storage by off-board regeneration motivates the research of hydride, and thus the cheap and effective way to regenerate such compounds becomes the most important for using them as hydrogen storage materials. This boosts the research stream of hydrides regeneration, but there is still a far way to cheaply produce hydrides for practical applications. (4) In our point of view, the reversibility of nanoconfined hydride system is still an exceptional important issue, although the off-board regeneration releases this reversibility criterion released by DOE. Importantly, the nanoconfinement of hydrides might enhance the possibility of hydride reversibility due to the restructuring of hydride inside nanopores and neutralizing the thermodynamics. This is only our hypothesis, which requires theoretical and/or experimental evidence, and could be an important topic of future study.

To shortly sum up, hydrides are highly potential to achieve the current, even ultimate, targets of onboard hydrogen applications, but the regeneration of the catalyst at 70 °C for PEM fuel cells is necessary between each cycle to recover the hydrides and regenerate the cavities of MOFs. It had been shown that strong caging effects, probably, are combined with polar and Lewis acid/base properties of the porous matrix, which are crucial for the use of mild reaction conditions. Nanoconfinement shows a significant effect on improving the kinetics and modifying the thermodynamics, but further evidences to support this conclusion are highly desirable. Compared with neat chemical hydrides, nanoconfined hydride@MOF materials show advantages for the improvement of hydrogen release thermodynamics and kinetics, lower operational temperature, and high purity of released hydrogen during pyrolysis. Up to now, the corresponding reports about hydride@MOFs are still scarce, which may be due to requirements for MOFs. For solid-state chemical hydride@MOF systems, the MOFs should have appropriate pore sizes, active metal sites, and especially thermal and chemical stability, which is the most important for application of MOFs in this field. In addition, even though the present system does not directly address the hydrogen storage challenge to fulfill every index, it is a step forward in realizing hydrogen storage materials that can operate below fuel cell operating temperature of less than 85 °C, which requires further efforts of study in the future.

15.3 MOF-Supported Metal Nanoparticle Catalysts for Hydrogen Generation from Hydrides

The physical and chemical characters of metal nanoparticles (M-NPs) are demonstrating huge advantages over those of bulk metals, such as thermal, magnetic, and electrical conductivities in virtue of the delocalization of free electrons. Since the high surface-area-to-volume ratio of M-NPs provides a great number of active sites, the size and shape control of M-NPs is crucial to achieve enhanced reactivity [229–234]. However, due to their high surface energy and large surface area, the stability of M-NPs is severely decreased, which obstructs the control of size and shape with high uniformity. M-NPs have poor cycling stability as a result of the agglomeration and coarsening of the nanoparticles [235–241]. The nucleation and growth of M-NPs can be averted if the metal or alloy nanoparticles are confined within highly porous scaffold materials, which could be used for the fabrication of metal nanoparticles (NPs) with controlled sizes inside the pores, thereby circumventing a common issue of nanoparticle aggregation [242–245]. In particular, the use of M-NPs in porous MOFs with confined void spaces has been proved to be an efficient way of preventing aggregation [1, 48, 246–252]. Recently, there have been extensive efforts to fabricate metal nanoparticles (M-NPs) in MOFs to elicit the properties that are hardly achieved by the individual material. MOFs have been utilized as supports for M-NPs since they could availably control the limited growth of M-NPs in the confined cavities and produce dispersed M-NPs [251, 253–266].

General synthetic methods to embed M-NPs in porous metal-organic framework matrix entail the impregnation of a metal precursor in a porous MOF, followed by reduction of the metal precursor to metal (0) atoms, which aggregate into M-NPs within the MOF. The precursor molecules that are most frequently used for the production of M-NPs are chloride or nitrate salts of the corresponding transition metal ions. In general, metal precursors included in MOFs are reduced with hydrogen gas [252, 255, 258–261], hydrazine [48, 262], oleylamine [251], or $NaBH_4$ [263–266] to generate M-NPs, and the reduction process is often performed at high temperature, followed by a washing step. In some cases, supercritical CO_2-methanolic solution has been used to load the precursor compound within a MOF, followed by a heating process [267]. The reaction conditions for fabricating M-NPs in MOFs should be determined, depending on the properties of the host matrices during the reduction process. Since the size of the NPs is significantly affected by the loading time of the metal precursor and the reduction conditions, these conditions should be carefully controlled.

Here we review recent progress on the MOF-supported metal nanoparticle catalysts for hydrogen generation from chemical hydrides, such as ammonia borane, dimethyl borane (DMAB), hydrazine(HZ), formic acid(FA), and so on. These are also proved to be one of the most convenient and effective ways to achieve property synergies of metal NPs and MOFs for dehydrogenation applications [257–261].

15.3.1 Ammonia Borane and Its Derivatives

Our group [268] reported the first example of water-stable MOF-supported catalysts for hydrogen generation from hydrolysis of AB. Highly dispersed Ni nanoparticles were successfully immobilized by metal-organic framework ZIF-8 via both chemical vapor deposition (CVD) [247, 269, 270] and chemical liquid deposition (CLD) [137, 271] approaches to obtain CVD-Ni@ZIF-8 and CLD-Ni@ZIF-8, respectively. The CVD-Ni@ZIF-8 and CLD-Ni@ZIF-8 showed high catalytic activity for hydrogen generation from hydrolysis of aqueous AB at room temperature. For CLD-Ni@ZIF-8, the hydrolysis reaction of the AB can be completed (Ni/AB = 0.019) in 19 min (H_2/AB = 3.0), giving a TOF value of 8.4 min^{-1}. The CVD-Ni@ZIF-8 sample showed a higher activity, with which the reaction can be completed (H_2/AB = 3.0) in 13 min (Ni/AB = 0.016), corresponding to a TOF value of 14.2 min^{-1}. As shown in Fig. 15.8, the experiments of durability/stability indicated that there was no significant decrease in catalytic activity even after 5 runs of hydrolysis reactions for both catalysts due to the fact that the highly dispersed Ni NPs have been effectively immobilized by the frameworks of ZIF-8.

By employing a “double solvents” method (DSM), we synthesized the highly active MOF-immobilized Pt nanocatalysts, Pt@MIL-101 [148]. The uniform three-dimensional distributed and ultrafine Pt NPs were successfully immobilized inside the pores throughout the interior cavities of MIL-101 without aggregation of Pt NPs

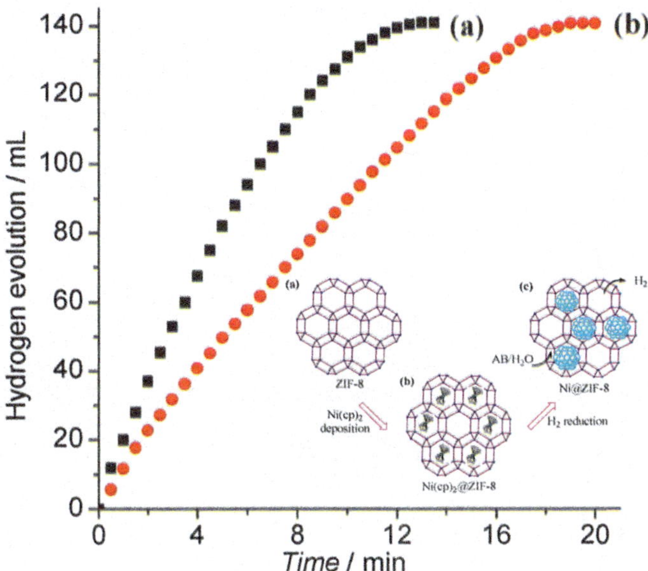

Fig. 15.8 Hydrogen generation from hydrolysis of aqueous AB (2 mmol in 2 mL water) in the presence of 10 mg (*a*) CVD-Ni@ZIF-8 (Ni/AB = 0.016) and (*b*) CLD-Ni@ZIF-8 (Ni/AB = 0.019) at room temperature (Reproduced from Ref. [268] by permission of The Royal Society of Chemistry)

on the external surface of the MOF. The "double solvents" method used in this work is based on a hydrophilic solvent (water) and a hydrophobic solvent (hexane), the former containing the metal precursor with a volume set equal to or less than the pore volume of the adsorbent (MIL-101), which can be absorbed within the hydrophilic adsorbent pores and, the latter, in a large amount, playing an important role to suspend the adsorbent and facilitate the impregnation process (Fig. 15.9). In the presence of 2 wt. % Pt@MIL-101 catalysts (Pt/AB = 0.0029 in molar ratio), the H_2 generation by hydrolysis of aqueous AB (ammonia borane) is completed within 2.5 min at room temperature, corresponding to a catalytic activity 2 times higher than that of 2 wt. % Pt/γ-Al_2O_3 [272, 273], the most active Pt catalyst reported for this reaction so far. Interestingly, the Pt@MIL-101 exhibited excellent catalytic activities for all three reactions in liquid-phase ammonia borane hydrolysis, solid-phase ammonia borane thermal dehydrogenation, and gas-phase CO oxidation.

Our group exploited a liquid-phase concentration-controlled reduction (CCR) strategy for the first time to control the size and location of the alloy AuNi NPs during reduction of the Au^{3+} and Ni^{2+} precursors which are introduced into the pores of MOF by using the double solvents method [274]. After an overwhelming reduction (OWR) approach with a high concentration reductant ($NaBH_4$) solution was employed, the ultrafine AuNi alloy NPs were successfully encapsulated into the MOF nanopores without aggregation on the external surface, which exhibited excellent catalytic performance in hydrolytic dehydrogenation of ammonia borane.

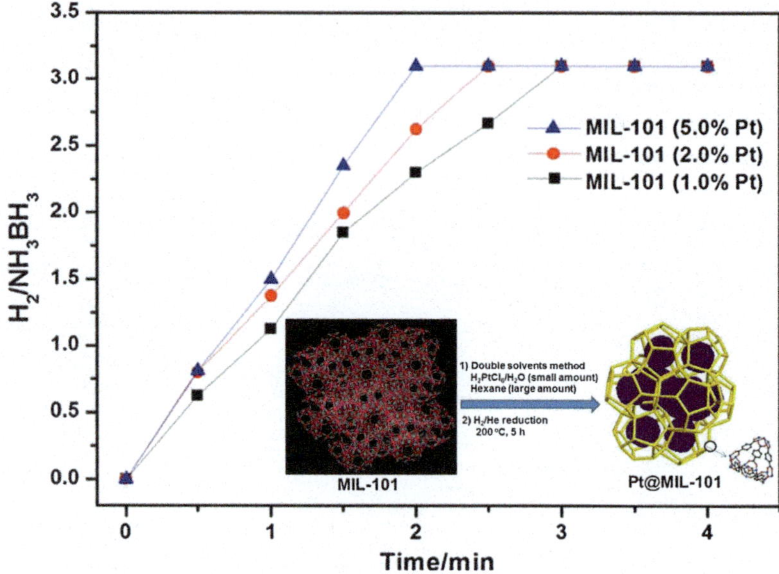

Fig. 15.9 Hydrogen generation from aqueous AB in the presence of Pt@MIL-101 catalysts at room temperature and representation of synthesis of Pt nanoparticles inside the MIL-101 matrix using double solvents method (*Inset*) (Reprinted with the permission from Ref. [148]. Copyright 2012 American Chemical Society)

As shown in Fig. 15.10, the AuNi@MIL-101 catalysts are more active than the monometallic counterparts, exhibiting synergistic effect between Au and Ni. The experimental results indicate that the AuNi@MIL-101 with the Au/Ni atomic ratio of 7:93 is the most active, in which the AB hydrolysis reaction is completed with a 70 mL H_2 release in 2.67 min ((Au + Ni)/AB = 0.017 in molar ratio), corresponding to a higher TOF value of 66.2 $\mathrm{mol_{H_2} \cdot mol_{cat}^{-1} \cdot min^{-1}}$ than those of the most active non-noble metal-based catalysts for this reaction reported so far and even higher than those of most Pt-, Rh-, and Ru-related catalysts [88, 274–278]. More importantly, the initial dehydrogenation rates indicate that the catalytic activity is severely decreased with the increase of AuNi alloy NPs size. As a result, a liquid-phase CCR strategy in combination with DSM could control the size and location of the MOF-immobilized metal NPs. Using the OWR approach, ultrafine AuNi alloy NPs would be fabricated inside the mesoporous MIL-101, which could serve as a high-performance catalyst for future development of ammonia borane into a practical hydrogen storage materials for clean energy applications. New avenues will be opened up for developing high-performance heterogeneous catalysts by using porous MOFs as hosts for ultrafine metal NPs, especially non-noble metal-based NPs.

Recently, Zahmakiran and coworkers reported the synthesis, characterization of Pd@$Cu_3(btc)_2$ with palladium(0) nanoparticles stabilized by the activated framework of HKUST-1 [134] or MOF-199 [279]. The in situ formation of well-dispersed palladium (0) nanoparticles with the size 4.3 ± 1.1 nm was supported

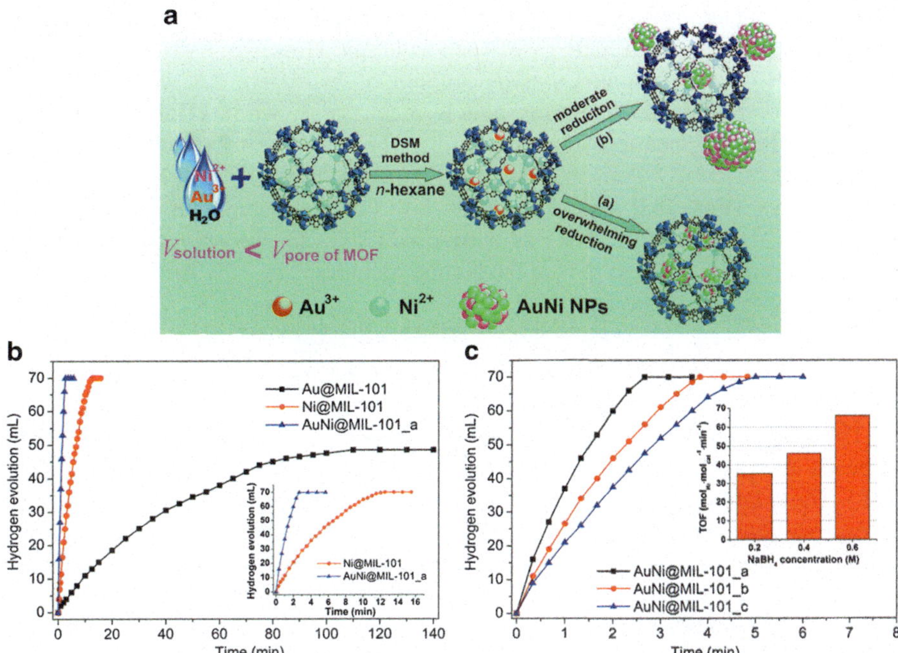

Fig. 15.10 (**a**) Representation of immobilization of the AuNis nanoparticles by the MIL-101 matrix using the DSM combined with a liquid-phase CCR strategy and plots of time vs. volume of hydrogen generated from AB (1 mmol in 5 mL water) hydrolysis at room temperature catalyzed by (**b**) the Au@MIL-101, Ni@MIL-101, and AuNi@MIL-101_a catalysts (50 mg, (Au + Ni)/AB (molar ratio) = 0.017) and (**c**)the AuNi@MIL-101_a – c catalysts (50 mg, (Au + Ni)/AB (molar ratio) = 0.017) prepared by reduction in $NaBH_4$ solution with different concentrations. *Inset*: the corresponding TOF values of the catalysts (Reprinted with the permission from ref. [274]. Copyright 2013 American Chemical Society)

on $Cu_3(btc)_2$ framework, which showed unprecedented catalytic activity, lifetime, and reusability for the dehydrogenation of DMAB at room temperature, providing an initial TOF value of 75 h^{-1} for complete conversion of DMAB to cyclic diborazane ($[Me_2NBH_2]_2$) and generation of 1 equiv. of H_2 per mole of DMAB [280]. In particular, the PdNPs@$Cu_3(btc)_2$ shows high stability against leaching and sintering throughout the catalytic runs, which make them a long-lived and reusable catalyst. Consequently, they provide a total turnover number (TTON) of 2100 and retain almost their inherent activity even at the fifth catalytic reuse in the dehydro genation of DMAB.

15.3.2 Hydrous Hydrazine

Anhydrous hydrazine (H_2NNH_2, HZ) has a hydrogen content as high as 12.5 wt. %, but it is hypergolic and explosively reacts upon exposure to a metal surface, which limits its application from safety point of view. Mostly the reactions of hydrazine must be highly diluted in inert gases such as argon. However, hydrous hydrazine, $H_2NNH_2 \cdot H_2O$, still contains a large amount of hydrogen, 8.8 wt.%, which is available for hydrogen generation and is much safer, while efforts need to be made from the engineering side to minimize the influence of toxicity. Hydrazine might be catalytically decomposed into ammonia, nitrogen, and hydrogen over some catalyst (Eqs. 15.7 and 15.8) [281–284]. For the incomplete decomposition reaction of hydrazine, the NH_3 as a by-product could not only complicate the separation process but also poison the Nafion membrane and the fuel cell catalysts.

$$H_2NNH_2 \rightarrow N_2(g) + 2H_2(g) \tag{15.7}$$

$$3H_2NNH_2 \rightarrow 4NH_3 + N_2(g) \tag{15.8}$$

The competition between decomposition reactions Eqs. (15.7) and (15.8) depends on the catalysts used and also on applied reaction conditions [285–288]. Thus, finding highly selective and active catalysts for immediate hydrogen release without ammonia is of great importance for practical applications. The only by-product of complete decomposition of hydrazine is N_2 that does not need to be collected for recycling. Moreover, N_2 can be transformed into ammonia by the Haber–Bosch process, homogeneous catalytic processes, or an electrolytic process [289–291] and subsequently to hydrazine on a large scale [292–294] or perhaps transformed directly to hydrazine by using an electrolytic process similar to that for ammonia synthesis. However, it is a thermodynamic unfavorable process for complete decomposition of hydrazine (Eq. 15.7) at ambient temperatures. Many attempts have been put forward to achieve this aim. Our group synthesized various kinds of monometallic as well as bimetallic nanoparticles [295–302].

Among them, Pt–Ni, Ir–Ni, and Rh–Ni bimetallic nanoparticles exhibited selectivity close to 100 % for hydrogen generation at room temperature [296–301]. Furthermore a highly efficient and low-cost catalyst of noble-metal-free Ni–Fe alloy nanoparticles exhibited excellent catalytic performance for the complete decomposition of hydrous hydrazine, for which the Ni–Fe nanocatalyst, with equimolar compositions of Ni and Fe, achieved 100 % hydrogen selectivity from hydrous hydrazine decomposition in an alkaline solution (0.5 M NaOH) at 70 °C [300]. The development of low-cost and high-performance catalysts may encourage the effective application of hydrous hydrazine as a promising hydrogen storage material. The present finding demonstrates the importance of combining different non-noble metals to develop active metal nanocatalysts whose distinct surface properties enable them to outperform their parent metals. These low-cost, high-performance catalysts may strongly encourage the effective application of hydrous hydrazine as a promising hydrogen storage material. Wang et al. [303] reported a supported catalyst by depositing Rh–Ni nanoparticles on graphene, which exhibited higher activity than the bare nanoparticle catalysts. Carbon-supported Ni_3Fe nanosphere catalysts were active, excreting 100 % selectivity for hydrogen from hydrous hydrazine at room temperature [304]. However, the aggregation of graphene nanosheets in aqueous solution and the weak interaction between carbon black and active metals would limit further improvement in the performance of the catalyst.

To improve the kinetic properties of hydrous hydrazine decomposition and decrease the noble metal loading, MOF-supported metal nanoparticles have contributed to the development of efficient heterogeneous catalysts. Recently, our group [305] reported highly dispersed Ni-Pt bimetallic NPs catalysts supported by ZIF-8 [306]. Four Ni-Pt@ZIF-8 composites were prepared from activated ZIF-8 (100 mg) with $NiCl_2$ and $K[PtCl_4]$ (total 0.2 mmol in molar rations of 95:5, 90:10, 80:20, 50:50) impregnated with deionized water (4 mL) at room temperature for 12 h, subsequently separated by centrifugation, dried, and reduced by $NaBH_4$ to yield catalysts I, II, III, and IV with 0.131, 0.133, 0.137 mmol metal loading (100 mg catalysts I, II, and III). For hydrogen generation from hydrous hydrazine, the four Ni-Pt@ZIF-8 composites exhibit high-performance catalytic activity, which strongly depends on the Ni-Pt composition, as shown in Fig. 15.11. The Ni-Pt@ZIF-8 is the first example of water-stable MOF-supported bimetallic Ni-Pt catalysts for hydrogen generation from selective decomposition of hydrous hydrazine. Similar dependences on temperature and base have been observed for the catalytic decomposition of hydrazine in aqueous solution over Ni-Pt@ZIF-8 catalysts. At room temperature, no hydrogen release in the absence of base. In the presence of base, hydrogen is generated and increases with increasing temperature.

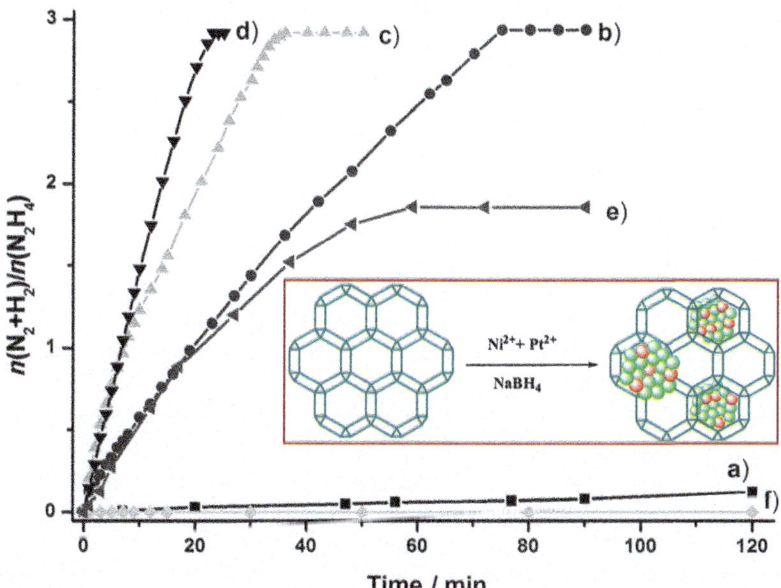

Fig. 15.11 Time course plots for the decomposition of hydrous hydrazine over ZIF-8 supported mono- and bimetallic catalysts (*a*) Ni@ZIF-8, (*b*) I, (*c*) II, (*d*) III, (*e*) IV, and (*f*) Pt@ZIF-8 in the presence of NaOH (0.5 M) at 323 K (catalyst = 0.100 g; $N_2H_4 \cdot H_2O$ = 0.1 mL). *Box*: Synthesis of ZIF-8 supported bimetallic nanocatalysts Ni-Pt@ZIF-8 (Reproduced with permission from Ref. [305] by permission of Jon Wiley & Sons Ltd)

At 50 °C, the hydrous hydrazine can be completely converted to N_2 and H_2 (nearly 2.9 equiv.), respectively, in 75, 36, and 26 min by catalysts I, II, and III, whereas catalyst IV shows only 11 % selectivity. Catalyst III shows the highest TOF value with 90 h^{-1} among all the catalysts. These highly efficient catalysts represent a promising step toward the practical applications of water-stable MOFs as effective matrices to immobilize metal NPs in the catalytic dehydrogenation reaction system and practical application of hydrous hydrazine as promising hydrogen storage material.

Interestingly, unlike bimetallic catalysts, monometallic catalyst Pt@ZIF-8 is catalytically inactive, and Ni@ZIF-8 has very low activity under the same reaction, which clearly indicates that alloying nickel with a small amount of platinum can tune the structure of the catalyst surface, resulting in high catalytic performance in decomposition of hydrazine in aqueous solution. In comparison with the previous Ni-Pt catalysts with surfactants as capping agents [296, 297], catalytic activities are improved for the MOF-supported Ni-Pt catalyst.

15.3.3 Formic Acid

Formic acid (FA), the simplest carboxylic acid with a density of 1.22 g cm^{-1}, a melting point of 8.4 °C and boiling point of 100.8 °C, is suitable for easy transportation, refueling, and handling [307]. Formic acid is considered as a convenient H_2 carrier because it is a liquid at room temperature, nontoxic, and contains a hydrogen content of 43 g kg^{-1}, corresponding to 4.4 wt. % hydrogen [308].

$$HCOOH \rightarrow H_2 + CO_2 \qquad \Delta G = -48.4 \text{ kJ mol}^{-1} \tag{15.9}$$

$$HCOOH \rightarrow H_2O + CO \qquad \Delta G = -28.5 \text{ kJ mol}^{-1} \tag{15.10}$$

The decomposition of FA can follow both dehydrogenation reaction producing CO_2 and H_2 (Eq. 15.9) and dehydration reaction producing CO and H_2O (Eq. 15.10). To maximize the efficacy of FA as a hydrogen storage material, the first decomposition pathway is regarded as a promising H_2-generating and desirable reaction process, which produces only gaseous products (H_2–CO_2), whose mixture can be easily separated under certain conditions. The selective decomposition of FA to CO_2 and H_2, which is reversible reaction of CO_2 hydrogenation, is crucial for formic acid-based hydrogen storage [309, 310]. In this case, a reduction of CO_2 emission is the use of CO_2 itself as a hydrogen carrier. However, the dehydrated pathway is not desirable because CO is toxic to fuel cell catalysts [311]. Recently, the selective dehydrogenation of FA in homogeneous reactions with organometallic complexes [312–314] and in heterogeneous reactions with noble metal nanoparticles (NPs) [315–320] has been reported.

In 2011, our group reported bimetallic Au–Pd NPs immobilized in mesoporous MIL-101 as efficient catalysts for the decomposition of formic acid for hydrogen

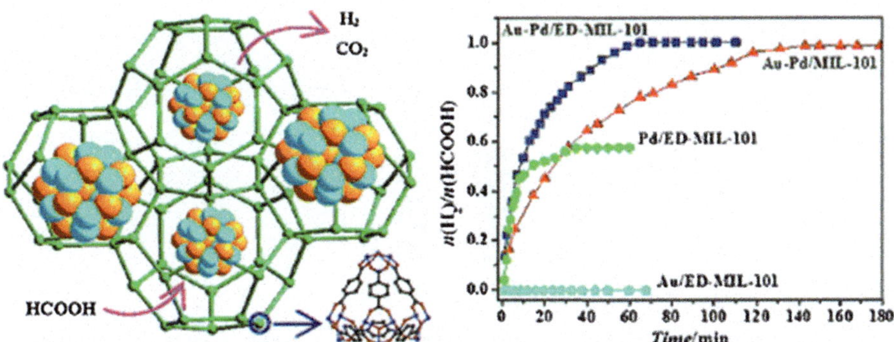

Fig. 15.12 Time course plots for hydrogen generation from formic acid in the presence of 20 mg of different metal NP catalysts, 140 mg of formic acid, 70 mg of sodium formate, and 1.0 mL of water at 90 °C (Reprinted with permission from Ref. [271]. Copyright 2011 American Chemical Society)

generation [271]. MIL-101 was chosen as a support because of its large pore sizes (2.9–3.4 nm), window sizes (1.2–1.4 nm), and its high pore surface, which facilitates the encapsulation of metal NPs and the adsorption of the substrate formic acid inside the pores. We grafted the electron-rich functional group ethylenediamine (ED) into MIL-101 (ED-MIL-101) for improving the interactions between the metal precursors and the MIL-101 support. By using a simple liquid impregnation method, the resulting bimetallic Au–Pd NPs immobilized in the MIL-101 and ED-MIL-101 (Au–Pd@MIL-101 and Au–Pd@ED-MIL-101) represent the first highly active MOF-immobilized metal catalysts for the complete conversion of formic acid to hydrogen. The study of its catalytic activity showed that formic acid (140 mg) can be completely converted to H_2 in 145 min ((Au + Pd) catalyst = 20 mg) at 90 °C, as shown in Fig. 15.12. Furthermore, Au–Pd@MIL-101 exhibited the enhanced kinetics with increasing temperature and 100 % selectivity for hydrogen generation from formic acid. The stability/durability test showed that the productivity of hydrogen remained almost unchanged after five runs in the decomposition of formic acid at 90 °C.

In another report, Martis et al. [321] used a photocatalytically active titanium-based MOF, MIL-125 [322], and its amine-functionalized-MOF, NH_2-MIL-125 [323], to immobilize Pd NPs by photoassisted and ion exchange deposition methods. Owing to the amine functional groups and small NP size, the Pd NPs embedded within the amine-functionalized NH_2-MIL-125 displayed higher catalytic activity for H_2 generation from FA at ambient temperature in comparison to MIL-125. Furthermore, the Pd@NH_2-MIL-125 is capable of generation of H_2 from FA without the unfavorable formation of CO (less than 5 ppm), giving a TOF value of 214 h^{-1} working at 32 °C. No H_2-producing activity was observed in the control experiment with the MIL-125, NH_2-125, TS-1 (Ti/Si = 57.7, prepared by photoassisted deposition method), and Ti-MCM-41 (5 mol % Ti, prepared by photoassisted deposition method) porous materials, but their Pd NPs-supported material exhibited some catalytic activity. As shown in Fig. 15.13, the Pd-TS-1

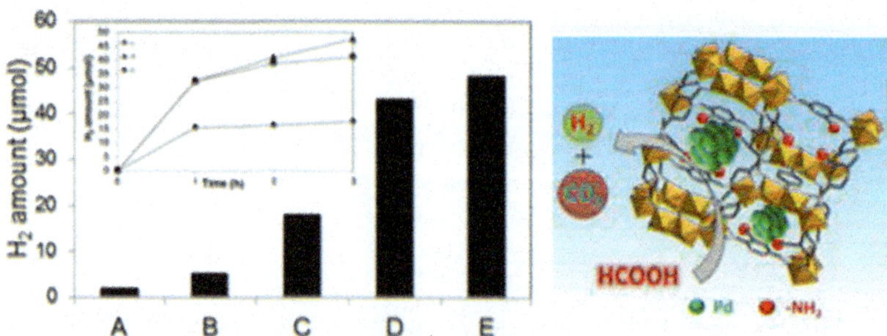

Fig. 15.13 Generation of H_2 from FA using different supports prepared by photoassisted deposition and the ion exchange method. (*A*) Pd-TS-1, (*B*) Pd – Ti-MCM-41, (*C*) Pd-MIL-125 (photoassisted), (*D*) Pd-NH2-MIL-125 (ion exchange), and (*E*) Pd-NH2-MIL-125 (photoassisted) (Reprinted with permission from Ref. [321]. Copyright 2013 American Chemical Society)

and Pd – Ti- MCM-41 were the least efficient catalysts, generating only 1.7 and 5.1 μmol of H_2 from 0.39 mL FA, respectively. Pd-NH_2-MIL-125 (photoassisted) displayed the best catalytic performance, generating 48.1 μmol, Pd-NH_2-MIL-125i (ion exchange) produced 43.1 μmol, and Pd-MIL-125 produced only 17.8 μmol from 0.39 mL FA. The ratio of H_2 and CO_2 generated during the reaction was approximately 1. The CO concentration during the dehydration reaction for Pd-NH_2-MIL-125 never exceeded 5 ppm. While the H_2 generation by Pd-MIL-125 nearly ceased after 1 h, the Pd-NH_2-MIL-125 continued to generate H_2 at a nearly linear rate after the initial period, but the H_2 production decreased for Pd-NH_2-MIL-125i.

From the experimental results, it is clear that the increased catalytic activity for FA decomposition was mainly caused by the –NH_2 functional groups within the nanoporous structure of the MOF, while the average size of Pd NPs is not the most important factor in attaining the high catalytic activity. High-performance monometallic gold nanoparticles were functionalized with amine and encapsulated in silica nanospheres. Due to the strong metal–molecular support interaction (SMMSI), amine-functionalized gold nanoparticles in the presence of amine in the silica sphere can make the gold nanoparticles highly active and 100 % selectivity for hydrogen generation from aqueous formic acid at a convenient temperature, although the unsupported or silica-supported gold NPs are inactive for this reaction. The amine-functionalized gold nanoparticles represented the highly active and stable monometallic nanocatalyst for hydrogen generation from aqueous formic acid decomposition [324]. The dehydrogenation mechanism of FA with Pd NPs supported on amine-functionalized MOF was deduced that the reaction intermediate of Pd formate undergoes β-hydride elimination to produce CO_2 and a Pd hydride species. The weakly basic –NH_2 group within the MOF with a positive effect on the O–H bond dissociation acts as a proton scavenger, forming –$^+NH_2$, which explained that the positive effect on the O–H bond dissociation and the cooperation of the amine functionality within the MOF structure and NPs is responsible for the high catalytic activity.

15.3.4 Conclusions

The wide range of adjustable metal centers and their surrounding environments as well as a variety of organic or pseudo-organic linkers provide multiple opportunities to create various MOFs with different catalytic properties. Loading of metal NPs inside the porous matrices of MOFs and their catalytic applications are of significant interest. The use of nanocluster catalysts appears to be an efficient way of preventing aggregation in mesoporous and microporous MOF, which could afford well-dispersed metal NP catalysts. It has been proved extremely significant that controllable integration of metal nanoparticles and metal-organic frameworks could obtain M-NP@MOF composite materials for improved kinetics of dehydrogenation of chemical hydrides; see Table 15.2.

Utilization of MOFs as supports for M-NPs has several advantages over other porous materials: their three-dimensional pore structures, the presence of organic linkers that stabilize M-NPs, robust structural property in some cases, and moderate thermal stability. Various MOFs with different pore sizes and shapes can be prepared from a wide range of metal ions and organic linkers, and thus an appropriate MOF can be selected as a host matrix. In choosing a MOF for loading M-NPs, the structural stability of the MOF upon the precursor loading or reduction procedure should be considered. Furthermore, to maintain a high surface area of the MOF, caution must be taken in the reduction process not to block the pores by the NPs or by unwanted by-products formed. Although the encapsulation of M-NPs in MOF pores is expected to limit the particle growth, the precursor compound and the product can actually diffuse out through the pores of the host to form M-NPs on the surface of the MOF crystal, instead of being inside the pores. Therefore, the location of the M-NPs, on the surface of the crystal or inside of the pores, should be verified by experimental evidence. Double solvent method (DSM) might be an effective way for the fabrication of M-NPs supported by MOFs. If MOFs have

Table 15.2 MOF-supported metal nanoparticle catalysts for hydrogen generation from chemical hydrides

MOF	M-NPs	Hydride[a]	TOF (h^{-1})[b]	Temperature (*K*)[c]	Ref.
ZIF-8	Ni	AB	852[d]	RT	[268]
			504[e]	RT	[268]
ZIF-8	Ni-Pt	HZ	90	323	[305]
MIL-101	Pt	AB	~24,490	RT	[148]
MIL-101	Au-Ni	AB	3,972	RT	[274]
HKUST-1	Pd	DMAB	75	RT	[280]
NH_2-MIL-125	Pd	FA	214	305	[321]

[a]*CH* chemical hydride, *AB* ammonia borane (NH_3BH_3), *DMAB* dimethylamine borane (H_3BNMe_2H), *FA* formic acid (HCOOH), *HZ* hydrazine(N_2H_4)
[b]mol H_2 per mol catalyst per hour
[c]At room temperature (298 K)
[d]Molar ratio Ni/AB = 0.016
[e]Molar ratio Ni/AB = 0.019

straight channels, they do not provide an adequate confinement effect, and thus there is a possibility that agglomeration may occur and M-NPs may escape through the pores. Therefore, it is necessary to develop appropriate novel porous MOFs that can be employed as supports for M-NPs for the improvement of efficiency.

M-NPs@MOFs have exhibited excellent catalytic activities in dehydrogenation of chemical hydrides due to the confinement effect of the M-NPs in a MOF as well as the limitation of the particles that remain constrained and do not grow further after catalytic reactions. In spite of being promising, the M-NPs@MOFs usually involved relatively cumbersome preparation processes and expensive organometallic precursor compounds. The development of a general, simple, and efficient route that can easily achieve a loading of MOFs with cavity-size-matched metal nanoparticles, and a homogeneous dispersion throughout the bulk MOF matrix is highly desirable and still remains challenging.

15.4 MOFs as Catalyst Precursors for Hydrogen Generation from Hydrides

Besides using the MOF framework itself as a catalyst, or using MOF as a host material for supporting NPs, MOFs can also act as precursors for forming catalysts by framework decomposition, which can be employed by hydrogen generation from chemical hydrides.

Li and coworkers reported in situ generated Ni NPs using a Ni-based MOF ([Ni(4,4′-bipy)(HBTC)], 4,4′-bipy = 4,4′-bipyridine) [325] as a precursor and AB as a reducing agent [275] in methanol. Due to the low activation energy and high reaction rate, the catalyst showed accelerated hydrogen generation kinetics, as shown in Fig. 15.14. With the MOF-based catalyst, the hydrogen generation from aqueous AB solution (0.32 M) was completed within 11, 7, 6, and 5 min at 298 K with Ni/AB molar ratios of 0.02, 0.05, 0.08, and 0.10, respectively. When the concentration of AB was 0.16 M, complete hydrolysis from aqueous NH_3BH_3 solution (1.0 mL) occurred in less than 2 min at 313 K in the presence of MOF-based catalyst (20 mg) and within 5 min at room temperature in the AB solution with the concentration of 0.32 M in the presence of MOF-based catalyst (MOF-based catalyst/NH_3BH_3 = 0.10), as shown in Fig. 15.13. More importantly, the catalyst can be recycled by centrifugal separation and reused for up to 20 cycles without obvious loss of activity.

Subsequently, they obtained other two catalysts from MOF precursors [Ni(pyz)][Ni(CN)4] (pyz = pyrazine) and $Ni_3[Fe(CN)_6]_2$ [326] by in situ reduction with AB in methanol (MOF/AB = 0.02). Using Ni^0-based catalysts (1.0 mol%) from the precursors of the two MOFs, the thermal decomposition of AB shows a significantly lower reaction onset temperature, remarkably accelerated kinetics, and decreased activation energy, as shown in Fig. 15.15. At 80 °C, AB with catalyst derived from [Ni(4,4′-bipy)(HBTC)] released 7.5 wt. % H_2 in 2 h without any

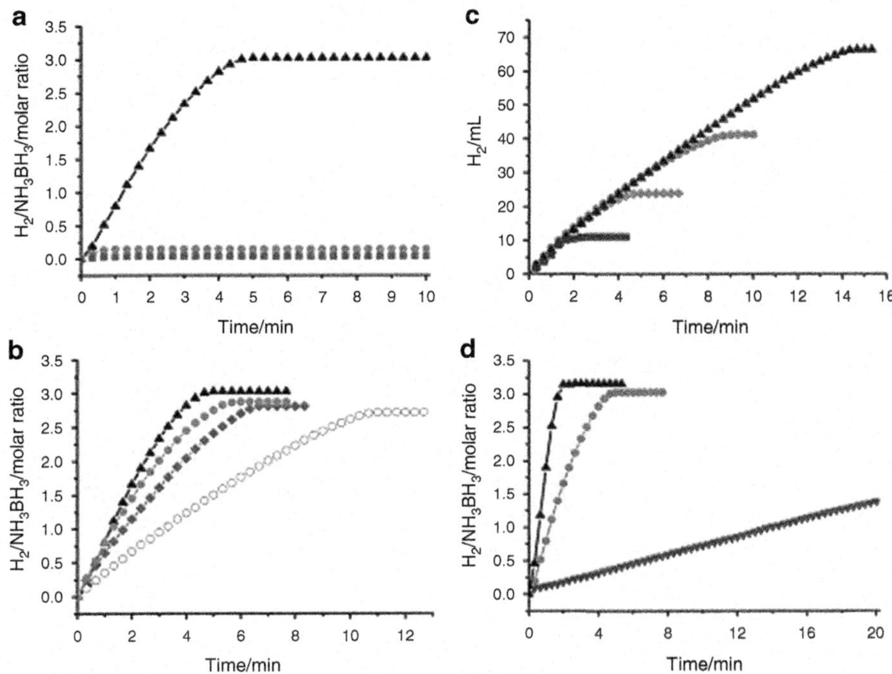

Fig. 15.14 (**a**) Hydrogen generation from the hydrolysis of NH_3BH_3 (0.32 M, 1.0 mL) in the presence of different catalysts at 298 K in air (MOF-based catalyst (▲), Ni powder mixed with MOF (○), no catalyst (■); (**b**) MOF-based catalyst with MOF-based catalyst/NH_3BH_3 in molar ratios of 0.02 (○), 0.05 (♦), 0.08 (·), and 0.10 (▲); (**c**) hydrogen release from aqueous NH_3BH_3 solution (1.0 mL) with different concentrations (0.16 (■), 0.32 (♦), 0.65 (♦), and 0.97 M (▲)) in the presence of MOF-based catalyst (20 mg); (**d**) hydrogen generation from aqueous NH_3BH_3 (0.32 M, 1.0 mL) in the presence of MOF-based catalyst (MOF-based catalyst/NH_3BH_3 = 0.10) at 273 (▼), 298 (·), and 313 (▲) K (Reproduced with permission from Ref. [325] by permission of John Wiley & Sons Ltd)

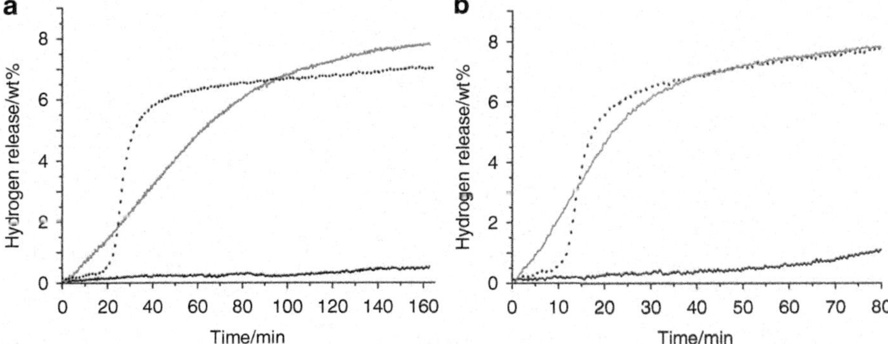

Fig. 15.15 (**a**) The volumetrically measured release of H_2 from AB/MOF1cat (*gray line*), AB/-MOF2cat (*dotted line*), and neat AB (*black line*) at 80 °C and (**b**) 90 °C (Reproduced with permission from Ref. [326] by permission of Jon Wiley & Sons Ltd)

induction period, and AB with catalyst derived from [Ni(pyz)][Ni(CN)$_4$] could generate 6.0 wt. % H_2 in 40 min. At 90 °C, 6.0 wt. % H_2 can be released from AB with the catalyst derived from [Ni(pyz)][Ni(CN)$_4$] within only 20 min. Therefore, for further development, both the frameworks and the metal sites in the MOFs can be well adjusted by crystal engineering to create more effective catalysts for various catalytic processes.

In another report, this group prepared a highly active Co(0) catalyst by the same in situ reduction of a MOF precursor $Co_2(bdc)_2(dabco)$ (bdc = 1, 4-benzenedicarboxylate; dabco = 1,4-diazabicyclo[2.2.2]-octane) [327]. It is interesting that the active Co(0) sites, generated by using $NaBH_4$ reduction in the micropores and channels, are surrounded by the organic linkers and stabilized in the residue of the framework. The hydrogen generation from aqueous AB solution (0.32 M) was completed within 1.4 min (MOF/$NaBH_4$/AB = 0.057:0.08:1) at room temperature.

Recently, Li and Kim used ZIF-9 ($Co(PhIm)_2 \cdot (DMF) \cdot (H_2O)$, PhIm = benzimidazolate) [148] framework as catalyst for the hydrolysis of $NaBH_4$ for the first time [328]. Unlike traditional Co-based catalysts [329–340], the experiments of the recycling and durability of ZIF-9 showed that the initial hydrogen generation rate is comparatively low and then linearly increase with the reaction time for the fresh ZIF-9, while the reused ZIF-9 obtains its rapid hydrogenation rate at the beginning, which is due to the formation of CoB active centers of ZIF-9 in the first cycle. Since ZIF-9 could consume a small amount of $NaBH_4$, the final volume of generated hydrogen decreases with the increase of ZIF-9 catalyst, although the same amount of $NaBH_4$, H_2O, and NaOH are used, as shown in Fig. 15.16. The hydrogen generation rate can reach 2,345.28 and

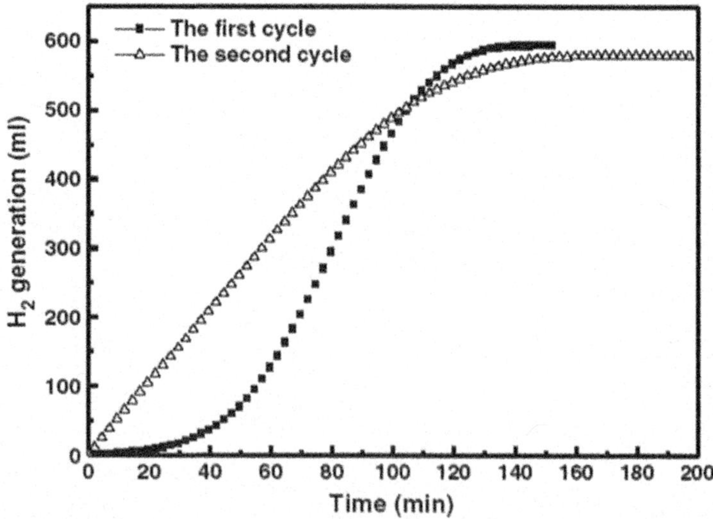

Fig. 15.16 The effect of catalyst amount on hydrogen production. $NaBH_4$:0.5 wt. %, NaOH 5 wt. %, H_2O 50 ml, temperature 30 °C (Reprinted from Ref. [328], Copyright 2012, with permission from Elsevier)

3,641.69 ml min^{-1} g^{-1} (Co) at 30 and 40 °C. After five cycles, ZIF-9 maintains its basic structure and crystallinity, but the long range order of ZIF-9 was destroyed to a certain extent during the operation. In this case, metal centers (Co) form CoB active centers firstly, which can promote the $NaBH_4$ hydrolysis. Thus, ZIF-9 plays both the roles of a support and a precursor for preparing the catalyst.

15.5 Conclusions and Future Outlook

The chapter reports recent developments of hydrogen production from hydrides with the help of MOFs and MOF-based materials. Chemical hydrides usually possess excellent hydrogen gravimetric storage capacities, and some of them show volumetric densities even surpassing that of liquid hydrogen. A group of hydrides stand as promising candidate for competitive hydrogen storage with reversible hydrogen for onboard applications. While many of these systems are commercially available, they often suffer from high thermal stability and poor reversibility under moderate conditions as well as sensitivity to oxygen and water. The chapter reports recent developments of MOFs as platforms for chemical hydrogen production from hydrides on properties including hydrogen release capacity, kinetics, cyclic behavior, toxicity, pressure, and thermal behavior. Many exciting developments emerged in the field of hydrogen generation from chemical hydrides, whereas the area of MOF-based catalysis for hydrogen generation reactions is still in its infancy. For most of the reactions, MOFs play important roles in controlling the reaction kinetics and product selectivity of hydrogen generation from chemical hydrides. Most reports have focused on simply documenting catalytic dehydrogenation behavior rather than on demonstrating catalytic chemistry that may prove useful to chemists working in areas other than porous materials, such as zeolites, mesoporous silica, and carbon nanotubes.

In addition, hydrogen is plentiful but is bound up primarily in water in nature. Economical and environmentally clean production of hydrogen from water is crucial to hydrogen-based power generation. Efficient conversion of sunlight to hydrogen by splitting water through photovoltaic cells driving electrolysis or through direct photocatalysis at energy costs competitive with fossil fuels is a major enabling milestone for a viable hydrogen economy. In the future, hydrogen is produced from water using sunlight and specialized microorganisms, such as green algae and cyanobacteria. Just as plants produce oxygen during photosynthesis, these microorganisms consume water and produce hydrogen as a by-product of their natural metabolic processes. Photobiological water splitting is also a long-term technology. Currently, the microbes split water too slowly to be used for efficient, commercial hydrogen production. Scientists are searching ways to modify the microorganisms and to identify other naturally occurring microbes that can produce hydrogen at higher rates. Photobiological water splitting is in the very early stages of research but offers long-term potential for sustainable hydrogen production with low environmental impact. Taking cues from these various natural processes, modeling and simulation approaches may be employed for rational design of hydride materials for unprecedented capabilities and more selective and high-performance catalysts for improved hydrogen production, storage, and application.

In short, MOFs as catalysts have been devoted to the dehydrogenation from hydrides to decrease their decomposition temperatures and enhance the kinetics and cycle life. However, the state-of-the-art hydrides are still far from meeting every aimed target for their transport applications. Therefore, further research work is needed to achieve the goal by practical application on hydrogenation and thermal and cyclic behavior of hydrides. At present hydride systems are not suitable for large-scale commercial applications. Due to thermodynamic constraints, hydrides are not reversibly hydrogenated under acceptable conditions. The known complex transition metal hydrides suffer from high-cost and low gravimetric loading. However, exceptions for specific hydrides are possible as hydrogen carrier for PEM fuel cells. Fundamental research is necessary to understand the mechanisms involved in the dehydrogenation and rehydrogenation of complex hydrides.

Acknowledgements We gratefully acknowledge the fine work of the talented and dedicated graduate students, postdoctoral fellows, and colleagues who have worked with us in this area and whose names can be found in the references. We are pleased to acknowledge AIST and METI of Japan for financial support. Y. Y. Zhao is grateful to the Project Grants 521 Talents Cultivation of Zhejiang Sci-Tech University and thanks to Zhejiang Provincial Top Key Academic Discipline of Chemical Engineering and Technology and National Natural Science Foundation of China (No. 21273202) for providing financial support for stay in AIST.

References

1. Jiang HL, Xu Q (2011) Porous metal-organic frameworks as platforms for functional applications. Chem Commun 47:3351–3370
2. Suh MP, Park HJ, Prasad TK et al (2012) Hydrogen storage in metal-organic frame works. Chem Rev 112:782–835
3. Murray LJ, Dinca M, Long JR (2009) Hydrogen storage in metal-organic frameworks. Chem Soc Rev 38:1294–1314
4. Ward MD (2013) Molecular fuel tanks. Science 300:1104–1105
5. Lim W-X, Thornton AW, Hill AJ (2013) High performance hydrogen storage from Be-BTB metal-organic framework at room temperature. Langmuir 29:8524–8533
6. Duan X, Yu J, Cai J et al (2013) A microporous metal-organic framework of a rare sty topology for high CH_4 storage at room temperature. Chem Commun 49:2043–2045
7. Nugent P, Belmabkhout Y, Burd SD et al (2013) Porous materials with optimal adsorption thermodynamics and kinetics for CO_2 separation. Nature 495:80–84
8. Geier SJ, Mason JA, Bloch ED et al (2013) Selective adsorption of ethylene over ethane and propylene over propane in the metal-organic frameworks M_2(dobdc) (M-Mg, Mn, Fe, Co, Ni, Zn). Chem Sci 4:2054–2061
9. Sumida K, Rogow D, Mason JA et al (2012) Carbon dioxide capture in metal-organic frameworks. Chem Rev 112:724–781
10. Herm ZR, Wiers BM, Mason JA et al (2013) Separation of hexane isomers in a metal-organic framework with triangular channels. Science 340:960–964

11. Bloch ED, Queen WL, Krishna R, Krishna R et al (2012) Hydrocarbon separations in a metal-organic framework with open iron (II) coordination sites. Science 335:1606–1610
12. Li J-R, Sculley J, Zhou H-C (2012) Metal-organic frameworks for separations. Chem Rev 112:869–932
13. Yang Q, Liu D, Zhong C et al (2013) Development of computational methodologies for metal-organic frameworks and their application in gas separations. Chem Rev 113:8261–8323
14. Li SL, Xu Q (2013) Metal-organic frameworks as platforms for clean energy. Energy Environ Sci 6:1656–1683
15. Pardo E, Train C, Gontard G et al (2011) High proton conduction in a chiral ferromagnetic metal-organic quartz-like framework. J Am Chem Soc 133:15328–15331
16. Ikezoe Y, Washino G, Uemura T et al (2012) Autonomous motors of a metal-organic framework powered by reorganization of self-assembled peptides at interfaces. Nat Mater 11:1081–1085
17. Cai D, Guo H, Wen L et al (2013) Fabrication of hierarchical architectures of Tb-MOF by a "green coordination modulation method" for the sensing of heavy metal ions. Cryst Eng Commun 15:6702–6708
18. Wright PA (2010) Opening the door to peptide-based porous solids. Science 329:1025–1026
19. Britt D, Tranchemontagne D, Yaghi OM (2008) Metal-organic frameworks with high capacity and selectivity for harmful gases. PNAS 105(33):11623–11627
20. Zou X, Goupil J-M, Thomas S et al (2012) Detection of Harmful gases by copper-containing metal-organic framework films. J Phys Chem C 116:16593–16600
21. Khan NA, Hasan Z, Jhung SH (2013) Adsorptive removal of hazardous materials using metal-organic frameworks (MOFs): a review. J Hazard Mater 244–245:444–456
22. Peterson GW, Rossin JA, DeCoste JB et al (2013) Zirconium hydroxide-metal-organic framework composites for toxic chemical removal. Int Eng Chem Res 52:5462–5469
23. McKinlay AC, Eubank JF, Wuttke S et al (2013) Nitric oxide adsorption and delivery in flexible MIL-88(Fe) metal organic frameworks. Chem Mater 25:1592–1599
24. Dhakshinamoorthy A, Alvaro M, Garcia H (2012) Commercial metal-organic frameworks as heterogeneous catalysts. Chem Commun 48:11275–11288
25. Seo JS, Whang D, Lee H et al (2000) A homochiral metal-organic porous material for enantioselective separation and catalysis. Nature 404:982–986
26. Yoon M, Srirambalaji R, Kim K et al (2012) Homochiral metal-organic frameworks for asymmetric heterogeneous catalysis. Chem Rev 112:1196–1231
27. Cirma A, García H, LIabrés I, Xamena FX (2010) Engineering metal organic frameworks for heterogeneous catalysis. Chem Rev 110:4606–4655
28. Cirujano FG, Leyva-Pérez A, Corna A et al (2013) MOFs as multifunctional catalysts: synthesis of secondary arylamines, quinolines, pyrroles, and arylpyrrolidines over bifunctional MIL-101. ChemCatChem 5:538–549
29. Dhakshinamoorthy A, Opanasenko M, Cejka J et al (2013) Metal organic frameworks as solid catalysts in condensation reactions of carbonyl groups. Adv Synth Catal 355:247–268
30. Lee JY, Farha OK, Roberts J et al (2009) Metal-organic framework materials as catalysts. Chem Soc Rev 38:1450–1459
31. Horcajada P, Gref R, Baati T et al (2012) Metal-organic frameworks in biomedicine. Chem Rev 112(2):1232–1268
32. Cui Y, Yue Y, Qian G et al (2012) Luminescent functional metal-organic frameworks. Chem Rev 112(2):1126–1162
33. Zhang W, Xiong R-G (2012) Ferroelectric metal-organic frameworks. Chem Rev 112:1163–1195
34. Kuppler RJ, Timmons DJ, Fang QR et al (2009) Potential applications of metal-organic frameworks. Coord Chem Rev 253:3042–3066
35. Furukawa H, Cordova KE, O'Keeffe M et al (2013) The chemistry and applications of metal-organic frameworks. Science 341:1230444

36. Liang Z, Du J, Sun L et al (2013) Design and synthesis of two porous metal-organic frameworks with nbo and agw topologies showing high CO_2 adsorption capacity. Inorg Chem 52:10720–10722
37. Latroche M, Surblé S, Serre C et al (2006) Hydrogen storage in the giant-pore metal-organic frameworks MIL-100 and MIL-101. Angew Chem Int Ed 45:8227–8231
38. Li T, Kozlowski MT, Doud EA et al (2013) Stepwise ligand exchange for the preparation of a family of mesoporous MOFs. J Am Chem Soc 135:11688–11691
39. Bew SP, Burrows AD, Düren T et al (2012) Calix[4]areene-based metal-organic frameworks: towards hierarchically porous materials. Chem Commun 48:4824–4826
40. Kim TK, Lee KJ, Cheon JY et al (2013) Nanoporous metal oxides with tunable and nanocrystalline frameworks via conversion of metal-organic frameworks. J Am Chem Soc 135:8940–8946
41. Grobler I, Smith VJ, Bhatt PM et al (2013) Tunable anisotropic thermal expansion of a porous zinc (II) metal-organic framework. J Am Chem Soc 135:6411–6414
42. Wu H, Chua YS, Krungleviciute V et al (2013) Unusual and highly tunable missing-linker defects in zirconium metal-organic framework UiO-66 and their important effects on gas adsorption. J Am Chem Soc 135:10525–10532
43. Cook T, Zheng YR, Stang PJ (2013) Metal-organic frameworks and self-Assembled supramolecular coordination complexes: comparing and contrasting the design, synthesis, and functionality of metal-organic materials. Chem Rev 113:734–777
44. Valtchev V, Tosheva L (2013) Porous nanosized particles: preparation, properties, and applications. Chem Rev 113:6734–6760
45. Jiang HL, Makal TA, Zhou HC (2013) Interpenetration control in metal-organic frameworks for functional applications. Coord Chem Rev 257:2232–2249
46. Wang C, Liu D, Lin W (2013) Metal-organic frameworks as a tunable platform for designing functional molecular materials. J Am Chem Soc 135:13222–13234
47. Eddaoudi M, Kim J, Rosi N et al (2002) Systematic design of pore size and functionality in isoreticular MOFs and their application in methane storage. Science 295:469–472
48. Moon HR, Lim DW, Suh MP (2013) Fabrication of metal nanoparticles in metal-organic frameworks. Chem Soc Rev 42:1807–1824
49. Huang YL, Gong YN, Jiang L et al (2013) A unique magnesium-based 3D MOF with nanoscale cages and temperature dependent selective gas sorption properties. Chem Commun 49:1753–1755
50. Mitra A, Hubley CT, Panda DK et al (2013) Anion-directed assembly of a non-interpenetrated square-grid metal-organic framework with nanoscale porosity. Chem Commun 49:6629–6631
51. Saito M, Toyao T, Ueda K et al (2013) Effect of pore sizes on catalytic activities of arenetricarbonyl metal complexes constructed within Zr-based MOFs. Dalton Trans 42:9444–9447
52. Chae HK, Siberio-Perez DY, Kim J et al (2004) A route to high surface area, porosity and inclusion of large molecules in crystals. Nature 427:23–527
53. Li H, Eddaoudi M, O'Keeffe M et al (1999) Design and synthesis of an exceptionally stable and highly porous metal-organic framework. Nature 402:276–279
54. Farha OK, Eryazici I, Jeong MC et al (2013) Metal-organic framework materials with ultrahigh surface areas: is the sky the limit? J Am Chem Soc 134:15016–15021
55. Rosi NL et al (2003) Hydrogen storage in microporous metal-organic frameworks. Science 300:1127–1129
56. Dalebrook AF, Gan WJ, Grasemann M et al (2013) Hydrogen storage: beyond conventional methods. Chem Commun 49:8735–8751
57. Krishna R, Titus E, Salimian M et al (2012) Hydrogen storage for energy application. In: Liu J (ed) Hydrogen storage. InTech, pp 252–254
58. Arora P, Zhang Z (2004) Battery separators. Chem Rev 104:4419–4462

59. Combelles C, Yahia MB, Pedesseau L et al (2011) Fe^{II}/Fe^{III} mixed-valence state induced by Li-insertion into the metal-organic-framework MIl-53(Fe): a DFT + U study. J Power Sources 196:3426–3432
60. de Combarieu G, Morcrette M, Millange F et al (2009) Influence of the Benzoquinone sorption on the structure and electrochemical performance of the MIL-53 (Fe) hybrid porous material in a lithium-ion battery. Chem Mater 21:1602–1611
61. Wiers BM, Foo M-L, Balsara NP et al (2011) A solid lithium electrolyte via addition of lithium isopropoxide to a metal-organic framework with open metal sites. J Am Chem Soc 133:14522–14525
62. Zhang L, Wu HB, Madhavi S, Hng HH et al (2012) Formation of Fe_2O_3 microboxes with hierarchical shell structures from metal-organic frameworks and their lithium storage properties. J Am Chem Soc 134:17388–17391
63. Shimizu GKH, Taylor JM, Kim SR (2013) Proton conduction with metal-organic frameworks. Science 341:354–356
64. Yoon M, Suh H, Natarajan S et al (2013) Proton conduction in metal-organic frameworks and related modularly built porous solids. Angew Chem Int Ed 52:2688–2700
65. Sen S, Nair NN, Tamada T et al (2012) High proton conductivity by a metal-organic framework incorporating Zn_8O clusters with aligned imidazolium groups decorating the channels. J Am Chem Soc 134:19432–19437
66. Kim SR, Dawson KW, Gelfand BS et al (2013) Enhancing proton conduction in a metal–organic framework by isomorphous ligand replacement. J Am Chem Soc 35:963–966
67. Taylor JM, Dawson KW, Shimizu GKH (2013) A water-stable metal-organic framework with highly acidic pores for proton-conducting applications. J Am Chem Soc 135:1193–1196
68. Morozan A, Jaouen F (2012) Metal organic frameworks for electrochemical applications. Energy Environ Sci 5:9269–9290
69. Díaz R, Orcajo MG, Botas J et al (2012) Co8-MOF-5 as electrode for supercapacitors. Mater Lett 68:126–128
70. Liu B, Shioyama H, Akita T et al (2008) Metal-organic framework as a template for porous carbon synthesis. J Am Chem Soc 130:5390–5391
71. Chaikittisilp W, Hu M, Wang H et al (2012) Nanoporous carbons through direct carbonization of a zeolitic imidazolate framework for supercapacitor electrodes. Chem Commun 48:7259–7261
72. Zhang F, Hao L, Zhang L et al (2011) Solid-state thermolysis preparation of Co_3O_4 nano/micro superstructures from metal-organic framework for supercapacitors. Int J Electrochem Sci 6:2943–2954
73. Stephan D (2013) A step closer to a methanol economy. Nature 495:54–55
74. Kopetz H (2013) Build a biomass energy market. Nature 494:29–31
75. Assadourian E, Prugh T (2013) State of the world 2013: is sustainability still possible? In: Murphy TW Jr (ed) Beyond fossil fuels: assessing energy alternatives, 15th edn. Springer, Heidelberg, pp 172–183
76. Steele BCH, Heinzel A (2001) Materials for fuel-cell technologies. Nature 414:345–352. doi:10.1038/35104620
77. DOE targets for onboard hydrogen storage systems for light-duty vehicles. http://www1.eere.energy.gov/hydrogenandfuelcells/storage/pdfs/targets_onboard_hydro_storage_explanation.pdf. Accessed 16 Nov 2011
78. Satyapal S, Read C, Ordaz G, Thomas G (2006) Annual DOE hydrogen program merit review: hydrogen storage, U.S. Department of Energy, Washington DC. Available at: http://www.hydrogen.energy.gov/pdfs/review06/2_storage_satyapal.pdf
79. Grochala W, Edwards PP (2004) Thermal decomposition of the non-interstitial hydrides for the storage and production of hydrogen. Chem Rev 104:1283–1316
80. Cundy CS, Cox PA (2003) The hydrothermal synthesis of Zeolites: history and development from the earliest days to the present times. Chem Rev 103:663–701

81. Zhou O, Shimoda H, Gao B et al (2002) Materials science of carbon nanotubes: fabrication, integration, and properties of macroscopic structures of carbon nanotubes. Acc Chem Res 35:1045–1053
82. Linares N, Serrano E, Rico M et al (2011) Incorporation of chemical functionalities in the framework of mesoporous silica. Chem Commun 47:9024–9035
83. Zhou C, Fang ZZ, Lu J et al (2013) Thermodynamic and kinetic destabilization of magnesium hydride using Mg–In solid solution alloy. J Am Chem Soc 135:10982–10985
84. Paskevicius M, Sheppard DA, Buckeley CE (2010) Thermodynamic changes in mechanochemically synthesized magnesium hydride nanoparticles. J Am Chem Soc 132:5077–5083
85. Lu J, Choi YJ, Fang ZZ et al (2010) Hydrogenation of nanocrystalline Mg at room temperature in the presence of TiH_2. J Am Chem Soc 132:6616–6617
86. Zhou C, Fang ZZ, Ren C et al (2013) Effect of Ti intermetallic catalysts on hydrogen storage properties of magnesium hydride. J Phys Chem C 117:12973–12980
87. Chen P, Xiong Z, Luo J et al (2002) Interaction of hydrogen with metal nitrides and imides. Nature 420:302–304
88. Yadav M, Xu Q (2012) Liquid-phase chemical hydrogen storage materials. Energy Environ Sci 5:9698–9725
89. Umegaki T, Yan JM, Zhang XB et al (2009) Boran- and nitrogen-based chemical hydrogen storage materials. Int J Hydrog Energy 34:2303–2311
90. Li L, Xu CC, Chen CC et al (2013) Sodium alanate systems for efficient hydrogen storage. Int J Hydrog Energy 38:8798–8812
91. Tan T, Yu X (2013) Chemical regeneration of hydrogen storage materials. RSC Adv 3:23879–23894
92. Moussa G, Moury R, Demirci UB et al (2013) Boron-based hydrides for chemical hydrogen storage. Int J Energy Res 37:825–842
93. Ravnsbak D, Filinchuk Y, Cerenius Y et al (2009) A series of mixed-metal borohydrides. Angew Chem Int Ed 48:6659–6663
94. Orimo S, Nakamori Y, Eliseo JR et al (2007) Complex hydrides for hydrogen storage. Chem Rev 107:4111–4132
95. Nicola CD, Karabach YY, Kirillov AM et al (2007) Supramolecular assemblies of trinuclear triangular copper(II) secondary building units through hydrogen bonds. Generation of different metal-organic frameworks, valuable catalysts for peroxidative oxidative of alkanes. Inorg Chem 46(1):221–230
96. Kida M, Sakagami H, Takahashi N et al (2013) Chemical shift changes and line narrowing in ^{13}C NMR spectra of hydrocarbon clathrate hydrates. J Phys Chem A 117:4108–4114
97. Struzhkin VV, Militzer B, Mao WL et al (2007) Hydrogen storage in molecular clathrates. Chem Rev 107:4133–4151
98. Koh DY, Juwoon HK, Shin W et al (2013) Atomic hydrogen production from semi-clathrate hydrates. J Am Chem Soc 134:5560–5562
99. Oldenhof S, de Bruin B, Lutz M et al (2013) Base-free production of H_2 by dehydrogenation of formic acid using an iridium-bisMETAMORPhos complex. Chem Eur J 19:11507–11511
100. Boddien A, Mellmann D, Gärtner F et al (2011) Efficient of dehydrogenation of formic acid using an iron catalyst. Science 333(23):1733–1736
101. Bi QY, Du XL, Liu YM et al (2012) Efficient subnanometric gold-catalyzed hydrogen generation via formic acid decomposition under ambient conditions. J Am Chem Soc 134:8926–8933
102. Grasemann M, Laurenczy G (2012) Formic acid as a hydrogen source – recent developments and future trends. Energy Environ Sci 5:8171–8181
103. Teichmann D, Arlt W, Wasserscheid P et al (2011) A future energy supply based on Liquid Organic Hydrogen Carriers (LOHC). Energy Environ Sci 4:2767–2773

104. Brückner N, Obesser K, Bösmann A et al (2013) Evaluation of industrially applied heat-transfer fluids as liquid organic hydrogen carrier systems. Chem Sus Chem 7(1):229–235. doi:10.1002/cssc.201300426
105. Teichmann D, Stark K, Müller K et al (2012) Energy storage in residential and commercial buildings via Liquid Organic Hydrogen Carriers (LOHC). Energy Environ Sci 5:9044–9054
106. Eblagon KM, Tam K, Tsang SCE (2012) Comparison of catalytic performance of supported ruthenium and rhodium for hydrogenation of 9-ethylcarbazole for hydrogen storage applications. Energy Environ Sci 5:8621–8630
107. Schlapbach L, Züttel A (2001) Hydrogen-storage materials for mobile applications. Nature 414:353–358
108. Klebanoff LE, Keller JO (2013) 5 years of hydrogen storage research in the U. S. DOE metal hydride center of excellence (MHCoE). Int J Hydrog Energy 38(11):4533–4576
109. Sutton AD, Burrell AK, Dixon DA et al (2011) Regeneration of ammonia borane spent fuel by direct reaction with hydrazine and liquid ammonia. Science 331(6023):1426–1429
110. Lu ZH, Xu Q (2011) Recent progress in boron- and nitrogen-based chemical hydrogen storage. Funct Mater Lett 5(1):1230001-1–1230001-9
111. Jiang HL, Xu Q (2011) Catalytic hydrolysis of ammonia borane for chemical hydrogen storage. Catal Today 170:56–63
112. Lu Z-H, Jiang H-L, Yadav M et al (2012) Synergistic catalysis of Au-Co@SiO2 nanospheres in hydrolytic dehydrogenation of ammonia borane for chemical hydrogen storage. J Mater Chem 22:5065–5071
113. Cohen SM (2012) Postsynthetic methods for the functionalization of metal-organic frameworks. Chem Rev 112(2):970–1000
114. Tanabe KK, Cohen SM (2011) Postsynthetic modification of metal-organic frameworks—a progress report. Chem Soc Rev 40:498–519
115. Wang Z, Cohen SM (2009) Postsynthetic modification of metal-organic frameworks. Chem Soc Rev 38:1315–1329
116. Kong GQ, Ou S, Zou C et al (2012) Assembly and post-modification of a metal-organic nanotube for highly efficient catalysis. J Am Chem Soc 134:19851–19857
117. Fei H, Cahill JF, Prather KA et al (2013) Tandem postsynthetic metal ion and ligand exchange in zeolitic imidazolate frameworks. Inorg Chem 52:4011–4016
118. Genna DT, Wong-Foy AG, Matzge AJ et al (2013) Heterogenization of homogeneous catalysts in metal-organic frameworks via cation exchange. J Am Chem Soc 135:10586–10589
119. Hoskins BF, Robson R (1990) Design and construction of a new class of scaffolding-like materials comprising infinite polymeric frameworks of 3D-linked molecular rods. A reappraisal of the zinc cyanide and cadmium cyanide structures and the synthesis and structure of the diamond-related frameworks $[N(CH_3)_4][Cu^IZn^{II}(CN)_4]$ and $Cu^I[4, 4', 4'', 4'''$-tetracyanotetraphenylmethane$]BF_4 \cdot xC_6H_5NO_2$. J Am Chem Soc 112:1546–1554
120. Fujita M, Kwon YJ, Washizu S et al (1994) Preparation, clathration ability, and catalysis of a two-dimensional square network material composed of cadmium (II) and 4,4′-bipyridine. J Am Chem Soc 116:1151–1152
121. Dhakshinamoorthy A, Opanasenko M, Cejka J et al (2013) Metal organic frameworks as heterogeneous catalysts for the production of fine chemicals. Catal Sci Technol 3:2509–2540
122. Gutowska A, Li L, Shin Y et al (2005) Nanoscaffold mediates hydrogen release and the reactivity of ammonia borane. Angew Chem Int Ed 44(23):3578–3582
123. Zhao J, Shi J, Zhang X et al (2010) A soft hydrogen storage material: poly(methylacrylate)-confined ammonia borane with controllable dehydrogenation. Adv Mater 22:394–397
124. Xia GL, Li L, Guo ZP et al (2013) Stabilization of $NaZn(BH_4)_3$ via nanoconfinement in SBA-15 towards enhanced hydrogen release. J Mater Chem A 1:250–257
125. Appelt C, Slootweg JC, Lammertsma K et al (2013) Reaction of a P/Al-based frustrated lewis pair with ammonia, borane, and amine–boranes: adduct formation and catalytic dehydrogenation. Angew Chem Int Ed 52:4256–4259

126. Tang Z, Chen H, Chen X et al (2012) Graphene oxide based recyclable dehydrogenation of ammonia borane within a hybrid nanostructure. J Am Chem Soc 134:5464–5467
127. Hu MG, Geanangel RA, Wendlandt WW (1978) The thermal decomposition of ammonia borane. Thermochim Acta 23:249–255
128. Guo XD, Zhu GS, Li ZY et al (2006) A lanthanide metal-organic framework with high thermal stability and available Lewis-acid metal sites. Chem Commun 30:3172–3174
129. Li Z, Zhu G, Lu G et al (2010) Ammonia borane confined by a metal-organic framework for chemical hydrogen storage: enhancing kinetics and eliminating ammonia. J Am Chem Soc 132:1490–1491
130. Wahab MA, Zhao H, Yao XD (2012) Nano-confined ammonia borane for chemical hydrogen storage. Front Chem Sci Eng 6:27–33
131. Cakey SR, Wong-Foy AG, Matzger AJ (2008) Dramatic tuning of carbon dioxide uptake via metal substitution in a coordination polymer with cylindrical pores. J Am Chem Soc 130:10870–10871
132. Remy T, Peter SA, Van der Perre S (2013) Selective dynamics CO_2 separations on Mg-MOF-74 at low pressures: a detailed comparison with 13X. J Phys Chem C 117:9301–9310
133. Gadipelli S, Ford J, Zhou W et al (2011) Nanoconfinement and catalytic dehydrogenation of ammonia borane by magnesium-metal-organic-framework-74. Chem Eur J 17:6043–6047
134. Chui SSY, Lo SMF, Charmant JPH et al (1999) A chemically functionalizable nanoporous material $[Cu_3(TMA)_2(H_2O)_3]_n$. Science 283:1148–1150
135. Yue Y, Qiao Z, Fulvio PF et al (2013) Template-free synthesis of hierarchical porous metal-organic frameworks. J Am Chem Soc 135:9572–9575
136. Wu H, Zhou W, Yildirim T (2009) High-capacity methane storage in metal-organic frameworks M_2(dhtp): the important role of open metal sites. J Am Chem Soc 131:4995–5000
137. Rowsell JLC, Yaghi OM (2006) Effects of functionalization, catenation, and variation of the metal oxide and organic linking units on the low-pressure hydrogen adsorption properties of metal-organic frameworks. J Am Chem Soc 128:1304–1315
138. Srinivas G, Ford J, Zhou W, Yildirim T (2012) Zn-MOF assisted dehydrogenation of ammonia borane: enhanced kinetics and clean hydrogen generation. Int J Hydrog Energy 37:3633–3638
139. Millange F, Guillou N, Walton RI et al (2008) Effect of the nature of the metal on the breathing steps in MOFs with dynamic frameworks. Chem Commun 39:4732–4734
140. Scherb C, Schöodel A, Bein T (2008) Directing the structure of metal-organic frameworks by oriented surface growth on an organic monolayer. Angew Chem 120:5861–5863
141. Srinivas G, Travis W, Ford J et al (2013) Nanoconfined ammonia borane in a flexible metal-organic framework Fe-MIL-53: clean hydrogen release with fast kinetics. J Mater Chem A 1:4167–4172
142. Staubitz A, Rbertson APM, Manners I (2010) Ammonia-Borane and related compounds as dihydrogen sources. Chem Rev 110:4079–4124
143. Peng Y, Ben T, Jia Y et al (2012) Dehydrogenation of ammonia borane confined by low density porous aromatic framework. J Phys Chem C 116:25694–25700
144. Férey G, Mellot-Draznieks C, Serre C et al (2005) A chromium terephthalate-based solid with unusually large pore volumes and surface area. Science 310:2040–2042
145. Si X-L, Sun L-X, Xu F, Jiao C-L, Li F, Liu S-S, Zhang J, Song L-F, Jiang C-H, Wang S, Liu Y-L, Sawada Y (2011) Improved hydrogen desorption properties of ammonia borane by Ni-modified metal-organic frameworks. Int J Hydrog Energy 36:6698–6704
146. Bernt S, Guillerm V, Serre C et al (2011) Direct covalent post-synthetic chemical modification of Cr-MIL-101 using nitrating acid. Chem Commun 47:2838–2840
147. Gao L, Li CYV, Yung H, Chan K-Y (2013) A functionalized MIL-101(Cr) metal-organic framework for enhanced hydrogen release from ammonia borane at low temperature. Chem Commun 49:10629–10631

148. Aijaz A, Karkamkar A, Choi YJ et al (2012) Immobilizing highly catalytically active Pt nanoparticles inside the pores of metal-organic framework: a double solvents approach. J Am Chem Soc 134:13926–13929
149. Phan A, Doonan CJ, Uribe-Romo FJ et al (2010) Synthesis, structure, and carbon dioxide capture properties of zeolitic imidazolate frameworks. Acc Chem Res 43(1):58–67
150. Banerjee R, Furukawa H, Britt D et al (2009) Control of pore size and functionality in isoreticular zeolitic imidazolate frameworks and their carbon dioxide selective capture properties. J Am Chem Soc 131:3875–3877
151. Fairen-Jimenez D, Moggach SA, Wharmby MF et al (2011) Opening the gate: framework flexibility in ZIF-8 explored by experiments and simulations. J Am Chem Soc 133:8900–8902
152. Lu G, Hupp JF (2010) Metal-organic frameworks as sensors: a ZIF-8 based Fabry-Pérot devices as a selective sensor for chemical vapors and gases. J Am Chem Soc 132:7832–7833
153. Zhong RQ, Zou RQ, Nakagawa T et al (2012) Improved hydrogen release from ammonia-borane with ZIF-8. Inorg Chem 51:2728–2730
154. Kalidindi SB, Esken D, Fischer RA (2011) B–N chemistry@ZIF-8: dehydrocoupling of dimethylamine borane at room temperature by size-confinement effects. Chem Eur J 17:6594–6597
155. Schüth F, Bogdanović B, Felderhoff M (2004) Light metal hydrides and complex hydrides for hydrogen storage. Chem Commun 20:2249–2258
156. Gilliard RJ Jr, Abraham MY, Wang Y et al (2012) Carbene-stabilized beryllium borohydride. J Am Chem Soc 134:9953–9955
157. Sun T, Wang H, Zhang QA et al (2011) Synergetic effects of hydrogenated Mg_3La and $TiCl_3$ on the dehydrogenation of $LiBH_4$. J Mater Chem 21:9179–9184
158. Wang FH, Liu YF, Gao MX et al (2009) Formation reactions and the thermodynamics and kinetics of dehydrogenation reaction of mixed alanate Na_2LiAlH_6. J Phys Chem C 113:7978–7984
159. Nakamori Y, Orimo S (2008) Borohydrides as hydrogen storage materials. In: Walker G (ed) Solid-state hydrogen storage. Woodhead Publishing, Cambridge, pp 420–449
160. Zhang Y, Ding H, Liu C et al (2013) Significant effects of graphite fragments on hydrogen storage performances of LiBH4: a first-principles approach. Int J Hydrog Energy 38:13717–13727
161. Rafieudedin QXH, Li P et al (2011) Hydrogen sorption improvement of $LiAlH_4$ catalyzed by Nb_2O_5 and Cr_2O_3 nanoparticles. J Phys Chem C 115:13088–13099
162. Muir SS, Yao X (2011) Progress in sodium borohydride as a hydrogen storage material: development of hydrolysis catalysts and reaction system. Int J Hydrog Energy 36:5983–5997
163. Xiong R, Sang G, Yan X et al (2013) Separation and characterization of the active species in Ti-doped $NaAlH_4$. Chem Commun 49:2046–2048
164. Michel KJ, Ozolin V (2011) Native defect concentrations in $NaAlH_4$ and Na_3AlH_6. J Phys Chem C 115:21443–21453
165. Kurban Z, Lovell A, Bennington SM et al (2010) A solution selection model for coaxial electro spinning and its application to nanostructured hydrogen storage materials. J Phys Chem C 114:21201–21213
166. Nielsen TK, Besenbacherb F, Jensen TR (2011) Nanoconfined hydrides for energy storage. Nanoscale 3:2086–2098
167. Baldé CP, Hereijgers BPC, Bitter JH et al (2008) Sodium alanate nanoparticle-linking size to hydrogen storage properties. J Am Chem Soc 130:6761–6765
168. Kumar LH, Rao CV, Viswanathan B (2013) Catalytic effects of nitrogen-doped grapheme and carbon nanotube additives on hydrogen storage properties of sodium alanate. J Mater Chem A 1:3355–3361
169. Hsu C-P, Jiang D-H, Lee S-L et al (2013) Buckyball-, carbon nanotube-, graphite-, and graphene-enhanced dehydrogenation of lithium aluminum hydride. Chem Commun 49:8845–8847

170. Hazrati E, Brpcls G, Wijsde GA (2012) First-principles study of $LiBH_4$ nanoclusters and their hydrogen storage properties. J Phys Chem C 116:18038–18047
171. Gao J, Ngene P, Lindemann I et al (2012) Enhanced reversibility of H2 sorption in nanoconfined complex metal hydrides by alkali metal addition. J Mater Chem 22:13209–13215
172. Xia G, Meng Q, Guo Z et al (2013) Nanoconfinement significantly improves the thermodynamics and kinetics of co-infiltrated 2LiBH4–LiAlH4 composites: stable reversibility of hydrogen absorption/resorption. Acta Mater 61:6882–6893
173. Nielsen TK, Javadian P, Polanski M et al (2014) Nanoconfined $NaAlH_4$: prolific effects from increased surface area and pore volume. Nanoscale 6:599–607
174. Vajeeston P, Sartori S, Ravindran P et al (2012) MgH_2 in carbon scaffolds: a combined experimental theoretical investigation. J Phys Chem C 116:21139–21147
175. Nielsen TK, Javadian P, Polanski M et al (2013) Nanoconfined $NaAlH_4$: determination of distinct prolific effects from pore size, crystallite size, and surface interactions. J Phys Chem C 116:21046–21051
176. Wahab MA, Jia Y, Jia D et al (2013) Enhanced hydrogen desorption from $Mg(BH_4)_2$ by combing nanoconfinement and a Ni catalyst. J Mater Chem A 1:3471–3478
177. Ward PA, Teprovich JA Jr, Peters B et al (2013) Reversible hydrogen storage in a $LiBH_4$-C_{60} nanocomposite. J Phys Chem C 117:22569–22575
178. Ngene P, Adelhelm P, Beale AM et al (2010) $LiBH_4$/SBA-15 nanocomposites prepared by melt infiltration under hydrogen pressure: synthesis and hydrogen sorption properties. J Phys Chem C 114:6163–6168
179. Guo L, Jiao L, Li L et al (2013) Enhanced desorption properties of $LiBH_4$ incorporated into mesoporous TiO_2. Int J Hydrog Energy 38:162–168
180. Schlesinger HI, Brown HC, Finholt AE et al (1940) Metallo borohydrides. III. lithium borohydride. J Am Chem Soc 62:3429–3435
181. Züttel A, Wenger P, Rentsch S et al (2003) $LiBH_4$ a new hydrogen storage material. J Power Sources 118:1–7
182. Mauron P, Buchter F, Friedrichs O et al (2008) Stability and reversibility of $LiBH_4$. J Phys Chem B 112:906–910
183. Yan Y, Remhof A, Hwang SJ et al (2012) Pressure and temperature dependence of the decomposition pathway of $LiBH_4$. Phys Chem Chem Phys 14:6514–6519
184. Vajo JJ, Olsen GL (2007) Tetrahydroborates as new hydrogen storage materials. Scr Mater 56:829–834
185. Züttel A, Rentsch S, Fischer P et al (2003) Hydrogen storage properties of $LiBH_4$. J Alloys Compd 356–357:515–520
186. Zhang Y, Zhang WS, Wang AQ et al (2007) $LiBH_4$ Nanoparticles supported by disordered mesoporous carbon: hydrogen storage performances and destabilization mechanisms. Int J Hydrog Energy 32:3976–3980
187. Yu XB, Wu Z, Chen QR, Huang TS et al (2007) Improved hydrogen storage properties of $LiBH_4$ destabilized by carbon. Appl Phys Lett 90:034106
188. Fang ZZ, Kang XD, Wang P et al (2008) Improved reversible dehydrogenation of lithium borohydride by milling with as-prepared dingle-walled carbon nanotubes. J Phys Chem C 112:17023–17029
189. Wang PJ, Fang ZZ, Ma LP et al (2008) Effect of SWNTs on the reversible hydrogen storage properties of $LiBH_4 - MgH_2$ composite. Int J Hydrog Energy 33:5611–5616
190. Gross AF, Vajo JJ, Van Atta SL et al (2008) Enhanced hydrogen storage kinetics of $LiBH_4$ in nanoporous carbon scaffolds. J Phys Chem C 112:5651–5657
191. Liu X, Peaslee D, Jost CZ et al (2010) Controlling the decomposition pathway of $LiBH_4$ via confinement in highly ordered nanoporous carbon. J Phys Chem C 114:14036–14041
192. Wang PJ, Fang ZZ, Ma LP et al (2010) Effect of carbon addition on hydrogen storage behaviors of Li–Mg–B–H system. Int J Hydrog Energy 35:3072–3075

193. Zhao-Karger Z, Witter R, Bardají EG et al (2013) Altered reaction pathways of eutectic $LiBH_4$–$Mg(BH_4)_2$ by nanoconfinement. J Mater Chem A 1:3379–3386
194. Liu X, Peaslee D, Jost CZ et al (2011) Systematic pore-size effects of nanoconfinement of $LiBH_4$: elimination of diborane release and tunable behavior for hydrogen storage applications. Chem Mater 23:1331–1336
195. Sun T, Liu J, Jia Y et al (2012) Confined $LiBH_4$: enabling fast hydrogen release at 100 °C. Int J Hydrog Energy 37:18920–18926
196. Fang ZZ, Wang P, Rufford TE et al (2008) Kinetic- and thermodynamic-based improvements of lithium borohydride incorporated into activated carbon. Acta Mater 56:6257–6263
197. Remhof A, Mauron P, Züttel A et al (2013) Hydrogen dynamics in nanoconfined lithiumborohydride. J Am Chem Soc 117:3789–3798
198. Nielsen TK, Bosenberg U, Dornheim M et al (2010) A reversible nanoconfined chemical reaction. ACS Nano 4:3903–3908
199. Li C, Peng P, Zhou D et al (2011) Research and progress in $LiBH_4$ for hydrogen storage: a review. Int J Hydrog Energy 36:14512–14526
200. Sun W, Li S, Mao J et al (2011) Nanoconfinement of lithium borohydride in Cu-MOFs towards low temperature dehydrogenation. Dalton Trans 40:5673–5676
201. Gross KJ, Majzoub EH, Spangler SW (2003) The effects of titanium precursors on hydriding properties of alanates. J Alloys Compd 356–357:423–428
202. Claudia W, Pommerin A, Felderhoff M et al (2003) On the state of the titanium and zirconium in Ti- or Zr-doped $NaAlH_4$ hydrogen storage material. Phys Chem Chem Phys 5:5149–5153
203. Bogdanović B, Schwickardi M (1997) Ti-doped alkali metal aluminium hydrides as potential novel reversible hydrogen storage materials. J Alloys Compd 253:1–9
204. Bogdanović B, Brand RA, Marjanović A et al (2000) Metal-doped sodium aluminium hydrides as potential new hydrogen storage materials. J Alloys Compd 302:36–58
205. Vegge T (2006) Equilibrium structure and Ti-catalyzed H_2 desorption in $NaAlH_4$ nanoparticles from density functional theory. Phys Chem Chem Phys 8:4853–4861
206. Zhao Y, Wang H, Guo J (2011) Energetics and structure of single Ti defects and their influence on the decomposition of $NaAlH_4$. Phys Chem Chem Phys 13:552–562
207. Frankcombe TJ (2012) Proposed mechanisms for the catalytic activity of Ti in $NaAlH_4$. Chem Rev 112:2164–2178
208. Wang P, Kang XD, Cheng HM (2005) Exploration of the nature of active Ti species in metallic Ti-doped $NaAlH_4$. J Phys Chem B 109:20131–20136
209. Frankcombe TJ (2013) Catalyzed Rehydrogenation of $NaAlH_4$: Ti and friends are active on NaH surfaces; Pt and friends are not. J Phys Chem C 117:8150–8155
210. Xiao X, Fan X, Yu K et al (2009) Catalytic mechanism of new TiC-doped sodium alanate for hydrogen storage. J Phys Chem C 113:20745–20751
211. Li L, Qiu F, Wang Y et al (2012) TiN catalyst for the reversible hydrogen storage performance of sodium alanate system. J Mater Chem 22:13782–13787
212. Suttisawat Y, Rangsunvigit P, Kitiyanan B et al (2007) Catalytic effect of Zr and Hf on hydrogen desorption/absorption of $NaAlH_4$ and $LiAlH_4$. Int J Hydrog Energy 32:1277–1285
213. Fan X, Xiao X, Chen L (2013) Significantly improved hydrogen storage properties of $NaAlH_4$ catalyzed by Ce-based nanoparticles. J Mater Chem A 1:9752–9759
214. Anton DL (2003) Hydrogen desorption kinetics in transition metal modified $NaAlH_4$. J Alloys Compd 356–357:400–404
215. Wang J, Ebner AD, Zidan R et al (2005) Synergistic effects of co-dopants on the dehydrogenation kinetics of sodium aluminum hydride. J Alloys Compd 391:245–255
216. Fan X, Xiao X, Chen L et al (2011) Enhanced hydriding–dehydriding performance of $CeAl_2$-doped $NaAlH_4$ and the evolvement of Ce-containing species in the cycling. J Phys Chem C 115:2537–2543

217. Fan X, Xiao X, Chen L et al (2011) Hydriding-dehydriding kinetics and the microstructure of La- and Sm-doped $NaAlH_4$ prepared via direct synthesis method. Int J Hydrog Energy 36:10861–10869
218. Gao J, Adelhelm P, Verkuijlen MHW et al (2010) Confinement of $NaAlH_4$ in nanoporous carbon: impact on H_2 release, reversibility, and thermodynamics. J Phys Chem C 114:4675–4682
219. Lohstroh W, Roth A, Hahn H et al (2010) Thermodynamic effects in nanoscale $NaAlH_4$. Chem Phys Chem 11:789–792
220. Stephens RD, Gross AF, Van Atta SL et al (2009) The kinetic enhancement of hydrogen cycling in $NaAlH_4$ by melt infusion into nanoporous carbon aerogel. Nanotechnology 20:204018
221. Majzoub EH, Zhou F, Ozoliņš V (2011) First-principles calculated phase diagram for nanoclusters in the Na–Al–H system: a single-step decomposition pathway for $NaAlH_4$. J Phys Chem C 115:2636–2643
222. Mueller T, Ceder G (2010) Effect of particle size on hydrogen release from sodium alanate nanoparticles. ACS Nano 4:5647–5656
223. Ngene P, van den Berg R, Verkuijlen MHW et al (2011) Reversibility of the hydrogen desorption from $NaBH_4$ confinement in nanoporous. Energy Environ Sci 4:4108–4115
224. Minella CB, Lindemann I, Nolis P et al (2013) $NaBH_4$ confined in ordered mesoporous carbon. Int J Hydrog Energy 38:8829–8837
225. Adelhelm P, de Jong KP, de Petra PE (2009) How intimate contact with nanoporous carbon benefits the reversible hydrogen desorption from NaH and $NaAlH_4$. Chem Commun 41:6261–6263
226. Bhakta RK, Herberg JL, Jacobs B et al (2009) Metal-organic frameworks as templates for nanoscale $NaAlH_4$. J Am Chem Soc 131:13198–13199
227. Nhakta RK, Maharrey S, Stavila V et al (2012) Thermodynamics and kinetics of $NaAlH_4$ nanocluster decomposition. Phys Chem Chem Phys 14:8160–8169
228. Stavila V, Bhakta RK, Alam TM et al (2012) Reversible hydrogen storage by $NaAlH_4$ confined within a titanium-functionalized MOF-74(Mg) nanoreactor. ACS Nano 6:9807–9817
229. Kelly KL, Coronado E, Zhao LL et al (2003) The optical properties of metal nanoparticles: the influence of size, shape, and dielectric environment. J Phys Chem B 107:668–677
230. Tao AR, Habas S, Yang P (2008) Shape control of colloidal metal nanocrystals. Small 4:310–325
231. Seo JK, Khetan A, Seo MH, Kim H et al (2013) First-principles thermodynamic study of the electrochemical stability of Pt nanoparticles in fuel cell applications. J Power Sources 238:137–143
232. Kim YH, Zhang L, Yu T et al (2013) Droplet-based microreactors for continuous production of palladium nanocrystals with controlled sizes and shapes. Small 9:3462–3467
233. Chen Y, Fernandes AA, Erbe A (2013) Control of shape and surface crystallography of gold nanocrystals for electrochemical applications. Electrochim Acta 113:810–816
234. Gong J, Zhou F, Li Z et al (2013) Controlled synthesis of non-epitaxially grown Pd@Ag core–shell nanocrystals of interesting optical performance. Chem Commun 49:4379–4381
235. Kang Y, Li M, Cai Y et al (2013) Heterogeneous catalysts need not be so "Heterogeneous": monodisperse Pt nanocrystals by combining shape-controlled synthesis and purification by colloidal recrystallization. J Am Chem Soc 135:2741–2747
236. Quan Z, Wang Y, Fang J (2013) High-index faceted noble metal nanocrystals. Acc Chem Res 46:191–202
237. Zhang Q, Le I, Joo JB (2013) Core-shell nanostructured catalysts. Acc Chem Res 46:1816–1824
238. Guo H, Chen Y, Ping H et al (2013) Facile synthesis of Cu and Cu@Cu–Ni nanocubes and nanowires in hydrophobic solution in the presence of nickel and chloride ions. Nanoscale 5:2394–2402

239. Sun Y (2013) Controlled synthesis of colloidal silver nanoparticles in organic solutions: empirical rules for nucleation engineering. Chem Soc Rev 42:2497–2511
240. Burda C, Chen X, Narayanan R et al (2005) Chemistry and properties of nanocrystals of different shapes. Chem Rev 105:1025–1102
241. Moghimi N, Abdellah M, Thomas JP et al (2013) Bimetallic FeNi concave nanocubes and nanocages. J Am Chem Soc 135:10958–10961
242. Aslan K, Wu M, Lakowicz JR et al (2007) Fluorescent core-shell Ag@SiO_2 nanocomposites for metal-enhanced fluorescence and single nanoparticle sensing platforms. J Am Chem Soc 129:1524–1525
243. Joo SH, Park JY, Tsung C-K et al (2009) Thermally stable Pt/mesoporous silica core-shell nanocatalysts for high-temperature reactions. Nat Mater 8:126–131
244. Joo SH, Choi SJ, Oh I et al (2001) Ordered nanoporous arrays of carbon supporting high dispersions of platinum nanoparticles. Nature 412:169–172
245. Song H, Rioux RM, Hoefelmeyer JD et al (2006) Hydrothermal growth of mesoporous SBA-15 silica in the presence of PVP-stabilized Pt nanoparticles: synthesis, characterization, and catalytic properties. J Am Chem Soc 128:3027–3037
246. Lim D-W, Yoon JW, Ryu KY et al (2012) Magnesium nanocrystals embedded in a metal-organic framework: hybrid hydrogen storage with synergistic effect on physic and chemisorption. Angew Chem Int Ed 51:9814–9817
247. Park YK, Choi SB, Nam HJ et al (2010) Catalytic nickel nanoparticles embedded in a mesoporous metal-organic framework. Chem Commun 46:3086–3088
248. Yadav M, Aijaz A, Xu Q (2012) Highly catalytically active palladium nanoparticles incorporated inside metal-organic framework pores by double solvents method. Funct Mater Lett 5:1250039–1250042
249. Hermannsdörfer J, Friedrich M, Kempe R (2013) Colloidal size effect and metal-particle migration in M@MOF/PCP catalysis. Chem Eur J 19:13652–13657
250. Li Z, Zeng HC (2013) Surface and bulk integrations of single-layered Au or Ag nanoparticles onto designated crystal planes {110} or {100} of ZIF-8. Chem Mater 25:1761–1768
251. Wu R, Qian X, Zhou K et al (2013) Highly dispersed Au nanoparticles immobilized on Zr-based metal-organic frameworks as heterostructured catalyst for CO oxidation. J Mater Chem A 1:14294–14299
252. Pan Y, Yuan B, Li Y et al (2010) Multifunctional catalysis by Pd@MIL-101: one-step synthesis of methyl isobutyl ketone over palladium nanoparticles deposited on a metal-organic framework. Chem Commun 46:2280–2282
253. Li H, Zhu Z, Zhang F et al (2011) Palladium nanoparticles confined in the cages of MIL-101: an efficient catalyst for the one-pot indole synthesis in water. ACS Catal 1:1604–1612
254. Shen L, Wu W, Liang R et al (2013) Highly dispersed palladium nanoparticles anchored on UiO-66(NH_2) metal-organic framework as a reusable and dual functional visible-light-driven photocatalyst. Nanoscale 5:9374–9382
255. Meilikhov M, Yusenko K, Esken D et al (2010) Selective palladium-loaded MIL-101 catalysts. Eur J Inorg Chem 2010:3701–3714
256. Cheon YE, Suh MP (2009) Enhanced hydrogen storage by palladium nanoparticles fabricated in a redox-active metal-organic framework. Angew Chem Int Ed 48:2899–2903
257. Aijaz A, Akita T, Tsumori N, Xu Q (2013) Metal-organic framework-immobilized polyhedral metal nanocrystals: reduction at solid–gas interface, metal segregation, core–shell structure, and high catalytic activity. J Am Chem Soc 135:16356–16359
258. Khajavi H, Stil HA, Kuipers PCEH et al (2013) Shape and transition state selective hydrogenations using egg-shell Pt-MIL-101(Cr) catalyst. ACS Catal 3:2617–2626
259. Huang Y, Lin Z, Cao R (2011) Palladium nanoparticles encapsulated in a metal-organic framework as efficient heterogeneous catalysts for direct C2 arylation of indoles. Chem Eur J 17:12706–12712
260. Yadav M, Xu Q (2013) Catalytic chromium reduction using formic acid and metal nanoparticles immobilized in a metal-organic framework. Chem Commun 49:3327–3329

261. Zlotea C, Campesi R, Cuevas F et al (2013) Pd nanoparticles embedded into a metal-organic framework: synthesis, structural characteristics, and hydrogen sorption properties. J Am Chem Soc 132:2991–2997
262. Wu F, Qiu L-G, Ke F, Jiang X (2013) Copper nanoparticles embedded in metal-organic framework MIL-101(Cr) as a high performance catalyst for reduction of aromatic nitro compounds. Inorg Chem Commun 32:5–8
263. Jiang HL, Akita T, Ishida T et al (2011) Synergistic catalysis of Au@Ag core-shell nanoparticles stabilized on metal-organic framework. J Am Chem Soc 133:1304–1313
264. Long J, Liu H, Liao S et al (2013) Selective oxidation of saturated hydrocarbons using Au–Pd alloy nanoparticles supported on metal-organic frameworks. ACS Catal 3:647–654
265. Chen G, Wu S, Liu H et al (2013) Palladium supported on an acidic metal-organic framework as an efficient catalyst in selective aerobic oxidation of alcohols. Green Chem 15:230–235
266. Antonels NC, Meijboom R (2013) Preparation of well-defined dendrimer Encapsulated ruthenium nanoparticles and their evaluation in the reduction of 4-nitrophenol according to the Langmuir–Hinshelwood approach. Langmuir 29:13433–13442
267. Zhao Y, Zhang J, Song J et al (2011) Ru nanoparticles immobilized on metal-organic framework nanorods by supercritical CO_2-methanol solution: highly efficient catalyst. Green Chem 13:2078–2082
268. Li PZ, Aranishi K, Xu Q (2012) ZIF-8 immobilized nickel nanoparticles: highly effective catalysts for hydrogen generation from hydrolysis of ammonia borane. Chem Commun 48:3173–3175
269. Huang XC, Lin YY, Zhang JP et al (2006) Ligand-directed strategy for zeolite-type metal-organic frameworks: Zinc(II) imidazolates with unusual zeolitic topologies. Angew Chem Int Ed 45:1557–1559
270. Hermes S, Schröter MK, Schmid R et al (2005) Metal@MOF: loading of highly porous coordination polymers host lattices by metal organic chemical vapor deposition. Angew Chem Int Ed 44:6237–6241
271. Gu X, Lu ZH, Jiang HL et al (2011) Synergistic catalysis of metal-organic framework-immobilized Au-Pd nanoparticles in dehydrogenation of formic acid for chemical hydrogen storage. J Am Chem Soc 133:11822–11825
272. Zahmakiran M, Ozkar S (2013) Transition metal nanoparticles in catalysis for the hydrogen generation from the hydrolysis of ammonia borane. Top Catal 56:1171–1183
273. Chandra M, Xu Q (2007) Room temperature hydrogen generation from aqueous ammonia-borane using noble metal nano-clusters as highly active catalysts. J Power Sources 168:135–142
274. Zhu Q-L, Li J, Xu Q (2013) Immobilizing metal nanoparticles to metal-organic frameworks with size and location control for optimizing catalytic performance. J Am Chem Soc 135:10210–10213
275. Yan J-M, Zhang X-B, Han S et al (2008) Iron-nanoparticle-catalyzed hydrolytic dehydrogenation of ammonia borane for chemical hydrogen storage. Angew Chem Int Ed 47:2287–2289
276. Yan J-M, Zhang X-B, Akita T et al (2010) One-step seeding growth of magnetically recyclable Au@Co core−shell nanoparticles: highly efficient catalyst for hydrolytic dehydrogenation of ammonia borane. J Am Chem Soc 132:5326–5327
277. Li PZ, Aijaz A, Xu Q (2012) Highly dispersed surfactant – free nickel nanoparticles and their remarkable catalytic activity in the hydrolysis of ammonia borane for hydrogen generation. Angew Chem Int Ed 47:6753–6756
278. Sanyal U, Demirci UB, Jagirdar BR et al (2011) Hydrolysis of ammonia borane as a hydrogen source: fundamental issues and potential solutions towards implementation. ChemSusChem 4:1731–1739
279. Tranchmontagne DJ, Hunt JR, Yaghi OM (2008) Room temperature synthesis of metal-organic frameworks: MOF-5, MOF-74, MOF-177, MOF-199, and IRMOF-0. Tetrahedron 64:8553–8557

280. Gucan M, Zahmakiran M, Özkar S (2014) Palladium(0) nanoparticles supported on metal organic framework as highly active and reusable nanocatalyst in dehydrogenation of dimethylamine-borane. Appl Catal B Environ 147:394–401
281. Zheng M, Cheng R, Chen X et al (2005) A novel approach for CO-free H2 production via catalytic decomposition of hydrazine. Int J Hydrog Energy 30:1081–1089
282. Chen X, Zhang T, Xia L et al (2002) Catalytic decomposition of hydrazine over supported molybdenum nitride catalysts in a monopropellant thruster. Catal Lett 79:21–25
283. Zheng M, Chen X, Cheng N et al (2005) Catalytic decomposition of hydrazine on iron nitride catalysts. Catal Commun 7:187–191
284. Chen X, Zhang T, Ying P et al (2002) A novel catalyst for hydrazine decomposition: molybdenum carbide supported on γ-Al_2O_3. Chem Commun 2:288–289
285. Chen X, Zhang T, Zheng M et al (2004) The reaction route and active site of catalytic decomposition of hydrazine over molybdenum nitride catalyst. J Catal 224:473–478
286. Al-Haydari YK, Saleh JM, Matloob MH (1985) Adsorption and decomposition of hydrazine on metal films of iron, nickel, and copper. J Phys Chem 89:3286–3290
287. Alberas DJ, Kiss J, Liu ZM et al (1992) Surface chemistry of hydrazine on Pt (111). Surf Sci 278:51–61
288. Prasad J, Gland JL (1991) Hydrazine decomposition on a clean rhodium surface: a temperature programmed reaction spectroscopy study. Langmuir 7:722–726
289. Hazari N (2010) Homogeneous iron complexes for the conversion of dinitrogen into ammonia and hydrazine. Chem Soc Rev 39:4044–4056
290. Chin JM, Schrock RR, Müller P (2010) Synthesis of diamidopyrrolyl molybdenum complexes relevant to reduction of dinitrogen to ammonia. Inorg Chem 49:7904–7916
291. Murakami T, Nishikiori T, Nohira T, Ito Y (2003) Electrolytic synthesis of ammonia in molten salts under atmospheric pressure. J Am Chem Soc 125:334–335
292. Sridhar S, Srinivasan T, Virendra U, Khan AA (2003) Pervaporation of ketazine aqueous layer in production of hydrazine hydrate by peroxide process. Chem Eng J 94:51–56
293. Hayashi H (1998) Hydrazine synthesis: commercial routes, catalysis and intermediates. Res Chem Intermed 24:183–196
294. Hayashi H (1990) Hydrazine synthesis by a catalytic oxidation process. Catal Rev 32:229–277
295. Singh KS, Zhang X-B, Xu Q (2009) Room-temperature hydrogen generation from hydrous hydrazine for chemical hydrogen storage. J Am Chem Soc 131:9894–9895
296. Singh KS, Xu Q (2010) Bimetallic Ni–Pt nanocatalysts for selective decomposition of hydrazine in aqueous solution to hydrogen at room temperature for chemical hydrogen storage. Inorg Chem 49:6148–6152
297. Singh KS, Xu Q (2010) Bimetallic nickel-iridium nanocatalysts for hydrogen generation by decomposition of hydrous hydrazine. Chem Commun 46:6545–6547
298. Singh KS, Iizuka Y, Xu Q (2011) Nickel-palladium nanoparticle catalyzed hydrogen generation from hydrous hydrazine for chemical hydrogen storage. Int J Hydrog Energy 36:11794–11801
299. Singh KS, Lu ZH, Xu Q (2011) Temperature-induced enhancement of catalytic performance in selective hydrogen generation from hydrous hydrazine with Ni-based nanocatalysts for chemical hydrogen storage. Eur J Inorg Chem 2011:2232–2237
300. Singh KS, Singh KA, Aranishi K, Xu Q (2011) Noble-metal-free bimetallic nanoparticle-catalyzed selective hydrogen generation from hydrous hydrazine for chemical hydrogen storage. J Am Chem Soc 133:19638–19641
301. Singh KS, Xu Q (2009) Complete conversion of hydrous hydrazine to hydrogen at room temperature for chemical hydrogen storage. J Am Chem Soc 131:18032–18033
302. Singh KA, Xu Q (2013) Synergistic catalysis over bimetallic alloy nanoparticles. ChemCatChem 5:652–676

303. Wang J, Zhang X, Wang Z, Wang L, Zhang Y (2012) Rhodium–nickel nanoparticles grown on graphene as highly efficient catalyst for complete decomposition of hydrous hydrazine at room temperature for chemical hydrogen storage. Energy Environ Sci 5:6885–6888
304. Tong DG, Tang DM, Chu W, Gua GF, Wu P (2013) Monodisperse Ni_3Fe single-crystalline nanospheres as a highly efficient catalyst for the complete conversion of hydrous hydrazine to hydrogen at room temperature. J Mater Chem A 1:6425–6432
305. Singh KA, Xu Q (2013) Metal-organic framework supported bimetallic Ni-Pt nanoparticles as high-performance catalysts for hydrogen generation from hydrazine in aqueous solution. Chem Cat Chem 5:3000–3004
306. Hayashi H, Côté AP, Furukawa H et al (2007) Zeolite imidazolate frameworks. Nat Mater 6:501–506
307. Johnson TC, Morris DJ, Wills M (2010) Hydrogen generation from formic acid and alcohols using homogeneous catalysts. Chem Soc Rev 39:81–88
308. Joó F (2008) Breakthroughs in hydrogen storage – formic acid as a sustainable storage material for hydrogen. ChemSusChem 1:805–808
309. Zhou X, Huang Y, Xing W, Liu C, Liao J, Lu T (2008) High-quality hydrogen from the catalyzed decomposition of formic acid by Pd–Au/C and Pd–Ag/C. Chem Commun 2008:3540–3542
310. Himeda Y, Miyazawa S, Horose T (2011) Interconversion between formic acid and H_2/CO_2 using rhodium and ruthenium catalysts for CO_2 fixation and H_2 storage. ChemSusChem 4:487–493
311. Faur Ghenciu A (2002) Review of fuel processing catalysts for hydrogen production in PEM fuel cell systems. Curr Opin Solid State Mater Sci 6:389–399
312. Morris DJ, Clarkson GJ, Wills M (2009) Insights into hydrogen generation from formic acid using ruthenium complexes. Organometallics 28:4133–4140
313. Gao Y, Kuncheria J, Puddephatt RJ, Yap GPA (1998) An efficient binuclear catalyst for decomposition of formic acid. Chem Commun 21:2365–2366
314. Fukuzumi S, Kobayashi T, Suenobu T (2010) Unusually large tunneling effect on highly efficient generation of hydrogen and hydrogen isotopes in pH-selective decomposition of formic acid catalyzed by a heterodinuclear iridium–ruthenium complex in water. J Am Chem Soc 132:1496–1497
315. Tedsree K, Li T, Jones S, Chan CWA, Yu KMK, Bagot PAJ, Marquis EA, Smith GDW, Tsang SCE (2011) Hydrogen production from formic acid decomposition at room temperature using a Ag–Pd core–shell nanocatalyst. Nat Nano 6:302–307
316. Gan W, Dyson PJ, Laurenczy G (2013) Heterogeneous silica-supported ruthenium phosphine catalysts for selective formic acid decomposition. ChemCatChem 10:3124–3130
317. Bulushev DA, Beloshapkin S, Ross JRH (2010) Hydrogen from formic acid decomposition over Pd and Au catalysts. Catal Today 154:7–12
318. Wang Z-L, Yan J-M, Wang H-L, Ping Y, Jiang Q (2012) Pd/C synthesized with citric acid: an efficient catalyst for hydrogen generation from formic acid/sodium formate. Sci Rep 2 (598):1–6
319. Yadav M, Singh AK, Tsumori N, Xu Q (2012) Palladium silica nanosphere-catalyzed decomposition of formic acid for chemical hydrogen storage. J Mater Chem 22:19146–19150
320. Zhou X, Huang Y, Liu C, Lioa J, Lu T, Xing W (2010) Available hydrogen from formic acid decomposed by rare earth elements promoted Pd-Au/C catalysts at low temperature. ChemSusChem 3:1379–1382
321. Martis M, Mori K, Fujiwara Ke et al (2013) Amine-functionalized MIL-125 with imbedded palladium nanoparticles as an efficient catalyst for dehydrogenation of formic acid at ambient temperature. J Phys Chem C 117:22805–22810
322. Dan-Hardi M, Serre C, Frot T et al (2009) New photoactive crystalline highly porous titanium (IV) dicarboxylate. J Am Chem Soc 131:10857–10859

323. Fu Y, Sun D, Chen Y et al (2012) An amine-functionalized titanium metal-organic framework photocatalyst with visible-light-induced activity for CO_2 reduction. Angew Chem Int Ed 124:3420–3423
324. Yadav M, Akita T, Tsumori N et al (2012) Strong metal-molecular support interaction (SMMSI): amine-functionalized gold nanoparticles encapsulated in silica nanospheres highly active for catalytic decomposition of formic acid. J Mater Chem 22:12582–12586
325. Li Y, Xie L, Li Y et al (2009) Metal-organic-framework-based catalyst for highly efficient H_2 generation from aqueous NH_3BH_3 solution. Chem Eur J 15:8951–8954
326. Li Y, Song P, Zheng J et al (2010) Promoted H_2 generation from NH_3BH_3 thermal dehydrogenation catalyzed by metal-organic framework based catalysts. Chem Eur J 16:10887–10892
327. Song P, Li Y, Li W et al (2011) A highly efficient Co(0) catalyst derived from metal-organic framework for the hydrolysis of ammonia borane. Int J Hydrog Energy 36:10468–10473
328. Li Q, Kim H (2012) Hydrogen production from $NaBH_4$ hydrolysis via Co-ZIF-9 catalyst. Fuel Process Technol 100:43–48
329. Schlesinger HI, Brown HC, Finholt AB, Gilbreath JR, Hockstra HR, Hydo EK (1953) Sodium borohydride as the hydrogen supplier for proton exchange membrane fuel cell systems. J Am Chem Soc 75:215–219
330. Brown HC, Brown CA (1962) New, highly active metal catalysts for the hydrolysis of borohydride. J Am Chem Soc 84:1493–1494
331. Demirci UB, Akdim O, Andrieux J, Hannauer J, Chamoun R, Miele P (2010) Sodium borohydride hydrolysis as hydrogen generator: issues, state of the art and applicability upstream from a fuel cell. Fuel Cells 10:335–350
332. Jeong SU, Kim RK, Cho EA, Kim HJ, Nam SW, Oh I, Hong SA, Kim SH (2005) A study on hydrogen generation from NaBH4 solution using the high-performance Co-B catalyst. J Power Sources 144:129–134
333. Wu C, Wu F, Bai Y, Yi B, Zhang H (2005) Cobalt boride catalysts for hydrogen generation from alkaline $NaBH_4$ solution. Mater Lett 59:1748–1751
334. Walter JC, Zurawski A, Montgomery D, Thornburg M, Revankar S (2008) Sodium borohydride hydrolysis kinetics comparison for nickel, cobalt, and ruthenium boride catalysts. J Power Sources 179:335–339
335. Patel N, Guella G, Kale A, Miotello A, Patton B, Zanchetta C, Mirenghi L, Rotolo P (2007) Thin films of Co–B prepared by pulsed laser deposition as efficient catalysts in hydrogen producing reactions. Appl Catal A 323:18–24
336. Patel N, Fernandes R, Guella G, Kale A, Miotello A, Patton B, Zanchetta C (2008) Structured and nanoparticle assembled Co-B thin films prepared by pulsed laser deposition: a very efficient catalyst for hydrogen production. J Phys Chem C 112:6968–6976
337. Gupta S, Patek N, Fernandes R, Kothari DC, Miotello KA (2013) Mesoporous Co–B nanocatalyst for efficient hydrogen production by hydrolysis of sodium borohydride. Int J Hydrog Energy 38:14685–14692
338. Figen AK (2013) Dehydrogenation characteristics of ammonia borane via boron-based catalysts (Co–B, Ni–B, Cu–B) under different hydrolysis conditions. Int J Hydrog Energy 38:9186–9197
339. Figen AK, Coşkuner B (2013) A novel perspective for hydrogen generation from ammonia borane (NH_3BH_3) with Co–B catalysts: "Ultrasonic Hydrolysis". Int J Hydrog Energy 38:2824–2835
340. Ma J, Xu L, Xu L, Wang H, Xu S, Li H, Xie S, Li H (2013) Highly dispersed Pd on Co–B amorphous alloy: facile synthesis via galvanic replacement reaction and synergetic effect between Pd and Co. ACS Catal 3:985–992

Chapter 16
Organometallics for Hydrogen Storage Applications

Torsten Beweries

Abstract In this chapter, the role of organometallic compounds for hydrogen storage applications is highlighted. In this context, the focus is on transition metal complex-catalysed dehydrogenations of amine-borane adducts as a special class of so-called chemical hydrides as well as dehydrogenation reactions of formic acid and alcohols, which are also discussed as possible fuel alternatives.

Keywords Alcohols • Formic acid • Homogeneous catalysis • Hydrogen storage • Main group compounds • Organometallic chemistry

16.1 Introduction

In light of the diminishing fossil resources and the increasing demand for energy owing to a rapidly growing population, much attention has been directed towards alternative ways of supplying food, water, housing and power. For natural scientists, especially the energy issue has attracted a great deal of interest. Apart from the problem of sustainable energy production, the development of a clean, workable, inexpensive and environmentally friendly way of storing energy is inevitable as all methods of producing "green" energy suffer discontinuous availability. Wind, solar and tidal power are nowadays well-established clean energy sources; however, energy must be stored in the absence of the primary source (e.g. at night or in times of variable wind). One option to overcome this problem can be the so-called hydrogen economy, a term which was first used by Bockris in 1972 [1]. This concept includes offshore (nuclear) power plants, which would electrolyse water into hydrogen and oxygen; the former could be "piped to distribution stations and thereafter sent to factory and homes", where fuel cells would produce electricity [1]. Other methods to produce hydrogen directly from aqueous media, which are currently under investigation, include photoelectrolysis, photocatalysis, electrocatalysis as well as photobiological techniques [2]. Apart from the fact that

T. Beweries (✉)
Leibniz-Institut für Katalyse e.V. an der Universität Rostock,
Albert-Einstein-Str. 29a, 18059 Rostock, Germany
e-mail: torsten.beweries@catalysis.de

W.-Y. Wong (ed.), *Organometallics and Related Molecules for Energy Conversion*, Green Chemistry and Sustainable Technology, DOI 10.1007/978-3-662-46054-2_16

nuclear power is currently discussed controversially in some parts of the world, transformation and adaption of Bockris' concept is part of today's models for a sustainable energy supply. Moreover, instead of "piping" the hydrogen gas to the electricity production site, direct transformation of the latter in to a convenient, transportable storage material is highly desirable. One approach in this context can be the methanisation of hydrogen, thus producing synthetic gas, which can be fed into the gas grid for industrial and domestic uses [3]. However, depending on the nature of the application site (size dimensions, energy demand, lifetime), a wide range of different solutions is needed. In order to be applicable as a hydrogen storage technique, the medium has to meet certain requirements such as high volumetric and gravimetric hydrogen density, a reasonably operable temperature range for charging and discharging, good stability as well as convenience of manipulation.

Well-established techniques of physical hydrogen storage include compressed hydrogen tanks where the energy carrier is contained as a gas under high pressure as well as liquid hydrogen tanks, which are operated at low temperatures. Apart from this, ways of storing hydrogen chemically were developed with one of the most considered approaches being based on metal-organic frameworks (MOFs), which consist of three-dimensional polymers of metal atoms bridged by organic ligands [4]. These can easily be prepared from wet chemical methods, and activation proceeds in most cases by the removal of solvent. At low temperatures and elevated pressures of 10–20 bar, hydrogen can be stored by physisorption in the cavities of the cage molecules or by chemisorption directly at the metal centres or at the organic linkers [5]. In metal hydrides as another potential hydrogen storage medium, hydrogen is bound more strongly to its binding partner, thus giving significantly higher operating temperatures. The disadvantage of relatively low gravimetric capacities due to the presence of heavy metals can be compensated by using lighter main-group hydrides as well as complex metal hydrides such as $NaAlH_4$, $LiBH_4$, $LiNH_2$ or MgH_2. In this contribution, an overview of current trends in hydrogen storage concepts that employ organometallic complexes either as catalysts for the hydrogenation (i.e. charging) or dehydrogenation (i.e. discharging) of a storage medium is presented [6].

16.2 Chemical Hydrides

The term "chemical hydride" is used to differentiate between the aforementioned metal hydrides and the multifaceted class of potential hydrogen storage compounds that display covalent element-hydrogen bonds. Typically, in these species, much lighter elements than in metal hydrides are used, thus giving higher gravimetric hydrogen capacities. Reversibility of hydrogen uptake and release is an important aspect that has to be considered when discussing the potential of a compound for hydrogen storage applications. In many cases, reaction enthalpies for either of the two half reactions are highly endothermic or exothermic, which results in reaction conditions for the respective reactions that are highly unpractical. Hence, an ideal hydrogen storage system based on chemical hydrides should have reaction enthalpies that are only slightly positive or negative in order to work under convenient conditions.

As an example for a purely organic hydrogen storage system, *N*-ethylcarbazole was studied in detail. Typically, dehydrogenation of organic cyclic molecules requires very harsh conditions; however, in this case, due to the presence of the nitrogen heteroatom in the fused ring system, dehydrogenation of the parent fully hydrogenated compound dodecahydro-*N*-ethylcarbazole proceeds at temperatures in the range 150–200 °C in the presence of supported noble-metal catalysts such as Pd/Al_2O_3 [7]. Re-hydrogenation takes place at moderately high temperatures of around 130 °C and hydrogen pressures of 70 bar over supported Ru catalysts [8]. As for the practical feasibility in automotive applications, *N*-ethylcarbazole would be a viable alternative to commonly used fossil energy carriers, as one could use existing infrastructures for distribution and fuelling. Discharged fuels could be removed from the automotive and be replaced by newly charged material.

Also on laboratory scale, molecularly defined model compounds for the reversible hydrogenation/dehydrogenation were reported, the most prominent being Stephan's system based on a so-called "frustrated Lewis pair" containing a perfluoroaryl bridged phosphine-borane moiety $R_2P\text{-}C_6F_4\text{-}B(C_6F_5)_2$ (**1a**) ($R = 2,4,6\text{-}Me_3\text{-}C_6H_2$) [9]. This species can add hydrogen at ambient conditions to give a zwitterionic compound $R_2P(H)\text{-}C_6F_4\text{-}B(H)(C_6F_5)_2$ (**1b**). Heating of the latter above 100 °C regenerates the starting material (Scheme 16.1).

Other model substances for hydrogenation/dehydrogenation of a hydrogen storage medium involve heavier main group elements such as gallium, germanium or tin. Power and co-workers reported on a series of reactions of dihydrogen with low-valent organoelement compounds (Scheme 16.2) [10, 11].

These examples show that apart from the well-established reactions of transition metal complexes with hydrogen to give metal hydrides and metal dihydrogen complexes, also low-valent main-group compounds can be very interesting for the activation of small molecules. The exploration of these species for catalytic applications is one of the main challenges for main group and organoelement chemists [12].

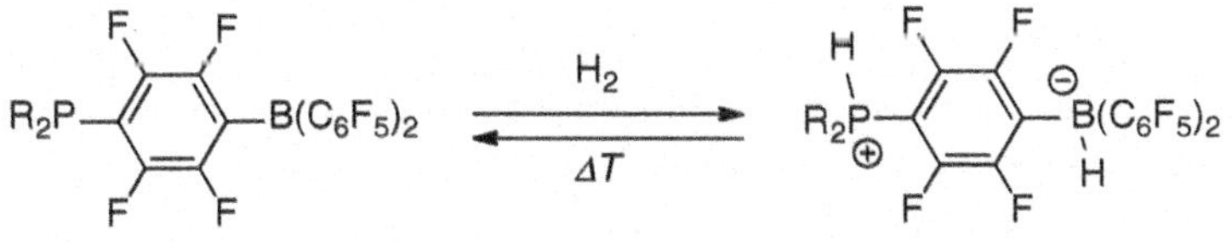

Scheme 16.1 Stephan's model system for hydrogenation/dehydrogenation

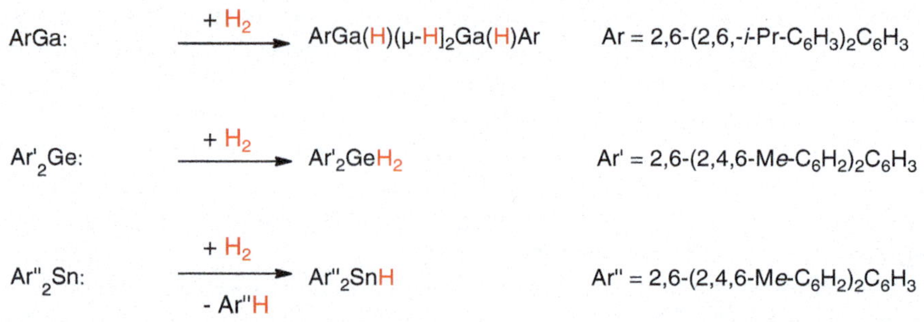

ArGa: $\xrightarrow{+ H_2}$ ArGa(H)(μ-H]$_2$Ga(H)Ar Ar = 2,6-(2,6,-*i*-Pr-C_6H_3)$_2$$C_6H_3$

Ar'$_2$Ge: $\xrightarrow{+ H_2}$ Ar'$_2$GeH_2 Ar' = 2,6-(2,4,6-*Me*-C_6H_2)$_2$$C_6H_3$

Ar''$_2$Sn: $\xrightarrow[- Ar''H]{+ H_2}$ Ar''$_2$SnH Ar'' = 2,6-(2,4,6-*Me*-C_6H_2)$_2$$C_6H_3$

Scheme 16.2 Reactions of group 13 and 14 organoelement compounds with dihydrogen

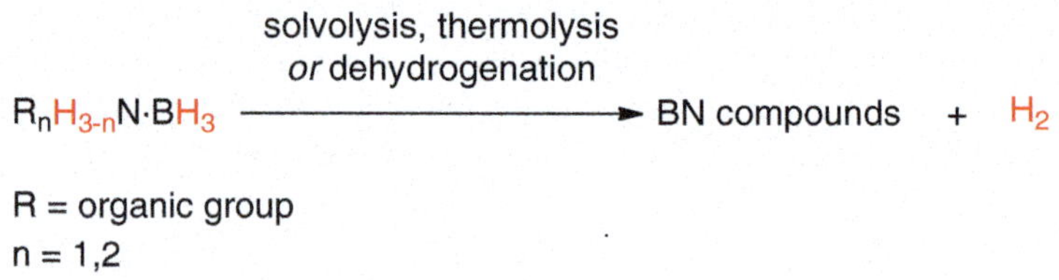

$R_nH_{3-n}N{\cdot}BH_3$ —(solvolysis, thermolysis *or* dehydrogenation)→ BN compounds + H_2

R = organic group
n = 1,2

Scheme 16.3 Hydrogen release from amine-borane adducts

16.3 Amine-Borane Adducts

16.3.1 Metal-Catalysed Hydrogen Release from Amine-Borane Adducts

The above mentioned systems featuring main group elements show great potential for the investigation of mechanisms of hydrogen activation and storage. However, in virtually all cases, gravimetric hydrogen capacities are very low, and most systems lack reversibility. The first aspect can be addressed by employing lightweight main group elements such as boron and nitrogen, which can both bind multiple hydrogen atoms and thus give compounds with high gravimetric hydrogen capacities. Also, due to the electronegativity difference, in corresponding element-hydrogen compounds (B: 2.0; N: 3.1; H: 2.2), B–H bonds are hydridic, whereas N–H bonds are typically protic, which turns out to be beneficial for the elimination of hydrogen from BN adducts, so-called amine-borane adducts (Scheme 16.3). Also, for the release of a maximum amount of hydrogen, a balanced ratio of protic and hydridic element-hydrogen bonds is important. In this section, a selection of examples of amine-borane compounds will be described with the focus on organometallic catalysts for dehydrogenation.

Ammonia borane (AB, $H_3N \cdot BH_3$) as the simplest of all amine-borane adducts has attracted a considerable amount of attention, as it contains the highest amount of hydrogen (gravimetric hydrogen capacity of 19.6 wt%). It is very stable at room temperature, solid and moderately soluble in water, making it very convenient to

$$H_3N{\cdot}BH_3 + H_2O \xrightarrow{\text{cat.}} 3H_2 + NH_4^+ + BO_2^-$$

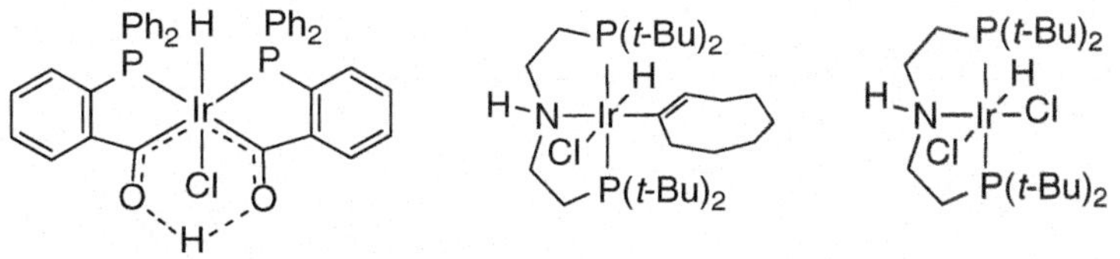

Fig. 16.1 Catalytic hydrolysis of AB and iridium complexes for this purpose

handle. Release of hydrogen from this compound is possible by solvolysis (thermal, acid- and metal-catalysed), thermolysis (solid state and solution) as well as dehydrogenation (acid and metal catalysed). A concise overview of these aspects was given recently by Baker and Manners [13]. Most of the examples known for the solvolysis of AB use metal nanoparticles as catalysts (e.g. Fe [14], Ni [15], Co@Au [16]); however, in recent reports, Garralda and later Amoroso and Abdur-Rashid describe very efficient systems based on molecularly defined iridium hydride complexes (**2**–**4**) (Fig. 16.1) [17, 18]. In the latter example, the catalyst is capable of dehydrogenating AB in different mixtures of alcohols and water as well as in the solid state in the presence of water vapour. Moreover, the catalysts' lifetime is very high, as evidenced by consecutive addition of fresh AB to the catalyst solution. Mechanistic details were not reported by the authors; a cooperative effect of the amine ligand is however likely to play a role, as evidenced by further studies on the dehydrogenation of amine-boranes (vide infra). Other examples were reported by the group of Djukic, who found that a series of ruthenium(II) dicarbonyls is active for the full hydrolytic release of hydrogen from AB; however, also in this case, no mechanistic insights were given [19].

The vast majority of transition metal-catalysed processes for the release of hydrogen from AB do not involve hydrolysis and formation of boronic acid and its derivatives but dehydrogenation with formation of oligomeric and polymeric BN compounds. Selected examples will be presented here; review articles that give a more detailed insight into this chemistry were published in the past [13, 20, 21]. Complexes that are active catalysts for the dehydrogenation of AB were established based on a broad range of early and late transition metals (Fig. 16.2).

So-called base metal catalysts were developed by the group of Baker, who used nickel complexes of the type $Ni(NHC)_2$ (**5**) (NHC = *N*-heterocyclic carbene) and found that the activity for AB dehydrogenation is strongly dependent on the substituents of the carbene ligand [22]. As for the mechanism, initial coordination of the B–H bond at the metal centre followed by protonation of a $C_{carbene}$ donor atom and formation of a Ni hydride is suggested. Recombination of the hydride and the carbon bound proton gives hydrogen, which can be eliminated from the complex to furnish the starting complex $Ni(NHC)_2$ and $[H_2NBH_2]$ [23], a species which

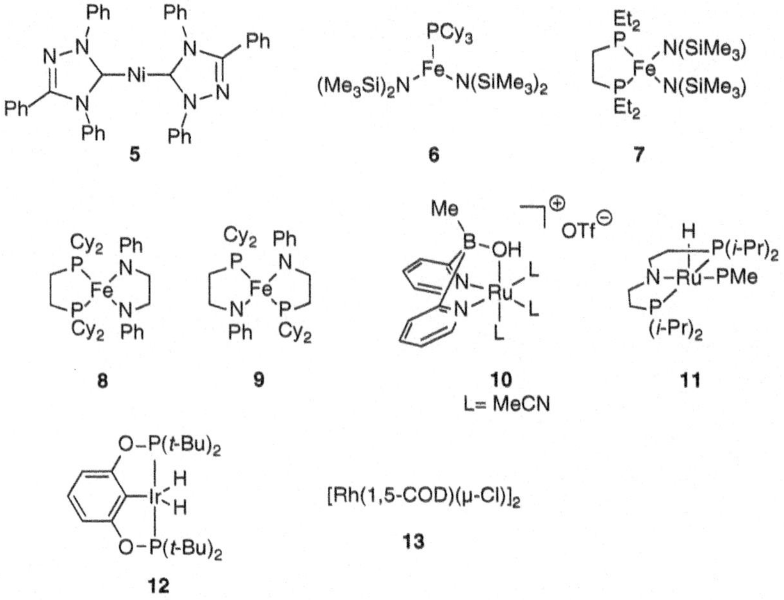

Fig. 16.2 Transition metal complexes referred to in this section used for the catalytic dehydrogenation of AB

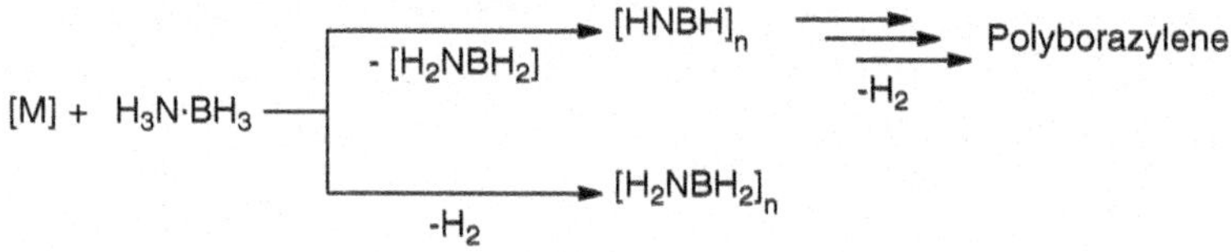

$$[M] + H_3N{\cdot}BH_3 \xrightarrow{-[H_2NBH_2]} [HNBH]_n \xrightarrow{-H_2} \text{Polyborazylene}$$

$$[M] + H_3N{\cdot}BH_3 \xrightarrow{-H_2} [H_2NBH_2]_n$$

Scheme 16.4 Different pathways for the dehydrogenation of AB

is further dehydrogenated to yield polyborazylene. Later, the same group also investigated iron complexes containing amido and phosphine supporting ligands (**6–9**) and found that the nature of the BN dehydrogenation products as well as the extent of hydrogen release (Scheme 16.4) is strongly dependent on the used ligands [24].

The authors postulate that in cases where the intermediate aminoborane $[H_2NBH_2]$ remains coordinated to the metal centre, formation of linear polyaminoborane (i.e. partial dehydrogenation with release of 1 eq. of H_2) is the favoured pathway over generation of polyborazylene (i.e. almost full dehydrogenation with release of up to 3 eq. of H_2). Also, it became evident that Fe(+2) fragments react much different than Fe(0) species, which are likely to be formed by dissociation of a labile amine ligand. Another example for AB dehydrogenation involving an iron complex was reported by Manners et al., who found that photoactivated $[CpFe(CO)_2]_2$ is a moderately active catalyst [25].

Complexes based on ruthenium were found to be among the most active catalysts for hydrogen release from AB. Williams and co-workers described a study of the Shvo complex [26], which was used before for a broad range of other catalytic applications and is capable of liberating up to two equivalents of hydrogen within two hours. In another study, the same group introduced a Lewis basic boron functionality into a Ru(II) complex (**10**) and found that the presence of this group is an important factor for its high activity. Additionally, this complex is very stable in air and water, thus making it very interesting for further optimisations [27]. Cooperative reactivity between a PNP pincer ligand and the metal centre was observed in AB dehydrogenation using a bifunctional ruthenium hydride complex (**11**) [28]. In this case, concerted transfer of a hydride (B–H) and a proton (N–H) from the substrate to a catalyst, similar to the Noyori-Morris mechanism for hydrogen transfer processes [29], is likely to be present. A molecularly defined iridium pincer dihydride complex (POCOP)IrH_2 (**12**) (POCOP = κ^3-C_6H_3-1,3-[OP$(tBu)_2]_2$) was used by Heinekey and Goldberg and co-workers for the catalytic dehydrogenation of AB [30]. As for the mechanism of hydrogen formation, two possibilities were discussed, including concerted activation of B–H and N–H bonds at a monovalent [(POCOP)Ir] centre as well as at a trivalent [(POCOP)IrH_2] species, respectively. Notably, this catalyst was also found to be active for the homogeneous dehydrogenation of alkanes, thus showing similarities between formally isoelectronic hydrocarbon and BN compounds (e.g. benzene/borazine, ethane/AB). Most importantly, in many cases during the catalytic cycle, the metal species exhibits structural and electronic changes, owing to the strongly reducing conditions present in the reaction mixture. As a consequence, the active species can strongly differ from the precatalyst, as demonstrated by Manners et al., who found that, upon addition to a solution containing AB, initial $[Rh(COD)(\mu\text{-}Cl)]_2$ (**13**) is reduced to form Rh nanoparticles, which serve as the dehydrogenation catalyst [31].

Despite being structurally very similar to AB, hydrazine borane ($N_2H_4 \cdot BH_3$, HB) and the parent hydrazine bis(borane) ($N_2H_4 \cdot (BH_3)_2$, HBB) are poorly explored in terms of chemical hydrogen storage. Both substances display comparably high gravimetric hydrogen capacities of 13.1 % and 13.4 %, respectively. Unfortunately, HBB is poorly suited for hydrogen storage applications owing to its shock sensitivity and its thermal lability in air. For HB, studies of thermolysis [32] as well as metal-catalysed hydrolysis [33] were described; however, to date, only one example is known for a catalytic dehydrogenation involving molecularly defined transition metal complexes. Group 4 metallocene alkyne and hydride complexes were found to be moderately active for the release of hydrogen from HB; however, this process was very slow, and substantial improvements are needed in order to make this approach attractive for applications [34]. Also, the study of the dehydrogenation mechanism turned out to be very difficult as the obtained HB dehydrogenation products were poorly soluble, even in highly polar solvents.

Further variations of the substituents at amine-borane adducts offer the possibility of adjusting physical properties such as solubility as well as melting and decomposition temperatures. Moreover, once introduced, organic groups at the

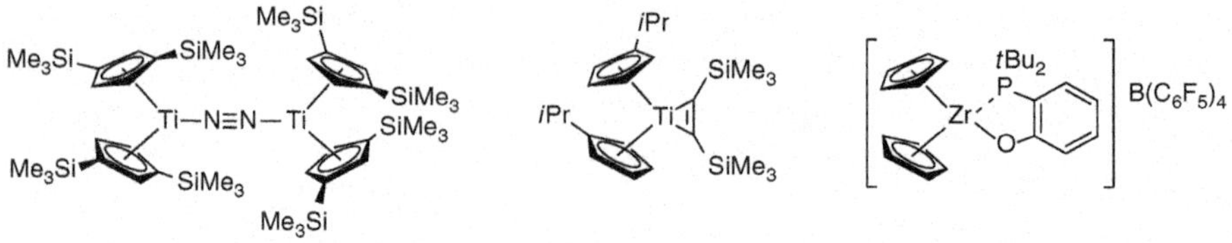

Fig. 16.3 Selected group 4 metallocene complexes used for the catalytic dehydrogenation of DMAB

heteroatoms can serve as a spectroscopic handle, which can be very useful in terms of mechanistic investigations. One of the simplest and most frequently used substituted amine-borane adducts is dimethylamine borane ($Me_2NH \cdot BH_3$, DMAB). Despite displaying a rather low gravimetric hydrogen capacity of only 3.4 %, the dehydrogenation chemistry of this compound is well studied for a broad range of main group and transition metal complexes.

A number of active catalysts were reported for group 4 metallocenes (Fig. 16.3). Chirik and co-workers found significant differences in reactivity when comparing the steric influence of the cyclopentadienyl substituents, with the sterically more demanding species being less active and complex **14** displaying good DMAB dehydrogenation efficiencies [35]. This effect was also observed for other titanocenes and zirconocenes that were formed from the corresponding metallocene bis(trimethylsilyl)acetylene complexes (e.g. **15**) [36, 37]. The mechanism of titanocene [Cp_2Ti]-catalysed DMAB dehydrogenation was investigated by Manners et al., who suggested the presence of titanocene(IV) hydrides as intermediates; however, the presence of these species was not confirmed [38]. Also, as the catalytically active component, a titanocene(II) species was proposed. Later, the same authors used EPR spectroscopic techniques to corroborate that Ti(III) species are present during DMAB dehydrogenation, thus pointing towards the importance of such paramagnetic species for the catalytic cycle [39]. In a study of DMAB dehydrogenation using cationic zirconocene-phosphinoaryloxide complexes such as **16** with frustrated Lewis pair character, Wass et al. demonstrated high activities (TOF up to 600 h^{-1}) and reaction times of only several minutes [40].

Similarly as for AB dehydrogenation described above, a variety of late transition metal complexes were investigated. Examples include amino olefin nickel(0) and nickel(I) complexes, the latter being very active compared to other noble-metal-free dehydrogenation catalysts (release of 1 eq. of H_2 within one minute) [41]. Mechanistic details were not reported; however, the presence of a hydridic intermediate is likely as a stoichiometric reaction of DMAB with the Ni(I) catalyst furnished a Ni (I) hydride fragment. Other examples described by Weller and co-workers highlight the importance of initial coordination of the DMAB substrate at the metal centre. For instance, in the dehydrogenation reaction using the Rh(III) complex Ru $H_2(PCy_3)_2Cl$ (**17**), hydrogen elimination from the complex was induced by oxidative addition of one of the B–H bonds of DMAB [42], probably by a σ-CAM [43]

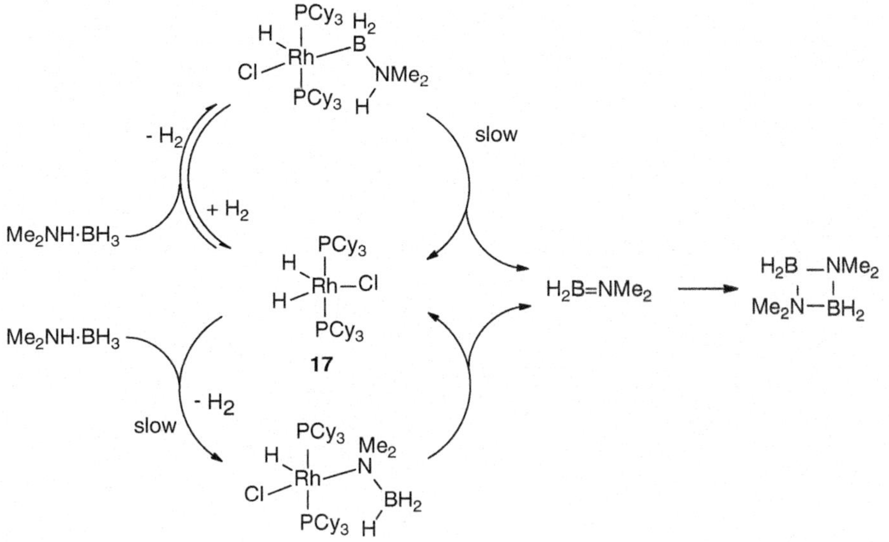

Scheme 16.5 Proposed mechanism for the dehydrogenation of DMAB using $RhH_2(PCy_3)_2Cl$ (**17**)

mechanism. As an alternative, yet much slower initial step, oxidative addition of the N–H bond to give an aminoborane complex was suggested. From either of the two intermediates, release of $H_2B{=}NMe_2$ takes place, which subsequently dimerises to yield the main dehydrocoupling product $[H_2B\text{-}NMe_2]_2$ (Scheme 16.5).

In another study, stoichiometric reaction of DMAB with an Ir(III) complex $[Ir(H_2)_2H_2(PCy_3)_2]BAr^F_4$ (**18**) furnished a κ^2-H-bound amine-borane σ-complex $[IrH_2(PCy_3)_2(\kappa^2\text{-}H\text{-}H_3B\text{-}NMe_2H)]BAr^F_4$ (**19**). Hydrogen release was found to take place from this species to give an aminoborane complex $[IrH_2(PCy_3)_2(\kappa^2\text{-}H_2BNMe_2)]BAr^F_4$ (**20**), which could release the fragment $H_2B{=}NMe_2$ in catalytic reactions to reform the catalytically active species (Scheme 16.6a) [44]. Also, in this case, the metal was found to be essential for the formation of the linear dimer $H_3B \cdot NMe_2BH_2 \cdot NMe_2H$, a species which is often observed spectroscopically in transition metal-catalysed dehydrogenations of DMAB (Scheme 16.6b). Generation of the cyclic dimer $[H_2B\text{-}NMe_2]_2$ is proposed to proceed via coupling of $H_2B{=}NMe_2$ with the aforementioned amine-borane σ-complex followed by elimination of one equivalent of hydrogen (Scheme 16.6c).

These reactivity patterns of the iridium centre demonstrate the multifaceted chemistry that should be considered when modelling and interpreting catalytic cycles in transition metal-catalysed amine-borane dehydrogenation. Further evidence for the essential role of such structural motifs was also given by Aldridge and co-workers, who isolated an 18-electron aminoborane adduct $[IrH_2(IMes)_2(\kappa^2\text{-}H_2BNMe_2)]BAr^F_4$ from reaction solutions containing DMAB and catalytic amounts of the catalyst precursor [45]. In the same study, the formation of a very unusual 14-electron rhodium(III) aminoboryl complex

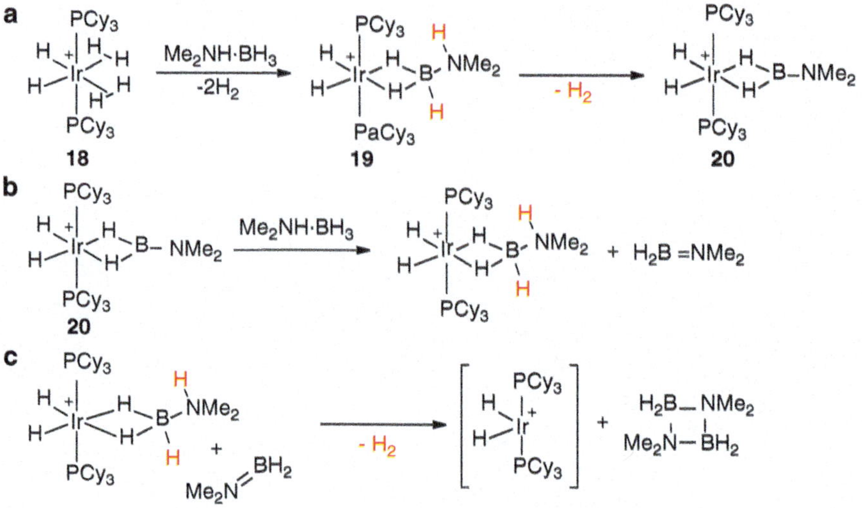

Scheme 16.6 Multiple reaction motifs in the dehydrogenation of DMAB with $Ir(H_2)_2H_2(PCy_3)_2$ (**18**). Note that BAr^F_4 anions are not shown

$\{Rh(IMes)_2(H)[B(H)NMe_2]\}BAr^F_4$ was observed from the similar reaction of the parent Rh precursor, thus pointing towards considerable differences in dehydrogenation reactivity when moving down one group.

As an alternative to transition metal catalysts, recently significant efforts to also develop systems based on main group metals were made. Hill and co-workers prepared a magnesium β-diketiminate complex by reaction of [HC{(Me)CN(Dipp)}$_2$Mg(*n*Bu)] with DMAB and found that this species [HC{(Me)CN(Dipp)}$_2$Mg($NMe_2BH_2NMe_2BH_3$)] (**21a**) plays a key role in the catalytic dehydrogenation of DMAB as elimination of the cyclic dimer $[H_2B\text{-}NMe_2]_2$ yields a magnesium hydride complex (**21b**), which can insert further DMAB with release of hydrogen and regeneration of the starting complex (Scheme 16.7) [46].

Also, a structurally similar calcium complex was investigated for dehydrogenation of DMAB; however, the rate of hydrogen evolution was much slower. In an attempt to rationalise this behaviour, the authors suggest that the efficiency of insertion of the intermediate $H_2B{=}NMe_2$ into the M–N bond and subsequent β- or δ-hydride elimination steps is dependent on the charge density and the polarising capability of the metal centre. Other d^0 dehydrogenation catalysts were developed by Wright et al., who investigated aluminium(III) and gallium(III) amides and found that similar insertion processes are present in these cases [47, 48]. Also, it should be noted that parent titanium(IV) and zirconium(IV) amides are moderately active for DMAB dehydrogenation, thus pointing towards the analogies of structurally similar d^0 complexes [36]. Organotin(IV) complexes $Cp^*_2SnCl_2$ and Ph_2SnCl_2 were tested for DMAB dehydrogenation; however, the catalytic activity was comparably low with both complexes requiring several days and high catalyst

Scheme 16.7 Hydrogen release from DMAB using a magnesium β-diketiminate complex

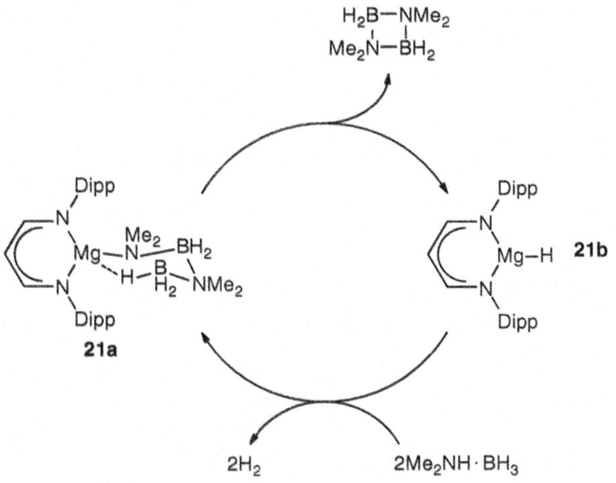

loadings in order to realise mentionable turnovers [49]. In contrast, in AB dehydrogenation experiments, these compounds have shown more promising activities.

16.3.2 Regeneration of Spent Amine-Borane Fuels

Hydrogen release from amine-borane compounds is exothermic and takes place at comparably low temperatures, which means that direct re-hydrogenation is impossible. Hence, a recycling protocol should consist of multiple steps in order to provide a thermodynamic driving force that facilitates the reformation of the hydrogen carrier (Scheme 16.8). For AB, Mertens et al. summarised these steps as (1) the digestion of the polymeric spent fuel by the formation of oxidised, or more specifically, halogenated boron species such as BX_3 (X = Cl, Br, I), (2) hydrogenation/hydrodehalogenation of these compounds and (3) replacement of the auxiliary reagents (bases) used as thermodynamic drivers in the hydrodehalogenation process by ammonia [50]. Alternatively, a strong reductant such as hydrazine can be used to hydrogenate polyborazylene, which is the BN-containing product when releasing more than two equivalents of hydrogen from AB [51]. Although being very simple from an experimental perspective, this procedure features the significant disadvantages of the toxic, highly energetic and unstable reagent hydrazine. Moreover, the described one-pot regeneration is carried out in liquid ammonia at 40 °C, thus requiring high-pressure equipment.

As for the first step of BN waste regeneration, digestion is required in order to solubilise the spent material. For AB, different reactions were described such as transformation of B–N bonds into B–Cl bonds by reaction with the superacidic system $AlCl_3/HCl/CS_2$. This is followed by hydrodehalogenation in the presence of the base NEt_3 with nickel boride as the catalyst. In the last step, substitution of the amine by NH_3 furnishes AB [50]. Main advantages of this reaction sequence are the

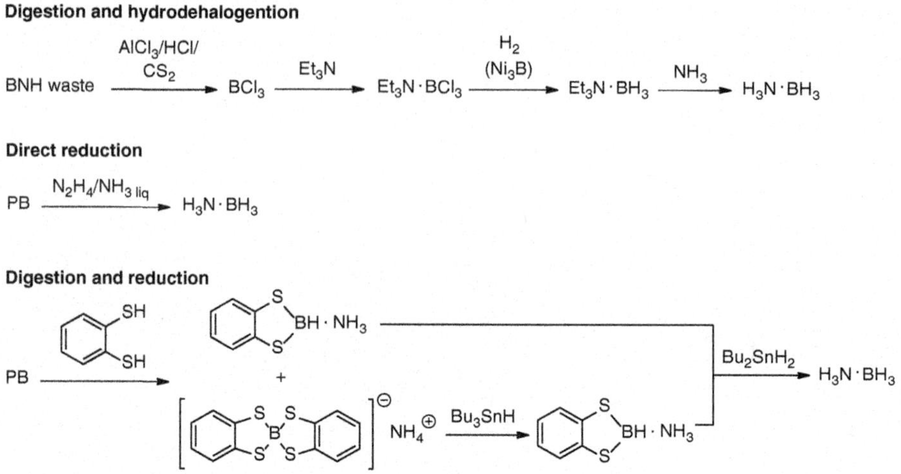

Digestion and hydrodehalogention

BNH waste $\xrightarrow{AlCl_3/HCl/CS_2}$ BCl_3 $\xrightarrow{Et_3N}$ $Et_3N \cdot BCl_3$ $\xrightarrow{H_2\ (Ni_3B)}$ $Et_3N \cdot BH_3$ $\xrightarrow{NH_3}$ $H_3N \cdot BH_3$

Direct reduction

PB $\xrightarrow{N_2H_4/NH_{3\ liq}}$ $H_3N \cdot BH_3$

Digestion and reduction

PB $\xrightarrow{C_6H_4(SH)_2}$ $C_6H_4S_2BH \cdot NH_3$ + $[B(S_2C_6H_4)_2]^{\ominus} NH_4^{\oplus}$ $\xrightarrow{Bu_3SnH}$ $C_6H_4S_2BH \cdot NH_3$ $\xrightarrow{Bu_2SnH_2}$ $H_3N \cdot BH_3$

Scheme 16.8 Examples for the recycling of spent AB material

use of inexpensive reagents which can theoretically be implemented into a closed regeneration scheme. Moreover, hydrogen gas serves as the reductant, a fact that is essential in terms of adapting this procedure for true hydrogen storage applications. In another report, Dixon, Gordon and co-workers reported on the digestion of polyborazylene with benzenedithiol to give the soluble adduct $C_6H_4S_2BH \cdot NH_3$ (Scheme 16.8) [52]. This can be hydrogenated using the organotin hydrides Bu_3SnH and Bu_2SnH_2. Similar to the hydrazine-based protocol described above, this approach uses equimolar amounts of a highly energetic reductant; moreover, it produces stoichiometric amounts of organotin-containing waste.

16.3.3 Transition Metal-Catalysed Dehydrogenation of Formic Acid

Formic acid (FA) is a very promising hydrogen storage material, owing to its low toxicity and the fact that it is formed by nothing else than reversible binding of hydrogen and carbon dioxide. The transition metal-catalysed dehydrogenation chemistry of FA was first investigated by Coffey, who used a series of noble-metal complexes based on iridium, palladium and platinum [53]. However, until the end of the twentieth century when hydrogen became more and more important in the context of energy storage, no significant improvements of catalysts were reported.

In 2000, Puddephatt and co-workers developed a binuclear, diphosphine-bridged diruthenium catalyst (**22a**) for the conversion of FA to H_2 and CO_2 [54]. As for the mechanism (Scheme 16.9), the authors propose that initial hydrogenation yields the active dihydride complex **22b**, which upon insertion of FA and elimination of

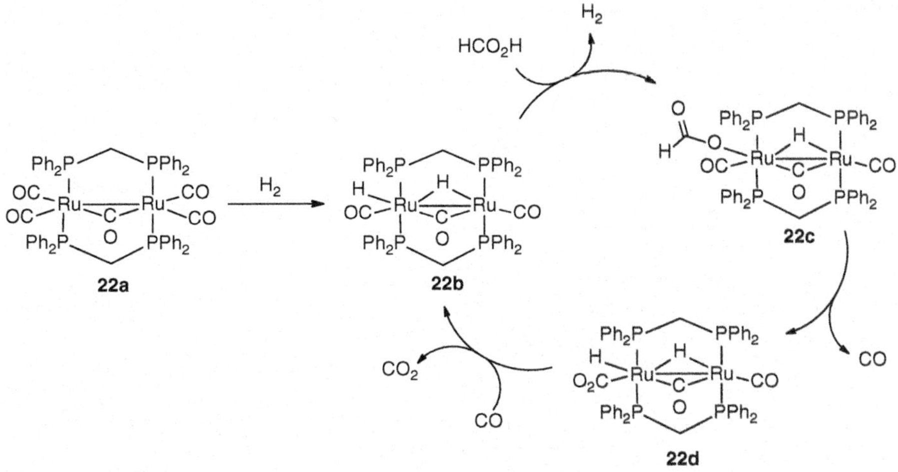

Scheme 16.9 Proposed mechanism for FA dehydrogenation with a binuclear Ru catalyst

hydrogen yields a formate monohydride complex (**22c**). Subsequent hydride transfer from the formate ligand to the ruthenium centre yields a dihydride complex (**22d**), from which CO_2 can dissociate to furnish the catalytically active species. Interestingly, the catalyst was found to catalyse the reverse reaction, i.e. hydrogenation of CO_2 to give FA, as well. In a mechanistic study, Wills et al. investigated the fate of molecularly simple mononuclear ruthenium complexes during dehydrogenation of a mixture of FA and Et_3N as a basic additive [55]. It became evident that during catalysis, chloride complexes such as $RuCl_2(DMSO)_4$, $RuCl_2(NH_3)_6$ and $RuCl_3$ form binuclear formate-bridged ruthenium species as addition of varying amounts of PPh_3 to reaction solutions furnished the corresponding binuclear isolable phosphine complexes.

Other examples for ruthenium-catalysed dehydrogenation of FA were reported by Beller and co-workers, who used commercially available complexes $[RuCl_2(p\text{-cymene})]_2$ and $RuCl_2(PPh_3)_3$ to dehydrogenate various FA/amine adducts [56]. The influence of the amine was found to be significant, as aliphatic dialkylmethylamines give higher activities with increasing chain length ($Oct_2NMe > Hex_2NMe > Bu_2NMe$), whereas the addition of sterically more demanding amines such as $PhNMe_2$ results in a decreased hydrogen yield. In a more detailed study, also the effect of a variation of the phosphine ligand was investigated, and $[RuCl_2(benzene)]_2$/dppe (dppe = 1,2-bis(diphenylphosphino)ethane) was found to be the most active catalyst system for hydrogen release from a 5:2 mixture of FA and Et_3N [57]. Notably, the released hydrogen gas was found to be virtually CO-free and could thus be used directly in a fuel cell setup for production of electricity. Visible light illumination was found to accelerate the dehydrogenation process. As for the influence of light, the authors suggest light-accelerated ligand dissociation steps to be the key factors for the large increase of catalytic activity (Scheme 16.10). It should be noted, that this effect was only observed for

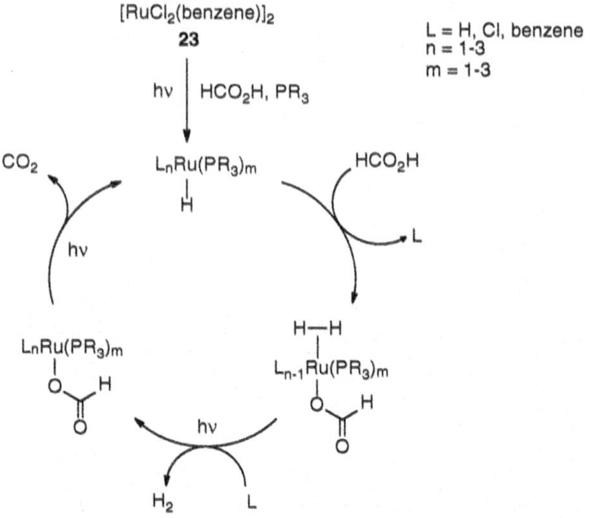

Scheme 16.10 Visible light-accelerated Ru catalysed dehydrogenation of FA

aryl phosphine ligands with the most significant improvements being found for the bidentate ligand dppe in combination with $[RuCl_2(benzene)]_2$ (**23**) (TON 2804, cf. 407 without illumination) [58].

A system that operates continuously in aqueous solutions was established by Laurenczy and co-workers [59]. Dehydrogenation of FA was investigated using the highly water-soluble ligand *meta*-trisulfonated triphenylphosphine (TPPTS) with either $[Ru(H_2O)_6]^{2+}$ or commercially available $RuCl_3$ with addition of a small amount of sodium formate, which is essential for activation of the catalyst. A study of the reaction mechanism implied that two competing catalytic cycles are present with a ruthenium monohydride aqua fragment $[Ru(TPPTS)_2(H_2O)_3H]^+$ being the key intermediate. Formate was found to be important for the generation of the active species, as replacement of a water ligand by formate induces further loss of water to generate free coordination sites which can be occupied by hydride ligands [60].

The first example of a catalyst that can reversibly store hydrogen using CO_2/FA in aqueous media was recently published by Hull, Himeda and Fujita et al. [61]. Most interestingly, the described binuclear iridium catalyst **24a/b** (Scheme 16.11) displays a proton-switchable ligand framework that – depending on the pH of the reaction system – is capable of either storing (high pH) or releasing hydrogen (low pH). Moreover, the catalyst is highly active for both reactions, operates at conditions that resemble those of common fuel cells and provides very pure, CO-free hydrogen gas. The reaction mechanism for the dehydrogenation was not reported to date; however, initial elimination of one of the aqua ligands followed by coordination of formate and loss of hydrogen is likely. In the hydrogen storage reaction, the deprotonated form of the catalyst can activate H_2 to give an

Scheme 16.11 An iridium catalyst that reversibly stores hydrogen in aqueous media. **24a′** denotes a partly protonated form of the ligand between **24a** and **24b**

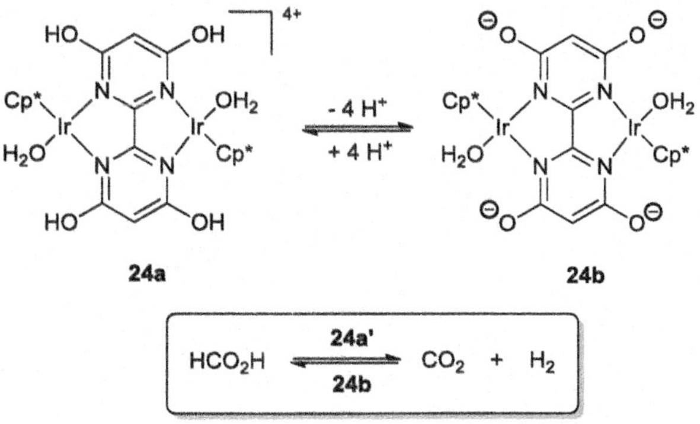

iridium hydride, which subsequently inserts CO_2 to furnish a formate complex. Further addition of H_2 induces elimination of FA and reforms the hydride species.

In order to lower the cost of the overall system for potential applications, Beller and co-workers introduced FA dehydrogenation catalysts based on less expensive, abundant iron complexes. In the first studies, multicomponent catalyst systems based on iron carbonyl complexes were presented [62, 63]. In isolated mononuclear molecularly defined complexes $Fe(CO)_3(PPh_2Bn)_2$ and $Fe(CO)_3(PBn_3)_2$ (Bn = Benzyl), moderate turnover numbers were observed in combination with terpyridine as a co-ligand [62]. The latter was found to be beneficial for the stability of the system, as it stabilises the iron centre during catalysis. Also, *ortho*-metalation of the Bn groups was found to be of importance for the long-term stability of the catalyst. Later, also in situ systems using a combination of $Fe_3(CO)_{12}$, PPh_3 and bidentate *N*-donor ligands such as phenanthroline and terpyridine were described [63]. Interestingly, also in these studies, visible light was found be relevant for the generation of the active catalyst as well as for driving the catalytic cycle, thus accelerating FA dehydrogenation significantly. More recently, an iron-based catalyst that operates in environmentally benign propylene carbonate was described [64]. The highly active, long-term, stable, in situ system consists of the precursor complex $Fe(BF_4)_2 \cdot 6\ H_2O$ and tris[(2-diphenylphosphino)ethyl]phosphine [P$(CH_2CH_2PPh_2)_3$, PP_3] and does not require additional base, thus giving significantly higher hydrogen capacities (4.4 wt%, cf. 2.3 % in commonly used 5:2 FA/Et_3N mixtures). Other examples for base-free FA dehydrogenation catalyst systems were reported by Reek [65], Himeda [66] and Puddephatt [54].

As mentioned above, the reversibility of FA decomposition into H_2 and CO_2 may offer the possibility of developing efficient hydrogen storage systems based on FA, formate and, most interestingly, cheap, readily available CO_2. The latter is also a greenhouse gas, and fixation would be a key element of a carbon-neutral energy storage cycle. Hence, besides significant research efforts addressing FA dehydrogenation, also transition metal complex-catalysed hydrogenation of CO_2 and carbonates was extensively studied in the past. Excellent review articles addressing the fundamentals of this conversion were published by Leitner [67], Noyori [68] as well as for more recent developments by Gong [69]. Examples for homogeneous

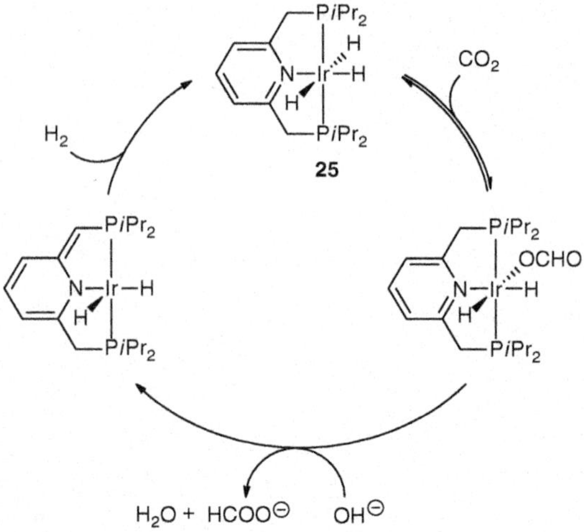

Scheme 16.12 Proposed catalytic cycle of CO_2 reduction to formate using an Ir-PNP pincer complex

transition metal complexes that catalyse hydrogenation of CO_2 or carbonates include an iridium trihydride PNP pincer (**25**) complex that is capable of selectively producing potassium formate from equimolar mixtures of H_2 and CO_2 in potassium hydroxide with a TON of up to $3.5 \cdot 10^6$ [70]. As for the mechanism, insertion of CO_2 into one of the Ir–H bonds is suggested, followed by base-induced deprotonation of the pyridine moiety and elimination of formate. Addition of hydrogen results in rearomatisation and closure of the catalytic cycle (Scheme 16.12).

Well-defined noble-metal-free catalysts for CO_2 hydrogenation were reported to display considerable activities. In a study of different iron precursor complexes, the combination of $Fe(BF_4)_2 \cdot 6\ H_2O$ and the tetradentate phosphine ligand PP_3 (vide supra) was identified to hydrogenate bicarbonate to formate at medium H_2 pressures of 60 bar and temperatures of 80 °C [71]. Mechanistic insights were gained after isolation of a hydride species $[FeH(PP_3)]BF_4$, which after reaction with CO_2 furnished the coordination compound $[FeH(CO_2)(PP_3)]BF_4$. ^{13}C NMR spectroscopic reaction control further indicated subsequent insertion of CO_2 into the Fe–H bond to give the corresponding formate complex $[Fe(HCO_2)(PP_3)]BF_4$. A non-precious metal-based system that was found to be competitive and even superior to noble-metal catalysts for hydrogenation of CO_2 or carbonates was identified in a study of different cobalt precursor complexes and the same phosphine ligand [72]. The combination $Co(BF_4)_2 \cdot 6\ H_2O/PP_3$ gave a maximum TON of 3,877. In contrast to the similar iron-based system described above, the corresponding cobalt monohydride was found to be inactive; in contrast, the dihydrogen complex $[Co(H_2)(PP_3)]BPh_4$ showed activities comparable to the in situ system, thus indicating significant differences in the hydrogenation mechanism. Also, in in situ NMR studies at higher hydrogen pressure, a signal corresponding to a dihydrogen complex could be identified.

Current efforts focus on the evolution of systems that can reversibly interconvert carbonates/CO_2 and H_2 and formates/FA (vide supra [61, 73]). For instance, the reversible coupling of formate dehydrogenation and bicarbonate hydrogenation using a commercially available ruthenium chloride complex $[RuCl_2(benzene)]_2$ in combination with dppm (dppm = 1,2-bis(diphenylphosphino)methane) was demonstrated. It should be noted that – in contrast to previous studies of FA dehydrogenation – the hydrogen release step does not require additional base [74]. Himeda and co-workers reported on well-defined ruthenium and rhodium catalysts for both reactions [75]. In this case, the nature of the used DHBP ligand (DHBP = 4,4-′-dihydroxy-2,2′-bipyridine) was found of special importance as the acid-base equilibrium involving the OH groups, and the aromatic framework is essential for driving catalysis. Hydrogenation of CO_2/bicarbonate was found to take place under basic conditions, whereas decomposition of FA/formate took place in acidic media. In the course of a mechanistic study of CO_2 hydrogenation using an Ir trihydride PNP pincer complex **25** (cf. Scheme 16.12), Nozaki et al. also developed a protocol for the catalytic dehydrogenation of FA in the presence of base using the same complex [76]. Observed activities were comparable to other FA dehydrogenation systems. The role of the pyridine ligand in the FA decomposition reaction was not elucidated; however, it was shown that reversible interconversion of CO_2 and FA is possible using this Ir catalyst.

First efforts to develop a hydrogen storage device based on the hydrogenation of bicarbonate and decomposition of formate in aqueous solution were made by Papp and Joo et al., who – taking into account the thermodynamic prerequisites – designed a simple setup that uses the water-soluble catalyst system consisting of $[RuCl_2(mTPPMS)_2]_2$ and mTPPMS (mTPPMS = sodium diphenylphosphinobenzene-3-sulfonate) [77]. Important features that need to be addressed in this context are the requirement for a closed system that can be switched between charging (i.e. CO_2 hydrogenation) and discharging (i.e. FA decomposition) simply by means of pressure and temperature variation as well as sufficient recyclability of the used components. Depending on the applied catalyst, the closed cycle system consisting of CO_2/bicarbonate and FA/formate fulfils these requirements and thus holds potential for further developments and future applications.

16.3.4 Transition Metal-Catalysed Dehydrogenation of Alcohols

Dehydrogenation of aromatic and aliphatic alcohols using transition metal complexes is a well-known reaction motif in homogeneous catalysis. Since the development of first catalysts – especially for transfer hydrogenation reactions – the knowledge in this field has been extended enormously. In this part of the chapter, selected systems that catalyse the dehydrogenation of alcohols without transfer of the produced hydrogen to an organic acceptor will be described. Also, as maximisation of the hydrogen content is one main target from the applicant's perspective, the focus will be on the dehydrogenation of low-molecular weight alcohols such as methanol, ethanol, propanol or isopropanol. Other examples referring to the dehydrogenation of alcohols with lower hydrogen content will not be discussed in this section [78].

One of the first studies of hydrogen production from alcohols by means of homogeneous catalysis was described by Dobson and Robinson, who used the complex $Ru(COOCF_3)_2(CO)(PPh_3)_2$ (**26a**) in combination with an excess of trifluoroacetic acid for the catalytic dehydrogenation of a range of alcohols [79]. As proposed, initial coordination of the alcohol at the ruthenium centre followed by proton shift to one of the acetate ligands and elimination of one equivalent of trifluoroacetic acid (via **26b** and **26c**) is followed by formation of a ruthenium hydride species (**26d**), which upon reaction with further acid eliminates hydrogen and the carbonyl product (Scheme 16.13). It should be noted that activities for low-molecular weight substrates were rather low; competitive results were however observed for benzyl alcohol. A reason for this difference in reactivity was later identified in the higher tendency of small primary alcohols to undergo decarbonylation reactions, thus producing the catalyst poison CO. Moreover, under the applied acidic conditions, competing aldol condensations are much more likely for primary alcohols [80].

An approach to avoid the use of strongly acidic additives was demonstrated by Hulshof and co-workers, who integrated tetrafluorosuccinic acid as a bidentate fluorinated ligand into a ruthenium catalyst for benzyl alcohol dehydrogenation, thus giving a cooperative effect between the ruthenium centre and the fluorinated succinate ligand [81]. As a consequence, the system operates at neutral pH and is less prone to side reactions.

The development of more efficient catalysts for the dehydrogenation of low-molecular weight alcohols proceeded as Morton and Cole-Hamilton used rhodium [82, 83] and ruthenium complexes [82] with addition of NaOH. For the more active ruthenium precatalyst $Ru(N_2)H_2(PPh_3)_3$ (**27a**), as for the beneficial influence of base, initial reaction of alkoxide (formed by reaction of NaOH with the corresponding alcohol substrate) with the fragment $[RuH_2(PPh_3)_3]$ (**27b**) is suggested [82]. Subsequent elimination of the carbonyl compound gives an anionic trihydride complex (**27d**), which reacts with alcohol to form an alcoholate and a dihydrogen complex (**27e**). The latter eliminates hydrogen to close the catalytic cycle (Scheme 16.14). Notably, also methanol and ethanol were converted using this catalyst. Light illumination was found to increase the activity of the investigated ruthenium-based systems significantly, most likely due to more facile release of hydrogen from the intermediate $[RuH_2(H_2)(PPh_3)_3]$ (**27e**).

In a detailed study of the catalytic dehydrogenation of isopropanol using different ruthenium precursor complexes in combination with different phosphine ligands and NaOH or Na as the base, Beller and Junge identified ruthenium chloride $RuCl_3 \cdot x\ H_2O$ with adamantylphosphines and biarylphosphines as the most active systems [84]. Notably, this combination was found to exceed the activity of the well-defined hydride complexes described before [82]. Later, further improvement

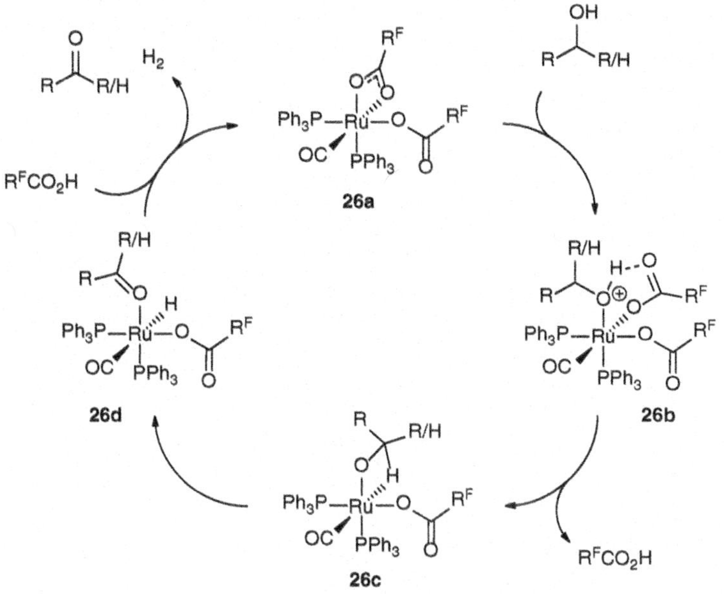

Scheme 16.13 Proposed catalytic cycle of alcohol dehydrogenation in acidic media

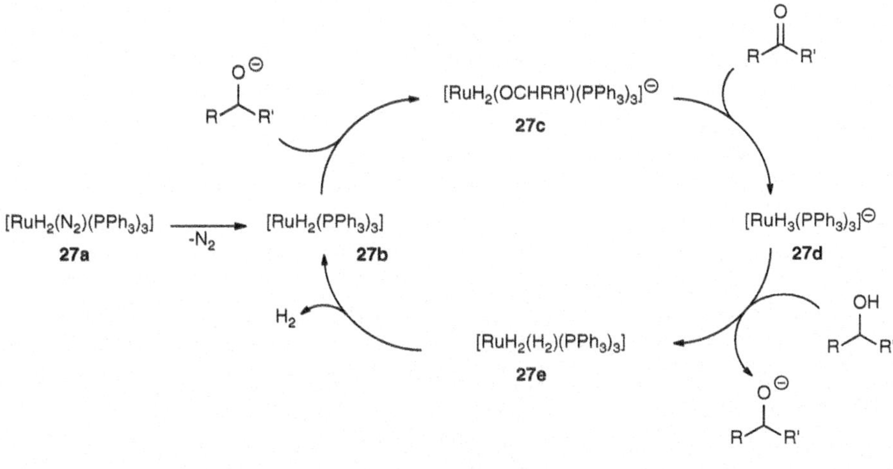

Scheme 16.14 Base-promoted alcohol dehydrogenation using a ruthenium hydride complex

of the catalyst system was achieved by using bidentate *N*-donor ligands to stabilise the metal precursor $[RuCl_2(p\text{-cymene})]_2$ [85]. Replacement of the phosphine ligands by amine ligands gives a significant increase of long-term stability. Moreover, the reaction temperatures were lowered to 90 °C (cf. 150 °C in reference 82). Combinations of ruthenium complexes with other chelating ligands based on pyridine were also found to be active for the dehydrogenation of alcohols. Milstein

Scheme 16.15 Outer-sphere mechanism in the catalytic dehydrogenation of alcohols using a *H*PNP pincer complex

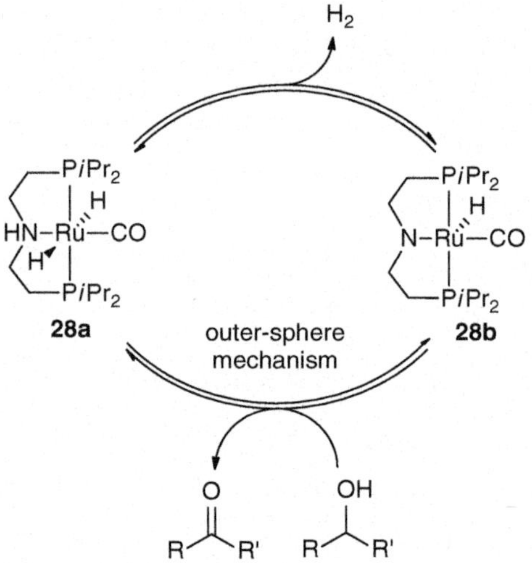

et al. developed PNN pincer complexes that – by cooperation between the metal centre and the ligand framework – are capable of coupling alcohol dehydrogenation to the synthesis of esters and amides [86, 87].

Significant improvements regarding the catalytic dehydrogenation of energetically interesting low-molecular weight primary and secondary alcohols were recently made as the efficient hydrogen release from ethanol and isopropanol at temperatures below 100 °C was reported [88]. In an evaluation of different pincer-type transition metal complexes and in situ combinations of ruthenium precursor complexes and pincer ligands, a system consisting of $RuH_2(PPh_3)_3CO$ and *H*N $(CH_2CH_2P\textit{i}Pr_2)_2$ was identified as the most active combination that – most interestingly – also dehydrogenates ethanol with high activities (TOF 1,483 h^{-1} after 2 h). The proposed outer-sphere mechanism includes loss of hydrogen from the ruthenium dihydride complex (**28a**) to give an amido species (**28b**), followed by addition of the alcohol to reform the amine functionality in the ligand and the metal dihydride (Scheme 16.15). In the case of ethanol being the hydrogen source, the process can also be used for the highly efficient and environmentally benign synthesis of ethyl acetate in the absence of an acceptor, thus lowering the amount of waste produced [89].

One of the most interesting substrates in terms of applications in an energy consuming device is however methanol, as it displays a very high hydrogen content of 12.6 wt% being liquid at ambient conditions. For this reason, the concept of a "methanol economy" as an alternative to the aforementioned "hydrogen economy" was intensively discussed [90]. The efficient reforming of methanol as an import aspect to realise this concept is however very challenging and requires high temperatures of more than 200 °C. Important contributions towards a homogeneous approach of methanol reforming were made by the group of Beller. Aqueous-phase methanol dehydrogenation and reforming virtually without formation of C1 residuals CH_4 and CO were observed using a molecularly defined ruthenium pincer

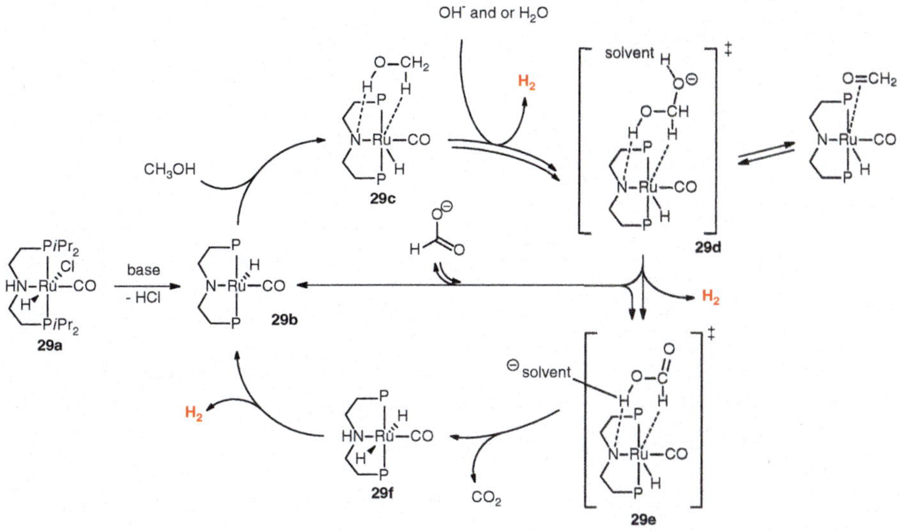

Scheme 16.16 Proposed catalytic cycle of homogeneous methanol reforming. Note that in the catalytic cycle the PNP pincer ligand is simplified for clarity

complex **29a** (Scheme 16.16) [91]. The described catalyst is highly active, is long-term stable and operates at temperatures around 90 °C. The proposed dehydrogenation mechanism involves initial base-assisted elimination of HCl to generate the active species **29b**, which can activate methanol in an outer-sphere mechanism that involves the amine/amide functionality of the ligand (**29c**). Elimination of hydrogen and interaction with hydroxide most likely furnish a *gem*-diolate species **29d** that can also be referred to as a masked formaldehyde complex. Another outer-sphere dehydrogenation step gives a formate complex **29e**, from which CO_2 can dissociate to give the amine group in the ligand as well as the ruthenium dihydride motif (**29f**). Finally, a third equivalent of hydrogen is generated to close the catalytic cycle.

A methanol-reforming system based on a molecularly defined nonnoble metal catalyst (*H*PNP)FeH(HBH_3)(CO) (*H*PNP = *H*N($CH_2CH_2PiPr_2$)$_2$) was found to produce hydrogen for a period of several days with a TON of up to 10,000, thus indicating that further developments in this field should pave the way to applications, e.g. in PEM fuel cells [92]. Similarly as for the above catalyst, base is needed to abstract the anionic ligand. Attempts to further optimise methanol reforming resulted in the development of base-free catalyst systems. Trincado, Grützmacher and co-workers demonstrated that full dehydrogenation of a methanol/water mixture is possible using a ruthenium hydride complex with a chelating bis(olefin) diazadiene ligand without using additional base (**30**) (Fig. 16.4a) [93]. The special feature of this compound is its ability to store up to two equivalents of hydrogen intramolecularly. Due to this, the oxidation states of the ruthenium centre change between 0 and +2. The use of a precatalyst without chloride ligand proved to be

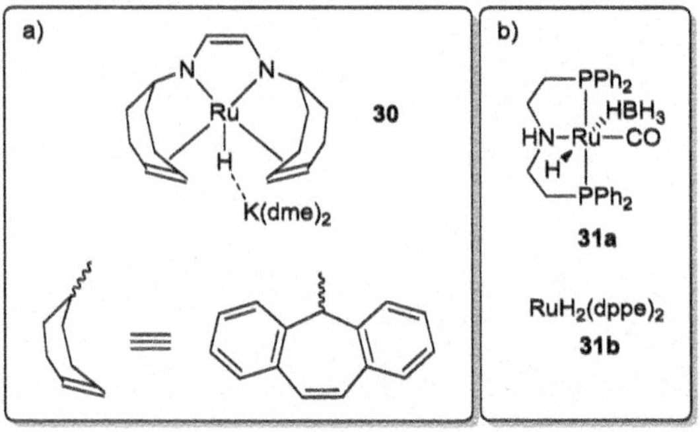

Fig. 16.4 Single component (**a**) and multicomponent (**b**) catalysts for the base-free catalytic reforming of methanol

necessary in order to observe catalytic activity for methanol dehydrogenation in the absence of base. For further improvements of the efficiency of ruthenium-based methanol-reforming systems, a bicatalytic approach can be applied, as demonstrated by Beller et al. [94]. The combination of two complexes Ru(MACHO)BH (**31a**) and $RuH_2(dppe)_2$ (**31b**) (Fig. 16.4b) gives significantly higher activities at mild conditions than one of the complexes alone. This effect can be rationalised, as formally, during methanol reforming, three separate dehydrogenation steps (methanol → formaldehyde → FA → CO_2) must be performed. Hence, it is sensible to add a second catalyst that can improve the rate of the methanol dehydrogenation by accelerating the decomposition of in situ formed FA.

16.4 Conclusions

The studies highlighted in this chapter demonstrate the vital importance of a fundamental understanding of organometallic complexes and its elementary reactivity for the development of new catalytic systems for hydrogen generation and storage in various storage media. Rational optimisations of catalyst systems and interpretation of side reactions and degradation pathways never go without deeper insights into elementary steps of the respective catalytic cycles. Some of the catalytic systems described herein display activities that already meet the requirements needed for potential applications. Further evolution of these as well as the transfer of such chemical solutions into applications designed by engineers is one main target of research in this field.

References

1. Bockris JOM (1972) A hydrogen economy. Science 176:1323
2. Riis T, Hagen EF, Vie PJS, Ulleberg O (2006) Hydrogen production and storage. R&D priorities and gaps. Available via https://www.iea.org/publications/freepublications. Accessed 12 Nov 2013

3. One widely discussed concept in this context is power-to-gas, see: Gahleitner G (2013) Hydrogen from renewable electricity: an international review of power-to-gas pilot plants for stationary applications. Int J Hydrog Energy 38:2039–2061
4. Collins DJ, Zhou H-C (2007) Hydrogen storage in metal–organic frameworks. J Mater Chem 17:3154–3160
5. Hügle T, Hartl M, Lentz D (2011) The route to a feasible hydrogen-storage material: MOFs versus ammonia borane. Chem Eur J 17:10184–10207
6. Note that only selected examples will be presented in this contributions as the main focus of this chapter will be on fundamental concepts of hydrogen storage using organometallic compounds
7. Sobota M, Nikiforidis I, Amende M, Zanón BS, Staudt T, Höfert O, Lykhach Y, Papp C, Hieringer W, Laurin M, Assenbaum D, Wasserscheid P, Steinrück H-P, Görling A, Libuda J (2011) Dehydrogenation of dodecahydro-*N*-ethylcarbazole on Pd/Al_2O_3 model catalysts. Chem Eur J 17:11542–11552
8. Eblagon KM, Rentsch D, Friedrichs O, Remhof A, Zuettel A, Ramirez-Cuesta AJ, Tsang SC (2010) Hydrogenation of 9-ethylcarbazole as a prototype of a liquid hydrogen carrier. Int J Hydrog Energy 35:11609–11621
9. Welch GC, San Juan RR, Masuda JD, Stephan DW (2006) Reversible, metal-free hydrogen activation. Science 314:1124–1126
10. Zhu Z, Wang X, Peng Y, Lei H, Fettinger JC, Rivard E, Power PP (2009) Addition of hydrogen or ammonia to a low-valent group 13 metal species at 25 °C and 1 atmosphere. Angew Chem Int Ed 48:2031–2034
11. Peng Y, Guo J-D, Ellis BD, Zhu Z, Fettinger J, Nagase S, Power PP (2009) Reaction of hydrogen or ammonia with unsaturated germanium or tin molecules under ambient conditions: oxidative addition versus arene elimination. J Am Chem Soc 131:16272–16282
12. Power PP (2010) Main-group elements as transition metals. Nature 463:171–177
13. Hamilton CW, Baker RT, Staubitz A, Manners I (2009) B-N compounds for chemical hydrogen storage. Chem Soc Rev 38:279–293
14. Yan J-M, Zhang X-B, Han S, Shioyama H, Xu Q (2008) Iron-nanoparticle-catalyzed hydrolytic dehydrogenation of ammonia borane for chemical hydrogen storage. Angew Chem Int Ed 47:2287–2289
15. Metin Ö, Mazumder V, Özkar S, Sun S (2010) Monodisperse nickel nanoparticles and their catalysis in hydrolytic dehydrogenation of ammonia borane. J Am Chem Soc 132:1468–1469
16. Yan J-M, Zhang X-B, Akita T, Haruta M, Xu Q (2010) One-step seeding growth of magnetically recyclable Au@Co core–shell nanoparticles: highly efficient catalyst for hydrolytic dehydrogenation of ammonia borane. J Am Chem Soc 132:5326–5327
17. Ciganda R, Garralda MA, Ibarlucea L, Pinilla E, Rosario Torres M (2010) A hydridoirida-β-diketone as an efficient and robust homogeneous catalyst for the hydrolysis of ammonia–borane or amine–borane adducts in air to produce hydrogen. Dalton Trans 39:7226–7229
18. Graham TW, Tsang C-W, Chen X, Guo R, Jia W, Lu S-M, Sui-Seng C, Ewart CB, Lough A, Amoroso D, Abdur-Rashid K (2010) Catalytic solvolysis of ammonia borane. Angew Chem Int Ed 49:8708–8711
19. Boulho C, Djukic J-P (2010) The dehydrogenation of ammonia–borane catalysed by dicarbonylruthenacyclic(II) complexes. Dalton Trans 39:8893–8905
20. Staubitz A, Robertson APM, Sloan MF, Manners I (2010) Amine- and phosphine-borane adducts: new interest in old molecules. Chem Rev 110:4023–4078
21. Staubitz A, Robertson APM, Manners I (2010) Ammonia-borane and related compounds as dihydrogen sources. Chem Rev 110:4079–4124
22. Keaton RJ, Blacquiere JM, Baker RT (2007) Base metal catalyzed dehydrogenation of ammonia-borane for chemical hydrogen storage. J Am Chem Soc 129:1844–1845
23. Yang X, Hall MB (2008) The catalytic dehydrogenation of ammonia-borane involving an unexpected hydrogen transfer to ligated carbene and subsequent carbon-hydrogen activation. J Am Chem Soc 130:1798–1799

24. Baker RT, Gordon JC, Hamilton CW, Henson NJ, Lin P-H, Maguire S, Murugesu M, Scott BL, Smythe NC (2012) Iron complex-catalyzed ammonia–borane dehydrogenation. A potential route toward B–N-containing polymer motifs using earth-abundant metal catalysts. J Am Chem Soc 134:5598–5609
25. Vance JR, Robertson APM, Lee K, Manners I (2011) Photoactivated, iron-catalyzed dehydrocoupling of amine–borane adducts: formation of boron–nitrogen oligomers and polymers. Chem Eur J 17:4099–4103
26. Conley BL, Williams TJ (2010) Dehydrogenation of ammonia-borane by Shvo's catalyst. Chem Commun 46:4815–4817
27. Conley BL, Guess D, Williams TJ (2011) A robust, air-stable, reusable ruthenium catalyst for dehydrogenation of ammonia borane. J Am Chem Soc 133:14212–14215
28. Käß M, Friedrich A, Drees M, Schneider S (2009) Ruthenium complexes with cooperative PNP ligands: bifunctional catalysts for the dehydrogenation of ammonia-borane. Angew Chem Int Ed 48:905–907
29. Noyori R, Ohkuma T (2001) Asymmetric catalysis by architectural and functional molecular engineering: practical chemo- and stereoselective hydrogenation of ketones. Angew Chem Int Ed 40:40–73
30. Denney MC, Pons V, Hebden TJ, Heinekey MD, Goldberg KI (2006) Efficient catalysis of ammonia borane dehydrogenation. J Am Chem Soc 128:12048–12049
31. Jaska CA, Temple K, Lough AJ, Manners I (2003) Transition metal-catalyzed formation of boron-nitrogen bonds: catalytic dehydrocoupling of amine-borane adducts to form aminoboranes and borazines. J Am Chem Soc 125:9424–9434
32. Hügle T, Kühnel MF, Lentz D (2009) Hydrazine borane: a promising hydrogen storage material. J Am Chem Soc 131:7444–7446
33. Hannauer J, Akdim O, Demirci UB, Geantet C, Herrmann J-M, Miele P, Xu Q (2011) High-extent dehydrogenation of hydrazine borane $N_2H_4BH_3$ by hydrolysis of BH_3 and decomposition of N_2H_4. Energy Environ Sci 4:3355–3358
34. Thomas J, Klahn M, Spannenberg A, Beweries T (2013) Group 4 metallocene catalysed full dehydrogenation of hydrazine borane. Dalton Trans 42:14668–14672
35. Pun D, Lobkovsky E, Chirik PJ (2007) Amineborane dehydrogenation promoted by isolable zirconium sandwich, titanium sandwich and N_2 complexes. Chem Commun 42:3297–3299
36. Beweries T, Hansen S, Kessler M, Klahn M, Rosenthal U (2011) Catalytic dehydrogenation of dimethylamine borane by group 4 metallocene alkyne complexes and homoleptic amido compounds. Dalton Trans 40:7689–7692
37. Beweries T, Thomas J, Klahn M, Schulz A, Heller D, Rosenthal U (2011) Catalytic and kinetic studies of the dehydrogenation of dimethylamine borane with an *i*Pr substituted titanocene catalyst. ChemCatChem 3:1865–1868
38. Sloan ME, Staubitz A, Clark TJ, Russell CA, Lloyd-Jones GC, Manners I (2010) Homogeneous catalytic dehydrocoupling/dehydrogenation of amine-borane adducts by early transition metal, group 4 metallocene complexes. J Am Chem Soc 132:3831–3841
39. Helten H, Dutta B, Vance JR, Sloan ME, Haddow MF, Sproules S, Collison D, Whitell GR, Lloyd-Jones GC, Manners I (2012) Paramagnetic titanium(III) and zirconium(III) metallocene complexes as precatalysts for the dehydrocoupling/dehydrogenation of amine-boranes. Angew Chem Int Ed 125:455–458
40. Chapman AM, Haddow MF, Wass DF (2011) Frustrated Lewis pairs beyond the main group: cationic zirconocene-phosphinoaryloxide complexes and their application in catalytic dehydrogenation of amine boranes. J Am Chem Soc 133:8826–8829
41. Vogt M, De Bruin B, Berke H, Trincado M, Grützmacher H (2011) Amino olefin nickel(I) and nickel(0) complexes as dehydrogenation catalysts for amine boranes. Chem Sci 2:723–727
42. Sewell LJ, Huertos MA, Dickinson ME, Weller AS (2013) Dehydrocoupling of dimethylamine borane catalyzed by $Rh(PCy_3)_2H_2Cl$. Inorg Chem 52:4509–4516
43. Perutz RN, Sabo-Etienne S (2007) The σ-CAM mechanism: σ complexes as the basis of σ-bond metathesis at late-transition-metal centers. Angew Chem Int Ed 46:2578–2592

44. Stevens CJ, Dallanegra R, Chaplin AB, Weller AS, Macgregor SA, Ward B, McKay D, Alcaraz G, Sabo-Etienne S (2011) $[Ir(PCy_3)_2(H)_2(H_2B{=}NMe_2)]^+$ as a latent source of aminoborane: probing the role of metal in the dehydrocoupling of $H_3B \cdot NMe_2H$ and retrodimerisation of $[H_2BNMe_2]_2$. Chem Eur J 17:3011–3020
45. Tang CY, Phillips N, Bates JI, Thompson AL, Gutmann MJ, Aldridge S (2012) Dimethylamine borane dehydrogenation chemistry: syntheses, X-ray and neutron diffraction studies of 18-electron aminoborane and 14-electron aminoboryl complexes. Chem Commun 48:8096–8098
46. Liptrot DJ, Hill MS, Mahon MF, MacDougall DJ (2011) Group 2 promoted hydrogen release from $NMe_2H \cdot BH_3$: intermediates and catalysis. Chem Eur J 16:8508–8515
47. Cowley HJ, Holt MS, Melen RL, Rawson JM, Wright DS (2011) Catalytic dehydrocoupling of Me_2NHBH_3 with $Al(NMe_2)_3$. Chem Commun 47:2682–2684
48. Hanssmann MM, Melen RL, Wright DS (2011) Group 13 BN dehydrocoupling reagents, similar to transition metal catalysts but with unique reactivity. Chem Sci 2:1554–1559
49. Erickson KA, Wright DS, Waterman R (2013) Dehydrocoupling of amine boranes via tin (IV) and tin(II) catalysts. J Organomet Chem 751:541–545
50. Reller C, Mertens FORL (2012) A self-contained regeneration scheme for spent ammonia borane based on the catalytic hydrodechlorination of BCl_3. Angew Chem Int Ed 51:11731–11735
51. Sutton AD, Burrell AK, Dixon DA, Garner EB III, Gordon JC, Nakagawa T, Ott KC, Robinson JP, Vasiliu M (2011) Regeneration of ammonia borane spent fuel by direct reaction with hydrazine and liquid ammonia. Science 331:1426–1429
52. Davis BL, Dixon DA, Garner EB, Gordon JC, Matus MH, Scott B, Stephens FH (2009) Efficient regeneration of partially spent ammonia borane fuel. Angew Chem Int Ed 48:6812–6816
53. Coffey RS (1967) The decomposition of formic acid catalysed by soluble metal complexes. Chem Commun 923–924
54. Gao Y, Kuncheria JK, Jenkins HA, Puddephatt RJ, Yap GPA (2000) The interconversion of formic acid and hydrogen/carbon dioxide using a binuclear ruthenium complex catalyst. J Chem Soc Dalton Trans 3212–3217
55. Morris DJ, Clarkson GJ, Wills M (2000) Insights into hydrogen generation from formic acid using ruthenium complexes. Organometallics 28:4133–4140
56. Loges B, Boddien A, Junge H, Beller M (2008) Controlled generation of hydrogen from formic acid amine adducts at room temperature and application in H_2/O_2 fuel cells. Angew Chem Int Ed 47:3962–3965
57. Boddien A, Loges B, Junge H, Beller M (2008) Hydrogen generation at ambient conditions: application in fuel cells. ChemSusChem 1:751–758
58. Loges B, Boddien A, Hunge H, Noyes JR, Baumann W, Beller M (2009) Chem Commun 45:4185-4187
59. Fellay C, Dyson PJ, Laurenczy G (2008) A viable hydrogen-storage system based on selective formic acid decomposition with a ruthenium catalyst. Angew Chem Int Ed 47:3966–3968
60. Fellay C, Yan N, Dyson PJ, Laurenczy G (2009) Selective formic acid decomposition for high-pressure hydrogen generation: a mechanistic study. Chem Eur J 15:3752–3760
61. Hull JF, Himeda Y, Wang W-H, Hashiguchi B, Periana R, Szalda DJ, Muckerman JT, Fujita E (2012) Reversible hydrogen storage using CO_2 and a proton-switchable iridium catalyst in aqueous media under mild temperatures and pressures. Nat Chem 4:383–388
62. Boddien A, Gärtner F, Jackstell R, Junge H, Spannenberg A, Baumann W, Ludwig R, Beller M (2010) Ortho-metalation of iron(0) tribenzylphosphine complexes: homogeneous catalysts for the generation of hydrogen from formic acid. Angew Chem Int Ed 49:8993–8996
63. Boddien A, Loges B, Gärtner F, Torborg C, Fumino K, Junge H, Ludwig R, Beller M (2010) Iron-catalyzed hydrogen production from formic acid. J Am Chem Soc 132:8924–8934

64. Boddien A, Mellmann D, Gärtner F, Jackstell R, Junge H, Dyson PJ, Laurenczy G, Ludwig R, Beller M (2012) Efficient dehydrogenation of formic acid using an iron catalyst. Science 333:1733–1736
65. Oldenhof S, de Bruin B, Lutz M, Siegler MA, Patureau FW, van der Vlugt JI, Reek J (2013) Base-free production of H_2 by dehydrogenation of formic acid using an iridium–bisMETAMORPhos complex. Chem Eur J 19:11507–11511
66. Himeda Y (2009) Highly efficient hydrogen evolution by decomposition of formic acid using an iridium catalyst with 4,4′-dihydroxy-2,2′-bipyridine. Green Chem 11:2018–2022
67. Leitner W (1995) Carbon dioxide as a raw material: the synthesis of formic acid and its derivatives from CO_2. Angew Chem Int Ed Engl 34:2207–2221
68. Jessop PG, Ikariya T, Noyori R (1996) Homogeneous hydrogenation of carbon dioxide. Chem Rev 95:259–272
69. Wang W, Wang S, Ma S, Gong J (2012) Recent advances in catalytic hydrogenation of carbon dioxide. Chem Soc Rev 40:3703–3727
70. Tanaka R, Yamashita M, Nozaki K (2009) Catalytic hydrogenation of carbon dioxide using Ir (III)-pincer complexes. J Am Chem Soc 131:14168–14169
71. Federsel C, Boddien A, Jackstell R, Jennerjahn R, Dyson PJ, Scopelliti R, Laurenczy G, Beller M (2010) A well-defined iron catalyst for the reduction of bicarbonates and carbon dioxide to formates, alkyl formates, and formamides. Angew Chem Int Ed 49:9777–9780
72. Federsel C, Ziebart C, Jackstell R, Baumann W, Beller M (2012) Catalytic hydrogenation of carbon dioxide and bicarbonates with a well-defined cobalt dihydrogen complex. Chem Eur J 18:72–75
73. Fujita E, Muckerman JT, Himeda Y (2013) Interconversion of CO_2 and formic acid by bio-inspired Ir complexes with pendent bases. Biochim Biophys Acta 1827:1031–1038
74. Boddien A, Federsel C, Gärtner F, Sponholz P, Mellmann D, Jackstell R, Junge H, Beller M (2011) CO_2-"neutral" hydrogen storage based on bicarbonates and formats. Angew Chem Int Ed 50:6411–6414
75. Himeda Y, Miyazawa S, Hirose T (2011) Interconversion between formic acid and H_2/CO_2 using rhodium and ruthenium catalysts for CO_2 fixation and H_2 storage. ChemSusChem 4:487–493
76. Tanaka R, Yamashita M, Chung LW, Morokuma K, Nozaki K (2011) Mechanistic studies on the reversible hydrogenation of carbon dioxide catalyzed by an Ir-PNP complex. Organometallics 30:6742–6750
77. Papp G, Csorba J, Laurenczy G, Joo F (2011) A charge/discharge device for chemical hydrogen storage and generation. Angew Chem Int Ed 50:10433–10435
78. Johnson TC, Morris DJ, Wills M (2010) Hydrogen generation from formic acid and alcohols using homogeneous catalysts. Chem Soc Rev 39:81–88
79. Dobson A, Robinson SD (1977) Complexes of the platinum metals. 7. Homogeneous ruthenium and osmium catalysts for the dehydrogenation of primary and secondary alcohols. Inorg Chem 16:137–142
80. Ligthart GBWL, Meijer RH, Donners MPJ, Meuldijk J, Vekemans JAJM, Hulshof LA (2003) Highly sustainable catalytic dehydrogenation of alcohols with evolution of hydrogen gas. Tetrahedron Lett 44:1507–1509
81. van Buijtenen J, Meuldijk J, Vekemans JAJM, Hulshof LA, Kooijman H, Spek AL (2006) Dinuclear ruthenium complexes bearing dicarboxylate and phosphine ligands. Acceptorless catalytic dehydrogenation of 1-phenylethanol. Organometallics 25:873–881
82. Morton D, Cole-Hamilton DJ (1987) Rapid thermal hydrogen production from alcohols catalysed by [Rh(2,2′-bipyridyl)$_2$]Cl. J Chem Soc Chem Commun 248–249
83. Morton D, Cole-Hamilton DJ (1988) Molecular hydrogen complexes in catalysis: highly efficient hydrogen production from alcoholic substrates catalysed by ruthenium complexes. J Chem Soc Chem Commun 1154–1156
84. Junge H, Beller M (2005) Ruthenium-catalyzed generation of hydrogen from *iso*-propanol. Tetrahedron Lett 46:1031–1034

85. Junge H, Loges B, Beller M (2007) Novel improved ruthenium catalysts for the generation of hydrogen from alcohols. Chem Commun 42:522–524
86. Zhang J, Leitus G, Ben-David Y, Milstein D (2005) Facile conversion of alcohols into esters and dihydrogen catalyzed by new ruthenium complexes. J Am Chem Soc 127:10840–10841
87. Gunanathan C, Ben-David Y, Milstein D (2007) Direct synthesis of amides from alcohols and amines with liberation of H_2. Science 317:790–792
88. Nielsen M, Kammer A, Cozzula D, Junge H, Gladiali S, Beller M (2011) Efficient hydrogen production from alcohols under mild reaction conditions. Angew Chem Int Ed 50:9593–9597
89. Nielsen M, Junge H, Kammer A, Beller M (2012) Towards a green process for bulk-scale synthesis of ethyl acetate: efficient acceptorless dehydrogenation of ethanol. Angew Chem Int Ed 51:5711–5713
90. Olah GA, Goeppert A, Surya Prakash GK (2009) Beyond oil and gas: the methanol economy, second updated and enlarged edition. Wiley-VCH Verlag GmbH & Co. KGaA, Weinheim
91. Nielsen M, Alberico E, Baumann W, Drexler H-J, Junge H, Gladiali S, Beller M (2013) Low-temperature aqueous-phase methanol dehydrogenation to hydrogen and carbon dioxide. Nature 495:85–90
92. Alberico E, Sponholz P, Cordes C, Nielsen M, Drexler H-J, Baumann W, Junge H, Beller M (2013) Selective hydrogen production from methanol with a defined iron pincer catalyst under mild conditions. Angew Chem Int Ed 52:14162–14166
93. Rodriguez-Lugo RE, Trincado M, Vogt M, Tewes F, Santiso-Quinones G, Grützmacher H (2013) A homogeneous transition metal complex for clean hydrogen production from methanol–water mixtures. Nat Chem 5:342–347
94. Monney A, Barsch E, Sponholz P, Junge H, Ludwig R, Beller M (2013) Base-free hydrogen generation from methanol using a bi-catalytic system. Chem Commun 50:707–709

Chapter 17
Hybrid Systems Consisting of Redox-Active π-Conjugated Polymers and Transition Metals or Nanoparticles

Toshiyuki Moriuchi, Toru Amaya, and Toshikazu Hirao

Abstract Nitrogen atoms of the quinonediimine (QD) moiety of the emeraldine base of poly(*o*-toluidine) (POT) are capable of participating in the complexation with metals (Pd, V, Cu, etc) to afford the single-strand or cross-linked network conjugated complexes. The complexation of the redox-active π-conjugated 1,4-benzoquinonediimines, unit molecules of the emeraldine base of polyanilines (PAn's), with the palladium(II) complexes affords a variety of conjugated complexes. The conjugated complexes can be used as electronic materials and catalysts, for example, the redox mediator in the Wacker reaction. The regulation of the coordination mode of the QD moiety permits controlled formation of the conjugated bimetallic or polymeric complexes. The introduction of the chiral complex into the QD nitrogens of POT via coordination interaction induces chirality into a π-conjugated backbone of POT, affording optically active conjugated complexes. Metal nanoparticles (NPs) such as Pd and Au are also assembled with PAn's to afford PAn's/metal NPs hybrids. Three representative approaches including direct reduction approach, template approach, and ligand exchange approach are developed here. The PAn's metal NPs hybrid can be utilized as a catalyst.

Keywords Polyaniline • Phenylenediamine • Quinonediimine • Redox active • Coordination mode • Conjugated complex • Chiral complexation • Metal nanoparticles

17.1 Introduction

π-Conjugated polymers and oligomers have attracted much attention in the application of electronic materials depending on their redox properties [1–5]. Polyanilines (PAn's) are promising conducting π-conjugated polymers with

T. Moriuchi • T. Amaya • T. Hirao (✉)
Department of Applied Chemistry, Graduate School of Engineering, Osaka University, Yamada-oka, Suita, Osaka 565-0871, Japan
e-mail: hirao@chem.eng.osaka-u.ac.jp

W.-Y. Wong (ed.), *Organometallics and Related Molecules for Energy Conversion*, Green Chemistry and Sustainable Technology, DOI 10.1007/978-3-662-46054-2_17

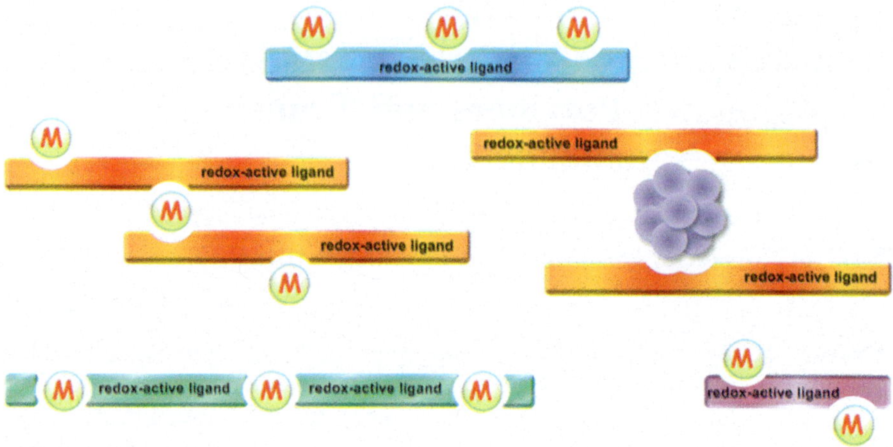

Fig. 17.1 Controlled design of the conjugated complexes with redox-active π-conjugated ligands

chemical stability. PAn has been extensively studied because of its unique electronic and redox properties as well as numerous potential use in a wide range of applications in many fields. PAn's exist in three different discrete redox forms, which include the fully reduced leucoemeraldine, the semioxidized emeraldine, and the fully oxidized pernigraniline base forms [1]. The neutral emeraldine base is composed of the quinonediimine (QD) and phenylenediamine (PD) moieties as oxidized and reduced forms, respectively. Another interesting function of PAn's is coordination properties of two nitrogen atoms of the QD moiety. The function of π-conjugated polymers is expected to be modified dramatically by incorporation of metal centers into the polymers [6–14]. π-Conjugated polymers and molecules, which possess redox-active properties and coordination sites, are allowed to serve as redox-active ligands to give the d,π-conjugated complexes. Depending on the number and geometry of the coordination sites, the systems with a variety of designed structures can be constructed as shown in Fig. 17.1. In the case of both metals and ligands having two coordination sites, these components are arrayed alternatively to give the corresponding polymer complexes. The multinuclear complex is derived by multi-coordination of the π-conjugated polymers. This can be extended to the complexes of metal nanoparticles (NPs) with the π-conjugated polymers. Such hybrids of metal NPs and π-conjugated polymers can be of their potential applicability as electronic devices, chemical sensors, and catalysts. This chapter summarizes the conjugated complexes with redox-active π-conjugated PAn's and 1,4-benzoquinonediimines.

17.2 Hybrid Systems Consisting of Redox-Active π-Conjugated Polymers and Transition Metals

17.2.1 Controlled Design of Conjugated Complexes with Redox-Active π-Conjugated Polyanilines and 1,4-Benzoquinonediimines

The complexation of the emeraldine base of poly(*o*-toluidine) (POT) with palladium(II) compounds is demonstrated in an organic solvent to afford the conjugated polymer complexes [15]. Two nitrogen atoms of the quinonediimine (QD) moiety are found to be available for complexation in the case of $Pd(OAc)_2$ and $PdCl_2(MeCN)_2$, which have two coordination sites, giving the cross-linked conjugated complexes as shown in Scheme 17.1. Poly(3-heptylpyrrole) is performed to serve as an efficient π-conjugated polymer ligand to afford the similar conjugated complex with $PdCl_2(MeCN)_2$. An organic light-emitting diode device with the thus-obtained conjugated complex film as a hole injection layer shows the maximum luminance of 11,000 cd/m^2 at 10 V, which is 2 V lower than a device

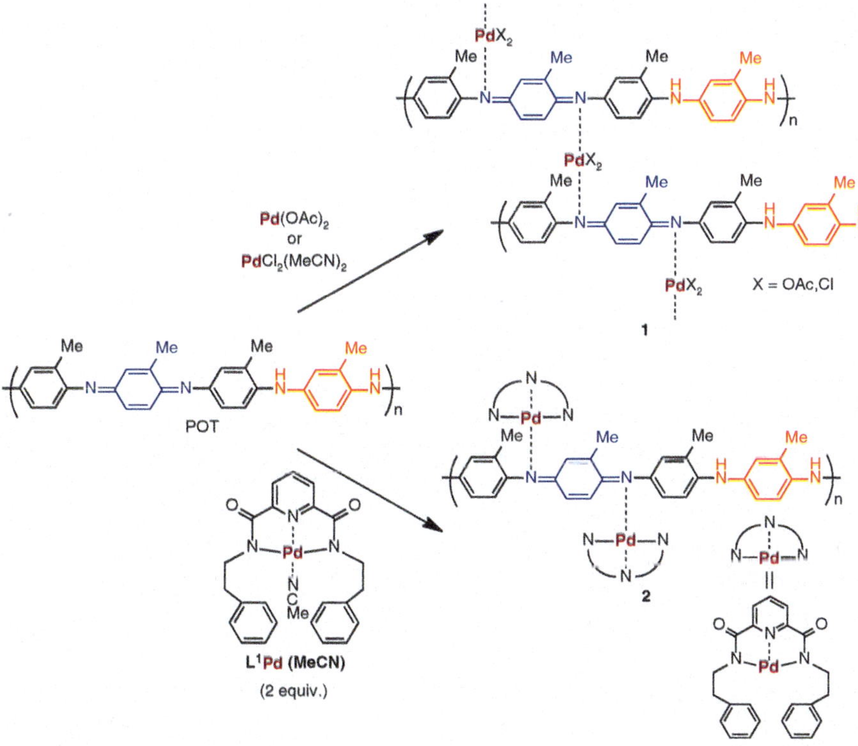

Scheme 17.1 Controlled formation of cross-linked conjugated complexes **1** and single-strand conjugated complex **2**

with the conventional copper phthalocyanine hole injection layer [16]. In contrast, the single-strand conjugated complex **2** is likely to be formed by using the palladium(II) complex [L^1Pd(MeCN)] [17, 18] bearing one interchangeable coordination site, which is obtained by treatment of the *N*-heterocyclic tridentate ligand, *N*, *N'*-bis(2-phenylethyl)-2,6-pyridinedicarboxamide (L^1H_2) [19, 20] with $Pd(OAc)_2$ (Scheme 17.1).

The conjugated polymer complexes with PAn's or polypyrroles are demonstrated to afford the redox systems depending on their structures and redox properties. For example, copper salts can affect the redox properties of PAn's to form the reversible redox cycle [21]. The conjugated polymer complex can be effectively employed as an oxidation catalyst [22–25], wherein the coordination of the QD moiety is considered to play an important role in reversible redox processes of the complex. In the Wacker reaction, PAn's or polypyrroles serve as a redox-active ligand.

The complexation behavior of the redox-active π-conjugated molecule, *N,N'*-bis (4'-dimethylaminophenyl)-1,4-benzoquinonediimine (L^2) [26], as a model molecule of PAn affords the further insight into the coordination and redox properties in the complexation with the QD moieties. Complexation of L^2 with [$PdCl_2(MeCN)_2$] having two interchangeable coordination sites in acetonitrile leads to the formation of the conjugated polymeric complex **3**, in which palladium centers are incorporated in the main chain (Scheme 17.2) [27].

Regulation of the coordination mode of the QD moiety is possible. The 1:2 conjugated homobimetallic palladium(II) complex [(L^1)Pd(L^2)Pd(L^1)] (**4**) is formed by the complexation of L^2 with two equimolar amounts of the palladium (II) complex [L^1Pd(MeCN)] as shown in Scheme 17.2 [28]. Variable temperature ^{1}H NMR studies on the conjugated complex **4** indicate that the *syn* configuration is enthalpically more favorable than the *anti* configuration in CD_2Cl_2, but entropically less favorable. The crystal structure of **4-*anti*** shows that the two [L^1Pd] units are bridged by the QD moiety of L^2 in *anti* conformation as depicted in Fig. 17.2. The steric repulsion between the hydrogen atoms at C(32) and at C(38) causes the phenylene ring of L^2 to rotate away from the orientation parallel to the QD moiety, resulting in a propeller twist between the planes of the two phenylene rings. The redox properties of the QD moiety are found to be modulated by complexation with the palladium(II) complex [L^1Pd(MeCN)]. The conjugated complex **4** in dichloromethane shows the successive one-electron reduction of the QD moiety to give the corresponding reduced species [28]. This result is in sharp contrast to the redox behavior of L^2 in dichloromethane, wherein an irreversible reduction wave is observed. Compared with the uncomplexed QD, the complexed QD is stabilized as an electron sink.

*17.2.2 Redox Complexation of Poly(*o*-toluidine)*

Electronic interaction of transition metals with the π-conjugated polymers through the coordination is envisaged to provide efficient redox systems. Vanadium compounds can exist in a variety of oxidation states and generally convert between the

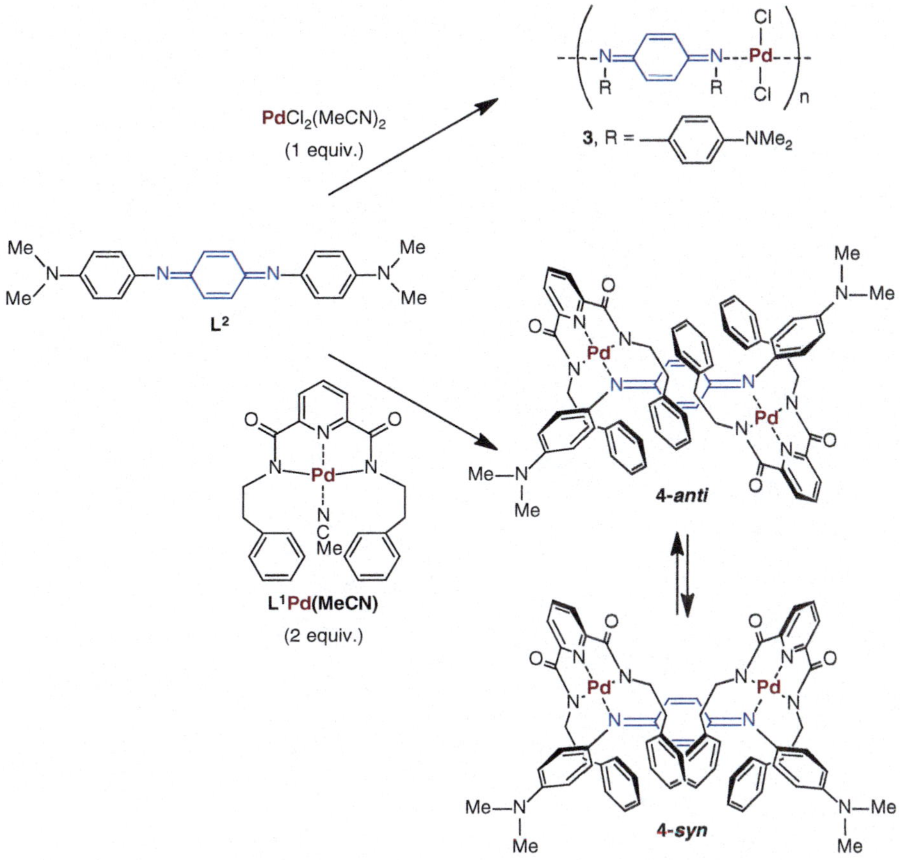

Scheme 17.2 Controlled formation of the conjugated polymeric complex **3** and the 1:2 conjugated homobimetallic palladium(II) complex **4**

states via a one-electron redox process [29]. Vanadium compounds in low oxidation states can serve as one-electron reductants, as exemplified by the redox process of V(III) to V(IV). The emeraldine base of POT is performed to undergo redox complexation with VCl_3, affording the corresponding conjugated complexes **5** (Scheme 17.3) [30]. The complexation proceeds via reduction of the QD moiety with oxidation of V(III) to V(IV), wherein the vanadium species is considered to play an important role in both the complexation and redox reactions.

17.2.3 Chirality Induction into the π-Conjugated Backbone of Polyanilines

Chirality induction into PAn's and oligomers has attracted much attention because of their potential use in diverse areas such as surface modified electrodes, molecular

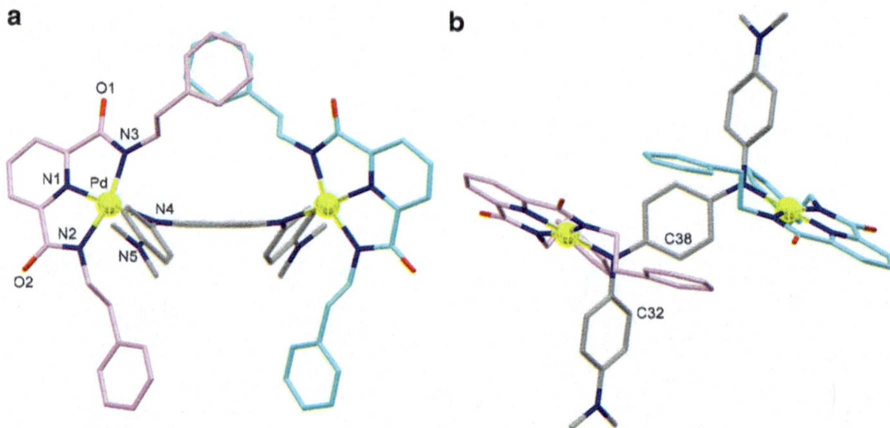

Fig. 17.2 (**a**) A top view and (**b**) a side view of the crystal structure of **4-*anti*** (hydrogen atoms are omitted for clarity)

Scheme 17.3 Complexation behavior of POT with VCl_3

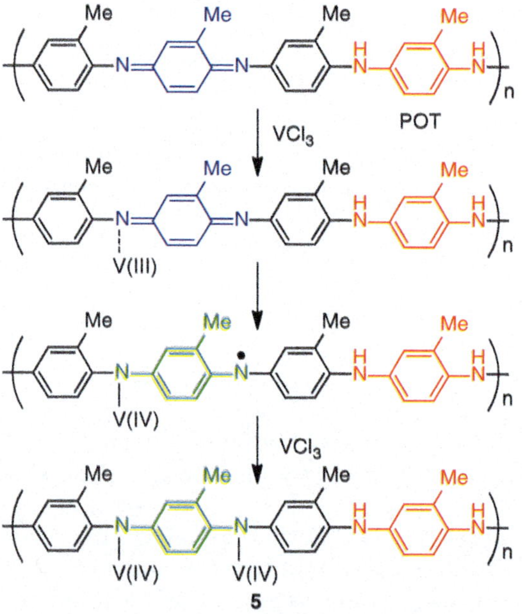

recognition, and chiral separation [31]. The reaction of the emeraldine base of POT with the chiral palladium(II) complex [(*S*,*S*)-L^3Pd(MeCN)] induces chirality to a π-conjugated backbone of POT, giving the corresponding optically active conjugated polymer complexes **(*S*,*S*)-6** (Scheme 17.4) [32, 33]. The CD spectrum of **(*S*,*S*)-6** shows an induced circular dichroism (ICD) at the absorbance region of the π-conjugated moiety around 500–800 nm (Fig. 17.3). The conjugated polymer complex **(*R*,*R*)-6** exhibits the mirror imaged ICD signal around 500–800 nm (Fig. 17.3), supporting the chirality induction into a π-conjugated backbone of

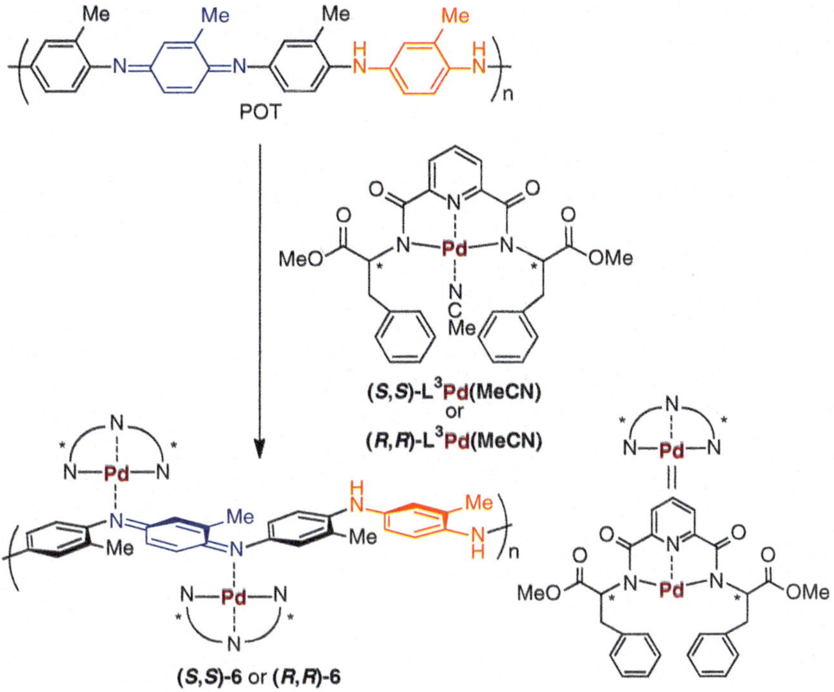

Scheme 17.4 Formation of optically active conjugated polymer complexes

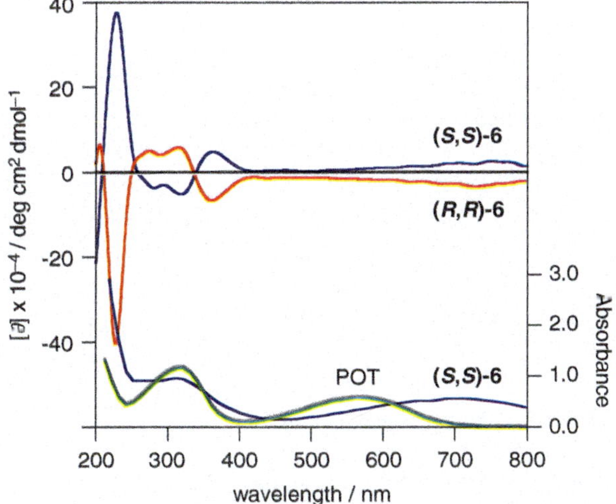

Fig. 17.3 CD spectra (top) of **(*S*,*S*)-6** and **(*R*,*R*)-6**, and UV-vis spectra (bottom) of **(*S*,*S*)-6** and POT in THF (1.3×10^{-3} M)

POT. The helical conformation with a predominant screw sense might be formed through chiral complexation.

17.3 Hybrid Systems Consisting of Redox-Active π-Conjugated Polymers and Transition Metal Nanoparticles

17.3.1 PAn/Pd Nanoparticles

Hybrid systems of metal NPs and π-conjugated polymers or molecules have become of great interest because of their potential applicability to electronic devices, chemical sensors, and catalysts [34]. In the catalytic applications, the NPs with small size and high surface-to-bulk ratio often exhibit advantages compared with the bulk materials [35]. Therefore, the smaller and well-dispersed NPs are desired. Metal NPs generally need a stabilizer (polymers are often used) to prevent metal NPs from the aggregation [36]. On the other hand, Pd NPs are known to catalyze various reactions [37]. Furthermore, their catalytic activity is often superior to that of the mononuclear Pd complexes [37]. In this section, the syntheses of PAn/Pd NPs are summarized, including three different synthetic approaches: (1) direct reduction approach, (2) template approach, and (3) ligand exchange approach.

17.3.1.1 Direct Reduction Approach

Direct reduction approach is the simple synthesis based on mixing of the metal ion and PAn followed by the reduction to form the metal NPs (Scheme 17.5a) [38].

Reductants such as hydroquinone, EtOH, ascorbic acid, or $NaBH_4$ are investigated, where ten-time excess amounts of PAn (emeraldine base) to $PdCl_2$ are employed (four units of the monomer are used to calculate the molar ratio). As a result of optimization, the procedure using $NaBH_4$ in the presence of $PdCl_2$ and PAn (emeraldine base) with 1/10 ratio affords small particle size, well dispersity, and better reproducibility [38]. The TEM image is shown in Scheme 17.5b (2–6 nm, average diameter = 3.4 nm). The particles are ascertained to be palladium in the EDX experiment.

17.3.1.2 Template Approach

Template approach involves two stages: (1) the complex formation of metal ion with PAn and (2) the reductive aggregation of metallic species of the coordinated complex [38, 39]. In the case of the direct reduction approach, generally, use

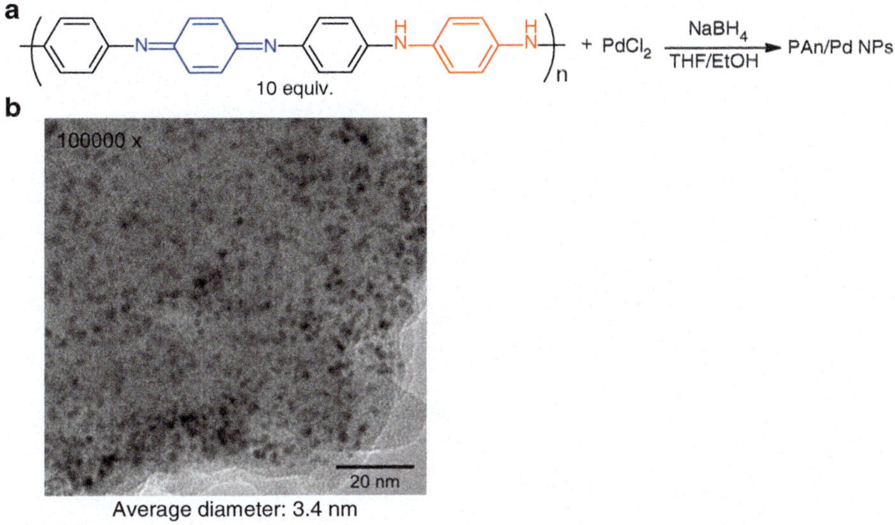

a

10 equiv.

+ $PdCl_2$ $\xrightarrow[\text{THF/EtOH}]{NaBH_4}$ PAn/Pd NPs

b

100000 x

20 nm

Average diameter: 3.4 nm

Scheme 17.5 (**a**) Synthesis of PAn/Pd NPs by direct reduction approach and (**b**) the TEM image

of excess amounts of the polymer to the metal ion is necessary to prepare small and well-dispersed NPs to inhibit the formation of largely aggregated metal colonies. The template approach is envisaged to provide small and well-dispersed NPs even in a high molar ratio of metal ion/polymer because the metal species are pre-organized on the polymer.

Scheme 17.6a shows template approach for the synthesis of PAn/Pd NPs [38, 39]. Preparation of Pd(II) pre-organized polymer is already described in Sect. 17.2.1; here $Pd(OAc)_2$ is employed. No NPs is observed by the TEM analysis of the thus-obtained complex. Reduction of the pre-organized complex gives small and dispersed NPs by reduction with refluxing EtOH, where EtOH works as a reductant. On the other hand, reduction not via pre-organization affords more aggregated particles. Thus, the effect of pre-organization is confirmed. Reduction of the complex with $NaBH_4$ also yields the PAn/Pd NPs (average diameter = 2.4 nm). The TEM image is shown in Scheme 17.6b. Notably, each particle is independent while the palladium density is high (16 wt%). The thus-obtained NPs catalyze the oxidative dimerization of 2,6-di-*t*-butylphenol to give the corresponding diphenoquinone compound in a good yield.

17.3.1.3 Ligand Exchange Approach

Direct reduction approach and template one cause partial or complete reduction of redox-active ligand PAn, whose redox state significantly contributes to the electronic, coordinating, and catalytic properties. Ligand exchange approach offers

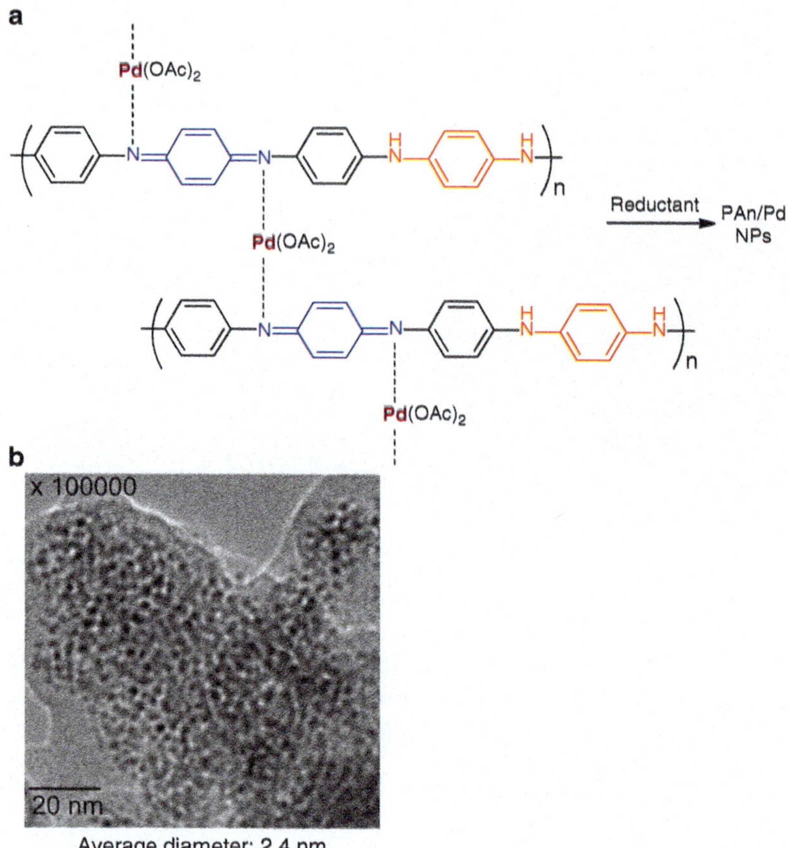

Scheme 17.6 (**a**) Synthesis of PAn/Pd NPs by template approach and (**b**) the TEM image

a reductant-less method to preserve the redox state of PAn. This method has two stages: the preparation of metal NPs with ligand A by addition of reductant and then ligand exchange reaction from ligand A to ligand B to provide metal NPs. The key of this method is the choice of ligand A, which is required to stabilize the metal NPs well and be easily removable by ligand exchange. Starch is selected as a ligand A based on the reasons that the coordination to the hydroxy groups in starch is weaker than that to the imino and amino groups of PAn, and the high water solubility permits to separate starch after ligand exchange by washing only with water. $Pd(OAc)_2$ and PAn (emeraldine base) are selected as a metal salt and a ligand B, respectively.

Preparation of the small and well-dispersed starch/Pd NPs (average diameter = 2.3 nm) is achieved by reduction of $Pd(OAc)_2$ with $NaBH_4$ in the presence of starch, followed by neutralization with 1 M HCl (Scheme 17.7a).

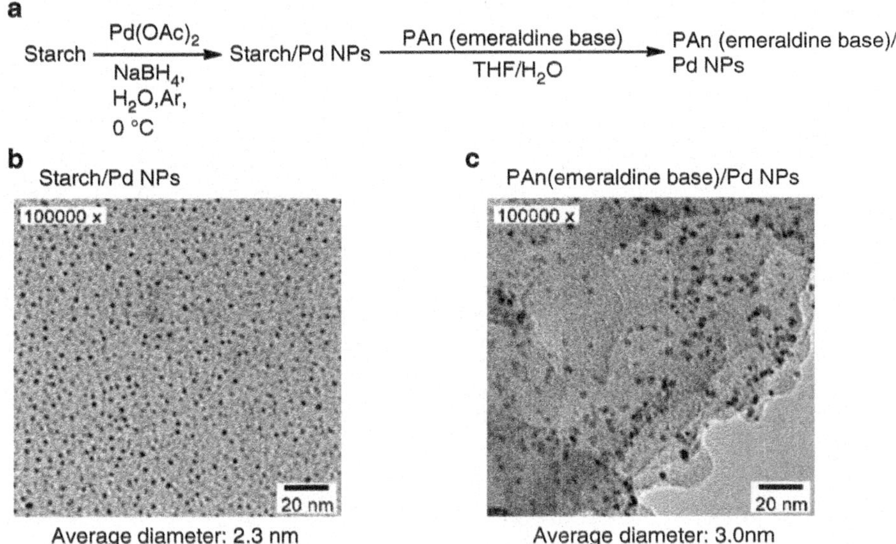

Scheme 17.7 (**a**) Synthesis of PAn/Pd NPs by ligand exchange approach and TEM images for (**b**) the starch/Pd NPs and (**c**) the PAn/Pd NPs

TEM image is shown in Scheme 17.7b [40]. The increasing size and aggregation of particles are not observed at least when the solution is kept at room temperature for 2 weeks.

The reaction procedure of ligand exchange is quite simple as follows: (1) mixing of PAn (emeraldine base) and starch/Pd NPs, (2) stirring for a day, (3) evaporation, (4) suspending in water, (5) filtration, and (6) washing the dark-blue residue thoroughly with water to remove starch. In order to dissolve both hydrophobic PAn (emeraldine base) and hydrophilic starch/Pd NPs, a cosolvent THF/H_2O (v/v =1:1) solution is used (Scheme 17.7a). The thus-obtained PAn (emeraldine base)/ Pd NPs are confirmed with IR and TEM. The IR spectrum of starch/Pd NPs exhibits the characteristic peak at around 3,300 cm^{-1} assignable to the stretching mode of the hydroxyl group. After ligand exchange, this characteristic peak disappears in the IR spectrum; instead the sharp peaks appear at 1,596 and 1,490 cm^{-1} assignable to the quinonediimine and the phenylenediamine of PAn (emeraldine base), respectively. The peak intensity ratio of two characteristic absorptions at 1,596 and 1,490 cm^{-1} is almost the same in both PAn (emeraldine base) and PAn (emeraldine base)/Pd NPs. These results clearly indicate that the ligand exchange reaction proceeds completely with preservation of the redox state of PANI (emeraldine base). The particle sizes are distributed in a range of diameter 2–7 nm (average diameter = 3.0 nm), showing a little growth of the particles during the ligand exchange reaction (Scheme 17.7c) [40].

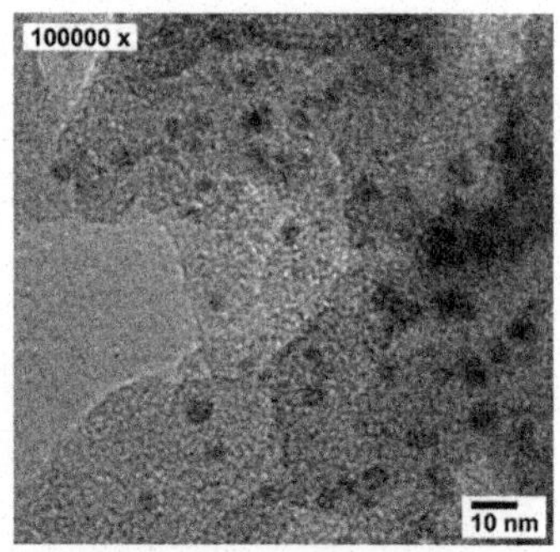

Fig. 17.4 TEM image for PAn (emeraldine base)/iron oxide NPs

17.3.2 PAn/Iron Oxide Nanoparticles

Hybrids of metal oxide NPs and π-conjugated polymers are also expected to create a novel class of electronic device materials and catalysts due to the characteristic properties as an electron mediator.

Synthesis of PAn (emeraldine base)/iron oxide NPs is demonstrated, and they are applied for the catalysis [41]. Fe(acac)$_3$ is employed as an iron source. The formation of iron oxide NPs depends on the thermal reaction under argon, which is carried out in the presence of PAn (emeraldine base) in DMF to give PAn (emeraldine base)/iron oxide NPs. The particle size is distributed in a range of diameter 3–9 nm as shown in the TEM image (Fig. 17.4). The presence of Fe is confirmed by EDX experiment. The characteristic absorptions for PAn (emeraldine base) are observed in the IR spectrum (1,596 cm^{-1} for the quinonediimine and 1,490 cm^{-1} for the phenylenediamine). There are no peaks derived from Fe(acac)$_3$. PAn (emeraldine base)/iron oxide NPs catalyzes the oxidative dimerization of 2,6-di-*t*-butylphenol to give the corresponding diphenoquinone compound.

17.3.3 PMAS/Au Nanoparticles

Synthesis of the water-soluble poly(2-methoxyaniline-5-sulfonic acid) (PMAS)/Au NPs is demonstrated for the catalytic aerobic dehydrogenative oxidation in aqueous solution [42]. The synthetic procedure is as follows: (1) treatment of PMAS (half ox) with $N_2H_4 \cdot H_2O$ and (2) addition of $NaAuCl_4$ to the reaction mixture (Scheme 17.8a), where the reduced PMAS can be a reductant to yield the Au NPs. TEM image of the PMAS/Au NPs is shown in Scheme 17.8b. The average diameter is 7.9 nm.

The thus-obtained PMAS/Au NPs can catalyze the aerobic dehydrogenative oxidative reactions of alcohols [42], 2-substituted indolines [43], *N*-phenyltetrahydroisoquinolines [44], and secondary amines [45] in an aqueous solution, where the redox mediating effect of PMAS is suggested.

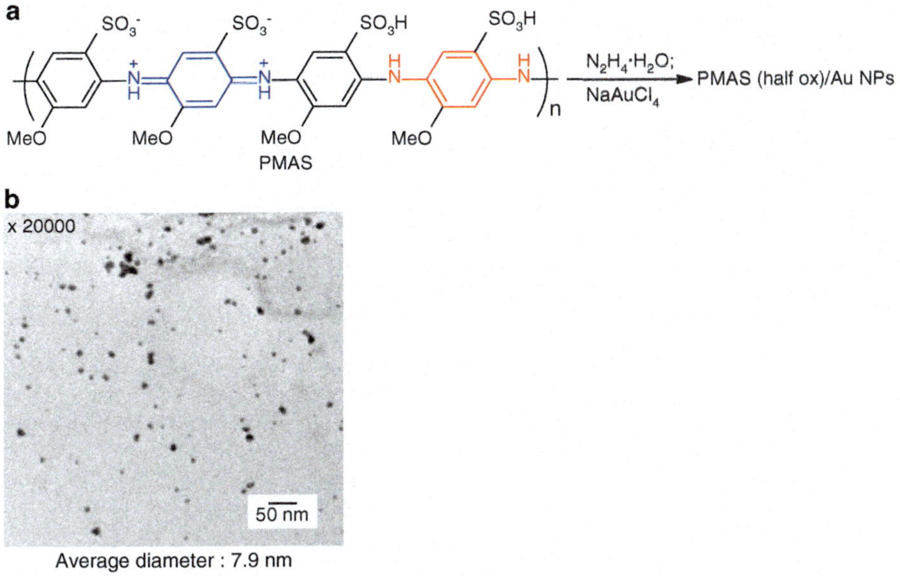

Scheme 17.8 (a) Synthesis of PMAS/Au NPs and (b) the TEM image

17.4 Conclusions

The π-conjugated PAn's and QD derivatives are demonstrated as a potential redox-active ligand based on the coordination ability of the QD nitrogen atoms, affording d,π-conjugated complexes. The combination of both redox properties provides an efficient redox system for electronic materials and catalysts. The controlled formation of the conjugated complexes with redox-active-π-conjugated PAn's and QD derivatives is achieved by the regulation of the coordination mode of the QD moiety and transition metal directed assembly. Incorporated metals play an important role as a metallic dopant, in which complexed QD becomes stabilized as an electron sink. Chirality induction into a π-conjugated backbone of PAn's is performed by chiral complexation through coordination interaction, affording optically active d,π-conjugated complexes. Metal NPs such as Pd and Au are also assembled with PAn's to afford PAn's/metal NPs hybrids. Three representative approaches including direct reduction approach, template approach, and ligand exchange approach are introduced here. The PAn's metal NPs hybrid can be utilized as a catalyst, in which PAn's serve as a redox mediator. Such conjugated complexes and NPs composed of the redox-active conjugated ligands are envisioned to provide future potential functional materials and redox catalysts.

References

1. Alcácer L (1987) Conducting polymers: special applications. Reidel, Dordrecht
2. Salaneck WR, Clark DT, Samuelsen EJ (1990) Science and application of conductive polymers. Adams Hilger, New York
3. Shirakawa H (2001) The discovery of polyacetylene film: the dawning of an era of conducting polymers (Nobel lecture). Angew Chem Int Ed 40:2574–2580
4. MacDiarmid AG (2001) "Synthetic metals": a novel role for organic polymers (Nobel lecture). Angew Chem Int Ed 40:2581–2590
5. Heeger AJ (2001) Semiconducting and metallic polymers: the fourth generation of polymeric materials (Nobel lecture). Angew Chem Int Ed 40:2591–2611
6. Holliday BJ, Swager TM (2005) Conducting metallopolymers: the roles of molecular architecture and redox matching. Chem Commun 23–36
7. Wolf MO (2006) Recent advances in conjugated transition metal-containing polymers and materials. J Inorg Organomet Polym Mater 16:189–199
8. Nguyen P, Gómez-Elipe P, Manners I (1999) Organometallic polymers with transition metals in the main chain. Chem Rev 99:1515–1548
9. Wong W-Y, Ho C-L (2006) Di-, oligo- and polymetallaynes: syntheses, photophysics, structures and applications. Coord Chem Rev 250:2627–2690
10. Hirao T (2002) Conjugated systems composed of transition metals and redox-active-π-conjugated ligands. Coord Chem Rev 226:81–91
11. Hirao T (2002) Conjugated complex system composed of quinonediimine unit. Macromol Symp 186:75–80
12. Hirao T (2006) Redox systems under nano-space control. Springer, Heidelberg
13. Amaya T, Hirao T (2011) Synthesis and application of redox-active hybrid catalytic systems consisting of polyanilines and transition metals. Synlett 435–448
14. Moriuchi T, Hirao T (2012) Design and redox function of conjugated complexes with polyanilines or quinonediimines. Acc Chem Res 45:347–360
15. Hirao T, Yamaguchi S, Fukuhara S (1999) Controlled formation of synthetic metal-transition metal conjugated complex systems. Tetrahedron Lett 40:3009–3012
16. Hirao T, Otomaru Y, Inoue Y, Moriuchi T, Ogata T, Sato Y (2006) Conjugated palladium complex with poly(3-heptylpyrrole) and its application. Synth Met 156:1378–1382
17. Moriuchi M, Bandoh S, Miyaji Y, Hirao T (2000) A novel heterobimetallic complex composed of the imide-bridged [3]ferrocenophane and the tridentate palladium(II) complex. J Organomet Chem 599:135–142
18. Moriuchi T, Kamikawa M, Bandoh S, Hirao T (2002) Architectural formation of a conjugated bimetallic Pd(II) complex via oxidative complexation and a tetracyclic Pd(II) complex via self-assembling complexation. Chem Commun 1476–1477
19. Hirao T, Moriuchi T, Mikami S, Ikeda I, Ohshiro Y (1993) A novel system for oxygenation. Effect of multidentate podand ligand in transition metal catalyzed epoxidation with molecular oxygen. Tetrahedron Lett 34:1031–1034
20. Hirao T, Moriuchi T, Ishikawa T, Nishimura K, Mikami S, Ohshiro Y, Ikeda I (1996) A novel catalytic system for oxygenation with molecular oxygen induced by transition metal complexes with a multidentate N-heterocyclic podand ligand. J Mol Catal A 113:117–130
21. Higuchi M, Imoda D, Hirao T (1996) Redox behavior of polyaniline-transition metal complexes in solution. Macromolecules 29:8277–8279
22. Hirao H, Higuchi M, Ikeda I, Ohshiro Y (1993) A novel synthetic metal catalytic system for dehydrogenative oxidation based on redox of polyaniline. J Chem Soc Chem Commun 194–195
23. Hirao H, Higuchi M, Hatano B, Ikeda I (1995) A novel redox system for the palladium(II)-catalyzed oxidation based on redox of polyanilines. Tetrahedron Lett 36:5925–5928
24. Higuchi M, Yamaguchi S, Hirao T (1996) Construction of palladium-polypyrrole catalytic system in the Wacker oxidation. Synlett 1213–1214

25. Higuchi M, Ikeda I, Hirao T (1997) A novel synthetic metal catalytic system. J Org Chem 62:1072–1078
26. Wei Y, Yang C, Ding T (1996) A one-step method to synthesize N, N'-bis(4′-aminophenyl)-1,4-quinonenediimine and its derivatives. Tetrahedron Lett 37:731–734
27. Moriuchi T, Miyaishi M, Hirao T (2001) Conjugated complexes composed of quinonediimine and palladium: controlled formation of a conjugated trimetallic macrocycle. Angew Chem Int Ed 40:3042–3045
28. Moriuchi T, Bandoh S, Miyaishi M, Hirao T (2001) A novel redox-active conjugated palladium homobimetallic complex. Eur J Inorg Chem 651–657
29. Hirao T (1997) Vanadium in modern organic synthesis. Chem Rev 97:2707–2724
30. Hirao T, Fukuhara S, Otomaru Y, Moriuchi T (2001) Conjugated complexes via oxidative complexation of polyaniline derivatives to vanadium(III). Synth Met 123:373–376
31. Kane-Maguire LAP, Wallace GG (2010) Chiral conducting polymers. Chem Soc Rev 39:2545–2576
32. Shen X, Moriuchi T, Hirao T (2004) Chirality induction of polyaniline derivatives through chiral complexation. Tetrahedron Lett 45:4733–4736
33. Moriuchi T, Shen X, Hirao T (2006) Chirality induction of π-conjugated chains through chiral complexation. Tetrahedron 62:12237–12246
34. Gangopadhyay R, De A (2000) Conducting polymer nanocomposites: a brief overview. Chem Mater 12:608–622
35. Fihri A, Bouhrara M, Nekoueishahraki B, Basset J-M, Polshettiwar V (2011) Nanocatalysts for Suzuki cross-coupling reactions. Chem Soc Rev 40:5181–5203
36. Roucoux A, Schulz J, Patin H (2002) Reduced transition metal colloids: a novel family of reusable catalysts? Chem Rev 102:3757–3778
37. Balanta A, Godard C, Claver C (2011) Pd nanoparticles for C–C coupling reactions. Chem Soc Rev 40:4973–4985
38. Amaya T, Saio D, Hirao T (2008) Versatile synthesis of polyaniline/Pd nanoparticles and catalytic application. Macromol Symp 270:88–94
39. Amaya T, Saio D, Hirao T (2007) Template synthesis of polyaniline/Pd nanoparticle and its catalytic application. Tetrahedron Lett 48:2729–2732
40. Saio D, Amaya T, Hirao T (2009) Synthesis of polyaniline/Pd nanoparticles via ligand exchange. J Inorg Organomet Polym 19:79–84
41. Amaya T, Nishina Y, Saio D, Hirao T (2008) Hybrid of polyaniline/iron oxide nanoparticles: facile preparation and catalytic application. Chem Lett 37:68–69
42. Saio D, Amaya T, Hirao T (2010) Redox-active catalyst based on poly(anilinesulfonic acid)-supported gold nanoparticles for aerobic alcohol oxidation in water. Adv Synth Catal 352:2177–2182
43. Amaya T, Ito T, Inada Y, Saio D, Hirao T (2012) Gold nanoparticles catalyst with redox-active poly(aniline sulfonic acid):application in aerobic dehydrogenative oxidation of cyclic amines in aqueous solution. Tetrahedron Lett 53:6144–6147
44. Amaya T, Ito T, Hirao T (2012) Selective cross-dehydrogenative coupling of *N*-phenyltetrahydroisoquinolines in aqueous media using poly(2-methoxyaniline-5-sulfonic acid)/gold nanoparticles. Heterocycles 86:927–932
45. Amaya T, Ito T, Hirao T (2013) Aerobic dehydrogenative imination in complete aqueous media catalyzed by poly(aniline sulfonic acid)/gold nanoparticles. Tetrahedron Lett 54:2409–2411

Chapter 18
Photochemical Solar Energy Conversion and Storage Using Cyclometalated Iridium Complexes

Zhen-Tao Yu

Abstract The appealing photophysical properties of cyclometalated iridium complexes, such as their intense and highly efficient luminescence and long-lived excited states, render them highly desirable for the conversion and storage of solar energy. In this chapter, we describe general considerations in terms of pursuing these applications and track the successful use of cyclometalated iridium complexes as photosensitizers in dye-sensitized solar cells (DSSCs) and light-driven hydrogen production, which have gained a large amount of attention in recent years. Particular emphasis is placed on the systematic elucidation of the correlation between the complex architectures, the corresponding photophysical behavior, and the aforementioned potential promising applications of cyclometalated iridium complexes, which are expected to contribute possible design implications to the development of superior light-harvesting components for efficiently driving photosynthesis.

Keywords Dye-sensitized solar cells • Iridium • Photochemistry • Photosensitizer • Water splitting

18.1 Introduction

The demand for sustainable energy is one of the most important challenges confronting the world today and far into the future [1]. As a long-term, carbon-neutral renewable energy source, the utilization of the sun's abundant energy is attracting increasingly great attention all over the world and has yet to fulfill its promise of providing an attractive alternative to traditional fuels [2]. One promising long-term solution for directly and efficiently harvesting and storing energy is to build artificial photosynthetic systems capable of solar energy conversion with concepts derived from natural photosynthesis [3]. Along these lines, one alternative

Z.-T. Yu (✉)
Collaborative Innovation Center of Advanced Microstructures, College of Engineering and Applied Sciences, Nanjing University, Nanjing, Jiangsu 210093, China
e-mail: yuzt@nju.edu.cn

W.-Y. Wong (ed.), *Organometallics and Related Molecules for Energy Conversion*, Green Chemistry and Sustainable Technology, DOI 10.1007/978-3-662-46054-2_18

approach is solar-driven hydrogen production via photoelectrochemical water splitting [4], which can store energy in the form of fuel. The research on photo-assisted hydrogen production from water began with the discovery of the TiO_2-based Honda-Fujishima effect in 1972, which demonstrated that hydrogen can be produced via the bandgap excitation of TiO_2 by ultraviolet light [5]. Instead of using semiconductor particles as the light absorber, in 1977, Lehn and Sauvage reported the first working artificial photosynthesis model for producing hydrogen using a Ru(II) polypyridine complex-based intermolecular system [6]. Another approach is the development of efficient solar cells, from which solar radiation can be directly converted into electrical power, unlike in photosynthesis. However, such cells cannot store energy. The landmark discovery in solar conversion research was a low-cost sandwich-type solar cell based on dye-sensitized colloidal TiO_2 films reported by O'Regan and Grätzel in 1991, which works quite similarly to a photosynthetic plant cell and does not produce fuel or use water as a source of electrons [7].

Light-induced charge separation is necessary for all of these energy conversion devices and is coupled to redox reactions [8]. The first requirement of such processes is a key photoactive component for capturing the sun's energy, giving rise to the efficient generation of a long-lived charge-separated state between adjacent donor or acceptor molecules and facilitating the extremely complex reduction of substrates at low overpotentials. In the past decade, both dye-sensitized solar cells and artificial photosynthesis have achieved photon absorption using synthetic photoactive and redox-active organometallic complexes of transition metals [9], especially since the first reports on solar conversion using ruthenium–polypyridyl complexes [10].

Transition metal complexes of d^6-configured metal ions with organic ligands are attracting great interest, as they generally strongly absorb light in the visible region due to their exceptional electronic structure to produce stable and long-lived excited states [11], which is of crucial importance for efficient charge generation and separation. The charge transfer transitions in metal complexes involve electron transfer between the metal and the ligands, including metal-to-ligand charge transfer (MLCT, for readily oxidized metal ions and ligands with low-lying acceptor orbitals) and ligand-to-metal charge transfer (LMCT, for readily reduced metal ions with strong donor ligands). The MLCT state, having an electron vacancy in the metal d-electron manifold and an excess electron in the acceptor ligand-based orbitals, is quite reactive and is probably the key to the photochemical behavior of the complexes [12]. Furthermore, these complexes can be designed to have readily accessible coordination sites for the binding of substrates in solution through coordination geometry changes and can be tuned to the desired optical absorption by careful ligand design and systematic variation of the metal center. In addition, as a successful sensitizer, the energy state of the electron upon excitation in the sensitizer should allow the thermodynamic driving force to be large enough for the charge transfer involved in the related processes [13]. After photoexcitation, the generating excited electron is appropriately energetic in terms of reduction potential to be capable of participating in redox reactions at the liquid/semiconductor interface [14].

To achieve these functions, the octahedral d^6ruthenium-tris(bipyridyl) complex has been extensively studied due to its unique combination of chemical stability and photophysical and acceptable redox properties [15]. However, in general, the hexacoordinate metal center bearing three bidentate ligands, such as bipyridine, exhibits notable chirality; as a result, these complexes are often available as racemic mixtures of enantiomers. From a synthetic perspective, product purification and analysis are often difficult. More importantly, the emission energy of this class of complexes cannot be tuned via the systematic chemical modification of the ligands, which suffer from the characteristics of a low-lying triplet metal-centered excited state [16].

As the d^6 congener of Ru(II), isoelectronic organometallic Ir(III) complexes have been highly appealing in attempts to extend the family of compounds with excited states capable of photochemical electron transfer reactions [17]. For these complexes, the occurrence of close-lying, ligand-centered transitions enables broader tuning of the electronic properties of the complex, such as its emission energy, emission lifetimes, and quantum yields, by modifying the metal coordination environment, potentially allowing control of the photoactivity and redox activity of the metal complex [18]. The stronger metal–ligand bonding between the Ir metal and ligands, which allows for more efficient MLCT to ligand π^* orbitals, provides the complexes with good stability and favorable photophysical and electrochemical properties well suited for solar conversion applications. In this section, we present the recent development of iridium complexes as sensitizers to facilitate the conversion of solar energy into electricity or chemical hydrogen fuel.

18.2 Basic Structure and Photophysical Properties of Iridium Complexes

Cyclometalation, which was discovered in the early 1960s, provides a direct pathway for accessing transition metal compounds that contain a covalent metal–carbon bond in place of a metal–nitrogen bond [19], allowing the tuning of orbital energies and thus the modification of the redox and photophysical properties of the resulting complex. Cyclometalated Ir(III) complexes have been designed and synthesized by the established chemistry [20] and show good light-harvesting capabilities. The family of cyclometalated iridium complexes, including one core Ir atom and three bidentate ligands in an octahedral geometry, most commonly applied are charged bis-cyclometalated complexes, usually with general formulation of $[Ir(C^\wedge N)_2(N^\wedge N)]^+$, and neutral tris-cyclometalated complexes, such as Ir $(C^\wedge N)_3$ [21]. In these complexes, there is sufficient charge compensation for metal-centered oxidation when the monoanionic cyclometalating ligand C^N, commonly a derivative of 2-phenylpyridine (ppy), in which the phenyl ring acts as the anionic

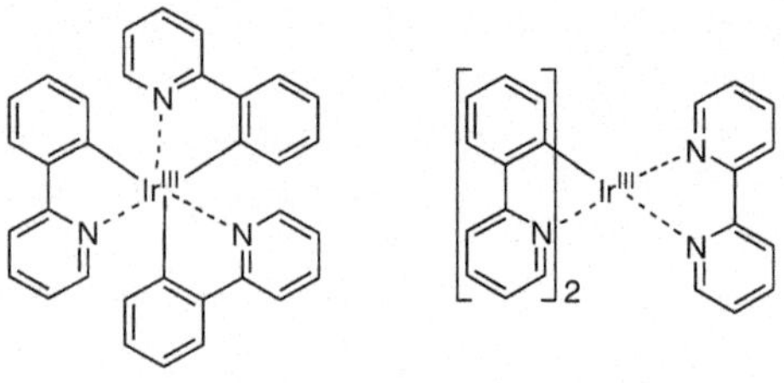

Fig. 18.1 General structure of cyclometalated iridium complexes (**1** and **2**) used as sensitizers

component, is used, binding the metal through an sp^2-hybridized C-atom to produce an ortho-fused five-membered ring through activation of the C–H bond of the phenyl ring in the ortho position. Neutral N^N ligands, such as oligopyridines, are used as ancillary ligands, which donate less charge density to the metal but endow the complexes with large molar extinction coefficients. In terms of the design of such complexes, it is especially important to select chelates with appropriate photophysical characteristics. The simplest cases of a prototypical Ir(III) system, $Ir(ppy)_3$ (**1**) and $[Ir(ppy)_2(bpy)]^+$ (**2**) (bpy = 2,2′-bipyridine), are illustrated in Fig. 18.1 and often serve as model systems.

Based on the Rehm–Weller correlation, the spectroscopic properties of the metal complex affect the redox character of the excited state. Iridium complexes are typically characterized by smaller Stokes shifts relative to similar ruthenium complexes [22]. As a result, excited state iridium complexes act as a much stronger potent reductant or oxidant than their ruthenium counterparts. To allow excited state electron transfer, the lifetime (τ) of the MLCT excited state of the chromophore must be sufficiently long ($\tau > 10^{-9}$ s) [23].

Consideration of the excited state properties, as given above, is important because the excited chromophore can be used to initiate an electron transfer process, and they play a vital role in the function of a light-harvesting complex [24]. Understanding the connection between the properties and structure not only provides a sound understanding of the role of the photoactive complexes in solar energy conversion but also provides information that can be used for developing successful photosynthetic strategies.

In an Ir(III) complex, due to the presence of remarkable delocalized molecular orbitals for the transition, there are at least four concurrent optical transitions: ^{1}MLCT, ^{3}MLCT, and ligand-centered (^{1}LC and ^{3}LC) transitions. These transitions follow the general energetic order of $^1\text{LC} > {}^1\text{MLCT} > {}^3\text{MLCT} > {}^3\text{LC}$ [25]. The cyclometalating ligand tends to be associated with the ^{3}LC transition and the ancillary ligand with the ^{3}MLCT transition. According to Kasha's rule, emission should be due to the lowest triplet state, which is a combination of the ^{3}MLCT and ^{3}LC transition states. Moreover, the strong spin–orbit coupling from the iridium center promotes the sufficiently fast intersystem crossing of the singlet excited state to the energetically triplet manifold that the excited state is usually a mixed excited state, leading to high phosphorescence quantum yields, even at room temperature [26].

Density functional theoretical (DFT) investigations have provided insight into the nature of the excited state in these iridium complexes [16]. In the case of **1**, the delocalized highest occupied molecular orbital (HOMO) primarily resides on the d-orbitals of the iridium atom and π orbitals of the phenyl rings of anionic ligands, while the lowest unoccupied molecular orbital (LUMO) preferentially resides at the formally neutral pyridine of the ligands. Similar results have been attained for the cationic Ir(III) complex **2** [27]. Thus, it appears feasible for the excited state of these complexes to contain mixed MLCT-LC character. Bis-cyclometalated complexes have appreciably lower quantum yields than the tris-cyclometalated family because of the lower LUMO orbitals of the 2,2′-bipyridine ligand [28]. The introduction of suitable electron-donating and electron-withdrawing substituents or changes in their positions at the ligand to decrease or increase the energy of the frontier orbitals (i.e., the HOMO and LUMO) is a promising method for attaining the desired excited state properties for iridium complexes. Generally, the introduction of electron-withdrawing F and/or CF_3 substituents on the phenyl of the C^N ligands, stabilizing the metal-based HOMO, or the introduction of donor substituents, such as dimethylamino groups on the bipyridine, destabilizing the LUMO, can result in high quantum yields. In this way, metal–ligand-based luminescence covering the whole visible spectrum [29], excited state lifetimes in the range of nanoseconds to several microseconds, and high phosphorescent yields close to 100 % can be realized [30]. In general, the properties of the lowest spin-forbidden^{3}MLCT state of the Ir(III) complex meet the requirement for light-harvesting systems in solar energy conversion. However, the iridium complexes absorb a smaller fraction of visible light than the ruthenium complexes, limiting their utilization as a light harvester.

18.3 Conversion of Light into Electricity

In the field of dye-sensitized solar cells (DSSCs), the properties of the sensitizers significantly influence the performance of the photovoltaic devices. The most effective DSSCs reported to date are based on a classical ruthenium–polypyridyl dye, such as cis-bis-(4,4′-dicarboxy-2,2′-bipyridine)dithiocyanato Ru(II) ([Ru $(dcbpy)_2(NCS)_2$], known as N3-dye) and its doubly protonated analog (N719), possessing a very high solar-to-electric power conversion efficiency in excess of 10 % [31]. The carboxylate groups of the dye ensure efficient adsorption of the dye on the surface of mesoporous semiconductor metal oxide (such as nanocrystalline TiO_2) films. The inclusion of two NCS groups in the dye improves visible-light absorption, leading to higher DSSC efficiencies [32]. However, concerning the future prospects of DSSC with these metal complexes, the long-term ability of dyes is greatly hampered by the liberation of the NCS^- ligands from the metal center and toxicity considerations [33].

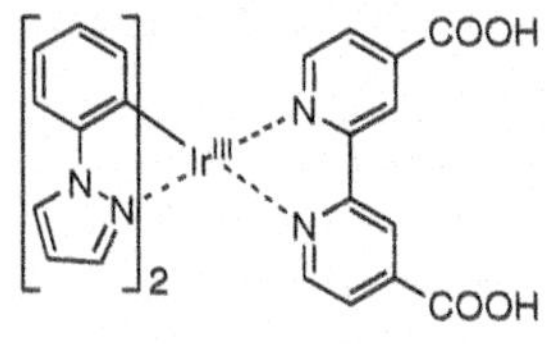

3 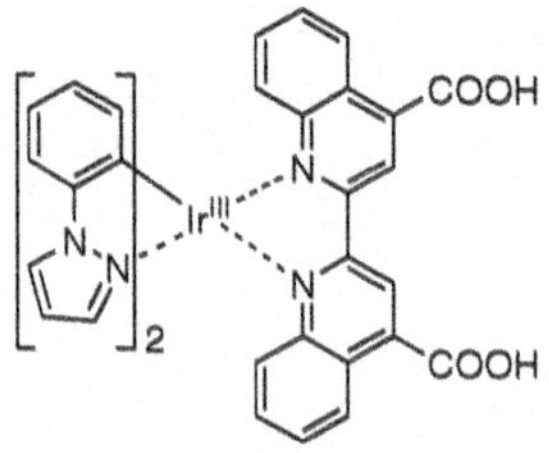 4 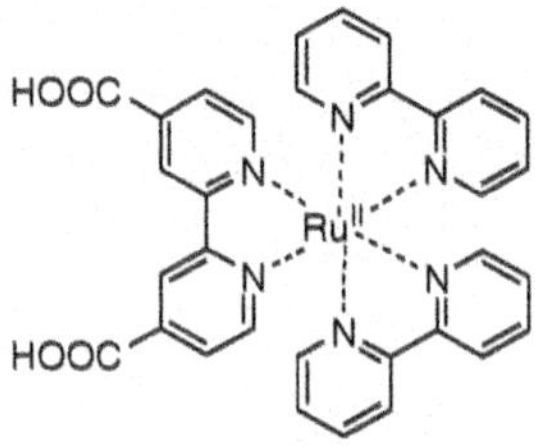 $[Ru(bpy)_2(dcbpy)]^{2+}$

In the solar cell device configuration, the use of cyclometalated iridium complexes as a dye has not been extensively explored. DSSC is fabricated on a fluorine-doped tin oxide (FTO) conductive-glass substrate using a dye-sensitized TiO_2 working electrode, a platinized FTO glass counter electrode, and a typical liquid electrolyte consisting of iodide/triiodide (I^-/I_3^-) ions as a redox couple. In related studies, Gray and coworkers pioneered the use of Ir(III) dyes in sensitizing TiO_2 as a functional solar cell in 2006 [34]. In particular, $[Ir(ppz)_2(dcbpy)]^+$ (**3**) and $[Ir(ppz)_2(dcbq)]^+$ (**4**) (ppz is phenylpyrazole and dcbq is 4,4-dicarboxy-2,2′-biquinoline; with PF_6^- counterions) complexes were reported. Based on DFT calculations, these complexes have absorption maxima at 455 and 495 nm and exhibit relatively weak absorption with a low molar extinction coefficient (700 M^{-1} cm^{-1}) in the visible region of the spectrum relative to a Ru(II)-bipyridyl analog, $[Ru(bpy)_2(dcbpy)]^{2+}$ (with PF_6^- counterions), due to transitions attributed to an LLCT transition from the cyclometalating ligand to the bpy or dcbq ligand. The efficiency (η) for the corresponding TiO_2-based DSSCs under simulated AM 1.0 sunlight was 0.65 for **3** and 0.50 for **4**, compared with 1.0 % for $[Ru(bpy)_2(dcbpy)]^{2+}$. The short-circuit current density (J_{sc}=1.99 and 2.24 mA cm^{-2}) induces a drop in efficiency, which is 3.35 mA cm^{-2} for the Ru (II) case. The report provides specific examples of the charge production based on injection from solely LLCT states [35], which is thought to be favorable to suppress the recombination of the photoinduced charges, rather than the typical MLCT states in ruthenium-based cells, in which the excitation of the dye involves the transfer of an electron from the metal to the π^* orbital of the surface anchoring carboxylated bipyridyl ligand. It is important to recognize that dual sensitization through LLCT and MLCT is expected to endow iridium complexes with better characteristics for DSSC devices.

These studies were subsequently extended to the complexes $[Ir(2\text{-phenyl-4-(2,5-dimethoxystyryl)pyridine})_2(4,4'\text{-dicarboxy(phenylethenyl)-2,2'-bipyridine})]^+$ (**5**), $[Ir(2\text{-phenyl-4-(2,5-dimethoxystyryl)pyridine})_2(dcbpy)]^+$ (**6**), $[Ir(2\text{-(2,4-difluorophenyl)-4-dimethylamino-pyridine})_2(dcbpy)]^+$ (**7**), and $[Ir(2\text{-phenyl-4-carboxylic acid-pyridine})_2(acetylacetonate)]$ (**8**) [36]. These complexes produced DSSCs with efficiencies of only 0.09 % for **5** due to the very low J_{sc} (0.27 mA cm^{-2}), 0.94 % for **6** (J_{sc}=2.70 mA cm^{-2}), 0.79 % for **7** (J_{sc}= 2.70 mA cm^{-2}), and 1.87 % for **8** (J_{sc}=4.30 mA cm^{-2}). The relatively low efficiencies were related to the incident photon-to-current conversion efficiency (IPCE) response of these sensitizers, which does not extend beyond 600 nm.

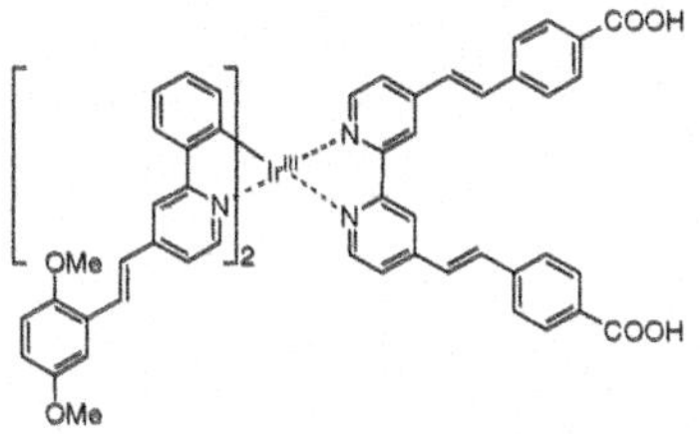

5 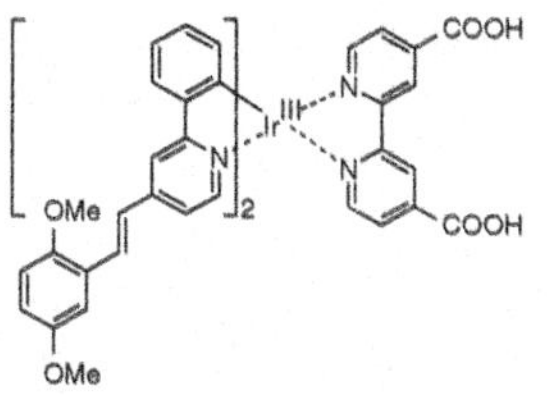 6 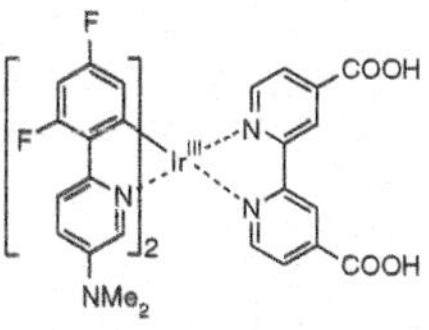 7

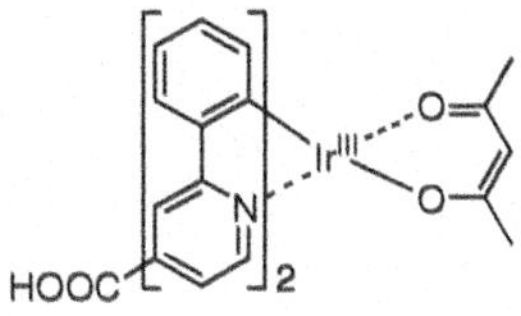

8 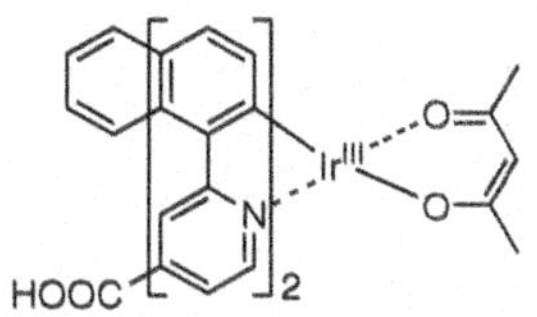 9 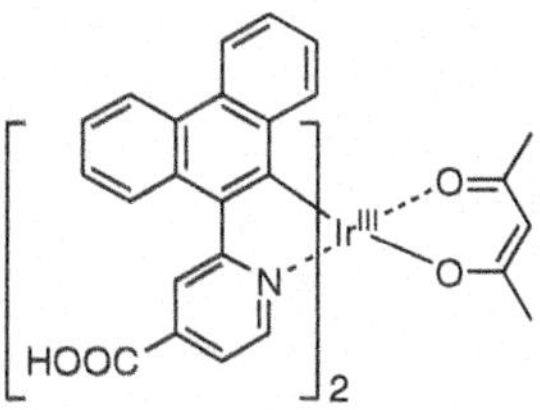 10

Encouraged by the promising performance of the simple neutral molecule **8**, in which the anchoring carboxyl group is directly connected to the main C^N ligands, this class of dyes for DSSCs was further studied [37]. In the visible region, compound **8** exhibits a band at 500 nm extending up to 590 nm with an absorption coefficient of approximately 2,200 M^{-1} cm^{-1}, while compounds **9** and **10** absorb at 472 nm extending up to 630 nm with absorption coefficients of 7,000 M^{-1} cm^{-1} and 4,700 M^{-1} cm^{-1}, respectively. These bands possess an MLCT character. DSSCs based on a 9-μm transparent TiO_2 film possessed J_{sc} values in the following sequence: dyes **10** (5.78 mA cm^{-2})>**9** (5.11 mA cm^{-2})> **8** (4.60 mA cm^{-2}). This ordering is consistent with the UV/Vis absorption results. The open-circuit voltage (V_{oc}) values of dyes **9** (539 mV) and **10** (491 mV) are lower than that of dye **8** (670 mV), which is probably due to the faster recombination of the former two dyes. As a result, the overall efficiency under standard global AM 1.5 solar conditions is 2.23 % for **8**, 1.96 % for **9**, and 2.04 % for **10**, respectively, with the standard electrolyte A6141 (0.6 M N-methyl-N-butyl imidazolium iodide (BMII), 0.03 M I_2, 0.1 M guanidinium thiocyanate, and 0.5 M tert-butylpyridine in acetonitrile/valeronitrile 85:15). However, due to the lack of a standard dye as a reference, the assessment of the benefits of the complexes is difficult. To increase the light-harvesting efficiency, a light-scattering TiO_2 film with approximately 400-nm TiO_2 particles was applied on the transparent TiO_2 layer. With 9+5-μm TiO_2 films, dye **8** shows the highest power conversion efficiency of 2.51 % under full sunlight.

A new type of efficient Ir(III) complex with carboxyl pyridine ligands was designed for the sensitization of TiO_2 injection solar cells [38]. In the visible region, dyes **11–13** show weak absorption assigned to ^{3}MLCT mixed with significant $^3\pi\pi$ transition character. The molar extinction coefficient of dye **12** is higher than those of dyes **11** and **13** due to the presence of the extended π-system conjugation in **12**. The photoluminescence quantum yield of **13** (0.38) is higher than those of **11** (0.0024) and **12** (0.0067), which might be correlated with the electron injection efficiency and therefore device performance of the DSSCs. It is notable that complex **13** was synthesized without anion metathesis using NH_4PF_6. In some cases, similar iridium complexes with dcbpy ligands are charge neutral because the terminal deprotonated carboxyl group can offer charge compensation for the highly charged centered metal ion. The weaker strength of the metal–ligand coordinative bond of the carboxyl pyridine group in these complexes may decrease the emission quantum yield. After adsorption on the TiO_2 film, dye **13** shows the highest absorbance in the long-wavelength region because of its relatively high dye loading and distinct molar extinction coefficient at similar wavelengths. The cell based on dye **13** achieves the highest conversion efficiency of 2.86 % (J_{sc}=9.59 mA cm^{-2}; V_{oc}= 0.55 V), compared with 6.8 % for N719 using MPII/I2 (MPII=1-methyl-3-propylimidazolium iodide) in acetonitrile/3-methyl-2-oxazolidinone 9:1 as an electrolyte. Using Br^-/Br_3^- (E_{ox}=1.09 V) as a redox electrolyte, an increase in the V_{oc} of DSSC was observed at 0.79 V due to the positive potential of the Ir complex (especially for **13**, 1.48 V).

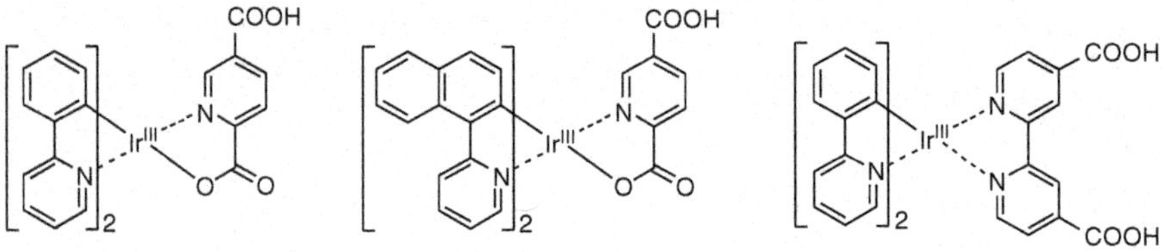

11 **12** **13**

To modify the light-harvesting effect, a series of the cyclometalated Ir(III) complexes **14–17** with tridentate ligands were prepared [39]. These compounds possess an absorption band between 400 and 550 nm, which is assigned to mixed MLCT and LLCT transitions. Importantly, due to their high molar extinction coefficients (10^4 M^{-1} cm^{-1}), which are superior to those of bidentate ligands, they are expected to be good sensitizers for DSSCs. The lifetimes for these complexes are approximately 200 ns, which are longer than that of N3 (50 ns at 77 K). Dye **17** in a nanocrystalline TiO_2-based solar cell shows a maximum of 63 % IPCE and the highest conversion efficiency of 2.16 % (J_{sc}=8.23 mA cm^{-2}; V_{oc}=0.46 V), compared with 3.32 % for N3 (J_{sc}=14.10 mA cm^{-2}; V_{oc}=0.46 V) with DMPII/I2 (DMPII=1,2-dimethyl-3-propylimidazolium iodide) in propionitrile/tetrahydrofuran (80:20, v/v) under simulated AM 1.5 sunlight. The results demonstrated that to improve the energy conversion efficiency of the photovoltaic devices, the light-harvesting capacity of Ir(III) complexes should be enhanced via the maximization of the spectral overlap with the solar spectrum.

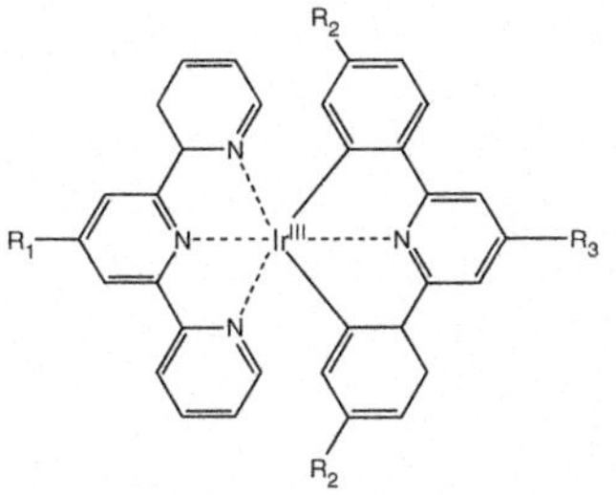

14: R_1 = -Ph-COOH, $R_2 = R_3$ = H
15: R_1 = -Ph-COOH, R_2 = H, R_3 = Ph
16: R_1 = -COOH, $R_2 = R_3$ = H
17: R_1 = -COOH, R_2 = H, R_3 = Me

For solar energy conversion, it is crucial to explore systems with high photon absorption efficiency, whereas the construction of structural complexity is simple from a synthesis viewpoint. Simple cationic cyclometalated Ir(III) complexes, such as **18** and **19**, were used for DSSC solar cells [40]. Compared with ppy in the similar complex **13**, an increase in the π-conjugation in the C^N ligand in **18** or **19** leads to a small redshift of the MLCT band of the absorption spectrum. The photovoltaic efficiencies are low (0.2 % for **18** and 0.3 % for **19**) compared to that of N719 with a standard electrolyte A6141 (7.0 %). However, cationic $[Ir(ppy)_2(dcbpy)]^+$ (with PF_6^- counterions) under the same conditions also exhibited a low efficiency (0.2 %), which is quite different from the unusual complex **13**, especially the emission quantum yield.

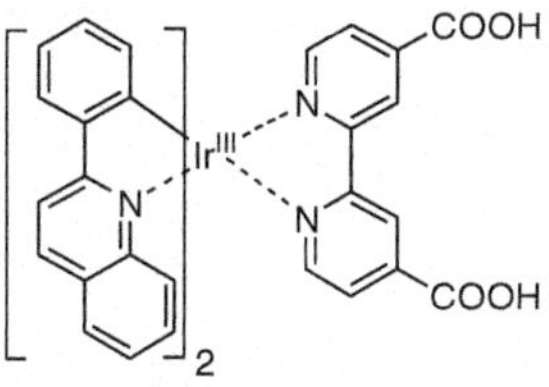

18 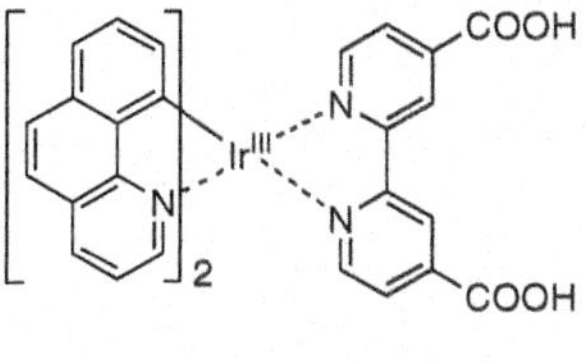 **19**

The use of rigid cyclohexenone derivatives as CN ligands extended the absorption response of the resulting iridium complexes **20** and **21** to a low-energy band near 550 nm [41]. Using these complexes as a sensitizer, the fabricated DSSCs exhibit a low but wide IPCE response in the range of 350–675 nm with peaks at 530 nm for **20** and 550 nm for **21**. These cells presented relatively low J_{sc} (2.1 mA cm^{-2} for **20** and 1.8 mA cm^{-2} for **21**) and low conversion efficiencies of 1.03 % for **20** and 0.83 % for **21** compared with 6.08 % for N719 with the standard electrolyte A6141 (V_{oc}=645 mV for **20** and 626 mV for **21**).

Using as similar structure to **21**, complexes **22–26** based on the 2-phenylbenzothiazole were tested for use in DSSCs [42]. In the visible region, all of the complexes except **26** exhibit weak absorption bands centered at ca. 408 nm (except for **23** at 415 nm) with a tail extending well into the visible region. The DFT calculations indicate that the electronic transition in low-energy regions for **22–25** is mainly attributed to the ^{3}MLCT [$d\pi(Ir) \rightarrow \pi^*_{N\wedge N}$] and/or ^{3}LLCT ($\pi_{N\wedge C} \rightarrow \pi^*_{N\wedge N}$). The absorption spectrum of **26** displays a particularly intense absorption band (438 nm) with a shoulder arising from an intraligand transition of the C^N ligand, probably mixing with the LLCT transition from the π orbital of the amino-substituted C^N to the π* orbital of the N^N moiety. Based on these dyes, dye-sensitized solar cells were constructed. For all five devices based on complexes **22–26**, the observed V_{oc} is approximately 0.5 V. The J_{sc} values vary from 2.2 to 3.7 mA cm^{-2}. The DSCC achieves a maximum overall efficiency (η) of 1.39 % for dye **24** compared with 7.8 % for N719.

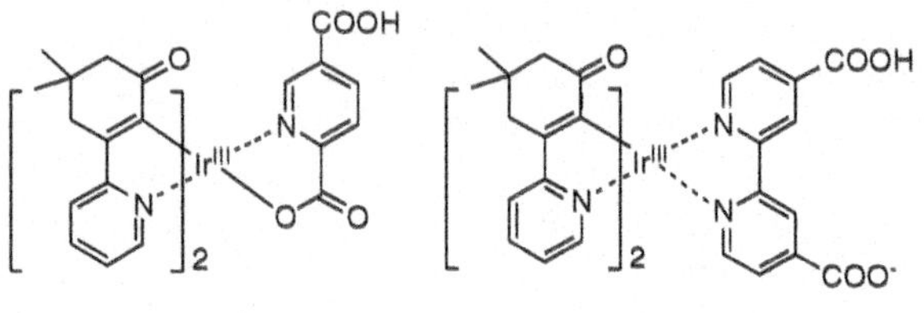

20 **21**

Organic photovoltaic (OPV) cells represent another approach pursuing for current generation. Iridium complexes are also increasing used for active layer formation in the fabrication of an efficient OPV cell [43, 44]; however, this topic exceeds the scope of this review.

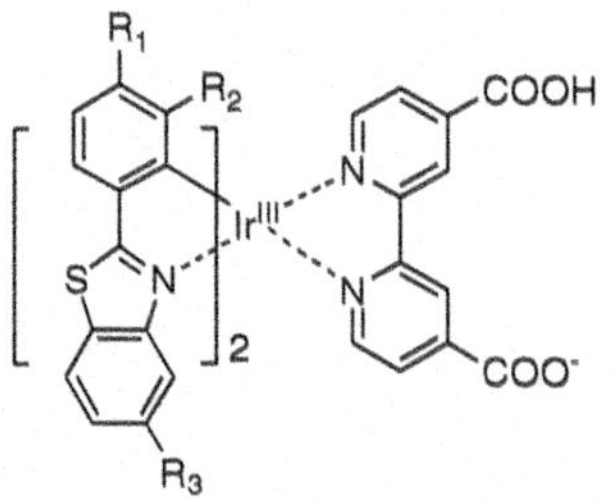

22: $R_1 = CF_3$, $R_2 = R_3 = H$
23: $R_1 = R_3 = H$, $R_2 = CF_3$
24: $R_1 = R_2 = R_3 = H$
25: $R_1 = R_2 = H$, $R_3 = CF_3$
26: $R_1 = NMe_2$, $R_2 = H$, $R_3 = CF_3$

18.4 Conversion of Light into Hydrogen

18.4.1 Bis-Cyclometalated Ir(III) Complexes

Parallel to the development of DSSCs based on iridium complexes described above, efforts have also been devoted toward photoinduced hydrogen production. The direct conversion of solar energy to hydrogen, in general, proceeds catalytically in a multicomponent system containing at least a light sensitizer (e.g., iridium species) for light-harvesting and a water reduction catalyst (WRC), such as colloidal platinum, in addition to a sacrificial electron donor such as triethylamine (TEA) or triethanolamine (TEOA). This field was initiated in the late 1970s. Lehn and coworkers discovered that the ortho-metalated complex $[Ir(bpy)_2(C^3,N'\text{-}bpy)]^{2+}$ (**27**) can be used as a sensitizer for the reduction of water to hydrogen at pH 7.8 [45]. Avnir and coworkers subsequently provided an early example of photoinduced electron transfer between compound **27** confined within an inert porous sol–gel SiO_2 matrix and 1,4-dimethoxybenzene (DMB) dissolved in a water phase in the pores of the glass [46]. From this heterogeneous system, hydrogen was produced from water reduction at acidic pH; the turnover number (TON) for Ir(III), which acts as both a photosensitizer (PS) and a catalyst, is estimated to be greater than 100. The TON is, by definition, being the number of one-electron proton reductions created by one molecule of photosensitizer in the initial reaction mixture over the system lifetime ($TON = n_H/n_{PS}$), reflecting a catalytic system's productivity and stability under the working conditions [47].

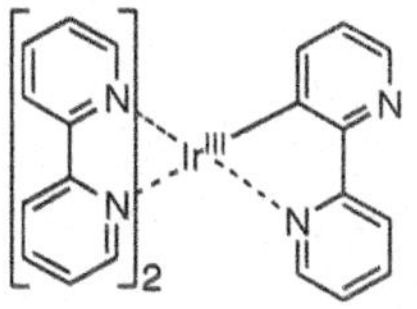

27

Significant improvements in system performance were realized by Bernhard and coworkers in 2005 [48]. They investigated a series of heteroleptic diimine iridium complexes, which were prepared using a combinatorial technique. These complexes **28–33** (with PF_6^- counterions) were tested for use as a photosensitizer for hydrogen production together with $Co(bpy)_3Cl_2$ (2.5 mM) as a WRC and TEOA (0.57 M) as a sacrificial reductant in a 20 mL 1:1 (v:v) water/MeCN mixture (the pH was adjusted by the addition of 0.4 mL of 37 % HCl). Under blue LED (500 mW; wavelength of center emission: approximately 465 nm) irradiation, these iridium complexes were proven superior to the classical ruthenium polypyridine derivatives such as $Ru(dmphen)_3^{2+}$ (dmphen=4,7-dimethy-1,10-phenanthroline) by a factor of 3–6. The author observed that the emission of iridium compounds can be quenched better than that of the standard ruthenium sensitizer, $Ru(dmphen)_3^{2+}$, by $Co(bpy)_3Cl_2$; moreover, it can also be quenched by TEOA, which does not quench the emission of $Ru(dmphen)_3^{2+}$. The dominant reductive quenching behaviors that occur for the iridium compounds are believed to be correlated with the increased hydrogen production. Furthermore, based on the DFT calculations, there is significant phenylpyridine ligand contribution to the HOMO levels of the iridium-based complexes, while that of the ruthenium complexes is almost exclusively metal centered. These differences in the HOMO between Ir- and Ru-based photosensitizers might be responsible for the variation in quenching behavior. The ppy-containing compounds **28–30**, with shorter excited state lifetimes varying from 390 to 1,033 ns, produce slightly less hydrogen than the corresponding F-containing compounds **31–33** ($\tau = 1{,}049$, 1,636, and 1,924 μs), which may be related to their quenching or electrochemical properties.

In the ground state, the iridium complexes are usually stable, the inner coordination sphere of their Ir cores being saturated. They become active in their photoexcitation-induced excited states; therefore, electron transfer involving direct interactions with Ir centers occurs. As a result, the complexes become unstable, and the weakly bound N^N ligand escapes from the complexes, providing an open coordination site for a solvent molecule to subsequently gain access. Mass spectroscopic analysis of the samples from the reaction media (9:3:1 MeCN/water/TEOA) containing complex **28** as the PS, in situ-formed Pt(0) catalyst and TEOA as a sacrificial reductant demonstrated that the photodecomposition of the photosensitizer to $[Ir(ppy)_2]^+$ and $[Ir(ppy)_2CH_3CN]^+$ during catalysis is the main cause of the deactivation of the catalytic process over time [49]. The lack of charge separation in the excited state of the inactive photoproducts, which involves the LC transitions on the cyclometalating ligands, leads to the loss of the driving force for forming hydrogen from water. The results also suggest that the choice of cosolvents is important because they are directly involved in the PS decomposition of the system. Therefore, the relatively weakly coordinating solvents, such as DMF and THF, are considered alternatives for enhancing the system stability. However, except using zwitterionic cyclometalated Ir(III) complexes, few photocatalytic systems display activity in pure water [50].

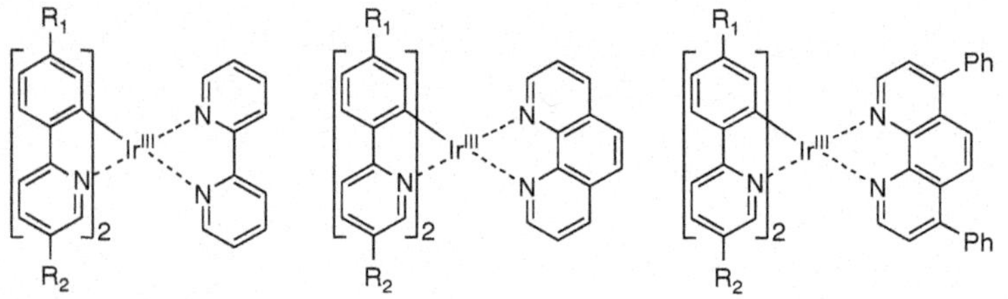

28: $R_1 = R_2 = H$
31: $R_1 = F$, $R_2 = Me$

29: $R_1 = R_2 = H$
32: $R_1 = F$, $R_2 = Me$

30: $R_1 = R_2 = H$
33: $R_1 = F$, $R_2 = Me$

Efficient iron-catalyzed hydrogen generation was found using an Fe(0) carbonyl complex such as $Fe_3(CO)_{12}$ as the WRC; the maximum TON was as high as 3,000 based on the Ir complex **28** in a solution containing 8:2:1 THF/TEA/H_2O with visible-light irradiation (420-nm cutoff filter) [51]. The process was investigated by in situ EPR spectroscopy [52]. When the solution in the absence of $Fe_3(CO)_{12}$ is irradiated, an intense isotropic signal of the reduced form of the iridium photosensitizer (Ir PS^-) formed by the sacrificial reductive quenching of the excited state of the iridium photosensitizer (Ir PS*) by TEA is observed. However, when a mixture containing all of the necessary components of the water reduction system is irradiated, no paramagnetic signal of Ir PS^- species is observed due to the rapid intermolecular electron transfer to the Fe catalyst. Using in situ FTIR techniques, the iron hydride $[HNEt_3][HFe_3(CO)_3(CO)_{11}]$ was identified as a catalytically active species involved in light-induced hydrogen generation. The result indicates that the one-electron transfer process in the catalytic cycles is initiated by the capture of light. The bioinspired diiron thiolate-bridged complex [Fe_2S_2] has also been used in further study of complex **28**, resulting in a TON of 660 in a 9:1 acetone–H_2O mixture [53]. In addition, through cation exchange, both complex **28** and $[Rh(bpy)_3]^{3+}$ can be entrapped within a macroreticular acidic resin [54], creating a recyclable heterogeneous photocatalyst and enabling hydrogen production in aqueous media (9:1 H_2O/MeCN).

The attachments of a CF_3 group and t-butyl groups to the difluorophenylpyridine ligand and the bpy ligand, respectively, significantly enhanced the lifetime and reducing power of the excited state of complex **34** with respect to those of other iridium complexes with the same N^N ligand [55]. Using the similar conditions as described previously for **28–33**, applying $Co(bpy)_3Cl_2$ as a WRC, with **34** (with PF_6^- counterions), the total hydrogen amount obtained was 1.8 times greater than that with the ruthenium-based standard. However, the amount of light absorbed by complex **34** is approximately 12.5 % that absorbed by $Ru(dmphen)_3^{2+}$. As a result, complex **34** is 14 times more efficient at converting light into hydrogen in terms of the relative quantum yield at 465 nm relative to $Ru(dmphen)_3^{2+}$. The improved H_2-evolving performance of **34** can be ascribed to a combination of the long lifetime ($\tau = 2.3$ μs) and significant reducing power (−1.21 V) of the excited state of complex **34**.

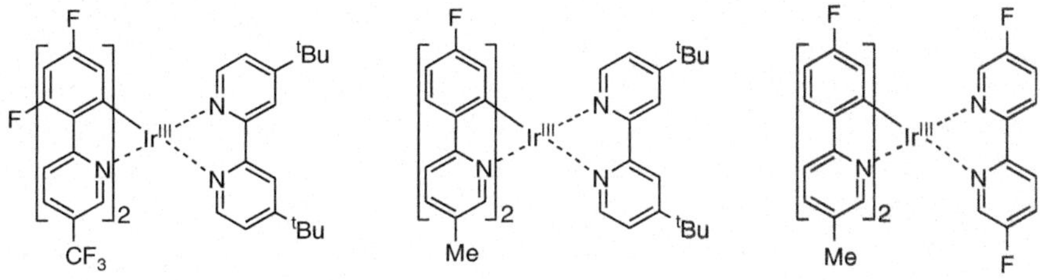

In addition to the optimized experimental conditions and the use of synthetic modification and catalyst screening, the combination of photosensitizer **35** (with PF_6^- counterions) with $[Rh(dtbbpy)_3]^{3+}$ (dtbbpy = 4,4′-di-*tert*-butyl-2,2′-dipyridyl) as a catalyst instead of $[Co(bpy)_3]^{2+}$ was a more effective H_2-evolving system than the other Ir(III) PSs and Rh WRCs and achieved over 5,000 turnovers for PS with quantum yields (1/2 H_2 per photon absorbed) exceeding 34 % [56]. The solvent properties can strongly affect the photophysical characteristics of the excited states of the PS and the electron transfer rates between the PS and WRC and thus affect the activity and stability of the catalysts. The use of 4:1 THF–H_2O as a reaction medium provided the maximum H_2 production, which may provide a good balance between the advantages of the lower dielectric constant at suitable THF concentrations and the need for water as a proton source for the formation of Rh–hydride species, which is believed to be the immediate precursor to H_2 formation. Isotope-labeling studies demonstrated that the hydrogen is predominantly formed from water reduction. Based on tests for system poisoning with mercury or quantitative CS_2, the system was identified as a true homogeneous system for hydrogen production. In studies with a Pd(0) catalyst, such as $[Pd(PPh)_3Cl_2]_2$, the system including complex **35** gave an incident photon-to-hydrogen efficiency of 12.3 % [57]. The addition of bpy considerably enhanced the durability and activity of the system (TON increased from 3,129 to 3,708). With a nickel–thiolate hexameric cluster, $Ni_6(SC_2H_4Ph)_{12}$, as a catalyst, using complex **35** as a PS also leads to a highly active water reduction system [58].

The substituents of the N^N ligand also have a dramatic effect on PS performance when the C^N ligand is held constant. For instance, the contact of the solvent with the iridium center is minimized when steric bulk ligands are located around the metal center, which presumably helps stabilize the PS by decreasing the possibilities for ligand substitution. Using a heterogeneous Pd catalyst generated in situ from K_2PdCl_4 in place of a platinum catalyst, the system with complex **36** (with PF_6^- counterions) [59], which is sterically encumbered with fluoro substituents on the bpy ligand, achieved turnover numbers over six times greater than those of the parent compound **28** under identical conditions. Additionally, when evaluated with the Ir-based PS, the Pd catalyst is much more active than the known Pt-based catalyst under the employed conditions. Steric bulk plays an important role in the photocatalytic activity of these Ir complexes. To further increase the overall photocatalytic performance of these systems, the stability of the Ir photosensitizer was increased by the utility of the improvement of the fundamental inertness of cationic Ir complexes [60]. The use of a triphenylsilyl groups introduced at the N^N ligand is attributed to sufficient steric protection through the site-isolation effect at the reaction site, thereby preventing the undesired photodissociation of the photosensitizer and substantially increasing the longevity. Indeed, significant improvements in the turnover numbers from several hundreds to over 10 thousand compared to the traditional complex **28** were observed in the cases of complexes **37** and **38**.

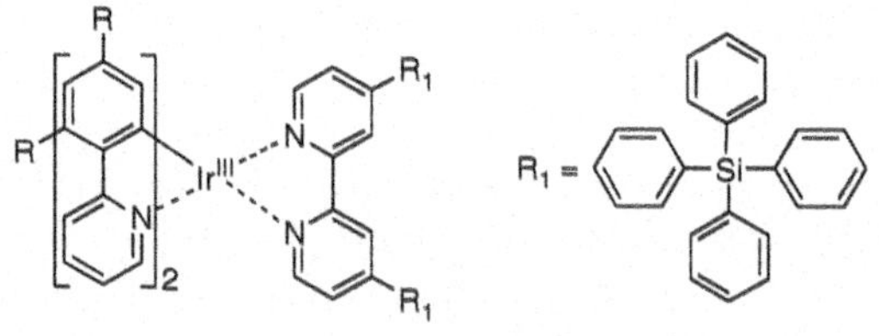

37: R = H
38: R = F

As described above, an ^{3}MLCT-dominant excited state in the tested Ir PSs proved necessary to harvest solar radiation. In principle, other complexes featuring non-^{3}MLCT, such as $[Ir(ppy)_2(dppe)]^+$ (dppe = 1,2-bis(diphenylphosphino)ethane), which exhibits a pure π–π^* emissive state [61], do not work. The development of a new efficient photosensitizer responsible for the initial absorption of a photon represents a critical issue for solar-to-fuel conversion. One efficient method is the judicious design of the ligand sphere of the luminescent iridium complex. Inspired by the basic structure of the classical iridium complex **28**, some iridium complexes were designed by the variation of the ligand sphere, especially the use of the N^N ligand with other neutral bidentate N^N ligands in place of the analogous bpy ligand, in which the basic 1,4-bisimine-coordinated Ir(III) structure remained unchanged [62]. For instance, complex **39**, bearing a 6-*i*Pr-substituted bpy ligand, was tested as a photosensitizer to promote hydrogen generation together with the iron catalyst generated in situ from $[HNEt_3][HFe_3(CO)_3(CO)_{11}]$ and PR_3 (R=3,5-bis(trifluoromethyl)phenyl) (1:1.5) in a mixture of 3:2:1 THF/TEA/H_2O. The electron-deficient monodentate phosphine can stabilize low-valent metal Fe(0) species or even the reduced form of the catalyst obtained from electron transfer from the Ir PS^-, improving hydrogen production by ca. 30 %. A similar strategy has been applied to improve the stability of a diimine–dioxime-based cobalt catalyst [63]. Using blue light irradiation (440 nm, 1.5 W), the catalytic activity of complex **39** proved to be more active than that of the basic structure with unsubstituted bipyridine. The incident photon-to-hydrogen yield reached 16.4 % and the maximum TONs reached 4,550 for the iridium photosensitizer [64]. In the same experiments, no significant activity was observed for complexes **40–44**, which was consistent with their photophysical properties.

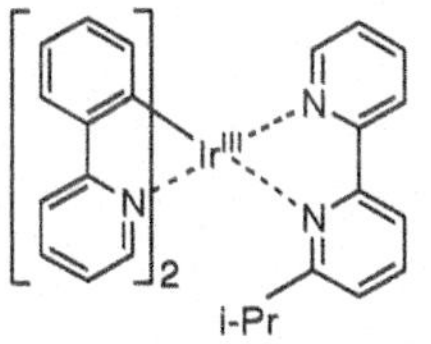

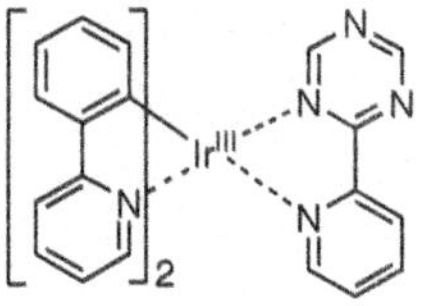

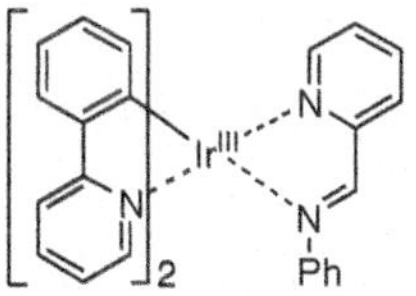

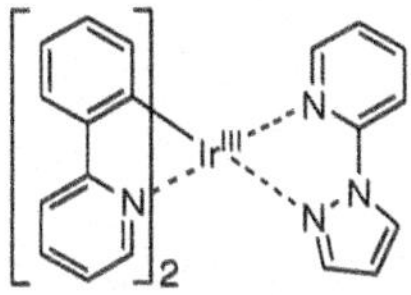

Using other cyclometalating ligands such as phenylazoles in place of phenylpyridine, the final iridium complexes **45** and **46** exhibited interesting photophysical properties [65], especially an unusually long excited state lifetime. Using an iron catalyst based on $[HNEt_3][HFe_3(CO)_3(CO)_{11}]$, complexes **45** and **46** exhibited better productivities than the standard photosensitizer **28** for hydrogen evolution from water. Using other complexes with similar structural motifs, reduced productivity was obtained despite their lowest triplet state also having a microsecond-scale lifetime. The report suggested that the long lifetime of the lowest triplet state does not significantly increase the catalytic efficiency of the system in the established experiments. Most cyclometalated iridium complexes meet the requirement of excited states longer than ca. 10^{-9} s, allowing the possibility of interacting with substrate molecules.

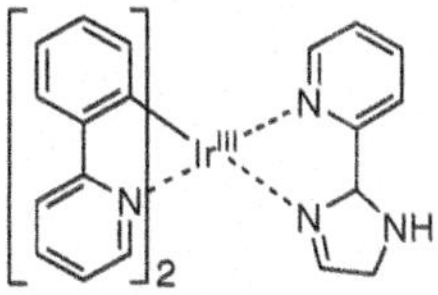

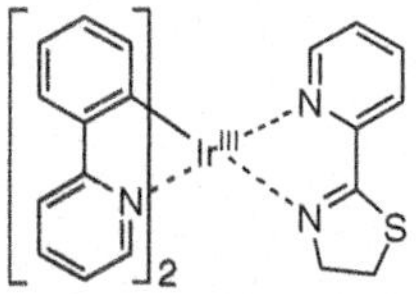

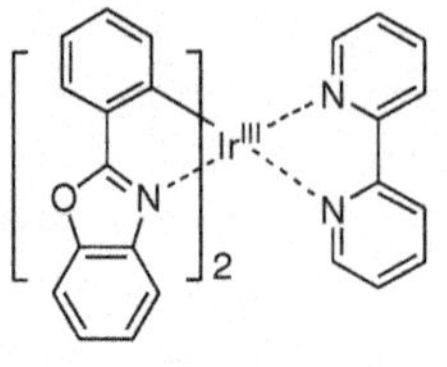

45 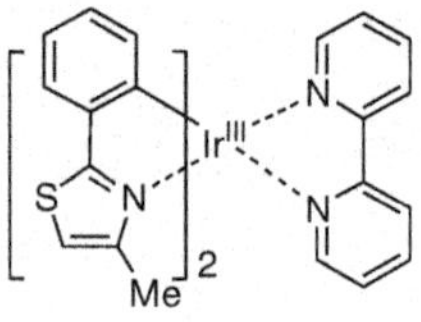 **46**

Although methods to create effective photosensitizers based on [Ir(C^N)$_2$(N^N)]$^+$ complexes in catalytic hydrogen-producing system have been developed, attention has turned to other structural complexes, such as [Ir(C^N^N)$_2$]$^+$ (**47**) (with PF_6^- counterions), carrying rigid tridentate 6-phenyl-2,2′-bipyridine ligands [66]. The tight binding of a tridentate ligand may create a distorted octahedral geometry of the Ir(III) coordination sphere, in which both the excited state lifetime and the fluorescence quantum yield will be affected. The absorptivity of complex **47** is higher than that of compound **28** for visible wavelengths. Moreover, due to the increased rigidity of complex **47**, which is based on C^N^N ligands, compared to that of complex **28**, which is based on C^N ligands, the former exhibits a longer excited state lifetime than that of the latter. When complex **47** is used as a photosensitizer, the lifetime of the catalytic activity of the system was approximately 3 times longer in a 4:1 MeCN/H_2O system than that of compound **28**; however, both catalytic systems have identical maximum hydrogen evolution rates of 85 μmol h^{-1}.

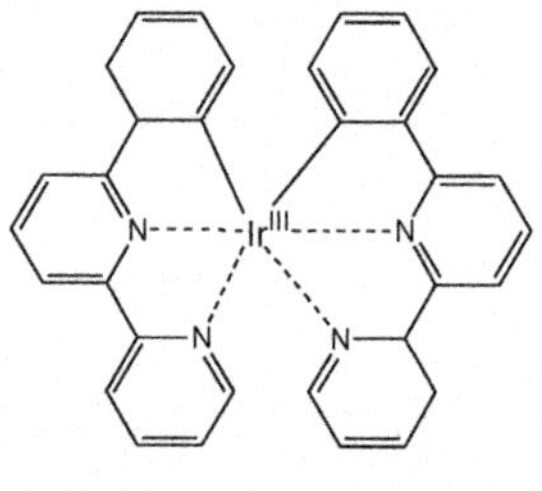

47

Various approaches have been explored and pursued to stabilize Ir-based PS to achieve robust photocatalytic water reduction. One successful strategy is the improvement of the interaction between photosensitizer and catalyst. Several studies have examined this effect using modified Ir(III) complexes. A very attractive example is the Ir(III) complex **48** (with PF_6^- counterions) containing vinyl moieties at the ligand backbone [67]. The use of vinyl groups can stabilize the colloidal metal nanoparticles and promote a faster electron transfer from the photosensitizer to the catalyst. Acting as a photosensitizer for the reduction of water, despite their similar initial photocatalytic activities, the turnover numbers of complex **48** and its vinyl-free derivative **49** are 8,500 and 1,300, respectively. The lifetime of the catalytic systems is significantly increased by the additional vinyl groups of the N^N ligand. The effect is explained by the possible interaction of the vinyl moieties with the in situ-formed colloidal Pt or Rh catalysts during this process. Similarly, a series of Ir(III) complexes containing pendant pyridyl moieties [68], such as complex **50**, which allow the complexes to adsorb on the surface of the colloidal metal nanoparticles, are used as a photosensitizer for the hydrogen-producing photoreaction. These complexes also have higher stability than photosensitizers that display similar photophysical properties but without adsorbing moieties.

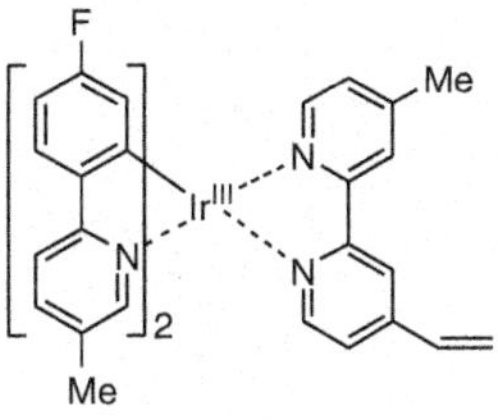

F Me Ir N N N 2 Me **48** F Me Ir N N N 2 Me Me **49** F Ir N N N 2 Me N N **50**

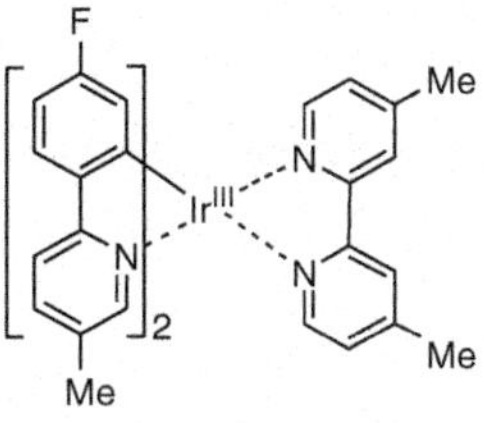

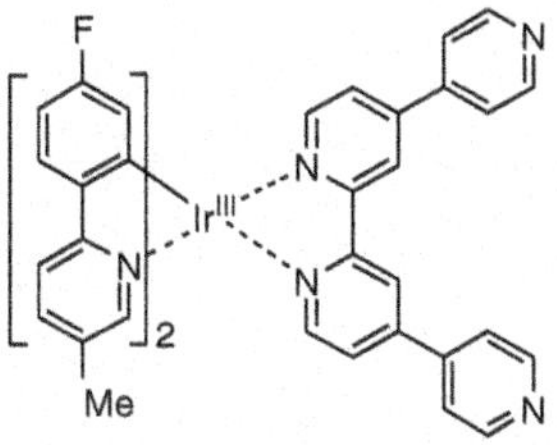

Water reduction by visible light in pure water without the use of organic cosolvents was carried out using the complexes **22–26** [42, 50]. The introduction of charged or polar substituents, such as carboxyl (COOH) groups, into the N^N ligand increases the aqueous solubility of its complexes because of the presence of the acid/base equilibrium of the ligand. The production of hydrogen from water using these complexes differs by the architecture of the complex and the electronic effect of the substituent of the cyclometalated ligand in particular. Among the systems, the better activity is achieved in 9:1 acetone/water medium based on complexes **23** or **24**, in which the strong electron-withdrawing CF_3 substituent is attached to a phenyl ring of the cyclometalated ligand. The directional nature of the carboxylate anchoring groups is shown to be suitable for grafting the complexes on the TiO_2 surface for efficient electron transfer, resulting in the enhancement of the hydrogen evolution compared to the homogeneous systems in the absence of TiO_2.

The utility of colloidal MoS_2 as a catalyst for photoinduced hydrogen production was developed with iridium sensitizers, such as complex **51** [69]. The system exhibited much higher performance in 1:1 MeOH/water medium than the K_2PtCl_4, $[Co(bpy)_3]Cl_2$, $[Rh(dtb\text{-}bpy)_3](PF_6)_3$, and $[Co(dmgH)_2(H_2O)_2]$ (dmgH = dimethylglyoximate) catalysts under the same conditions. The conversion efficiency with complex **51** reached 12.4 % at 400 nm in the optimum MoS_2-TEOA system. Other Ir(III) complexes, such as **52** and **53** (with PF_6^- counterions) with other substituents such as CH_2OH or NH_2, were less active in the reduction of protons with colloidal MoS_2. The introduction of adsorbing moieties of carboxylate groups allows the species to interact with the MoS_2 nanoparticles, leading to the promotion of the electron transfer, positively affecting the activity of the catalytic system.

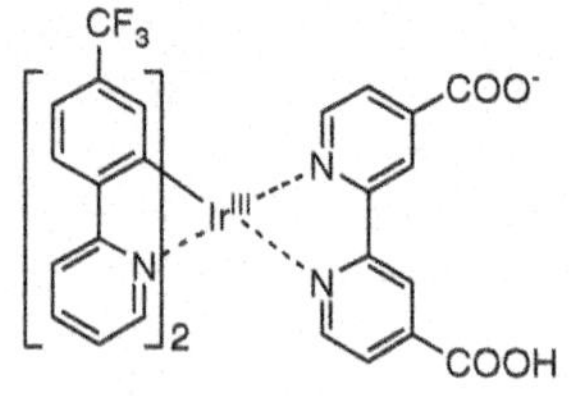

51 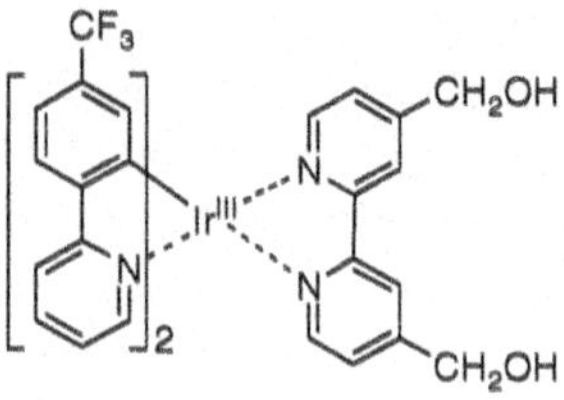 52 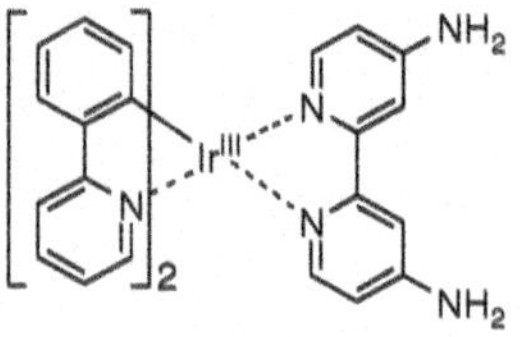 53

18.4.2 *Tris-Cyclometalated Ir(III) Complexes*

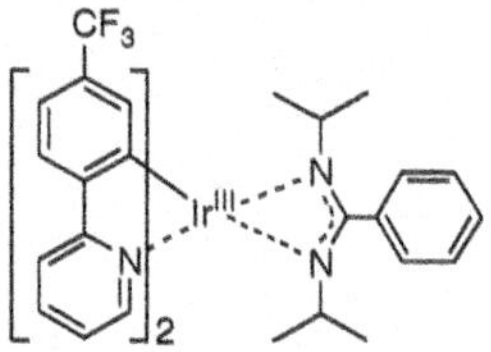

54 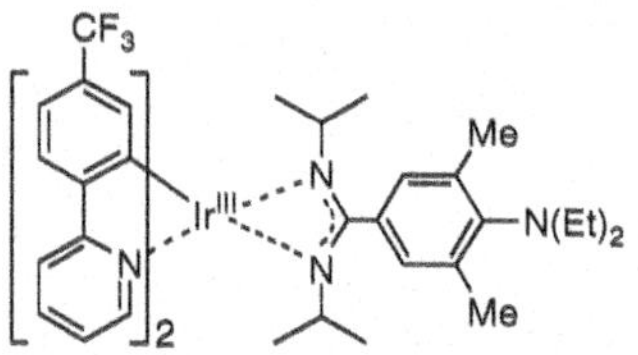 55

The capabilities of such bis-cyclometalated Ir(III) complexes have been demonstrated in photocatalytic hydrogen production. These complexes usually carry bpy moieties, which dissociate from the complex during catalysis, leading to the termination of catalysis [70]. The exclusively reductive quenching step in the photochemical system, in which PS^- is generated by electron transfer to the excited photosensitizer from the sacrificial reductant, was thought to be responsible for this decomposition [71]. In contrast with the success of the bis-cyclometalated Ir(III) complexes in photocatalytic hydrogen production described in the previous section, only a very limited number of tris-cyclometalated Ir(III) complexes have been reported so far in this field. Monoanionic N,N-bis-imine-type ligands, such as amidinates (RNR′CNR), were used instead of neutral bipyridines for increased stability for the resulting charge-neutral complexes **54** and **55** [72]. Using a cobalt catalyst, the complexes display long-term photostability for photoinduced hydrogen production under constant illumination (xenon-arc lamp, 300 W) over 72 h; complex **54** achieved a relatively high turnover number of 1,880 in 4:1 acetone/water medium during this reaction period. The improved utility of the photocatalytic systems likely benefits from an oxidative quenching pathway in the photochemically driven step that occurs for the iridium-based photosensitizers, in which PS^+ is generated by electron transfer from the excited photosensitizer to the catalyst. This argument was supported by fluorescence-quenching experiments. The observed photoluminescence spectra suggest that the excited states of the neutral complexes **54** and **55** cannot be quenched upon the addition of TEOA but can be quenched upon the addition of $[Co(bpy)_3]^{2+}$. This is an obvious difference from the behaviors of the bis-cyclometalated Ir(III) complexes.

56: $R_1 = R_2 = R_3 = H$
57: $R_1 = CF_3$, $R_2 = R_3 = H$
58: $R_1 = R_3 = H$, $R_2 = CF_3$
59: $R_1 = NMe_2$, $R_2 = H$, $R_3 = CF_3$

To further increase the stability of the iridium-based photosensitizers, a series of charge-neutral, heteroleptic tris-cyclometalated Ir(III) complexes $Ir(thpy)_2(bt)$ (**56**–**59**, where thpy=2,2′-thienylpyridine; bt=2-phenylbenzothiazole and its derivatives) were described for producing hydrogen over an extended period of time in the presence of $Co(bpy)_3^{2+}$ and triethanolamine (TEOA) [73]. Similar to complexes **54** and **55**, the PS excited state can be quenched through an oxidative quenching mechanism only. With $[Co(bpy)_3Cl_2]$ as a WRC and TEOA as a sacrificial electron donor, a system for light-induced hydrogen production from water was obtained with a TON of approximately 300 based on complex **56** after 72 h of illumination. The maximal amount hydrogen is obtained under constant irradiation (300 W Xenon lamp; $\lambda > 420$ nm) over 72 h, and the system can regain its activity upon the addition of a cobalt-based catalyst after hydrogen evolution ceases. When hydrogen production ceased, the decomposition products were determined after the photolysis experiments using **56** by GC-MS. The results demonstrate that in the photocatalytic systems, the complete loss in activity of the catalytic system was linked with a loss of the active species of the $Co(bpy)_3^{2+}$ catalyst, limiting the long-term catalytic duration of the systems.

60: R = H
61: $R = CF_3$
62: R = n-Bu

The neutral homoleptic tris-cyclometalated compounds of *fac*-$Ir(ppy)_3$ are known to feature high phosphorescence efficiencies and relatively long excited state lifetimes [74]. When the charge-neutral homoleptic tris-cyclometalated iridium complexes chelated by 2-phenylpyridine and its derivatives as cyclometalating ligands, $Ir(C^\wedge N)_3$**60**–**62** (HC^N=2-phenylpyridine and its derivatives) [75], were used as PSs for photocatalytic hydrogen formation in the presence of $[Rh(dtbbpy)_3]^{3+}$ as a proton reduction catalyst and TEOA as a sacrificial donor in an acetone/water (4:1) solvent mixture, these systems displayed a long lifetime far exceeding 72 h, maximum H_2 turnovers of up to 3,040 (for **60**) and a maximum apparent quantum yield of 2.59 % at 380 nm with visible-light irradiation by a 300 W Xenon lamp equipped with a cut-off filter ($\lambda > 420$ nm).

18.4.3 Iridium-Based Supramolecular Systems

The intermolecular reaction used in the above strategy is a diffusion-controlled process and depends critically on the efficiency of the intermolecular collision between the species involved [76]. In contrast, only several studies have been conducted on the intermolecularly linked component systems in which the Ir PS is combined with the catalyst for hydrogen evolution in one molecule via coordinative interactions or covalent linkages [77].

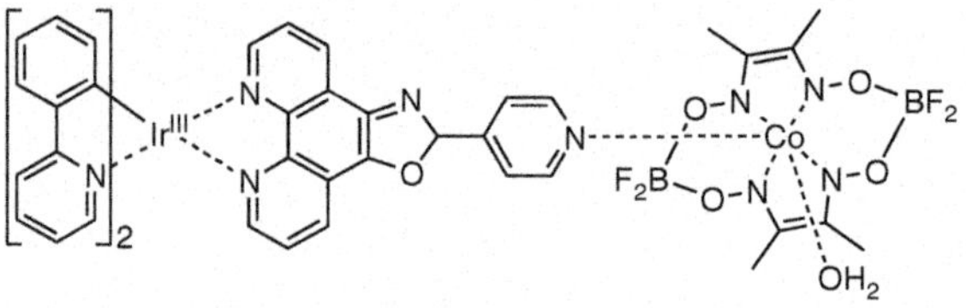

63

In 2008, Artero and coworkers reported a single photocatalytic system for hydrogen production based on cyclometalated iridium as a photosensitizer and cobaloxime H_2-evolving catalytic moiety (complex **63**, with PF_6^- counterions) [78]. A turnover number of 210 and a quantum efficiency for H_2 production of 0.10 were obtained in a 15-h experiment; moreover, the replacement of Ir cyclometalated with Ru(dmphen) in the photosensitizer subunits was found to significantly decrease the activity under the same conditions. The author suggested that the reaction might be determined by the reductive quenching of the excited photosensitizer by TEA because the first electron transfer process between the photosensitizer and the catalyst was assumed not to be a rate-determining step for this type of catalysis based on the initial turnover frequencies of the supramolecular architecture and the reference bicomponent system of $[Ir(ppy)_2(phen)]^+$ and $[Co(dmgBF_2)_2(OH_2)_2]$. Under the same experimental conditions, the supramolecular system was shown to be more stable than the bicomponent catalytic system, being capable of 165 TON.

Another system based on $[Ir(ppy)(bpy)]^+$ and $[Co(bpy)_3]^{2+}$ cores was developed through a self-assembled supramolecular approach [79]. Cyclometalated iridium complexes bearing a pendant bipyridine spontaneously coordinated with a cobalt ion to afford self-assembled species, such as compound **64** (with PF_6^- counterions), in which the two components were brought together by a nonconjugated bridge. Hydrogen was evolved in a 1:1 H_2O/MeCN solution with a TON of up to 20, which is double compared with that of the corresponding multicomponent system, where $[Ir(ppy)(bpy)]^+$ and $[Co(bpy)_3]^{2+}$ were used under the conditions of identical component concentrations.

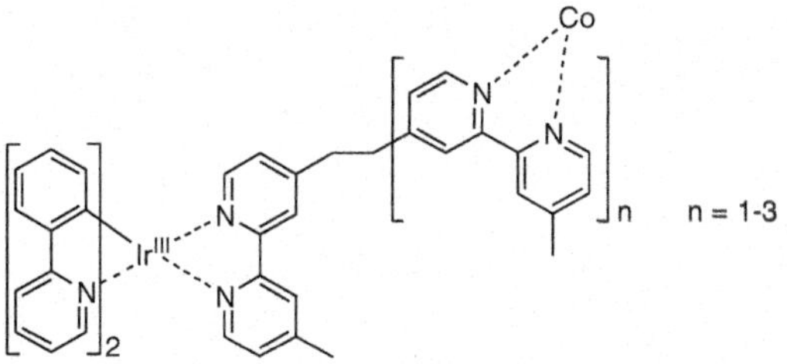

n = 1-3

64

Using the bridging ligand 2,2′:5′,2″-terpyridine, iridium–Pt/Pd dinuclear complexes **65** and **66** were designed (with PF_6^- counterions) [80]. These complexes exhibited good activity for hydrogen production with TONs with visible excitation that are comparatively higher than those produced in the UV region, which resulted from the interplay between two independent excited states. The DFT method was used to understand the wavelength-dependent photocatalytic behaviors. The results obtained from the calculations indicated that MLCT transitions directly involving the Pd centers occur in the UV region of the spectrum, which was believed to be inactive, acting as a photocatalyst for hydrogen formation.

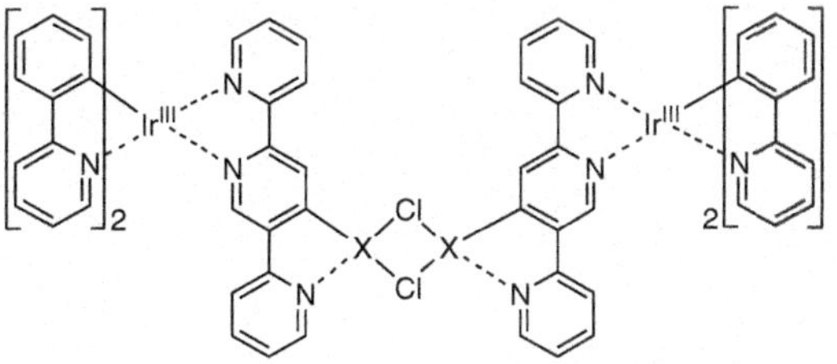

65: X = Pt
66: X = Pd

18.5 Conclusions and Perspectives

The construction of artificial synthetic systems for generating clean and sustainable fuel cycles at high efficiencies has remained a central academic issue for many years. As outlined above, cyclometalated iridium complexes have gained importance in this context and have represented significant advances in terms of the efficiency and operating lifetime of molecular-based photochemical conversion systems. However, solar conversion efficiencies have remained too low to implement these compounds in large-scale practical devices in real-world applications due to issues with the durability, ability, and cost of such photosensitizers and the lack of a deep understanding of the structural role of the metal complexes in their reactive properties.

In the future, significant developments are expected in the important desired application of iridium complexes as the photoactive constituent of solar-to-energy devices by an increasingly precise understanding of the theoretical fundamentals of the complicated photochemical processes involving the interplay between the photoexcitation, energy or electron transfer, charge separation and transport, and catalytic activation. Furthermore, the further improvement of their light-harvesting and charge-generation abilities would be enabled by finding new members of this family through systematic investigation and various structural modifications to continuously improve the emission efficiency and increase the molar extinction coefficient as well as long-term chemical and electrochemical stability.

References

1. Lewis NS, Nocera DG (2006) Powering the planet: chemical challenges in solar energy utilization. Proc Natl Acad Sci U S A 103:15729–15735
2. Armaroli N, Balzani V (2007) The future of energy supply: challenges and opportunities. Angew Chem Int Ed 46:52–66
3. Walter MG, Warren EL, McKone JR et al (2010) Solar water splitting cells. Chem Rev 110:6446–6473
4. Crabtree GW, Dresselhaus MS, Buchanan MV (2004) The hydrogen economy. Phys Today 57:39–44
5. Fujishima A, Honda K (1972) Electrochemical photolysis of water at a semiconductor electrode. Nature 238:37–38
6. Lehn JM, Sauvage JP (1977) Chemical storage of light energy: catalytic generation of hydrogen by visible light or sunlight irradiation of neutral aqueous solutions. Nouv J Chim 1:449–451
7. O'Regan B, Grätzel M (1991) A low-cost, high-efficiency solar cell based on dye-sensitized colloidal TiO_2 films. Nature 353:737–740
8. McConnell I, Li G, Brudvig GW (2010) Energy conversion in natural and artificial photosynthesis. Chem Biol 17:434–447
9. Frischmann PD, Mahata K, Würthner F (2013) Powering the future of molecular artificial photosynthesis with light-harvesting metallosupramolecular dye assemblies. Chem Soc Rev 42:1847–1870
10. Sutin N, Creutz C (1980) Light induced electron transfer reactions of metal complexes. Pure Appl Chem 52:2717–2738
11. Prier CK, Rankic DA, MacMillan DWC (2013) Visible light photoredox catalysis with transition metal complexes: applications in organic synthesis. Chem Rev 113:5322–5363
12. Tucker JW, Stephenson CRJ (2012) Shining light on photoredox catalysis: theory and synthetic applications. J Org Chem 77:1617–1622
13. Krassen H, Ott S, Heberle J (2011) 'In vitro' hydrogen production – using energy from the sun. Phys Chem Chem Phys 13:47–57
14. Hirata N, Lagref JJ, Palomares EJ et al (2004) Supramolecular control of charge-transfer dynamics on dye-sensitized nanocrystalline TiO_2 films. Chem Eur J 10:595–602
15. Medlycott EA, Hanan GS (2005) Designing tridentate ligands for ruthenium(II) complexes with prolonged room temperature luminescence lifetimes. Chem Soc Rev 34:133–142
16. Lowry MS, Bernhard S (2006) Synthetically tailored excited states: phosphorescent, cyclometalated iridium(III) complexes and their applications. Chem Eur J 12:7970–7977
17. Ladouceur S, Zysman-Colman E (2013) A comprehensive survey of cationic iridium(III) complexes bearing nontraditional ligand chelation motifs. Eur J Inorg Chem 17:2985–3007
18. Flamigni L, Barbieri A, Sabatini C, Ventura B, Barigelletti F (2007) Photochemistry and photophysics of coordination compounds: iridium. Top Curr Chem 281:143–203

19. Chi Y, Chou PT (2010) Transition-metal phosphors with cyclometalating ligands: fundamentals and applications. Chem Soc Rev 39:638–655
20. Holder E, Langeveld BMW, Schubert US (2005) New trends in the use of transition metal–ligand complexes for applications in electroluminescent devices. Adv Mater 17:1109–1121
21. Dixon IM, Collin JP, Sauvage JP, Flamigni L, Encinas S, Barigelletti F (2000) A family of luminescent coordination compounds: iridium(III) polyimine complexes. Chem Soc Rev 29:385–391
22. Lamansky S, Djurovich PI, Murphy D et al (2001) Highly phosphorescent bis-cyclometalated iridium complexes: synthesis, photophysical characterization, and use in organic light emitting diodes. J Am Chem Soc 123:4304–4312
23. Happ B, Winter A, Hager MD, Schubert US (2012) Photogenerated avenues in macromolecules containing Re(I), Ru(II), Os(II), and Ir(III) metal complexes of pyridine-based ligands. Chem Soc Rev 41:2222–2255
24. Balzani V, Juris A, Venturi M (1996) Luminescent and redox-active polynuclear transition metal complexes. Chem Rev 96:759–833
25. You Y, Park SY (2009) Phosphorescent iridium(III) complexes: toward high phosphorescence quantum efficiency through ligand control. Dalton Trans 8:1267–1282
26. You Y, Nam W (2012) Photofunctional triplet excited states of cyclometalated Ir(III) complexes: beyond electroluminescence. Chem Soc Rev 41:7061–7084
27. Hay PJ (2002) Theoretical studies of the ground and excited electronic states in cyclometalated phenylpyridine Ir(III) complexes using density functional theory. J Phys Chem A 106:1634–1641
28. Slinker JD, Gorodetsky AA, Lowry MS et al (2004) Efficient yellow electroluminescence from a single layer of a cyclometalated iridium complex. J Am Chem Soc 126:2763–2767
29. Tamayo AB, Garon S, Sajoto T et al (2005) Cationic bis-cyclometalated iridium(III) diimine complexes and their use in efficient blue, green, and red electroluminescent devices. Inorg Chem 44:8723–8732
30. Lowry MS, Hudson WR, Pascal RA et al (2004) Accelerated luminophore discovery through combinatorial synthesis. J Am Chem Soc 126:14129–14135
31. Nazeeruddin MK, Angelis FD, Fantacci S et al (2005) Combined experimental and DFT-TDDFT computational study of photoelectrochemical cell ruthenium sensitizers. J Am Chem Soc 127:16835–16847
32. Young KJ, Martini LA, Milot RL et al (2012) Light-driven water oxidation for solar fuels. Coord Chem Rev 256:2503–2520
33. Bessho T, Yoneda E, Yum JH et al (2009) New paradigm in molecular engineering of sensitizers for solar cell applications. J Am Chem Soc 131:5930–5934
34. Mayo EI, Kilså K, Tirrell T et al (2006) Cyclometalated iridium(III)-sensitized titanium dioxide solar cells. Photochem Photobiol Sci 5:871–873
35. Geary EAM, Yellowlees LJ, Jack LA et al (2005) Synthesis, structure and properties of [Pt(II)(diimine)(dithiolate)] dyes with 3,3′, 4,4′ and 5,5′–disubstituted bipyridyl: applications in dye-sensitized solar cells. Inorg Chem 44:242–250
36. Baranoff E, Yum JH, Graetzel M et al (2009) Cyclometallated iridium complexes for conversion of light into electricity and electricity into light. J Organomet Chem 694:2661–2670
37. Baranoff E, Yum JH, Jung II et al (2010) Cyclometallated iridium complexes as sensitizers for dye-sensitized solar cells. Chem Asian J 5:496–499
38. Ning Z, Zhang Q, Wu W et al (2009) Novel iridium complex with carboxyl pyridyl ligand for dye-sensitized solar cells: high fluorescence intensity, high electron injection efficiency? J Organomet Chem 694:2705–2711
39. Shinpuku Y, Inui F, Nakai M et al (2011) Synthesis and characterization of novel cyclometalated iridium(III) complexes for nanocrystalline TiO_2-based dye-sensitized solar cells. J Photochem Photobiol A Chem 222:203–209

40. Dragonetti C, Valore A, Colombo A et al (2012) Simple novel cyclometallated iridium complexes for potential application in dye-sensitized solar cells. Inorg Chim Acta 388:163–167
41. Wang D, Wu Y, Dong H et al (2013) Iridium (III) complexes with 5,5-dimethyl-3-(pyridin-2-yl)cyclohex-2-enone ligands as sensitizer for dye-sensitized solar cells. Org Electron 14:3297–3305
42. Yuan YJ, Zhang JY, Yu ZT et al (2012) Impact of ligand modification on hydrogen photogeneration and light-harvesting applications using cyclometalated iridium complexes. Inorg Chem 51:4123–4133
43. Yang CM, Wu CH, Liao HH et al (2007) Enhanced photovoltaic response of organic solar cell by singlet-to-triplet exciton conversion. Appl Phys Lett 90:133509–133509-3
44. Lee W, Kwon TH, Kwon J, Kim J, Lee C, Hong JI (2011) Effect of main ligands on organic photovoltaic performance of Ir(III) complexes. New J Chem 35:2557–2563
45. Kirch M, Lehn LM, Sauvage LP (1979) Hydrogen generation by visible light irradiation of aqueous solutions of metal complexes. An approach to the photochemical conversion and storage of solar energy. Helv Chim Acta 62:1345–1384
46. Slama-Schwok A, Avnir D, Ottolengbi M (1989) Photoinduced charge separation across the solid-liquid interface of porous sol-gel glasses: catalyzed hydrogen generation from water. J Phys Chem 93:7544–7547
47. Tinker LL, McDaniel ND, Bernhard S (2009) Progress towards solar-powered homogeneous water photolysis. J Mater Chem 19:3328–3337
48. Goldsmith JI, Hudson WR, Lowry MS, Anderson TH, Bernhard S (2005) Discovery and high-throughput screening of heteroleptic iridium complexes for photoinduced hydrogen production. J Am Chem Soc 127:7502–7510
49. Tinker LL, McDaniel ND, Curtin PN, Smith CK, Ireland MJ, Bernhard S (2007) Visible light induced catalytic water reduction without an electron relay. Chem Eur J 13:8726–8732
50. Yuan YJ, Yu ZT, Chen XY, Zhang JY, Zou ZG (2011) Visible-light-driven H_2 generation from water and CO_2 conversion by using a zwitterionic cyclometalated iridium(III) complex. Chem Eur J 17:12891–12895
51. Gärtner F, Sundararaju B, Surkus AE et al (2009) Light-driven hydrogen generation: efficient iron-based water reduction catalysts. Angew Chem Int Ed 48:9962–9965
52. Hollmann D, Gärtner F, Ludwig R et al (2011) Insights into the mechanism of photocatalytic water reduction by DFT-supported in situ EPR/Raman spectroscopy. Angew Chem Int Ed 50:10246–10250
53. Zhang P, Wang M, Na Y, Li X, Jiang Y, Sun L (2010) Homogeneous photocatalytic production of hydrogen from water by a bioinspired [Fe_2S_2] catalyst with high turnover numbers. Dalton Trans 39:1204–1206
54. Mori K, Kubota Y, Yamashita H (2013) Iridium and rhodium complexes within a macroreticular acidic resin: a heterogeneous photocatalyst for visible-light driven H_2 production without an electron mediator. Chem Asian J 8:3207–3213
55. Lowry MS, Goldsmith JI, Slinker JD et al (2005) Single-layer electroluminescent devices and photoinduced hydrogen production from an ionic iridium(III) complex. Chem Mater 17:5712–5719
56. Cline ED, Adamson SE, Bernhard S (2008) Homogeneous catalytic system for photoinduced hydrogen production utilizing iridium and rhodium complexes. Inorg Chem 47:10378–10388
57. Hansen S, Pohl MM, Klahn M, Spannenberg A, Beweries T (2013) Investigation and enhancement of the stability and performance of water reduction systems based on cyclometalated iridium(III) complexes. ChemSusChem 6:92–101
58. Kagalwala HN, Gottlieb E, Li G, Li T, Jin R, Bernhard S (2013) Photocatalytic hydrogen generation system using a nickel-thiolate hexameric cluster. Inorg Chem 52:9094–9101
59. Curtin PN, Tinker LL, Burgess CM, Cline ED, Bernhard S (2009) Structure-activity correlations among iridium(III) photosensitizers in a robust water-reducing system. Inorg Chem 48:10498–10506

60. Whang DR, Sakai K, Park SY (2013) Highly efficient photocatalytic water reduction with robust iridium(III) photosensitizers containing arylsilyl substituents. Angew Chem Int Ed 52:11612–11615
61. Brooks AC, Basore K, Bernhard S (2013) Photon-driven reduction of Zn^{2+} to Zn metal. Inorg Chem 52:5794–5800
62. Gärtner F, Boddien A, Barsch E et al (2011) Photocatalytic hydrogen generation from water with iron carbonyl phosphine complexes: improved water reduction catalysts and mechanistic insights. Chem Eur J 17:6425–6436
63. Zhang P, Jacques PA, Chavarot-Kerlidou M et al (2012) Phosphine coordination to a cobalt diimine–dioxime catalyst increases stability during light-driven H_2 production. Inorg Chem 51:2115–2120
64. Gärtner F, Cozzula D, Losse S et al (2011) Synthesis, characterisation and application of iridium(III) photosensitisers for catalytic water reduction. Chem Eur J 17:6998–7006
65. Gärtner F, Denurra S, Losse S et al (2012) Synthesis and characterization of new iridium photosensitizers for catalytic hydrogen generation from water. Chem Eur J 18:3220–3225
66. Tinker LL, Bernhard S (2009) Photon-driven catalytic proton reduction with a robust homoleptic iridium(III) 6-phenyl-2,2′-bipyridine complex ($[Ir(C^\wedge N^\wedge N)_2]^+$). Inorg Chem 48:10507–10511
67. Metz S, Bernhard S (2010) Robust photocatalytic water reduction with cyclometalated Ir(III) 4-vinyl-2,2′-bipyridine complexes. Chem Commun 46:7551–7553
68. DiSalle B, Bernhard S (2011) Orchestrated photocatalytic water reduction using surface-adsorbing iridium photosensitizers. J Am Chem Soc 133:11819–11821
69. Yuan YJ, Yu ZT, Liu XJ, Cai JG, Guan ZJ, Zou ZG (2014) Hydrogen photogeneration promoted by efficient electron transfer from iridium sensitizers to colloidal MoS_2 catalysts. Sci Rep 4:4045–4045-9
70. Bokarev SI, Hollmann D, Pazidis A et al (2014) Spin density distribution after electron transfer from triethylamine to an $[Ir(ppy)_2(bpy)]^+$ photosensitizer during photocatalytic water reduction. Phys Chem Chem Phys 16:4789–4796
71. Han Z, McNamara WR, Eum MS, Holland PL, Eisenberg R (2012) A nickel thiolate catalyst for the long-lived photocatalytic production of hydrogen in a noble-metal-free system. Angew Chem Int Ed 51:1667–1670
72. Yu ZT, Yuan YJ, Cai JG, Zou ZG (2013) Charge-neutral, amidinate-containing iridium complexes capable of efficient photocatalytic water reduction. Chem Eur J 19:1303–1310
73. Yuan YJ, Yu ZT, Gao HL, Zou ZG, Zheng C, Huang W (2013) Tricyclometalated iridium complexes as highly stable photosensitizers for light-induced hydrogen evolution. Chem Eur J 19:6340–6349
74. Tamayo AB, Alleyne BD, Djurovich PI et al (2003) Synthesis and characterization of facial and meridional tris-cyclometalated iridium(III) complexes. J Am Chem Soc 125:7377–7387
75. Yuan YJ, Yu ZT, Cai JG, Zheng C, Huang W, Zou ZG (2013) Highly efficient water reduction systems associated with homoleptic cyclometalated iridium complexes of various 2-phenylpyridines. ChemSusChem 6:1357–1365
76. Tschierlei S, Presselt M, Kuhnt C et al (2009) Photophysics of an intramolecular hydrogen-evolving Ru–Pd photocatalyst. Chem Eur J 15:7678–7688
77. Andreiadis ES, Chavarot-Kerlidou M, Fontecave M, Artero V (2011) Artificial photosynthesis: from molecular catalysts for light-driven water splitting to photoelectrochemical cells. Photochem Photobiol 87:946–964
78. Fihri A, Artero V, Pereira A, Fontecave M (2008) Efficient H_2-producing photocatalytic systems based on cyclometalated iridium- and tricarbonylrhenium-diimine photosensitizers and cobaloxime catalysts. Dalton Trans 41:5567–5569
79. Jasimuddin S, Yamada T, Fukuju K, Otsuki J, Sakai K (2010) Photocatalytic hydrogen production from water in self-assembled supramolecular iridium–cobalt systems. Chem Commun 46:8466–8468
80. Soman S, Bindra GS, Paul A et al (2012) Wavelength dependent photocatalytic H_2 generation using iridium–Pt/Pd complexes. Dalton Trans 41:12678–12680

The manufacturer's authorised representative in the EU is Springer Nature Customer Service Centre GmbH, Europaplatz 3, 69115 Heidelberg, Germany. If you have any concerns regarding our products, please contact ProductSafety@springernature.com

Printed and bound by CPI Group (UK) Ltd, Croydon, CR0 4YY
15/07/2026
02167634-0007